String Theory Methods for Condensed Matter Physics

The discovery of a duality between Anti–de Sitter spaces (AdS) and Conformal Field Theories (CFT) has led to major advances in our understanding of quantum field theory and quantum gravity. String theory methods and AdS/CFT correspondence maps provide new ways to think about difficult condensed matter problems. String theory methods based on the AdS/CFT correspondence allow us to transform problems so they have weak interactions and can be solved more easily. They can also help map problems to different descriptions, for instance, mapping the description of a fluid using the Navier-Stokes equations to the description of an event horizon of a black hole using Einstein's equations. This textbook covers the applications of string theory methods and the mathematics of AdS/CFT to areas of condensed matter physics. Bridging the gap between string theory and condensed matter, this is a valuable textbook for students and researchers in both fields.

Horaţiu Năstase is a Researcher at the Institute for Theoretical Physics at the State University of São Paulo, Brazil. To date, his career has spanned four continents. As an undergraduate he studied at the University of Bucharest and Copenhagen University. He later completed his Ph.D. at the State University of New York, Stony Brook, before moving to the Institute for Advanced Study, Princeton, where his collaboration with David Berenstein and Juan Maldacena defined the pp-wave correspondence. He has also held research and teaching positions at Brown University and the Tokyo Institute of Technology.

String Theory Methods for Condensed Matter Physics

HORAŢIU NĂSTASE

Universidade Estadual Paulista, São Paulo

CAMBRIDGE
UNIVERSITY PRESS

CAMBRIDGE
UNIVERSITY PRESS

University Printing House, Cambridge CB2 8BS, United Kingdom

One Liberty Plaza, 20th Floor, New York, NY 10006, USA

477 Williamstown Road, Port Melbourne, VIC 3207, Australia

314-321, 3rd Floor, Plot 3, Splendor Forum, Jasola District Centre, New Delhi - 110025, India

79 Anson Road, #06-04/06, Singapore 079906

Cambridge University Press is part of the University of Cambridge.

It furthers the University's mission by disseminating knowledge in the pursuit of education, learning and research at the highest international levels of excellence.

www.cambridge.org
Information on this title: www.cambridge.org/9781107180383
DOI: 10.1017/9781316847978

© Horaţiu Năstase 2017

First published 2017

A catalogue record for this publication is available from the British Library

ISBN 978-1-107-18038-3 Hardback

To the memory of my mother,
who inspired me to become a physicist

Contents

Preface

Over the past 20 years, string theory has started to branch out and has tried to tackle difficult problems in several areas, mostly through the advent of the AdS/CFT correspondence. While at the beginning the sought-for applications were mostly in particle physics and cosmology, over the last 15 years or so we have seen a gradual increase in the number of applications to condensed matter theory, to the point that now various string theorists work completely in condensed matter. The purpose of this book is to provide an introduction to the various methods that have been developed in string theory for condensed matter applications, and to be accessible to graduate students just beginning to learn about either string theory or condensed matter theory, with the aim of leading them to where they can start research in the field. I assume a solid working knowledge of Quantum Field Theory, as can be obtained from a two-semester graduate course, and an advanced undergraduate course on Solid State physics, as well as some basic elements of General Relativity (but not necessarily a full course). Familiarity with string theory or modern condensed matter theory is helpful, but not necessary, since I try to be as self-consistent as possible. To that end, in Part I, I give an introduction to modern topics in condensed matter from the perspective of a string theorist. In Part II, I give a very basic introduction to general relativity and string theory, mostly for the benefit of people who haven't seen them before, as they are not very detailed. Parts III and IV then describe the string theory applications to the condensed matter problems of Part I. The goal is to give an introduction to the various tools available, but I will not try to be extensive in my treatment of any of them. Instead, my aim is to have a fair overview of all the methods currently available in the field. Part III describes tools that are by now standard: the pp wave correspondence, spin chains and integrability, AdS/CFT phenomenology ("AdS/CMT"), and the fluid-gravity correspondence. Part IV focuses on more advanced topics that are still being developed, like Fermi liquids and non-Fermi liquids, insulators, the quantum Hall effect, nonstandard statistics, etc.

Acknowledgments

I have to start by thanking all the people who have guided me along the path of becoming a physicist, without whom it would have been impossible to even think about writing this book. My mother, Ligia, who first inspired me in the love of physics, through her example as a physicist. My high school physics teacher, Iosif Sever Georgescu, who showed me that I could be successful at physics and make a career out of it. Poul Olesen at the Niels Bohr Institute, who during a student exchange period under his supervision introduced me to string theory. My Ph.D. advisor, Peter van Nieuwenhuizen, from whom I learned not just about supergravity and string theory, but about the rigor and beauty of theoretical physics, and the value of perseverance in calculations. Juan Maldacena at IAS from whom I learned, during my postdoc years, more about the AdS/CFT correspondence, the basis for most of the methods in this book.

This book is an expanded version of a course I gave at the IFT in São Paulo, so I would like to thank all the students who participated in the course for their input that helped shape the material.

I have to thank of course all my collaborators with whom I developed my research which led to my being able to write this book. I have also to thank my students and postdocs for their patience in dealing with my reduced time to work with them. A special thanks to my Ph.D. student Heliudson de Oliveira Bernardo, who helped me get rid of typos and errors in equations by carefully checking a previous version of the book. To my editor at Cambridge University Press, Simon Capelin, thank you for believing in me and helping me to get the book published, and to all the staff at CUP for making sure the book is up to standard.

Introduction

Before learning about string theory methods for condensed matter physics, we must define the problems we want to tackle and what string theory is. The interesting and difficult problems in condensed matter theory are in general strong coupling ones, though not always. In the first part of the book we will therefore learn about a variety of problems of topical interest, like Fermi liquids, spin chains, integrable systems, conformal field theories, quantum phase transitions, superconductivity, the quantum Hall effect, the Kondo problem, and hydrodynamics.

We then need to define string theory and its main low-energy tool, general relativity. After defining general relativity, we will focus on the objects of most interest, black holes and their extensions, that will be used extensively in the book. String theory is then defined, mostly as a perturbative theory of strings, a field theory on the 2-dimensional "worldsheets" spanned by the string. But we will also learn that there are various nonperturbative "dualities," where one description is mapped to another, in terms of other variables and valid in other regions of parameter space. We will also learn about the most important tool available for condensed matter applications, the AdS/CFT correspondence.

The AdS/CFT correspondence relates string theory, usually in its low-energy version of *supergravity*, a supersymmetric theory of gravity, and living in a curved background spacetime, to field theory in a flat spacetime of lower dimensionality. The correspondence is *holographic*, which means that in some sense, the physics in the higher dimension is projected onto a flat surface without losing information, at the price of encoding the information nonlocally.

The original incarnation of the AdS/CFT correspondence defined by Juan Maldacena related a certain field theory, 4-dimensional "$\mathcal{N} = 4$ SYM," to string theory in a curved background, $AdS_5 \times S^5$. In that case, although Maldacena's presentation emphasized it is a conjecture, the overwhelming number of checks that were performed since then makes it a pedantic point to still call it a conjecture; to all intents and purposes it is proven. Similar comments apply to the first (to my knowledge) application to condensed matter, the "pp wave correspondence" and the generalization to spin chains, started by the work of Berenstein, Maldacena, and myself, as an off-shoot of the original correspondence.

However, more common applications to condensed matter usually are of two types: "top-down" constructions, where one starts with a duality defined in string theory, with fewer checks than in the original case, but nevertheless defined beyond the level of conjecture. In that case, however, it is usually less clear why the field theory can be applied to the condensed matter problem of interest. Or "bottom-up" constructions, where one starts with a field theory we would like to describe, for condensed matter reasons, but then one guesses a possible holographic dual gravitational (string) theory, in which case the duality itself is less than clear. The point of view most commonly stated is that either way, one finds nice gravitational ways to parametrize the condensed matter problem, which makes it less important whether we can actually rigorously prove that the gravitational description is rigorously equivalent to the condensed matter problem.

There are other cases as well. Sometimes it is possible to view the condensed problem in a different way by simply embedding it into string theory. This is the case of Fermi gas constructions, particle-vortex duality, and Chern-Simons and Quantum Hall Effect constructions pre-dating holographic models. Yet in other cases like the fluid/gravity correspondence, which includes a mapping of the Navier-Stokes equations of viscous fluid hydrodynamics to the Einstein equations near the event horizon of a black hole, there are ways to argue for the map, but it is not clear that the new description is better than the original one, just different.

There are a number of reviews on various string theory applications to condensed matter, but they each mostly focus on one particular aspect and assume a lot of background. These include [1], [2], [3], and [4]. The book [5] presents many, but not all, of the applications described here and spends less time on the background material needed to understand it.

PART I

CONDENSED MATTER MODELS AND PROBLEMS

1 Lightning Review of Statistical Mechanics, Thermodynamics, Phases, and Phase Transitions

In Part I, I will describe various condensed matter models and problems of interest. In other words, we will study the condensed matter issues that can be described in one way or another using string theory methods.

To set up the notation, in Chapter 1, I will make a lightning review of thermodynamics, phase transitions, and statistical mechanics. These are issues that are supposed to be known, but we will review them in a way that will be useful for us later and in order to have a common starting point.

1.1 Note on Conventions

In most of this book, I will use field theorists, conventions, with $\hbar = c = 1$, unless needed to emphasize some quantum or (non)relativistic issues. We can always reintroduce $\hbar$ and c by dimensional analysis, if needed. In these conventions, there is only one dimensionful unit, namely $mass = 1/length = energy = 1/time = \cdots$. When I speak of dimension of a quantity, I refer to mass dimension.

For the Minkowski metric $\eta^{\mu\nu}$ I use the mostly plus signature convention, so in the most relevant case of 3+1 dimensions the signature is $(-+++)$, for $\eta^{\mu\nu} = diag(-1, +1, +1, +1)$.

I also use the Einstein summation convention, i.e. repeated indices are summed over. The repeated indices will be one up and one down, unless we are in Euclidean space, when it doesn't matter, so we can put all indices down.

1.2 Thermodynamics

In thermodynamics, we use two types of quantities:

• *Intensive quantities*, which are quantities that are independent of the size of the system. The relevant examples for us are

$$T, P, \vec{E}, \vec{H}, \mu_\alpha, \{P_j\}. \tag{1.1}$$

Here T is the temperature, P is the pressure, $\vec{E}$ is the electric field, $\vec{H}$ is the magnetic field, μ_α are chemical potentials for the particle species α, i.e. the increase in energy required to add one particle to the system, and P_j are generalized pressures (such that $P_0 = P$).

- *Extensive quantities*, which are quantities which increase with the size of the system, usually with the volume (though it can also be with the surface area, for example). The relevant examples for us are

$$S, V, V\vec{D}, V\vec{B}, N_\alpha, \{X_j\}, \tag{1.2}$$

respectively; i.e. these are the conjugate quantities corresponding to the intensive quantities above. Here S is the entropy, V the volume, $\vec{D}$ is the electric induction, $\vec{B}$ is the magnetic induction, N_α are numbers of particles of species α, and X_j are generalized volumes (such that $X_0 = -V$).

1. The *first law of thermodynamics* is then the following statement about the energy differential:

$$dU = TdS - PdV + \sum_j P_j dX_j + \vec{E} \cdot d(V\vec{D}) + \vec{H} \cdot d(V\vec{B}) + \sum_\alpha \mu_\alpha N_\alpha. \tag{1.3}$$

For the electro-magnetic quantities, we need to add the *material relations* that relate the inductions to the fields:

- Electric case:

$$\vec{D} = \epsilon_0 \vec{E} + \vec{P} = \epsilon \vec{E}, \tag{1.4}$$

where ϵ_0 is the vacuum electric permittivity, ϵ is the electric permittivity in the material, and $\vec{P}$ is the polarization.

- Magnetic case:

$$\vec{B} = \mu_0(\vec{H} + \vec{M}) = \mu\vec{H}, \tag{1.5}$$

where μ_0 is the vacuum magnetic permeability, μ is the magnetic permeability in the material, and $\vec{M}$ is the magnetization.

From the first law and the material relations we can deduce the electric and magnetic energy densities (or reversely from the energy densities and material relations we can deduce the first law for the electro-magnetic case):

- in the vacuum:

$$\rho_e = \frac{\epsilon_0 \vec{E}^2}{2}$$
$$\rho_m = \frac{\mu_0 \vec{H}^2}{2}. \tag{1.6}$$

- in the material:

$$\rho_e = \frac{\epsilon \vec{E}^2}{2} = \frac{\vec{D}^2}{2\epsilon}$$
$$\rho_m = \frac{\mu \vec{H}^2}{2} = \frac{\vec{B}^2}{2\mu}. \tag{1.7}$$

2. The *second law of thermodynamics* is the statement that the entropy always increases in a process, i.e. its variation is positive or zero:

$$\Delta S \geq 0. \tag{1.8}$$

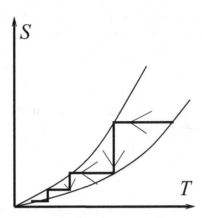

Fig. 1.1 Third law of thermodynamics.

3. The *third law of thermodynamics* is usually less known in its correct form. The statement is that as the temperature goes to zero, $T \to 0$, then the entropy goes to a constant, $S \to S_0$. Often it is stated as the fact that $S \to 0$, but there are in fact examples with nonzero entropy $S_0 \neq 0$, i.e. with a degenerate ground state, at zero temperature. One very important example for us will be the case of black holes, which later will be embedded in string theory.

An equivalent statement of the third law is that *It is impossible to reach $T = 0$ in a finite number of steps from $T \neq 0$.* To understand the equivalence, consider the (S, T) diagram, and two curves for the system, reaching the same $(S_0, 0)$ point, as in Figure 1.1. The most efficient cooling process is the Carnot process, i.e. an isothermal line ($T = \text{const.}$), followed by an isentropic line ($S = \text{const.}$), i.e. a vertical line, followed by a horizontal line. To reach $T = 0$ by moving on verticals and horizontals between the two curves, we can see that we need an infinite number of steps.

For expediency, we will include formally $\{\vec{E}, \vec{H}, \mu_\alpha\}$ into $\{P_j\}$ and $\{V\vec{D}, V\vec{B}, N_\alpha\}$ into $\{X_j\}$.

• Then the *Euler equation* (obtained from the homogeneity properties of T and S) says that for a system with extensive quantities S and $\{X_j\}$, for $j = 1, \ldots, r$, we have

$$U = TS + \sum_{j=1}^{r} P_j X_j. \tag{1.9}$$

• Together with the first law, the Euler equation implies the *Gibbs-Duhem equation*:

$$S dT + \sum_{j=1}^{r} X_j dP_j = 0. \tag{1.10}$$

Thermodynamic Potentials

For many systems, it is useful to work with variables that include only part of the extensive quantities and the rest of the intensive quantities, i.e. with $\{P_l\}$, $l = 1, \ldots, m$ and

$\{X_{l'}\}$, $l' = m + 1, \ldots, r$. As always, that is done by performing a *Legendre transform* over X_l; i.e. one uses the following *thermodynamic potentials*:

$$\bar{U} = \bar{U}[P_1, \ldots, P_m] = U - \sum_{l=1}^{m} P_l X_l. \qquad (1.11)$$

Their differentials, i.e. the first law of thermodynamics for systems at constant $P_1, \ldots, P_m$, are given by

$$d\bar{U} = -\sum_{l=1}^{m} X_l dP_l + \sum_{l'=m+1}^{r} P_{l'} dX_{l'}. \qquad (1.12)$$

This is the general formalism, but there are special cases of thermodynamic potentials that are important and have been named before:

- The *free energy*, or *Helmholtz potential*,

$$F = U - TS. \qquad (1.13)$$

- The *enthalpy*,

$$H = U + PV, \qquad (1.14)$$

and we can also define a *generalized enthalpy*,

$$H^* = U - P_j X_j, \qquad (1.15)$$

where $j \neq 0$ and $j \neq$ chemical potential.

- The *free enthalpy*, or *Gibbs potential*,

$$G = U - TS + PV, \qquad (1.16)$$

and we can also define a *generalized Gibbs potential*,

$$G^* = U - TS - \sum_{l=1}^{m} P_l X_l. \qquad (1.17)$$

- The *grand-canonical potential*,

$$\Omega = U - TS - \sum_{\alpha=1}^{n} \mu_\alpha N_\alpha. \qquad (1.18)$$

The reason for considering the thermodynamic potentials is that *at equilibrium, for a system in contact with a reservoir that fixes $T, P_1, \ldots, P_n$*, i.e. the situation that we will denote by

$$\mathcal{S} \cup \mathcal{R}_{T, P_1, \ldots, P_n}, \qquad (1.19)$$

the equilibrium is obtained for the minimum of the thermodynamic potential $\bar{U}[T, P_1, \ldots, P_n]$.

1.3 Phase Transitions

One of the quantities that we are interested in, in order to describe a phase transition, is the dimensionality of the phase manifold. In order to obtain it, we have the *Gibbs phase rule*:

Consider a system with f phases, n chemical (nonreacting) components, and q nonchemical degrees of freedom, whose intensive parameters we fix (e.g., T, P). Then the *number of effective degrees of freedom = dimensionality of the phase manifold* is

$$v = n + q - f. \tag{1.20}$$

Example To understand it, consider the simplest case, of a system with two phases ($f = 2$), one component ($n = 1$), and we fix T and P, i.e. $q = 2$, then $v = 1 + 2 - 2 = 1$, i.e. the separation between the two phases is a line (1-dimensional) that can be, for instance, drawn in the (T, P) plane.

Phase transition types

There are two important classifications of phase transitions, the first being developed by **Ehrenfest**. It is based on the thermodynamic potential $\bar{U}$ for the system under consideration. According to it, we have phase transitions of

• *first order*, which means that the thermodynamic potential $\bar{U}$ is constant (continuous) across the phase transition, but its first derivatives are not, i.e. they are discontinuous across the phase transition, $\partial \bar{U} / \partial a_j \neq$ const.

• *second order*, which means that the thermodynamic potential $\bar{U}$ is constant across the phase transition, as are its first derivatives $\partial \bar{U} / \partial a_j =$ const., but its second derivatives are discontinuous across the phase transition, $\partial^2 \bar{U} / \partial a_i \partial a_j \neq$ const.

In principle, we could go on and define third order (only the third order derivative of the thermodynamic potential being discontinuous), etc., but no such phase transition was found until now, so most likely there is no other type of phase transition.

The other important classification is due to **Landau**, who showed that the first and second order phase transitions have a different interpretation, and the difference between them can be described as follows.

• *first order* is when the phases differ only quantitatively, but not qualitatively.

• *second order* is when the phases also differ qualitatively. Specifically, there exists a so-called *order parameter* that we will denote by ψ, such that $\psi = 0$ in one phase that is symmetric under some symmetry, and $\psi \neq 0$ in the other phase, where we have no such symmetry, i.e. in the asymmetric phase.

Example of first order phase transition

Consider a system in contact with a (T, P) reservoir, i.e. $\mathcal{S} \cup \mathcal{R}_{T,P}$. Then the thermodynamic potential is the Gibbs potential $G = G(T, P) = U - TS + PV$, with $dG = -S dT + V dP + \cdots$. If the phase transition is first order, $G(T, P)$ is continuous across the phase transition, but its first derivatives,

$$S = -\left(\frac{\partial G}{\partial T}\right)_P \quad \text{and} \quad V = \left(\frac{\partial G}{\partial P}\right)_T, \tag{1.21}$$

are discontinuous, i.e. we have nonzero ΔV and ΔS, thus nonzero $\Lambda_{12} \equiv T\Delta S$. They are in fact related by the *Clausius-Clapeyron relation*: Across the coexistence line of the two phases (i.e. for $V_1 \leq V \leq V_2$), we have

$$\left.\frac{\partial P}{\partial T}\right|_{\text{phase trans.}} = \frac{\Delta S}{\Delta V} = \frac{\Lambda_{12}}{T(V_2 - V_1)}. \tag{1.22}$$

Example of second order phase transition

For the same system in contact with a (T, P) reservoir, the Gibbs potential $G(T, P)$ is again the relevant thermodynamic potential, and it and its first derivatives, S and V, are continuous, but its second derivatives are discontinuous, thus the *specific heat*,

$$C_P = \frac{T}{\nu}\left(\frac{\partial S}{\partial T}\right)_P = -\frac{T}{\nu}\left(\frac{\partial^2 G}{\partial T^2}\right)_P \tag{1.23}$$

is discontinuous. Here ν is the number of moles, actually defining the molar specific heat.

We can also consider a system in contact with a reservoir of magnetic and/or electric field $\vec{H}, \vec{E}$. In that case, the second derivatives of the thermodynamic potential are the following susceptibilities:

- the *magnetic susceptibility* χ_m, defined as

$$\chi_m = \frac{\partial M}{\partial H} = \frac{1}{\mu_0}\frac{\partial B}{\partial H} - 1 = -\frac{1}{\mu_0}\left(\frac{\partial^2 \bar{U}}{\partial H^2}\right) \tag{1.24}$$

is discontinuous, and

- the *electric susceptibility*

$$\kappa_e = \frac{1}{\epsilon_0}\frac{\partial P}{\partial E} = \frac{1}{\epsilon_0}\frac{\partial D}{\partial E} = -\frac{1}{\epsilon_0}\left(\frac{\partial^2 \bar{U}}{\partial E^2}\right) \tag{1.25}$$

is also discontinuous.

1.4 Statistical Mechanics and Ensembles

Statistical Mechanics

We now review the basics of statistical mechanics. We consider first the classical case, when there is a distribution function in the N-particle phase space. The one-particle phase space is $\{\vec{r}_k, \vec{p}_k\}$, and the N-particle phase space is shortened as $\vec{r}^N \equiv \{\vec{r}_1\vec{r}_2, \ldots, \vec{r}_N\}$.

The infinitesimal probability to be in phase space (around the point $(\vec{r}^N, \vec{p}^N)$) at time t is

$$dP(\vec{r}^N, \vec{p}^N, t) = \mathcal{P}(\vec{r}^N, \vec{p}^N, t)d\Gamma_N = \frac{d\mathcal{N}}{\mathcal{N}}, \tag{1.26}$$

where $\mathcal{N}$ is the number of particles (out of which $d\mathcal{N}$ are in the given state), $\mathcal{P}$ (or also ρ later) is the *distribution function of the statistical ensemble*, and $d\Gamma_N = d\vec{r}^N d\vec{p}^N$ is the infinitesimal volume element of phase space.

We also define the quantities:

- The phase space volume for maximum energy E,

$$\Gamma_N(E) = \int_0^E d\Gamma_N(E),   \tag{1.27}$$

- the phase space volume between E and $E + \Delta E$,

$$\Omega_N(E, \Delta E) = \Gamma_N(E + \Delta E) - \Gamma_N(E),   \tag{1.28}$$

- and the density of states,

$$\omega(E) = \frac{\partial \Gamma(E)}{\partial E}.   \tag{1.29}$$

Actually, to obtain the correct *number of states*, we should divide the phase space volume by $(N!h^{3N})$, the quantum unit of phase space times the symmetry factor for N particles.

In fact, above, when we called $\mathcal{P}$ the distribution function of the statistical ensemble, we have implicitly used the *ergodic hypothesis*, stating that the temporal average should equal the ensemble average,

$$\langle f \rangle_{\text{temporal}} = \langle f \rangle_{\text{ensemble}}.   \tag{1.30}$$

This is an ergodic postulate, associated with Gibbs and Tollman (who formalized statistical mechanics by basing it on postulates from which one can find everything else).

The next important assumption is the one of equilibrium, when there is no time dependence. While there are many interesting things about systems out of equilibrium, like in the case of heavy ion collisions, for instance, in this course we will stick with the assumption of equilibrium. We then have the *postulate of a priori equal probabilities*, which states that: The probability density is constant in the allowed domain $\mathcal{D}$ in phase space and zero outside it.

From the above postulates, we can obtain that the distribution function depends explicitly only on the total energy of the system E, and depends on phase space only implicitly, i.e. that

$$\mathcal{P}(\vec{r}^N, \vec{p}^N) = f\left(E(\vec{r}^N, \vec{p}^N)\right).   \tag{1.31}$$

Statistical Ensembles

We now review the most important statistical ensembles.

Microcanonical

The first one can be immediately derived from the above statement that the distribution function depends explicitly only on E. We consider then a constant energy $E = E_0$, thus having a probability $P(E(\dots))$ constant for $E = E_0$ and zero for $E \neq E_0$. It is in fact correct to consider a *quasi-microcanonical ensemble*, with energy $E \in (E_0, E_0 + \Delta E)$.

Boltzmann Formula

The next step is in some sense a postulate, meaning that it cannot be truly derived, but must be postulated, and in fact there might be alternatives to it for some systems. The postulate is Boltzmann's formula relating entropy with the distribution function. We can relate S to the logarithm of the number of states (remembering that, up to the factor of $1/(N!h^{3N})$, the total volume of phase space for the quasi-microcanonical case, Ω, is the number of states). The result is

$$S = k_B \ln \Omega, \tag{1.32}$$

where k_B is Boltzmann's constant. In the thermodynamical large N limit, we have also the equivalent formulas

$$S = k_B \ln \omega = k_B \ln \Gamma(E) = -k_B \langle \ln \rho \rangle. \tag{1.33}$$

The entropy is $S = S(E, V, N, \ldots)$. Note that the above equality is a bit counter-intuitive, but correct, since in the large N limit, the volume of the $3N$-dimensional space bounded by E is approximately equal to the volume of the space between E and $E + dE$, as we can check.

Since the first law is written as (renaming U as E)

$$dS = \frac{1}{T}dE + \frac{P}{T}dV - \frac{\mu}{T}dN + \cdots \tag{1.34}$$

by combining it with the Boltzmann formula, we obtain the relations ($\beta \equiv 1/(k_B T)$):

$$\beta = \left(\frac{\partial \ln \Omega}{\partial E}\right)_{V,N} ; \quad \beta P = \left(\frac{\partial \ln \Omega}{\partial V}\right)_{E,N} ; \quad -\beta \mu = \left(\frac{\partial \ln \Omega}{\partial N}\right)_{E,V}. \tag{1.35}$$

Canonical

Consider the system in contact with a reservoir of temperature, $\mathcal{S} \cup \mathcal{R}_T$, in which case the distribution function is

$$\rho(\mathcal{H}) = \frac{e^{-\beta \mathcal{H}}}{Z(\beta, V, N, \ldots)}, \tag{1.36}$$

where the $\mathcal{H}$ is the Hamiltonian and the *partition function Z* is simply the sum over states of the numerator,

$$Z = \sum_n g_n e^{-\beta E_n}, \tag{1.37}$$

where g_n is the degeneracy of the energy E_n.

The thermodynamical potential is the free energy and is given in terms of the partition function as

$$F(\beta, V, N) = -k_B T \ln Z. \tag{1.38}$$

Grand-Canonical

Consider a system in contact with a temperature and a particle (chemical potential) reservoir, $\mathcal{S} \cup \mathcal{R}_{T,\mu}$. The distribution function is then

$$\rho(\mathcal{H}) = \frac{e^{-\beta(\mathcal{H}-\mu N)}}{Z(\beta, \beta\mu, \ldots)}, \tag{1.39}$$

and the partition function is

$$Z(\beta, \beta\mu, V, \ldots) = \sum_{N \geq 0} e^{-\beta(\mathcal{H}-\mu N)} \frac{Z_1^N}{f}. \tag{1.40}$$

Here Z_1 is a single-particle partition function, and $f = N!$ for free particles and $f = 1$ for interacting particles.

Isothermal-Isobaric

For the case $\mathcal{S} \cup \mathcal{R}_{T,P}$, the distribution function is

$$\rho = \frac{e^{-\beta(\mathcal{H}+PV)}}{Z(\beta, \beta P, \ldots)} \tag{1.41}$$

and the partition function is given as an integration over the volume (which is a continuous variable, thus integrated instead of summed)

$$Z(\beta, \beta P) = \int_0^\infty dV e^{-\beta PV} Z_{\text{can}}(\beta, V, N), \tag{1.42}$$

and $Z_{\text{can}}(\beta, V, N)$ is the canonical partition function.

The thermodynamic potential is the Gibbs potential and is again given in terms of the partition function as

$$G(T, P, N) = -k_B T \ln Z. \tag{1.43}$$

Generalized Canonical

The distribution function is

$$\rho = \frac{e^{-\beta\mathcal{H}+\sum_{j=1}^n \beta P_j X_j (-\beta PV + \beta\mu N)}}{Z}, \tag{1.44}$$

where in brackets we have considered the case that we write explicitly the P and μ variables, and the partition function is

$$Z = \left(\int_0^\infty dV \sum_{N \geq 0} e^{-\beta PV + \beta\mu N} \right) \int d\Gamma_{V,N} e^{-\beta\mathcal{H}_{V,N} + \sum_{j=1}^n \beta P_j X_j}$$
$$= Z(\beta, (\beta P, \beta\mu), \beta P_1, \ldots, \beta P_n, X_{n+1}, \ldots, X_r). \tag{1.45}$$

Note that if we don't put the P and μ in the reservoir, Z depends on V and N instead.

Then the thermodynamic potential is again given in terms of the partition function as before:

$$\bar{U} = -k_B T \ln Z(\beta, \beta P_1, \ldots, \beta P_n, X_{n+1}, \ldots, X_r). \tag{1.46}$$

The total energy is the statistical average of the sum of the energies, $U = \langle E \rangle$, and by writing it explicitly in terms of the distribution function, we see that we can write it in terms of the derivative of $\ln Z$:

$$U = \langle E \rangle = -\frac{\partial \ln Z}{d\beta}. \tag{1.47}$$

Similarly, the extensive quantity X_j is the average of some A_j, $X_j = \langle A_j \rangle$, and by writing it explicitly in terms of the distribution function, we see that we can write is as a derivative,

$$X_j = \langle A_j \rangle = \frac{\partial \ln Z}{\partial \beta P_j}, \tag{1.48}$$

for $j = 1, \ldots, n$. For the other values, $l = n+1, \ldots, r$, we obtain

$$P_l = -\frac{1}{\beta} \frac{\partial \ln Z}{\partial X_l}. \tag{1.49}$$

1.5 Distributions

Classical Distribution: Maxwell-Boltzmann

We now analyze the standard distributions, starting with the classical distribution of Maxwell-Boltzmann. The total probability is the product of the individual probabilities, so the same holds for the distributions functions $\mathcal{P}$ (or ρ),

$$\mathcal{P} = \prod_i \mathcal{P}(\vec{r}_i, \vec{p}_i), \tag{1.50}$$

and the partition function is $Z = Z_1^N/N!$.

The distribution function for a single particle is

$$\mathcal{P}(\vec{r}, \vec{p}) = \frac{e^{-\beta E}}{\int \ldots \int d\vec{r} d\vec{p} e^{-\beta E}}. \tag{1.51}$$

Quantum Distributions

Consider one-particle states α, with occupation number n_α. Then the energy is

$$E = \sum_\alpha \epsilon_\alpha n_\alpha \tag{1.52}$$

and the number of particles is

$$N = \sum_\alpha n_\alpha. \tag{1.53}$$

The quantum distribution function is then $\langle n_\alpha \rangle$. The partition function is

$$Z = \prod_\alpha Z_\alpha, \tag{1.54}$$

and the single-state partition function is

$$Z_\alpha = \sum_{n_\alpha} e^{-\beta n_\alpha(\epsilon_\alpha - \mu)}. \tag{1.55}$$

We can now specialize for bosons and fermions.

Bose-Einstein Distribution

For bosons, we can have any occupation number for a state, i.e. $n_\alpha \in (0, \infty)$. In that case we obtain

$$Z_\alpha = \frac{1}{1 - e^{-\beta(\epsilon_\alpha - \mu)}}. \tag{1.56}$$

The thermodynamic potential is the grand-canonical potential, which equals

$$\Omega = -k_B T \ln Z. \tag{1.57}$$

From the general relations derived earlier, we have

$$\langle N \rangle = -\frac{\partial \Omega}{\partial \mu}$$
$$\equiv \sum_\alpha \langle n_\alpha \rangle. \tag{1.58}$$

Doing the calculation, we obtain

$$\langle n_\alpha \rangle = \frac{1}{e^{-\beta(\epsilon_\alpha - \mu)} - 1} \equiv f_{BE}(\epsilon_\alpha). \tag{1.59}$$

Fermi-Dirac Distribution

For fermions, a state can be either occupied or unoccupied, i.e. $n_\alpha = 0, 1$. Then we get

$$Z_\alpha = 1 + e^{-\beta(\epsilon_\alpha - \mu)}. \tag{1.60}$$

Using the same formula (1.58) as above for this new Z_α, we get

$$\langle n_\alpha \rangle = \frac{1}{e^{\beta(\epsilon_\alpha - \mu)} + 1} \equiv f_{FD}(\epsilon_\alpha). \tag{1.61}$$

We note that at $\beta \to \infty$, both the Bose-Einstein and the Fermi-Dirac distributions tend to the Maxwell-Boltzmann one.

We also note that at $T \to 0$, the Fermi-Dirac distribution becomes

$$\langle n_\alpha \rangle = 1, \quad \epsilon_\alpha < \mu$$
$$= 0, \quad \epsilon_\alpha > \mu, \tag{1.62}$$

and then this zero temperature chemical potential, $\mu(T = 0)$, is called ϵ_F, the Fermi energy. In other words, we get

$$\langle n_\alpha \rangle = \theta(\epsilon_F - \epsilon_\alpha). \tag{1.63}$$

Important Concepts to Remember

- All of the concepts in this chapter are important, since it is a very condensed review.

Further Reading

Any thermodynamics and statistical mechanics textbook.

Exercises

(1) Calculate S and U for the Fermi-Dirac and the Bose-Einstein distributions, as a function of $\{\epsilon_\alpha\}$.

(2) Consider a relativistic free particle gas, with $\epsilon^2 = \vec{p}^2 + m^2$. Calculate the number density $n = N/V$ and the energy density $\rho = E/V$ as a 1-dimensional integral. Solve it for the ultrarelativistic limit, $T \gg m, \mu$, for the BE and FD distributions.

(3) Consider that the endpoints V_1 and V_2 of a first order phase transition line lie on the curve

$$V = V_0 - \alpha(T - T_c)^2. \tag{1.64}$$

Calculate $P(T)$ for the phase transition.

(4) Consider a classical system $\mathcal{S} \cup \mathcal{R}_{T,P}$ and $E_n = k \cdot n$. Calculate S and V using the natural ensemble for the system.

(5) Consider an ideal nonrelativistic monoatomic gas, with

$$\mathcal{H}_1 = \frac{\vec{p}^2}{2m} + U(\vec{R}), \tag{1.65}$$

and the density of states

$$\Gamma(E) = \frac{1}{N! h^{3N}} \mathcal{V}(E), \tag{1.66}$$

and $\mathcal{V}(E)$ the N-atom phase space volume of energy E. Calculate the entropy $S(E, V, N)$ and the energy U.

(6) Consider a system of harmonic independent isotropic oscillators, with

$$\mathcal{H}_1 = \frac{\vec{p}^2}{2m} + \frac{m\omega^2 \vec{r}^2}{2}, \tag{1.67}$$

and the density of states defined via a quantum unit of energy $\hbar\omega$. Prove that

$$U = 3Nk_B T. \tag{1.68}$$

(7) Consider the system of electric dipoles with

$$\mathcal{H}_1 = \frac{\vec{p}^2}{2m} + \frac{1}{2I}\left(p_\theta^2 + \frac{p_\phi^2}{\sin^2\theta}\right) \tag{1.69}$$

and polarization along the electric field

$$\mathcal{P}_z = \epsilon\delta\cos\theta_i. \tag{1.70}$$

Show that the electric susceptibility satisfies the "Curie law"

$$\chi_e = \frac{N}{V}\frac{\delta^2}{3\epsilon_0 k_B T}. \tag{1.71}$$

2

Magnetism in Solids

In this chapter I will describe the various types of magnetism in solids and the theories for them. We start by remembering the terminology. The relation between the magnetic induction $\vec{B}$ and the magnetic field $\vec{H}$ is given by

$$\vec{B} = \mu_0(\vec{H} + \vec{M}) = \mu\vec{H}. \tag{2.1}$$

The magnetization $\vec{M}$ is related to the magnetic field $\vec{H}$ by the relation

$$\vec{M} = \chi\vec{H}, \tag{2.2}$$

which in general is matriceal, i.e. χ is in general a matrix, but here we consider only the scalar case. χ is called the magnetic susceptibility. On the other hand, we have

$$\frac{\mu}{\mu_0} = 1 + \chi \equiv \mu_r \quad \text{or} \quad \kappa_m. \tag{2.3}$$

The magnetic field is vorticity-free, $\vec{\nabla} \times \vec{H} = 0$.

The mean field (magnetic induction) in vacuum, or the "applied magnetic field" is

$$\vec{B}_0 = \mu_0\vec{H}. \tag{2.4}$$

Note that Kittel [6] uses the definition for the magnetic susceptibility

$$\tilde{\chi} = \chi_p = \frac{M}{B/\mu_0} = \frac{M}{H + M} = \frac{\chi}{1 + \chi}. \tag{2.5}$$

If χ is very small, then the difference between the two definitions is negligible.

For comparison, we remind the reader what happens in the electric case. There one has the electric induction

$$\vec{D} = \epsilon_0\vec{E} + \vec{P} = \epsilon\vec{E}, \tag{2.6}$$

where $\vec{P}$ is the polarization vector, and

$$\epsilon = \epsilon_0\epsilon_r; \quad \epsilon_r - 1 = \kappa_e \tag{2.7}$$

is the dielectric susceptibility.

2.1 Types of Magnetism: Diamagnetism, Paramagnetism, (Anti)Ferromagnetism, Ferrimagnetism

The magnetism can be divided in three types, which divide into further subtypes.

1. Diamagnetism

It has $\chi < 0$, is independent of $\vec{H}$, and is very small: $\chi_{\text{dia}} \sim 10^{-5} - 10^{-6}$. $\chi < 0$ is somewhat counterintuitive, since it means that the magnetization opposes the original field.

It can have two different origins, which we will consider separately:

(a) *Diamagnetism of bound electrons (in an atom)*. It is also called *Langevin, or orbital, diamagnetism*.

- It is present in all materials, but since it is very small, it can be overwhelmed by other phenomena.
- It is caused by the precession of the orbital $\vec{L}$ or spin $\vec{S}$ angular momentum around the axis of the field $\vec{B}$.
- One can treat the system classically, as Langevin did, or quantum mechanically.

(b) *Landau diamagnetism, of free electrons in metals*. It is due to quantum effects.

2. Paramagnetism

It has $\chi > 0$, is independent of the magnetic field, and again is very small: $\chi_{\text{para}} \sim 10^{-3} - 10^{-5}$. This is somewhat more intuitive, since it means the magnetic moments align themselves with the magnetic field.

(a) *Paramagnetism of bound electrons*, described by *Brillouin*. It is due to the orienting of intrinsic magnetic moments $\vec{\mu}$ along the magnetic field $\vec{B}$. Initially the magnetic moments are oriented randomly (chaotically), because of thermal fluctuations, and of having weak interactions among themselves, and the magnetic field aligns them.

The paramagnetic χ obeys the "Curie law"

$$\chi_{\text{para}} = \frac{C}{T},\tag{2.8}$$

where C is a constant, and paramagnetism can be described both classically (as Langevin did) or quantum mechanically (as Brillouin did).

(b) *Paramagnetism of free electrons* in metals, due to spin, or *Pauli paramagnetism*. It is of an intrinsically quantum nature, being due to spin effects, and gives

$$\chi_{\text{para}}(T) \simeq \text{const.}\tag{2.9}$$

(c) *Paramagnetism of polarization (van Vleck paramagnetism)*. It arises for materials with asymmetric external shells (orbitals), and it also gives

$$\chi_{\text{para}}(T) \simeq \text{const.}\tag{2.10}$$

3. Magnetism of Materials with Magnetic Order

It is for materials that have a intrinsic (proper) magnetic moments $\vec{\mu}$ with strong interactions between them, predominating over the disorder that would tend to randomize them.

(a) Ferromagnetism

In this case, the magnetic moments $\vec{\mu}$ are ordered even in the absence of a magnetic field, leading to a *spontaneous magnetization M_s*. The magnetic susceptibility χ is still positive, like for paramagnetism, but now it depends on the magnetic field H and temperature T. Moreover, instead of being very small, it is now very large, $\chi_{\text{ferro}} \sim 10^3 - 10^1$.

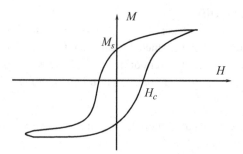

Hysteresis curve: the value of the magnetization $M(H)$ depends on the history.

The first, phenomenological, model of spontaneous magnetization was due to Weiss in 1907, and we will describe it in this chapter.

There is a hysteresis curve for the variation of M with H, i.e. the variation in the (M, H) plane depends on its history, as in Figure 2.1. It passes through points of zero field $\vec{H} = 0$, but nonzero magnetization M_s, and no magnetization, but nonzero field H_c.

For temperatures larger than a critical temperature $T > T_C$, called the Curie temperature, the ferromagnet becomes a paramagnet obeying the "Curie-Weiss law,"

$$\chi_{\text{C-W}} = \frac{C}{T - T_C},\tag{2.11}$$

as in Figure 2.2.

(b) Antiferromagnetism

In this case, again χ depends on the field $\vec{H}$ and the temperature T. However, now at $T = 0$, the exchange interactions orient the magnetic moments sequentially antiparallel, so in effect there are two subnetworks of opposite moments, with no total magnetization $M = 0$. But at nonzero T, we have a nonzero spontaneous magnetization $M_s \neq 0$.

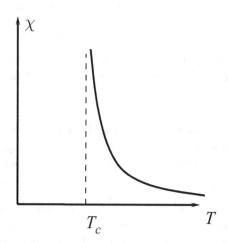

Curie law for the ferromagnet.

For temperatures higher than a critical one, $T > T_N$, called the *Neél temperature*, we have a paramagnet, obeying the "Curie-Neél law":

$$\chi_{C-N} = \frac{C}{T + T_N}.$$ (2.12)

(c) Ferrimagnetism

Finally, we have ferrimagnetism, which is a sort of uncompensated antiferromagnetism. That is, we have two subnetworks of different moments, that do not cancel each other, so at $T = 0$ we still have a spontaneous magnetization.

2.2 Langevin Diamagnetism

We now describe the theory of the various types of magnetism described above, starting with Langevin diamagnetism.

Langevin Diamagnetism for Bound Electrons in Atoms

The first information we need is the fact that a rotating electron has a unit of magnetic moment called the Bohr magneton μ_B. For an electron rotating in a circle of radius r with area πr^2 the magnetic moment is

$$\mu = IA = \frac{ev}{2\pi r}\pi r^2 = \frac{evr}{2} = \frac{e}{2m}L.$$ (2.13)

But since the unit of angular momentum is $L = \hbar$, we have the unit of magnetic moment being the Bohr magneton

$$\mu_B = \frac{e\hbar}{2m}.$$ (2.14)

The second information needed is that in a magnetic field $\vec{B}$, an electron has a precession (on a helical motion) with the *Larmor frequency*:

$$\omega = \frac{eB}{2m}.$$ (2.15)

The current generated by the rotating electron is $I = e\omega/(2\pi)$, and the area enclosed by the circle is $A = \pi\langle\rho^2\rangle$, where ρ is the radius *transverse* to the magnetic field $\vec{B}$, so

$$\langle\rho^2\rangle = \langle x^2\rangle + \langle y^2\rangle = \frac{2}{3}\langle r^2\rangle,$$ (2.16)

where $\langle r^2\rangle = \langle x^2\rangle + \langle y^2\rangle + \langle z^2\rangle$. Here we have used that in quantum mechanics, the electron distribution is isotropic. The total moment of the material is obtained by multiplying with the number of atoms N and the charge of the atom Z,

$$\mu = (NZ)\frac{e^2}{6m}\langle r^2\rangle B.$$ (2.17)

From first principles we can calculate only up to here. For specific atoms, we can use quantum mechanics to calculate $\langle r^2\rangle$.

Landau Diamagnetism of Free Electrons in Metals

The electrons in metal are effectively in a situation of contact with a reservoir of temperature and particles, $S \cup R_{T,\mu}$, i.e. we are in the grand-canonical ensemble. Since this is a quantum effect due to the spin of the electrons, we must use the Fermi-Dirac distribution, and then the grand-canonical potential is (as seen in Chapter 1)

$$\Omega = -k_B T \sum_k 2 \ln \left[1 + e^{\beta(\mu - \epsilon_k)} \right]. \tag{2.18}$$

In the presence of a magnetic field $\vec{H}$, the electron forms *Landau levels*, with energy levels

$$\epsilon_n = (n + 1/2)\hbar\omega_c, \tag{2.19}$$

where $\omega_c = eB/m$. We will describe them further later in Section 10.3.

The grand-canonical potential Ω is written as sum over k states, but we can use the density of states $dN/d\epsilon = Am/(\pi\hbar^2)$ to write it as a sum over n, by a "discretized approximation":

$$\Omega = -k_B T \frac{Am}{\pi\hbar^2} \hbar\omega_c \sum_n \ln \left[1 + e^{\beta(\mu - \epsilon_n)} \right]. \tag{2.20}$$

One can evaluate the sum by approximating it in terms of integrals, and find that

$$\Omega(H) = \Omega(0) + \frac{e^2}{24\pi mc^2} H^2 + \frac{\partial\Omega}{\partial\mu} \delta\mu(H), \tag{2.21}$$

where the last term comes from the fact that the chemical potential μ itself depends on H, but can be ignored.

We can then derive M as $-\partial\Omega/\partial H$, and equate with χH to derive finally

$$\chi_{\text{dia}} = -\frac{e^2}{12\pi mc^2}. \tag{2.22}$$

2.3 Paramagnetism of Bound Electrons (Langevin-Brillouin)

We will use a quantum description of the magnetic moment (since, as we know, the classical magnetic moment of the electron is off by a factor of g, the Landé factor, with respect to the quantum moment) but the classical statistics of Maxwell-Boltzmann.

The interaction Hamiltonian between the magnetic moment and the magnetic field is $\mathcal{H}_m = -\vec{\mu} \cdot \vec{B}$.

The magnetization of a single electron is given by

$$\vec{M} = -\mu_B(\vec{L} + 2\vec{S}) = -g\mu_B\vec{J}, \tag{2.23}$$

where $\vec{L}$ is the orbital angular momentum, $\vec{J} = \vec{L} + \vec{S}$ is the total angular momentum, and the Landé g-factor is

$$g = 1 + \frac{J(J+1) + S(S+1) - L(L+1)}{2J(J+1)}. \tag{2.24}$$

The projection of the magnetization on the z axis is then $M_z = g\mu_B m_J$.

The energy levels ϵ are shifted by $B_0 M_z$, where $B_0 = \mu_0 H$ (the applied magnetic field, in vacuum):

$$\epsilon = \epsilon_0 - \mu_0 H g \mu_B m_J. \tag{2.25}$$

Using Maxwell-Boltzmann statistics, with $Z = Z_1^N$ (Z_1 being the partition function of a single electron), after some calculations, we get

$$M = N g J \mu_B B_J(x), \tag{2.26}$$

where

$$x = \frac{g \mu_B B_0}{k_B T} J \tag{2.27}$$

and $B_J(x)$ is a complicated function (given by an integral), called the *Brillouin function*. Two relevant limits are $J = 1/2$:

$$B_{1/2}(x) = \tanh x, \tag{2.28}$$

and for $x \ll 1$,

$$B_J(x) \simeq \frac{x}{3J}(J + 1). \tag{2.29}$$

From this limit we get

$$\chi = \frac{N g^2 \mu_B^2 \mu_0 J(J + 1)}{3 k_B} \frac{1}{T} \equiv \frac{C}{T}, \tag{2.30}$$

i.e. the Curie law.

2.4 Pauli Paramagnetism

Pauli paramagnetism is a paramagnetism of free, i.e. conduction, electrons in metals, and it is due to the spin of the electrons, so we must use the quantum Fermi-Dirac statistics.

The magnetization of a system of $n_\uparrow$ up spins and $n_\downarrow$ down spins is

$$M = \mu_B \left(\frac{1}{2} n_\uparrow - \frac{1}{2} n_\downarrow \right), \tag{2.31}$$

and the energy is (the free energy of the electrons plus interaction Hamiltonian given by $B_0 = \mu_0 H$)

$$\epsilon_{\uparrow, \downarrow} = \frac{\hbar^2 k^2}{2m} \pm \mu_B B_0. \tag{2.32}$$

The number of spins up and down is given by the zero-temperature FD distribution $f_0^{FD}(\epsilon)$ and the density of states (orbitals) $g_{\uparrow, \downarrow}(\epsilon)$, by

$$n_{\uparrow, \downarrow} = \int d\epsilon f_0^{FD}(\epsilon) g_{\uparrow, \downarrow}(\epsilon). \tag{2.33}$$

At zero temperature $T = 0$, the FD distribution is (as we saw in the last chapter) $f_0^{DF}(\epsilon_0) = \theta(\epsilon_F - \epsilon_0)$, so we obtain

$$n_\uparrow = \int_{-\mu_B B}^{\epsilon_F} d\epsilon_0 g_\uparrow(\epsilon_0 + \mu_B B) \simeq \int_0^{\epsilon_F} d\epsilon g_\uparrow(\epsilon) + \mu_B B g_\uparrow(\epsilon_F)$$

$$n_\downarrow = \int_{\mu_B B}^{\epsilon_F} d\epsilon_0 g_\downarrow(\epsilon_0 + \mu_B B) \simeq \int_0^{\epsilon_F} d\epsilon g_\downarrow(\epsilon) - \mu_B B g_\downarrow(\epsilon_F). \quad (2.34)$$

We will see in Chapter 3 that $g(\epsilon_F) = 3N/(2\epsilon_F)$, so we obtain $M = \mu_B/2(2\mu_B B g(\epsilon_F))$, giving

$$\chi = \frac{M}{H} = \frac{\mu_0 M}{B} = \frac{\mu_0 \mu_B}{B}[\mu_B B g(\epsilon_F)] = \frac{3N}{2}\frac{\mu_0 \mu_B^2}{k_B T_F}. \quad (2.35)$$

This was the zero temperature result $\chi_{\text{para}}^{T=0}$, but we can calculate more precisely and find that the result doesn't change much at nonzero temperature.

2.5 Van Vleck Paramagnetism

This type of paramagnetism is independent of temperature and arises in the case of asymmetrical external atomic shells in the material. In this case there are nondiagonal elements of μ_z (the projection onto the magnetic field of the magnetic moment) $\langle s|\mu_z|0\rangle$, where $|0\rangle$ is the ground state, and $|s\rangle$ is an excited state, with an energy gap $\Delta = E_s - E_0$.

Then, first order perturbation theory means that if the perturbation in energy due to the magnetic field is small, $\mu B \ll \Delta$, we have a rotation of (ψ_0, ψ_s) with the angle $\mu B/\Delta$:

$$\psi_0' \simeq \psi_0 + \frac{B}{\Delta}\langle s|\mu_z|0\rangle\psi_s$$

$$\psi_s' \simeq \psi_s - \frac{B}{\Delta}\langle 0|\mu_z|s\rangle\psi_0, \quad (2.36)$$

from which we find that ($\psi_0(x) = \langle x|0\rangle$ and $\psi_s(x) = \langle x|s\rangle$)

$$\langle 0'|\mu_z|0'\rangle \simeq \frac{2B}{\Delta}|\langle s|\mu_z|0\rangle|^2$$

$$\langle s'|\mu_z|s'\rangle \simeq -\frac{2B}{\Delta}|\langle s|\mu_z|0\rangle|^2. \quad (2.37)$$

Then, if we are at small temperatures, $k_B T \ll \Delta$, nearly all particles are in the ground state, so the total magnetization is given by N times (for N atoms) the matrix element of μ_z in the perturbed vacuum,

$$M \simeq N\langle 0'|\mu_z|0'\rangle, \quad (2.38)$$

so that finally

$$\chi = \frac{M}{B_0} \simeq \frac{2N}{\Delta}|\langle s|\mu_z|0\rangle|^2. \quad (2.39)$$

As we wanted, it is independent of temperature (for small temperatures).

2.6 Ferromagnetism

The first theory of ferromagnetism was the Weiss theory, which is a phenomenological theory, i.e. it does not involve a microscopic understanding, only some macroscopic input.
  Indeed, the Weiss theory starts with two hypotheses:

- There are domains of spontaneous magnetization, within which the magnetic moments are parallel and constant.
- There is a *molecular (or atomic) field* proportional to the magnetization, $\vec{H}_m = \lambda \vec{M}$.

This is an internal effective field, due to exchange interactions. Note that normally we have $\vec{M} = \chi \vec{H}$, but now we say that *at small scales* there is also an opposite effect, the magnetization generating a magnetic field. In turn this gives a magnetic induction $B_e = \mu_0 H_m = \mu_0 \lambda M$. Since also $\vec{M} = \chi \vec{H}$, so $\mu_0 \vec{H} = \chi \vec{B}_0$, where $\vec{B}_0 = \mu_0 \vec{H}$, now we write a similar relation in terms of the total microscopic magnetic induction,

$$\mu_0 M = \chi_p (B_a + B_e), \tag{2.40}$$

where χ_p is the paramagnetic susceptibility, $\vec{B}_a = \mu_0 \vec{H}$ is an *applied* magnetic induction, and $\vec{B}_e$ is the molecular field. We have for the usual susceptibility $\chi = \mu_0 M / B_a = M/H$, which means that we obtain

$$\chi_p = \frac{M}{H + H_e}. \tag{2.41}$$

The effective experienced magnetic field is

$$\vec{H}_{\text{effective}} = \vec{H} + \lambda \vec{M}, \tag{2.42}$$

and the applied magnetic induction at the microscopic level is

$$\vec{B}_{\text{micro}} = \vec{B}_a + \vec{B}_e = \mu_0 \vec{H}_{\text{effective}}. \tag{2.43}$$

For this quantity we can use paramagnetic theory, but *with the effective magnetic field* $\vec{H}_{\text{effective}}$, which is the one experienced by the magnetic moments. Therefore the magnetization is as before

$$M = N g \mu_B J B_J(a), \tag{2.44}$$

but where a is like x, but with $H_{\text{effective}}$ instead of H:

$$a = \frac{g \mu_B J}{k_B T} \mu_0 H_{\text{effective}}. \tag{2.45}$$

For $J = 1/2$, as we said, we have $B_{1/2}(a) = \tanh a$, and $\mu = \mu_B$.
  Since $H_{\text{effective}}$ has M in it, we have a transcendental equation for M. For no magnetic field, $H = 0$, we still obtain a nonzero magnetization, for $T < T_c$, which is

$$T_c = \frac{N \mu_B^2 \lambda}{k_B} \mu_0. \tag{2.46}$$

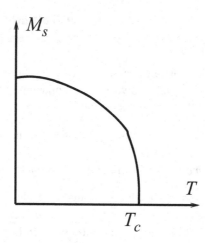

Fig. 2.3   Spontaneous magnetization at $H = 0$ as a function of temperature.

We leave the proof of this statement as an exercise. Nonzero magnetization at $H = 0$ means we obtain the required spontaneous magnetization M_s.

Also, in the case $J = 1/2$, as $x \to \infty$, $\tanh x \to 1$, so we obtain the maximum magnetization

$$M_{max} = N\mu_B. \tag{2.47}$$

On the other hand, for $T > T_c$, at $H = 0$ we have $M_s = 0$ (no spontaneous magnetization), and the $M_s(T)$ curve is represented in Figure 2.3. But if we introduce a nonzero magnetic field H, we obtain a paramagnetic behavior.

Then

$$\frac{M}{H_{\text{effective}}} = \chi_{\text{para}} = \frac{C}{T}. \tag{2.48}$$

Replacing $H_{\text{effective}} = H + \lambda M$, we obtain

$$\chi = \frac{M}{H} = \frac{\chi_p}{1 - \lambda \chi_p} = \frac{C}{T - T_c}, \tag{2.49}$$

where

$$T_c = C\lambda. \tag{2.50}$$

(For $J = 1/2$, $C = N\mu_B^2/k_B$.) This is the Curie-Weiss law.

The next model for ferromagnetism is a microscopic one, due to Heisenberg.

The Heisenberg model is very important, and we will come back to it later to describe it in much detail. For the moment we give its general form, as a spin-spin interaction with a coupling that depends on the distance between the spins, together with the interaction of the spins with the magnetic field:

$$\mathcal{H} = -\sum_{m,n} J(|\vec{R}_n - \vec{R}_m|)\vec{S}_n \cdot \vec{S}_m - g\mu_0\mu_B\vec{H} \cdot \sum_n \vec{S}_n. \tag{2.51}$$

We can rewrite this as

$$\mathcal{H} = - \left(\sum_n \vec{S}_n \cdot \vec{H}_{\text{effective}} \right) g\mu_0\mu_B, \tag{2.52}$$

where

$$\vec{H}_{\text{effective}} = \vec{H} + \frac{A\langle\vec{S}\rangle}{g\mu_0\mu_B} \tag{2.53}$$

and

$$A = \sum_m J(|\vec{R}_n - \vec{R}_m|). \tag{2.54}$$

The implicit assumption is that A is independent of n (if it isn't, the rewriting of the Hamiltonian is an identity).

2.7 Ferrimagnetism and Antiferromagnetism

We continue with the ferrimagnetic case, since it is more general than the antiferromagnetic one.

In this case there are two subnetworks of opposite spins, each with its own magnetization M. The interaction is of the antiferromagnetic type, i.e. the magnetization opposes the applied field (it comes with the opposite sign).

We can write the magnetic field in the most general case as a result of the magnetizations in the two subnetworks:

$$H_{m,A} = -\lambda M_A - \mu M_B$$
$$H_{m,B} = -\mu M_A - \nu M_B. \tag{2.55}$$

Considering only the molecular field due to the interaction of the two subnetworks (considering that there is no self-interaction), $\lambda = \nu = 0$, we have

$$M_{m,A} = -\mu M_B$$
$$M_{m,B} = -\mu M_A. \tag{2.56}$$

Since $M = \chi_p H_{\text{effective}}$, we obtain

$$M_A = \chi_{p,A}(H - \mu M_B)$$
$$M_B = \chi_{p,B}(H - \mu M_A). \tag{2.57}$$

At $H = 0$, having nonzero M_A and M_B means that the system of equations is degenerate:

$$\det \begin{pmatrix} 1 & \mu\chi_{p,A} \\ \mu\chi_{p,B} & 1 \end{pmatrix} = 0 \Rightarrow \mu^2\chi_{p,A}\chi_{p,B} = 1. \tag{2.58}$$

Since the paramagnetic susceptibilities are $\chi_{p,A} = C_A/T$ and $\chi_{p,B} = C_B/T$, we obtain

$$T = T_c = \mu\sqrt{C_A C_B}. \tag{2.59}$$

Note that this means that nonzero M_A and M_B is possible only at $H = 0$ for $T = T_c$. For $T > T_c$, we solve (2.57) for H and M_B in terms of M_A, and then we get

$$\chi = \frac{M_A + M_B}{H} = \frac{(C_A + C_B)T - 2\mu C_A C_B}{T^2 - T_c^2}. \tag{2.60}$$

Antiferromagnetism

This is a particular case of the antiferromagnetic case, with $C_A = C_B \equiv C$, giving a Néel temperature,

$$T_c = T_N = \mu C, \tag{2.61}$$

and a susceptibility obeying the Curie-Néel law,

$$\chi = \frac{2C}{T + T_N}. \tag{2.62}$$

The antiferromagnetic interaction has as a microscopic origin the Heisenberg Hamiltonian with $J < 0$, so that antiparallel spins ($\vec{S}_n \cdot \vec{S}_m < 0$) are energetically preferred (minimum energy).

Important Concepts to Remember

- Diamagnetism has $\chi < 0$ and is very small, paramagnetism has $\chi > 0$ and is small, and ferromagnetism has very large $\chi > 0$, a spontaneous magnetization, and a χ that depends on H and T.
- Antiferromagnetism is understood as having two subnetworks with oppositely oriented spins, and ferrimagnetism is uncompensated antiferromagnetism.
- Diamagnetism is of two types: of bound electrons in atoms (Langevin), due to precession of angular momenta around B, and of free electrons in metals (Landau), due to Landau level precession.
- Paramagnetism is of three types: of bound electrons in atoms (Brillouin), due to orienting of magnetic moments along $\vec{B}$ and satisfying the Curie law $\chi = C/T$; of free conduction electrons in metals (Pauli), due to spins; and of polarization (van Vleck), due to asymmetrical external shells.
- Ferromagnetism is phenomenologically described by the Weiss theory, in terms of spontaneous magnetization domains, and a molecular field $\vec{H}_m = \lambda \vec{M}$; in terms of $H_{\text{effective}} = H + H_m$, we have a paramagnetic theory, but with $\chi = C/(T - T_c)$ (the Curie-Weiss law).
- Ferromagnetism is microscopically described by the Heisenberg model, of spins interacting with a coupling depending on the distance.
- Ferrimagnetism has a Néel law $\chi = C/(T + T_N)$.
- Antiferromagnetism corresponds microscopically to the Heisenberg model with $J < 0$.

Further Reading

Chapters 11 and 12 in the book by Kittel [6].

Exercises

(1) Calculate the *Landau levels* for an electron in a uniform magnetic field B on the Oz axis, with $\vec{A} = -By\vec{e}_x$. Use the ansatz $\psi = \chi(y)e^{ik_x x}$ in the Schrödinger equation.

(2) Calculate the thermodynamic potential for Langevin-Brillouin paramagnetism.

(3) Prove that for the Weiss theory of ferromagnetism we have

$$T_c = \frac{N\mu_B^2 \lambda}{k_B}\mu_0. \tag{2.63}$$

(4) Calculate the first correction in T to Pauli paramagnetism (for χ_{para}).

(5) For Heisenberg's model used in Weiss's theory of ferromagnetism show that

$$T_c = \frac{AS(S+1)}{3k_B}, \tag{2.64}$$

where

$$A = \sum_m J(|\vec{R}_n - \vec{R}_m|). \tag{2.65}$$

(6) Show that at normal temperatures,

$$\frac{\chi_{\text{Curie}}}{\chi_{\text{Pauli}}} \sim 100. \tag{2.66}$$

(7) For a Weiss ferromagnet, show that at small T,

$$\frac{M_{spont.}}{M_{max}} \sim 1 - \frac{1}{J}e^{-\frac{3}{JH}\frac{T_c}{T}}, \tag{2.67}$$

and at large T,

$$\left(\frac{M_{spont}}{M_{max}}\right)^2 = \frac{10}{3}\frac{(J+1)^2}{(J+1)^2 + J^2}\left(\frac{T}{T_c}\right)^2\left(1 - \frac{T}{T_c}\right). \tag{2.68}$$

Electrons in Solids: Fermi Gas vs. Fermi Liquid

In this chapter, we will describe the Fermi gas and Fermi liquid theory, with applications for metals and semiconductors.

3.1 Free Fermi Gas

We start by considering only free electrons, which is the case relevant to (most) metals, specifically their conduction electrons.

For free fermions, we have seen that we have the Fermi-Dirac distribution:

$$f_{FD}(\epsilon) = \frac{1}{e^{\frac{\epsilon - \epsilon_F}{k_B T}} + 1}. \tag{3.1}$$

At $T = 0$, the chemical potential becomes the *Fermi energy*, $\epsilon_F = \mu(T = 0, n)$.

For semiconductors, besides the electrons with distribution $f_{0n}(\epsilon)$, we are interested in "holes," which are excitations given by the absence of one occupied Fermi state in the Fermi sea, the propagating "hole." As such, its distribution is the complement of the FD one:

$$f_{0p}(\epsilon) = 1 - f_{0n}(\epsilon) = \frac{1}{e^{\frac{\epsilon_F - \epsilon}{k_B T}} + 1}. \tag{3.2}$$

$T = 0$

Also at $T = 0$, the FD distribution becomes the Heaviside function $f_{FD}(\epsilon) \rightarrow \theta(\epsilon_F - \epsilon)$, and its derivative the delta function, $df_{FD}(\epsilon)/d\epsilon \rightarrow -\delta(\epsilon - \epsilon_F)$. In the following, we will therefore consider that the states follow this step function, with equal probability until ϵ_F, and zero thereafter.

The simplest case, which we consider first, is of the *free electron gas in one dimension*, on the interval $(0, L)$. We have the wavefunction $\psi_n = A \sin(k_n x)$, giving the nonrelativistic energy (the electrons in a metal are nonrelativistic)

$$\epsilon_n = \frac{\hbar^2 k_n^2}{2m}, \tag{3.3}$$

where the allowed momenta with the given boundary conditions are $k_n = n\pi/L$.

Electrons obey FD statistics, so according to the Pauli exclusion principle there is only one per state (that includes the two spin orientations $s = \pm 1/2$), so the Fermi wavenumber n_F (of electron k_n states) is related to the total number of electron states N by $n_F = N/2$.

The maximum wavevector k is given by $n = n_F$ and is

$$k_F = \frac{n_F \pi}{L}. \tag{3.4}$$

The allowed k's are $0 \leq k \leq k_F$. The Fermi energy as a function of k_F is then

$$\epsilon_F = \frac{\hbar^2 k_F^2}{2m}. \tag{3.5}$$

Three Dimensions

We now move to the 3-dimensional case, of a free electron system in the form of a cube with side L, with periodic boundary conditions and wavefunction $\psi_{\vec{k}}(\vec{r}) = e^{i\vec{k}\cdot\vec{r}}$. We still have the same relation between the Fermi energy and Fermi momentum,

$$\epsilon_F = \frac{\hbar^2 k_F^2}{2m}. \tag{3.6}$$

Since we are in three dimensions, the allowed k's live inside a sphere of radius k_F, $|\vec{k}| \leq k_F$. Thus the boundary of the occupied zone in k space is a *spherical Fermi surface*. We can also define the Fermi velocity and Fermi temperature, in the obvious way:

$$v_F = \frac{\hbar k_F}{m}; \quad T_F = \frac{\epsilon_F}{k_B}. \tag{3.7}$$

The periodic boundary conditions mean that the wavevector depends on three integers as

$$k_{n_i} = \frac{2\pi n_i}{L}. \tag{3.8}$$

Then the element of volume in the k-space (the volume that encloses a single electron state) is $(2\pi/L)^3$, so the number of electron states is given by two (for the two spins) times the total volume divided by the element of volume:

$$N = 2\frac{\frac{4\pi}{3}k_F^3}{(2\pi/L)^3} = \frac{V}{3\pi^2}k_F^3. \tag{3.9}$$

Inverting the relation, we obtain

$$k_F = \left(\frac{3\pi^2 N}{V}\right)^{1/3}, \tag{3.10}$$

which in turn gives the Fermi energy

$$\epsilon_F = \frac{\hbar^2}{2m}(3\pi^2 n)^{2/3}, \tag{3.11}$$

where $n = N/V$ is the density of occupied electron states or the total density of electrons. This result was derived for an electron gas living in an empty cubical slice of space at zero temperature.

We can define the density of states (with respect to energy) per unit volume:

$$g(\epsilon) = \frac{dN}{V d\epsilon}. \tag{3.12}$$

Considering that there is a shift in the zero for the energy, thus replacing ϵ with $\epsilon - \epsilon_0$ (thus being able to treat metals and semiconductors together; see Section 3.2), and replacing m by an effective mass m^*, which we will see happens due to electron-electron interactions, and still keeping a spherical Fermi surface (the surface of Fermi energy is spherical in k space). We obtain for a general energy (since the derivation above was valid for any energy, not just for the Fermi one)

$$\epsilon - \epsilon_0 = \frac{\hbar^2 k^2}{2m^*} = \frac{\hbar^2}{2m^*}\left(3\pi^2\frac{N}{V}\right)^{2/3}, \tag{3.13}$$

where N is the number of states with energy up to this value. Inverting and differentiating, we obtain

$$g(\epsilon) = \frac{(2m^*)^{3/2}}{2\pi^2\hbar^3}(\epsilon - \epsilon_0)^{1/2}, \tag{3.14}$$

a more general result of Fermi theory.

One can in fact show that the most general result for an isoenergetic surface Σ of arbitrary shape (including the Fermi surface) is

$$g(\epsilon) = \frac{1}{(2\pi)^3}\int_\Sigma \frac{dS}{|\nabla_{\vec{k}}\epsilon|}. \tag{3.15}$$

Nonzero Temperature and Fermi Integrals

For nonzero temperatures, $T > 0$, the free fermion results are given by Fermi integrals. To exemplify them, we first consider the integral giving the number density of electrons: the integral of the density of states with respect to energy times the Fermi-Dirac distribution giving the probability to have a state with a given energy:

$$n = I_F(\mu) = \int_0^\infty g(\epsilon)f_{FD}(\epsilon)d\epsilon. \tag{3.16}$$

At small temperatures $T \ll T_F$, we can do a Taylor expansion of the result.

We first define

$$G(\epsilon) = \int g(\epsilon)d\epsilon. \tag{3.17}$$

Note that at $T = 0$, ϵ in $g(\epsilon)$ goes only up to ϵ_F.

By a partial integration in $I_F(\mu)$, we can write

$$I_F(\mu) = -\int_0^\infty G(\epsilon)f'_{FD}(\epsilon), \tag{3.18}$$

where the prime denotes differentiation with respect to ϵ. Considering the Taylor expansion of this result in T, and the fact that $f'_{FD}(\epsilon, T = 0) \to -\delta(\epsilon - \epsilon_F)$, we get

$$I_F(\mu) = G(\mu) + \frac{\pi^2}{6}(k_BT)^2G''(\mu) + \cdots \tag{3.19}$$

In fact, the general formula is

$$I(\epsilon_F) = \sum_{l \geq 0} 2(k_B T)^l (1 - 2^{1-l}) \zeta(l) G^{(l)}(\epsilon = \epsilon_F), \tag{3.20}$$

where $\zeta(x)$ is the Riemann zeta function and $G^{(l)}(\epsilon)$ stands for the l-th derivative with respect to ϵ.

The general Fermi integrals, allowing us to calculate all quantities for the free Fermi gas, are the functions

$$F_\alpha(y) \equiv \int_0^\infty \frac{x^\alpha dx}{1 + e^{x-y}}, \tag{3.21}$$

where

$$y = \frac{\epsilon_F - \epsilon_0}{k_B T}; \quad x = \frac{\epsilon - \epsilon_0}{k_B T}. \tag{3.22}$$

Then the density of electrons is

$$n = I_F(\mu) = \int_0^\infty g(\epsilon) f_{FD}(\epsilon) d\epsilon = \frac{(2m^* k_B T)^{3/2}}{2\pi^2 \hbar^3} F_{1/2}(y). \tag{3.23}$$

Similarly, the total energy is

$$U = \int_0^\infty f_{FD}(\epsilon) g(\epsilon) (\epsilon - \epsilon_0) d\epsilon = \frac{(2m^* k_B T)^{3/2}}{2\pi^2 \hbar^3} k_B T F_{3/2}(y). \tag{3.24}$$

We can rewrite this as

$$U = n K_B T \frac{F_{3/2}(y)}{F_{1/2}(y)}. \tag{3.25}$$

A Taylor expansion around $T = 0$, left as an exercise, gives

$$U \simeq U_0 \left[1 + \frac{5}{12} \pi^2 \left(\frac{k_B T}{\epsilon_F} \right)^2 \right] + \cdots, \tag{3.26}$$

with

$$U_0 = \frac{3}{5} N \epsilon_F, \tag{3.27}$$

leading to the *specific heat capacity of the free electron gas*,

$$C_V^{(el)} = \left. \frac{\partial U}{\partial T} \right|_{V,N} = \frac{N\pi^2}{2\epsilon_F} k_B^2 T = \frac{N\pi^2}{2} k_B \frac{T}{T_F}. \tag{3.28}$$

Experimentally, in metals one finds

$$C_{exp} = \gamma T + A T^3, \tag{3.29}$$

where the first term is due to the free electrons and the second is due to phonons (quanta of atom lattice oscillations). So at small T, the free electron gas-specific heat C dominates.

However, we can see that the ratio of the experimental and theoretical (free) values is a constant, and use it to *define* the effective electron mass m^*, by

$$\frac{\gamma_{exp}}{\gamma_{free}} \equiv \frac{m^*}{m}. \tag{3.30}$$

Since $C \propto \frac{1}{T_F}$ and

$$\frac{1}{T_F} \propto \frac{1}{\epsilon_F} \propto m$$

then $\gamma \propto m$. $\tag{3.31}$

Electrical Conductivity

Consider electron scattering in a metal. The electrons are not really free, basically because of two effects: scattering by impurities, and scattering by phonons (quanta of oscillations of the atom lattice), i.e. not directly from scattering by the individual atoms in the lattice.

Under an external electric field $\vec{E}$, the electrons experience a force

$$\vec{F} = m\frac{d\vec{v}}{dt} = \hbar\frac{d\vec{k}}{dt} = -e\vec{E}. \tag{3.32}$$

That means that during a time t, the electron gains a momentum

$$\Delta\vec{k} = -e\vec{E}\frac{t}{\hbar}, \tag{3.33}$$

and so a velocity

$$\Delta\vec{v} = \frac{\hbar\Delta\vec{k}}{m} = -e\vec{E}\frac{t}{m}. \tag{3.34}$$

Now consider that the electrons are accelerated from zero to a velocity v and then lose it completely in a collision. So instead of $\Delta\vec{v}$ we write $\vec{v}$ (the average velocity), if we replace t by τ, the "lifetime," or time between collisions. Then for a concentration n of electrons, each of which has a charge $q = -e$, we get

$$\vec{j} = nq\vec{v} = \frac{ne^2\tau}{m}\vec{E}, \tag{3.35}$$

and so the coefficient is the conductivity σ, whose inverse, the resistivity, becomes

$$\rho = \frac{1}{\sigma} = \frac{m}{ne^2\tau}. \tag{3.36}$$

As we said, the collisions are mainly due to collisions with impurities, with τ_i, and with phonons, with τ_L, and moreover

$$\frac{1}{\tau} = \frac{1}{\tau_i} + \frac{1}{\tau_L}, \tag{3.37}$$

leading to

$$\rho = \rho_L + \rho_i. \tag{3.38}$$

We will generalize this in Chapter 13 to the Drude model, which involves a real and imaginary part of σ.

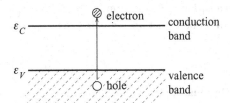

Fig. 3.1 Valence and conduction bands in metals and semiconductors.

3.2 Fermi Surfaces

Fermi Surface in Materials

In materials, the Fermi surface in $\vec{k}$ space is not necessarily a sphere, but rather a complicated surface.

But we are also interested in position space $\vec{x}$, where the Fermi surface is also a complicated surface in general. If for individual atoms electron states are described by fixed orbitals (s, p, d, f, etc.), which are the same for all identical atoms, when we create an atom lattice by putting these atoms very close together, because of the Pauli exclusion principle, the orbitals coalesce, and the energies corresponding to various atoms split finely, creating *energy bands*. In particular, the orbitals of valence electrons coalesce into the *valence band*, and the orbitals of conduction electrons coalesce into the *conduction band*, as in Figure 3.1.

Denoting by ϵ_C the lowest energy of the conduction band, for metals we have in general that the Fermi energy is larger, $\epsilon_F > \epsilon_C$, so we have a fraction of the conduction electron states being filled and giving us charge carriers.

Semiconductors

Semiconductors can also be characterized by the structure of their electron bands. We start with statements valid for general (nondegenerate) semiconductors.

In semiconductors, we have $\epsilon_C > \epsilon_V$, where ϵ_V is the highest energy in the valence band. Moreover, $\epsilon_C > \epsilon_F > \epsilon_V$. In fact, as we will see shortly, at $T = 0$ we have $\epsilon_F = (\epsilon_C + \epsilon_V)/2$. Since the Fermi energy falls between the conductance and valence bands, one electron can jump from the valence to the conduction band, leaving a "hole" behind. The hole, the absence of an electron in an occupied Fermi sea, acts as a positive charge carrier. Its effective mass m_p^* is in general different (larger) than m_n^*, the effective electron mass.

As we mentioned, the electron obeys the FD distribution, so

$$f_{0n}(\epsilon) = \frac{1}{e^{\frac{\epsilon - \epsilon_F}{k_B T}} + 1}, \tag{3.39}$$

whereas the hole has the complement of the FD distribution:

$$f_{0p}(\epsilon) = 1 - f_{0n}(\epsilon) = \frac{1}{e^{\frac{\epsilon_F - \epsilon}{k_B T}} + 1}. \tag{3.40}$$

We can adapt the general formula (3.14) for the density of states of free fermions to the electrons in semiconductors, replacing ϵ_0 by ϵ_C, so

$$g_n(\epsilon) = \frac{(2m_n^*)^{3/2}}{4\pi^2\hbar^3}(\epsilon - \epsilon_C)^{1/2}. \tag{3.41}$$

A similar formula applies for the holes, with $\epsilon_V - \epsilon$ replacing $\epsilon - \epsilon_C$, giving

$$g_p(\epsilon) = \frac{(2m_p^*)^{3/2}}{4\pi^2\hbar^3}(\epsilon_V - \epsilon)^{1/2}. \tag{3.42}$$

We also can write formulas for the electron and hole densities in the same way,

$$n = \frac{(2m_n^*k_BT)^{3/2}}{2\pi^2\hbar^3}F_{1/2}(y_n)$$

$$p = \frac{(2m_p^*k_BT)^{3/2}}{2\pi^2\hbar^3}F_{1/2}(y_p), \tag{3.43}$$

where

$$y_n = \frac{\epsilon_F - \epsilon_C}{k_BT}; \quad x_n = \frac{\epsilon - \epsilon_C}{k_BT}$$

$$y_p = \frac{\epsilon_V - \epsilon_F}{k_BT}; \quad x_p = \frac{\epsilon_V - \epsilon}{k_BT}. \tag{3.44}$$

Since $\epsilon_F - \epsilon_C < 0$ and $\epsilon_V - \epsilon_F < 0$, it means that $y_n < 0$ and $y_p < 0$, so

$$F_{1/2}(y < 0) = \int_0^\infty \frac{x^{1/2}dx}{1 + e^{x-y}} = e^y\int_0^\infty \frac{x^{1/2}dx}{e^y + e^x} \simeq e^y\int_0^\infty x^{1/2}e^{-x}dx = e^y\Gamma(1/2 + 1). \tag{3.45}$$

Replacing the y's from the above formulas, we get

$$n = N_Ce^{\frac{\epsilon_F - \epsilon_C}{k_BT}}$$

$$p = N_Ve^{\frac{\epsilon_V - \epsilon_F}{k_BT}}, \tag{3.46}$$

where

$$N_C = \frac{(2\pi m_n^*k_BT)^{3/2}}{4\pi^3\hbar^3}$$

$$N_V = \frac{(2\pi m_p^*k_BT)^{3/2}}{4\pi^3\hbar^3}. \tag{3.47}$$

Specializing to the case of an *intrinsic* semiconductor, i.e. a semiconductor that is neutral in terms of its charge carriers, $n = p$, and replacing the Formulas for n and p, we get a formula for ϵ_F as a function of temperature T:

$$\epsilon_F = \frac{\epsilon_C + \epsilon_V}{2} + \frac{3}{4}k_BT \ln\frac{m_p^*}{m_n^*}. \tag{3.48}$$

As promised, we see that at $T = 0$, $\epsilon_F = (\epsilon_C + \epsilon_V)/2$, and since $m_p^* > m_n^*$, at $T > 0$ the Fermi energy ϵ_F moves up, closer to ϵ_C, as in Figure 3.2.

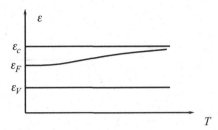

Fig. 3.2 The Fermi energy as a function of temperature in semiconductors.

Nondegeneracy, or classicality of the free fermions, is equivalent to the statement that the exponential in the FD distribution is very small (using the formula for n):

$$e^{\frac{\epsilon_F - \epsilon_C}{k_B T}} = \frac{4\pi^3 \hbar^3 n}{(2\pi m_n^* k_B T)^{3/2}} \ll 1. \tag{3.49}$$

Extrinsic Semiconductors

These are semiconductors without charge carrier balance (neutrality). In them we have impurities doping the material, which can be of two kinds: donors of electrons and acceptors of electrons. These impurities give the charge carriers.

The total concentration of donors N_d is made up of un-ionized impurities (n_d) and ionized impurities (N_d^+),

$$N_d = n_d + N_d^+, \tag{3.50}$$

and the total concentration of acceptors N_a is made up of un-ionized impurities (p_a) and ionized impurities (N_a^-),

$$N_a = p_a + N_a^-. \tag{3.51}$$

Denote by n the total concentration of carriers in the conduction band and p the total concentration of carriers in the valence band. Electric neutrality of the material then implies that the total positive charge, made up of carriers plus ionized positive impurities, equals the total negative charge, made up of carriers plus ionized negative impurities:

$$p + N_d^+ = n + N_a^-. \tag{3.52}$$

Replacing N_d^+ and N_a^-, we get

$$p + p_a + N_d = n + n_d + N_a. \tag{3.53}$$

Consider now a semiconductor with only one type of impurity, like an n semiconductor with no acceptors, $N_a = p_a = 0$. Replacing in the above, we obtain

$$n + n_d = p + N_d. \tag{3.54}$$

Note that here

$$n = N_C e^{\frac{\epsilon_F - \epsilon_C}{k_B T}}$$
$$p = N_V e^{\frac{\epsilon_V - \epsilon_F}{k_B T}}, \tag{3.55}$$

just like in the case of the intrinsic semiconductor, since there is no difference with respect to n and p whether we have an extrinsic or intrinsic semiconductor.

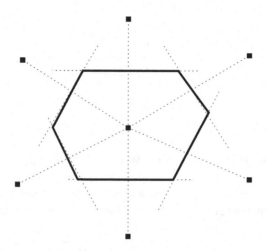

Fig. 3.3   The first Brillouin zone.

Fermi Surfaces in Metals

Now that we described Fermi surfaces in position space, we remember that Fermi surfaces are actually firstly defined in momentum space, so we will try to understand a bit about that.

The Fermi surface separates unfilled orbitals from filled orbitals. The electrical properties of metals depend on the volume and shape of the Fermi surface. But to understand all the properties of the metal, it is in fact enough to talk about the Fermi surface for the first Brillouin zone. Therefore we need to understand first what it means to draw the first Brillouin zone.

The electron density (and all other quantities) is periodic on the coordinate space lattice. As such, when writing a Fourier expansion for it,

$$n(\vec{r}) = \sum_{\vec{B}} n_{\vec{B}} e^{i\vec{B}\cdot\vec{r}}, \tag{3.56}$$

invariance under all the translations of the crystalline structure means that the Fourier transformed coordinate $\vec{B}$ is written in terms of integers times unit vectors,

$$\vec{B} = n_1 \vec{b}_1 + n_2 \vec{b}_2 + n_3 \vec{b}_3, \tag{3.57}$$

where $n_i \in \mathbb{Z}$, and the unit vectors are

$$\vec{b}_1 = 2\pi \frac{\vec{a}_2 \times \vec{a}_3}{\vec{a}_1 \cdot (\vec{a}_2 \times \vec{a}_3)}; \quad \vec{b}_2 = 2\pi \frac{\vec{a}_3 \times \vec{a}_1}{\vec{a}_1 \cdot (\vec{a}_2 \times \vec{a}_3)}; \quad \vec{b}_3 = 2\pi \frac{\vec{a}_1 \times \vec{a}_2}{\vec{a}_1 \cdot (\vec{a}_2 \times \vec{a}_3)}, \tag{3.58}$$

i.e. the vectors of the *reciprocal lattice*.

The reciprocal lattice is relevant for Bragg-type diffraction. The diffraction condition is

$$\vec{k}' - \vec{k} = \vec{B}, \tag{3.59}$$

which means that $k'^2 = k^2 + B^2 + 2\vec{k} \cdot \vec{B}$, and for elastic scattering, $|\vec{k}'| = |\vec{k}|$, so we have

$$2\vec{k} \cdot \vec{B} + \vec{B}^2 \Rightarrow \vec{k} \cdot \left(\frac{1}{2}\vec{B}\right) = -\left(\frac{1}{2}\vec{B}\right)^2. \tag{3.60}$$

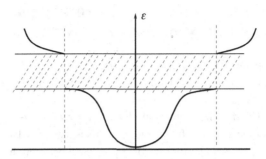

Energy bands and the modified Fermi surface.

We see that this condition defines the first Brillouin zone as follows: we consider a center of the Brillouin zone (a given atom), and around it we consider the neighboring atoms. At the midpoints from the center to each neighbor we consider the perpendicular. The figure formed by all these perpendiculars is the first Brillouin zone, as in Figure 3.3. It is a figure that is in some sense a discretized form of a sphere.

Using this first Brillouin zone, and adding copies of it by symmetry ad infinitum in all directions of the reciprocal lattice, we can describe in this auxiliary space all the properties of the material.

Interactions modify the spherical isoenergetic Fermi surface of the free fermions to a more complicated form. For instance, if in one k dimension, the free surface is a parabola $\epsilon = (\hbar^2/(2m))k^2$, and it is an approximation for the real surface in the first and second energy bands, interrupted during the energy gap between the bands, as in Figure 3.4.

Let us now review the occupation types of the conduction and valence bands for the three cases of interests for solids, as in Figure 3.5:

- Insulator: Both the valence and conduction bands are completely filled, so there are no charge carriers that can move.
- Metal: The valence band is filled, and the conduction band is partially (comparable with half) filled, so that there are many conduction channels open in the conduction band.
- Semiconductor: These can be of two types: The conduction band is only a very little filled (leading to electron conduction) or almost completely filled (leading to hole conduction).

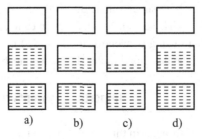

Energy bands in solids, valence, conduction, plus another for (a) Insulator; (b) Metal; (c) Semiconductor (n);
(d) Semiconductor (p).

3.3 Fermi Liquid

In most metals, the electrons are *interacting*, instead of being free. For most of them, we have *Landau's Fermi liquid theory*. Landau developed this theory based on two assumptions:

(1) The dominant effect of electron interactions is simply to renormalize the effective mass of the electron, from m to m^*. This is usually a 10%–50% effect.

(2) There is a one-to-one correspondence between the excited states of the normal state of a metal and the states of a noninteracting electron gas. That means that we can think of the metal as a modification of the states of the free Fermi gas, and instead of the electrons we have "quasi-particles," which are stable only within the metal, for a lifetime τ. These particles can be composites of the real particles.

Since the states at the edge of allowed region (Fermi surface) are stable, as the energy $\epsilon \to \epsilon_F$, the lifetime goes to infinity, $\tau \to \infty$.

We can now consider electron scattering. To be able to scatter them, the first particle has to be above ϵ_F, $\epsilon_1 > \epsilon_F$, and it can scatter against an electron with $\epsilon_2 < \epsilon_F$. But then, after scattering, both states must be outside the Fermi surface (so they can be mobile), i.e. $\epsilon_3, \epsilon_4 > \epsilon_F$. On the other hand, energy conservation gives

$$\epsilon_1 + \epsilon_2 = \epsilon_3 + \epsilon_4. \tag{3.61}$$

That means that if $\epsilon_1 = \epsilon_F$, then also $\epsilon_2 = \epsilon_3 = \epsilon_4 = \epsilon_F$. That in turn means that the number of states into which it can scatter is zero, thus that the scattering rate, proportional to the density of scattering states, goes to zero, so the lifetime goes to infinity, $\tau \to \infty$.

If $\epsilon_1 > \epsilon_F$, then there is a $\epsilon_1 - \epsilon_F$ window to choose from for ϵ_2, ϵ_3, and $\epsilon_4 = \epsilon_1 + \epsilon_2 - \epsilon_3$ is fixed by energy conservation. The scattering rate is then proportional to the density of scattering states, i.e. with $(\epsilon_1 - \epsilon_F)^2$ (one factor for each of ϵ_2 and ϵ_3). And if the initial excited electron was excited thermally, i.e. if $\epsilon_1 - \epsilon_F \propto k_B T$, then it follows that the scattering rate ($\propto 1/\tau$) is proportional to T^2, and therefore the resistivity is

$$\rho \propto T^2. \tag{3.62}$$

This result was derived under very general assumptions put forward in Landau's theory, and indeed for most materials with strong interactions and an effective description, which therefore should obey Fermi liquid theory, we find $\rho \propto T^2$. However, it was found that for cuprate high T_c superconductors in the normal state, we have $\rho \propto T$ instead, all the way down to T_c. These materials have been called *non-Fermi liquids*, and their description is very difficult using conventional condensed matter theory. It is one of the cases where AdS/CMT (to be described in Part IV) can contribute best.

Important Concepts to Remember

- The free electron gas in three dimensions at $T = 0$ has a spherical Fermi surface $k \leq k_F$.
- The free electron gas in three dimensions at $T = 0$ has $\epsilon_F = \hbar^2/(2m)[3\pi^2 n]^{2/3}$.

- The density of energy states of the (possibly interacting, thus $m \to m^*$) fermions is $g(\epsilon) = [(2m^*)^{3/2}/(2\pi^2\hbar^3)](\epsilon - \epsilon_0)^{1/2}$, or in general $1/(2\pi^3)\int_\Sigma dS/|\nabla_{\vec{k}}\epsilon|$.
- At nonzero temperatures, we have Fermi integrals, $F_\alpha(y) = \int_0^\infty x^\alpha d\alpha/(1 + e^{x-y})$, where $y = (\epsilon_F - \epsilon_0)/(k_B T)$ and $x = (\epsilon - \epsilon_0)/(k_B T)$.
- Examples of use are $n = I_F(\mu) = \int_0^\infty g(\epsilon)f_{FD}(\epsilon)d\epsilon$ and $U = nK_BTF_{3/2}(y)/F_{1/2}(y)$.
- The heat capacity of the free electron gas is $C_V = N\pi^2/2[k_B T/T_F]$, and experimentally $C_V = \gamma T + AT^3$, so at small temperatures, the free electron contribution dominates. We can define m^* from $\gamma_{exp}/\gamma_{free} = m^*/m$.
- The electrical conductivity in metals is due to scattering from impurities and phonons, $\rho = m/(ne^2\tau)$, with $1/\tau = 1/\tau_i + 1/\tau_L$.
- The semiconductors have as charge carriers electrons and holes, with density of states $g_n \propto (\epsilon - \epsilon_C)^{1/2}$ and $g_p \propto (\epsilon_V - \epsilon)^{1/2}$.
- In an intrinsic semiconductor, $\epsilon_F = (\epsilon_C + \epsilon_V)/2 + 3k_B T/4 \ln(m_p^*/m_n^*)$.
- In metals, we can describe the k-space Fermi surface only in the first Brillouin zone.
- In an insulator, the valence and conduction bands are completely filled, in a metal the conduction band is partially filled, and in a semiconductor the conduction band can be only a little filled (electron conduction) or almost completely filled (hole conduction).
- For a Fermi liquid, according to Landau's theory, the dominant effect of interactions is to change $m \to m^*$, and there is a one-to-one correspondence between the states of the Fermi liquid and the states of the free Fermi gas.
- For a Fermi liquid, $\rho \propto T^2$, but there are strongly interacting Fermion systems that have $\rho \propto T$ instead, called non-Fermi liquids, like cuprate high T_c superconductors in the normal state.

Further Reading

Chapters 6, 2, 9, and 14 in Kittel's book [6] and chapter 11.5 in [7].

Where It will be Addressed in String Theory

In Chapter 40 of Part IV, we will describe how to use the AdS/CFT correspondence to describe the Fermi gas and liquid.

Exercises

(1) Complete the missing steps from the text to find

$$U = U_0\left[1 + \frac{5}{12}\pi^2\left(\frac{k_B T}{\epsilon_F}\right)^2\right], \tag{3.63}$$

where

$$U_0 = \frac{3}{5}N\epsilon_F, \tag{3.64}$$

for a free electron gas.

(2) Using the result of the last exercise, prove that the pressure of the electron gas is

$$p = \frac{2}{3}\frac{U}{V}.$$ (3.65)

(3) Show that for a free Fermi gas

$$P = g_s \frac{k_B T}{\Lambda^3}\frac{F_{3/2}(\zeta)}{\Gamma(5/2)}$$ (3.66)

and

$$n = \frac{g_s}{\Lambda^3}\frac{F_{1/2}(\zeta)}{\Gamma(1/2)},$$ (3.67)

where the *fugacity* is $\zeta = e^{\beta\mu}$, $g_s = 2$, and

$$n = \left(\frac{2\pi m}{\beta h^2}\right)^{-1/2}.$$ (3.68)

Then show that if $\zeta(n)$ admits a Taylor expansion,

$$P(n, T) = nk_B T \left[1 + \frac{\Lambda^3}{g_s 2^{5/2}}n + \frac{\Lambda^6}{g_s^2}\left(\frac{1}{8} - \frac{2}{3^{5/2}}\right)n^2 + \cdots\right].$$ (3.69)

(4) Consider the generalized Fermi gas in spatial d dimensions, with dispersion relation $\epsilon_k = ak^x$ instead of $\epsilon = \frac{\hbar^2}{2m}k^2$. Show that the density of states is

$$\omega(E) = g_s \left(\frac{L}{2\pi}\right)^d \frac{\Omega_d}{xa^{d/x}}\epsilon^{\frac{d}{x}-1},$$ (3.70)

where Ω_d is the volume of the unit sphere, L is the size of the system, and then that

$$U = U_0 \left[1 + \frac{\pi^2}{6}\left(\frac{d}{x} + 1\right)\left(\frac{k_B T}{\epsilon_F}\right)^2\right]$$ (3.71)

and

$$U_0 = N\epsilon_F \frac{d}{d+x}.$$ (3.72)

(5) Show that for a general (nonspherical) Fermi surface

$$n = 2\int_0^{\epsilon_F} g(\epsilon)d\epsilon + \frac{\pi^2}{3}(k_B T)^2 \frac{dg}{d\epsilon_F}(\epsilon_F)$$ (3.73)

and from it,

$$\epsilon_F = \epsilon_F(T = 0) - \frac{\pi^2}{6}(k_B T)^2 \frac{dg(\epsilon_F)}{g(\epsilon_F)d\epsilon_F}\bigg|_{\epsilon_F = \epsilon_F(T=0)}.$$ (3.74)

(6) For an extrinsic semiconductor ($n \neq p$) use the electric neutrality relation $p + N_d^+ = n + N_a^-$ and the relations

$$n_d = \frac{N_d}{1 + \frac{1}{2}e^{\frac{\epsilon_d - \epsilon_F}{k_B T}}}; \quad p_a = \frac{N_a}{1 + \frac{1}{2}e^{\frac{\epsilon_F - \epsilon_a}{k_B T}}} \tag{3.75}$$

to find the equation for $\epsilon_F(T)$.

(7) For a semiconductor type n, apply the equation at Exercise 6 above to find (for $p \ll N_d$)

$$\epsilon_F = \epsilon_d + k_B T \ln \frac{1}{4}\left[\sqrt{1 + \frac{8N_d}{N_C}e^{\frac{\epsilon_C - \epsilon_D}{k_B T}}} - 1\right]. \tag{3.76}$$

4 Bosonic Quasi-Particles: Phonons and Plasmons

In this chapter we will study bosonic quasi-particles in materials, which are (composite) particles that live only inside the material and may have a finite lifetime (though usual particles also can have a finite lifetime). The two such quasi-particles that we will analyze are phonons, the quanta of oscillation of the crystalline network (the ions, without the conduction electrons), and the plasmons, which are the quanta of oscillation of the plasma of free (conduction) electrons.

So these are the *collective excitations* of the solids, the ones of the ions being the phonons, and the ones of the conduction electrons being the plasmons.

4.1 Phonons in $1+1$ and $3+1$ Dimensions

1-Dimensional Phonon Model

We start the analysis with the phonons, and the simplest model for them is a 1-dimensional chain of N atoms joined by harmonic springs (that link the nearest neighbors), which is a good 1-dimensional approximation to a solid.

Considering this chain of atoms linked by springs, the equation of motion for it is a discrete set of equations for each site, with a spring of length $x_{i+1} - x_i$ pulling the atom one way, and another of length $x_i - x_{i-1}$ pulling the other way. This gives

$$M\frac{d^2 x_n}{dt^2} = -k_{el}(2x_n - x_{n-1} - x_{n+1}),\tag{4.1}$$

where M is the mass and k_{el} the elastic constant of the springs. Integrating these equations and summing, we get the action

$$S = \sum_{n=1}^{N} \int dt \left[\frac{M}{2}\left(\frac{dx_n}{dt}\right)^2 - \frac{k_{el}}{2}(2x_n^2 - x_{n-1}x_n - x_n x_{n+1}) \right],\tag{4.2}$$

and in turn this leads to the Hamiltonian

$$H = \sum_n \left[\frac{P_n^2}{2M} + \frac{M\omega^2}{2}(x_n - x_{n+1})^2 \right],\tag{4.3}$$

where

$$\omega^2 = k_{el}/M.\tag{4.4}$$

We next diagonalize this Hamiltonian. First, we do a discrete Fourier transform,

$$P_n = \frac{1}{\sqrt{N}} \sum_k e^{ikna} P_k$$

$$x_n = \frac{1}{\sqrt{N}} \sum_k e^{ikna} x_k, \tag{4.5}$$

which leads to

$$\sum_n P_n^2 = \sum_k P_k P_{-k}; \quad \sum_n x_n^2 = \sum_k x_k x_{-k}; \quad \sum_n x_n x_{n+1} = \sum_k x_k x_{-k} e^{-ika}. \tag{4.6}$$

Thus the Hamiltonian becomes

$$H = \frac{1}{2M} \sum_k P_k P_{-k} + \frac{M\omega^2}{2} \sum_k (2x_k x_{-k} - x_k x_{-k} e^{ika} - x_k x_{-k} e^{-ika})$$

$$= \frac{1}{2M} \sum_k P_k P_{-k} + \frac{M}{2} \sum_k \omega_k^2 x_k x_{-k}, \tag{4.7}$$

where

$$\omega_k^2 = 2\omega^2 (1 - \cos ka) = 4\omega^2 \sin^2 \frac{ka}{2}. \tag{4.8}$$

Now we can do the usual rescaling and mixing of the x's and the P's, as

$$\tilde{Q}_k = x_k \sqrt{\frac{M\omega_k}{2}}; \quad \tilde{P}_k = \frac{P_k}{\sqrt{2M\omega_k}}, \tag{4.9}$$

followed by

$$b_k = \tilde{Q}_k - i\tilde{P}_{-k}; \quad b_k^\dagger = \tilde{Q}_{-k} + i\tilde{P}_k. \tag{4.10}$$

Then the oscillators obey the usual algebra,

$$[b_k, b_{k'}^\dagger] = \delta_{kk'}, \tag{4.11}$$

and the Hamiltonian is simply the free decoupled sum of usual oscillators with frequencies ω_k:

$$H = \sum (\hbar) \omega_k \left(b_k^\dagger b_k + \frac{1}{2} \right). \tag{4.12}$$

These are the phonon modes in the 1-dimensional solid, also called acoustic phonons.

The $b_k(t)$ have the usual time dependence,

$$b_k(t) = b_k(t = 0) e^{-i\omega_k t}, \tag{4.13}$$

and the coordinates are found to be simply

$$x_n(t) = \sum_k \frac{1}{\sqrt{2MN\omega_k}} \left(b_k e^{-i\omega_k t} + b_{-k}^\dagger e^{+i\omega_k t} \right) e^{ikna}. \tag{4.14}$$

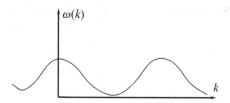

Fig. 4.1   Dispersion relation $\omega(k)$ of the monoatomic solid.

Phonon Modes in 3-Dimensional Solid

In the solid, the ion potential is a 2-point interaction:

$$V_{\text{ion}} = \sum_{i<j} V(\vec{R}_i - \vec{R}_j). \tag{4.15}$$

There is a stable set of equilibrium points, the crystalline structure $\vec{R}_i^0$. Expanding around this stable set as

$$\vec{R}_i = \vec{R}_i^0 + \vec{Q}_i, \tag{4.16}$$

it follows that we can approximate up to quadratic fluctuations by

$$V_{\text{ion}} \simeq \sum_{i<j} V(\vec{R}_i^0 - \vec{R}_j^0) + \frac{1}{2}\sum_{i<j}(\vec{Q}_i - \vec{Q}_j)_\mu (\vec{Q}_i - \vec{Q}_j)_\nu \frac{\partial^2}{\partial R_\mu \partial R_\nu} V(\vec{R}_i^0 - \vec{R}_j^0). \tag{4.17}$$

This is of basically the same form as in the 1-dimensional case, except it now has three directions. Therefore we can already import the result from one dimension, with the only modification being the addition of a polarization vector $\lambda_{\vec{k}}$, which contains one longitudinal polarization (to the momentum $\vec{k}$) and two transversal polarizations, and therefore we can write

$$\vec{Q}_i(t) = \sum_{\vec{k},\vec{\lambda}} \frac{1}{\sqrt{2MN\omega_{\vec{k},\vec{\lambda}}}} \left(b_{\vec{k},\vec{\lambda}}\vec{\lambda}_{\vec{k}} e^{-i\omega_{\vec{k},\lambda}t} + b^\dagger_{-\vec{k},\vec{\lambda}}\vec{\lambda}^*_{-\vec{k}} e^{i\omega_{\vec{k},\lambda}t} \right) e^{i\vec{k}\cdot\vec{R}_i^0}. \tag{4.18}$$

Also similarly to the 1-dimensional case, we can write the ion potential in terms of the phonon harmonic modes:

$$V_{\text{ion}} = \sum_{i<j} V(\vec{R}_i^0 - \vec{R}_j^0) + \frac{M}{2}\sum_{\vec{k},\vec{\lambda}}\omega^2_{\vec{k},\vec{\lambda}}\vec{Q}_{\vec{k},\vec{\lambda}}\vec{Q}_{\vec{k},\vec{\lambda}}. \tag{4.19}$$

Dispersion Relations

As we saw in the 1-dimensional case, the dispersion relation one obtains in this case is

$$\omega^2(k) = 2\omega^2(1 - \cos ka), \tag{4.20}$$

where $\omega^2 = k_{el}/M$, as in Figure 4.1.

For a propagating wave with a dispersion relation $\omega(k)$, there are two velocities of interest: the phase velocity $v_{\text{phase}} = \omega/k$ describes the propagation of the phase of the wave, whereas the *group velocity* $v_{\text{group}} = d\omega/dk$, or rather

$$\vec{v}_g = \vec{\nabla}_{\vec{k}}\,\omega \tag{4.21}$$

describes the propagation of a group of waves, carrying *information*. In our case we obtain

$$v_{\text{phase}} = \frac{2\omega}{k}\sin\frac{ka}{2}; \quad v_{\text{group}} = \omega a \cos\frac{ka}{2}. \tag{4.22}$$

In the continuum limit, $ka \ll 1$ (lattice unit is much smaller than the wavelength), the dispersion relation becomes the usual one,

$$\omega^2(k) = \frac{k_{el}}{M}k^2 a^2, \tag{4.23}$$

and the phase and group velocities coalesce to the same value,

$$v_p = v_g = \sqrt{\frac{k_{el}}{M}}\,a. \tag{4.24}$$

4.2 Optical and Acoustic Branches; Diatomic Lattices

Crystals with two or more atoms per primitive basis (cell), for instance, salt (NaCl) crystals, have at least two branches for $\omega(k)$, (1) an optical branch and (2) an acoustic branch.

In reality, counting the three polarizations (one longitudinal and two transverse), we have three acoustic and three optical branches. More generally, for a crystal with p atoms per primitive cell, one can show that we have three acoustic and $3p - 3$ optical branches.

The optical branch is defined by a decreasing $\omega(k)$, and $\omega(0) \neq 0$. The acoustic branch is defined by an increasing $\omega(k)$ and $\omega(0) = 0$.

Diatomic Structure (e.g. NaCl)

Consider a crystalline structure with two alternating types of atoms. For simplicity, we will consider the 1-dimensional case as in the monoatomic case. We denote by u_s (with mass M_1) the positions of the chain of one type of atoms, and with v_s (with mass M_2) the positions of the chain of the other type of atoms. Thus, we have, in order, for instance, $u_{s-1}, v_{s-1}, u_s, v_s, u_{s+1}, v_{s+1}$.

Generalizing the monoatomic case, the equations of motion in this diatomic case are found easily to be

$$M_1 \frac{d^2 u_s}{dt^2} = -k_{el}(2u_s - v_s - v_{s+1})$$
$$M_2 \frac{d^2 v_s}{dt^2} = -k_{el}(2v_s - u_s - u_{s+1}). \tag{4.25}$$

We write an ansatz for the (normal mode) solution:

$$u_s = u e^{iska} e^{-i\omega t}$$
$$v_s = v e^{iska} e^{-i\omega t}. \tag{4.26}$$

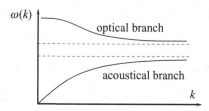

Fig. 4.2   Dispersion relations $\omega(k)$ for the biatomic solid: optical and acoustical branches.

Substituting the ansatz in the equations of motion, we find a system of equations that needs, to find a solution, the determinant of the matrix acting on the (u, v) pair to be equal to zero (since there are two linear relations between u and v, with a zero constant term):

$$\det \begin{pmatrix} 2k_{el} - M_1\omega^2 & -k_{el}(1 + e^{ika}) \\ -k_{el}(1 + e^{ika}) & 2k_{el} - M_2\omega^2 \end{pmatrix} = 0. \tag{4.27}$$

If $ka \ll 1$, there are two solutions for $\omega(k)$ coming from this equation:

$$\omega^2 = 2k_{el}\left(\frac{1}{M_1} + \frac{1}{M_2}\right); \quad \text{optical branch}$$

$$\omega^2 = \frac{k_{el}/2}{M_1 + M_2}k^2a^2; \quad \text{acoustic branch.} \tag{4.28}$$

In general, for ka arbitrary, the optical branch goes down until (asymptotically) $\sqrt{2k_{el}/M_2}$, and the acoustic branch goes up to (asymptotically) $\sqrt{2k_{el}/M_1}$, as in Figure 4.2.

4.3 Electron-Phonon Interaction

We now describe the interaction between the conduction electrons and the phonons (quanta of oscillation of the ion lattice). This is a type of collective interaction of the whole lattice with the electrons.

The potential of interaction between the electrons and the ions is

$$V_{ei} = \sum_{ij} V_{ei}(\vec{r}_j - \vec{R}_i) \simeq \sum_{ij} V_{ei}(\vec{r}_j - \vec{R}_i^0) - \sum_{ij} \vec{Q}_i \cdot \vec{\nabla}_j V_{ei}(\vec{r}_i - \vec{R}_j^0) + \mathcal{O}(Q^2). \tag{4.29}$$

Here we have expanded the interaction between the electrons $\vec{r}_j$ and the ions $\vec{R}_i$ around the stable points $\vec{R}_i^0$ of the ion lattice (whereas for the free – conduction – electrons there is no stable point, so nothing to expand around). The first term is a "constant" (with respect to $\vec{Q}$) V_0, interesting in its own right, but not containing the phonon interaction, which comes from $\vec{Q}$, as we saw before.

We write the electron-ion interaction potential in Fourier components through

$$V_{ei}(r) \simeq \frac{1}{N}\sum_{\vec{k}} V_{ei}(\vec{k})e^{i\vec{k}\cdot\vec{r}}. \tag{4.30}$$

Replacing $\vec{Q}_i$ with its phonon expansion (4.18) and the Fourier expansion in the above (4.29) results in

$$V_{ei} \simeq V_0 - \frac{i}{N} \sum_{\vec{k},i,j} V_{ei}(\vec{k}) e^{i\vec{k}\cdot\vec{r}_j} \vec{k} \cdot \sum_{\vec{q},\lambda} \frac{1}{\sqrt{2MN\omega_{\vec{q},\lambda}}} \vec{\lambda}_{\vec{q}} \left(b_{\vec{q},\lambda} + b^\dagger_{-\vec{q},\lambda} \right) e^{i(\vec{q}-\vec{k})\cdot\vec{R}^0_i} \equiv V_0 + H_{e-ph}.$$

(4.31)

The second term (H_{e-ph}) is the electron-phonon interaction. We now rewrite it, and go from the classical expression above to a second quantized one.

We can perform the sum over i (where $\vec{q}$ and $\vec{k}$ are in the first Brillouin zone) and write it as a sum over the reciprocal lattice vectors $\vec{L}$ as

$$\frac{1}{N} \sum_i e^{i(\vec{q}-\vec{k})\cdot\vec{R}^0_i} = \sum_{\vec{L}} \delta_{\vec{k},\vec{q}+\vec{L}}.$$

(4.32)

The momentum-space electron density

$$\rho_{\vec{k}} = \sum_j e^{i\vec{k}\cdot\vec{r}_j}$$

(4.33)

can be replaced with a second quantized form,

$$\hat{\rho}_{\vec{k}} = \left(\sum_{\vec{r}} \right) \sum_{\vec{k}_1,\vec{k}_2} \langle \vec{k}_1 | e^{i\vec{k}\cdot\vec{r}} | \vec{k}_2 \rangle a^\dagger_{\vec{k}_1} a_{\vec{k}_2}.$$

(4.34)

But for free waves,

$$\frac{1}{V} \sum_{\vec{r}} \langle \vec{k}_1 | e^{i\vec{k}\cdot\vec{r}} | \vec{k}_2 \rangle = \delta_{\vec{k}_1,\vec{k}_2+\vec{k}}.$$

(4.35)

For more general electron wavefunctions, we can define the overlap of two states,

$$\alpha_{\vec{q}_1,\vec{q}_2} \equiv \langle \vec{q}_1 | \vec{q}_2 \rangle,$$

(4.36)

from which we obtain

$$\frac{1}{V} \sum_{\vec{k}_1,\vec{k}_2} \sum_{\vec{r}} \langle \vec{k}_1 | e^{i\vec{k}\cdot\vec{r}} | \vec{k}_2 \rangle a^\dagger_{\vec{k}_1} a_{\vec{k}_2} = \sum_{\vec{k}_2} \alpha_{\vec{k}_2+\vec{k},\vec{k}_2} a^\dagger_{\vec{k}_2+\vec{k}} a_{\vec{k}_2}.$$

(4.37)

Finally, substituting the above equations, we can write the electron-phonon Hamiltonian as

$$\hat{H}_{e-ph} = \sum_{\vec{q},\vec{L},\lambda} \left(\frac{-i}{\sqrt{2MN\omega_{\vec{q},\lambda}}} \right) (\vec{q}+\vec{L}) \cdot \vec{\lambda}_{\vec{q}} V_{ei}(\vec{q}+\vec{L}) \left(b_{\vec{q},\lambda} + b^\dagger_{-\vec{q},\lambda} \right) \alpha_{\vec{k}+\vec{q}+\vec{L},\vec{k}} a^\dagger_{\vec{q}+\vec{L}+\vec{k}} a_{\vec{k}}.$$

(4.38)

The general electron-phonon interaction process described by this Hamiltonian is an "Umklapp" process. In German, *umklappen* means to "flip over." So this process is one of nonconservation of momentum, or conservation only modulo a vector $\vec{L}$ of the reciprocal lattice, i.e. $\vec{k}' = \vec{k} + \vec{L}$. Physically this corresponds to a scattering off the lattice, when the lattice absorbs part of the momentum via the phonon oscillation and part as a whole (being very heavy).

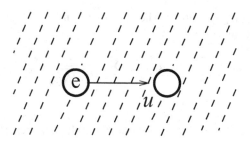

　Displacement of an electron in the plasma generates a plasmon mode.

We can have two kinds of inelastic electron scattering processes by the phonon of vector $\vec{k}$:

- It can be *absorbed*,

$$\vec{k} + \vec{q} + \vec{L} = \vec{k}',\tag{4.39}$$

where $\vec{k}$ is the initial electron momentum, $\vec{q}$ the phonon momentum and $\vec{k}'$ is the final electron momentum.

- It can be *created*,

$$\vec{k} + \vec{L} = \vec{k}' - \vec{q}.\tag{4.40}$$

The interaction between the phonons and photons (photons interacting with the ion lattice by making it oscillate) can be treated similarly to the electron-phonon interaction. It also has the same Umklapp (inelastic) scattering.

4.4 Plasmons

Plasmons are collective excitations of the electron gas (plasma). We first derive their frequency (see Figure 4.3).

Consider a plasma of electron density n (conduction electrons) inside the material (ion lattice), so that it doesn't dissipate due to the electrostatic interaction between the electrons. Consider a 1-dimensional situation, when a small charge of linear size u gets displaced by the same u (toward the boundary of the interval), as in Figure 4.3, generating a surface charge density $\sigma = neu$, and correspondingly an electric field

$$E = \frac{\sigma}{\epsilon} = \frac{neu}{\epsilon_0}.\tag{4.41}$$

Then the restoring force, that tries to revert the fluctuation, is $\vec{F} = q\vec{E}$, with the charge density being $q/V = -ne$, leading to the equation of motion

$$nm\frac{du^2}{dt^2} = -neE = -\frac{n^2 e^2 u}{\epsilon_0},\tag{4.42}$$

which in turn implies the harmonic motion

$$\frac{du^2}{dt^2} + \omega_p^2 u = 0, \tag{4.43}$$

where the *plasma frequency* is

$$\omega_p = \sqrt{\frac{ne^2}{\epsilon_0 m}}. \tag{4.44}$$

Note, however, that this result (harmonic oscillation with ω_p) is an approximation. A more rigorous classical calculation gives the result

$$\ddot{u}_k + \omega_p^2 u_k = \sum_i e^{i\vec{k}\cdot\vec{r}_i} \left(\frac{\vec{k}\cdot\vec{p}_i}{m} + \frac{\hbar^2 k^2}{2m}\right)^2 - \frac{e^2}{\epsilon_0 m} \sum_{\vec{q}\neq\vec{k}} \vec{k}\cdot\vec{q}\, \frac{n_{\vec{k}-\vec{q}} n_{\vec{q}}}{q^2}. \tag{4.45}$$

But in the *Random Phase Approximation* (RPA), due to the random phase, most of the oscillatory sum cancels, and the right-hand side becomes negligible. At the quantum level, the same RPA appears as the statement that for a Fermi state $|F\rangle$, the expectation value of a product equals the product of expectation values:

$$\left\langle F|\rho_{\vec{q}} a^\dagger_{\vec{k}+\vec{q}} a_{\vec{q}+\vec{q}}|F\right\rangle = \langle F|\rho_{\vec{q}}|F\rangle \left\langle F|a^\dagger_{\vec{k}+\vec{q}} a_{\vec{k}+\vec{q}}|F\right\rangle. \tag{4.46}$$

4.5 Thomas-Fermi Screening

Finally, we describe screening of charges in the plasma (electron gas), using the *Thomas-Fermi approximation*.

When we introduce a charge Q in the electron gas, there is an induced screening charge $-e\delta n(\vec{r})$ that tries to compensate for it, so the true electric potential V_{eff} due to Q in the electron gas is given by

$$\vec{\nabla}^2 V_{\text{eff}} = \frac{Q\delta(\vec{r}) - e\delta n(\vec{r})}{\epsilon_0}. \tag{4.47}$$

In Fourier space, the potential energy $U_{\text{eff}} = eV_{\text{eff}}$ becomes

$$k^2 U_{\text{eff}}(\vec{k}) = -\frac{eQ}{\epsilon_0} + \frac{e^2}{\epsilon_0}\delta n(\vec{k}). \tag{4.48}$$

To calculate it, the approximation due to Thomas and Fermi is to assume that the response is as for a locally free electron gas, namely as a simple shifting of the chemical potential to a *local* one, $\mu \to \mu - U_{\text{eff}}(\vec{r})$. More precisely, because of the shift in energy of the states,

$$\epsilon_{\vec{k}}(\vec{r}) = \frac{\hbar^2 k^2}{2m} + U_{\text{eff}}(\vec{r}), \tag{4.49}$$

the FD distribution gets modified to

$$n_{\vec{k}}(\vec{r}) = \frac{1}{1 + e^{\beta(\epsilon_{\vec{k}}(\vec{r}) - \mu)}}. \tag{4.50}$$

Then the electron density (as the average of the FD distribution) can be approximated by

$$\langle n(\vec{r}) \rangle = 2 \int \frac{d^3 k}{(2\pi)^3} n_{\vec{k}}(\vec{r}) \simeq n_e - \frac{\partial n_e}{\partial \mu} U_{\text{eff}}(\vec{r}) + \cdots. \tag{4.51}$$

Therefore the variation in electron density is

$$\delta n_e(\vec{r}) = \langle n(\vec{r}) \rangle - n_e \simeq -\frac{\partial n_e}{\partial \mu} U_{\text{eff}}(\vec{r}). \tag{4.52}$$

Substituting this in the equation for U_{eff}, we obtain

$$U_{\text{eff}}(\vec{k}) \simeq -\frac{eQ/\epsilon_0}{k^2 + \frac{e^2}{\epsilon_0} \frac{\partial n_e}{\partial \mu}} \equiv -\frac{eQ/\epsilon_0}{k^2 + k_{TF}^2}, \tag{4.53}$$

which in coordinate space gives (by the usual Fourier transform)

$$U_{\text{eff}}(r) = -\frac{eQ e^{-k_{TF} r}}{\epsilon_0 r}. \tag{4.54}$$

As a first approximation, we can use the $n_e(\mu)$ derived in Chapter 3 for free fermions,

$$n_e(\mu) \simeq \left(\frac{2m\mu}{\hbar^2} \right)^{3/2} \frac{1}{3\pi^2}, \tag{4.55}$$

leading to the Thomas-Fermi scale

$$k_{TF}^2 = \frac{e^2}{\epsilon_0} \frac{3n_e}{2\mu} = \frac{e^2}{\epsilon_0} \frac{3n_e}{m v_F^2} = \frac{3\omega_p^2}{v_F^2}. \tag{4.56}$$

To have a plasma frequency of oscillations, we need $\omega_p^2 \gg k^2 v_F^2$, i.e. $k \ll k_{TF}$. Otherwise, the screening effect overwhelms the plasma oscillations before they can begin.

Important Concepts to Remember

- Bosonic quasi-particles in materials are composite particles that exist only inside the material and have a finite lifetime: phonons are quanta of the oscillation of the crystal ion lattice, and plasmons are quanta of the oscillations of the (conduction) electron plasma.
- The 1-dimensional phonon model has dispersion relation $\omega_k^2 = 4\omega^2 \sin^2(ka/2)$, corresponding to an acoustic branch.
- In a 3-dimensional solid, small fluctuations of the ions in the network obey a 3-dimensional version of the 1-dimensional model.
- In general, dispersion relations have optical branches, where $\omega(k = 0) \neq 0$ and decreasing with k, and acoustic branches, where $\omega(k = 0) = 0$ and increasing with k.
- For a 3-dimensional system with p atoms per primitive cell, we have three acoustic branches and $3p - 3$ optical branches.
- For a diatomic lattice (like NaCl), at small k, the acoustic branch has $\omega \propto k^2$ and $\omega \simeq$ constant.

- In the interaction between the phonons and electrons, we have Umklapp interactions, where momentum is conserved only up to arbitrary elements $\vec{L}$ of the reciprocal lattice, $\vec{k} + \vec{q} + \vec{L} = \vec{k}'$.
- The plasma frequency is $\omega_p = \sqrt{ne^2/(\epsilon_0 m)}$, but harmonic plasma oscillations are valid only in the random phase approximation (RPA).
- The Thomas-Fermi approximation for screening of a charge in a plasma corresponds to assuming that the response of the plasma is as for a locally free electron gas, shifting the chemical potential to a local one, by the full effective potential, $\mu \to \mu - U_{eff}(\vec{r})$. It leads to an effective potential with a "mass" $k_{TF} = \sqrt{(e^2/\epsilon_0)(\partial n_e/\partial \mu)}$. For a free electron gas, $k_{TF} = \sqrt{3}\omega_p/v_F$.
- To have a plasma frequency, we need $\omega_p \gg k v_F$, i.e. $k \ll k_{TF}$.

Further Reading

Chapters 4 and 14 in the book by Kittel [6] and chapters 8.1 and 8.2 in the book by Phillips [7].

Where It will be Addressed in String Theory

The phonons, i.e. ion lattice oscillations, will be implicitly discussed in Chapter 47, where we discuss a holographic lattice. The plasmons, i.e. electron plasma oscillations, will be implicitly discussed in Chapters 40 and 47, where we discuss a strongly coupled electron plasma.

Exercises

(1) For optical phonons (bosons), $\omega \simeq \omega_0 = \text{const.}(k)$, calculate the heat capacity $C_V(T)$ (Einstein model).
(2) For acoustic phonons, $\omega = v_s k$ varying from 0 to $\omega_{\text{Debye}} = v_s(6\pi^2 n)^{1/2}$, derive this formula, and calculate the energy U and heat capacity.
(3) Find the general solutions to the phonon equation for (u, v) and $\omega(k)$ (no approximations).
(4) Calculate the continuum limit of the equation for lattice oscillations (phonons).
(5) Consider the lattice of identical ions of charge e and oscillations around the equilibrium points. Consider the number density of ions n, defined as one ion within a sphere of radius R. Calculate the value of ω ($= k_{el}/M$) as a function of e, R, and M.
(6) Consider a cubic lattice of ions of lattice spacing a. Write the form of the general Umklapp elastic photon scattering process as a function of a and arbitrary integers.
(7) Consider the Thomas-Fermi approximation for electrons in a type n semiconductor. Calculate k_{TF} and the screened conduction electron potential $U_{\text{eff}}(\vec{r})$.

Spin-Charge Separation in 1+1–Dimensional Solids: Spinons and Holons

In the last chapter, we saw that plasma oscillations with frequency ω_p are quantized, giving plasmons, bosonic quasi-particles. We also have interacting electrons, which are, according to Fermi liquid theory, in one-to-one correspondence with the states of the free electrons.

In this chapter, we will study a phenomenon that happens only in 1+1 dimensions. We will first see that instead of Fermi liquid theory, in 1+1 dimensions we have a *Luttinger liquid*, for which the number of degrees of freedom is different than for free electrons. We will prove that by using a fact that is peculiar to 1+1 dimensions: Fermions and bosons cannot be distinguished from their general properties, existing a nonlinear (and nonlocal) map between them, called *bosonization*. Using bosonization of the Luttinger liquid, we will see that the electrons break apart into separate spin and charge degrees of freedom, whose quanta of oscillation are called spinons and holons, respectively.

This mapping between fermions and bosons happens only in one spatial dimension, since one way to distinguish fermions is to do a rotation by 2π around an axis, after which a fermion picks up a minus sign (so it comes back to itself only by a 4π rotation). But in one spatial dimension, this is not available, so there is no physical difference between bosons and fermions, which is why we can map one into another.

5.1 One-Dimensional (Fermionic) Hubbard Model and the Luttinger Liquid

The interaction of 1 dimensional fermions on a discrete (atomic) lattice, or rather *chain* (since we are in one dimension), is described by the (fermionic, since there is also a bosonic version of it) Hubbard model. Its Hamiltonian is

$$H_L = -\frac{t}{2}\sum_{n,\sigma}[\psi_{n,\sigma}^{\dagger}\psi_{n+1,\sigma} + h.c.] + \mu\sum_{n}\psi_{n,\sigma}^{\dagger}\psi_{n,\sigma} + U\sum_{n,\sigma}\rho_{n,\sigma}\rho_{n,-\sigma}$$
$$= H_{\text{free}} + H_{\text{int}}. \tag{5.1}$$

Here $\psi_{n,\sigma}$ annihilates electrons at site n with spin $\sigma = (\uparrow, \downarrow)$, so $\rho_{n,\sigma} = \psi_{n,\sigma}^{\dagger}\psi_{n,\sigma}$ is the electron density. The first term is a hopping term between the site n and the site $n + 1$, annihilating an electron from site $n + 1$ and creating one at site n, thus t is called a hopping matrix element. The second term is a chemical potential term, μ being the chemical potential.

By substituting $\psi_n \sim e^{inka}$, we see that in momentum (k) space, the energy of the free term is $\epsilon_k = -t \cos ka + \mu$, where a is the lattice spacing, so to have zero energy at $k = k_F$, we must have $\mu = t \cos k_F a$. This is done by analogy: For a system of free fermions, we have

$$\epsilon_k = \frac{\hbar^2}{2m} k^2 - \epsilon_F, \qquad (5.2)$$

which does have $\epsilon_k = 0$ for $k = k_F$. For interacting fermions,

$$\epsilon_k = -2t \cos ka - \epsilon_F, \qquad (5.3)$$

again obeying the same relation, and being the same as the free fermion energy band at small k.

The interaction term has an energy U as the cost to have both spin states (up and down) at the same site.

The field splits into left-moving and right-moving degrees of freedom on the discrete chain labeled by n. In the vicinity of the Fermi surface, $k \simeq k_F$, we get, isolating the left-moving and right-moving modes with momentum $\simeq k_F$ but allowing still for an x-dependence that comes with a small momentum variation $\delta k = k - k_F \to \tilde{k}$,

$$\psi_{n,\sigma} = e^{+ink_F a} \psi_{n,\sigma+} + e^{-ik_F na} \psi_{n,\sigma-} \equiv R_\sigma(n) + L_\sigma(n). \qquad (5.4)$$

We can then expand the fields $\psi_{n+1,\sigma\pm}$ in R and L in a continuum limit as (the first nontrivial term in the Taylor expansion only)

$$\psi_{n+1,\sigma\pm} = \psi_{n,\sigma\pm} + a\partial_x \psi_{n,\sigma\pm}. \qquad (5.5)$$

Because of $\mu = t \cos k_F a$, the $\psi_n^\dagger \psi_n$ terms in H_{free} cancel, and we are left only with terms with derivatives. Actually, we would have also obtained oscillating terms, with $e^{\pm 2ik_F na}$ and a product of left and right fields, but that product gives zero when summed over n (e.g. $\sum_n e^{+2ik_F na} \psi_{n\sigma-}^\dagger \psi_{n\sigma+} \simeq 0$), so in the linearized limit (only the first Taylor term) we get

$$H_{\text{free}} \simeq -\hbar v_F \sum_\sigma \int_{-L/2}^{+L/2} dx [\psi_{\sigma+}^\dagger(x)(i\partial_x)\psi_{\sigma+}(x) + \psi_{\sigma-}^\dagger(x)(-i\partial_x)\psi_{\sigma-}(x)], \qquad (5.6)$$

where

$$\hbar v_F = at \sin k_F a. \qquad (5.7)$$

Writing a Fourier (momentum space) representation for the R (right) and L (left) fields ($\psi_{n\sigma\pm}$), which had a $e^{ik_F(na)}$ factored out, but which still allows for a small momentum $\delta k = k - k_F \to \tilde{k}$,

$$\psi_{\sigma\pm}(x) = \int_{-\infty}^{+\infty} \frac{d\tilde{k}}{2\pi} e^{i\tilde{k}x} \psi_{\sigma\pm}(\tilde{k}), \qquad (5.8)$$

we obtain from the above free linearized Hamiltonian the linearized dispersion relation

$$E_\pm(k) = \pm \hbar v_F \tilde{k} = \pm \hbar v_F (k - k_F). \qquad (5.9)$$

That means that there are infinitely negative energy states, an analogue of the Dirac sea (positrons) of the relativistic fermions, so the system has no ground state (since we can always find a state of a lower energy).

To make sense of this system, we need to find a way to fill up the Dirac sea. Luttinger's solution was to do a canonical transformation that effectively does this:

$$\begin{aligned}
\psi_{\sigma+}(k) &\rightarrow b_{k\sigma}, && k \geq 0 \\
&\quad\;\; c^\dagger_{k\sigma}, && k < 0 \\
\psi_{\sigma-}(k) &\rightarrow b_{k\sigma}, && k < 0 \\
&\quad\;\; c^\dagger_{k\sigma}, && k \geq 0.
\end{aligned} \tag{5.10}$$

Here the b and c oscillators obey the usual commutation relations. This transformation is effectively a Bogoliubov-type transformation, mixing up the creation and annihilation operators, and thus exchanging an excited state for one set into a ground state for another.

Replacing in the free Hamiltonian, one gets

$$H_{\text{free}} = \hbar v_F \int_{-\infty}^{+\infty} dp|p|[b^\dagger_p b_p + c^\dagger_p c_p] + W, \tag{5.11}$$

where W is a constant,

$$W = -\hbar v_F \int_0^\infty dpp + \hbar v_F \int_{-\infty}^0 pdp. \tag{5.12}$$

We see then that the energy is now positive definite, except for an (negative) infinite constant, which can be dropped. Since the remaining Hamiltonian is normal ordered, the procedure is equivalent to normal ordering and eliminates the infinite energy of the filled Dirac sea.

Normal ordering here means the usual procedure of writing $a^\dagger$'s to the left and a's to the right, e.g. $: a^\dagger a := : aa^\dagger := a^\dagger a$.

The resulting system in the effective description of the fermionic Hubbard model in 1+1 dimensions is the Luttinger liquid.

5.2 Bosonization

Within the context of the fermionic Hubbard model, we can moreover map the fermionic system to a bosonic one that will be easier to understand, by a process specific to 1+1 dimensions called bosonization. Bosonization in 1+1 dimensions is more general than the specific context we use it in, but we will not describe other applications here.

We consider the components of the $U(1)$ (charge) current for the fermion fields:

$$\begin{aligned}
j_0^\sigma &= \psi^\dagger_{\sigma+}\psi_{\sigma+}(x) + \psi^\dagger_{\sigma-}\psi_{\sigma-}(x) \\
j_1^\sigma &= \psi^\dagger_{\sigma+}\psi_{\sigma+}(x) - \psi^\dagger_{\sigma-}\psi_{\sigma-}(x).
\end{aligned} \tag{5.13}$$

These are operators, so if we consider them in the Heisenberg representation (as appropriate for free quantum fields), the Heisenberg equation of motion for j_0^σ is

$$-i\partial_t j_0^\sigma = [H_{\text{free}}, j_0^\sigma]. \tag{5.14}$$

Using the commutation relations for $\psi_{n,\sigma\pm}$ and their dagger conjugates to calculate the commutator above, we easily find

$$- i\partial_t j_0^\sigma = i\partial_x j_1^\sigma, \tag{5.15}$$

or, using relativistic notation,

$$\partial^\mu j_\mu^\sigma(x) = 0, \tag{5.16}$$

which is the conservation relation for currents. The currents are bosons (being products of two fermions), so we can trivially solve the conservation relation in terms of derivatives on a *real* boson field $\Phi_\sigma(x)$,

$$j_0^\sigma(x) \equiv \frac{1}{\sqrt{\pi}}\partial_x \Phi_\sigma(x)$$

$$j_1^\sigma(x) \equiv -\frac{1}{\sqrt{\pi}}\partial_t \Phi_\sigma(x) = -\frac{1}{\sqrt{\pi}}\Pi_\sigma(x), \tag{5.17}$$

where Π_σ is the momentum canonically conjugate to Φ_σ, so we have the canonical commutation relations

$$[\Phi_\sigma(x), \Pi_{\sigma'}(y)] = i\delta_{\sigma\sigma'}\delta(x-y). \tag{5.18}$$

Then the Heisenberg evolution equation for $\partial_x\Phi_\sigma$ in the boson representation, using the Hamiltonian

$$H_{\text{free}} = \frac{\hbar v_F}{2}\sum_\sigma \int dx [\Pi_\sigma^2(x) + (\partial_x \Phi_\sigma(x))^2], \tag{5.19}$$

is the same as the Heisenberg evolution equation (5.14) for $j_0^\sigma(x)$, and therefore the boson free Hamiltonian (5.19) is equivalent with the fermion free Hamiltonian (5.6).

This is the relation describing the boson in terms of fermions, and so is the *fermionization* relation.

Bosonization Map
The bosonization map is the inverse one, writing the fermion in terms of bosons:

$$\psi_{\sigma\pm}(x) = \frac{1}{\sqrt{2\pi a}} : e^{-i\sqrt{\pi}\left[\int_{-\infty}^x dx' \Pi_\sigma(x')\pm\Phi_\sigma(x)\right]} := \frac{1}{\sqrt{2\pi a}} : e^{\pm i\sqrt{\pi}\Phi_{\sigma\pm}(x)} :, \tag{5.20}$$

where

$$\Phi_{\sigma\pm}(x) = \Phi_\sigma(x) \mp \int_{-\infty}^x dx' \Pi_\sigma(x'). \tag{5.21}$$

It is in fact the inverse map to the above, as we will shortly see. Moreover, we can show that the usual commutation relation are preserved (the boson commutation relations imply the fermionic ones, and vice versa). Then the free part of the Hamiltonian for the bosons was found as above, so we need to look at the interaction part.

Including the factors of a coming from $\sum_n = \int dx/a$ and the normalization between R, L and $\psi_{n\sigma\pm}$, the interaction Hamiltonian is

$$H_{int} = aU \int_{-L/2}^{+L/2} dx (R_\uparrow^\dagger(x) + L_\uparrow^\dagger(x))(R_\uparrow(x) + L_\uparrow(x))(R_\downarrow^\dagger(x)$$
$$+ L_\downarrow^\dagger(x))(R_\downarrow(x) + L_\downarrow(x)). \tag{5.22}$$

Again, as in the case of the free Hamiltonian, and for the same reason, we consider only the nonoscillatory terms in H_{int}. Substituting R and L, we find

$$H_{int} = aU \int_{-L/2}^{+L/2} dx (\psi_{\uparrow+}^\dagger(x)e^{-ik_F na} + \psi_{\uparrow-}^\dagger(x)e^{+ik_F a})(\psi_{\uparrow+}(x)e^{+ik_F na} + \psi_{\uparrow-}(x)e^{-ik_f na})$$
$$\times (\psi_{\downarrow+}^\dagger(x)e^{-ik_F na} + \psi_{\downarrow-}^\dagger(x)e^{+ik_F na})(\psi_{\downarrow+}(x)e^{+ik_F na} + \psi_{\downarrow-}(x)e^{-ik_F na}). \tag{5.23}$$

Keeping only the nonoscillatory terms (with no exponential factors), we find

$$H_{int} = aU \int_{-L/2}^{+L/2} dx \Big[(\psi_{\uparrow+}^\dagger \psi_{\uparrow+} + \psi_{\uparrow-}^\dagger \psi_{\uparrow-})(\psi_{\downarrow+}^\dagger \psi_{\downarrow+} + \psi_{\downarrow-}^\dagger \psi_{\downarrow-})$$
$$+ (\psi_{\downarrow+}^\dagger \psi_{\uparrow-} \psi_{\downarrow-}^\dagger \psi_{\downarrow+} + H.c.) \Big]. \tag{5.24}$$

We see that the normal ordered Hamiltonian can be written in terms of products of normal ordered factors containing two fermions each, specifically

$$H_{int} = aU \int_{-L/2}^{+L/2} dx \Big[(: \psi_{\uparrow+}^\dagger \psi_{\uparrow+} : + : \psi_{\uparrow-}^\dagger \psi_{\uparrow-} :)(: \psi_{\downarrow+}^\dagger \psi_{\downarrow+} : + : \psi_{\downarrow-}^\dagger \psi_{\downarrow-} :)$$
$$+ (: \psi_{\downarrow+}^\dagger \psi_{\uparrow-} : : \psi_{\downarrow-}^\dagger \psi_{\downarrow+} : +h.c.) \Big]. \tag{5.25}$$

We can now write this Hamiltonian in terms of the bosons Φ_σ. To do that, we need to use the relation

$$e^A e^B = e^{A+B} e^{\frac{1}{2}[A,B]} \tag{5.26}$$

Applying this for $\psi^\dagger = e^A$ and $\psi = e^B$ written in terms of Φ_σ and Π, and using the commutation relations for Φ_σ and Π_σ (the canonical ones), we get

$$\psi_{\sigma+}^\dagger \psi_{\sigma-} = \frac{1}{2\pi a} e^{-i\sqrt{4\pi}\Phi_\sigma(x)} e^{-i\pi/2} = \frac{-i}{2\pi a} e^{-u\sqrt{\pi}\Phi_\sigma(x)}. \tag{5.27}$$

We also have that, if $C = A + B$, where A contains $A^\dagger$ and B contains A's, then

$$: e^C := e^A e^B = e^{\frac{1}{2}[A,B]} e^C. \tag{5.28}$$

Then we also obtain that, if $A = A^+ + A^-$ and $B = B^+ + B^-$, we have

$$: e^A : : e^B := e^{[A^-, B^+]} : e^{A+B} :. \tag{5.29}$$

The details are left as an exercise.

We want to apply this formula for $A = -i\sqrt{4\pi}\,\Phi^\uparrow(x)$, $B = +i\sqrt{4\pi}\,\Phi_\downarrow(x)$. Since from (5.27) we have

$$\psi^\dagger_{\uparrow+}\psi_{\downarrow-} = \frac{-i}{2\pi a}e^{-i\sqrt{4\pi}\,\Phi_\uparrow(x)}$$

$$\psi^\dagger_{\downarrow-}\psi_{\downarrow+} = \frac{i}{2\pi a}e^{+i\sqrt{4\pi}\,\Phi_\downarrow(x)}, \tag{5.30}$$

and since $[\Phi^-_\uparrow, \Phi^+_\downarrow] = 0$, we obtain

$$: \psi^\dagger_{\uparrow+}\psi_{\uparrow-} : : \psi^\dagger_{\downarrow-}\psi_{\downarrow+} : + h.c. = \frac{1}{(2\pi a)^2} : \left[e^{-i\sqrt{4\pi}(\Phi_\uparrow - \Phi_\downarrow)} + h.c. \right] :$$

$$= \frac{1}{2\pi^2 a^2} : \cos[\sqrt{4\pi}(\Phi_\uparrow(x) - \Phi_\downarrow(x)] : . \tag{5.31}$$

This takes care of one of the terms in H_{int}, and for the other types of terms, we need to calculate $: \psi^\dagger_{\sigma\pm}(x)\psi_{\sigma\pm}(x) :$. This, however, needs to be regularized as

$$: \psi^\dagger_{\sigma\pm}(x+\epsilon)\psi_{\sigma\pm}(x-\epsilon) :\, \sim \frac{1}{2\pi a} : e^{\mp i\sqrt{\pi}(\Phi_{\sigma\pm}(x+\epsilon) - \Phi_{\sigma\pm}(x-\epsilon))} : \tag{5.32}$$

But we will calculate it more precisely using the 2-point correlator. Going to momentum space, we have

$$\langle \psi^\dagger_{\sigma+}(x)\psi_{\sigma+}(0) \rangle = \int_{-\infty}^{+\infty} \frac{dk}{2\pi} \int_{-\infty}^{+\infty} \frac{dq}{2\pi} e^{iqx} \langle \psi^\dagger_{\sigma+}(q)\psi_{\sigma+}(k) \rangle, \tag{5.33}$$

but we introduce a regulator e^{-ak} inside the integral, and then using $\psi_{\sigma+}(k) = b_{k\sigma}$ for $k \geq 0$ and $c^\dagger_{k\sigma}$ for $k < 0$, we have $\langle 0|c^\dagger_{k\sigma}c_{k'\sigma'}|0\rangle = 2\pi\delta_{\sigma\sigma'}\delta(k - k')$. Finally that leads to

$$\langle \psi^\dagger_{\sigma+}(x)\psi_{\sigma+}(0) \rangle = \frac{1}{2\pi(a - ix)}, \tag{5.34}$$

and we use it for $x = \epsilon$.

Then we realize that taking the VEV of (5.29), and taking into account that $\langle : e^A : \rangle = 1$, as $\langle 0| : a^\dagger \ldots a : |0\rangle = 0$, we have

$$\langle 0| : e^A : : e^B : |0\rangle = e^{[A^-, B^+]}, \tag{5.35}$$

when the commutator is a c-number, so replacing in (5.29), we get

$$: e^A : : e^B : =: e^{A+B} : \langle 0| : e^A : : e^B : |0\rangle, \tag{5.36}$$

which means that

$$: \psi^\dagger_{\sigma\pm}(x)\psi_{\sigma\pm}(x) := \lim_{\epsilon \to 0}\lim_{a \to 0} : e^{\mp i\sqrt{\pi}(\Phi_{\sigma\pm}(x+\epsilon) - \Phi_{\sigma\pm}(x-\epsilon))} : \langle 0|\psi^\dagger_{\sigma\pm}(x+\epsilon)\psi_{\sigma\pm}(x)|0\rangle.$$
$$\tag{5.37}$$

Note the order of limits (we must first take a to zero, then ϵ to zero). Using the results we calculated, we obtain

$$
\begin{aligned}
: \psi_{\sigma\pm}^{\dagger}(x)\psi_{\sigma\pm}(x) : &= \lim_{\epsilon\to 0}\lim_{a\to 0}\frac{1}{2\pi a} : e^{\mp 2i\epsilon\sqrt{\pi}\partial_x\Phi_{\sigma\pm}(x)} : \left(\frac{a}{a \mp i\epsilon}\right) \\
&= \lim_{\epsilon\to 0}\lim_{a\to 0}\frac{1}{2\pi a} : [1 \mp 2i\sqrt{\pi}\epsilon\partial_x\Phi_{\sigma\pm} + \mathcal{O}(\epsilon^2)] : \left(\frac{a}{a \mp i\epsilon}\right) \\
&= \lim_{\epsilon\to 0}\left(\frac{1}{\mp 2\pi i\epsilon} + : \frac{1}{\sqrt{\pi}}\partial_x\Phi_{\sigma\pm} : + \mathcal{O}(\epsilon)\right).
\end{aligned}
\tag{5.38}
$$

We drop the infinite constant and keep only

$$
: \psi_{\sigma\pm}^{\dagger}(x)\psi_{\sigma\pm}(x) :=: \frac{1}{\sqrt{\pi}}\partial_x\Phi_{\sigma\pm} : .
\tag{5.39}
$$

Now we are finally in a position to calculate H_{int}, using the above relation and (5.31), obtaining

$$
: H_{int} := aU \int dx \sum_{\sigma}\left\{\frac{: \partial_x\Phi_{\uparrow}\partial_x\Phi_{\downarrow} :}{\pi} + \frac{1}{2\pi^2 a^2} : \cos\left[\sqrt{4\pi}(\Phi_{\uparrow}(x) - \Phi_{\downarrow}(x))\right]\right\}.
\tag{5.40}
$$

5.3 Spin-Charge Separation: Spinons and Holons

We now can find that the bosonized electron splits into two degrees of freedom, one describing charge transport and one spin transport. Specifically, we define

$$
\Phi_C = \frac{\Phi_{\uparrow} + \Phi_{\downarrow}}{\sqrt{2}}; \quad \Phi_S = \frac{\Phi_{\uparrow} - \Phi_{\downarrow}}{\sqrt{2}},
\tag{5.41}
$$

where the bosonic field Φ_C is the charge carrier, the "holon," and the bosonic field Φ_S is the spin carrier, the "spinon."

With these redefinitions, we easily see that the Hamiltonian splits as

$$
\begin{aligned}
H &= H_C + H_S \\
H_C &= \frac{\hbar v_F}{2}\int dx\left[\Pi_C^2 + g_c^2(\partial_x\Phi_C)^2\right] \\
H_S &= \frac{\hbar v_F}{2}\int dx\left[\Pi_S^2 + g_s^2(\partial_x\Phi_S)^2\right] + \frac{U}{2\pi^2 a^2}\int dx : \cos[\sqrt{8\pi}\Phi_S] :,
\end{aligned}
\tag{5.42}
$$

where

$$
g_c^2 = 1 + \frac{aU}{2\pi\hbar v_F}; \quad g_s^2 = 1 - \frac{aU}{2\pi\hbar v_F}.
\tag{5.43}
$$

Rescaling

$$
\Phi_C\sqrt{g_c} = \tilde{\Phi}_C, \quad \Phi_S\sqrt{g_s} = \tilde{\Phi}_S, \quad \frac{\Pi_C}{\sqrt{g_c}} = \tilde{\Pi}_C, \quad \frac{\Pi_S}{\sqrt{g_s}} = \tilde{\Pi}_S,
\tag{5.44}
$$

the Hamiltonians admit a canonical form,

$$H_C = \frac{\hbar v_F^C}{2} \int dx [\tilde{\Pi}_C^2 + (\partial_x \tilde{\Phi}_C)^2]$$

$$H_S = \frac{\hbar v_F^S}{2} \int dx \left[\tilde{\Pi}_S^2 + (\partial_x \tilde{\Phi}_S)^2\right] + \frac{U}{2\pi^2 a^2} \int dx : \cos\left(\sqrt{\frac{8\pi}{g_s}} \tilde{\Phi}_S\right) :, \qquad (5.45)$$

where

$$v_F^C = v_F g_c; \quad v_F^S = v_F g_s. \qquad (5.46)$$

In this form, we see clearly that we have obtained separation of the two degrees of freedom, the (bosonic) charge carrier, called a holon, and the (bosonic) spin carrier, called a spinon. Moreover, as we can see, the two degrees of freedom move at different velocities.

The calculation that we performed in this chapter means that the electron falls apart into the holon and spinon quasi-particles. This is a departure from the Fermi liquid behavior, since one degree of freedom splits into two. This behavior occurs only in one spatial dimension; it is unclear if there is an analogue in higher dimensions (perhaps in the case of the non-Fermi liquids). The resulting system is called *Luttinger liquid*, as we mentioned.

Note that the bosonization map relates the terms

$$\frac{1}{\pi a} \cos[\sqrt{4\pi} \Phi] \leftrightarrow i(R^\dagger L - L^\dagger R), \qquad (5.47)$$

and the right-hand side corresponds to a mass term for the fermion.

5.4 Sine-Gordon vs. Massive Thirring Model Duality

In fact, the bosonization relates the so-called *sine-Gordon model*, a bosonic model with the canonical kinetic term and a cos potential, like we obtained, with a *massive Thirring model*, a fermionic model with a mass term, as we also obtained. This duality was first obtained by Coleman and is the quintessential example of *duality*, which is a nonlocal map that relates fundamental charges (currents) with topological charges (currents), and weak coupling with strong coupling.

Indeed, in relativistic notation, we obtain the duality between the bosonic action (sine-Gordon)

$$S = \int d^2x \left[\frac{(\partial_\mu \Phi)^2}{2} + \frac{\alpha}{\beta^2}(\cos \beta \Phi - 1)\right], \qquad (5.48)$$

and the fermionic action (massive Thirring)

$$S = \int d^2x \left[\bar{\psi} i \slashed{\partial} \psi - m_F \bar{\psi}\psi - \frac{g}{2}(\bar{\psi}\gamma^\mu \psi)^2\right]. \qquad (5.49)$$

For this action, the bosonization map is a slight generalization (with a parameter β) of the map we derived,

$$\psi_{\pm}(x) = C_{\pm} : \exp\left[\frac{2\pi}{i\beta}\int_{-\infty}^{x}\frac{\partial\Phi(x')}{\partial t}dx' \mp \frac{i\beta}{2}\Phi(x)\right] :, \tag{5.50}$$

so as we said, it is a nonlocal map. Its inverse is

$$\frac{m_0^2}{\beta^2}\cos[\beta\Phi] = -m_F\bar{\psi}\psi. \tag{5.51}$$

The coupling relation maps weak to strong coupling:

$$\frac{\beta}{4\pi^2} = \frac{1}{1+g\pi}. \tag{5.52}$$

Moreover, it maps a fundamental (electric) fermion current to a topological (monopole, i.e. solitonic) current:

$$\bar{\psi}\gamma^{\mu}\psi = -\frac{\beta}{2\pi}\epsilon^{\mu\nu}\partial_{\nu}\Phi. \tag{5.53}$$

Indeed, whereas the conservation of the left-hand side is a dynamical statement (on-shell), the conservation of the right-hand side is a topological statement, i.e. it is trivially true. It is left as an exercise to show that this map is obtained in the relativistic case from the nonrelativistic maps written in this chapter.

Important Concepts to Remember

- In 1+1 dimensions, fermions are mapped into bosons by the bosonization map.
- In 1+1 dimensions, interacting electrons are described by a Luttinger liquid instead of a Fermi liquid, and they break apart into charge carrier (bosonic) quasi-particles called holons and spin carrier (bosonic) quasi-particles called spinons, moving at different velocities.
- The fermionic Hubbard model has a hopping term, a chemical potential term, and an interaction term between electrons of opposite spins at the same site.
- We need to make a canonical transformation, equivalent to a normal ordering, to fill up the Dirac sea of the fermionic Hubbard model.
- The bosonization map relates j_0^{σ} with $\partial_x\Phi_{\sigma}$ and j_1^{σ} with $-\partial_t\Phi_{\sigma}$, and $\psi_{\sigma\pm}$ with $: e^{\pm\sqrt{\pi}\Phi_{\sigma\pm}} :$.
- The charge and spin carriers are the sum and difference of the bosonized fields Φ_{σ}.
- Bosonization relates the bosonic sine-Gordon model with the fermionic massive Thirring model, mapping weak to strong coupling and fundamental (electric) particles with topological (solitonic) particles.

Further Reading

Chapter 9.1 in the book by Phillips [7].

Where It will be Addressed in String Theory

This chapter lays the groundwork for further elaborations. The fermionic Hubbard model is relevant for insulators, described in Chapter 13, and which will be addressed in string theory via the holography in Chapter 47. The spinons, or spin waves, or magnons (in the context of the Heisenberg model) will be developed further in the next chapter and will be addressed in string theory, holographically, in Chapters 28 and 29.

Exercises

(1) Calculate the equal time commutation relations for $j_0^\sigma(x)$ and $j_1^\sigma(x)$.

(2) Prove that

$$: e^A :: e^B := : e^{A+B} : e^{\frac{1}{2}[A^+, B^-]}, \tag{5.54}$$

where $A = A^+ + A^-$ and $B = B^+ + B^-$.

(3) Prove that the relativistic form of the $\psi\psi \to \phi$ map is

$$-\frac{\beta}{2\pi}\epsilon^{\mu\nu}\partial_\nu\phi = \bar{\psi}\gamma^\mu\psi, \tag{5.55}$$

given the nonrelativistic relations in the text.

(4) Consider

$$\Phi_L^\mu = \sum_n \alpha_n^\mu a_n e^{-in(t+x)} + \sum_n \alpha_n^{*\mu} a_n^\dagger e^{+in(t+x)}, \tag{5.56}$$

and similarly for Φ_R^μ, with $[a_n, a_m^\dagger] = \delta_{n,m}$, etc. Calculate

$$: e^{ik_\mu^1(\Phi_L^\mu + \Phi_R^\mu)} :: e^{ik_\mu^2(\Phi_L^\mu + \Phi_R^\mu)} : \tag{5.57}$$

in terms of normal ordered terms.

6   The Ising Model and the Heisenberg Spin Chain

We saw in the last chapter that in 1+1–dimensional systems, spin and charge degrees of freedom separate (the electron breaks up into separate degrees of freedom). That means that we can treat the spin degrees of freedom separately and consider a purely spin model. Moreover, there are indeed materials that are best described in terms of simply the spin degrees of freedom.

6.1 The Ising Model in $1+1$ and $2+1$ Dimensions

The first and most famous model is a classical spin model called the Ising model. Wilhelm Lenz gave this problem to his Ph.D. student Ernst Ising, who solved it in 1+1 dimensions, where he found no phase transition. In 2+1 dimensions, however we will see that we do have a phase transition.

The model is the simplest thing we can imagine, a classical nearest-neighbor interaction between the spins 1/2, described by elements σ_i taking the values $+$ or $-$ (up or down), so

$$H = -\sum_{\langle ij \rangle} J_{ij}\sigma_i\sigma_j - \sum_j h_j\sigma_j. \tag{6.1}$$

Here $\langle ij \rangle$ means the nearest neighbors ij, and h_j is an external magnetic field. If $J_{ij} > 0$, we are in the ferromagnetic case, if $J_{ij} < 0$ we are in the antiferromagnetic case.

The probability distribution is

$$P_\beta(\sigma) = \frac{e^{-\beta H}}{Z(\beta)}, \tag{6.2}$$

and the partition function is

$$Z(\beta) = \sum_{\sigma=\pm} e^{-\beta H(\sigma)}. \tag{6.3}$$

We usually consider the case of constant coupling, $J_{ij} = J$, and we consider a (hyper)cubic lattice, $\Lambda = \mathbb{Z}^d$.

If we also have $h = 0$ (no external field), we get

$$H(\sigma) = -J\sum_{\langle ij \rangle} \sigma_i\sigma_j. \tag{6.4}$$

6.1.1 $d = 1$ Case

The 1+1–dimensional case was solved by Ising in his 1924 Ph.D. thesis. He found that there is no phase transition, which he did by calculating that the two-point function is exponentially suppressed,

$$\langle \sigma_i \sigma_j \rangle_\beta \le C \exp[-C(\beta)|i - j|], \tag{6.5}$$

which means a disordered system, independently of temperature.

We will instead solve the problem of the 1-dimensional Hamiltonian (spin chain):

$$H(\sigma) = -J \sum_{i=1}^{L} \sigma_i \sigma_{i+1} - h \sum_i \sigma_i. \tag{6.6}$$

Then we find that the free energy (Legendre transformed over the magnetic field h) per particle (i.e. per spin, so we divide by the length L of the chain) is

$$f^*(\beta, h) = -\lim_{L \to \infty} \frac{1}{\beta L} \ln Z(\beta)$$

$$= -\frac{1}{\beta} \ln \left[e^{\beta J} \cosh \beta h + \sqrt{e^{2\beta J}(\sinh \beta h)^2 + e^{-2\beta J}} \right]. \tag{6.7}$$

$h = 0$ Case

We first solve the case of zero field. We define $\sigma_j' = \sigma_j \sigma_{j-1}$ for $j \le 2$. That means that we can turn the sum over σ_i into a sum over σ_i' times a factor of 2, since $\sigma_i' = +$ if $(\sigma_i, \sigma_{i-1}) = (++)$ or $(--)$, and $\sigma_i' = -$ if $(\sigma_i, \sigma_i') = (+-)$ or $(-+)$. Thus

$$Z(\beta) = \sum_{\sigma_1, \ldots, \sigma_L = \pm 1} e^{\beta J \sigma_1 \sigma_2} e^{\beta J \sigma_2 \sigma_3} \ldots e^{\beta J \sigma_{L-1} \sigma_L}$$

$$= 2 \prod_{j=2}^{L} \sum_{\sigma_j' = \pm} e^{\beta J \sigma_j'}$$

$$= 2[e^{\beta J} + e^{-\beta J}]^{L-1}. \tag{6.8}$$

Then we find

$$f^*(\beta, 0) = f(\beta, 0) = -\frac{1}{\beta} \ln[e^{\beta J} + e^{-\beta J}]. \tag{6.9}$$

$h \ne 0$ Case

We solve the case of nonzero field using the *transfer matrix method*. Now we can write the partition function as

$$Z(\beta) = \sum_{\sigma_1, \ldots, \sigma_L} V_{\sigma_1 \sigma_2} V_{\sigma_2 \sigma_3} \ldots V_{\sigma_L \sigma_1}; , \tag{6.10}$$

where we used $L + 1 \equiv 1$ (periodic boundary conditions), and we defined the *transfer matrix* (in the spin σ_i, $\sigma_{i+1} = \pm 1$ space)

$$V_{\sigma\sigma'} = e^{\frac{\beta h}{2}\sigma}\, e^{\beta J\sigma\sigma'}\, e^{\frac{\beta h}{2}\sigma'}. \tag{6.11}$$

As a matrix then, this becomes

$$V = \begin{pmatrix} e^{\beta(h+J)} & e^{-\beta J} \\ e^{-\beta J} & e^{-\beta(h-J)} \end{pmatrix}. \tag{6.12}$$

Then the partition function (in the thermodynamic, $L \to \infty$ limit) is

$$Z(\beta) = \text{Tr}[V^L] = \lambda_1^L + \lambda_2^L = \lambda_1^L\left(1 + \left(\frac{\lambda_1}{\lambda_2}\right)^L\right) \simeq \lambda_1^L, \tag{6.13}$$

where λ_1 and λ_2 are the eigenvalues of the transfer matrix V, with λ_1 being the largest.

To find the eigenvalues, we solve the equation $\det(V - \lambda\mathbf{1}) = 0$, which gives

$$\lambda_{1,2} = e^{\beta J}\cosh(\beta h) \pm \sqrt{e^{2\beta J}(\sinh\beta h)^2 + e^{-2\beta J}}. \tag{6.14}$$

Then finally, the free energy of the system is

$$f^*(\beta, h) = -k_B T \ln\lambda_1 = -J - k_B T \ln\left[\cosh(\beta h) + \sqrt{\sinh^2(\beta h) + e^{-4\beta J}}\right]. \tag{6.15}$$

We can now calculate the magnetization

$$\mathcal{M} = \frac{\partial f^*}{\partial h} = \frac{\sinh(\beta h) + \frac{\sinh(\beta h)\cosh(\beta h)}{\sqrt{\sinh^2(\beta h) + e^{-4\beta J}}}}{\cosh(\beta h) + \sqrt{\sinh^2(\beta h) + e^{-4\beta J}}}. \tag{6.16}$$

We notice from the above that at a zero magnetic field ($h = 0$), there is no magnetization, independent of temperature, which means that, as advertised, *there is no phase transition in one spatial dimension.*

For small βh, $\beta h \ll 1$, the magnetization becomes

$$\mathcal{M} \simeq -\frac{\beta h + \beta h e^{2\beta J}}{1 + e^{-2\beta J}} = -\beta h e^{2\beta J}. \tag{6.17}$$

Therefore the magnetic susceptibility at small βh is

$$\chi = \frac{\partial\mathcal{M}}{\partial h} \simeq \frac{e^{2\beta J}}{k_B T}. \tag{6.18}$$

To calculate the energy per particle (per spin), we note that $f = -s\,dT + \cdots$, so $s = -\partial f/\partial T$, and since $u = f + Ts$, we find

$$u = f - T\frac{\partial f}{\partial T} = f + \beta\frac{\partial f}{\partial\beta} \simeq -J\tanh(\beta J) \tag{6.19}$$

at small βh. That means that the volumic specific heat is

$$c_V = \left.\frac{\partial u}{\partial T}\right|_V = \frac{J^2}{k_B T^2}\frac{1}{\cosh^2\beta J}. \tag{6.20}$$

6.1.2 The $d = 2$ Case

In two spatial dimensions (or higher), in finite volume, the partition function cannot have a singularity (since it is a discrete sum of analytic functions, it means itself is analytic). Therefore, one couldn't have a phase transition from these partition functions coming from sums, it was thought. But in infinite volume, we can have a singularity. In 1933, Rudolph Peierls gave an argument for the existence of a phase transition in the 2-dimensional Ising model. However, the solution of the model was written only in 1944 by Onsager, and only in 1952 did we have the first published proof, by C. N. Yang, of the formula.

We will not give the derivation here, since it is difficult. We will say only that Onsager's solution has the magnetization at $h = 0$,

$$\mathcal{M} = \{1 - [\sinh(2\beta J)]^{-4}\}^{1/8}, \tag{6.21}$$

for $T < T_c$. From the above, we can calculate the critical temperature, by equating $\mathcal{M}$ with zero, obtaining

$$T_c = \frac{2J}{k_B \log(1 + \sqrt{2})}. \tag{6.22}$$

6.2 Mean Field Approximation

Instead of showing the full solution of Onsager, we will calculate the magnetization, and the resulting T_c, from a simple mean field approximation that we basically already did (without the full details) in Chapter 2, for the ferromagnetism case. We should note that there is no small parameter for a mean field approximation: It is an *uncontrolled* approximation, and one only hopes (and gets, as we will see) to obtain qualitative information, but quantitatively the result will be off by a factor of order 1.

The average spin is related to the magnetization by

$$\langle \vec{S} \rangle = \frac{1}{L} \left(\sum_{i=1}^{L} S_i \right) = \frac{m}{g\mu_B}. \tag{6.23}$$

The Hamiltonian of the system,

$$\mathcal{H} = -J \sum_{\langle ij \rangle} \vec{S}_i \cdot \vec{S}_j - g\mu_0\mu_B \vec{H} \cdot \sum_i \vec{S}_i, \tag{6.24}$$

can be rewritten as

$$\mathcal{H} = -g\mu_0\mu_B \sum_i \vec{S}_i \cdot \vec{H}_{eff}, \tag{6.25}$$

where the effective field is

$$\vec{H}_{eff} = \vec{H} + J \frac{\langle \vec{S} \rangle}{g\mu_0\mu_B}. \tag{6.26}$$

Then we can follow the same steps as in the paramagnetic case and define the variable

$$x = \beta g \mu_0 \mu_B H_{eff} S. \tag{6.27}$$

We calculate the total magnetization, and as usual we find

$$\mathcal{M} = L g S \mu_B B_S(x), \tag{6.28}$$

where $B_S(x)$ is the Brillouin function. On the other hand, we have (by definition)

$$\mathcal{M} = L g \mu_B \langle \vec{S} \rangle. \tag{6.29}$$

In the case of $S = 1/2$, we have σ_i instead of the general $\vec{S}_i$, and we can rewrite the Hamiltonian as

$$\mathcal{H} = -\frac{1}{2} L J m^2 + (J m + g \mu_0 \mu_B H) \sum_i \sigma_i. \tag{6.30}$$

Indeed, we can check that if we eliminate m by its equation of motion, we get the original Hamiltonian, with the replacement

$$\sum_{\langle ij \rangle} \vec{S}_i \cdot \vec{S}_j \rightarrow \left(\sum_i \vec{S}_i \right)^2. \tag{6.31}$$

This is the mean field approximation.

The partition function is obtained then as

$$\begin{aligned} Z &= e^{-\frac{\beta L J m^2}{2}} \sum_{\sigma_i} \cdots \sum_{\sigma_N} e^{\beta (J m + g \mu_0 \mu_B H) \sum_i \sigma_i} \\ &= e^{-\frac{\beta L J m^2}{2}} \left[\sum_{\sigma = \pm 1} e^{\beta (J m + g \mu_0 \mu_B H) \sigma} \right]^L \\ &= e^{-\frac{\beta L J m^2}{2}} \left[2 \cosh \beta (J m + g \mu_0 \mu_B H) \right]^L. \end{aligned} \tag{6.32}$$

From this, we can calculate the free energy as

$$f = -\frac{1}{\beta L} \ln Z, \tag{6.33}$$

and then the magnetization

$$m = \frac{\mathcal{M}}{L} = -\frac{\partial f}{\partial H} = \tanh[\beta (J m + g \mu_0 \mu_B H)], \tag{6.34}$$

where tanh is the Brillouin function $B_{1/2}$.

At zero magnetic field, $H = 0$, we have a transcendental equation,

$$m = \tanh(\beta J m), \tag{6.35}$$

which has a nontrivial solution ($m \neq 0$) for $T < T_c$, and we can find the critical temperature as $\beta_C J = 1$, giving

$$T_C = \frac{J}{k_B}, \tag{6.36}$$

so we obtain the correct parametric behavior, but the coefficient is wrong by a factor of $2/\log(1 + \sqrt{2})$.

6.3 Kramers-Wannier Duality at $H = 0$

Kramers-Wannier duality is an important property of the Ising model in two dimensions that is an example of a *nonperturbative (weak coupling–strong coupling) duality*.

The Ising model in two dimensions at $H = 0$ can be written as

$$Z = \sum_{\sigma_i=\pm 1} e^{\beta J \sum_{(ij)} \sigma_i \sigma_j} = \sum_{\sigma_i=\pm 1} \prod_{\langle ij \rangle} e^{\beta J \sigma_i \sigma_j}$$

$$= e^{\beta J L} \sum_{\sigma_i=\pm 1} e^{-2\beta J \hat{L}}, \tag{6.37}$$

where $\hat{L}$ is the perimeter (number of sites) of the domain walls between the islands of $\sigma = +1$ and $\sigma = -1$. This formula appears because only on these domain walls do we have $e^{-\beta J \hat{L}}$, since $\sigma_i \sigma_j = -1$, whereas in the rest we have $e^{+\beta J \hat{L}}$, since $\sigma_i \sigma_j = +1$. This formula is defined on the dual lattice.

Low-Temperature Expansion (βJ Large)
We can expand this formula in the case βJ is large. Then, writing

$$Z = \sum_{\hat{L}} n(\hat{L}) e^{-2\beta J}, \tag{6.38}$$

where $n(\hat{L})$ is the number of configurations of closed loops of perimeter $\hat{L}$, we have that in the large βJ limit,

$$n(\hat{L}) \propto k^{\hat{L}}, \tag{6.39}$$

where k is some constant (the result must be an exponential in $\hat{L}$). In turn, that means that

$$Z \propto \sum_{\hat{L}} e^{\hat{L}(\ln k - 2\beta J)}, \tag{6.40}$$

so we have a phase transition (transition from convergent to divergent, or rather from a phase dominated by short loops to a phase dominated by long loops) at

$$T_C = \frac{2J}{\ln k}. \tag{6.41}$$

As we see, we obtain the correct form of the transition temperature, without being able to calculate k, which is in fact $(1 + \sqrt{2})$.

High T_C Expansion (βJ Small)
On the other hand, we can do an expansion in βJ small of the formula for $Z(\beta)$. Using the fact that $e^{\pm x} = \cosh x \pm \sinh x$, we can write ($\sigma_i \sigma_j = \pm$)

$$e^{\beta J \sigma_i \sigma_j} = \cosh(\beta J) + \sigma_i \sigma_j \sinh(\beta J). \tag{6.42}$$

To calculate

$$Z = \sum_{\sigma_i = \pm 1} \prod_{\langle ij \rangle} e^{\beta \sigma_i \sigma_j} = \sum_{\sigma_i = \pm 1} \prod_{\langle ij \rangle} [\cosh(\beta J) + \sigma_i \sigma_j \sinh(\beta J)], \qquad (6.43)$$

we first expand this product into a sum of 2^{N_l} terms, where N_l is the number of links of the dual lattice, some with $\cosh(\beta J)$ and some with $\sigma_i \sigma_j \sinh(\beta J)$. Then we define

$$d_{\langle ij \rangle} = 0, \quad \text{if} \quad \cosh(\beta J) \in \text{link}$$
$$= 1, \quad \text{if} \quad \sigma_i \sigma_j \sinh(\beta J) \in \text{link}. \qquad (6.44)$$

Then, similarly to the original formula in terms of $\hat{L}$, we write Z as

$$Z = \cosh(\beta J)^{N_l} \sum_{\sigma_i = \pm 1} \sum_{d_{\langle ij \rangle} = 0,1} \prod_{\langle ij \rangle} (\sigma_i \sigma_j \tanh(\beta J))^{d_{\langle ij \rangle}}. \qquad (6.45)$$

On the other hand,

$$\sum_{\sigma_i = \pm 1} \prod_{\langle ij \rangle} (\sigma_i \sigma_j)^{d_{\langle ij \rangle}} = \sum_{\sigma_i = \pm 1} \prod_i (\sigma_i)^{b_i}, \qquad (6.46)$$

where $b_i = \sum_{j \text{ next to } i} d_{\langle ij \rangle}$. Then, since $\sum_{\sigma_j = \pm 1} \sigma_j = 0$, the above formula gives 0 if b_i is odd (when $\sigma_i^{b_i} = \sigma_i$) and 2^{N_l} if b_i is even (when $\sigma_i^{b_i} = 1$).

Then

$$Z = (2 \cosh(\beta J))^{N_l} \sum_{d_{\langle ij \rangle} = 0,1; \, b_i = 0,2,4,\dots} (\tanh(\beta J))^{\sum_{\langle ij \rangle} d_{\langle ij \rangle}}$$
$$= (2 \cosh(\beta J))^{N_l} \sum_{\text{closed loops}} (\tanh(\beta J))^{\hat{L}} \propto [\tanh(\beta J)]^{\hat{L}}, \qquad (6.47)$$

since $\hat{L} = \sum_{\langle ij \rangle} d_{\langle ij \rangle}$.

Now calling βJ in this high T expansion $\widehat{\beta J}$, to distinguish it from βJ in the low T expansion, we can equate the high T expansion of a system with $\widehat{\beta J}$ with the small T expansion of a system with βJ, and obtain

$$e^{-2\beta J} = \tanh(\widehat{\beta J}). \qquad (6.48)$$

This is the example of nonperturbative duality that we explained, between a system with small coupling $\widehat{\beta J}$ and a system with large coupling βJ. It is a concrete example in condensed matter physics of duality, perhaps the only one that one can actually observe (other dualities, in quantum field theories, are useful mostly as theoretical tools).

6.4 The Heisenberg Model and Coordinate Bethe Ansatz

The Heisenberg Model

We now move on to a quantum version of the classical Ising model. To do that, we can simply promote the σ_i taking values ± 1 to the Pauli matrices σ^i describing the quantum system. That in particular means that we think of the spin as a 3-dimensional quantity, defined, however, only on the 1-dimensional spin chain. Therefore the Hamiltonian of the model is

$$H = J \sum_{j=1}^{L} \sum_{\alpha=1}^{3} \sigma_j^\alpha \sigma_{j+1}^\alpha. \tag{6.49}$$

We considered here the opposite sign for J for comparison with the literature, such that $J < 0$ corresponds to the ferromagnetic case and $J > 0$ to the antiferromagnetic case. The model above has the name of the Heisenberg XXX spin 1/2 chain. The name comes from the fact that the coupling is the same for all the directions, $J_x = J_y = J_z$. A general case, with different couplings for the different components, would be

$$H = J \sum_{j=1}^{L} \sum_{\alpha=1}^{3} J_\alpha \sigma_j^\alpha \sigma_{j+1}^\alpha, \tag{6.50}$$

which goes under the name of the Heisenberg XYZ spin 1/2 chain. One also talks about the XXZ chain if $J_x = J_y \neq J_z$.

This model was introduced by Heisenberg in 1928 and was "solved" by Bethe in 1931 by the method of the "coordinate Bethe ansatz." More precisely, he wrote an ansatz, and the resulting Bethe ansatz equations that the ansatz must satisfy, though of course solving these equations is a difficult problem that is still the subject of research and will be addressed using string theory methods.

Coordinate Bethe Ansatz

A basis of eigenstates for the Hamiltonian is written in terms of states of definite $\vec{S}^2$ and S_z at each site j, i.e. spin up $|\uparrow\rangle$ and spin down $|\downarrow\rangle$. We denote by $|x_1, \ldots, x_N\rangle$ the spin chain state with N spins up, called *magnons* at sites $\{x_i\} = \{x_1, \ldots, x_N\}$ along a chain of spins down, e.g.

$$|1, 3, 4\rangle_{L=5} \equiv |\uparrow\downarrow\uparrow\uparrow\downarrow\rangle. \tag{6.51}$$

We also define the permutation operator of sites i and j, P_{ij}, that permutes the value of the spins at sites i and j, i.e.:

$$P_{ij}|\uparrow_i\uparrow_j\rangle = |\uparrow_i\uparrow_j\rangle; \quad P_{ij}|\downarrow_i\downarrow_j\rangle = |\downarrow_i\downarrow_j\rangle$$
$$P_{ij}|\uparrow_i\downarrow_j\rangle = |\downarrow_i\uparrow_j\rangle; \quad P_{ij}|\downarrow_i\uparrow_j\rangle = |\uparrow_i\downarrow_j\rangle. \tag{6.52}$$

Then we can easily check that we can write P_{ij} as

$$P_{ij} = \frac{1}{2} + \frac{1}{2}\vec{\sigma}_i \cdot \vec{\sigma}_j = \frac{1}{2} + \frac{1}{2}\sigma_i^3\sigma_j^3 + \sigma_i^+\sigma_j^- + \sigma_i^-\sigma_j^+. \tag{6.53}$$

Using this formula, the Heisenberg Hamiltonian becomes

$$H = J \sum_{j=1}^{L} (2P_{j,j+1} - 1). \tag{6.54}$$

We consider periodic boundary conditions for the spin chain: $L + 1 \equiv 1$.

It is easier to subtract a constant factor JL from the energy, so that the total Hamiltonian is

$$H = 2J \sum_{i=1}^{L} (P_{j,j+1} - 1). \tag{6.55}$$

Note that

$$(P_{j,j+1} - 1)|x\rangle = |x - 1\rangle - |x\rangle, \quad \text{for} \quad j + 1 = x$$
$$= |x + 1\rangle - |x\rangle, \quad \text{for} \quad j = x. \tag{6.56}$$

The simplest state, a "*one-magnon*" *state* (single quasi-particle state, with a single spin up on the chain), is one that diagnonalizes trivially the Hamiltonian, by a discrete Fourier transform (on the chain):

$$|\psi(p_1)\rangle = \sum_{x=1}^{L} e^{ip_1 x} |x\rangle. \tag{6.57}$$

Indeed, substituting in the Schrödinger equation,

$$H|\psi(p_1)\rangle = 2J \sum_{x=1}^{L} e^{ip_1 x} (|x - 1\rangle + |x + 1\rangle - 2|x\rangle)$$

$$= 2J(e^{ip_1} + e^{-ip_1} - 2) \sum_{x=1}^{L} e^{ip_1 x} |x\rangle$$

$$= -8J \sin^2 \frac{p_1}{2} |\psi(p_1)\rangle, \tag{6.58}$$

so the energy of a single magnon is

$$E = -8J \sin^2 \frac{p}{2}. \tag{6.59}$$

Two-Magnon State

The Bethe ansatz becomes interesting in the case of two magnons (two quasi-particles). We want to find a state that is an eigenstate of the Hamiltonian:

$$H|\psi(p_1, p_2)\rangle = E|\psi(p_1, p_2)\rangle. \tag{6.60}$$

The Bethe ansatz is now that the state is a superposition of the incoming plane wave and the outgoing plane wave, with an S-matrix $S(p_1, p_2)$ for the scattered particles. Specifically,

$$\psi(x_1, x_2) = e^{i(p_1 x_1 + p_2 x_2)} + S(p_2, p_1) e^{i(p_2 x_1 + p_1 x_2)}. \tag{6.61}$$

The idea is that particles exchange momenta and scatter by the matrix S. The eigenstate is

$$|\psi(p_1, p_2)\rangle = \sum_{1 \leq x_1 < x_2 \leq L} \psi(x_1, x_2)|x_1, x_2\rangle. \tag{6.62}$$

Substituting the ansatz in the Schrödinger equation, one first gets the rather surprising result that the total energy of the system is simply the sum of the energies of the individual magnons:

$$E = -8J \left(\sin^2 \frac{p_1}{2} + \sin^2 \frac{p_2}{2} \right), \tag{6.63}$$

instead of having an interaction energy as well; and second, one obtains the form of the S-matrix,

$$S(p_1, p_2) = \frac{\phi(p_1) - \phi(p_2) + i}{\phi(p_1) - \phi(p_2) - i} = S^{-1}(p_2, p_1), \tag{6.64}$$

where

$$\phi(p) = \frac{1}{2} \cot \frac{p}{2} \tag{6.65}$$

is called the "Bethe root," for reasons to be explained shortly. The details of the calculation are left as an exercise (Exercise 1).

Note that this S-matrix has poles at $\phi_{12} \equiv \phi_1 - \phi_2 = i$ in the complex plane, which means, according to S-matrix theory, that there is a bound state of two magnons. S-matrix theory was thought in the 1960s to be able to completely constrain the physics of a system. That program was overly ambitious, yet one can still gain a lot of information from letting momenta and angular momenta (in spatial dimensions higher than one) be complex and studying the analyticity properties of the S-matrix. Here we *must* in general consider complex momenta for p_i, as we will describe in more detail in Chapter 7, since the only constraint on measurable physics is that the *total* energy of the magnons be real, but individual ones can be complex.

So the Schrödinger equation fixes the form of the S-matrix and calculates the total energy. But there is one more constraint on the system that fixes the parameters of the solution, the momenta, to specific values. More precisely, it fixes $\phi(p_i)$ as the *solutions* of equations known as the Bethe ansatz equations (BAE), or the *roots* of these equations, hence the name Bethe roots.

The condition stems from the periodicity of the spin chain, $L + 1 \equiv 1$, together with the symmetry properties of the wavefunction. We have seen that we must have $x_1 < x_2$ in the sum, so if we interchange the coordinates, we must add by periodicity an L to x_1 to restore the condition, thus we impose

$$\psi(x_1, x_2) = \psi(x_2, x_1 + L). \tag{6.66}$$

This leads to the *Bethe ansatz equations*, as we can easily see (the second one comes from interchanging indices 1 and 2):

$$\begin{aligned} e^{ip_1 L} &= S(p_1, p_2) \\ e^{ip_2 L} &= S(p_2, p_1). \end{aligned} \tag{6.67}$$

These equations restrict the possible values of p_1, p_2. In particular, multiplying the two equations and remembering that $S(p_2, p_1) = S^{-1}(p_1, p_2)$, we get

$$e^{i(p_1 + p_2)} = 1, \tag{6.68}$$

which implies that

$$p_1 + p_2 = \frac{2\pi n}{L}; \quad n = 0, 1, \ldots, L - 1. \tag{6.69}$$

Note that as we said, p_1 and p_2 could be complex, only their sum is constrained to be the real value above. But in particular, we have a solution with $p_1 = -p_2$ and real (so $n = 0$), in which case the two Bethe ansatz equations reduce to the condition

$$e^{ip_1(L-1)} = 1, \tag{6.70}$$

solved by

$$p_1 = \frac{2\pi n}{L - 1}, \tag{6.71}$$

and substituting $p_1 = -p_2$ in the S-matrix we get

$$S(p_1, p_2) = \frac{\cot p_1/2 + i}{\cot p_1/2 - i} = e^{ip_1}. \tag{6.72}$$

Substituting in the Bethe ansatz, we get

$$\psi(x_1, x_2) = e^{ip_1(x_1-x_2)} + e^{-ip_1 L} e^{-ip_1(x_1-x_2)} = e^{-i\frac{p_1}{2}} \left(e^{ip_1(l+1/2)} + e^{-ip_1(l+1/2)} \right), \tag{6.73}$$

where $x_1 - x_2 = l$. All in all, we get for the eigenstate (depending on a single integer n from p_1)

$$|\psi(n)\rangle = C_n \cos\left(\pi n \frac{2l + 1}{L - 1} \right) |x_2 + l'; x_2\rangle, \tag{6.74}$$

where

$$C_n = 2e^{-\frac{i\pi n}{L-1}}. \tag{6.75}$$

Note that we cannot put $p_1 = p_2$, they must be different numbers, because of the fermionic nature of the spin chain that implies a Pauli exclusion principle, the absence of two quasiparticles in the same state. Indeed, for $p_1 = p_2$, we obtain $S(p_1, p_2) = -1$, which in turn leads to $\psi = 0$.

In the next chapter we will generalize the analysis started here to spin chains and integrable systems.

Important Concepts to Remember

- The Ising model is a classical nearest-neighbor interaction between spins 1/2, with values $\sigma_i = \pm 1$ (up or down).
- The Ising model in 1+1 dimensions is solved by the transfer matrix method, and one finds no phase transition, as there is no spontaneous magnetization at $h = 0$, independent of the temperature.
- In 2+1 dimensions, Onsager's solution to the Ising model has a phase transition temperature $k_B T_C = 2J/\ln(1 + \sqrt{2})$, between a state with spontaneous magnetization at $h = 0$ for $T < T_C$ and a state with no magnetization at $h = 0$ for $T > T_C$.

- The mean field approximation to the Ising model in 2+1 dimensions gives $k_B T_C = J$, thus correct parametrical dependence, but off by a numerical factor of $2/\log(1+\sqrt{2})$.
- The Ising model in 2+1 dimensions has Kramers-Wannier duality, a nonperturbative (weak coupling vs. strong coupling) duality between the low T expansion of a system vs. the high T expansion of another similar system, with $e^{-2\beta J} = \tanh(\widehat{\beta J})$.
- The Heisenberg model is a quantum version of the Ising model in 1+1 dimension, i.e. on a *spin chain*, replacing the σ_i with the Pauli matrices $\vec{\sigma}_i$, thus with the spin living in three spatial dimensions. The XXX model has equal couplings in all three spin directions, the XYZ has different couplings, and the XXZ has two couplings equal.
- The coordinate Bethe ansatz is an ansatz in terms of quasi-particles called magnons, which correspond to particles with a given momentum (discrete Fourier transform) on the chain.
- The two-magnon solution is written in terms of a wavefunction composed of an incoming wave, and a scattered wave with interchanged momenta and a scattering matrix $S(p_1, p_2)$.
- The Bethe ansatz equations are obtained from periodicity and symmetry properties of the chain, and restrict the momenta to belong to discrete values, called Bethe roots.
- The momenta of the magnons cannot be equal, due to the fermionic nature of the spin chain, which means we cannot have two identical particles in the same state.

Further Reading

See also chapter 18 in [8].

Where It will be Addressed in String Theory

The Heisenberg spin chain will be addressed in Chapters 28 and 29, using a certain limit (pp wave, or Penrose, limit) of the AdS/CFT correspondence.

Exercises

(1) Check that the Bethe ansatz for two magnons satisfies the Schrödinger equation for the Heisenberg Hamiltonian.
(2) Write the Bethe ansatz and Bethe ansatz equations for four magnons.
(3) Show that the total sum of solution sets of M magnons for the spin chain with L sites must be

$$\sum_{M=1}^{L} n_M = 2^L. \tag{6.76}$$

(4) The result of Onsager's solution for the 2-dimensional Ising model at $h = 0$ is

$$g(T) = -k_B T \ln[2\cosh(2\beta J)] - \frac{k_B T}{2\pi} \int_0^\pi d\phi \ln\left[\frac{1}{2}(1 + \sqrt{1 - c^2 \sin^2 \phi})\right], \tag{6.77}$$

where

$$C = \frac{2}{\cosh(2\beta J)\coth(2\beta J)}.$$

(6.78)

Calculate the energy per spin, $\epsilon(T)$, and the heat capacity per spin near $T = T_C$.

(5) Calculate the self-duality temperature for the Kramers-Wannier duality and compare with the transition temperature.

(6) Write a classical Ising model for spin 1 and a quantum generalization similar to the Heisenberg model, and discuss its properties.

7 Spin Chains and Integrable Systems

In the last chapter we studied the Heisenberg spin chain in the coordinate Bethe ansatz. Now we will try to generalize the lessons learned there and define spin chains in more generality and, even more generally, to define integrable systems.

7.1 Classical Integrable Systems

We begin with a definition of classically integrable systems, where everything is easy to understand. More particularly, we are interested in system with a finite number n of degrees of freedom. Consider a system with Hamiltonian $H(x_i, p_j)$ and Poisson brackets

$$\{x_i, p_j\}_{P.B.} = \delta_{ij}. \tag{7.1}$$

Then the Hamiltonian equations of motion are

$$\dot{x}_i = \{x_i, H\}_{P.B.} = \frac{\partial H}{\partial p_i}; \quad \dot{p}_i = \{p_i, H\}_{P.B.} = -\frac{\partial H}{\partial x_i}. \tag{7.2}$$

We then say that *the system is integrable if and only if there are n independent integrals of motion $I_i(x, p)$, $i = 1, \ldots, n$,* i.e. such that

$$\dot{I}_i = \{I_i, H\}_{P.B.} = 0 \tag{7.3}$$

(integrals of motion must be constant in time) and

$$\{I_i, I_j\}_{P.B.} = 0 \tag{7.4}$$

(integrals of motion are independent).

For integrable systems, sometimes we have a *Lax pair*, which is a pair of $N \times N$ matrices $L(x, p)$ and $M(x, p)$, such that the *Lax equation*

$$\dot{L} = [L, M] \tag{7.5}$$

is equivalent with the Hamiltonian equations of motion of the system (7.2). Note that N is unknown a priori, and in fact there is no algorithm for finding out if there is a Lax pair (much less find the Lax pair, we have to work by trial and error), we know only that $N \geq n$, since we need to get at least n (independent) integrals of motion. Indeed, we can check that

$$I_i = \text{Tr}[L^{n_i}], \tag{7.6}$$

where n_i is some integer, is an integral of motion, since

$$\dot{I}_i = n_i \operatorname{Tr}[L^{n_i-1}\dot{L}_i] = n_i \operatorname{Tr}[L^{n_i-1}[L, M]] = 0. \tag{7.7}$$

Moreover, for $n_i \geq N$, the I_i's are functionally dependent on the traces of the lower powers of L, so we have at most N independent quantities, and thus we need $N \geq n$.

Note that the Lax pair is by no means unique. In fact, there is a sort of "gauge transformation" acting on them that leaves the Lax equation invariant:

$$L \rightarrow S^{-1}LS; \quad M \rightarrow S^{-1}MS - S^{-1}\dot{S}. \tag{7.8}$$

In fact, we can make this analogy more precise by introducing a spurious coordinate σ, so that L, M are independent on it, and we define $L_\alpha : (L_0 \equiv M, L_1 \equiv L)$ for $\alpha = 0, 1$, and $\alpha = 0$ is time t, whereas $\alpha = 1$ is σ. Then the Lax equation can be written as a zero curvature (zero field strength, or flat connection) equation,

$$\partial_\alpha L_\beta - \partial_\beta L_\alpha + [L_\alpha, L_\beta] = 0, \tag{7.9}$$

which immediately gives the gauge invariance acting on L_α (and leaving the zero field strength equation invariant)

$$L_\alpha \rightarrow hL_\alpha h^{-1} + (\partial_\alpha h)h^{-1}. \tag{7.10}$$

In the context of the string, we can generalize this notion of integrability as zero curvature to a nontrivial dependence on σ, now understood as the spacelike coordinate along the string. We can then fix a gauge using a generalized form of the Lorenz gauge,

$$\partial^\alpha L_\alpha = 0 \Rightarrow -\partial_0 L_0 + \partial_1 L_1 = 0 \Rightarrow \dot{M} = \partial_1 L, \tag{7.11}$$

but as usual, this doesn't completely fix the gauge invariance (there is residual gauge invariance).

A stronger form of integrability appears if there are Lax pairs depending on an additional complex parameter z called a *spectral parameter*, i.e. there is a $(L(z), M(z))$ pair that is a Lax pair for all z. Moreover, we can define the *spectral curve* as the set of pairs of spectral parameter z for which $L(z)$ is diagonalizable and its corresponding eigenvalue k:

$$\Gamma = \{(u, z) \in \mathbb{C} \times \mathbb{C} | \det(k \mathbb{1} - L(z)) = 0\}. \tag{7.12}$$

With this spectral curve we associate a one-form $d\lambda = kdz$. Then the integrals of motion I_i can be recovered from the (Taylor) series expansion of

$$\operatorname{Tr}L(z) = \sum_n L_n z^n. \tag{7.13}$$

7.2 Quantum Mechanical Integrable Systems in $1+1$ Dimensions

We continue with the case of quantum mechanical integrable systems, for which the general definition seems less clear. However, there is one case that is easy enough to understand and will serve as a guide. That is the case of scattering of massive particles in $1+1$ dimensions.

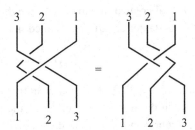

Diagrammatic representation of the Yang-Baxter equation: Different order for the S-matrix interactions should not matter if the endpoint is the same.

For particles in 1+1 dimensions, we can solve the relativistic relation $E^2 - p^2 = m^2$ in terms of the *rapidity* μ by

$$E = m\cosh\mu, \quad p = m\sinh\mu. \tag{7.14}$$

Then the relativistic invariant for the 2-body scattering becomes

$$(p_1 - p_2)^2 = -m_1^2 - m_2^2 + 2m_1 m_2 \cosh(\mu_1 - \mu_2). \tag{7.15}$$

This S-matrix should depend on this relativistic invariant, therefore it is a function of the difference of rapidities:

$$S = S((p_1 - p_2)^2) = S(\mu_1 - \mu_2). \tag{7.16}$$

In general, the 2-body S-matrix for scattering of particles in states $\alpha = 1, \ldots, n$ will be $S_{\alpha\beta}^{\alpha'\beta'}(\lambda - \mu)$.

Now we can define an integrable system as a system whose general S-matrix (for N-body scattering) is factorizable in terms of just S-matrices for 2-body scattering. In particular, it is enough to have that the 3-body S-matrix factorizes into 2-body S-matrices, since then by induction we can factorize any S-matrix into 2-body ones.

The factorization of the 3-body S-matrix into 2-body scatterings can be done in two independent ways, for which we can write a diagrammatic representation (the 2-body S-matrix interchanges two lines, then the 3-body S-matrix is composed of three interchanges of two lines, which can be done in two orders), as in Figure 7.1. The equivalence of the two ways of factorization is the famous *Yang-Baxter equation*, found by C. N. Yang and R. J. Baxter (first by Baxter in the context of a lattice in two spatial dimensions at finite temperature, then generalized by Yang):

$$S_{123}^{(3)} = S_{12}(\lambda_1 - \lambda_2)S_{13}(\lambda_1 - \lambda_3)S_{23}(\lambda_2 - \lambda_3) = S_{23}(\lambda_2 - \lambda_3)S_{13}(\lambda_1 - \lambda_3)S_{12}(\lambda_1 - \lambda_2). \tag{7.17}$$

Although the equation was derived in the context of massive particles in 1+1 dimensions, it is valid in more general contexts, due to its algebraic nature. Moreover, often λ, μ are not rapidities per se, but other things, and in general can be considered to be complex parameters.

In a more general context, the matrix $\mathcal{R}$ is defined in a tensor product of algebras, $\mathcal{A} \otimes \mathcal{A}$. Then the Yang-Baxter equation is defined in $\mathcal{A} \otimes \mathcal{A} \otimes \mathcal{A}$:

$$\mathcal{R}_{12}\mathcal{R}_{13}\mathcal{R}_{23} = \mathcal{R}_{23}\mathcal{R}_{13}\mathcal{R}_{12}. \tag{7.18}$$

In this equation, $\mathcal{R}_{12} = \mathcal{R} \otimes \mathbb{1}$ (R-matrix in the first two and identity in the third), $\mathcal{R}_{23} = \mathbb{1} \otimes \mathcal{R}$ (identity in the first, and R-matrix in the last two), and $\mathcal{R}_{13}$ is an R-matrix in the first and third algebra, and identity in the second.

7.3 Algebraic and Coordinate Bethe Ansatz

Then, unlike the case of the Heisenberg spin chain in the coordinate Bethe ansatz, where one starts with a Hamiltonian that is integrable and then solves the system via an ansatz, now integrable systems can be defined as *representations of the Yang-Baxter equation*. This is the essence of the *algebraic Bethe ansatz*. Thus one does not input anything about the dynamics (Hamiltonian, etc.) of the theory, only about the symmetries. The symmetries of the theory define the representation of the Yang-Baxter equation, and we solve the theory entirely on their basis, under the assumption of integrability.

Then even the Hamiltonian is not essential, appearing as simply one of the various integrals of motion of the theory, which are found from the Lax operator $L(z)$. For spin chains (in 1+1 dimensions), one usually finds integrals of motion I_k that contain interaction over k (neighboring) sites, so in that case we could define the Hamiltonian as the one that has only nearest-neighbor interactions, but other than that there is nothing special about it to distinguish it from the other integrals of motion.

For example, for the spin 1/2 XXX Heisenberg model, one can check that the operator

$$R_{ij}(u, v) \equiv u - v + P_{ij} \tag{7.19}$$

satisfies the Yang-Baxter equation, which is left as an exercise. To solve the spin chain, one defines a total Hilbert space for the chain $\mathcal{H} = h \otimes \cdots \otimes h$, and a Lax operator, with an algebra on it. One finds the Hamiltonian among the integrals of motion, which are obtained as follows. One finds operators $L_j(u)$, for $j = 1, \ldots, L$ indexing the sites, which are matrices in the total Hilbert space $\mathcal{H}$ and in an auxiliary space. From them one can construct the *monodromy matrix* $T(u)$,

$$T(u) = L_L(u) \cdots L_2(u)L_1(u). \tag{7.20}$$

Here $L_j(u)$ are Lax operators and act as a "connection," that defines parallel transport along a curve. Indeed, for instance, on a curved spacetime, the "spin connection" defines parallel transport, and one can argue that for gauge theories the same can be said about the gauge field (connection). The parallel transport is defined by the *Lax equation*,

$$\psi_{j+1} = L_j\psi_j, \tag{7.21}$$

where ψ_j is the wavefunction at site j. The above is a discretized version of the (time independent) Dirac equation, which in one spatial dimension would be

$$(\partial_x + A_x(x))\psi(x) = 0 \Rightarrow \psi(x + dx) = (1 + A_x(x))\psi(x). \tag{7.22}$$

One can prove that the traces of the monodromy matrix commute,

$$[\text{Tr}\, T(u), \text{Tr}\, T(v)] = 0, \tag{7.23}$$

and one can expand it in Fourier modes,

$$T(u) = \sum_{n=0}^{L} t_n u^n, \tag{7.24}$$

in which case the traces of the modes are the commuting integrals of motion, like in the classical case,

$$[\text{Tr}\, t_n, \text{Tr}\, t_m] = 0. \tag{7.25}$$

Among these integrals of motion we find the Hamiltonian $\hat{H}$ and the discrete lattice momentum $\hat{P}$, so now we have a usual definition of the system.

We can then solve the Yang-Baxter equation and define a Fock space using the *ansatz* (assumption) that in the monodromy matrix (2×2 matrix) $T(u)$, written as

$$T(u) = \begin{pmatrix} A(u) & B(u) \\ C(u) & D(u) \end{pmatrix}, \tag{7.26}$$

(so that $\text{Tr}\, T(u) = A(u) + D(u)$), the $B(u)$ is a creation operator, and $C(u)$ is an annihilation operator. The Fock space is thus

$$|\vec{u}\rangle = B(u_1) \cdots B(u_M)|\Omega\rangle. \tag{7.27}$$

One can derive the Bethe ansatz equations algebraically, from the same formalism, though it is a bit complicated, so we will not do it here.

Instead, for the Heisenberg spin 1/2 XXX system, we describe the general case in the coordinate Bethe ansatz, like in Chapter 6, since it is easier to understand.

The M-body problem is defined from the 2-body problem, because of integrability. The M-body wavefunction is

$$\psi(x_1, \ldots, x_M) = \sum_{P \in Perm(M)} \exp\left\{ i \sum_{i=1}^{M} p_{P(i)} x_i + \frac{i}{2} \sum_{i<j} \delta_{P(i)P(j)} \right\}, \tag{7.28}$$

where the *phase shifts* $\delta_{ij} = -\delta_{ji}$ are given by $S(p_i, p_j) = e^{i\delta_{ij}}$. We can check that we obtain the wavefunction from last chapter for $M = 2$, except for an overall phase $e^{i\delta_{12}}$.

Plugging the ansatz into the Schrödinger equation, like in the $M = 2$ case, we first find that the total energy is the sum of the individual magnon energies:

$$E = 2J \sum_{j=1}^{M} 2(\cos p_j - 1) = -2J \sum_{j=1}^{M} \frac{1}{u_j^2 + 1/4}. \tag{7.29}$$

Also as before, from the periodicities, we get the M Bethe ansatz equations:

$$e^{ip_k L} = \prod_{i \neq k, i=1}^{M} S(p_k, p_i).$$ (7.30)

Since $S(p_k, p_j) = S^{-1}(p_j, p_k)$, by taking the product of these equations, we obtain the identity on the right-hand side, so

$$e^{i(\sum_{k=1}^{M} p_k)L} = 1,$$ (7.31)

which gives

$$\hat{P} = \sum_{i} p_i = \frac{2\pi n}{L}(mod\ 2\pi), n = 0, 1, \ldots, L-1.$$ (7.32)

As we showed in Chapter 6, the p_i's must be different, since $S(p, p) = -1$, and we get a zero wavefunction because of the fermionic nature of the chain.

The Bethe ansatz equations have both real and complex solutions for fixed M and L, and in fact all of them can be acceptable physically, since we need only to have real total energy, the individual energies or momenta are not measurable.

We obtain sets of solutions $\{p_1, \ldots, p_M\}_n$, where n labels the solution set. We can prove that for $M = 1$ we have $n_1 = L - 1$ sets, for $M = 2$ we have $n_2 = L(L-3)/2$, etc. (for higher M it is harder to find the formula). But it is clear that the sum is

$$\sum_{M=1}^{L} n_M = 2^L$$ (7.33)

for the total number of sets of solutions, which equals the dimension of the total Hilbert space (there are two states at each site).

Each solution $\{p_i\}$ is associated with a set of integers $\{n_i\}$ that appear from taking the log of the Bethe equation (7.30), which is

$$p_k L = \sum_{i \neq k, i=1}^{M} \delta_{ki} + 2\pi n_k,$$ (7.34)

and the n_k's appear because of the branches (Riemann sheets) of the log function.

In the case of real solutions $\{p_i\}$, we must have sets of different u_i's, as we showed. However, in the case of complex $\{p_i\}$'s, we can have the same real parts.

Denoting $\phi(p_k) = 1/2 \cot p/2$ as u_k, we first note that

$$\frac{u_k + i/2}{u_k - i/2} = \frac{\cos p_k/2 + i \sin p_k/2}{\cos p_k/2 - i \sin p_k/2} = e^{ip_k},$$ (7.35)

and thus we can write the Bethe ansatz equations as

$$\left(\frac{u_k + i/2}{u_k - i/2}\right)^L = \prod_{j \neq k, j-1}^{M} \left(\frac{u_k - u_j + i}{u_k - u_j - i}\right), \quad k = 1, \ldots, M.$$ (7.36)

For $M = 1$, we have seen that the wavefunction is simply the free wave, and the Bethe equation is $e^{ip_k L} = 1$, solved by $p_n = 2\pi n/L$, for $n = 1, \ldots, L$. That means that there are

$L - 1$ states, with a single spin flipped traveling around the chain with momentum p_n, like for a free particle with periodic boundary conditions. That means that we have a quasi-particle interpretation, as in Chapter 6, where the quasi-particles are called *magnons* and have energy

$$\epsilon_j = 4J(\cos p_j - 1). \tag{7.37}$$

We note that for these magnons, since $u = 1/2 \cot p/2$, we have

$$\frac{dp}{du} = -\frac{1}{u^2 + 1/4} = \epsilon(u). \tag{7.38}$$

Note that the magnons are a nonrelativistic system.

This is formally the same relation as for a massive relativistic particle in 1+1 dimensions, for which in terms of the rapidity μ we have

$$E = m \cosh \mu, \quad p = m \sinh \mu \Rightarrow$$
$$E = \frac{dp}{d\mu}, \tag{7.39}$$

so the u is the nonrelativistic analog of the relativistic rapidity of a massive particle. That is why the u_i's, the Bethe roots (roots of the Bethe equation), are also called rapidities, by extension.

We noted that the Bethe roots (and the associated momenta) need not be real, they can be complex. The only constraint is to have real total energies, which is the only thing we can measure. But we note that if u_k is a Bethe root, so is u_k^*, thus we always have pairs of complex conjugate Bethe roots, leading to a real total energy! So we can consider the complex roots as well.

An important limit of the spin chain is the thermodynamic limit, $L, M \to \infty$, with $M/L =$ fixed. As we mentioned, two magnons can have the same real part, but because of the fermionic nature of the chain, the imaginary parts must be different. In fact, from the Bethe equations we see that the S-matrix has poles when $u_k = u_j \pm i$, so the imaginary parts are different by integers times i. We can have then discretized straight lines, $u_k = \text{Re}(u) + ik$. But in the thermodynamic limit, we must divide the u's by L as well, so the differences between the roots become continuous, and moreover the above argument is modified, and instead of straight lines in complex u space, we have curved lines called *Bethe strings*.

Solving the Bethe equations, we find Bethe roots and Bethe strings, but it is in general very difficult to do. In fact, we will see that string theory helps here, and actually Bethe strings do correspond to physical strings.

7.4 Generalizations

One can generalize the discussion of the Heisenberg spin chain to other integrable spin chains. We obtain them as other representations of the Yang-Baxter equations, with other symmetries. However, the requirement of integrability turns out to be quite restrictive, and there are only a few examples of integrable spin chains. One such example is the case of the Heisenberg XXX spin chain with a spin S at each site. However, from the algebraic Bethe ansatz, one finds the *integrable* Hamiltonian for arbitrary spin S to be

$$H = \sum_{k=1}^{L} \sum_{\alpha} \left[S_k^\alpha S_{k+1}^\alpha - (S_k^\alpha S_{k+1}^\alpha)^2 \right]. \tag{7.40}$$

One can also find the Hamiltonian for the spin 1/2 XXZ Heisenberg model as an integrable one.

Moreover, there is a priori no reason to restrict to spin S representations of the group $SU(2)$ as being the degrees of freedom at each site, other than this physical spin can be made from electrons. But as a matter of principle, we could instead consider a general representation $\mathcal{R}$ of a group G instead. Only for specific representations of specific groups does one find integrable models. One such example will be mentioned in Part III, as it comes from string theory; it is a rather complicated $SO(6) = SU(4)$ model.

Important Concepts to Remember

- An integrable system is a system with as many integrals of motion as there are degrees of freedom.
- A integrable system sometimes has a Lax pair of $N \times N$ matrices (L, M), which means that the Hamiltonian equations of motion are equivalent with the matrix equation $\dot{L} = [L, M]$.
- If we have a Lax pair that depends on a spectral parameter z, $L(z)$, $M(z)$, then the integrals of motion are obtained from the Taylor expansion (in z) of $L(z)$.
- Quantum integrable systems, at least in the case of scattering, can be defined as systems for which the 3-body scattering is factorizable into 2-body scatterings, and as such obeys the Yang-Baxter equation $\mathcal{R}_{12}\mathcal{R}_{13}\mathcal{R}_{23} = \mathcal{R}_{23}\mathcal{R}_{13}\mathcal{R}_{12}$.
- The algebraic Bethe ansatz consists of solving for representations of the Yang-Baxter equation based on symmetries, and deriving the wavefunctions, energies, and even the Hamiltonian from consistency of the solutions.
- For spin chains, we have Lax operators $L_j(u)$ defined at each site, defining a Lax equation $\psi_{j+1} = L_j \psi_j$ and a monodromy matrix $T(u) = L_L(u) \cdots L_2(u) L_1(u)$. The traces of its Taylor expansion, $\text{Tr}\, t_m$, with $T(z) = \sum_m t_m z^m$, give the integrals of motion.
- The solutions of the Bethe ansatz equations are sets of Bethe roots $\{u_k\}$, which for the spin 1/2 XXX Heisenberg case are different from each other, complex, and in the thermodynamic limit $L, M \to \infty$, L/M fixed lie on Bethe strings in the complex Bethe root plane.
- The rapidities of the solutions are nonrelativistic analogs of the relativistic rapidities of massive particles.

Further Reading

Chapter 18 in [8] and the (original) review by Faddeev [9].

Where It will be Addressed in String Theory

The Heisenberg spin chain will be addressed in Chapters 28 and 29. Bethe strings will be obtained from classical strings in Chapter 30, and general integrability issues using the AdS/CFT correspondence will be addressed in Chapter 31.

Exercises

(1) Prove that a free particle is an integrable system.
(2) Prove that for the KdV integrable system, with equation of motion

$$u_t = 6u\partial_x u - \partial_x^3 u, \qquad (7.41)$$

the pair

$$L = -\partial_x^2 + u$$
$$M = 4\partial_x^3 - 3(2u\partial_x + \partial_x u) \qquad (7.42)$$

is a Lax pair.
(3) Check that the Heisenberg spin chain R-matrix

$$R_{ij}(u, v) \equiv u - v + P_{ij} \qquad (7.43)$$

satisfies the Yang-Baxter equation. Denoting 1 by a, 2 by b, 3 by j, setting the third variable (w) to 1/2 and defining

$$L_j(u) = u - \frac{i}{2}\sigma^\alpha \sigma_j^\alpha, \qquad (7.44)$$

prove that we can rewrite the Yang-Baxter equation as

$$R_{12}(u - v)(L_j(u) \otimes I)(I \otimes L_j(v)) = (I \otimes L_j(v))(L_j(u) \otimes I)R_{12}(u - v), \qquad (7.45)$$

or symbolically

$$R_{12}(u, v)L_j^1(u)L_j^2(v) = L_j^2(v)L_j^1(u)R_{12}(u, v), \qquad (7.46)$$

where the action is in the quantum Hilbert space of spin j and *auxiliary* Hilbert spaces a and b.
(4) Using that

$$L_j^a(-i/2) = -P_{j,a}, \quad \frac{d}{du}L_j^a(a) = \mathbb{1}, \qquad (7.47)$$

show that the shift operator is

$$U = P_{12}P_{23}\dots P_{N-1,N} = i^N \operatorname{Tr} T(-i/2) = e^{i\hat{P}} \qquad (7.48)$$

and the Hamiltonian is

$$\hat{H} = -J\left(L + 2i\frac{d}{du}\ln \operatorname{Tr} T(-i/2)\right). \qquad (7.49)$$

(5) Check that the 3-magnon Bethe ansatz satisfies the Schrödinger equation.

8 The Thermodynamic Bethe Ansatz

In the previous chapter, we have defined integrable systems by generalizing lessons from spin chains. We were led to the Bethe ansatz, in particular the *algebraic Bethe ansatz*, as a way to define them, and we found that there we could obtain an ansatz in terms of "Bethe roots" with "rapidities," nonrelativistic analogs of quantities for massive relativistic systems.

In this chapter we will see that we can actually derive thermodynamical properties of real, massive relativistic particles, in the case of integrable 1+1–dimensional theories, from the large volume limit of their factorizable S matrices. This is known as the *Thermodynamic Bethe Ansatz (TBA)*, and it was developed by A. B. Zamolodchikov in 1989, based on a famous paper by Yang and Yang from 1969.

The procedure will be as follows. We will first write Bethe ansatz equations, then take their thermodynamic limit, defining densities of states, and integral equations, like in Chapter 7. But now we will apply the procedure to real rapidities for relativistic systems, as opposed to the complex ones in Chapter 7, where for the poles of the S-matrix, $u_{ik} = u_i - u_k$ takes the imaginary values $\pm i$, corresponding to bound states.

8.1 Massive Relativistic 1+1–Dimensional Integrable Systems and Bethe Ansatz Equations

For a massive relativistic particle in 1+1 dimensions of rapidity θ, as we saw before, we have

$$E = m \cosh \theta; \quad p = m \sinh \theta. \tag{8.1}$$

We will consider *purely elastic scattering* theory in 1+1 dimensions, which means a quantum field theory with an S-matrix that is factorizable and diagonal. Factorizability means, as before, that the S-matrix can be written as the product of 2-body S-matrices, and implies that the Yang-Baxter equation (YBE) is automatically satisfied. Indeed, this was also the case for the coordinate Bethe ansatz, where we assumed factorizability, and in fact to deduce the YBE we had assumed factorizability. Therefore, factorizability does not mean triviality, we can have a more complicated bound state structure.

Defining the relative rapidity

$$\theta_{ab} = \theta_a - \theta_b, \tag{8.2}$$

the 2-body S-matrix will depend only on it and is defined as usual through

$$|a(\theta_a), b(\theta_b)\rangle_{\text{in}} = S_{ab}(\theta_{ab})|a(\theta_a), b(\theta_b)\rangle_{\text{out}}. \tag{8.3}$$

But one can impose various physical constraints on the possible form of the S-matrix. It has to be real analytic, unitary, and crossing symmetric, giving respectively

$$S_{ab}^*(\theta) = S_{ab}(-\theta^*)$$
$$S_{ab}(\theta)S_{ab}(-\theta) = 1$$
$$S_{ab}(\theta) = S_{ab}(i\pi - \theta). \tag{8.4}$$

If moreover one imposes meromorphicity in θ and being bounded by polynomials in momenta, it was proven that one restricts the possible form of the S-matrix to be a product of given factors,

$$S_{ab}(\theta) = \prod_{\alpha \in A_{ab}} f_\alpha(\theta), \tag{8.5}$$

where the basic component is

$$f_\alpha(\theta) = \frac{\sinh((\theta + i\pi\alpha)/2)}{\sinh((\theta - i\pi\alpha)/2)}, \tag{8.6}$$

and the only things that depend on the theory are the sets A_{ab} of α's.

We consider here *minimal theories*, which are massive perturbations of nontrivial conformal field theories (conformal field theories that are not free bosons). Then the simple pole of $S_{ab}(\theta)$ at the imaginary relative rapidity $\theta_{ab} = iu_{ab}^c$ indicates a bound state c of a and b with mass

$$m_c^2 = m_a^2 + m_b^2 + 2m_a m_b \cos u_{ab}^c. \tag{8.7}$$

The same is true for any odd order pole. Note that these statements are not true for non-minimal theories.

We then consider an elastic, minimal scattering theory, defined on a circle of length L that will be taken to infinity, $L \to \infty$, with N particles, N_a of species a, at positions $x_1, \ldots, x_N$. Then asymptotically, when the particles are well separated, we have an ansatz similar to the coordinate Bethe ansatz for the Heisenberg spin chain:

$$\psi(x_1, \ldots x_N) = \exp\left(i \sum_j p_j x_j\right) \sum_{Q \in S_N} A(Q)\Theta(x_Q). \tag{8.8}$$

Here Q are permutations, $\Theta(x_Q) = 1$ only if $x < Q_1 < \cdots x_{Q_N}$ and is zero otherwise, and if Q and Q' differ just by permuting i and j we have

$$A(Q') = S_{ij}(\theta_i - \theta_j)A(Q). \tag{8.9}$$

We can define the phase shifts as usual, $S_{ij} = e^{i\delta_{ij}}$, or

$$\delta_{ij}(\theta_i - \theta_j) = -i \ln S_{ij}(\theta_i - \theta_j), \tag{8.10}$$

and then the expression for the wave function becomes essentially the one in the coordinate Bethe ansatz, though note that in that case one considered "scattering" of quasi-particles (magnons) in the nonrelativistic spin chain system, whereas here we are considering scattering of real particles in a relativistic 1+1–dimensional system.

Like in the case studied by Bethe, we impose periodicity on the asymptotic wavefunction, and we get the exact analog of the Bethe equations:

$$e^{iLm_i \sinh \theta_i} \prod_{j \neq i; j} S_{ij}(\theta_i - \theta_j) = (-1)^{F_i}. \tag{8.11}$$

Since we are dealing with both bosons and fermions, we consider periodic boundary conditions for the bosons and antiperiodic boundary conditions for the fermions, leading to the $(-1)^{F_i}$ term on the right-hand side. Taking the log of the above equation gives the Bethe ansatz equation in the case we are considering:

$$Lm_i \sinh \theta_i + \sum_{j \neq i, j} \delta_{ij}(\theta_i - \theta_j) = 2\pi n_i, \quad i = 1, \ldots, N, \tag{8.12}$$

where $\{n_i\}$ can be taken to be the quantum number of the state of the multi-particle system, and n_i is integer for bosons and half integer for fermions.

8.2 Thermodynamic Limit

As in Chapter 7, from the Bethe ansatz equations we can calculate the momenta (or the rapidities θ_i) of a multiparticle system in the box of size L. For a general theory, the Bethe ansatz will be exactly valid only asymptotically for $L \to \infty$, i.e. an *asymptotic Bethe ansatz*, and at finite L it will be valid only up to $1/L$ corrections. Since we are interested in real particles, we consider only real rapidities θ_i here.

We are therefore led to consider the thermodynamic limit, $L \to \infty$, for $N \to \infty$ particles, for each species $N_a \to \infty$, with $N_a/L \to$ fixed. We take the thermodynamic limit of the Bethe ansatz equations (8.12).

We define the *rapidity density* $\rho_a^{(r)}(\theta)$ in a similar manner to the Bethe root density from Chapter 7, as the number of particles of species a with rapidities in an interval $\Delta\theta$, divided by $L\Delta\theta$.

We moreover define the functions

$$J_a(\theta) = m \sinh \theta + 2\pi \sum_{b=1}^{n} \left(\delta_{ab} * \rho_b^{(r)} \right)(\theta), \tag{8.13}$$

where $*$ denotes the convolution, defined as

$$(f * g)(\theta) \equiv \int_{-\infty}^{+\infty} \frac{d\theta'}{2\pi} f(\theta - \theta') g(\theta'), \tag{8.14}$$

such that the Bethe equations (8.12) become simply

$$J_a(\theta) = \frac{2\pi n_{a,i}}{L}. \tag{8.15}$$

It can be proven that the functions $J_a(\theta)$ are monotonically increasing. As before, the solutions to these equations are *Bethe roots*, i.e. sets of $\theta = \theta_{a,i}$.

Consider the Bethe roots that actually occur in the system as real particles, giving the root (particle) density $\rho_a^{(r)}(\theta)$, and consider also the Bethe roots that do not occur in the system as real particles, i.e. the corresponding integers $n_{a,i}$ don't occur, called *holes* by analogy with the case of the electron band structure, with hole density $\rho_a^{(h)}(\theta)$. Then the Bethe equations (8.15) imply that the total density of roots and holes is

$$
\begin{aligned}
\rho_a(\theta) &= \rho_a^{(r)}(\theta) + \rho_a^{(h)}(\theta) \\
&= \frac{1}{2\pi} \frac{d}{d\theta} J_a(\theta) \\
&= \frac{m_a}{2\pi} \cosh\theta + \sum_{b=1}^{n} \left(\phi_{ab} * \rho_b^{(r)} \right)(\theta),
\end{aligned}
\tag{8.16}
$$

where

$$
\phi_{ab}(\theta) \equiv \frac{d}{d\theta} \delta_{ab}(\theta).
\tag{8.17}
$$

8.3 Thermodynamics and TBA Equations

We are finally ready to find the thermodynamics of the system. Define, in analogy with the familiar relation for the fraction of occupied states of bosonic species a (Bose-Einstein distribution) with chemical potential μ_a at energy E_a,

$$
\frac{\rho_a^{(r)}(\theta)}{\rho_a(\theta)} = \frac{1}{e^{(E_a(\theta) - \mu_a)/T} + 1},
\tag{8.18}
$$

the quantity $\epsilon_a(\theta)$ (energy divided by temperature for species a) by

$$
\frac{\rho_a^{(r)}(\theta)}{\rho_a(\theta)} = \frac{1}{e^{\epsilon_a} + 1},
\tag{8.19}
$$

such that $E_a(\theta) = T\epsilon_a(\theta) + \mu_a$ is the "dressed" one-particle excitation energy.

From quantum statistical mechanics, for the system of particles and holes, the entropy density is given by $s = \rho \ln \rho - \rho^{(r)} \ln \rho^{(r)} - \rho^{(h)} \ln \rho^{(h)}$, giving in total

$$
\begin{aligned}
s(\rho, \rho^{(r)}) &= \sum_{a=1}^{n} s_a(\rho_a, \rho_a^{(r)}) \\
&= \sum_{a=1}^{n} \int_{-\infty}^{\infty} d\theta \left[\rho_a \ln \rho_a - \rho_a^{(r)} \ln \rho_a^{(r)} - (\rho_a - \rho_a^{(r)}) \ln(\rho_a - \rho_a^{(r)}) \right].
\end{aligned}
\tag{8.20}
$$

The energy per unit length is given by the particle (root) density, times the energy of individual particles:

$$
h(\rho^{(r)}) = \sum_{a=1}^{n} \int_{-\infty}^{+\infty} d\theta \, \rho_a^{(r)}(\theta) m_a \cosh\theta.
\tag{8.21}
$$

The equilibrium thermodynamics is found by minimizing the free energy per unit length,

$$f(\rho) = h(\rho^{(r)}) - Ts(\rho, \rho^{(r)}), \tag{8.22}$$

subject to the constraint of fixed particle densities,

$$D_a \equiv \frac{N_a}{L} = \int_{-\infty}^{+\infty} d\theta \, \rho_a^{(r)}(\theta), \tag{8.23}$$

which are introduced with Lagrange multipliers that are the chemical potentials μ_a, as usual.

Defining the quantity

$$L_a(\theta) \equiv \ln\left(1 + \exp(-\epsilon_a(\theta))\right), \tag{8.24}$$

and ratios with respect to the smallest mass m_1,

$$\hat{\mu} \equiv \frac{\mu}{m_1}; \quad \hat{m}_a \equiv \frac{m_a}{m_1}; \quad r \equiv Rm_1; \quad R \equiv \frac{1}{T}; \Rightarrow$$
$$\hat{\mu}_a r = \frac{\mu_a}{T}; \quad \hat{m}_a r = \frac{m_a}{T}, \tag{8.25}$$

one can show that the extremization of $f(\rho)$ leads to the following *thermodynamic Bethe ansatz (TBA) equations*:

$$\epsilon_a(\theta) = -\hat{\mu}_a r + \hat{m}_a r \cosh\theta - \sum_{b=1}^{n} (\phi_{ab} * L_b)(\theta); \quad a = 1, \ldots, n. \tag{8.26}$$

The proof is left as an exercise.

These TBA equations are equations for $\epsilon(\theta, T, \mu_a)$, which from their definition (8.19) together with the Bethe particle density equations (derived from the Bethe equations) (8.16) give all the densities $\rho_a(\theta_a)$, $\rho_a^{(r)}(\theta)$.

We can calculate the extremized free energy by using (8.26) and (8.16) in the definition of the free energy to find

$$f(R, \mu) = -\frac{1}{2\pi R} \sum_{a=1}^{n} \int_{-\infty}^{+\infty} d\theta L_a(\theta, r, \hat{\mu}) m_a \cosh\theta + \sum_{a=1}^{n} \mu_a D_a(R, \hat{\mu}). \tag{8.27}$$

Moreover, because of the thermodynamical relation $f = -P + \sum_a \mu_a D_a$, we find for the pressure

$$P(T, \mu) = \frac{T}{2\pi} \sum_{a=1}^{n} \int_{-\infty}^{+\infty} d\theta L_a(\theta, m/T, \mu) m_a \cosh\theta, \tag{8.28}$$

and the thermodynamics relation

$$dP = s \, dT + \sum_a D_a \, d\mu_a \tag{8.29}$$

allows us to calculate further quantities.

We see that we have obtained a complete description of the thermodynamics, based solely on the factorized S-matrix.

Important Concepts to Remember

- The thermodynamic Bethe ansatz refers to obtaining the thermodynamics of an integrable system of 1+1–dimensional massive relativistic particles, based solely on its factorized S-matrix.
- The S-matrix is described by the 2-body S-matrix composed of given factors, depending on the rapidity θ and a parameter α.
- One imposes asymptotically a Bethe ansatz similar to the coordinate Bethe ansatz in the Heisenberg case.
- The solutions of the Bethe ansatz equations are given in terms of particles (occupied roots) and holes (unoccupied roots), giving particle and hole densities in the thermodynamic limit.
- The thermodynamics is calculated by extremizing the free energy for constant particle densities and leads to the TBA equations.

Further Reading

The paper [10] describes the Thermodynamic Bethe Ansatz in more detail.

Where It will be Addressed in String Theory

While it was not yet addressed, some speculative comments will be made in Chapter 30.

Exercises

(1) Prove that the extremization of $f(\rho)$ leads to the TBA equations (8.26).
(2) Calculate C_V for the system described by the TBA equations.
(3) Check that $f_{\alpha\beta}$ in (8.6) satisfies the requirements of real analyticity, unitarity, and crossing symmetry.
(4) The *sine-Gordon model*, a scalar field with potential $V(\phi) = -m_0^2/\beta^2 \cos(\beta\phi)$, is integrable, and one can calculate its S-matrices exactly. The 2-body S-matrix for scattering of "elementary" particles is

$$S = \frac{\sinh\theta + i\sin(\gamma/8)}{\sinh\theta - i\sin(\gamma/8)},\qquad (8.30)$$

which is of the form (8.6). Here $\gamma = \beta^2/(1 - \beta^2/(8\pi))$. Find its poles, and calculate the resulting Mandelstam variables $s = (p_1 + p_2)^2$ and $u = (p_1 + p_4)^2$ for these particles (the convention for s and u is: all momenta are in) as a function of the mass m of the elementary particles. Remember that θ stands for θ_{12}, the relative rapidity of incoming particles.

(5) For the same sine-Gordon model, there are solitons ("kinks") that have masses m. The scattering of a soliton and antisoliton, at $\gamma = 8\pi/n$, has an S-matrix

$$S_T(\theta) = e^{in\pi} \prod_{k=1}^{n-1} \frac{e^{\theta - i(\pi k/n)} + 1}{e^{\theta} + e^{-i(\pi k/n)}}. \tag{8.31}$$

From it, find the masses of the bound states of soliton and antisoliton apparent in this S-matrix.

9 Conformal Field Theories and Quantum Phase Transitions

In this chapter we will treat together conformal field theories and quantum phase transitions. Conformal field theories arise near a phase transition, when there are no scales involved. One usually talks about thermal phase transitions, but a lot of the recent interest in condensed matter physics has been around quantum phase transitions, which are phase transitions that happen as a coupling parameter (like, for instance, doping) is varied. In that case, it turns out that the phase transition, as well as a new phase arising from it, the quantum critical phase, can be described in terms of modifications of a conformal field theory, so in this chapter we will treat the two together.

9.1 Thermal Phase Transitions and Conformal Field Theory

We consider first the usual type of phase transitions, namely thermal ones, which occur at a critical temperature T_C. Near this critical point, for $T \to T_C$, we have *critical behavior* for thermodynamic quantities that characterize the material. For a second order phase transition, the second derivatives of the thermodynamic potential, C, χ_m, κ_e for example, blow up in a specific way, namely as power laws in $(T - T_C)/T_C$:

$$
\begin{aligned}
C &\propto \left(\frac{T - T_C}{T_C} \right)^{-\alpha} \\
\chi_m &\propto \left(\frac{T - T_C}{T_C} \right)^{-\gamma_m} \\
\kappa_e &\propto \left(\frac{T - T_C}{T_C} \right)^{-\gamma_e} .
\end{aligned}
\tag{9.1}
$$

This behavior is due to the fact that the *correlation length*, i.e. the distance over which correlation functions of physical quantities is nontrivial, blows up in a similar way:

$$
\xi \propto \left(\frac{T - T_C}{T_C} \right)^{-\nu} .
\tag{9.2}
$$

The 2-point correlation functions at small distances go like

$$
G(r) \propto \frac{1}{r^{d-2+\eta}} e^{-r/\xi},
\tag{9.3}
$$

but ξ will eventually become much larger than the lattice spacing a, in which case we can adopt a (continuum) field theory description.

For instance, for the Ising model on a (hyper)square lattice $\mathbb{Z}^d$, we can define correlation functions by

$$\langle \phi_1^{lat}(r_1)\phi_2^{lat}(r_2)\ldots\phi_n^{lat}(r_n)\rangle = Z^{-1} \sum_{\{s\}} \phi_1^{lat}(r_1)\ldots\phi_n^{lat}(r_n)W(\{s\}), \qquad (9.4)$$

where

$$W(\{s\}) = e^{-\beta H(\{s\})} \qquad (9.5)$$

is the usual Boltzmann factor, $\{s\}$ are lattice variables, usually "spins," but can be defined more generally, and the fields $\phi_j^{lat}(r)$ are (in the spin case) sums of products of nearby spins over a region of size of order a (just a few sites), e.g. the energy density

$$\mathcal{H} \propto \sum_{r'} J(|r - r'|)S(r)S(r'). \qquad (9.6)$$

Then we can define the correlation function for the *scaling fields*, which will take the role of quantum fields in the continuum theory, renormalized while we take a scaling limit, $a \to 0$ with ξ fixed, such that

$$\lim_{a\to 0}\left(\prod_{i=1}^{n}\frac{1}{a^{\Delta_i}}\right)\langle\phi_1^{lat}(r_1)\ldots\phi_n^{lat}(r_n)\rangle = \langle\phi_1(r_1)\ldots\phi_n(r_n)\rangle. \qquad (9.7)$$

The correlation functions are thought of as VEVs with respect to some path integral measure, modulo some renormalization subtleties. The *scaling dimensions* Δ_i are defined such that under scale transformations, we have

$$\langle\phi_1(br_1)\ldots\phi_n(br_n)\rangle = b^{-\sum_j \Delta_j}\langle\phi_1(r_1)\ldots\phi_n(r_n)\rangle, \qquad (9.8)$$

where the fields ϕ are normalized such that

$$\langle\phi_j(r_1)\phi_j(r_2)\rangle = \frac{1}{|r_1 - r_2|^{2\Delta_j}}. \qquad (9.9)$$

In a general quantum field theory we can define an *operator product expansion* (OPE), that appears when we have the product of two operators at nearby points. The expansion is in terms of operators defined at one of the points (or the midpoint), with coefficients that depend on the distance between operators,

$$\mathcal{O}_i(x)\mathcal{O}_j(x') = \sum_k C_{ijk}(|x - x'|)\mathcal{O}_k\left(\frac{x + x'}{2}\right), \qquad (9.10)$$

and this is meant to be an operator relation, and thus valid inside any correlation function.

Thus, for example, in our case of fields obtained as limits of products of spins at a site, we have

$$\langle\phi_i(r_i)\phi_j(r_j)\ldots\rangle = \sum_k \frac{C_{ijk}}{|r_i - r_j|^{\Delta_i + \Delta_j - \Delta_k}}\left\langle\phi_k\left(\frac{r_i + r_j}{2}\right)\ldots\right\rangle. \qquad (9.11)$$

Since the field theory that appears near the phase transition T_C has an infinite correlation length, all correlators will be power laws (since there are no dimensional parameters to form exponentials), and the theory near T_C will be a *conformal field theory*.

Conformal Field Theory

We now turn to defining this conformal field theory. Conformal field theories can be defined as above, as the Euclidean theories (on the spatial dimensions) near a phase transition in condensed matter systems, but their applications extend beyond that, to string theory, which is defined as a conformal field theory on the (Minkowskian) 2-dimensional worldsheet of the string.

A conformal field theory is a theory *in flat space* invariant under a conformal transformation, which is a generalization of the scale transformation. A scale transformation multiplies the metric with a constant,

$$ds^2 \rightarrow ds'^2 = \lambda^2 ds^2, \qquad (9.12)$$

whereas a conformal transformation is a transformation $x^\mu \rightarrow x'^\mu$ *still of flat space* that multiplies the metric with a conformal factor, i.e. an overall spacetime-dependent function:

$$ds^2 \rightarrow ds'^2 = \Omega^2(x) ds^2. \qquad (9.13)$$

Note that conformal transformations are transformations *on flat space* (even though the metric after the transformation might not look flat). So even though the transformations can be thought of as being a subset of general coordinate transformations, conformal invariance is not an invariance of curved space (for which general coordinate transformations put it in an equivalent form in terms of the transformed metric), but a transformation (invariance) of flat space theory.

9.2 Conformal Field Theory in Two and d Dimensions

In dimensions $d > 2$, the solution to the condition (9.13) that defines conformal transformations gives the symmetry group $SO(d+1, 1)$ for d Euclidean dimensions, and $SO(d, 2)$ for $(d-1, 1)$ (Minkowski) dimensions. While this is very interesting, and conformal field theories in $d > 2$ will be described using AdS/CFT (thus string theory) in Part III, here we will focus on $d = 2$ dimensions, which is relevant as the theory on the worldsheet of the string.

$d = 2$ **Dimensions**

In general, we would define general relativity tensors with covariant indices by their transformation law, which in two dimensions would be

$$T_{i_1 \ldots i_j}(z_1, z_2) = T'_{j_1 \ldots j_n}(z'_1, z'_2) \frac{\partial z'_{j_1}}{\partial z_{i_1}} \ldots \frac{\partial z'_{j_n}}{\partial z_{i_n}}. \qquad (9.14)$$

But in two dimensions we can form complex coordinates, $z = z_1 + iz_2, \bar{z} = z_1 - iz_2$, and express everything in terms of them. Then it is easy to see that a conformal transformation would be simply a holomorphic transformation,

$$z' = f(z); \quad \bar{z}' = \bar{f}(\bar{z}), \qquad (9.15)$$

since then

$$ds^2 = dz' d\bar{z}' = |f'(z)|^2 dz d\bar{z} \equiv \Omega^2(z, \bar{z}) dz d\bar{z}. \tag{9.16}$$

In two dimensions expressed in complex coordinates, a general relativity tensor would be an object with an integer number of z and of $\bar{z}$ indices. But since conformal transformations can be embedded in general coordinate transformations, we should take advantage of the definition of general relativity tensors and define a notion of tensor for conformal transformations only.

Such a tensor is called a *primary field* of dimensions (h, h'), and it transforms as a general relativity tensor with h indices z and h' indices $\bar{z}$:

$$T_{z...z\bar{z}...\bar{z}} = T'_{z...z\bar{z}...\bar{z}} \left(\frac{dz'}{dz}\right)^h \left(\frac{d\bar{z}'}{d\bar{z}}\right)^{h'}. \tag{9.17}$$

But unlike the case of general relativity tensors, we don't need to have integer values for h and h', since they are not associated with a physical number of indices. The primary fields are denoted by $\phi^{(h, h')}(z, \bar{z})$.

The infinitesimal transformation of a primary field under the transformation $z' = z + \epsilon(z)$ is

$$\delta_\epsilon \phi(z) = \epsilon(z) \partial \phi(z) + h \partial \epsilon(z) \phi(z) + H.c. \tag{9.18}$$

9.3 Quantum Conformal Field Theory

Until now we defined a classical field theory, but the interesting behavior of conformal field theory occurs at the quantum level. For that, we need to consider a model. Consider the free scalar field model (relevant for string theory in conformal gauge, as we will see later),

$$S[\phi] = \int d^2 z \, \partial \phi \bar{\partial} \phi. \tag{9.19}$$

For this free field action, we can invert the kinetic term to find the propagator, or free 2-point function, as

$$\langle \phi(z) \phi(w) \rangle = -\partial^{-2}(z, w) = -\log|z - w|^2 = -\log(z - w) - \log(\bar{z} - \bar{w}). \tag{9.20}$$

This is a familiar result for a free propagator, but in the unfamiliar complex notation, so the proof is left as an exercise.

Taking derivatives $\partial_z \partial_w$ on the above relation, we find

$$\langle \partial \phi(z) \partial \phi(w) \rangle = -\frac{1}{(z - w)^2}. \tag{9.21}$$

Note that the dimensions of the above operators are the classical ones, $[\phi] = 0$ and $[\partial \phi] = 1$.

For general fields, of dimensions $(h, \bar{h})$, we have

$$\langle \phi_{h,\bar{h}}(z, \bar{z}) \phi_{h,\bar{h}}(w, \bar{w}) \rangle = \frac{1}{(z - w)^{2h}} \frac{1}{(\bar{z} - \bar{w})^{2\bar{h}}}. \tag{9.22}$$

An important quantity in any theory, but especially in conformal field theory, is the energy momentum tensor. Here $T_{\mu\nu}$ is conserved, $\partial^\mu T_{\mu\nu} = 0$, symmetric $T_{\mu\nu} = T_{\nu\mu}$ and it is traceless. The last condition (tracelessness) comes from conformal invariance. Indeed, the usual energy momentum tensor (Belinfante tensor) is defined as

$$T_{\mu\nu} = \frac{2}{\sqrt{g}} \frac{\delta S}{\delta g^{\mu\nu}}. \tag{9.23}$$

But conformal invariance is the invariance under $g^{\mu\nu} \to e^{2\omega} g^{\mu\nu}$ in flat space, so we should have

$$\eta^{\mu\nu} \frac{\delta S}{\delta g^{\mu\nu}} = 0. \tag{9.24}$$

That translates into tracelessness of the energy momentum tensor, $\eta^{\mu\nu} T_{\mu\nu} = 0$. In light-cone or complex coordinates, the metric is $\eta^{+-} = 2$, so tracelessness means $T_{z\bar{z}} + T_{\bar{z}z} = 0$. When coupled with the symmetric property, $T_{z\bar{z}} = T_{\bar{z}z}$, we obtain $T_{z\bar{z}} = T_{\bar{z}z} = 0$. Finally, conservation of $T_{\mu\nu}$ means holomorphicity:

$$\partial^{\bar{z}} T_{zz} = 0 \Rightarrow T_{zz} \equiv T = T(z)$$
$$\partial^z T_{\bar{z}\bar{z}} = 0 \Rightarrow T_{\bar{z}\bar{z}} \equiv \bar{T} = \bar{T}(\bar{z}). \tag{9.25}$$

So we have only a holomorphic function $T(z)$ and an antiholomorphic one $\bar{T}(z)$ defining the energy-momentum tensor.

One can prove that the general OPE of the energy momentum tensor $T(z)$ with the primary fields Φ_i gives

$$T(z)\Phi_i(z_i) = \frac{h_i}{(z - z_i)^2} \Phi_i(z_i) + \frac{1}{z - z_i} \partial \Phi_i(z_i) + \text{non-singular}, \tag{9.26}$$

where the nonsingular terms are descendants (to be seen shortly what that means) of Φ_i.

As for any holomorphic function, $T(z)$ can be expanded in a Laurent series:

$$T(z) = \sum_{n \in \mathbb{Z}} L_n z^{-n-2}. \tag{9.27}$$

Then the L_n satisfy the *Virasoro algebra*:

$$[L_m, L_n] = (m - n)L_{m+n} + \frac{c}{12} m(m^2 - 1)\delta_{m+n,0}. \tag{9.28}$$

This is related to the fact that two $T(z)$'s satisfy the OPE:

$$T(z)T(z') = \frac{c}{2(z - z')^4} + \frac{2}{(z - z')^2} T(z') + \frac{1}{z - z'} \partial T(z') + \text{non-singular}. \tag{9.29}$$

From this, we note that $T(z)$ itself is only a primary field if $c = 0$. As we can see, the Virasoro is a type of an infinite dimensional algebra, with a *central charge* c.

By unitarity, we must have $L_{-n} = L_n^\dagger$, a fact that will become clearer when we define representations.

Representations are constructed by analogy with the case of spin j representations of $SU(2)$, by starting with a *highest weight state* $|h\rangle$. In conformal field theory, there is a correspondence between states and operators. For the case of the highest weight state, this state is mapped to a primary field by the construction

$$|h\rangle = \lim_{z \to 0} \phi_h(z)|0\rangle. \tag{9.30}$$

The state is defined by the conditions

$$L_0|h\rangle = h|h\rangle, L_n|h\rangle = 0, \quad n > 0, \tag{9.31}$$

which are the analogs of the usual creation and annihilation operator definitions of the vacuum by $H|0\rangle = E_0|0\rangle$ and $a_n|0\rangle = 0$. Therefore we see that L_0 acts as a sort of energy (Hamiltonian), whereas L_n for $n > 0$ are annihilation operators.

Then the states $L_{-n}|h\rangle$ are called descendants and form the representation of the conformal algebra, since L_{-n} acts as an $a_n^\dagger$ operator. Indeed, from the Virasoro algebra we have $[L_0, L_{-n}] = nL_{-n}$, or $L_0 L_{-n} = L_{-n} L_0 + nL_{-n}$, so

$$L_0(L_{-n}|h\rangle) = L_{-n}L_0|h\rangle + nL_{-n}|h\rangle = (h + n)(L_{-n}|h\rangle). \tag{9.32}$$

So we see that L_{-n} increases the energy L_0, hence really $|h\rangle$ should be called the "lowest weight" state, but it is called "highest weight state" for historical reasons.

The representation of the conformal (Virasoro) algebra is thus constructed as $|h\rangle, L_{-1}|h\rangle$, $(L_{-1})^2|h\rangle$, $L_{-2}|h\rangle$, etc., and the *sum* of these representations (each defined by a highest weight state) is called the *Verma module*.

Minimal Unitary Series
It is left as an exercise to prove that for unitarity we must have $h \geq 0$ and $c \geq 0$.

A particular series of representations, labeled by an integer $m \in \mathbb{Z}$, is called the *minimal unitary series*. The representations have central charges

$$c(m) = 1 - \frac{6}{m(m + 1)}, \tag{9.33}$$

and the dimensions in the representation are defined by

$$h_{pq}(m) = \frac{[(m + 1)p - mq]^2 - 1}{4m(m + 1)}, \tag{9.34}$$

where $p, q \in \mathbb{Z}$, with $p = 1, \ldots, m - 1$ and $q = 1, \ldots, m$ defining the primary field members of the representation.

Note that $m = 1$ would give $c < 0$ so is excluded, whereas $m = 2$ would give $h = 0$ for all allowed primary fields, so it is also excluded.

The first nontrivial case, $m = 3$, which gives $c = 1/2$, corresponds to the Ising model. The next case, $m = 4$, which gives $c = 7/10$, corresponds to the tricritical Ising model.

Note that for all solutions (in p, q) of the equation $(m + 1)p - mq = $ constant, we have the same $h_{p,q}(m)$.

In the case $c = 1/2$, i.e. the Ising model, the primary fields are found to be $\phi_{1,1}$, with $h_{1,1} = 1/16$ and $\phi_{1,3} = \phi_{2,1}$, with $h_{1,3} = h_{2,1} = 1/2$. We should have in general

$$\langle \phi_{h,\bar{h}}(z_1)\phi_{h\bar{h}}(z_2) \rangle = \frac{1}{(z_1 - z_2)^{2h}(\bar{z}_1 - \bar{z}_2)^{2\bar{h}}}. \tag{9.35}$$

But the two obvious quantities formed at each site based on spin are the spin σ_n itself, and the energy density (energy per site), $\epsilon_n \sim \sigma_n\sigma_{n+1}$ (a single term in the sum in the Heisenberg Hamiltonian). It is found that they satisfy

$$\langle \sigma(z_1)\sigma(z_2) \rangle = \frac{1}{|z_1 - z_2|^{1/4}},$$

$$\langle \epsilon(z_1)\epsilon(z_2) \rangle = \frac{1}{|z_1 - z_2|^2}, \tag{9.36}$$

so they correspond to the values of $(h, \bar{h})$ equal to $(1/16, 1/16)$ and $(1/2, 1/2)$ respectively, as they should, since we saw that the only primary fields are the ones with these dimensions.

9.4 Quantum Phase Transitions

Up to now we have described normal phase transitions, occurring in the temperature, and thus where the relevant fluctuations responsible for it are thermal.

But there is another kind of phase transition, which happens as we tune a parameter (the "coupling") g, perhaps at $T = 0$, and at a finite $g = g_C$ we have the transition. That means that the phase transition is instead governed by quantum fluctuations and thus is called a *quantum phase transition*. In this case, generically the phases are different topologically, or the phase transitions are of type II (second order). Again the system in the vicinity of the phase transition is described by a conformal field theory, since there is no dimensional scale available.

A classic example of such a transition is the insulator–type II superconductor phase transition that occurs as we change the chemical doping g. Note that of course we also have in this case the usual phase transition occurring as the temperature is changed (the thermal superconductor/insulator phase transition), but here we are talking about a transition that occurs in g, at $T = 0$.

As we move to nonzero temperature, the phase transition point $g = g_C$ opens up into an entire phase, bounded to the left (small g) by the *Kosterlitz-Thouless phase transition* and at the right by a crossover, as in Figure 9.1 (a crossover is a smooth change in the thermodynamical potential, unlike a phase transition, which means a discontinuity in an n-th derivative of the thermodynamical potential). This phase is called the *quantum critical phase* and thus corresponds to putting the conformal field theory (that appears at $T = 0$ near the phase transition) at finite temperature. This phase is conformal, strongly coupled (since the coupling is close to g_C, which is large), and thus is poorly understood using conventional condensed matter theory methods, and is ideal for a treatment by AdS/CFT, which will be described in Part III.

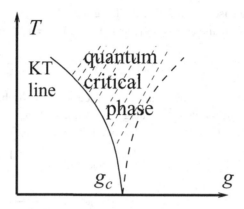

Quantum critical phase bounded from the left by the Kosterlitz-Thouless phase transition line and from the right by a crossover.

Kosterlitz-Thouless (KT) Phase Transition

The KT transition *in 2+1 dimensions* is a transition between a state with bound vortex-antivortex pairs (when single vortices disappear) at low temperature, and a state with unpaired (single) vortices at high temperature, but as seen in Figure 9.1, the transition is also defined at constant temperature, as a transition in the coupling g, i.e. a *quantum phase transition*, governed by quantum fluctuations, not temperature fluctuations.

The transition can be realized physically in systems modeled by the rotor model, known also as the $O(2)$ model, or the XY model. For instance, Josephson junction arrays, disordered superconductors, or granular flows are examples of such systems.

The low-temperature or low-coupling phase is an ordered phase, described by power law correlations, and at high temperature or large couplings, we can have exponential correlations, with a correlation length $\xi \propto 1/\sqrt{T - T_C}$. However, the order referred to here is not the usual order in type II phase transitions (occurring in temperature only), but rather is *topological order*, defined by the presence (or not) of vortices.

There is a simple semiquantitative argument for the existence of a phase transition. For a 2+1–dimensional system of size R, with vortices of vortex core size a, the energy of a vortex is found to be

$$E \sim \kappa \ln \frac{R}{a}, \tag{9.37}$$

for $R \gg a$, where κ is a constant that depends on the system. On the other hand, the number of possible positions for the vortex are

$$N \sim \frac{R^2}{a^2}, \tag{9.38}$$

which leads to a Boltzmann entropy

$$S = k_B \ln N = 2k_B \ln \frac{R}{a}. \tag{9.39}$$

Then the free energy of the system is

$$F = U - TS = (\kappa - 2k_B T) \ln \frac{R}{a}. \tag{9.40}$$

We see that for low temperature we have $F > 0$, which means that the single vortices are not preferred, so indeed the low-temperature phase has only bound vortex pairs. Physically, we understand that the energy of a single vortex diverges as $\ln(R/a)$, but bound pairs have finite energy, as the vortex and antivortex attract each other, as if they are particles of charge $+1$ and -1 with a $1/r$ force (log potential), but the total pair is neutral. On the other hand, for high temperatures we have $F < 0$, which means that the single vortices are stable, and we have a gas of unbound vortices.

The phase transition temperature, or Kosterlitz-Thouless temperature T_{KT}, between the two phases is

$$T_C = T_{KT} = \frac{\kappa}{2k_B}. \tag{9.41}$$

Example As an example of the KT transition, consider the Ising model in a transverse magnetic field, i.e. the H field to be along the x direction perpendicular to z. Specifically,

$$H = -J \sum_i (g\sigma_i^x + \sigma_j^z \sigma_{j+1}^z). \tag{9.42}$$

Then we see that at $g = 0$ we have the usual ferromagnetic theory, whereas at $g = \infty$ we have all spins oriented along the x direction. That means that somewhere in between there must be a phase transition. In fact, the phase transition happens at $g = 1$.

9.5 Insulator-Superconducting Phase Transition

We now try to describe the quantum phase transition in the relevant case of the insulator to superconductor transition in g. For instance, consider an array of superconducting islands ("grains") on a 2-dimensional lattice: the electrons in the grains are locked into Cooper pairs described by a bosonic field ($\phi = \psi_1 \psi_2$) with the same phase throughout the grain. That is, in the jth grain, we have amplitude $|\phi_j|$ and phase θ_j:

$$\phi_j = |\phi_j| e^{i\theta_j}. \tag{9.43}$$

In this case, the boson that corresponds to the order parameter is the composite boson for the Cooper pairs (two electrons of opposite spin locked together). But there are also truly bosonic examples that realize the same transition, specifically a system of ^{87}Rb cold atoms on an optical lattice. In this case, we have a superfluid to normal transition (basically the same as the superconducting to insulator transition, except for the existence of charge).

The phase transition is described by the *Bose Hubbard model* (note that the usual Hubbard model is fermionic, and this is a bosonic version of the same). The Hamiltonian for it is

$$H = E_C \sum_i (\hat{n}_i - n_0)^2 - t \sum_{\langle ij \rangle} (b_i^\dagger b_j + b_j^\dagger b_i).$$ (9.44)

Here as usual $\langle ij \rangle$ means nearest neighbor and $\hat{n}_i = b_i^\dagger b_i$. E_C is the energy required to remove a boson (or Cooper pair) from the system to infinity, and the minimum energy corresponds to $\hat{n}_i = n_0$ (constant boson density); t is a hopping energy, since it is the energy corresponding to removing a boson from site j and creating one at the nearest neighbor i, i.e. hopping.

We can rewrite the Hamiltonian as

$$H = \frac{U}{2} \sum_i n_i(n_i - 1) - \mu \sum_i n_i - t \sum_{\langle ij \rangle} (b_i^\dagger b_j + b_j^\dagger b_i),$$ (9.45)

where $E_C = U/2$ and $2E_C n_0 = U/2 + \mu$, and μ is a chemical potential.

We can also rewrite it in terms of commuting coordinates and their derivatives. We can represent the relation

$$[\hat{n}_i, b_i^\dagger] = b_i^\dagger$$ (9.46)

as

$$\hat{n}_i \equiv \frac{\partial}{i \partial \theta_i}; \quad b_i^\dagger = \sqrt{n_0} e^{i\theta_i}.$$ (9.47)

Then substituting in the Hamiltonian, we find

$$H = E_C \sum_i \left(\frac{\partial}{i \partial \theta_i} \right)^2 - J \sum_{\langle ij \rangle} \cos(\theta_i - \theta_j),$$ (9.48)

where $J = 2t$. This is known as the quantum rotor model. Its mean field solution generates the qualitative KT transition, as we have shown above.

In the ordered (superconducting) phase at low g, we have vortices and antivortices, and the Cooper pairs condense with rigid phase. In that case, we can describe the physics in terms of vortices. In the disordered (insulator) phase at high g, we have Cooper pairs localized, and vortices condense. In this case, we can describe the physics in terms of particles. So on one side of the transition we have particles, and on the other vortices.

At the phase transition temperature (criticality), vortices unbind, and there is a duality between the vortex and the particle descriptions. The phase transition temperature is written in terms of the effective mass of the charge carriers m^* and the superfluid density $n_S(T)$ as

$$k_B T_C = \frac{\pi \hbar^2 n_S(T)}{2m^*}.$$ (9.49)

Landau-Ginzburg Theory

We can write an effective field theory for the boson Hubbard model. Consider the ground state of the system, with an equal number of bosons at each site, $n_i = n_0$. Then introduce a creation operator $a_i^\dagger$ to produce an extra particle at each site, and a creation operator $h_i^\dagger$ to produce an extra "hole" (corresponding to an antiparticle in an usual relativistic QFT description) at each site.

Then define a "relativistic" quantum field as

$$\phi_i \sim \alpha_i a_i + \beta_i h_i^\dagger, \tag{9.50}$$

a discretized version of the usual definition. Then "time" appears from temperature, as an analytical continuation from Euclidean time.

All in all, we can write an effective "relativistic" Lagrangean for the corresponding continuum quantum field theory, which takes the form of a relativistic version of the Landau-Ginzburg Lagrangean, with

$$S = \int d^3x \left[-(\partial_t \phi)^2 + v^2 |\vec{\nabla}\phi|^2 + (g - g_c)|\phi|^2 + u|\phi|^4 \right], \tag{9.51}$$

where $u = U/2$.

In conclusion, we see that we can describe the quantum critical phase through a continuum quantum field theory of the relativistic Landau-Ginzburg type.

Important Concepts to Remember

- Near usual, thermal, phase transitions, thermodynamic quantities characterizing the material (second order derivatives for second order phase transitions) diverge as power laws, giving critical behavior. It is due to the divergence of the correlation length ξ.
- At the phase transition, we have an effective continuum field theory with conformal invariance, or conformal field theory.
- The conformal field theory is determined by the operator product expansions (OPEs).
- A CFT primary field is the analog of general relativity tensors with a continuous number of indices.
- In $d > 2$, the conformal group is $SO(d + 1, 1)$ in the Euclidean case and $SO(d, 2)$ in the Minkowski case. In two dimensions, the conformal group is infinite, and conformal transformations are holomorphic ones.
- The energy-momentum tensor in two dimensions is characterized by a holomorphic $T(z)$, whose moments satisfy the Virasoro algebra, an infinite dimensional algebra with a central charge.
- Representations of the Virasoro algebra are found from highest weight states, corresponding to primary operators, acted upon by L_{-n}, giving descendants fields.
- The minimal unitary series is labeled by an m, with $c = c(m)$ and the conformal dimensions $h_{p,q}(m)$, and starts with the Ising model.
- Quantum phase transitions are phase transitions occurring in a coupling g as opposed to temperature, thus being governed by quantum fluctuations instead of thermal fluctuations. A conformal field theory describes the transition point.

- At nonzero T, the quantum phase transition point opens up into a quantum critical phase, strongly coupled and poorly understood, that is a $T \neq 0$ version of the conformal field theory. It is bounded on one side (low T, low g) by the Kosterlitz-Thouless phase transition line, and on the other by a crossover.
- The KT phase transition corresponds to a transition in T between bound vortex-antivortex pairs at low T and single vortices for high T.
- The quantum insulator-superconducting phase transition in g is described by the boson Hubbard model or the quantum rotor model, reduces to a "relativistic" Landau-Ginzburg continuum effective field theory, and is a transition between vortices at low g and particles at high g, with particle-vortex duality and vortex unbinding at the transition point.

Further Reading

For conformal field theory, see chapter 2 in [11], and for quantum phase transitions, see [12].

Where It will be Addressed in String Theory

Conformal field theory is relevant to the description of the string theory itself, as we will see in Chapter 23. Moreover, the AdS/CFT correspondence itself, the main tool to describe condensed matter systems in string theory, is mostly defined for conformal field theories, as we will first see in Chapter 27. General quantum phase transitions and quantum critical systems will be addressed holographically in Chapter 42, while specific examples like strange metals will be addressed in Chapters 47 and 48.

Exercises

(1) Check that

$$\partial \bar{\partial} \ln |z|^2 = \partial \frac{1}{\bar{z}} = \bar{\partial} \frac{1}{z} = 2\pi \delta^2(z, \bar{z}). \tag{9.52}$$

(2) Prove that unitarity of the CFT (implying $L_m^\dagger = L_{-m}$) and the Virasoro algebra imply $h \geq 0, c \geq 0$.

(3) Consider a *classical* energy-momentum tensor with Laurent modes

$$L_m = \frac{1}{2} \sum_{m \in \mathbb{Z}} \alpha_{m-n}^\mu \alpha_n^\mu, \tag{9.53}$$

where α_m^μ satisfy the Poisson bracket algebra

$$[\alpha_m^\mu, \alpha_n^\nu]_{P.B.} = im\delta_{m+n,0}\eta^{\mu\nu}. \tag{9.54}$$

Calculate $[L_m, L_n]_{P.B.}$ and show that it is given by the classical version of the Virasoro algebra, for $[,] \rightarrow +i[,]_{P.B.}$ and $c = 0$.

(4) For $d > 2$ Euclidean dimensions, calculate the most general infinitesimal transformation solution $\delta\sigma^1 = v^a(\sigma)$ to the conformal transformation inducing (by general coordinate transformations)

$$\delta g_{ab} = -\partial_a v_b - \partial_b v_a. \tag{9.55}$$

As explained in the text, the various components form the $SO(1, d + 1)$ algebra.

(5) Check that the transformations $z \to z + \epsilon(z)$, $\bar{z} \to \bar{z} + \bar{\epsilon}(\bar{z})$, with

$$z \to z - a_n z^{n+1}, \quad \bar{z} \to \bar{z} - \bar{a}_n \bar{z}^{n+1}, \quad n \in \mathbb{Z}, \tag{9.56}$$

generated by the operators

$$L_n = -z^{n+1}\frac{d}{dz}; \quad \bar{L}_n = -\bar{z}^{n+1}\frac{d}{d\bar{z}}, \tag{9.57}$$

satisfy the 2-dimensional local conformal algebra, i.e. two copies of the Virasoro algebra for $c = \bar{c} = 0$:

$$[L_n, L_m] = (n - m)L_{n+m}; \quad [\bar{L}, \bar{L}_m] = (n - m)\bar{L}_{n+m}; \quad [L_n, \bar{L}_m] = 0. \tag{9.58}$$

(6) Consider the expansion of the energy-momentum tensor

$$T(z) = \sum_{n\in\mathbb{Z}} \frac{\hat{L}_n(w)}{(z - w)^{n+2}}, \tag{9.59}$$

where

$$\hat{L}_{-n}(w)\Phi(w, \bar{w}) \equiv \oint_w \frac{dz}{2\pi i}(z - w)^{-n+1}T(z)\Phi(w, \bar{w}). \tag{9.60}$$

From the OPE of $T(w)$ with $\Phi(z)$, find $\hat{L}_n\Phi(z, \bar{z})$, $n \in \mathbb{Z}$.

(7) Using the mean field approximation, show explicitly that at g very large in the Heisenberg model in H field, all spins are along x, i.e. there is a (very large) magnetization.

Classical vs. Quantum Hall Effect

In this chapter we will study the Hall effect, starting with the classical one, and moving on to the integer quantum Hall effect, which will take most of the chapter, and end with the fractional quantum Hall effect.

10.1 Classical Hall Effect

Consider a quasi-2-dimensional sample of conducting material (metal), with length L in the x direction, and "width" W in the y direction, and negligible extension in the third direction. Consider applying a voltage V_L along the x direction, with an electric field E_x, such that $V_L = LE_x$, and a magnetic field B in the transverse z direction, as in Figure 10.1. Then the conduction electrons will move under the Lorentz force in the y direction, until one generates a compensating electric field E_y, with a potential difference $V_H = WE_y$, which is called the *Hall voltage*. This is the classical Hall effect.

Now let us be more quantitative. The Lorentz force acting on the electrons is

$$\vec{F} = -e(\vec{E} + \vec{v} \times \vec{B}). \tag{10.1}$$

At equilibrium, the generated electric field on E_y compensates for the magnetic force, so

$$E_y = v_x B_z. \tag{10.2}$$

As usual, in the conductor there is a relaxation time for the collisions with impurities or phonons, giving the average velocity for conduction in the presence of an electric field E_x as

$$v_x = -\frac{eE_x\tau}{m}. \tag{10.3}$$

On the other hand, consider the current density j_x (normally per unit transverse area, but considering that one of the transverse directions is very small, per unit transverse length). Then, for the density of electrons per unit sample area n_e (also normally it would be per unit volume, but since the z direction has very small extension, only per unit area (WL)), we have

$$j_x = n_e e v_x = \frac{e^2 \tau n_e E_x}{m}. \tag{10.4}$$

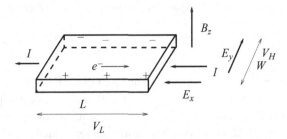

Fig. 10.1 The classical Hall effect: Current in the x direction and magnetic field in the z direction result in Hall potential in the y direction.

An electron in a transverse magnetic field rotates in a circle with the *cyclotron frequency* ω_c calculated from equating the Lorentz force with the centrifugal force:

$$F_{Lorentz} = evB = F_{centrif.} = mv\omega \Rightarrow$$

$$\omega_c = \frac{eB}{m}. \qquad (10.5)$$

Then

$$E_y = v_x B_z = -\frac{eB}{m}\tau E_x = -\omega_c \tau E_x. \qquad (10.6)$$

We define the transverse, or *Hall resistance* R_H, as the ratio of the Hall voltage (in the y direction) and the applied current (in the x direction):

$$R_H = \frac{V_H}{I} = \frac{W E_y}{W j_x} = -\frac{m\omega_c \tau}{e^2 \tau n_e} = -\frac{m\omega_c}{n_e e^2} = -\frac{B}{n_e e}. \qquad (10.7)$$

Note that the minus sign is conventional, just to remind that the Hall voltage is a response to compensate for the effect of the applied current.

We can also define the inverse, the *Hall conductance*:

$$\sigma_H = \frac{1}{R_H} = -\frac{n_e e}{B}. \qquad (10.8)$$

10.2 Hall Effect Experimental Results

That classical picture was everything that there was until 1980, when Klaus von Klitzing made a remarkable experimental discovery. He measured the Hall effect on a 2-dimensional electron gas confined at the interface between an oxide, SiO_2 and the semiconductor Si in a Si MOSFET (Metal-Oxide-Semiconductor Field-Effect Transistor, a common component in electronics). He found that the Hall voltage as a function of B does not increase simply linearly, as implied by the classical relation (10.7), but rather has *plateaus*, whose position is independent of the sample. He calculated that the Hall conductance σ_H where the plateaus are located is quantized in units of e^2/h, where h is the Planck constant,

$$\sigma_H = -\frac{ne^2}{h}, \tag{10.9}$$

where n is a positive integer (to one part in 10^7). For these values, the common (longitudinal) resistance vanishes, $R_L = V_L/I = 0$, i.e. we have dissipationless transport. This is the *Integer Quantum Hall Effect (IQHE)*. For its discovery, von Klitzing won the 1985 Nobel Prize in Physics (an extremely short time for modern standards between discovery and the Nobel Prize).

Very shortly after the integer quantum Hall effect was discovered, in 1982, when considering much higher magnetic fields B and lower temperatures T, Tsui, Stormer, and Gossard found Hall plateaus at fractional values of the conductance,

$$\sigma_H = \nu \frac{e^2}{h}, \quad \nu = \frac{k}{2p+1}, \tag{10.10}$$

where ν is called a *filling fraction* for reasons to be explained shortly, and k is a positive integer and $2p+1$ a positive and odd integer. This is called the *Fractional Quantum Hall Effect (FQHE)*. It was "explained" in 1983 by Robert Laughlin. The quotation marks are because only the filling fractions $\nu = 1/(2p+1)$ were explained this way, and moreover, the explanation came in the form of an ansatz for a wavefunction for interacting electrons, the "Laughlin wavefunction," but it is not clear how to obtain a microscopical derivation for this wavefunction. For this theoretical explanation, Laughlin won the 1998 Nobel Prize in Physics.

10.3 Integer Quantum Hall Effect

The understanding of the integer quantum Hall effect is based on the concept of Landau levels, so we will understand them first.

Landau Levels

We want to write the Schrödinger equation for electron motion in a plane, with constant transverse magnetic field B, just like in the case of the Hall effect. For a magnetic field B in the z direction, we can choose a gauge where the vector potential in the y direction is $A_y = Bx$, and $A_x = 0$. Since the momentum in the y direction is now $i(\partial_y - eA_y/\hbar)$, we have

$$-\frac{\hbar^2}{2m} \left[\partial_x^2 + (\partial_y - \frac{ie}{\hbar} A_y)^2 \right] \psi(x, y) = E\psi(x, y). \tag{10.11}$$

Because of translational invariance in the y direction, we can choose the ansatz

$$\psi_{n,k}(x, y) = e^{iky} f_n(y). \tag{10.12}$$

Replacing the ansatz in the Schrödinger equation, and defining the *magnetic length*

$$l = \sqrt{\frac{\hbar}{eB}} = \frac{250\text{Å}}{B(\text{T})}, \tag{10.13}$$

we obtain

$$\frac{\hbar\omega_c}{2}\left[-l^2\partial_x^2 + \left(\frac{x}{l} - lk\right)^2\right]f_n(x) = E_n f_n(x). \tag{10.14}$$

This is simply the equation for the harmonic oscillator of variable $x/l - lk$, i.e. shifted by

$$x_k = l^2 k. \tag{10.15}$$

Therefore the solution for the wavefunction is

$$\psi_{n,k}(x, y) = e^{iky}H_n\left(\frac{x}{l} - lk\right)e^{-\frac{(x-x_k)^2}{2l^2}}, \tag{10.16}$$

where H_n are the Hermite polynomials, and the energy levels are simply the energy levels of the harmonic oscillator,

$$E_n = \hbar\omega_c\left(n + \frac{1}{2}\right). \tag{10.17}$$

Here the harmonic oscillator index n is called the *Landau level*, and the motion is harmonic oscillator motion, but at a position shifted by x_k (and rescaled by l). Each Landau level n contains many electron states, however, since there are many possible k's, thus x_k's. We can calculate the degeneracy of a given Landau level n from the fact that the momentum k is quantized, as the direction y has length W, thus

$$k_m = \frac{2\pi m}{W}. \tag{10.18}$$

However, each such k corresponds to a shift by x_k, hence k can be up to a maximum value that corresponds to a shift equal to the length of the material, $x_{k_{max}} = L$, leading to a maximum value of the electron state number m (for fixed Landau level):

$$N_{\text{max}} \equiv m_{\text{max}} = \frac{LW}{2\pi l^2} = \frac{LWeB}{h}. \tag{10.19}$$

On the other hand, BLW is the magnetic flux Φ passing through the material (of area LW). Therefore we find

$$N_{\text{max}} = \frac{\Phi}{\Phi_0}, \tag{10.20}$$

where

$$\Phi_0 = \frac{h}{e} \tag{10.21}$$

is a *flux quantum*, since N_{max} is an integer, corresponding to the total (maximum) number of electron states in each Landau level. Note that this number is independent of n (the level).

Then we find that the (maximum) number of electron states per unit area of a single Landau level is

$$n_B = \frac{N_{max}}{LW} = \frac{1}{2\pi l^2} = \frac{eB}{m}. \tag{10.22}$$

If we have an integer total number of totally filled Landau levels,

$$\nu = \frac{n_e}{n_B} = n \in \mathbb{N}, \tag{10.23}$$

the Hall conductance is

$$\sigma_H = -\frac{n_e e}{B} = -\frac{n n_B e}{B} = -\frac{ne^2}{h}. \tag{10.24}$$

That means that ν is the filling fraction, and we have obtained the integer quantum Hall effect with the quantum number n being the number of filled Landau levels.

Alternative Derivation

We can make an alternative derivation of the integer quantum Hall effect, one that will also obtain the other important consequence of the integer quantum Hall effect, the zero normal resistance.

In the equilibrium situation, the total Lorentz force is zero, so $\vec{v} \times \vec{B} = -\vec{E}$, in components:

$$v_i B_k = E_j \epsilon_{ijk}. \tag{10.25}$$

For fixed k, corresponding specifically to z, since the magnetic field is only in the z direction, we can write

$$v_i = \frac{\epsilon_{ij} E_j}{B_z}. \tag{10.26}$$

Since the number of fully occupied Landau levels is n, each with N_{max} states, we have

$$Q = n N_{max} \tag{10.27}$$

electrons in the system, so the current density is

$$j_i = \frac{eQ}{LW} v_i = \frac{eQ}{BLW} \epsilon_{ij} E_j \equiv \sigma_{ij} E_j. \tag{10.28}$$

That means that the conductance *tensor* is

$$\sigma_{ij} = \frac{eQ}{BLW} \epsilon_{ij} = \frac{en N_{max}}{N_{max} \frac{h}{e}} \epsilon_{ij} = \frac{ne^2}{h} \epsilon_{ij}, \tag{10.29}$$

i.e. as a matrix,

$$\sigma = \begin{pmatrix} 0 & \frac{ne^2}{h} \\ -\frac{ne^2}{h} & 0 \end{pmatrix}. \tag{10.30}$$

In particular, that means that the normal conductance, $\sigma_{xx} = 0$. If the conductance would be a scalar, then this would mean we have an insulator. However, since it is a matrix, inverting it we obtain the resistance matrix:

$$R = \sigma^{-1} = \begin{pmatrix} 0 & -\frac{1}{\sigma_{xy}} \\ +\frac{1}{\sigma_{xy}} & 0 \end{pmatrix}. \tag{10.31}$$

That means that $R_{xx} = 0$, i.e. zero longitudinal (normal) resistance, or dissipationless transport, the opposite of an insulator (which would have been the conclusion drawn simply from $\sigma_{xx} = 0$).

Yet the explanation of the effect is indeed related to the explanation for the insulator. In the case of an insulator, the valence band is completely filled, so there are no conduction (electrons or) holes in it, and the conduction band is empty, so there are no electrons in it, and the energy gap between the bands is sufficiently large so that it cannot be excited thermally. The Fermi level is between the two energy bands (at midpoint at $T = 0$), i.e. in the middle of the gap.

In the Hall case, the situation is similar: The highest occupied Landau level n is completely filled (the Landau level splits slightly due to the interaction between the electrons, becoming an energy band), and there is an energy gap until the lowest (completely) unoccupied level $n + 1$, and for the same reason as for the insulator, the Fermi level falls in the middle of the energy gap, leading to σ_{xx} like for the insulator.

The difference is that now $\sigma_{xx} = 0$, coupled to $\sigma_{ij} \propto \epsilon_{ij}$, leads to $R_{xx} = 0$, explaining the dissipationless transport of the integer quantum Hall effect. Normally, in a conductor, disorder is important, being the leading influence in the calculation of resistance. Now disorder is important, but it does not affect the Hall conductance σ_H, as we saw.

10.4 Edge Currents and Laughlin's Gauge Principle

We have seen in two ways how to derive the integer quantum Hall effect. But there is another derivation that emphasizes the role of topology, namely that the quantization of σ_H is due to a topological number for the gauge field, the *first Chern class*. This has been first described by Laughlin and goes under the name of Laughlin's gauge principle.

We want to see the role of topology, in this case, of boundaries. Therefore let us consider a planar circular disk geometry instead of the rectangular geometry used until now. If we are in a transverse magnetic field, the conduction electrons in the disk move on small circles in the bulk of the disk, so there is no net current, as we can understand if we draw diagrams of the electron motions close to each other (for the magnetic length much larger than the Fermi length, $l \gg \hbar/p_F$), as in Figure 10.2(a). However, in the presence of the circular boundary, there is an effective *edge current* I_2, since on the edge, the electrons can do only tiny semicircles, after which they are reflected by the boundary, and the process of hopping in semicircles leads to an effective current around the boundary; see Figure 10.2(a).

If moreover we introduce a circular hole in the middle of the disk, then the semicircles are on the other side of the circle, and drawing this we can convince ourselves that the effective edge current I_1 flows around this new boundary in the opposite direction to the outside boundary; see Figure 10.2(b). Since moreover the current doesn't depend on

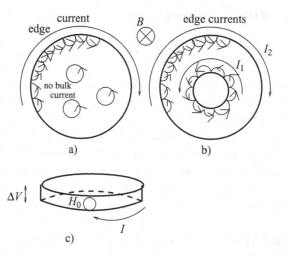

Fig. 10.2 (a) Edge currents form when we have a magnetic field perpendicular to the planar sample. There are no bulk currents. (b) If there is a hole, a current opposite to the one outside forms, for a total vanishing edge current. (c) Equivalent form for the circular disk with a hole: deformed to a thin circular band.

the size of the hole, only on the properties of the material and the magnetic field, if both edges are at the same chemical potential μ (the chemical potential is the energy required to take an electron to infinity, so a difference in μ can be achieved by a potential energy difference), then the current is the same, and $I_1 - I_2 = 0$, so there is no net current flowing around the disk.

But if we introduce a difference in Fermi energies (chemical potentials at $T = 0$) between the outside and inside boundary, for instance, by introducing an electric field transverse to the boundaries E_0, such that $\epsilon_{F1} - \epsilon_{F2} = eE_0$, and if we have n fully occupied Landau levels, we will have a total energy difference of $\Delta U = neE_0$, and then we can use the same argument as above (the second derivation) to prove that $\sigma_{xy} = ne^2/h$.

However, it is useful to modify the geometry in a another way: make the circular strip very thin (the inside boundary close to the outside one), and then pull out the inner boundary to make a very thin cylinder (of very small height, compared to the radius), as in Figure 10.2(c). For this geometry, we can arrange to have the magnetic field H_0 be everywhere perpendicular to the cylinder (strip)'s surface. The current will now flow around the cylinder, so L is the circumference now, and the Hall effect will be in the direction of the height (the thin width), W. The electric field E_0 now will be in this direction, so $\Delta V = E_0 W$.

By the Stokes theorem, we have

$$\int_S \vec{B} \cdot d\vec{A} = \int_{\partial S} \vec{A} \cdot d\vec{l}. \tag{10.32}$$

In our case then, we have that the integral of the vector potential along the circumference of the strip (actually, the difference of the integrals on the two sides, but we can consider a gauge where the vector potential is zero on one boundary and nonzero on the other) equals the magnetic flux flowing through the strip:

$$\oint_L \vec{A} \cdot d\vec{l} = \Phi_{\text{magn}} = n\Phi_0 = n\frac{h}{e}. \tag{10.33}$$

This is a topological quantity called in mathematics the first Chern class, and as we can see it is quantized, and the quantization is the same as the quantization of the magnetic flux explained above. We want to see that the quantization of σ_H is due to the quantization of this topological quantity.

We can solve this for the vector potential, assumed to be constant on the boundary where it is nonzero:

$$A = \frac{nh}{eL}. \tag{10.34}$$

If we modify the vector potential A by a small ΔA, then we shift the centers of the Landau levels by $\Delta x_k = \Delta A/B$ (since $\Delta A \simeq B\Delta x_k$).

Consider now two neighboring states of index m and $m+1$ in the same Landau level n, separated by

$$\Delta x_k = x_{k_{m+1}} - x_{k_m} = l^2(k_{m+1} - k_m) = \frac{2\pi l^2}{L} = \frac{h}{eBL}\left(=\frac{\Delta A}{B}\right). \tag{10.35}$$

If instead one adds a single unit of flux quantum, giving a $\Delta A = h/(eL)$, then the positions of *the same state* x_{k_m} shift by

$$\Delta x_{k_m} = \frac{\Delta A}{B} = \frac{h}{eBL} = x_{k_m} - x_{k_{m-1}}, \tag{10.36}$$

so the positions of the mth state in the Landau level shift by one unit, and since this happens for all states, effectively one electron from one side of the strip moves to the other side.

The net effect is that quantization of the gauge potential (through the topological first Chern class) leads to quantization of charge transfer, as we advertised, and this is called Laughlin's gauge principle.

Single-Particle Energy Change

The single-particle (electron) energy of the conduction electron on the nontrivial boundary, where we have vector potential and electric current flowing, is

$$\epsilon_\alpha = \langle\psi_\alpha|H|\psi_\alpha\rangle = \langle\psi_\alpha|\frac{1}{2m}(\vec{p} - e\vec{A})^2|\psi_\alpha\rangle. \tag{10.37}$$

Then

$$\frac{\partial\epsilon_\alpha}{\partial A} = -\frac{e}{m}\langle\psi_\alpha|\vec{p} - e\vec{A}|\psi_\alpha\rangle. \tag{10.38}$$

But, from Maxwell's equations, the variation of the potential is given by the time variation of the magnetic flux,

$$\Delta V = \frac{d\Phi_{\text{magn}}}{dt}, \tag{10.39}$$

and the energy stored in the current is

$$E = \int dt\, I \Delta V = \int I d\Phi_{\text{magn}},$$

(10.40)

so we can also write for a single particle

$$d\epsilon_\alpha = I_\alpha d\Phi_{\text{magn}} = I_\alpha dA\, L \Rightarrow \frac{\partial \epsilon_\alpha}{\partial A} = L I_\alpha,$$

(10.41)

where I_α is the single-particle current.

On the other hand, the electrons in Landau level n will move on circles situated at position x_k in the width of the strip, so the single-particle energies will scale linearly with x_k because of the presence of the compensating electric field E_0 in the x direction:

$$\Delta \epsilon_\alpha = e E_0 x_k.$$

(10.42)

Under the modification of the vector potential A by a small ΔA, which shifts the centers of the Landau levels by $\Delta x_k = \Delta A / B$, the variation in the single-particle energy is

$$\Delta \epsilon_\alpha = e E_0 \frac{\Delta A}{B} \Rightarrow \frac{\partial \epsilon_\alpha}{\partial A} = \frac{e E_0}{B}$$

(10.43)

and is (approximately) independent of the state. We can finally write, independently of α,

$$\frac{\partial \epsilon_\alpha}{\partial A} = \frac{e E_0}{B} = L I_\alpha = -\frac{e}{m} \langle \psi_\alpha | \vec{p} - e\vec{A} | \psi_\alpha \rangle.$$

(10.44)

10.5 The Fractional Quantum Hall Effect

As we mentioned, the fractional quantum Hall effect happens at much higher B and lower T, and for filling fractions $\nu = k/(2p+1)$, specifically observed values are $\nu = 1/3, 4/3, 5/3, 7/3, 1/5, 2/5, 3/5, 7/5, 8/5, \ldots$. However, only in the case $\nu = 1/(2p+1)$ do we have a simple theoretical explanation, due to Laughlin, though not a microscopic derivation. There are of course theories for other ν's as well, though they are more involved and less standard.

Laughlin postulated the *Laughlin wavefunction* for *interacting electrons* (a system called a "Laughlin liquid") in a solid. Unlike the integer quantum Hall effect, the fractional case cannot be obtained from free electrons in a solid.

The wavefunction is

$$\Psi(\vec{r}_1, \ldots, \vec{r}_N) = K \prod_{1 \le i < j \le N} f(z_i - z_j) \exp\left(-\sum_{i=1}^{N} \frac{|z_i|^2}{4l^2}\right),$$

(10.45)

where $z = x + iy$ is a complex coordinates for the two space coordinates, and $f(z_i - z_j)$ is a polynomial function (think about why it needs to be so; a hint is in one of the exercises). Then, since the wavefunction for electrons (fermions) needs to be antisymmetric in the interchange of z_i and z_j, we are restricted to odd powers:

$$f(z_1 - z_2) = (z_1 - z_2)^{2p+1}.$$

(10.46)

Therefore the wavefunction for integer $m = 2p + 1$ is

$$\Psi_m = K \prod_{1 \leq i < j \leq N} (z_i - z_j)^m \exp\left(-\sum_{i=1}^{N} \frac{|z_i|^2}{4l^2}\right). \tag{10.47}$$

We can check that the wavefunction is an eigenvalue of the angular momentum operator $L_{\text{total},z}$, with eigenvalue

$$L = \frac{N(N-1)m}{2}. \tag{10.48}$$

The probability density is

$$|\Psi_m|^2 = K^2 e^{-\beta \Phi}, \tag{10.49}$$

i.e. a Boltzmann-type factor for $1/(k_B T) \equiv \beta = m$, where the potential Φ is

$$\Phi(z_1, \ldots, z_N) = -2 \sum_{1 \leq i < j \leq N} \ln |z_i - z_j| + \frac{1}{2ml^2} \sum_{i=1}^{N} |z_i|^2. \tag{10.50}$$

We could rescale $\Phi \rightarrow \Phi/(4\pi \epsilon_0)$ and $\beta \rightarrow 4\pi \epsilon_0 \beta$ to simulate an electric potential better. Indeed, note that

$$-\frac{2}{4\pi \epsilon_0} \sum_{i<j} \ln |z_i - z_j| \tag{10.51}$$

is the 2 dimensional electric potential of charges $\sqrt{2}$ situated at positions z_i. Then

$$U_b(z_i) = \frac{1}{4\pi \epsilon_0} \frac{|z_i|^2}{2ml^2}, \tag{10.52}$$

satisfying

$$\vec{\nabla}^2 U_b = 4\partial \bar{\partial} U_b = \frac{1}{4\pi \epsilon_0} \frac{2}{ml^2}, \tag{10.53}$$

can be thought of as a compensating background for the delta function charges at z_i, so it needs to satisfy

$$\vec{\nabla}^2 U_b = \frac{n_e}{\epsilon_0} = \frac{\nu n_B}{\epsilon_0}, \tag{10.54}$$

which leads to (since $n_B = 1/(2\pi l^2)$)

$$\frac{\nu}{2\pi l^2} = n_e = \frac{1}{2\pi l^2 m} \Rightarrow \nu = \frac{1}{m} = \frac{1}{2p+1}. \tag{10.55}$$

From this simple physical picture, we have obtained a fractional filling fraction ν, specifically $\nu = 1/(2p+1)$.

The normalization K of the Laughlin wavefunction can be calculated to be

$$K = \left[\frac{1}{2l^2 2\pi (\sqrt{2})^{2m} \sqrt{m!}(\sqrt{N-1})^{m+1}} \right]^{\frac{N(N-1)}{2}}. \tag{10.56}$$

This Laughlin state accurately describes the ground state of the N-electron system, as verified experimentally in solids, up to 99% precision. But it still needs a microscopic derivation and also needs to be extended to $\nu = k/(2p+1)$ for $k \neq 1$. Since the problem is a strong coupling one, it is ideal for treatment using the AdS/CFT duality, as we will see in Part IV.

On top of the ground state, to create a "quasi-hole" at the origin, viewed as an excitation of the system, we add a factor of $\prod_i z_i$:

$$\Psi = \left(\prod_i z_i\right)\Psi_m = \left(\prod_i |z_i|\right)e^{i\sum_i \phi_i}\Psi_m. \tag{10.57}$$

That means that in fact we can understand the Laughlin ground state as a collection of m "quasi-holes" at each position. We see then that the quasi-hole, besides carrying a quantum of vorticity (it is a vortex, having a phase equal to the angle in polar coordinates), has an electric charge of $\nu = 1/m$ (putting m of them together we get the charge of the electron).

Important Concepts to Remember

- The Hall effect is the appearance of a voltage transverse to the electron current in a 2-dimensional sample with magnetic field transverse to it, due to the Lorentz force.
- Classically, the Hall resistance and thus the Hall voltage, is proportional to the magnetic field.
- The integer quantum Hall effect is the fact that the Hall voltage has plateaus in B (instead of the simple linear relation), where the Hall conductance is quantized in terms of e^2/h: $\sigma_H = ne^2/h$. The voltage plateaus reflect the fact that the normal resistance vanishes at these values, $R_{xx} = 0$.
- For the fractional quantum Hall effect, the integer is replaced by a rational filling fraction $\nu = k/(2p+1)$.
- An electron in a transverse magnetic field has shifted harmonic oscillator states called Landau levels, with centers at $x_k = l^2 k$ and k a longitudinal momentum, and the level n being the state of the oscillator.
- The integer quantum Hall effect can be explained by a quantization of the magnetic flux, in terms of the unit h/e, with the flux quantum number being the number of states in a single Landau level. The quantization of the conductance is in terms of the number of completely filled Landau levels.
- One can explain the zero normal resistance in terms of the Fermi level being between the highest fully occupied Landau level and the lowest fully unoccupied Landau level, leading to the insulator-like behavior $\sigma_{xx} = 0$. Only since $\sigma_{ij} \propto \epsilon_{ij}$, that also means $R_{xx} = 0$, where as a matrix $R = \sigma^{-1}$.
- In the quantum Hall effect, there are edge currents, leading to a topological behavior. The quantization of charge transport in the Hall effect is derived from the quantization of a topological number, the first Chern class. This is Laughlin's gauge principle. The mechanism is that, adding one unit of magnetic flux, all the Landau level states shift by one unit, leading to a net charge transport.

- The fractional quantum Hall effect is explained for $\nu = 1/m = 1/(2p+1)$ by Laughlin's wavefunction, which is postulated for the interacting electron gas in a solid (Laughlin liquid).
- In Laughlin's wavefunction, the "charges" of individual particles at positions z_i are compensated by a constant background, leading to $\nu = 1/(2p+1)$.

Further Reading

Chapter 14 in [7].

Where It will be Addresssed in String Theory

The integer and fractional quantum Hall effect in string theory will be addressed in Chapter 41. In Chapter 12 we will see that we can have nonstandard statistics (anyons and nonabelions) in the quantum Hall effect, which will be addressed via AdS/CFT in Chapter 44.

Exercises

(1) Write the Lorentz + drift (relaxation) force in a static magnetic field B on Oz, and show that at equilibrium, we can write the tensor relation

$$\begin{pmatrix} j_x \\ j_y \\ j_z \end{pmatrix} = \frac{\sigma_0}{1+(\omega_c\tau)^2} \begin{pmatrix} 1 & -\omega_c\tau & 0 \\ \omega_c\tau & 1 & 0 \\ 0 & 0 & 1+(\omega_c\tau)^2 \end{pmatrix} \begin{pmatrix} E_x \\ E_y \\ E_x \end{pmatrix}. \qquad (10.58)$$

Show that for $B \to \infty$ ($\omega_c\tau \ll 1$),

$$\sigma_{yx} = -\sigma_{xy} = \sigma_H. \qquad (10.59)$$

(2) For a system with two charge carriers like a semiconductor, with concentrations n and p, relaxation times τ_e, τ_h and masses m_e, m_h, defining the mobility

$$\mu = \frac{|v|}{E} \left(= \frac{e\tau}{m} \right), \qquad (10.60)$$

show that

$$R_H = \frac{1}{e} \frac{p - nb^2}{(p+nb)^2}, \qquad (10.61)$$

where

$$b = \frac{\mu_e}{\mu_h}. \qquad (10.62)$$

(3) Consider a conducting disk of radius R, with a magnetic field B perpendicular to it, giving 100 units of flux. Cut a concentric hole at radius $R/2$ in the disk. How does the Hall resistance change?

(4) Consider the Laughlin wavefunction for three electrons. Derive the result for the normalization constant K.

(5) Show that the Laughlin wavefunction at filling fraction $\nu = 1/m$ has m vortices at each particle position.

(6) Consider a conducting sphere of radius R with a monopole of 10 units of magnetic charge in the middle and two holes of angular size $\Delta\Omega = 4\pi/10$ each, situated at opposite ends (North and South poles in the sphere). Calculate the first Chern class and the Hall conductance.

11 Superconductivity: Landau-Ginzburg, London, and BCS

In this chapter we will learn about superconductivity, and we will describe it in increasing detail, starting with an experimental characterization, the phenomenological characterization of the Landau-Ginzburg model, the semiphenomenological one of the London equations, and finally the BCS microscopic theory.

11.1 Superconductivity Properties

Superconductivity was discovered in Leiden in 1911 by Heike Kamerlingh-Onnes. Specifically, it was found that for a temperature $T < T_c$, the resistance falls to approximately zero. Its most important properties are as follows:

(1) For $T < T_c$, the resistance is extremely small. That means that one can have a finite current I at almost zero potential V. We can create a current I induced by a variable magnetic field $\vec{B}(t)$, and then the current will remain constant. In fact, since in general the decay of a current is dominated by its inductance L, with $I(t) = I(0)e^{-\frac{R}{L}t}$ (meaning the decay constant is $\tau = L/R$), experimentally this is found to be larger than about 10^5 years.

(2) Isotopic effect: For isotopes of masses M_i of the same element, the critical temperature varies according to the law

$$T_c M_i^{\alpha} \simeq \text{const.}, \tag{11.1}$$

where $\alpha \simeq 1/2$. The reason for this is that for superconductivity, the electron interaction with the lattice, described as an electron-phonon-electron interaction, are important. And for such a process, in the BCS theory, one finds $T_c \propto \theta_{\text{Debye}} \propto M_i^{-1/2}$.

(3) Electronic specific heat: For a normal state of matter, we have found in Chapter 3 that the electronic specific heat is

$$C_{e,n} = \gamma T; \quad \gamma = \frac{\pi^2}{2} \frac{n_0 k_B^2}{\epsilon_F}. \tag{11.2}$$

Instead, in the superconducting state, the electronic specific heat changes to

$$C_{e,s} = 3\gamma T_c \left(\frac{T}{T_c}\right)^3 \propto T^3 \quad T < T_c, \tag{11.3}$$

or

$$C_{e,s} - C_{e,n} = \gamma T_c \left[3\left(\frac{T}{T_c}\right)^3 - \frac{T}{T_c}\right], \tag{11.4}$$

known as Keeson's law.

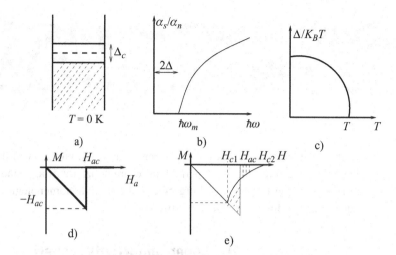

Fig. 11.1 (a) Gap at zero temperature. (b) Absorption as a function of the frequency, normalized to the normal state. (c) Gap as a function of the temperature. (d) Magnetization as a function of the applied magnetic field H_a for type I superconductors. (e) Magnetization as a function of the applied magnetic field H_a for type II superconductors.

At even smaller T, the behavior changes instead to

$$C_{e,s} = \alpha \gamma T_c e^{-\frac{\beta T_c}{T}} \ll T^3. \tag{11.5}$$

(4) There is an *energy gap* Δ that characterizes superconductivity, which will be found to be the energy required to break up a Cooper pair inside the material; see Figure 11.1(a). Thus in the energy bands, there is the fully occupied band, and the free valence band is separated from it by a gap Δ (or Δ_c). Correspondingly, one finds that the absorption spectrum of the superconductor, the optical absorption coefficient α_s as a function of $\hbar\omega$, has a gap 2Δ, i.e. the curve (normalized to the absorption of the normal state, α_s/α_n) starts at $\hbar\omega_m = 2\Delta$ and curves upwards and to the right, as in Figure 11.1(b).

The gap Δ depends on temperature, such that $\Delta(T)/(k_BT)$ makes a quarter circle arc, starting horizontally at $(\Delta(T)/(k_BT))(0) = 1.76$, and ending at $\Delta(T_c) = 0$, as in Figure 11.1(c).

(5) There is a critical magnetic field B_{ac}, above which, at $B > B_{ac}(T)$, the superconductivity disappears. Moreover, experimentally one finds

$$B_{ac} = B_{a0}\left(1 - \frac{T^2}{T_c^2}\right). \tag{11.6}$$

(6) The Meissner(-Ochsenfeld) effect. The effect is that there is no magnetic field inside the semiconductor. Since $B = \mu_0(H + M)$, that means that $M = -H = B_a/\mu_0$ (B_a is the applied B), i.e. the resulting magnetization is completely opposite to the applied H, thus we have an *ideal diamagnet*, of $\chi = M/H = -1$ (remember that a normal diamagnet has χ of about 10^{-5}).

(7) We can in fact classify the superconductors according to the magnetization M vs. H curve they have.

Type I, or "soft," superconductors obey exactly the laws at points 5 and 6, i.e. we have $M = -H$ until an $H = H_{ac}$, after which superconductivity disappears, and we are back in a normal state with $\chi \sim 10^{-5}$, so M drops effectively to zero, as in Figure 11.1(d). Most metallic elements and some alloys behave this way.

Type II, or "hard," superconductors instead have a transition region, with a continuous $M(H)$, not the abrupt drop of type I. For $H < H_{c1}$, we have the same behavior of point 6, of the perfect diamagnet. For $H_{c1} < H < H_{c2}$, we have a paramagnetic component to the susceptibility χ, i.e. now the magnetization increases with H, $\partial M / \partial H > 0$, from the negative value $-H_{c1}$ to $M = 0$ at $H = H_{c2}$, as in Figure 11.1(e). This behavior, and the paramagnetic component, is due to the presence of vortices in the material in this region. For $H > H_{c2}$, we have the normal state. These type II superconductors are more interesting theoretically, especially because of the presence of vortices.

The vortices are 2+1–dimensional solutions, but since the material is 3+1–dimensional, they appear as strings (parallel to the magnetic field) inside the superconductor. So the superconducting material is threaded by these vortex strings, which form a lattice in the two transverse dimensions, known as an *Abrikosov vortex lattice*.

Most materials of type II have low T_c. However, some cuprate materials have high T_c but are still of type II. These are the most interesting of all superconductors, since they are very poorly understood using normal condensed matter models, and string theory methods have a chance of describing them.

The phase transition between the normal and the superconducting state is of the second order, being a transition between a highly ordered state, the superconducting state, and a disordered state, the normal state. The superconducting state is highly ordered since its negligible resistance is indicative of negligible disorder for conduction, and of a stable ground state, thus low entropy. Therefore the two phases differ qualitatively, and there must exist an order parameter.

11.2 Landau's Theory of Second Order Phase Transitions

We can thus describe the semiconductor purely phenomenologically, without invoking anything about the microscopics, entirely in terms of the thermodynamic theory of second order phase transitions, developed by Landau. We have reviewed it to some degree in Chapter 1, but here we will present more details.

We consider an order parameter ψ, which must be an extensive quantity, that is nonzero $\psi \neq 0$ in the asymmetric, ordered, phase, and $\psi = 0$ in the symmetric, disordered, phase. Such ψ will be Legendre conjugate to an external field h. Consider moreover a natural thermodynamic potential for the system $\bar{U}$, then

$$\psi = -\frac{\partial \bar{U}}{\partial h}. \tag{11.7}$$

Consider also an intensive quantity P, then

$$\frac{\partial \psi}{\partial P} = \frac{\partial^2 \bar{U}}{\partial P \partial h} \tag{11.8}$$

is discontinuous across the phase transition (for second order phase transitions, the second derivatives of the thermodynamic potential are discontinuous).

No External Field, $h = 0$

We start with the case $h = 0$. Define the order parameter per particle,

$$\eta = \frac{\psi}{N}. \tag{11.9}$$

Consider also the case of a system at fixed T and P, $\mathcal{R}_{T,P}$. Then the natural thermodynamic potential is the Gibbs potential G, and consider the Gibbs potential per particle:

$$g(T, P, \eta) = \frac{G(T, P, \psi)}{N}. \tag{11.10}$$

The ordered state must have nonzero ψ, or η, as a solution, and the order must reflect in a choice of ψ. The simplest case is a discrete choice, between ψ and $-\psi$, that is, we must have a Gibbs potential that has a $\mathbb{Z}_2$ reflection symmetry in ψ broken by the vacuum. Hence we consider a Taylor expansion in η for g, and consider only the first three terms (constant, quadratic, quartic). Moreover, to have an ordered state $\eta \neq 0$ only for $T < T_c$, and a disordered state $\psi = 0$ for $T > T_c$, the coefficient of the quadratic piece must be an odd function of $T - T_c$, and keeping only the first term in a Taylor expansion in $T - T_c$, we finally get

$$g(T, P, \eta) = g_0(T, P) + a(P)[T - T_c(P)]\eta^2 + b(P)\eta^4. \tag{11.11}$$

The minimum of this function in terms of η is found for

$$\eta(T, P) = 0; \quad T > T_c(P)$$
$$= \sqrt{\frac{a(P)}{2b(P)}[T_c(P) - T]}, \quad T < T_c(P), \tag{11.12}$$

and substituting back in the Gibbs potential we find at the minimum that

$$g(T, P) = g_0(T, P), \quad T > T_c(P)$$
$$= g_0(T, P) - \frac{a^2(P)}{4b(P)}[T - T_c(P)]^2; \quad T < T_c(P). \tag{11.13}$$

But since $G = -SdT + \cdots$, we have for the entropy per particle

$$s = -\left(\frac{\partial g}{\partial T}\right)_P, \tag{11.14}$$

which leads to

$$s(T, P) = s_0(T, P); \quad T > T_c(P)$$
$$= s_0(T, P) - \frac{a^2}{2b}(T_c - T); \quad T < T_c(P). \tag{11.15}$$

Moreover, since the (pressure) specific heat is

$$C_P = T \left(\frac{\partial S}{\partial T} \right)_P,$$
(11.16)

we find that the specific heat jumps across the transition by

$$\Delta C_P = \left. \left(C_P^{(n)} - C_P^{(s)} \right) \right|_{T \to T_c} = \frac{a^2}{2b} T_c.$$
(11.17)

Nonzero External Field, $h \neq 0$

Consider now the thermodynamic potential $g^* = g - \eta h$. Then we can assume that as a first approximation, g is independent (indirectly, through some other quantities, since directly is already independent, as depending on η instead), then

$$g^* \simeq g(h = 0) - \eta h.$$
(11.18)

Then minimizing the potential with respect to η, $\delta g^*/\delta \eta = 0$, gives the external field as

$$h = 2a(T - T_c)\eta + 4b\eta^3.$$
(11.19)

This leads to the order parameter at zero external field, $h = 0$, which implies that

$$\eta_0 = \sqrt{\frac{a(T_c - T)}{2b}}$$
(11.20)

for $T < T_c$. If instead we calculate the local extrema of $h(\eta)$, we find

$$\eta = \pm \eta_M = \pm \sqrt{\frac{a(T_c - T)}{6b}}.$$
(11.21)

The curve $h(\eta)$ for $T < T_c$ grows from negative η for negative h until $h = 0$ at $\eta = -\eta_0$, then until a local maximum h_M at $\eta = -\eta_M$, then drops to $h = 0$ at $\eta = 0$, and a local minimum $h = -h_M$ for $\eta = +\eta_M$, and finally grows to $h = 0$ at $\eta = +\eta_0$, and then on to $+\infty$. Since really η is the response to the external field h, it is more logical to represent it as $\eta(h)$, in which case it would seem like we have a multivalued $\eta(h)$ in the region $(-h_M, +h_M)$; see Figure 11.2. For $T > T_c$, the curve for $h(\eta)$ increases monotonically, passing with $\partial h/\partial \eta > 0$ at $\eta = 0$.

However, for stability, we must have $\chi \equiv \partial \eta/\partial h \geq 0$ (in fact, for stability, all second order derivatives of the thermodynamic potential must be positive), and in fact one can prove that the behavior for $T < T_c$ in the quadrants II and IV (with the sign of η and the sign of h being opposite) must be avoided (even though we could have a metastable case of reaching the local minimum and local maximum while still $\partial \eta/\partial h > 0$). We can then quantify the condition of stability simply as $sgn(\eta) = sgn(\chi)$, leading to a jump at $h = 0$ from $\eta = -\eta_0$ to $\eta = +\eta_0$. Then, since for $T < T_c$, we have

$$\left. \frac{\partial h}{\partial \eta} \right|_{h=0} = 2a(T - T_c) + 12b\eta_0^2 = 4a(T_c - T),$$
(11.22)

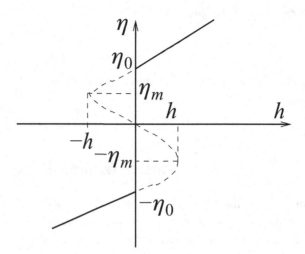

Fig. 11.2 $\eta(h)$ curve, from inverting $h(\eta)$ (dotted), and making it single valued (full line).

and for $T > T_c$, we have

$$\left.\frac{\partial h}{\partial \eta}\right|_{h=0} = 2a(T - T_c) + 12b \times 0 = 2a(T - T_c), \tag{11.23}$$

the *susceptibility* χ is discontinuous and divergent across the transition, specifically

$$\chi \equiv \left.\left(\frac{\partial \eta}{\partial h}\right)\right|_{h=0} = \frac{1}{2a(T - T_c)}; \quad T > T_c$$

$$= \frac{1}{4a(T_c - T)}; \quad T < T_c. \tag{11.24}$$

This can be written as

$$\chi \propto \frac{1}{|T - T_c|} \propto (T - T_c)^{-\nu}, \tag{11.25}$$

which means the susceptibility has a critical exponent $\nu = 1$ at the phase transition, which is a conformal point.

11.3 Thermodynamics of the Superconducting Phase Transition

Until now, we have described only issues that depend only on the type II character of the phase transition, but we did not input anything about the superconductivity itself. Now we turn to something that depends specifically on the superconducting nature of the transition.

Consider a system at fixed temperature and magnetic field H, $\mathcal{R}_{T,H}$, i.e. a system described in terms of the free energy F with

$$dF = -SdT - MdB_a + \cdots, \tag{11.26}$$

where $B_a \equiv \mu_0 H$ is the applied magnetic field. The magnetization is

$$M = \left(\frac{\mu}{\mu_0} - 1\right)\frac{B_a}{\mu_0},$$ (11.27)

so we can integrate it to write for the free energy

$$F(B_a) = F(0) - \left(\frac{\mu}{\mu_0} - 1\right)\frac{B_a^2}{2\mu_0}.$$ (11.28)

In a normal state, the normal diamagnetic component is negligible, so $\mu \simeq \mu_0$, and thus

$$F_n(B_a) \simeq F_n(0).$$ (11.29)

In a superconducting state, $B \simeq 0$ instead, i.e. $\mu \simeq 0$, which gives

$$F_s(B_a) \simeq F_s(0) + \frac{B_a^2}{2\mu_0}.$$ (11.30)

But moreover, the free energy itself must be continuous across the superconductor to normal phase transition happening at $B_a = B_{ac}$ (even though its second order derivatives are discontinuous), so

$$F_n(B_{ac}) = F_s(B_{ac}) \Rightarrow \Delta F \equiv F_n(0) - F_s(0) = \frac{B_{ac}^2}{2\mu_0}.$$ (11.31)

Moreover, since the entropy is given by the derivative with respect to temperature by

$$S = -\left.\frac{\partial F}{\partial T}\right|_{B_{ac}},$$ (11.32)

and inputting the experimental result about the superconducting transition that

$$B_{ac}(T) = B_{a0}\left(1 - \frac{T^2}{T_c^2}\right),$$ (11.33)

we obtain

$$S_n(0) - S_s(0) = -\frac{B_{ac}}{\mu_0}\frac{dB_{ac}}{dT} > 0,$$ (11.34)

so $S_s < S_n$, which was to be expected, since the superconducting state is more ordered, and preferred at zero magnetic field. The latent heat necessary to perform the transition between the two states is

$$\Lambda_{n\leftrightarrow s} = T(S_n - S_s) = -T\frac{B_{ac}}{\mu_0}\frac{dB_{ac}}{dT}.$$ (11.35)

The specific heat is given by

$$C = \frac{dQ}{dT} = \frac{T\,dS}{dT},$$ (11.36)

so we obtain the difference in (electronic) specific heats of the normal and superconducting states

$$C_{e,s} - C_{e,n} = \frac{T}{\mu_0}\left[\left(\frac{dB_{ac}}{dT}\right)^2 + B_{ac}\frac{d^2 B_{ac}}{dT^2}\right] = \frac{2TB_{a0}^2}{T_c^2\mu_0}\left[3\left(\frac{T}{T_c}\right)^2 - 1\right].$$ (11.37)

11.4 Landau-Ginzburg Theory and Gross-Pitaevskii Equation

We move on to a description of the superconducting state that is more precisely suited to the superconductor. Specifically, we now identify the order parameter ψ of the general theory of Landau for second order phase transitions as a bosonic position-dependent quantity (field) $\psi(\vec{r})$, such that

$$|\psi(\vec{r})|^2 = \psi^*(\vec{r})\psi(\vec{r}) = n_s(\vec{r}) \tag{11.38}$$

is the local concentration of superconducting electrons. Then it must follow that $\psi(\vec{r})$ can be interpreted as a wavefunction in the superconducting state. In fact, we will see later, in the BCS theory, that this bosonic quantity can be thought of as the composite field for Cooper pairs, pairs of electrons of opposite momenta and spin.

Then the free energy density for the superconducting state is a position-dependent generalization of the general theory above, with the usual spatial gradient term giving the expected contribution to the energy:

$$F_s(\vec{r}) = F_n - \alpha|\psi|^2 + \frac{\beta}{2}|\psi|^4 + \frac{1}{2m_*}|(-i\hbar\vec{\nabla} - q\vec{A})\psi|^2 - \int_0^{B_a} \vec{M} \cdot d\vec{B}_a. \tag{11.39}$$

The total free energy is the integral of this density, $\int dV F_s(\vec{r})$, so varying with respect to the order parameter we obtain

$$\frac{\delta}{\delta\psi^*(\vec{r})} \int dV F_s(\vec{r}) = 0 \Rightarrow$$

$$\left[-\alpha + \beta|\psi|^2 - \frac{1}{2m_*}(-i\hbar\vec{\nabla} - q\vec{A})(+i\hbar\vec{\nabla} - q\vec{A}) \right]\psi = 0. \tag{11.40}$$

Thus the resulting equation, called the *Landau-Ginzburg (LG) equation*,

$$\left[-\frac{1}{2m_*}| -i\hbar\vec{\nabla} - q\vec{A}|^2 - \alpha + \beta|\psi|^2 \right]\psi = 0, \tag{11.41}$$

is a Schrödinger-like equation, more precisely a gauged, nonlinear version of the Schrödinger equation, but one that makes sense, since the nonlinear term is proportional to the electron density $|\psi|^2$.

We can define a *coherence length* ξ in the following way.

Consider the system in the absence of a vector potential, $\vec{A} = 0$, and for $\beta|\psi|^2 \ll \alpha$, thus negligible nonlinear term. Then the LG equation becomes

$$\frac{\hbar^2}{2m_*}\frac{d^2\psi}{dx^2} = \alpha\psi, \tag{11.42}$$

with the solution

$$\psi \propto e^{-\frac{x}{\xi}}; \quad \xi \equiv \sqrt{\frac{\hbar^2}{2m_*\alpha}}. \tag{11.43}$$

Here ξ is called the *coherence length*.

If we keep also β, we can consider a solution that has $\psi(x = 0) = 0$, i.e. $x = 0$ is the line where the material ends. Then at $x \to \infty$, we are deep within the superconductor, where the wavefunction is in the vacuum, $\psi = \psi_0$, where ψ_0 is the constant solution of the LG equation, $|\psi_0| = \sqrt{\alpha/\beta}$. The solution to the 1+1–dimensional problem is then

$$\psi(x) = \psi_0 \tanh\left(\frac{\sqrt{2}x}{\xi}\right). \tag{11.44}$$

We can calculate the critical field H_c by imposing that the free energy of the superconducting state, $F_s = F_n - \alpha^2/(2\beta)$, equals the free energy of the normal state in the presence of H_c,

$$F_s = F_n - \frac{\alpha^2}{2\beta} = F_n(H_c) = F_n - \mu_0 \frac{H_c^2}{2}, \tag{11.45}$$

which leads to

$$H_c = \frac{\alpha}{\sqrt{\mu_0\beta}}. \tag{11.46}$$

We can also define the superconducting current by varying the free energy with respect to the gauge potential,

$$-\vec{j}_s(\vec{r}) = \frac{\delta}{\delta\vec{A}(\vec{r})} \int F_s(\vec{r}), \tag{11.47}$$

which leads to

$$\vec{j}_s(\vec{r}) = -\frac{iq}{2m_*}(\psi^*\vec{\nabla}\psi - \psi\vec{\nabla}\psi^*) - \frac{q^2}{m_*}\psi^*\psi\vec{A}. \tag{11.48}$$

The current deep inside the superconductor, i.e. near the minimum, $\psi \simeq \psi_0$, is then (the gradient terms are negligible)

$$\vec{j}_s(\vec{r}) \simeq -\frac{q^2}{m_*}|\psi_0|^2\vec{A} = -\frac{q^2}{m_*}\frac{\alpha}{\beta}\vec{A}. \tag{11.49}$$

As a function of temperature, the free energy of the superconductor is

$$F = F_n + \int dV\left[\frac{\hbar^2}{2m_*}|\vec{\nabla}\psi|^2 + a(T)|\psi|^2 + b(T)|\psi|^4\right], \tag{11.50}$$

and $a(T) < 0$ for $T < T_c$ (in the superconducting state) and $a(T) > 0$ for $T > T_c$ (in the normal state). In a Taylor expansion in T, we can approximate $a(T) \simeq a_1(T - T_c)$, with $a_1 > 0$.

For the description of type II superconductors, it is important to consider the vortex solutions of the Landau-Ginzburg theory. The solutions in this condensed matter context (of the LG theory) were found by Abrikosov, but in the particle physics context (for an abelian theory) were found by Nielsen and Olesen, so the vortices are called Abrikosov-Nielsen-Olesen (ANO) vortices. We can use the 1+1–dimensional solution as a guide to construct the topological vortex solutions. The solution will be similar to the 1+1–dimensional solution, just in the radial direction, i.e. as in Figure 11.3, and it will have a nontrivial gauge field.

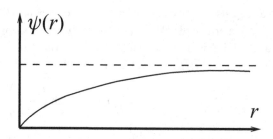

Fig. 11.3 Vortex profile.

Since at infinity we must have the superconductor in the vacuum state, it means that $|\psi| \to \psi_0$ at $r \to \infty$. On the other hand, at $r \to 0$, we find $|\psi(r)| \sim r$. To have a topological solution, however, with a conserved topological number, we have

$$\psi(\vec{r}) = |\psi(r)|e^{i\theta}, \tag{11.51}$$

where θ is the polar angle in two spatial dimensions. Such a solution carries a unit of topological charge for the gauge field, specifically

$$\int_{\mathbb{R}^2} F_{12}dx^1 dx^2 = \frac{2\pi\hbar}{e}. \tag{11.52}$$

Of course, the gauge field will decay exponentially away from the vortex, since in the superconductor that lives at infinity, the photon has a mass, thus the gauge field cannot penetrate it. The vortex solution is extended in 3+1 dimensions trivially by translational invariance, to obtain strings, as we said. Then the topological charge becomes

$$\int \vec{B} \cdot d\vec{S} = \Phi = \frac{2\pi\hbar}{e} = \Phi_0. \tag{11.53}$$

We have obtained therefore the quantization of magnetic flux, again in terms of the fluxon unit.

Gross-Pitaevskii Equation

The above Landau-Ginzburg equation can be re-derived in a related context, where it comes under the name Gross-Pitaevskii equation. Consider the Hamiltonian for a system of interacting bosons described by a bosonic field operator $\hat{\psi}(\vec{r}, t)$, under an external potential V_{ext} and an interaction potential V_{int}. In a second-quantized treatment, we have

$$\hat{H} = \int d^3r\hat{\psi}^\dagger(\vec{r})\left[\left(-\frac{\hbar^2\nabla^2}{2m}\right) + V_{\text{ext}}(\vec{r})\right]\hat{\psi}(\vec{r})$$
$$+ \int d^3r \int d^3r' \hat{\psi}^\dagger(\vec{r})\hat{\psi}^\dagger(\vec{r}')V_{\text{int}}(\vec{r} - \vec{r}')\psi(\vec{r})\psi(\vec{r}'). \tag{11.54}$$

For a sufficiently dilute gas of bosons, we have a delta function interaction,

$$V_{\text{int}}(\vec{r} - \vec{r}') = g\delta^3(\vec{r} - \vec{r}'), \tag{11.55}$$

so the *time-dependent* Schrödinger equation coming from it is

$$i\hbar\frac{\partial}{\partial t}\psi(\vec{r},t) = \left[\left(-\frac{\hbar^2\nabla^2}{2m}\right) + V_{ext}(\vec{r}) + g\psi^\dagger(\vec{r},t)\psi(\vec{r},t)\right]\psi(\vec{r},t). \qquad (11.56)$$

In the superconductor, the bosonic field is $\psi(\vec{r},t) \simeq \psi_0(\vec{r},t) + \delta\psi(\vec{r},t)$, where ψ_0 is the (Cooper pair) condensate, so finally we arrive at

$$i\hbar\frac{\partial}{\partial t}\psi_0(\vec{r},t) = \left[\left(-\frac{\hbar^2\nabla^2}{2m}\right) + V_{ext}(\vec{r}) + g|\psi_0(\vec{r},t)|^2\right]\psi_0(\vec{r},t). \qquad (11.57)$$

This is the *time-dependent Gross-Pitaevskii equation*. We can write a time-independent equation by replacing $i\hbar\partial\psi_0/\partial t$ with $\mu\psi$ (factoring out $e^{-i\frac{\mu t}{\hbar}}$), where μ is the chemical potential, obtaining

$$\mu\psi(\vec{r}) = \left[\left(-\frac{\hbar^2\nabla^2}{2m}\right) + V_{ext}(\vec{r}) + g|\psi(\vec{r})|^2\right]\psi(\vec{r}). \qquad (11.58)$$

This is the *time-independent Gross-Pitaevskii equation*, here appearing as a nonlinear Schrödinger equation, and it is really the same Landau-Ginzburg equation (11.41), just without the gauge coupling.

11.5 Electrodynamics of the Superconductor: London Equation

We now describe a phenomenological model found by the brothers H. London and F. London, the London-London model. While still a phenomenological model like the ones before, it contains more microscopic information. The model is a bifluid model, where we assume that inside the superconductor we have a fluid of normal electrons and a fluid of superconducting electrons. As the temperature varies, one fluid changes into another: at $T = 0$, we have only superconducting fluid, and at $T = T_c$, we have only a normal state.

The sum of the densities equals the total electron density, i.e. $n_n + n_s = n$, and a similar situation for the currents, i.e. $\vec{j} = \vec{j}_s + \vec{j}_n$, where

$$\vec{j}_n = -en_n\vec{v}_n; \qquad \vec{j}_s = -en_s\vec{v}_s. \qquad (11.59)$$

For the normal state, disorder is essential, and we learned that we have an average (drift) motion, since the velocity increases from zero to the maximal value, after which a collision drives it back to zero. That results in $\vec{j} = \sigma\vec{E}$, but for the superconducting state $\sigma = \infty$ ($R = 0$), so there are no collisions, and the electric field simply accelerates the electron according to

$$-e\vec{E} = m_*\frac{d\vec{v}_s}{dt}. \qquad (11.60)$$

In turn, that now leads to

$$\frac{d\vec{j}_s}{dt} = +\frac{n_s e^2}{m_*}\vec{E} = -\frac{n_s e^2}{m_*}\frac{d\vec{A}}{dt}. \qquad (11.61)$$

This the first London equation and is really postulated. The second London equation is also postulated:

$$\vec{\nabla} \times \vec{j_s} = -\frac{n_s e^2}{m_*} \vec{B} = -\frac{n_s e^2}{m_*} \vec{\nabla} \times \vec{A}. \qquad (11.62)$$

Thus these two equations are argued for based on experimental information, like in the Maxwell equations, but like in that case must be ultimately postulated.

Moreover, we must also postulate that $\vec{j}$ is proportional to $\vec{A}$, with a negative proportionality constant, *for a simply connected superconductor* (we will shortly see that otherwise we get a very interesting situation). From the London equations, it immediately follows that we have

$$\vec{j} = -\frac{n_s e^2}{m_*} \vec{A} \equiv -\frac{1}{\mu_0 \lambda_L^2} \vec{A}. \qquad (11.63)$$

Here we have defined the *penetration depth*

$$\lambda_L = \sqrt{\frac{m_*}{n_s e^2 \mu_0}}. \qquad (11.64)$$

From the Maxwell equation, $\vec{\nabla} \times \vec{B} = \mu_0 \vec{j}$, we obtain

$$\vec{\nabla}^2 \vec{B} = \vec{\nabla} \times (\vec{\nabla} \times \vec{B}) = \mu_0 \vec{\nabla} \times \vec{j} = +\frac{\mu_0 n_s e^2}{m_*} \vec{B} \equiv \frac{1}{\lambda_L^2} \vec{B}. \qquad (11.65)$$

The solution is an exponentially decaying magnetic field,

$$\vec{B}(r) = \vec{B}(0) e^{-\frac{r}{\lambda_L}}, \qquad (11.66)$$

so we see that indeed λ_L is the depth that B penetrates inside the superconductor. We find that $\lambda_L \sim 10$–100 nm, so for all intents and purposes the superconductor has no magnetic field inside.

Multiply Connected Sample

For a multiply connected sample, like a ring or a cylinder, $\vec{j}$ is not proportional to $\vec{A}$, yet still we have $\vec{\nabla} \times \vec{j} \propto \vec{B}$.

In that case, since $|\psi|^2 = n$, we have $\psi = \sqrt{n} e^{i\theta}$, where $e^{i\theta}$ is a phase. Then, defining

$$\vec{v} = \frac{1}{m}(\vec{p} - q\vec{A}) = \frac{1}{m}(-i\hbar\vec{\nabla} - e\vec{A}), \qquad (11.67)$$

the current is

$$\vec{j} = q(\psi^* \vec{v} \psi - \psi \vec{v} \psi^*) = \frac{nq}{m_*}(\hbar\vec{\nabla}\theta - q\vec{A}). \qquad (11.68)$$

In the case of a superconducting ring threaded by a magnetic field, we have $\vec{j} = 0$, which leads to

$$\hbar\vec{\nabla}\theta = q\vec{A}. \qquad (11.69)$$

Doing the integral over the ring of the above equation, $\oint_C d\vec{l}$, the left-hand side becomes

$$\hbar \oint_C d\vec{l} \cdot \vec{\nabla}\theta = \hbar(\theta_2 - \theta_1), \tag{11.70}$$

where θ_2 is the phase of ψ after rotating around the ring and θ_1 is the phase before the rotation. Since ψ must be single valued, we must have $e^{i\Delta\theta} = 1$, thus the left-hand side is $\hbar(\theta_2 - \theta_1) = 2\pi\hbar k$. The right-hand side is

$$q \oint_{C=\partial S} \vec{A} \cdot d\vec{l} = q \int_S \vec{B} \cdot d\vec{S} = q\Phi, \tag{11.71}$$

i.e. the magnetic flux times the electron charge. Equating the two sides, we obtain again the magnetic flux quantization condition in terms of the fluxon:

$$\Phi = \frac{\hbar(\theta_2 - \theta_1)}{q} = \frac{2\pi\hbar}{q}k = \frac{h}{e}k. \tag{11.72}$$

11.6 BCS Theory

We now finally describe a more microscopic theory for superconductivity, the standard description given by John Bardeen, Leon Cooper, and John Robert Schriefer (BCS), who received the 1972 Nobel Prize in Physics for this theory (John Bardeen had already been awarded a Nobel Prize in 1956 for the discovery of the transistor, together with William Shockley and Walter Brattain, the only person to get two Nobel Prizes in Physics). The first relevant idea for the BCS theory was by Fröhlich in 1950, specifically the idea that there must be an electron-electron coupling via phonons (the quanta of oscillation of the crystal lattice). Then, in 1955, Cooper described the binding of the two electrons (via the phonon interaction), of opposite momenta and opposite spins, into what is now known as a Cooper pair. Finally, in 1956, Bardeen, Cooper, and Schriefer wrote the multi-electron BCS theory.

BCS theory is complicated, and people still write papers about it, so we will sketch only some of the relevant steps for the most important effects. The electron-phonon interaction is, as we have proven in Chapter 4 (except ignoring the Umklapp processes)

$$H = H_0 + H_{e-ph}$$
$$H_0 = \sum_{\vec{q}} \hbar\omega_{\vec{q}} b_{\vec{q}}^\dagger b_{\vec{q}} + \sum_{\vec{k}} \epsilon_{\vec{k}} a_{\vec{k}}^\dagger a_{\vec{k}}$$
$$H_{e-ph} = \sum_{\vec{k},\vec{q}} M_{\vec{q}} a_{\vec{k}+\vec{q}}^\dagger a_{\vec{k}} (b_{\vec{q}} + b_{-\vec{q}}^\dagger), \tag{11.73}$$

where $\omega_{\vec{q}}$ is the frequency of the phonons created by $b_{\vec{q}}^\dagger$, $\epsilon_{\vec{k}}$ is the energy of the electrons created by $a_{\vec{k}}^\dagger$, and $M_{\vec{q}}$ is a matrix element whose form was written in Chapter 4.

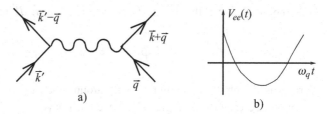

(a) Feynman diagram for the electron-electron interaction via a phonon. (b) The time dependence of the potential, $V_{ee}(t)$.

Electron-Phonon Binding Interaction

We can now present Frölich's idea of the electron-electron binding interaction via phonons.

Making a unitary transformation on the Hamiltonian of the type

$$\hat{H} = e^{-S} H e^{S}, \tag{11.74}$$

with a complicated S, specifically

$$S = \sum_{\vec{k},\vec{q}} \left[\frac{b^{\dagger}_{-\vec{q}}}{\epsilon_{\vec{k}} - \epsilon_{\vec{k}+\vec{q}} - \hbar\omega_{-\vec{q}}} + \frac{b_{\vec{q}}}{\epsilon_{\vec{k}} - \epsilon_{\vec{k}+\vec{q}} + \hbar\omega_{\vec{q}}} \right] M_{\vec{q}} a^{\dagger}_{\vec{k}+\vec{q}} a_{\vec{k}}, \tag{11.75}$$

and after a calculation that will not be done here, one arrives at the new Hamiltonian

$$\hat{H} = H_0 + \sum_{\vec{k},\vec{k}',\vec{q}} |M_{\vec{q}}|^2 a^{\dagger}_{\vec{k}+\vec{q}} a^{\dagger}_{\vec{k}'-\vec{q}} a_{\vec{k}'} a_{\vec{k}} \frac{\hbar\omega_{\vec{q}}}{(\epsilon_{\vec{k}'} - \epsilon_{\vec{k}'-\vec{q}})^2 - (\hbar\omega_{\vec{q}})^2}, \tag{11.76}$$

which describes an interaction between two electrons of momenta $\vec{k}$ and $\vec{k}'$, via an exchanged phonon of momentum $\vec{q}$ and frequency $\omega_{\vec{q}}$. The Feynman diagram implied by this interaction Hamiltonian is: The two electrons of momenta $\vec{k}$ and $\vec{k}'$ come, exchange the phonon, and then out go electrons of modified momenta $\vec{k} + \vec{q}$ and $\vec{k}' - \vec{q}$, as in Figure 11.4.

Then the total electron-electron interaction in momentum space is given by the usual Coulomb repulsion, divided by a dynamical screening dielectric constant for the material (since $\epsilon_0 \to \epsilon_0 \epsilon(\vec{q})$), plus the electron-electron interaction via the phonon written above, so

$$V_{ee}(\vec{k},\vec{q}) = \frac{V_e(\vec{q})}{\epsilon(\vec{q})} + \frac{\hbar\omega_{\vec{q}}|M_q|^2}{(\epsilon_{\vec{k}} - \epsilon_{\vec{k}-\vec{q}})^2 - (\hbar\omega_{\vec{q}})^2} \equiv \frac{V_e(\vec{q})}{\epsilon(\vec{q})} + \Delta V_{\vec{k},\vec{q}}. \tag{11.77}$$

The Fourier transform over ω of this potential (replacing the $(\epsilon_{\vec{k}} - \epsilon_{\vec{k}-\vec{q}})$ with ω) becomes an instantaneous ($\delta(t)$) Coulomb potential, plus a sinusoidal contribution:

$$V_{ee}(t) = \int_{-\infty}^{+\infty} \frac{d\omega}{2\pi} e^{-i\omega t} V_{ee} = V_{ee}^{\text{Coulomb}} \delta(t) - |M_{\vec{q}}|^2 \sin\omega_{\vec{q}} t. \tag{11.78}$$

This potential is a positive contribution, which can be smeared a bit (since there are no delta functions in real life), followed by an oscillatory contribution, that becomes negative (attractive) on a certain time scale, but then goes back to being repulsive, etc.

Cooper Pairs

We can now show that Cooper pairs form in the system. Pair binding of electrons becomes possible, since the attractive interaction is defined over a definite time scale. One can compute the matrix element to bind two electrons of opposite spins and find that it is nonzero only for opposite spins. Close to the Fermi surface, one finds that the matrix element of the interaction is approximately of the type

$$\langle \vec{p}_{4\uparrow}\vec{p}_{3\downarrow} | V_{ee} | \vec{p}_{1\uparrow}\vec{p}_{2\downarrow} \rangle = -V_0 \delta_{\vec{p}_1+\vec{p}_2, \vec{p}_3+\vec{p}_4}. \tag{11.79}$$

More precisely, the potential is a negative constant in a vicinity of the Fermi surface defined by ω_D, and zero otherwise:

$$V_{ee} = -V_0, \quad k, k' > p_F, \quad \text{and} \quad |\epsilon_{\vec{p}} - \epsilon_{\vec{p}_F}| < \hbar\omega_D$$

$$= 0, \quad \text{otherwise.} \tag{11.80}$$

Then we can consider the (toy) model of the full Fermi sea populated by N noninteracting electrons, and two electrons outside the sea. One can write the Schrödinger equation for the wavefunction, which is antisymmetric with respect to spins, and symmetric with respect to coordinates:

$$\left[-\frac{\hbar^2}{2m}(\vec{\nabla}_1^2 + \vec{\nabla}_2^2) + V(\vec{r}_1 - \vec{r}_2) - E \right] \psi(\vec{r}_1, \vec{r}_2, \uparrow_1, \downarrow_2) = 0. \tag{11.81}$$

It can be solved via an ansatz factorized in a spin part and a coordinate part,

$$\psi(\vec{r}_1, \vec{r}_2, \uparrow_1, \downarrow_2) = \tilde{\psi}(\vec{r}_1, \vec{r}_2)\chi_{\text{singlet}}(\uparrow_1, \downarrow_2), \tag{11.82}$$

where we further split the coordinate part into a free wave part (with momentum $\vec{Q}$) for the center of mass motion, and a wavefunction for the relative motion,

$$\tilde{\psi}(\vec{r}_1, \vec{r}_2) = \phi(\vec{r})e^{\frac{i\vec{Q}\cdot\vec{R}}{\hbar}}, \tag{11.83}$$

with $\vec{r} = \vec{r}_1 - \vec{r}_2$ and $\vec{R} = (\vec{r}_1 + \vec{r}_2)/2$. Since $\vec{k}\cdot\vec{r} = \vec{k}\cdot\vec{r}_1 - \vec{k}\cdot\vec{r}_2$, expanding $\phi(\vec{r})$ into free waves for the relative motion, $e^{i\vec{k}\cdot\vec{r}} = e^{i\vec{k}\cdot\vec{r}_1}e^{-i\vec{k}\cdot\vec{r}_2}$ amounts to considering a pair state, for one electron with momentum $\vec{k}$, and one with momentum $-\vec{k}$.

After a calculation, one finds that the eigenenergy is

$$E \simeq 2\epsilon_F - 2\hbar\omega_D e^{-\frac{2}{V_0 N(\epsilon_F)}}, \tag{11.84}$$

where $N(\epsilon_F)$ is the density of states at the Fermi surface $\epsilon = \epsilon_F$. This means that we have a *Cooper instability* toward forming a *bound* (Cooper) pair, of electrons of opposite spins and momenta, since the energy of the state is smaller than the sums of the individual energies (which by assumption were a little bit larger than ϵ_F).

Note that all this was in the assumption of two single electrons near a noninteracting Fermi sea. To describe things more precisely, we would need to formulate the (BCS theory of) multi-electrons in the presence of Cooper pairs.

BCS Ground State and Its Properties

We can now define the BCS ground state, since we have proven that the electrons form Cooper pairs. The ground state will be a combination of the vacuum and the action of the

Cooper pair creation operator,

$$b_{\vec{p}}^{\dagger} = a_{\vec{p}\uparrow}^{\dagger} a_{-\vec{p}\downarrow}^{\dagger}, \tag{11.85}$$

with coefficients u and v called coherence factors:

$$|\psi_{BCS}\rangle = \prod_{\vec{p}} (u_{\vec{p}} + v_{\vec{p}} b_{\vec{p}}^{\dagger})|0\rangle. \tag{11.86}$$

The $u_{\vec{p}}$, $v_{\vec{p}}$ coefficients are found from a variational problem (minimizing the energy with respect to them) as follows.

Consider the average energy, with the constraint that the average number of particles is fixed, a condition introduced by a Lagrange multiplier, the chemical potential, i.e. consider

$$E' = \langle H \rangle - \mu \langle N \rangle, \tag{11.87}$$

and that the free single-particle energies are redefined by including the chemial potential, $\epsilon_{\vec{p}} = \vec{p}^2/2m - \mu$. A calculation in the BCS ground state leads to the result

$$E' = 2 \sum_{\vec{p}} |v_{\vec{p}}|^2 \epsilon_{\vec{p}} + \sum_{\vec{p},\vec{p}'} V_{\vec{p}\vec{p}'} u_{\vec{p}} u_{\vec{p}}^* v_{\vec{p}}^* v_{\vec{p}'}. \tag{11.88}$$

One can then define

$$\Delta_{\vec{p}} \equiv - \sum_{\vec{p}'} V_{\vec{p}\vec{p}'} u_{\vec{p}'}^* v_{\vec{p}'}, \tag{11.89}$$

which will play the role of energy gap, and the quantity

$$\tilde{\epsilon}_{\vec{p}} \equiv \sqrt{\epsilon_{\vec{p}}^2 + |\Delta_{\vec{p}}|^2}, \tag{11.90}$$

which will be the quasi-particle energy in the superconductor. With these definitions, and varying E' with respect to $u_{\vec{p}}$ and $v_{\vec{p}}$ (for a given normalization constant), one finds

$$u_{\vec{p}} = \sqrt{\frac{\epsilon_{\vec{p}} + \tilde{\epsilon}_{\vec{p}}}{2\tilde{\epsilon}_{\vec{p}}}}$$

$$v_{\vec{p}} = \sqrt{\frac{\Delta_{\vec{p}}}{\Delta_{\vec{p}}^*} \left(\frac{\tilde{\epsilon}_{\vec{p}} - \epsilon_{\vec{p}}}{2\tilde{\epsilon}_{\vec{p}}} \right)} = \frac{\Delta_{\vec{p}}}{\sqrt{2\tilde{\epsilon}_{\vec{p}}(\epsilon_{\vec{p}} + \tilde{\epsilon}_{\vec{p}})}}, \tag{11.91}$$

and replacing in the energy, one finally finds the difference in energy between the superconducting and normal states in terms of $\Delta = \Delta_{\vec{p}}$ as

$$E_s - E_n = -\frac{N(\epsilon_F - \mu)}{2} \Delta^2. \tag{11.92}$$

Here as before $N(x)$ is a density of states, and appears because of converting sums into integrals over the energy. Then indeed, Δ plays the role of energy gap. It satisfies an equation found by replacing the variational values for $u_{\vec{p}}$ and $v_{\vec{p}}$ in the definition of $\Delta_{\vec{p}}$ as

$$\Delta_{\vec{p}} = - \sum_{\vec{p}'} V_{\vec{p}\vec{p}'} u_{\vec{p}'}^* v_{\vec{p}'} = - \sum_{\vec{p}'} V_{\vec{p}\vec{p}'} \frac{\Delta_{\vec{p}'}}{2\sqrt{\epsilon_{\vec{p}'}^2 + |\Delta_{\vec{p}'}|^2}}, \tag{11.93}$$

known as the gap equation.

One can look for physical solutions of the gap equation, and, besides the trivial $\Delta = 0$, one finds the nontrivial solution with

$$\Delta(T = 0) \simeq 2\hbar\omega_D \exp\left[-\frac{1}{N(0)V_0}\right]. \qquad (11.94)$$

Note that compared to the single-pair result (11.84), this exact result has half the exponent in the exponential, because of inclusion of pairing of particles below the Fermi sea (neglected in the single Cooper pair calculation before), which effectively replaces the density of states $N(0)$ with twice its value, $2N(0)$.

One can also calculate the critical temperature in BCS theory, though I will not explain that here, and find

$$k_B T_c \simeq 1.14\hbar\omega_D \exp\left[-\frac{1}{N(0)V_0}\right], \qquad (11.95)$$

leading finally to the result

$$\frac{2\Delta}{k_B T_c} \simeq \frac{4}{1.14} \simeq 3.52, \qquad (11.96)$$

a robust prediction of BCS theory. This is verified experimentally very well in normal superconductors, as mentioned at the beginning of the chapter, but high T_c superconductors violate this result, the ratio being larger by a factor of 2 or so, and this *strong coupling result* indicates that one needs a new theory to describe them. String theory will offer some help in that respect.

11.7 Vortices in Type II Superconductors

Let us say a few words about how vortices explain the magnetic effects in the type II superconductors. We have two critical magnetic fields, H_{c1} and H_{c2}.

The first field corresponds to when the first vortex manages to enter the material. That happens because one flux quantum can now fit inside the penetration depth of the material λ, i.e. the vortex can "enter from the outside." It leads to

$$(\pi\lambda^2)(\mu_0 H_{c1}) = \Phi_0 \Rightarrow H_{c1} = \frac{\Phi_0}{\mu_0\pi\lambda^2} = \frac{h}{e\mu_0\pi\lambda^2}. \qquad (11.97)$$

On the other hand, H_{c2} appears when the vortices are maximally packed inside the superconductor, and the introduction of one more fluxon will break the superconductor. That means that the vortices are situated at the coherence length ξ between them (since the coherence length is the "size of the vortex"):

$$(\pi\xi^2)(\mu_0 H_{c2}) = \Phi_0 \Rightarrow H_{c2} = \frac{\Phi_0}{\mu_0\pi\xi^2} = \frac{h}{e\mu_0\pi\xi^2}. \qquad (11.98)$$

It remains to calculate ξ. We have seen that in the phenomenological LG model, $\xi = \sqrt{\hbar^2/(2m_*\alpha)}$, but m_* and α are phenomenological parameters. We want instead to write

ξ in terms of something we can measure. In BCS theory, one finds the exact result

$$\xi_0 = \frac{2\hbar v_F}{\pi \Delta_g}. \tag{11.99}$$

But there is a simple argument to estimate it as follows. Consider a modulated plane wave acting on the material, with k_F modulated by a $q_0 = 1/\xi_0$. Then the wave is

$$\frac{e^{i(k+q)x} + e^{ikx}}{\sqrt{2}}, \tag{11.100}$$

and the energy of the system is

$$\frac{\hbar^2}{2m}\left[\frac{(k+q)^2 + k^2}{2}\right] \simeq \frac{\hbar^2}{2m}k^2 + \frac{\hbar^2}{2m}kq. \tag{11.101}$$

Consider then that the extra energy of the modulated wave is enough to destroy the superconductivity, and so equals the energy gap Δ_g,

$$\frac{\hbar^2}{2m}k_F\frac{1}{\xi_0} = \Delta_g \Rightarrow \xi_0 = \frac{\hbar^2 k_F}{2m\Delta_g} = \frac{\hbar v_F}{2\Delta_g}, \tag{11.102}$$

which is just a factor of $\pi/4$ different than the correct result.

11.8 Josephson Junctions

Consider the situation of two superconductors separated by a thin layer of insulator, what we will call a *Josephson junction*. Classically, there should be no current passing through the insulator, but we will see that quantum mechanically there is.

There are two main effects related to it: the *DC Josephson effect*, which is that there is a DC electric current flowing though the insulator gate even when there is no applied field, and the *AC Josephson effect*, that there is an AC current flowing through the insulator gate in the case of an applied DC voltage.

The DC Josephson Effect

Since superconductors are intrinsically quantum mechanical, one considers the quantum mechanical time-dependent Schrödinger equation for the two (coupled by the thin insulator) superconductors. Denoting by ψ_1 and ψ_2 the wavefunctions for the superconductors and by $\hbar T$ the quantum mechanical interaction Hamiltonian between them, we split $i\hbar\partial\psi/\partial t = \mathcal{H}\psi$ into

$$i\hbar\frac{\partial\psi_1}{\partial t} = \hbar T\psi_2; \quad i\hbar\frac{\partial\psi_2}{\partial t} = \hbar T\psi_1. \tag{11.103}$$

The presence of $\hbar$ on the right-hand side signals that this is a quantum effect, and of course T decreases as the width of the insulator is increased, but we are not interested in that at this time.

As we saw, inside a superconductor, the wavefunction is macroscopically coherent, with a single phase θ and the absolute value related to the number density:

$$\psi_1 = \sqrt{n_1}e^{i\theta_1}; \quad \psi_1 = \sqrt{n_2}e^{i\theta_2}. \tag{11.104}$$

Multiplying the Schrödinger equation for ψ_1 with $-i\sqrt{n_1}e^{-i\theta_1}$ and the equation for ψ_2 with $-i\sqrt{n_2}e^{-i\theta_2}$, and defining $\delta = \theta_2 - \theta_1$, we obtain

$$\frac{1}{2}\frac{\partial n_1}{\partial t} + in_1\frac{\partial\theta_1}{\partial t} = -iT\sqrt{n_1 n_2}e^{i\delta}$$

$$\frac{1}{2}\frac{\partial n_2}{\partial t} + in_2\frac{\partial\theta_2}{\partial t} = -iT\sqrt{n_1 n_2}e^{-i\delta}, \tag{11.105}$$

or, splitting the equations into real and imaginary parts,

$$\frac{\partial n_1}{\partial t} = 2T\sqrt{n_1 n_2}\sin\delta$$

$$\frac{\partial n_2}{\partial t} = -2T\sqrt{n_1 n_2}\sin\delta$$

$$\frac{\partial\theta_1}{\partial t} = -T\sqrt{\frac{n_2}{n_1}}\cos\delta$$

$$\frac{\partial\theta_2}{\partial t} = -T\sqrt{\frac{n_1}{n_2}}\cos\delta. \tag{11.106}$$

The last two equations give, for identical superconductors 1 and 2, i.e. for $n_1 = n_2$,

$$\frac{\partial\delta}{\partial t} = 0; \tag{11.107}$$

the phases of the two superconductors are relatively locked. The first two equations give first

$$\frac{\partial n_2}{\partial t} = -\frac{\partial n_1}{\partial t}, \tag{11.108}$$

which means that there is a constant current flowing from 1 to 2, proportional to this quantity, and the proportionality with $\sin\delta$ means that

$$J = J_0\sin\delta = J_0\sin(\theta_2 - \theta_1). \tag{11.109}$$

We have thus proven the DC Josephson effect.

AC Josephson Effect

We now add also a voltage V to the junction, which means that the Hamiltonian is shifted by $-eV$ for superconductor 1 and $+eV$ for superconductor 2, since a Cooper pair, of charge $q = -2e$, with wavefunction ψ, will experience an energy difference of $-2eV$ across the gate.

The Schrödinger equations are thus modified to

$$i\hbar\frac{\partial\psi_1}{\partial t} = \hbar T\psi_2 - eV\psi_1; \quad i\hbar\frac{\partial\psi_2}{\partial t} = \hbar T\psi_1 + eV\psi_2, \tag{11.110}$$

which means that with the definitions (11.104) we obtain

$$\frac{1}{2}\frac{\partial n_1}{\partial t} + in_1\frac{\partial \theta_1}{\partial t} = -iT\sqrt{n_1 n_2}e^{i\delta} + \frac{ieV}{\hbar}n_1$$
$$\frac{1}{2}\frac{\partial n_2}{\partial t} + in_2\frac{\partial \theta_2}{\partial t} = -iT\sqrt{n_1 n_2}e^{-i\delta} - \frac{ieV}{\hbar}n_2, \tag{11.111}$$

and splitting them into real and imaginary parts:

$$\frac{\partial n_1}{\partial t} = 2T\sqrt{n_1 n_2}\sin\delta$$
$$\frac{\partial n_2}{\partial t} = -2T\sqrt{n_1 n_2}\sin\delta$$
$$\frac{\partial \theta_1}{\partial t} = -T\sqrt{\frac{n_2}{n_1}}\cos\delta + \frac{eV}{\hbar}$$
$$\frac{\partial \theta_2}{\partial t} = -T\sqrt{\frac{n_1}{n_2}}\cos\delta - \frac{eV}{\hbar}. \tag{11.112}$$

Now the first two equations give $\partial n_2/\partial t = -\partial n_1/\partial t$, which means a common current, and moreover a current proportional to $\sin\delta$, whereas the last two equations now give the time variation:

$$\frac{\partial \delta}{\partial t} = -\frac{2eV}{\hbar} \Rightarrow \delta(t) = \delta(0) - \frac{2eV}{\hbar}t. \tag{11.113}$$

Substituting in the current, we obtain the sinusoidal (AC) current:

$$J(t) = J_0 \sin\left[\delta(0) - \frac{2eV}{\hbar}t\right]. \tag{11.114}$$

We have thus proven the AC Josephson effect as well.

Josephson Loop and Quantum Interference

If we consider two Josephson junctions a and b connected in parallel, so that we have a single piece of superconductor on the left, and a single one on the right, connected in two places by the insulator junctions, we obtain a superconducting loop. Consider next having a magnetic flux Φ pass through it. Then, according to (11.72), we have an extra difference in the *total* phase around the loop:

$$(\theta_2 - \theta_1)_b - (\theta_2 - \theta_1)_a \equiv \delta_b - \delta_a = q\frac{\Phi}{\hbar} = 2e\frac{\Phi}{\hbar}. \tag{11.115}$$

We can interpret this as a modification of δ_a and δ_b as

$$\delta_b = \delta(\Phi = 0) + \frac{e\Phi}{\hbar}; \quad \delta_a = \delta(\Phi = 0) - \frac{e\Phi}{\hbar}. \tag{11.116}$$

Then the total current passing through the parallel junctions is

$$J_{\text{total}} = J_a + J_b = J_0\left[\sin\left(\delta(\Phi = 0) + \frac{e\Phi}{\hbar}\right) + \sin\left(\delta(\Phi = 0) - \frac{e\Phi}{\hbar}\right)\right]$$
$$= 2J_0\sin(\delta(\Phi = 0))\cos\frac{e\Phi}{\hbar}. \tag{11.117}$$

This is an interference pattern as a function of the magnetic flux Φ, which has been experimentally observed.

This is the kind of effect depending on topology that will be considered in the next chapter.

11.9 High T_c Superconductors

Since in fact the most interesting superconductors are the high T_c ones, here we will collect a few important facts about them.

The high T_c superconductors have critical temperatures above those of liquid nitrogen, so that they can be formed easily. The vast majority of them are *cuprate* superconductors (with copper oxides), and are layered, so that they can be thought of as 2+1–dimensional. There are also some organic superconductors.

An important fact that makes them very amenable to treatment using AdS/CFT (which will be described in the last part of the book), but not by usual condensed matter methods, is the fact that they are intrinsically strongly coupled. One manifestation of that is the fact that the ratio $2\Delta/(k_B T_c)$ is not $\simeq 3.52$ like in BCS theory, valid for usual semiconductors, but is rather bigger by a factor of 2 or so, indicating that there are strong coupling effects, and one needs a new theory for them.

The other important observation is that, as a function of doping x, viewed as a coupling parameter of the theory, the (T, x) phase diagram contains a "strange metal" region, thought to be a quantum critical phase. The phase diagram contains a superconducting dome centered around a x_m and $T = 0$, separated from a low-x antiferromagnetic phase by a pseudogap phase, as in Figure 11.5. On top of the superconducting dome there is an open (V-shaped) "strange metal" phase, thought to be quantum critical. On the left there is the pseudogap phase, and on the right a Fermi liquid phase. The "strange metal" is "strange" because it doesn't obey Fermi liquid theory, hence it is also called "non-Fermi liquid."

Note that "heavy fermion" compounds, where the electron effective mass (due to interactions) is up to thousands of times larger than the free electron mass, have a similar phase diagram, just that without the "pseudogap" phase (so the antiferromagnetic phase continues until the superconducting dome), and is also believed to be a quantum critical phase.

The most important characteristic of the strange metal phase of high T_c superconductors, and the hardest to reproduce, is the non-Fermi liquid behavior. Fermi liquid theory seems to be almost independent of the microscopic details of the material and predicts a linear low-temperature specific heat,

$$C_e \sim \gamma T + \cdots ; \quad \gamma \propto m_*, \tag{11.118}$$

as we saw, and a low-temperature resistivity that is quadratic,

$$\rho_e = \rho_0 + AT^2 + \cdots , \tag{11.119}$$

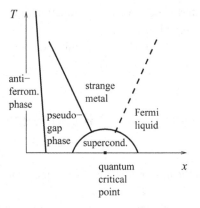

Phase diagram including a strange metal phase: superconducting dome, with strange metal on top. To the left, pseudogap phase and antiferromagnetic phase. To the right, Fermi liquid phase, separated by a crossover.

whereas for the strange metal phase we have a linear resistivity above a critical temperature,

$$\rho_e = B\,T + \cdots\,; \quad T > T_c. \tag{11.120}$$

The way this is obtained is via an electron spectral function $G_R(\omega, k)$ that, as one approaches a Fermi surface (which still exists for the non-Fermi liquid, but excitations have a much broader peak than for a Fermi liquid), is still of the type

$$G_R(\omega, k) \simeq \frac{h}{\omega - v_F(k - k_F) + \Sigma(\omega, k)}, \tag{11.121}$$

but, instead of the Fermi liquid prediction of

$$\Sigma(\omega) = i\frac{\Gamma}{2} \sim i\omega^2, \tag{11.122}$$

now obeys a modified law,

$$\Sigma(\omega) \simeq c\omega \log \omega + d\omega, \tag{11.123}$$

with c real and d complex. Then the decay rate, equal to the imaginary part of Σ, is now linear in ω instead of quadratic and is too large, and the particle decays before it can propagate enough, meaning that the quasi-particle interpretation breaks down. Therefore a non-Fermi liquid has a Fermi surface without quasi-particles.

Important Concepts to Remember

- Superconductivity has negligible resistance for $T < T_c$, persistent currents, isotopic effect $T_c \propto M_i^{-1/2}$, Keesom's law for the specific heat, energy gap Δ observed in the absorption coefficient, a critical magnetic field above which superconductivity disappears, and the Meissner-Ochesenfeld effect that the magnetic field is expulsed from the material, except for a penetration depth λ.

- Superconductors are type I, which abruptly end the perfect diamagnetic behavior $\chi = -1$ to normal state $\chi \simeq 0$ at a H_c, and type II, which end it gradually, from a H_{c1} the magnetization starting to increase (paramagnetic component) until it reaches 0 at H_{c2}.
- Type II superconductors form vortices, extended to strings parallel to the applied magnetic field, organized in an Abrikosov vortex lattice.
- One can use Landau's theory of second order phase transitions to describe the superconductor, and find a ΔC_P and a critical exponent -1 for $\chi \sim (T - T_c)^{-1}$.
- Landau-Ginzburg theory describes the superconductor, where the order parameter is a bosonic wavefunction for the superconductor, and find a Landau-Ginzburg equation that takes the form of a (nonrelativistic) nonlinear Schrödinger equation.
- In a 1+1–dimensional LG model, one finds that the wavefunction changes over a coherence length ξ from the zero value outside the material to the ψ_0 value inside the superconductor.
- In a 2+1–dimensional LG model, one finds (ANO) vortex solutions, carrying topological charge (magnetic charge) and with ψ varying from ψ_0 outside the vortex to $\psi = 0$ inside the core of the vortex.
- The London-London phenomenological model has a normal and a superconducting fluid in the material, changing from one to the other as the temperature is varied between T_c and 0. In a simply connected superconductor, $\vec{j} \propto \vec{A}$.
- In the microscopic BCS theory, Cooper pairs form via the electron-electron-phonon interaction. Cooper pairs are electrons of opposite spins and momenta and are bound, i.e. there is a binding energy Δ.
- In the BCS ground state, $2\Delta / k_B T_c \simeq 3.52$.
- In type II superconductors, H_{c1} corresponds to one vortex penetrating over the penetration depth λ, and H_{c2} corresponds to vortices being maximally packed, i.e. with the penetration depth ξ between them.
- In a Josephson junction, separating two superconductors by a thin insulator, there is a DC effect, for a current in the absence of voltage, and an AC effect, for an AC current in the presence of an applied DC voltage.
- One can create a quantum interference loop out of two parallel Josephson junctions.
- High T_c superconductors are mostly cuprates (copper oxides) and have a "strange metal" phase that is most likely a quantum critical phase.
- Non-Fermi liquids like the strange metals have linear resistivity and Fermi surfaces without quasi-particles.

Further Reading

Chapter 10 in [6] and chapter 11 in [7].

Where It will be Addressed in String Theory

Superconductivity will be addressed using phenomenological AdS/CFT in Chapter 37. High T_c type II superconductors, as an example of quantum critical phase, will be addressed in Chapter 42, and the "strange metal" phase will be addressed in Chapters 47 and 48.

Exercises

(1) Explain why a small permanent magnet brought down along the (vertical) axis of a superconducting ring levitates (doesn't fall under gravity).

(2) From the LG free energy derive the expression for the superconducting current used in the proof of flux quantization and show that for n vortices in a superconductor we have n units of flux.

(3) In the London theory, show that by writing the energy of the magnetic field and the kinetic term for the superconducting electrons, we obtain the free energy

$$f = \frac{1}{2\mu_0} \int_0^\infty (\vec{B}^2 + \lambda_L^2 \mu_0^2 \vec{j}^2) 2\pi r dr \tag{11.124}$$

and from it we obtain the London equation

$$\lambda_L^2 \vec{\nabla}^2 \vec{B} = \vec{B}. \tag{11.125}$$

(4) Show that for N vortices we have $|\psi(r)| \sim r^N$ and $\psi(\vec{r}) = |\psi(r)|e^{iN\theta}$.

(5) Writing the LG wavefunction equation at nonzero $\vec{A}$ in magnetic field in the presence of a vortex, assuming that there is almost no magnetization ($M \simeq 0 \Rightarrow B \simeq H$) near $H \simeq H_{c2}$, calculate from the resulting equation of motion that the maximum H that allows for vortices, thus for superconductivity, is

$$H_{c2} = \frac{\Phi_0}{e\pi\mu_0\xi^2}. \tag{11.126}$$

12

Topology and Statistics: Berry and Chern-Simons, Anyons, and Nonabelions

In this chapter, we will present several interesting properties related to topology and the statistics of particles that are of interest for current research. We will see that the topology defines an interesting observable called the Berry phase. Then we will define particles with nonstandard statistics, anyons, and nonabelions. Finally, we will describe topological superconductors, which are materials whose properties depend only on topological quantities.

12.1 Berry Phase and Connection

We start with the modern concept of Berry phase and Berry connection, which arose out of a way to understand and generalize the Aharonov-Bohm effect.

Aharonov-Bohm Effect

Something one can ask about gauge theories like electromagnetism is the following. Given that there is gauge invariance, and that the physically measurable (observable) fields are the field strengths, is there any need for the gauge field itself, or is it just a convenient mathematical construct? At the classical level, it is just a convenient mathematical construct, but at the quantum level, we can experimentally measure its effects in the Aharonov-Bohm effect.

The fact that the momentum operator acts on the wavefunction of a particle as $\vec{p}\psi = -i\hbar\frac{\partial}{\partial x}\psi$ means that a particle that doesn't interact with an electromagnetic field has a wavefunction with a phase factor of the type

$$e^{\frac{i}{\hbar}\int_P \vec{p}\cdot d\vec{x}},\tag{12.1}$$

where P is the path of the particle. But if we introduce the interaction with an electromagnetic field, the effect is to shift $\vec{p} \to \vec{p} - q\vec{A}$, so the phase factor is

$$e^{\frac{i}{\hbar}\int_P (\vec{p}-q\vec{A})\cdot d\vec{x}} \equiv e^{i\delta}.\tag{12.2}$$

But if we consider a *closed* path C for the particle, we see that there is a phase for the loop:

$$\delta = \frac{q}{\hbar}\oint_{C=\partial S} \vec{A}\cdot d\vec{x} = \frac{q}{\hbar}\int_S \vec{B}\cdot d\vec{S} = \frac{q\Phi}{\hbar}.\tag{12.3}$$

In fact, we have seen this fact in the case of superconductors at the end of Chapter 11, when discussing the quantum interference of two Josephson junctions in parallel. Actually, that case could be set up like an example of the Aharonov-Bohm effect.

The Aharonov-Bohm effect is the fact that this change in the quantum phase, which can be observed in interference experiments, depends only on the flux *enclosed* by the loop of particle paths, but we could set things up such that $\vec{B} \simeq 0$ along the particle paths and is only nonzero in the middle of the loop. This means that the particles are really measuring $\int_P \vec{A} \cdot d\vec{x}$.

Berry Phase

Consider a quantum system in an eigenstate of the Hamiltonian $|n\rangle$, and consider an adiabatic evolution of the Hamiltonian, such that the system remains in the state $|n\rangle$, up to a phase factor. There is of course a phase due to the time evolution of the state according to the Schrödinger equation, $i\hbar\partial_t\psi_n = H\psi_n$, but there is also a phase factor due to the change in the eigenstate $|n\rangle$ itself, due to the change in H. The latter is the *Berry phase*, and for a *cyclical* motion the phase along the cyclical path is physical, invariant, and observable,

$$\Delta\theta_n[C] = i \oint_C \langle n[k(t)]|\nabla_k|n[k(t)]\rangle dk, \tag{12.4}$$

where $k(t)$ is the adiabatic parameter, parametrized by t. On an open path, we would write

$$\Delta\theta_n[t] = i \int_{k(0)}^{k(t)} \langle n[k]|\nabla_k|n[k]\rangle dk = i \int_0^t \left\langle n[k(t')]\left|\frac{d}{dt'}\right|n[k(t')]\right\rangle dt', \tag{12.5}$$

which makes it clear that it is an extra contribution coming from the Schrödinger equation.

Since the wavefunction is not an observable in itself, there is a gauge invariance defined by the transformation

$$|\tilde{n}[k]\rangle = e^{i\beta[k]}|n[k]\rangle, \tag{12.6}$$

which implies that on an open path, the Berry phase changes to

$$\Delta\tilde{\theta}_n[t] = \Delta\theta_n[t] + \beta[k(t)] - \beta[k(0)]. \tag{12.7}$$

On a closed path C, requiring $\beta[k(T)] - \beta[k(0)] = 2\pi m$, we obtain that the Berry phase is invariant modulo 2π under the gauge transformations.

Berry Connection and Curvature

We can now define the *Berry connection* by writing the Berry phase as the holonomy of a connection,

$$\Delta\theta_n[C] = i \oint_C A_n(k) \cdot dk, \tag{12.8}$$

where we have defined

$$A_n(k) = i\langle n(k)|\nabla_k|n(k)\rangle. \tag{12.9}$$

Considering the gauge invariance just defined, we see that it transforms this connection in the usual way:

$$\tilde{A}_n(k) = A_n(k) + \nabla_k\beta(k). \tag{12.10}$$

We can also define the *Berry curvature* in the usual way, as the field strength associated with the Berry connection,

$$F_n(k) = \nabla_k \times A_n(k).\tag{12.11}$$

We now observe that if the Berry connection is a real connection (gauge field), the Berry phase turns into an Aharonov-Bohm phase, thus making the former a generalization of the latter.

Note that we have intentionally denoted the adiabatic parameter by k, since in insulators with an electronic band structure, we can consider the Berry connection in the first Brillouin zone (in $\vec{k}$ space):

$$A^n_{k_i} = i\left\langle n, \vec{k} \left| \frac{\partial}{\partial k_i} \right| n, \vec{k} \right\rangle.\tag{12.12}$$

Nonabelian Generalization

We can consider a generalization to N occupied states $|n\rangle$, $n = 1, \dots, N$ instead of a single one or, in the case of band structure, to several occupied bands, in which case

$$A(k) = i\sum_n \langle n, k | \nabla_k | n, k \rangle\tag{12.13}$$

is still an abelian connection. However, in that case, the above is only the abelian (diagonal $U(1)$) component of a *nonabelian Berry connection*,

$$A^{(nm)}(k) = i\langle n, k | \nabla_k | m, k \rangle,\tag{12.14}$$

which takes values in $U(N)$.

12.2 Effective Chern-Simons Theories in Solids

In a solid, as possible gauge fields with which to build up a theory of the material we have the electromagnetic field; we have seen that we can also have the Berry connection, but there is one more possible gauge field, which has no dynamics, the Chern-Simons or statistical gauge field.

Consider a generic microscopic Hamiltonian for electrons interacting via a potential, in an external electromagnetic potential $A_\mu = (A_0, \vec{A})$:

$$H_e = \sum_{j=1}^N \frac{|\vec{p}_j - e\vec{A}(\vec{r}_j)|^2}{2m_b} + \sum_{i<j} v(\vec{r}_i - \vec{r}_j) + \sum_{i=1}^N eA_0(\vec{r}_i).\tag{12.15}$$

The wavefunction for the multielectron state is called $\Psi_e(\vec{r}_1, \dots, \vec{r}_N)$, so the Schrödinger equation is

$$H_e\Psi_e(\vec{r}_1, \dots, \vec{r}_N) = E\Psi_e(\vec{r}_1, \dots, \vec{r}_N).\tag{12.16}$$

We now make a canonical transformation from Ψ_e to $\Phi_e = U\Psi_e$ as

$$\Phi_e(\vec{r}_1, \ldots, \vec{r}_N) = U\Psi_e(\vec{r}_1, \ldots, \vec{r}_N) = \left[\prod_{i<j} e^{-i\frac{\theta}{\pi}\alpha(\vec{r}_i - \vec{r}_j)} \right] \Psi_e(\vec{r}_1, \ldots, \vec{r}_k). \qquad (12.17)$$

Here $\alpha(\vec{r}_i - \vec{r}_j)$ is the angle made by the relative distance between the electrons, $\vec{r}_{ij} = \vec{r}_i - \vec{r}_j$, with a fixed axis.

The corresponding change in the Hamiltonian due to the canonical transformation is obtained by $H_e \to U^{-1}H_eU$. But we can easily check that

$$U^{-1}(\vec{p}_i - e\vec{A}(\vec{r}_i))U = \vec{p}_i - e\vec{A}(\vec{r}_i) - e\vec{a}(\vec{r}), \qquad (12.18)$$

where we have introduced an effective, statistical, gauge field $\vec{a}(\vec{r})$ defined by

$$e\vec{a}(\vec{r}_i) = \frac{\theta}{\pi} \sum_{j\neq i} \vec{\nabla}_i\alpha(\vec{r}_i - \vec{r}_j) = \frac{\theta}{\pi} \sum_{j\neq i} \frac{\hat{z} \times (\vec{r}_i - \vec{r}_j)}{|\vec{r}_i - \vec{r}_j|^2}, \qquad (12.19)$$

where $\hat{z}$ is a fixed direction, the third spatial direction. Note that, since $\alpha_{ij} \equiv \alpha(\vec{r}_i - \vec{r}_j)$ is the angle made with a fixed direction, then

$$\alpha_{ij} = \alpha_{ji} + \pi. \qquad (12.20)$$

Then we obtain that the Hamiltonian is changed to

$$H_e' = \sum_{j=1}^{N} \frac{|\vec{p}_j - e\vec{A}(\vec{r}_j) - e\vec{a}(\vec{r}_j)|^2}{2m_b} + \sum_{i<j} v(\vec{r}_i - \vec{r}_j) + \sum_{i=1}^{N} eA_0(\vec{r}_i). \qquad (12.21)$$

Moreover, since $\alpha_{ij} = \alpha_{ji} + \pi$, under the interchange of two electrons, $\vec{r}_i$ with $\vec{r}_j$, the new wavefunction Φ_e picks up an extra factor of $e^{i\theta}$ besides the minus from the original fermionic wavefunction, so in total

$$\Phi_e(\ldots, \vec{r}_j, \ldots, \vec{r}_i, \ldots) = -e^{i\theta}\Phi_e(\ldots, \vec{r}_i, \ldots, \vec{r}_j, \ldots), \qquad (12.22)$$

which means that the statistics is changed from Fermi to Bose if $\theta = (2k+1)\pi$. Moreover, now we have the possibility that θ is arbitrary, which leads to *fractional, or anyonic, statistics*. We will describe that in more detail in the next section.

Finally, note that the new Hamiltonian, if we are in 2+1 dimensions, is obtained from an action of the *Chern-Simons type*, since the magnetic field associated with $\vec{a}$ is proportional to the charge density:

$$f_{12}(\vec{r}_i) \equiv b(\vec{r}_i) = \vec{\nabla} \times \vec{a}(\vec{r}_i) = \frac{2\theta}{e} \sum_{j\neq i} \delta(\vec{r}_i - \vec{r}_j) = \frac{2\theta}{e^2} \rho_{\text{charge}}. \qquad (12.23)$$

But this would be the equation of motion for the action:

$$S = \int d^{2+1}x \left[\frac{e^2}{4\theta}\epsilon^{\mu\nu\rho}a_\mu\partial_\nu a_\rho - J^\mu a_\mu \right]. \qquad (12.24)$$

The first term in the action is of the *Chern-Simons* type and is the action for the statistical gauge field a_μ.

The Chern-Simons action with a source is usually written as

$$S = S_{CS} + S_{\text{source}} = \int d^{2+1}x \left[\frac{k}{2\pi} \epsilon^{\mu\nu\rho} A_\mu \partial_\nu A_\rho - J^\mu A_\mu \right], \qquad (12.25)$$

and its equation of motion is

$$\frac{k}{2\pi} \epsilon^{\mu\nu\rho} F_{\nu\rho} = J^\mu \Rightarrow F_{\mu\nu} = \frac{\pi}{k} \epsilon_{\mu\nu\rho} J^\rho. \qquad (12.26)$$

The Chern-Simons action is topological in nature, in the sense that it was written in a way that is independent of the metric $g_{\mu\nu}$ on the 3-dimensional space, and it can be derived as the result on the boundary of a 4-dimensional space, for what would be in the nonabelian case a known topological invariant, the *instanton number* or *Pontryagin index*:

$$S_{CS} = \frac{k}{2\pi} \int_{S=\partial M} d^3x \epsilon^{\mu\nu\rho} A_\mu \partial_\nu A_\rho = \frac{k}{8\pi} \int_M d^4x \epsilon^{\mu\nu\rho\sigma} F_{\mu\nu} F_{\rho\sigma}. \qquad (12.27)$$

Another observation is that the Chern-Simons term is of mass dimension 3, whereas the Maxwell kinetic term is of dimension 4 (higher), which means that at low enough energies we can always neglect the Maxwell term, even if there is one, and keep only the Chern-Simons term.

Then at the quantum level, the coefficient k, called the *level* of the Chern-Simons theory, needs to be an integer, $k \in \mathbb{Z}$, for consistency of e^{iS} in the presence of Dirac monopoles (singular magnetic sources). Under a gauge transformation $\delta A_\mu = \partial_\mu \lambda$, the CS action changes by

$$\delta S_{CS} = \frac{k}{\pi} \int d^{2+1}x \epsilon^{\mu\nu\rho} (\partial_\mu \lambda) \partial_\nu A_\rho = \frac{k}{\pi} \int_S dS^\mu \lambda \epsilon_{\mu\nu\rho} \partial^\nu A^\rho, \qquad (12.28)$$

which is zero in the absence of magnetic monopoles, but in their presence consistency requires $k \in \mathbb{Z}$, since otherwise we would obtain noninteger magnetic fluxes.

12.3 Anyons and Fractional Statistics

The construction in the previous subsection was in fact taylored for the description of *anyons*, which are objects of a general statistics, neither bosons, which are symmetric under exchange, nor fermions, which are antisymmetric. Rather, under the permutation of two anyons, the wavefunction would change by a general phase factor $e^{i\alpha}$. Note that for bosons and fermions, two permutations cancel each other, and we are back to the original wavefunction, but if we have an anyonic exchange phase factor instead, after two interchanges we get something nontrivial, $e^{2i\alpha}$.

The simplest construction of anyons is to attach magnetic flux *only at the positions of the anyons*, as was done in the last subsection, in (12.23). Such a delta function magnetic field implies that there is a magnetic flux associated with the particle:

$$\Phi = \int_S f_{12} dx^1 \wedge dx^2 = \frac{2\theta}{e}. \qquad (12.29)$$

In turn, that means that there is an Aharonov-Bohm phase if we move one of the anyons of charge e around the other:

$$\exp\left(ie\oint_C \vec{A}\cdot d\vec{x}\right) = \exp(ie\Phi[C]) = \exp(2i\theta). \tag{12.30}$$

But this process, of rotating one anyon around another, which gives back the original position, is interpreted as a double interchange, thus the anyonic phase factor for a single interchange is

$$e^{i\theta}, \tag{12.31}$$

as we saw in the last subsection as well. Hence we see that θ has the interpretation of *anyonic phase*, and hence the Chern-Simons term

$$\frac{4e^2}{\theta}\int d^{2+1}x \epsilon^{\mu\nu\rho}a_\mu\partial_\nu a_\rho \tag{12.32}$$

has the interpretation of kinetic term for the statistical gauge field a_μ.

Ultimately, we see that any field theory with a Chern-Simons term will lead to anyons if we add delta function sources for the gauge field, since then we find the delta function magnetic field (12.23).

Another relevant example is the Abelian-Higgs model with a CS term, with action

$$S = \int d^3x\left[-\frac{1}{4}F_{\mu\nu}F^{\mu\nu} + \frac{k}{2\pi}\epsilon^{\mu\nu\rho}A_\mu\partial_\nu A_\rho - \frac{1}{2}|D_\mu\phi|^2 - \lambda\left(|\phi|^2 - \frac{\mu^2}{2\lambda}\right)^2\right], \tag{12.33}$$

which has a vortex (soliton) solution that is anyonic, due to the CS term.

Anyons in the Fractional Quantum Hall Effect

In the Fractional Quantum Hall effect, one can describe part of the physics in terms of a simple effective action for the electromagnetic field A_μ coupled to an emergent Chern-Simons gauge field a_μ, with

$$S_{\text{eff}} = \int_{M_3} d^3x\left[\frac{1}{2\pi}\epsilon^{\mu\nu\rho}A_\mu\partial_\nu a_\rho - \frac{r}{4\pi}\epsilon^{\mu\nu\rho}a_\mu\partial_\nu a_\rho\right]. \tag{12.34}$$

The coupling of the electromagnetic gauge field A_μ with the emergent gauge field a_μ is done again through a Chern-Simons term. If we would integrate out a_μ via its equation of motion,

$$f_{\mu\nu} = \frac{1}{r}F_{\mu\nu}, \tag{12.35}$$

which implies that up to a gauge transformation, $a_\mu = A_\mu/r$, we would obtain by replacing it back in the action

$$S'_{\text{eff}} = \frac{1}{r}\frac{1}{4\pi}\int_{M_3} d^3x \epsilon^{\mu\nu\rho}A_\mu\partial_\nu A_\rho. \tag{12.36}$$

In it, $1/r$ appears instead of the integer k, which suggests the fact (which is actually true) that the Hall conductivity is $1/r$, a fractional value instead of an integer one.

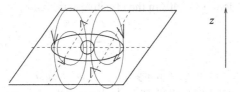

Fig. 12.1 A small localized flux tube must be surrounded by a larger opposite flux tube (passing through the plane in the opposite direction), because of flux conservation.

If we introduce "quasi-particles" with charge q under the emergent field a_μ, by adding to the effective action the source term

$$\int_{M_3} J^\mu a_\mu = q \int dt \, a_0(x_0, t), \tag{12.37}$$

the equation of motion for a_0 is now

$$\frac{F_{12}}{2\pi} - \frac{r f_{12}}{2\pi} + q\delta(x - x_0) = 0, \tag{12.38}$$

which means that there is a delta function flux associated with the quasi-particle, either in the electromagnetic field F_{12} or in the emergent field f_{12}. We can thus solve it by either

$$\frac{F_{12}}{2\pi} = -q\delta(x - x_0) \tag{12.39}$$

or

$$\frac{f_{12}}{2\pi} = \frac{q}{r}\delta(x - x_0), \tag{12.40}$$

but either way, we have a flux attached to the quasi-particle that leads, as we have explained before, to anyons. The first possibility (electromagnetic flux) leads to

$$\theta = \pi\frac{q}{e}, \tag{12.41}$$

whereas the second (emergent field flux) leads to

$$\theta = \frac{\pi}{r}\frac{q}{e}. \tag{12.42}$$

Note, however, that a delta function flux for F_{12} is harder to accept. It could be thought of as an idealization of a tiny flux tube, extending in the third spatial dimension $\hat{z}$, sort of like a permanent magnet in $\hat{z}$. But then we would expect the flux lines to return to the 2+1 dimensions piercing them in the opposite direction, surrounding the flux with the opposite-sign flux, such that when course graining, we would have $\int F = 0$, as in Figure 12.1. So the emergent field flux possibility makes more sense, and we generically obtain a fractional anyonic phase (for $q = e$ and $r > 1$). Note that it does not matter what kind of gauge field we have, as long as it has an Aharonov-Bohm or Berry phase, and that is true of both A_μ and a_μ.

Considering (12.40) in the effective action S_{eff}, we obtain the electric charge

$$J_0 = \frac{\delta S_{eff}}{\delta A_0} = \frac{f_{12}}{2\pi} = \frac{q}{r} \delta(x - x_0),$$
(12.43)

so it is a fractional ordinary electric charge, q/r.

An alternative explanation of why we have anyons in the Fractional Quantum Hall Effect is based on the Laughlin wavefunction (10.47) and the wavefunction to add a "quasi-hole," $\left(\prod_i z_i\right) \Psi_m$. As we saw in Chapter 10, this meant that the quasi-hole has a quantum of vorticity, and a fractional electric charge, $\nu = 1/m$. Considering two such quasi-holes, and taking one in a loop around the other, leads to an Aharonov-Bohm phase of $2\pi\nu$, as we saw at the beginning of the section, and moreover, this is interpreted as two consecutive interchanges of position of the quasi-holes. Therefore the quasi-holes are anyons with anyonic phase $\pi\nu$.

12.4 Nonabelian Anyons (Nonabelions) and Statistics

In the previous section we have considered anyons, i.e. particles that under interchange modify the wavefunction not by ± 1 as for the Bose/Fermi case, but by a more general phase $e^{i\alpha}$. But the general possibilities for the wavefunctions under interchanges of the particles are classified by representations of the *braid group* $\mathcal{B}_N$, as opposed to representations of the permutation group $\mathbb{Z}_N$. Anyons, characterized by a phase under the interchange, can be thus said to be *abelian* representations of the braid group $\mathcal{B}_n$ but we can also have *nonabelian* representations. This was first pointed out, in the context of the Fractional Quantum Hall Effect, by Moore and Read, and the objects that carry these representations were called nonabelian anyons, or nonabelions.

Specifically, this nonabelions transform under the interchange of two excitations r and s as

$$\psi_{p:\{i_1,...,i_r,...,i_s,...,i_N\}}(z_1, \ldots, z_{i_r}, \ldots, z_{i_s}, \ldots, z_N) = \sum_q B_{pq}[i_1, \ldots, i_N] \psi_{q:\{i_1,...,i_N\}}(z_1, \ldots, z_N).$$
(12.44)

Here p is a shorthand for the set of indices $\{i_1, \ldots, i_N\}$.

To construct wavefunctions for states obeying nonabelian statistics, Moore and Read took advantage of the fact that in a conformal field theory there is an operation that can be used to bring together two excitations to create a new one, namely the operator product of "fusion," symbolically

$$\phi_j \times \phi_k = \sum_k N^i_{jk} \phi_i.$$
(12.45)

The fusing procedure must be commutative and associative, so

$$N^i_{jk} = N^i_{kj}; \quad \phi_i \times (\phi_j \times \phi_k) = (\phi_i \times \phi_j) \times \phi_k \Rightarrow \sum_l N^m_{il} N^l_{jk} = \sum_n N^n_{ij} N^m_{nk}. \quad (12.46)$$

The nonabelian braiding appears when in the conformal field theory we decompose correlation functions (n-point functions) in terms of *conformal block functions* $\mathcal{F}_p$ as

$$\left\langle \prod_{a=1}^{n} \phi_{i_a}(z_a, \bar{z}_a) \right\rangle = \sum_{p} |\mathcal{F}_{p:\{i_1,\ldots,i_n\}}(z_1,\ldots z_n)|^2, \qquad (12.47)$$

where $\mathcal{F}_p$ are holomorphic, and p (where p stands for the set of indices $\{i_1,\ldots,i_n\}$) labels a basis of functions $\mathcal{F}_p$ and $\mathcal{F}_p$ transforms like ψ_p above. Roughly speaking, we factorize $\phi_i(z,\bar{z}) = \phi_i(z)\bar{\phi}_i(\bar{z})$ and use the fusion operators $\phi_i(z)$ as above.

The natural correspondence was to assume that *holomorphic wave functions of systems in 2+1 dimensions are conformal blocks of some CFT in two Euclidean (spatial) dimensions.*

For the case of the Fractional Quantum Hall Effect, the identification was done to a Chern-Simons theory in 2+1 dimensions that can be mapped to a 1+1–dimensional "rational torus" conformal field theory. This allowed for the interpretation of the Laughlin ground state,

$$\Psi_{\text{Laughlin}}(z_1,\ldots,z_N) = \prod_{i<j}(z_i - z_j)^m \exp\left[-\frac{1}{4}\sum_k |z_k|^2\right], \qquad (12.48)$$

as a conformal field theory correlator in the thermodynamic limit $N \to \infty$,

$$\Psi_{\text{Laughlin}} = \lim_{N\to\infty} \left\langle \prod_{i=1}^{N} e^{i\sqrt{m}\phi(z_i)} \exp\left[-i\int d^2z' \sqrt{m}\rho_0\phi(z')\right]\right\rangle_{\text{CFT}}, \qquad (12.49)$$

where $\rho_0 = 1/(2\pi m)$ is the electron density, the exponentials are normal ordered and ϕ is a free massless scalar in two dimensions, satisfying

$$\langle \phi(z)\phi(w)\rangle = -\log(z-w). \qquad (12.50)$$

The wavefunction for a quasi-hole at w is similarly interpreted,

$$\Psi_{\text{quasi-hole}}(z_1,\ldots,z_N;w) = \left\langle e^{i\frac{\phi(w)}{\sqrt{m}}} \prod_{i=1}^{N} e^{i\sqrt{m}\phi(z_i)} \exp\left[-i\int d^2z' \sqrt{m}\rho_0\phi(z')\right]\right\rangle$$

$$= \prod_i(z_i - w)\prod_{i<j}(z_i - z_j)^m \exp\left[-\frac{1}{4}\sum_k |z_k|^2 - \frac{1}{4m}|w|^2\right], \qquad (12.51)$$

and several quasi-holes are found in a similar way:

$$\Psi_{\text{quasi-holes}}(z_1,\ldots,z_N;w_1,\ldots,w_M)$$

$$\sim \left\langle \sum_{j=1}^{M} e^{i\frac{\phi(w_j)}{\sqrt{m}}} \prod_{i=1}^{N} e^{i\sqrt{m}\phi(z_i)} \exp\left[-i\int d^2z' \sqrt{m}\rho_0\phi(z')\right]\right\rangle. \qquad (12.52)$$

One can also construct a series of states with filling fractions that are continued fractions,

$$\nu = \cfrac{1}{m + \cfrac{1}{2p_2 + \cdots + \cfrac{1}{2p_{2s-1} + \cfrac{1}{2p_{2s}}}}} \equiv \frac{p}{q}, \tag{12.53}$$

and statistics that are also continued fractions (for $n = 2s$):

$$\frac{\theta}{\pi} = \cfrac{(-1)^{n-1}}{2p_n + \cfrac{1}{2p_{n-1} + \cdots + \cfrac{1}{2p_2 + \frac{1}{m}}}} \equiv \frac{p'}{q'}. \tag{12.54}$$

The details of these constructions can be found in the original paper.

Then to construct the nonabelian statistics states, one takes the opposite viewpoint, and starts with some known conformal blocks and interprets them as electron systems with nonabelian excitations.

The "Pfaffian state" for N electrons without spin,

$$\Psi_{Pf}(z_1, \ldots, z_N) = \text{Pfaff}\left(\frac{1}{z_i - z_j}\right) \prod_{i<j}(z_i - z_j)^m \exp\left[-\frac{1}{4}|z_j|^2\right], \tag{12.55}$$

is interpreted as a correlator of energy operators in the Ising model CFT times a rational torus correlator,

$$\Psi_{Pf} = \left\langle \prod_{i=1}^{N} \psi(z_i) \right\rangle_{\text{Ising}} \left\langle \prod_i e^{i\sqrt{m}\phi(z_i)} \exp\left[-i\int d^2z \sqrt{m}\rho_0\phi(z')\right] \right\rangle_{\text{rat.torus}}, \tag{12.56}$$

and then states with four quasi-holes around it obey nonabelian statistics,

$$\Psi_{4\,\text{quasi-hole}} = \left\langle \psi(z_i)\exp[i\sqrt{m}\phi(z_i)] \times \prod_{i=1}^{4} \sigma(v_i)\exp\left[\frac{i}{2\sqrt{m}}\phi(v_i)\right] \right.$$
$$\left. \times \exp\left[-i\int d^2z' \sqrt{m}\rho_0\phi(z')\right] \right\rangle, \tag{12.57}$$

as one can prove from the monodromy of the conformal blocks.

12.5 Topological Superconductors

A topological superconductor is a material that has fully gapped quasi-particle excitations in the bulk, i.e. the Cooper pairs responsible for superconductivity, but has topologically protected gapless quasi-particle states propagating on the boundary.

The first kind of topological superconductors to be predicted preserve time-reversal symmetry (T-invariant). The states propagating on the boundary are Majorana fermions (i.e., in QFT terms, the particle is its own antiparticle).

One interesting version of topological superconductors breaks time-reversal symmetry, and there the (zero-mode) states on the boundary are (odd numbers of) Majorana fermions and obey nonabelian statistics (nonabelions).

Topological Field Theory of $3+1$–**Dimensional T-Invariant**
Topological Superconductors

A simple topological theory is obtained for $3+1$–dimensional T-invariant topological superconductors. Consider such a system with several Fermi surfaces (2-dimensional surfaces), and for each of them consider the abelian Berry connection

$$a_j = -i \left\langle n\vec{k} \left| \frac{\partial}{\partial k_j} \right| n\vec{k} \right\rangle, \qquad (12.58)$$

and its corresponding *first Chern number*, the flux number defined by integration over this
2-dimensional Fermi surface Σ_n, by

$$C_{1n} = \frac{1}{2\pi} \int_{\Sigma_n} d\Omega^{ij} (\partial_i a_j - \partial_j a_i). \qquad (12.59)$$

There is a topological invariant that equals the sum of the first Chern numbers of all the
Fermi surfaces, weighted by the sign of a quantity called the pairing,

$$\Delta_{n\vec{k}} = T(\langle n\vec{k}|)\Delta_{\vec{k}}|n\vec{k}\rangle, \qquad (12.60)$$

where T is the time-reversal operator and $\Delta_{\vec{k}}$ is the pairing (off-diagonal) Hamiltonian
coupling momentum $\vec{k}$ with momentum $-\vec{k}$. Thus the topological invariant is

$$N = \frac{1}{2} \sum_n C_{1n} sgn(\Delta_{n\vec{k}}). \qquad (12.61)$$

There is a theory of the topological superconductor that can be derived from $4+1$–
dimensional Chern-Simons theory, on a spatial manifold $M_3 \times I$, where I is an interval,
with two $3+1$–dimensional boundaries that are topological s-wave (usual) *superconductors*, with phases for superconductors θ_L and θ_R (left and right). The effective field theory
for the topological superconductor is for the bosonic field: $\theta_L(x_\mu)$, $\theta_R(x_\mu)$ defined on the
boundaries ($\mu = 0, 1, 2, 3$) and $A_a = (A_\mu, A_4 = 0)$, with A_μ the physical electromagnetic
field. There are also fermions $c, \bar{c}$ living on the $3+1$–dimensional boundaries, which are
integrated out. There is a gauge invariance

$$c(x_a) \to c(x_a)e^{i\varphi(x_a)}; \qquad \delta A_a(x_b) = \partial_a \varphi;$$
$$\delta\theta_R(x_\mu) = 2\varphi(x_\mu, x_4 = 0); \qquad \delta\theta_L(x_\mu) = 2\varphi(x_\mu, x_4 = L), \qquad (12.62)$$

which can be fixed by the gauge transformation

$$\varphi(x_a) = -\frac{1}{2L_4}[\theta_L(L_4 - x_4) + \theta_R x_4], \qquad (12.63)$$

such that the fields θ_L, θ_R are canceled, and $\tilde{A}_a = A_a + \partial_a \varphi$. Therefore now

$$A_4 = 0 \to \tilde{A}_4 = \partial_4 \varphi = \frac{\theta_L - \theta_R}{2L_4}. \qquad (12.64)$$

The 4+1–dimensional Chern-Simons term reduces on the 3+1–dimensional boundary (left plus right) to the effective field theory,

$$S_{eff,3D} = S_{CS}[\tilde{A}_a] = \frac{1}{24\pi^2} \int d^{4+1}x \epsilon^{abcde} \tilde{A}_a \partial_b \tilde{A}_c \partial_d \tilde{A}_e$$

$$= \frac{1}{32\pi^2} \int d^{3+1}x \epsilon^{\mu\nu\rho\sigma} \frac{\theta_L - \theta_R}{2} F_{\mu\nu} F_{\rho\sigma}, \tag{12.65}$$

which is an "axion field theory" (using particle physics jargon) with "axion" $\theta_L - \theta_R$.

Besides this we have the usual Maxwell term, as well as the Higgs coupling to the electromagnetic field. One can also introduce a Josephson-type coupling $\cos(\theta_L - \theta_R)$, for a total effective action for the topological superconductor:

$$S_{eff,3D} = \int d^{3+1}x \left[\epsilon^{\mu\nu\rho\sigma} \frac{\theta_L - \theta_R}{64\pi^2} F_{\mu\nu} F_{\rho\sigma} - \frac{1}{4e^2} F_{\mu\nu} F^{\mu\nu} + \frac{1}{2}\rho_L(\partial_\mu \theta_L - 2A_\mu)^2 \right.$$

$$\left. + \frac{1}{2}\rho_R(\partial_\mu \theta_R - 2A_\mu)^2 + J\cos(\theta_L - \theta_R) \right]. \tag{12.66}$$

Important Concepts to Remember

- The Aharonov-Bohm effect corresponds to gauge fields at negligible field strength along a closed path having nontrivial effects due to $\oint_{C=\partial M} \vec{A} \cdot d\vec{x} = \int_M \vec{B} \cdot d\vec{S} \neq 0$.
- The Berry phase appears due to the adiabatic change of an eigenstate of the Hamiltonian on a cyclical path, as $i \oint_C \langle n[k(t)]|\nabla_k|n[k(t)]\rangle$.
- The Berry phase is the line integral of the Berry connection, with Berry curvature.
- In a solid, a canonical transformation on the system leads to an effective, statistical gauge field of Chern-Simons type, which changes the statistical properties of the wavefunction.
- Anyons are objects that under interchange multiply the wavefunction by an arbitrary phase $e^{i\theta}$ and appear in the presence of Chern-Simons gauge fields, which associate a delta function magnetic flux to the particles.
- In the Fractional Quantum Hall Effect we have quasi-particles with fractional electric charge and anyonic phase. The quasi-holes have charge $\nu = 1/m$ and anyonic phase $\pi\nu$, besides vorticity one.
- One can have particles with nonabelian statistics (nonabelian representation of the braiding group), called nonabelian anyons or nonabelions.
- Holomorphic wavefunctions of systems in 2+1 dimensions can be associated with conformal blocks of some CFT in two Euclidean (spatial) dimensions.
- One can construct the Laughlin wavefunction, its quasi-hole excitation, as well as abelian and nonabelian anyonic wavefunctions.
- A topological superconductor has fully gapped quasi-particle excitations in the bulk, but topologically protected gapless quasi-particle states propagating on the boundary.
- We can have boundary states be Majorana fermions, and also have nonabelian anyonic statistics.

- The topological field theory of a simple T-invariant superconductor has an axionic term (theta term), for the "axion," equal to the difference of the phases of the two superconducting phases $\theta_L - \theta_R$.

Further Reading

For an introduction to Chern-Simons theory, see the review [13]. For an introduction to anyons, see the reviews [14] and [15]. For a modern treatment of the fractional quantum Hall effect see E. Witten's lectures [16]. Nonabelian anyons in the context of the quantum Hall effect were defined in [17]. For a review of topological superconductors, see [18]. The topological field theory described here was defined in [19].

Where It will be Addressed in String Theory

The Berry phase and anyons will be addressed using the AdS/CFT correspondence in Chapter 44. The topological superconductor will be addressed in Chapter 41.

Exercises

(1) Consider a ring made of a regular superconductor, threaded by n fluxons. What is the Aharonov-Bohm phase around the ring? What about if the material shows fractional quantum Hall effect with quasi-holes of electric charge $1/r$?

(2) Consider a spin $1/2$ particle with magnetic moment μ in a magnetic field B. Calculate its eigenstates and the Berry connection that results from identifying $k(t)$ with angles.

(3) Show that for a system with statistical gauge field action (12.24), and with only electronic states (no fractional quasi-particles), we can have only bosonic or fermionic wavefunction descriptions.

(4) Consider the Abelian-Higgs model with a CS term (12.33). Show that the vortex solution is anyonic.

(5) Consider a topological superconductor with the effective action (12.66). Dimensionally reduce it to find the action on its 2+1–dimensional boundary.

13

Insulators

We have seen that the important mechanism for transport is momentum relaxation, i.e. how do the electrons lose their momentum, in the presence of an external electric field? In this chapter we ask the question: What turns a material into an insulator, i.e. with a very small conductivity? We know there is a transition from superconductor to insulator, which we have explored in Chapter 11, but the more critical question is how does the transition between a metal and an insulator happen? And how does one even characterize an insulator?

We will try to answer these questions in this chapter. There are two basic mechanisms for obtaining a metal-insulator phase transition. One is the mechanism described by Mott, leading to what is known as a Mott insulator, which mechanism is an extension of the analysis of the fermionic Hubbard model, hence we will start our discussion with this model. The other is the idea of Anderson localization due to scattering from a random potential generated by impurities. Finally, we will discuss material of important current research, topological insulators, with properties dictated by topology.

Band Insulator

Of course, the simplest example of an insulator would be a *band insulator*. As we described in Chapter 3, we can have a partially filled conduction band, allowing for metallic transport (the electrons are free to move onto unoccupied levels and hop between atoms), or we can have a fully filled conduction band, separated by a gap from the next band, not allowing for electronic transport, thus obtaining an insulator. In this case however, the system is always an insulator and does not present a phase transition.

Electric Transport: Drude Model

To characterize insulators, we must first understand electric transport. This is in general defined by a matrix relation between the current and the applied electric field, valid at each momentum and frequency:

$$j_\alpha(\omega, \vec{k}) = \sigma_{\alpha\beta}(\omega, \vec{k}) E_\beta(\omega, \vec{k}). \tag{13.1}$$

For a metal, the usual transport theory (Drude model) gives (from $d\vec{p}/dt = e\vec{E} - \vec{p}/\tau$ in the presence of a time $-= \vec{E}_0 e^{i\omega t}$, a generalization of the description in Section 3.1)

$$\sigma_{\alpha\beta}(\omega) = \frac{\sigma_{DC}}{1 - i\omega\tau}\delta_{\alpha\beta}, \tag{13.2}$$

where

$$\sigma_{DC} = \frac{ne^2\tau}{m^*}. \tag{13.3}$$

That then gives for the real part of the zero temperature, DC ($\omega \to 0$) conductivity

$$\text{Re}\left[\sigma_{\alpha\beta}(T=0, \omega \to 0)\right] = (D_c)_{\alpha\beta}\frac{\tau}{\pi(1+\omega^2\tau^2)}, \tag{13.4}$$

where from the above D_c is given as a function of the effective mass m_* of the electron by

$$(D_c)_{\alpha\beta} = \frac{\pi n e^2}{m_*}\delta_{\alpha\beta}, \tag{13.5}$$

and as the relaxation time τ becomes large (for a highly conductive metal), we obtain the *Drude peak* at zero frequency:

$$\text{Re}\left[\sigma_{\alpha\beta}(T=0, \omega \to 0, \tau^{-1} \to 0)\right] = (D_c)_{\alpha\beta}\delta(\omega). \tag{13.6}$$

Another way of explaining this result is that $\sigma = \sigma_{\text{DC}}/(1-i\omega\tau) \to i\tau^{-1}/\omega$ as $1/\tau \to 0$, and since for any variable

$$\frac{1}{x} = P\left(\frac{1}{x}\right) - i\pi\delta(x), \tag{13.7}$$

where P is the principal part, the imaginary part of σ having a $1/\omega$ pole implying the real part has a $\delta(\omega)$ piece.

This Drude peak is a result of the translational invariance in the extreme IR (for $\omega = 0$), leading to an infinite conductivity, and is a characteristic of metals. In fact, we can say that we could *define* metals by its presence.

We see that the key to the finiteness of σ is *momentum dissipation* (the Drude peak is obtained in the limit of $1/\tau \to 0$, when there is no momentum dissipation).

Then on the contrary, an insulator would be an object of negligible zero temperature DC conductivity:

$$\lim_{T\to 0}\lim_{\omega\to 0}\lim_{\vec{q}\to 0}\text{Re}\left[\sigma_{\alpha\beta}(T, \omega, \vec{q})\right] = 0. \tag{13.8}$$

To obtain it, we must break the translational invariance of the system even in the extreme IR, i.e. obtain momentum dissipation at all scales.

13.1 Fermionic Hubbard Model and Metal-Insulator Transition

The fermionic Hubbard model describes the features of the transition from metal to insulator. In this case, the metal will be the weakly coupled state, which is disordered, whereas the insulator will be the strongly coupled state, which will be ordered.

One considers the tight-binding approximation, when the electrons belong to each atom (site on the lattice), but they can "hop" between sites. The model has

$$H = -t\sum_{\langle ij\rangle,\sigma} c_{i\sigma}^{\dagger} c_{j\sigma} + U\sum_i n_{i\uparrow}n_{i\downarrow} - \mu\sum_i n_i, \tag{13.9}$$

where $n_{i\sigma} = c_{i\sigma}^{\dagger}c_{i\sigma}$ and $n_i = n_{i\uparrow} + n_{i\downarrow}$. Here $\langle ij\rangle$ signifies that the sum is taken only over nearest neighbors; the first term is then the "hopping" term for an electron to hop from site

j to neighboring site i, U is an on-site interaction to have two electrons (of opposite spins, thus obeying the Fermi exclusion principle) at the same site, and μ is a chemical potential.

We are mostly interested in the model at half-filling, that is $\langle n_i \rangle = 1$ (on the average, one electron per site instead of the allowed two, of opposite spins). It should also be at $\mu = 0$, since it should take no energy to add another electron.

Weak Coupling

Consider the $U \to 0$ limit ($U \ll t$). Go to the momentum basis (reciprocal space):

$$c_{\vec{k}\sigma} = \frac{1}{\sqrt{N}} \sum_l e^{i\vec{k}\cdot\vec{l}} c_{\vec{l}\sigma}. \tag{13.10}$$

Then the Hamiltonian becomes diagonal,

$$
\begin{aligned}
H &= -\frac{t}{N} \sum_{\vec{k},\vec{k}',\sigma} c_{\vec{k}\sigma}^\dagger c_{\vec{k}'\sigma} \left(e^{i\vec{k}\cdot\vec{j} - i\vec{k}'\cdot(\vec{j}+1)} + e^{i\vec{k}\cdot\vec{j} - i\vec{k}'\cdot(\vec{j}-1)} \right) \\
&= -t \sum_{\vec{k},\vec{k}',\sigma} c_{\vec{k}\sigma}^\dagger c_{\vec{k}'\sigma} (2\cos k_x' + 2\cos k_y') \delta_{\vec{k}\vec{k}'} \equiv \sum_{\vec{k}} \epsilon_{\vec{k}} c_{\vec{k}\sigma}^\dagger c_{\vec{k}\sigma},
\end{aligned} \tag{13.11}
$$

where the eigenenergies are

$$\epsilon_{\vec{k}\sigma} = -2t(\cos k_x + \cos k_y) - \mu, \tag{13.12}$$

and we have added for completeness a chemical potential. Note that in one dimension we have $k = 2\pi n/N$, and in general the momenta are discrete. So we obtain an electron energy band of width $4t$, and the electrons are approximately free.

One can analyze more precisely the Hamiltonian and find that the true ground state has antiferromagnetic order.

Strong Coupling

Consider now instead the strong coupling $U \gg t$ limit (or $t \to 0$). Since we are at small t/U, we can do perturbation theory in it. Consider an effective Heisenberg Hamiltonian for the model,

$$H_{\text{eff}} = J \sum_{\langle ij \rangle} \vec{S}_i \cdot \vec{S}_j. \tag{13.13}$$

If we have two electrons in a singlet state ($\uparrow\downarrow$) and an unoccupied neighbor, *virtual* hopping (tunneling), followed by return to the original state will give an energy of $-2t^2/U$, where the 2 comes from the two possible configurations ($\uparrow\downarrow$ and $\downarrow\uparrow$), $-t$ from the energy of the hopping and t/U from the probability. This energy is to be compared with $JS_1^- S_2^+/2$, which fixes

$$J = 4t^2/U. \tag{13.14}$$

One can do a much more rigorous analysis, and obtain

$$H_{\text{eff}} = \frac{4t^2}{U} \sum_{\langle ij \rangle} \left(\vec{S}_i \cdot \vec{S}_j - \frac{n_i n_j}{4} \right), \tag{13.15}$$

for an antiferromagnetic coupling Heisenberg Hamiltonian.

But since the Heisenberg Hamiltonian for $J > 0$ has an antiferromagnetic ground state, the strong coupling limit has an antiferromagnetic behavior. This would be an insulator, which in fact will be identified with a "Mott insulator" later on, since there is little hopping, i.e. transport, allowed. In the middle there will be a phase transition, which will be called a "Mott transition" between metal and insulator.

Mean Field Approximation in $1+1$ Dimensions
We want now to find a mean field description of the 1-dimensional model. Consider that each electron feels an average interaction from the others, and expand

$$n_{i\uparrow} = \langle n_i \rangle + \delta n_{i\uparrow}, \tag{13.16}$$

where δn is considered small, which means that the interaction term is

$$n_{l\uparrow} n_{l\downarrow} \simeq \langle n_{l\uparrow} \rangle \langle n_{l\downarrow} \rangle + \delta n_{l\uparrow} \langle n_{l\downarrow} \rangle + \delta n_{l\downarrow} \langle n_{l\uparrow} \rangle \simeq n_{l\uparrow} \langle n_{l\downarrow} \rangle + n_{l\downarrow} \langle n_{l\uparrow} \rangle - \langle n_{l\downarrow} \rangle \langle n_{l\uparrow} \rangle. \tag{13.17}$$

Substituting in the Hamiltonian, and going to momentum space, we obtain

$$\begin{aligned} H &= \sum_k \left[-t \cos k (n_{k\downarrow} + n_{k\uparrow}) + U(n_{k\uparrow} \langle n_{k\downarrow} \rangle + n_{k\downarrow} \langle n_{k\uparrow} \rangle) \right] - NU \langle n_\uparrow \rangle \langle n_\downarrow \rangle \\ &= \sum_k [\epsilon_{k\uparrow} n_{k\uparrow} + \epsilon_{k\downarrow} n_{k\downarrow}] - NU \langle n_\uparrow \rangle \langle n_\downarrow \rangle, \end{aligned} \tag{13.18}$$

where the eigenenergies are

$$\epsilon_{k\uparrow\downarrow} = -2t \cos k + U \langle n_{\uparrow\downarrow} \rangle. \tag{13.19}$$

Doing the sums using $\sum_k = (N/\pi) \int^{k_F} dk$, so that $k_F = \pi \langle n \rangle / N$, we obtain the total energy of the system:

$$E = -\frac{2Nt}{\pi} \left[\sin(\pi \langle n_\downarrow \rangle) + \sin(\pi \langle n_\uparrow \rangle) \right] + NU \langle n_\uparrow \rangle \langle n_\downarrow \rangle. \tag{13.20}$$

Since we are at half-filling, $\langle n \rangle = \langle n_\downarrow \rangle + \langle n_\uparrow \rangle = 1$, so

$$E(\langle n_\uparrow \rangle) = \frac{4Nt}{\pi} \left[-\sin(\pi \langle n_\uparrow \rangle) + \frac{U}{t} \frac{\pi}{4} \langle n_\uparrow \rangle (1 - \langle n_\uparrow \rangle) \right]. \tag{13.21}$$

This would imply the system goes from a paramagnet at small U/t (with $\langle n_\uparrow \rangle = \langle n_\downarrow \rangle = 1/2$ energetically preferred) to a ferromagnet at large U/t, with $\langle n_\uparrow \rangle = 0$ or 1 preferred. But this is wrong: In fact, there is an exact solution to the 1-dimensional model, and at small U/t an antiferromagnetic state forms, and as we saw also at large U/t we have an antiferromagnetic state. No exact solution is known in $d > 1$ dimensions.

So the phase diagram of the half-filled Hubbard model, for the temperature vs. U/t, has a kind of dome covering an antiferromagnetic state that extends from small U/t to very large U/t, and at large U/t, down to very small temperatures, we have a "Mott insulator state."

13.2 Defects and Anderson Localization

The insight of Anderson was to see that disorder, in the form of a random potential, induces an Anderson transition from delocalization to localization, in any spatial dimension higher than 2. In $d = 1$ and $d = 2$ spatial dimensions, *any amount of disorder localizes all the electronic states.* We will see that the Anderson transition occurs in the strong-disorder limit, so perturbation theory (weak-localization) is not very useful.

For the random potential, the model Anderson took is a tight binding model, with electrons at site i in electronic (orbital) states, with a probability for transition (hopping) given by V:

$$H = \sum_n \epsilon_n a_n^\dagger a_n + V \sum_{\langle nm \rangle} a_n^\dagger a_m. \tag{13.22}$$

In general, we don't need to take only nearest-neighbor hopping, and V depends on the distance, but it needs to decay fast enough at infinity. This Hamiltonian corresponds to conduction in an impurity band where the site energy ϵ_n is attributed randomly. Randomness in the value of V does not change the problem qualitatively. One can extend the site labels j to site and orbital levels $(j\mu)$.

The random site energies are distributed with some continuous probability, which in the original Anderson model was taken to be a simple Heaviside function, i.e. a uniform distribution of width W:

$$P(\epsilon_j) = \frac{1}{W} \theta \left(\frac{W}{2} - |\epsilon_j| \right). \tag{13.23}$$

Note that if $W = 0$, all states have the same energy, and transport is ballistic (unchallenged) in this regime. But when $V = 0$, the sites are not connected, so transport stops. Therefore, the localization/delocalization transition should depend on the ratio W/V. In $d > 2$, as expected, there is a critical W/V, $(W/V)_c$, at which we have a change between localization and delocalization. But in $d = 1$ and $d = 2$, a single impurity implies localization.

However, there is no rigorous proof of these statements, only various arguments and numerical simulations.

Transport

Transport is determined by the transition probability of an electron of energy E from position x to position x',

$$P(x, x'; E) = |\langle x|G(E)|x' \rangle|^2 = |c_{xx'}(E)|^2, \tag{13.24}$$

where $c_{xx'}$ is the probability amplitude, $G(E)$ is the Green's function,

$$G(E) = (E - H)^{-1} = \sum_n \frac{|n\rangle \langle n|}{E - \epsilon_n}, \tag{13.25}$$

satisfying the Dyson equation,

$$(E - \epsilon_i)G_{ij} = \delta_{ij} + \sum_{i \neq k} V_{ik} G_{kj}(E), \tag{13.26}$$

which is solved perturbatively for the diagonal elements as

$$G_{ii}(E) = \frac{1}{E - \epsilon_i} + \sum_{k \neq i} \frac{V_{il}V_{1j}}{(E - \epsilon_i)(E - \epsilon_l)} + \cdots$$

$$\equiv \frac{1}{E - \epsilon_i - \Sigma_i(E)}, \tag{13.27}$$

where $\Sigma_i(E)$ is a self-energy.

Then we can define the electron wavefunction in the material,

$$|\psi\rangle = \sum_i c_{ij}|j\rangle, \tag{13.28}$$

and its localization with localization length λ means that $\psi(r) = f(r)e^{-r/\lambda}$, which is to say that

$$c_{ij}(E) \equiv \langle i|G(E)|j\rangle \propto e^{-\frac{|i-j|}{\lambda}}. \tag{13.29}$$

In numerical simulations one indeed obtains such a behavior.

Scaling Arguments

Consider the conductance (inverse resistance) $G = 1/R$, a function of the linear dimension L of the sample, and the rescaled conductance

$$g(L) = \frac{2\hbar}{e^2}G(L). \tag{13.30}$$

In d spatial dimensions, Ohm's law is written as $G(L) = \sigma L^{d-2}$, where σ is the conductivity, so that for $d = 3$, we have the well-known linear relation. Considering $g(L)$ as a coupling constant, we can define a beta function for it:

$$\beta = \frac{d \ln g(L)}{d \ln L}. \tag{13.31}$$

It will have zeroes $g = g_c$, $\beta(g_c) = 0$, which will correspond to transition points, and we can linearize it around them,

$$\beta(g) \simeq \frac{g - g_c}{\nu g_c}, \tag{13.32}$$

which stands as a definition of ν. This can then be integrated to give

$$\frac{L}{L_0} = \left(\frac{g - g_c}{g_0 - g_c}\right)^\nu, \tag{13.33}$$

with $g_0 = g(L_0)$.

Defining the length scale

$$\xi = L_0 \left(\frac{g_c}{g_0 - g_c}\right)^\nu \equiv L_0|\epsilon|^{-\nu}, \tag{13.34}$$

which defines ϵ as well, we can write the scaling relation for the conductance near the transition point g_c:

$$g = g_c \left(1 - \left(\frac{L}{\xi} \right)^{\frac{1}{\nu}} \right). \tag{13.35}$$

As we see from its definition, ξ diverges as $g_0 \to g_c$, i.e. at the transition point *approached from the insulating side*, and thus can be identified with the localization length λ, which also diverges at the transition point. This is a hypothesis at the basis of this scaling theory.

At long enough length scales *from the metallic (conducting) side of the transition*, $g = \sigma L^{d-2}$, as we said (Ohm's law), but more precisely as we saw from the (13.35) at $L \ll \xi$, it is a function of L/ξ, so $\sigma \propto \xi^{2-d}$, and since $\xi \equiv L_0 |\epsilon|^{-\nu}$, we have

$$\sigma \propto |\epsilon|^{(d-2)\nu}, \tag{13.36}$$

which means the conductivity on the metallic side vanishes smoothly at the transition point (since $\epsilon \to 0$ as $g \to g_c$).

13.3 Mott Insulators

As we saw in the first section, the Hubbard model predicts the existence of a type of insulator at strong coupling, via an antiferromagnetic state, that was called a "Mott insulator." Therefore a Mott insulator has a strong electron coupling (the electrons are strongly correlated) and is difficult to describe analytically.

This state can be described independently of the Hubbard model, and is in fact described by further properties ("Mottness"). Most of these, however, are difficult to describe here, so we will not attempt to do so. What we can say in generality is that the state stops conducting at low temperature or high pressure, despite classical (weak coupling) theory predicting conductance. Thus band gap theory (weakly coupled, using only Fermi exclusion to define the bands) predicts a metal, but we have an insulator instead. The localization of the electrons is induced by interaction. When the distance between the atoms is larger than the "Bohr radius" of the orbitals, the electrons cannot "hop," and we have an insulator.

Under the name Mott insulator we can find objects with the presence or absence of a spontaneously broken symmetry (like the antiferromagnetic state in the Hubbard model), gapped or gapless neutral particle low-energy excitations, and the presence or absence of topological order or fractional charge states.

The Mott phase transition between metal and insulator is a quantum phase transition, since it occurs at $T = 0$. It is generally first order, though second order examples exist. In a (T, p) phase diagram, we have a phase transition line that goes all the way to $T = 0$, $p = p_c$. Mott proposed a pseudo-order parameter ϵ, as the difference between the ionization energy I of an electron in the crystal and the electron affinity E of *one* atom in the crystal:

$$\epsilon = I - E. \tag{13.37}$$

If there are electron-electron interactions, we have by definition $I > E \to \epsilon > 0$, otherwise $I = E$, $\epsilon = 0$. This will be a function of the lattice spacing d. As a function of $1/d$, ϵ drops toward small values, and can:

- Have a finite value at the transition, then drop to zero for $1/d < 1/d_c$, giving a first order phase transition.
- Go to zero smoothly, obtaining a second order phase transition.

To calculate analytically the strong interaction between electrons is notoriously difficult, though there exist techniques that are used. One such technique is the density functional theory (in a Hartree-Fock approximation). One considers the full energy of the multielectron system, kinetic plus potential, due to the crystal (external), interelectron interaction, and (Fermi) exchange energy,

$$E[\rho(\vec{r})] = E_{\text{kinetic}}[\rho(\vec{r})] + \int d^3\vec{r} V_{\text{ext}}(\vec{r})\rho(\vec{r}) + \frac{1}{2}\int d^3\vec{r}d^3\vec{r}' \frac{\rho(\vec{r})\rho(\vec{r}')}{|\vec{r}-\vec{r}'|} + E_{\text{exchange}}[\rho(\vec{r})],$$

(13.38)

and one writes an effective theory for a single electron in an effective potential, with Schrödinger equation

$$\left[\frac{\hbar^2\vec{\nabla}^2}{2m} + V_{\text{eff}}(\vec{r})\right]\Psi_i(\vec{r}) = \epsilon_i\Psi(\vec{r}),$$

(13.39)

where

$$V_{\text{eff}}(\vec{r}) = V_{\text{ext}}(\vec{r}) + \int d^3\vec{r}' \frac{\rho(\vec{r}')}{|\vec{r}-\vec{r}'|} + \frac{\delta E_{\text{exchange}}[\rho(\vec{r})]}{\delta\rho(\vec{r})}.$$

(13.40)

The density functional $\rho(\vec{r})$ is defined by

$$\rho(\vec{r}) = \sum_i f(\epsilon_i)|\Psi_i(\vec{r})|^2,$$

(13.41)

and one extremizes over the functions $f(\epsilon_i)$.

13.4 Topological Insulators

Topological insulators are solids for which the bulk behaves like an insulator, having fully gapped states, but the surface contains conducting states. In the bulk, we can have a band insulator, i.e. of the usual (not Mott or Anderson) type, where strong interactions are not necessary to describe it; however, we must have the surface states be protected by symmetry, especially time-reversal invariance. They are characterized by a topologically invariant index ($\mathbb{Z}_2$ index). Topological insulators were experimentally discovered (after being predicted) and have potential application in quantum computation.

As a function of momentum, the valence and conduction bands of insulators are oppositely curved, being closest at $k = 0$. The surface states form a double cone tangent to the bands, which therefore makes them touch each other at the tip of the cones, allowing transport. The Fermi surface, which is situated between the valence and conduction bands in the insulator, crosses the surface state bands.

The time-reversal symmetry operator T is anti-unitary, i.e. it acts on $c\psi$ as $c^*T\psi$. In a time-reversal invariant insulator, for every Bloch state $\psi_n(\vec{k}, \sigma)$, there will be an energy-degenerate state:

$$T\psi_n(\vec{k}, \sigma) = \psi_n'(-\vec{k}, -\sigma). \tag{13.42}$$

If however, $-\vec{k} = \vec{k} + \vec{G}$, with $\vec{G}$ in the dual lattice, i.e. at T-invariant states, then energy levels are doubly-degenerate, i.e. surface states must cross there.

Like in the case of topological superconductors, there are various possible types of topological insulators. We will describe some cases where the topology is simple to understand.

The 2-dimensional topological insulator, also called the *quantum spin Hall insulator*, can be realized, though in a very weak sense, in graphene. It can also be realized in some *quantum well* systems, like HgTe/CdTe, which are made by a layer of a material sandwiched between two slabs of a different material, with a wider bandgap.

Three-dimensional topological insulators have surface states that are 2-dimensional massless Dirac fermions, and whose dispersion relation ($\omega(k)$) forms a Dirac cone, on which a single massless Dirac fermion lives.

2+1–Dimensional Topological Field Theory

For a band insulator with M occupied bands, we consider the abelian (diagonal) part of the Berry connection,

$$a_j = -i \sum_{\alpha \in occupied} \left\langle \alpha\vec{k} \left| \frac{\partial}{\partial k_j} \right| \alpha\vec{k} \right\rangle; \quad j = x, y. \tag{13.43}$$

For the insulator, the longitudinal conductivity vanishes, $\sigma_{xx} = 0$, but the off-diagonal Hall conductivity is given by the first Chern number of the Berry connection,

$$\sigma_H = \sigma_{xy} = \frac{e^2}{2\pi h} \int dk_x dk_y f_{xy}(\vec{k}) = \frac{e^2}{h} C_1, \tag{13.44}$$

where C_1 is the first Chern number:

$$C_1 = \frac{1}{2\pi} \int dk_x dk_y f_{xy}(\vec{k}) \in \mathbb{Z}. \tag{13.45}$$

Therefore the Hall conductivity is quantized in units of $\sigma_0 = e^2/h$, i.e. we have an integer quantum Hall effect.

The Hall response is written more completely as

$$j_i = \sigma_H \epsilon^{ij} E_j, \tag{13.46}$$

but moreover, it also implies (using current conservation and the Maxwell equations)

$$\partial_0 \rho = -\partial_i j^i = -\sigma_H \epsilon^{ij} \partial_i E_j = \sigma_H \partial_0 B \Rightarrow \rho(B) - \rho_0 = \sigma_H B. \tag{13.47}$$

Note that in 2+1 dimensions, we have $F_{0i} = E_i$ and $F_{ij} = B\epsilon_{ij}$. Therefore the equations for j^i and $\rho(B)$ fit together into j^μ as the Lorentz covariant equation:

$$j^\mu = \frac{e^2}{\hbar}\frac{C_1}{2\pi}\epsilon^{\mu\nu\rho}\partial_\nu A_\rho. \tag{13.48}$$

This is a response equation (i.e., the current following from a gauge field input), and considering that in general we can define the current as $j^\mu = \delta S/\delta A_\mu$, the response is encoded in the topological Chern-Simons field theory:

$$S_{\text{eff}} = \frac{e^2}{\hbar}\frac{C_1}{4\pi}\int d^{2+1}x\epsilon^{\mu\nu\rho}A_\mu\partial_\nu A_\rho. \tag{13.49}$$

All topological responses of the Quantum Hall state are encoded in this topological field theory, which is the product of the first Chern number (a topological object, an integer) in the two spatial directions of the Berry connection a_i of the state, with the CS action for the electromagnetic field A_μ.

4+1–Dimensional Topological Field Theory

One can then do something similar and define a fictitious 4+1–dimensional topological insulator, and then dimensionally reduce to 3+1 dimensions to find the properties of a real topological insulator.

One can define a band Hamiltonian in 4+1–dimensions, though the details are not important for us, and similarly to the above find the effective action that encodes the topological response of the 4+1–dimensional theory. Not surprisingly, it will simply be the 4+1–dimensional Chern-Simons action:

$$S_{eff} = \frac{C_2}{24\pi^2}\int d^{4+1}dx\epsilon^{\mu\nu\rho\sigma\tau}A_\mu\partial_\nu A_\rho\partial_\sigma A_\tau. \tag{13.50}$$

Here C_2 is the second Chern number of the now *nonabelian* Berry connection:

$$a_j^{\alpha\beta} = -i\left\langle\alpha,\vec{k}\left|\frac{\partial}{\partial k_j}\right|\beta,\vec{k}\right\rangle. \tag{13.51}$$

The second Chern number C_2 is the integral over the four spatial dimensions,

$$C_2 = \frac{1}{32\pi^2}\int d^4x\epsilon^{ijkl}\,\text{Tr}[f_{ij}f_{kl}], \tag{13.52}$$

where $f_{ij}^{\alpha\beta} = \partial_i a_j^{\alpha\beta} - \partial_j a_i^{\alpha\beta} + i[a_i, a_j]^{\alpha\beta}$ is the field strength of the nonabelian Berry connection.

The topological response of the system to the electromagnetic field is then

$$j^\mu = \frac{C_2}{8\pi^2}\epsilon^{\mu\nu\rho\sigma\tau}\partial_\nu A_\rho\partial_\sigma A_\tau. \tag{13.53}$$

Dimensional Reduction to 3+1–Dimensional Time Reversal Invariant Insulators

We can now do a dimensional reduction to a physical 3+1–dimensional topological insulator. For that, consider a system that is independent on the fourth spatial coordinate, which we denote by θ_0, and consider an A_4 component of the gauge field that is now replaced by the *field*

$$\delta\theta(\vec{x}, t) = \theta(\vec{x}, t) - \theta_0, \tag{13.54}$$

so the effective action becomes

$$S_{\text{eff,3D}} = \frac{G_3(\theta_0)}{4\pi} \int d^{3+1}x\epsilon^{\mu\nu\rho\sigma} \delta\theta \partial_\mu A_\nu \partial_\rho A_\sigma. \tag{13.55}$$

The $G_3(\theta_0)$ coefficient is determined from the Berry curvature as

$$G_3(\theta_0) = \frac{1}{8\pi^2} \int d^3k\epsilon^{ijkl} \text{Tr}[f_{ij}f_{kl}], \tag{13.56}$$

where the Berry connection is defined in the *four space* dimensions $(k_x, k_y, k_z, \theta_0)$,

$$a_i^{\alpha\beta} = -i\left\langle \vec{k}, \theta_0; \alpha \left| \frac{\partial}{\partial k_i} \right| \vec{k}, \theta_0; \beta \right\rangle; \quad a_\theta^{\alpha\beta} = -i\left\langle \vec{k}, \theta_0; \alpha \left| \frac{\partial}{\partial \theta_0} \right| \vec{k}, \theta_0; \beta \right\rangle, \tag{13.57}$$

and now

$$\int d\theta_0 G_3(\theta_0) = C_2 \in \mathbb{Z}. \tag{13.58}$$

Define a generalized charge polarization of the 3+1–dimensional system $P_3(\theta_0)$, such that

$$G_3(\theta_0) = \frac{\partial P_3(\theta_0)}{\partial \theta_0}. \tag{13.59}$$

Substituting this formula inside the 3+1–dimensional integral in the effective action and integrating by parts to put the ∂_ν derivative on $\delta\theta$, we obtain

$$S_{\text{eff,3D}} = \frac{1}{4\pi} \int d^{3+1}x\epsilon^{\mu\nu\rho\sigma} A_\mu \frac{\partial P_3}{\partial\theta} \partial_\nu \delta\theta \partial_\rho A_\sigma, \tag{13.60}$$

and noting that

$$\frac{\partial P_3}{\partial\theta} \partial_\nu \delta\theta = \partial_\nu P_3(\theta(\vec{x}, t)), \tag{13.61}$$

and then partially integrating back the derivative on P_3 onto A_μ, we finally get

$$S_{\text{eff,3D}} = \frac{1}{4\pi} \int d^{3+1}x\epsilon^{\mu\nu\rho\sigma} P_3(\vec{x}, t)\partial_\mu A_\nu \partial_\rho A_\sigma. \tag{13.62}$$

In an analogy with particle physics (specifically, the theta term of QCD), this would be a *theta term (axionic) in the effective action for electromagnetism inside the topological insulator*, with P_3 playing the role of the axion.

To understand the usefulness of the above term in the effective action (13.62), among its physical consequences are the following:

1. The Hall effect induced by a spatial gradient of P_3 (the axion). Consider $P_3 = P_3(z)$ and calculate the current response from this effective action. This leads to

$$j^\mu = \frac{\delta S_{\text{eff,3D}}}{\delta A_\mu} = \frac{\partial_z P_3}{2\pi}\epsilon^{\mu\nu\rho}\partial_\nu A_\rho; \quad \mu, \nu, \rho = t, x, y. \tag{13.63}$$

This implies a Hall conductivity

$$\sigma_{xy} = \frac{j^x}{E_y} = \frac{\partial_z P_3}{2\pi}.$$
(13.64)

2. Topological magneto-electric effect. Consider instead $P_3 = P_3(t)$. Then similarly,

$$j^i = -\frac{\partial_t P_3}{2\pi}\epsilon^{ijk}\partial_j A_k; \quad i,j,k = x,y,z,$$
(13.65)

or in other words

$$\vec{j} = -\frac{\partial_t P_3}{2\pi}\vec{B}.$$
(13.66)

Since $\vec{j} = \partial_t \vec{P}$, the relation above integrates to

$$\vec{P} = -\frac{\vec{B}}{2\pi}(P_3 + \text{const.}).$$
(13.67)

That means that the magnetic field induces a charge polarization, i.e. we have a magneto-electric effect.

Important Concepts to Remember

- Regular, perturbative insulators are band insulators, due to a fully occupied band.
- A metal-insulator phase transition is described by the Hubbard model, with a hopping term t and an interaction term U, as a function of U/t. It has an antiferromagnetic states, and transitions between a metallic state with a band structure at small coupling to a "Mott insulator state" at large coupling.
- Another mechanism for a phase transition between metal and insulator is the Anderson localization due to impurities. In $d = 1$ and $d = 2$ spatial dimensions, any impurity leads to Anderson localization, and in $d > 2$ we have a phase transition, though there are no rigorous proofs of these statements.
- The localization length, defined as the scale of the exponential decay of the transition amplitude, diverges at the phase transition point, approached from the insulator side.
- When approaching from the metallic side of the Anderson transition, the conductivity vanishes smoothly.
- A Mott insulator can be defined independently of the Hubbard model, as a zero temperature (quantum) phase transition between metal and insulator, and is due to a strong interaction between electrons. It is generally first order, though it can also be second order.
- The Mott phase transition is difficult to obtain, though approximation methods like functional density theory can be used.
- A topological insulator is an insulator in the bulk, with fully gapped states, but having surface conducting states protected by topology and symmetries, especially time-reversal invariance. They are also characterized by a $\mathbb{Z}_2$ topological invariant index.
- The surface states are cones, extending out of the bulk bands, touching each other's tips, and thus allowing transport (the Fermi energy crosses them).

- A 2+1–dimensional (time-reversal invariant) topological insulator has an electromagnetic Chern-Simons topological field theory encoding its response.
- A 3+1–dimensional (time-reversal invariant) insulator is obtained from dimensional reduction of a 4+1–dimensional insulator with a Chern-Simons effective action encoding its response. One obtains an axionic (theta term) topological field theory encoding its response.

Further Reading

For Anderson localization, see, for instance, chapter 12 in [7]. A field theory of topological insulators was developed in [20].

Where It will be Addressed in String Theory

Insulators will be addressed holographically in Chapter 47, and strange metals in Chapter 48. Topological insulators will be addressed in string theory in Chapter 41, specifically Subsection 41.1.1.

Exercises

(1) Add a small magnetic field to the Fermionic Hubbard model at strong and weak coupling and argue whether there are any qualitative physics changes.

(2) Write the formal expression for the self-energy $\Sigma_i(E)$ for the conduction electrons in the Anderson model, resulting from the Dyson equation.

(3) Why would $f(\epsilon_i) = \delta(\epsilon - \epsilon_0)$ (pure ground state) not be the needed solution for the density functional minimization in (13.41)?

(4) Consider the 2+1–dimensional topological insulator effective action and reduce it onto a 1+1–dimensional boundary. Derive physical consequences of the resulting action.

(5) Consider the dimensional reduction onto a 2+1–dimensional boundary of the effective action (13.62) for the 3+1–dimensional topological insulator and compare with the 2+1–dimensional topological insulator. What are the differences?

14 The Kondo Effect and the Kondo Problem

An important question about transport is the effect of local magnetic moments, i.e. magnetic impurities, on electric transport, specifically the resistivity.

It is known that for certain metals with magnetic impurities, the resistivity decreases as the temperature decreases, but then reaches a minimum, and after that it increases as $-\ln T$ as the temperature decreases further. Moreover, the temperature dependence doesn't increase indefinitely, but disappears below a characteristic temperature called the Kondo temperature T_K.

Also the magnetic and spin properties change in the neighborhood of the Kondo temperature. The magnetic susceptibility follows a Curie $1/T$ law above T_K, and tends to a constant below T_K. Moreover, at T_K the impurity and conduction electron spins condense into singlet states, thus having a vanishing of the local magnetic moment, the opposite of ferromagnetic behavior.

The Kondo model (Jun Kondo, 1964) was to describe the above behavior in a perturbation theory to second order in J, an exchange interaction between the local impurity spin and the conduction electrons. It will obtain a minimum of the resistivity as a function of temperature, but as $T \to 0$, we have a diverging resistivity (as $-\ln T$). The perturbation theory breaks down at T_K, so Kondo's solution is valid only for $T \gg T_K$. Finding a solution to this divergence is known as the *Kondo problem*.

Anderson introduced in 1970 a scaling hypothesis ("poor man's scaling") showing that the exchange interaction increases in magnitude as one includes the effects of more high-energy excitations. Finally in 1975 Wilson used his (then recently found) renormalization group procedure to understand the Kondo problem, though a full analytical understanding is still lacking.

14.1 The Kondo Hamiltonian and Impurity-Conduction Spin Interaction

The Kondo model is based on an interaction between the spin of the conduction electrons and the spin of the impurities:

$$H_K = \sum_{\vec{k},\sigma} \epsilon_{\vec{k}} n_{\vec{k}\sigma} - \sum_{\vec{k},\vec{k}'} \frac{J_{\vec{k},\vec{k}'}}{\hbar^2} (\psi_{\vec{k}'}^\dagger \vec{S} \psi_{\vec{k}}) \cdot (\psi_i^\dagger \vec{S} \psi_i). \tag{14.1}$$

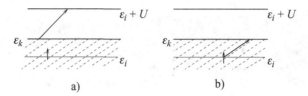

Impurity electron at ϵ_i, conduction electron at $\epsilon_{\vec{k}}$ and excited state at $\epsilon_i + U$. (a) Process where first a conduction electron jumps to an excited state. (b) Process where first the impurity electron jumps to a conduction state.

Here

$$\psi_{\vec{k}} = \begin{pmatrix} a_{\vec{k}\uparrow} \\ a_{\vec{k}\downarrow} \end{pmatrix} \tag{14.2}$$

is a doublet for the conduction electrons, with $\psi_{\vec{k}'}^{\dagger} \vec{S} \psi_{\vec{k}}$ the transition spin operator between conduction states $\vec{k}$ and $\vec{k}'$, and

$$\psi_i = \begin{pmatrix} a_{i\uparrow} \\ a_{i\downarrow} \end{pmatrix} \tag{14.3}$$

is a doublet for the impurity electrons, with $\psi_i^{\dagger} \vec{S} \psi_i$ the spin operator for impurity state i. Here as usual $\vec{S} = \hbar \vec{\sigma}/2$, with $\vec{\sigma}$ being the Pauli matrices. We note the similarity with Heisenberg's spin chain model, and note that $J_{\vec{k}\vec{k}'} < 0$ corresponds to an antiferromagnetic interaction.

We actually have $J_{\vec{k}\vec{k}'} < 0$, as we now show. The coefficient is related to the amplitude for spin flip between a conduction electron and an impurity electron,

$$M_{(\vec{k},\sigma)(i,-\sigma) \to (\vec{k}',-\sigma)(i,-\sigma)} = -\frac{1}{2} J_{\vec{k}\vec{k}'}, \tag{14.4}$$

as we can easily see by considering $\sigma_x = \begin{pmatrix} 1 & 0 \\ 0 & 1 \end{pmatrix}$ and $\sigma_y = \begin{pmatrix} 0 & -i \\ i & 0 \end{pmatrix}$ in the Kondo Hamiltonian (14.1) and considering only the components $a_{\vec{k}\downarrow}$, $a_{\vec{k}'\uparrow}$, $a_{i\uparrow}$, $a_{i\downarrow}$. More precisely, write the interacting part of the Hamiltonian H' as

$$H' = -\sum_{\vec{k},\vec{k}',\sigma} \frac{J_{\vec{k}\vec{k}'}}{2\hbar} [\psi_{i\sigma}^{\dagger} S_i^z \psi_{i\sigma} (a_{\vec{k}'\uparrow}^{\dagger} a_{\vec{k}\uparrow} - a_{\vec{k}'\downarrow}^{\dagger} a_{\vec{k}\downarrow}) + \psi_{i\sigma}^{\dagger} S_i^+ \psi_{i\sigma} (a_{\vec{k}'\downarrow}^{\dagger} a_{\vec{k}\uparrow}) + \psi_{i\sigma}^{\dagger} S_i^- \psi_{i\sigma} (a_{\vec{k}'\uparrow}^{\dagger} a_{\vec{k}\downarrow})], \tag{14.5}$$

where $S_i^{\pm} = S_i^x \pm i S_i^y$.

But this spin flip process is obtained in second order perturbation theory, through an intermediate state, in which one has generically

$$M \sim V_{\vec{k}i} \frac{1}{E_{\text{initial}} - E_{\text{intermediate}}} V_{i\vec{k}}. \tag{14.6}$$

We apply this to the case that the impurity energy ϵ_i is situated in the middle of the conduction band, the conduction states $\epsilon_{\vec{k}}$ and $\epsilon_{\vec{k}'}$ are at the upper side of it, and the excited impurity state is above the conduction band, at $\epsilon_i + U$, as in Figure 14.1.

Then consider the two possible processes: first the conduction electron jumps to the (opposite spin) excited impurity state, then the impurity electron jumps to the conduction electron state, as in Figure 14.1(a); or, first the impurity electron jumps to the (opposite spin) conduction electron state, and then the conduction electron jumps to the excited impurity state, as in Figure 14.1(b). In the first process, $E_{\text{intermediate}} = 2\epsilon_i + U$, in the second $E_{\text{intermediate}} = \epsilon_{\vec{k}} + \epsilon_{\vec{k}'}$, and in both $E_{\text{initial}} = \epsilon_i + \epsilon_{\vec{k}}$. In both cases, there is a minus sign in the amplitude due to the exchange (in the final state) of the two electrons (fermions) between the conduction and the impurity. Adding up the two processes, we get

$$
M_{(\vec{k},\sigma)(i,-\sigma)\to(\vec{k}',-\sigma)(i,-\sigma)} = -V_{\vec{k}i}V_{i\vec{k}'}\left(\frac{1}{\epsilon_{\vec{k}} - \epsilon_i - U} + \frac{1}{\epsilon_i - \epsilon_{\vec{k}'}}\right), \qquad (14.7)
$$

meaning

$$
J_{\vec{k}\vec{k}'} = 2V_{\vec{k}i}V_{i\vec{k}'}\left(\frac{1}{\epsilon_{\vec{k}} - \epsilon_i - U} + \frac{1}{\epsilon_i - \epsilon_{\vec{k}'}}\right). \qquad (14.8)
$$

Consider that both conduction electron states are around the Fermi energy, $\epsilon_{\vec{k}} \simeq \epsilon_{\vec{k}'} \simeq \epsilon_F$, and the impurity states ϵ_i and $\epsilon_i + U$ are symmetric around ϵ_F, and since then $V_{\vec{k}i} = (V_{i\vec{k}'})^*$, we obtain

$$
J_{\vec{k}\vec{k}'} = -8\frac{|V_{\vec{k}i}|^2}{U} < 0. \qquad (14.9)
$$

In the following we will assume that it is momentum independent: $J_{\vec{k}\vec{k}'} = J < 0$.

14.2 Electron-Impurity Scattering and Resistivity Minimum; Kondo Temperature

To calculate the resistivity of the material, we calculate the scattering of the conduction electrons off impurities.

We are interested in the process where only the momentum of the conduction electron changes, but not the spins or the state of the impurity, i.e. the process $(\vec{k}\uparrow)(i\downarrow) \to (\vec{k}'\uparrow)(i\downarrow)$. Note that because of the antiferromagnetic interaction, we pair the conduction and impurity spins to be opposite, since this configuration gives the smallest energy. There is also another process, where the spin of the conduction electrons is flipped, but we will not be interested in it. It will in fact lead to the same temperature behavior.

Consider first the zeroth order process. The amplitude is

$$
M^{(0)}_{(\vec{k}\uparrow)(i\downarrow)\to(\vec{k}'\uparrow)(i\downarrow)} = -\frac{J}{2}m_s \equiv M, \qquad (14.10)
$$

as we explained in the previous section, where $m_s = 1/2$ is the spin projection for the impurity. The scattering rate is obtained by summing the amplitude squared over the final states:

$$\Gamma^{(0)}_{(\vec{k}\uparrow)(i\downarrow)\to(\vec{k}'\uparrow)(i\downarrow)} = \frac{2\pi}{\hbar}\sum_{\vec{k}'}\delta(\epsilon_{\vec{k}}-\epsilon_{\vec{k}'})|M^{(0)}_{(\vec{k})(i\downarrow)\to(\vec{k}'\uparrow)(i\downarrow)}|^2 = \frac{2\pi}{\hbar}\sum_{\vec{k}'}\delta(\epsilon_{\vec{k}}-\epsilon_{\vec{k}'})\left(\frac{J}{2}\right)^2 m_s^2.$$

$$(14.11)$$

The sum over final states, for $\vec{k}$ close to the Fermi energy, gives $\rho_0 V$, where $\rho_0 = mk_F/(2\pi^2\hbar^3) = 3n_e/(4\epsilon_F)$ is the *single-spin* electron density of states at the Fermi surfaces from Chapter 3, $g(\epsilon)/2$ [see (3.14)], so that

$$\Gamma^{(0)}_{(\vec{k}\uparrow)(i\downarrow)\to(\vec{k}'\uparrow)(i\downarrow)} = \frac{\pi}{2\hbar}\rho_0 V J^2 m_s^2. \qquad (14.12)$$

But we need to consider also the second-order process of the generic form (14.6), since it will lead to a divergence.

There are four possible processes at second order. (1) An electron in $\vec{k}$ can scatter to unoccupied state with $\vec{q}$ and from there to $\vec{k}'$, without spin flip, i.e. $(\vec{k}\uparrow)(i\downarrow)\to(\vec{q}\uparrow)(i\downarrow)$ followed by $(\vec{q}\uparrow)(i\downarrow)\to(\vec{k}'\uparrow)(i\downarrow)$. (2) Reversely, first an electron in occupied state $\vec{q}$ scatters to $\vec{k}'$, then the initial electron $\vec{k}$ scatters in the vacated $\vec{q}$, i.e. first $(\vec{q}\uparrow)(i\downarrow)\to(\vec{k}'\uparrow)(i\downarrow)$, followed by $(\vec{k}\uparrow)(i\downarrow)\to(\vec{q}\uparrow)(i\downarrow)$. (3) The same as (1), but with a spin flip for the intermediate electron $\vec{q}$ and a corresponding (opposite) spin flip for the impurity, i.e. first $(\vec{k}\uparrow)(i\downarrow)\to(\vec{q}\downarrow)(i\uparrow)$, followed by $(\vec{q}\downarrow)(i\uparrow)\to(\vec{k}'\uparrow)(i\downarrow)$. (4) The same as (2), but with a spin flip for the intermediate electron $\vec{q}$ and a corresponding (opposite) spin flip for the impurity, i.e. first $(\vec{q}\downarrow)(i\uparrow)\to(\vec{k}'\uparrow)(i\downarrow)$, followed by $(\vec{k}\uparrow)(i\downarrow)\to(\vec{q}\downarrow)(i\uparrow)$.

Process 1. Apply (14.6) for $V_{\vec{k}i} = M$, and obtain the contribution

$$|M|^2 I_1, \qquad (14.13)$$

where the sum over intermediate states is

$$I_1 = \sum_{\vec{q}} \frac{1-f_{\vec{q}}}{\epsilon_{\vec{k}}-\epsilon_{\vec{q}}+i\eta}. \qquad (14.14)$$

Here $1-f_{\vec{q}}$, with $f_{\vec{q}}$ the Fermi-Dirac distribution, guarantees that we sum only over states $\vec{q}$ that are empty at $T=0$ (when the FD distribution is simply a Heaviside function), and the infinitesimal $i\eta$ is added to obtain only outgoing waves.

Process 2. Entirely similarly, apply (14.6) for $V_{\vec{k}i} = M$, obtaining the contribution

$$|M|^2 I_2, \qquad (14.15)$$

where now the sum over intermediate states is different,

$$I_2 = -\sum_{\vec{q}} \frac{f_{\vec{q}}}{\epsilon_{\vec{q}}-\epsilon_{\vec{k}'}+i\eta}, \qquad (14.16)$$

and the factor $f_{\vec{q}}$ now guarantees that we sum only over states $\vec{q}$ that are *occupied*. We also have a minus sign because of the interchange in the order of fermionic creation and annihilation with respect to process 1, since $a_{\vec{k}}a^{\dagger}_{\vec{k}'} = -a^{\dagger}_{\vec{k}'}a_{\vec{k}}$.

In the sum of these two processes, actually the Fermi-Dirac distribution cancels for an energy-conserving process, $\epsilon_{\vec{k}} = \epsilon_{\vec{k}'}$, obtaining a contribution

$$|M|^2(I_1 + I_2) = \left(-\frac{Jm_s}{2}\right)^2 \sum_{\vec{q}} \frac{1}{\epsilon_{\vec{k}} - \epsilon_{\vec{q}}}. \tag{14.17}$$

Since the FD distribution cancels, these terms will not lead to a temperature dependence. These two processes will actually be subleading and will not lead to the stated divergence.

Process 3. We obtain the same sum I_1, but the matrix element is modified with respect to process 1, since we now have

$$M_{(\vec{k}\uparrow)(i\downarrow)\to(\vec{k}'\downarrow)(i\uparrow)} = \frac{M}{m_s} M_{\uparrow\downarrow} = \frac{M}{m_s}\left\langle s, m_s + 1 \left| \frac{S_i^+}{\hbar} \right| s, m_s \right\rangle, \tag{14.18}$$

leading to a contribution

$$\frac{|M|^2}{m_s^2} I_1 \frac{1}{\hbar^2} \left|\left\langle s, m_s + 1 \left| \frac{S_i^+}{\hbar} \right| s, m_s \right\rangle\right|^2 = I_1 \left(-\frac{J}{2}\right)^2 [s(s+1) - m_s(m_s + 1)]. \tag{14.19}$$

Process 4. We obtain the same sum I_2, but the matrix element is modified with respect to process 2, since we now have

$$\frac{M}{m_s} M_{\downarrow\uparrow} = \frac{M}{m_s}\left\langle s, m_s - 1 \left| \frac{S_i^-}{\hbar} \right| s, m_s \right\rangle, \tag{14.20}$$

leading to a contribution

$$\frac{|M|^2}{m_s^2} I_2 \frac{1}{\hbar^2} \left|\left\langle s, m_s - 1 \left| \frac{S_i^-}{\hbar} \right| s, m_s \right\rangle\right|^2 = I_2 \left(-\frac{J}{2}\right)^2 [s(s+1) - m_s(m_s - 1)]. \tag{14.21}$$

Then the sum of the contributions from processes 3 and 4 is

$$\left(\frac{J}{2}\right)^2 [M_{\uparrow\downarrow}I_1 + M_{\downarrow\uparrow}I_2] = \left(\frac{J}{2}\right)^2 [M_{\uparrow\downarrow}(I_1 + I_2) + 2m_s I_2] = \left(\frac{J}{2}\right)^2 [M_{\downarrow\uparrow}(I_1 + I_2) - 2m_s I_1]. \tag{14.22}$$

Now the Fermi-Dirac distributions do not cancel between the two terms, and we can obtain a temperature dependence. But they do cancel in the $I_1 + I_2$ terms, leaving the relevant contribution either as $+2m_s I_2$, or as $-2m_s I_1$.

All in all, at second order in perturbation we have

$$M^{(2)}_{(\vec{k}\uparrow)(i\downarrow)\to(\vec{k}'\uparrow)(i\downarrow)} = \left(\frac{J}{2}\right)^2 [2m_s I_2 + (s(s+1) - m_s)(I_1 + I_2)]$$

$$= \left(\frac{J}{2}\right)^2 [-2m_s I_1 + (s(s+1) + m_s)(I_1 + I_2)]. \tag{14.23}$$

We can ignore the temperature-independent terms, and write

$$M^{(2)}_{(\vec{k}\uparrow)(i\downarrow)\to(\vec{k}'\uparrow)(i\downarrow)} \simeq \pm\left(\frac{J}{2}\right)^2 2m_s I_{2/1}, \tag{14.24}$$

with indices 2/1 corresponding to the $\pm$ signs, respectively.

We now convert the sums I_2 and I_1 into integrals using the *total density* of states at momentum k, $\rho = mk/(\pi^2 \hbar^3)$ and $\epsilon_{\vec{k}} = k^2/(2m)$:

$$I_2 = \int_0^{\epsilon_F} d\epsilon_q \frac{\rho}{\epsilon_{\vec{k}} - \epsilon_{\vec{q}} - i\eta} = \frac{m}{\pi^2 \hbar^3} \int_0^{k_F} \frac{q^2 dq}{k^2 - q^2 - i\eta} \tag{14.25}$$

and

$$I_1 = \int_{\epsilon_F}^{D} d\epsilon \frac{\rho}{\epsilon_{\vec{k}} - \epsilon_{\vec{q}} + i\eta} = \frac{m}{\pi^2 \hbar^3} \int_{k_F}^{k_D} \frac{q^2 dq}{k^2 - q^2 + i\eta}, \tag{14.26}$$

where D is the maximum energy in the conduction band for the electrons.

We can do the k (and $\epsilon_{\vec{k}}$) integrals, obtaining

$$\int_0^{k_F} \frac{q^2 dq}{k^2 - q^2 - i\eta} = -k_F - \frac{k}{2} \ln \left| \frac{k - k_F}{k + k_F} \right| \simeq -\frac{k}{2} \ln \left| \frac{k - k_F}{k + k_F} \right|$$

$$\int_{\epsilon_F}^{D} d\epsilon \frac{\rho}{\epsilon_{\vec{k}} - \epsilon_{\vec{q}} + i\eta} \simeq \frac{k_F}{2} \ln \left| \frac{\epsilon_k - \epsilon_F}{\epsilon_k - \epsilon_D} \right|. \tag{14.27}$$

The proof is left as an exercise. But at $T \neq 0$, yet $T \ll T_F$, the range of thermally excited electrons is given by $|\epsilon_k - \epsilon_F| \leq k_B T$, and thus we can approximate $|\epsilon_k - \epsilon_F|$ by $k_B T$, and $|k - k_F|$ by $m k_B T/k_F$. In the other terms, we replace $\epsilon_k \simeq \epsilon_F$ and $k \simeq k_F$.

Finally, we obtain approximately

$$M^{(2)}_{(\vec{k}\uparrow)(i\downarrow) \to (\vec{k}'\uparrow)(i\downarrow)} \simeq 2m_s \left(\frac{J}{2} \right)^2 \rho_0 \ln \frac{T_F}{T} \tag{14.28}$$

in one way, and in the other we replace T_F by $(D - \epsilon_F)/K_B$. Adding this expression to the zeroth order one, we obtain

$$M^{(2)}_{(\vec{k}\uparrow)(i\downarrow) \to (\vec{k}'\uparrow)(i\downarrow)} \simeq -\frac{J m_s}{2} \left(1 - J \rho_0 \ln \frac{T_F}{T} + \cdots \right), \tag{14.29}$$

leading to a scattering rate of (since the scattering rate is proportional to $|M|^2$, as seen in (14.11))

$$\Gamma \simeq \Gamma^{(0)} \left(1 - 2J \rho_0 \ln \frac{T_F}{T} + \cdots \right). \tag{14.30}$$

This would lead to a resistivity (since $\Gamma \propto 1/\tau$ and in the Drude model $\rho = m^*/(ne^2\tau)$)

$$\rho_R(T) = \rho_R(0) \left(1 - 2J \rho_0 \ln \frac{T_F}{T} + \cdots \right), \tag{14.31}$$

(or with T_F replaced by $(D - \epsilon_F)/k_B$) where $J < 0$, so the resistivity decreases with temperature. However, note that we actually have a leading contribution to the resistivity that comes from phonons and scales as T^5, so the correct dependence of the total resistivity is of the form

$$\rho_{R,\text{total}}(T) = aT^5 - b \ln \frac{T}{T_F}, \tag{14.32}$$

where $a, b > 0$, and it has a minimum at a temperature

$$T_{\min} \simeq \left(\frac{b}{5a}\right)^{1/5}. \tag{14.33}$$

Note that the difference between T/T_F and $k_B T/(D - \epsilon_F)$ in the result is negligible to logarithmic accuracy, which is why we have obtained both by different approximations.

However, note that when the divergent log term becomes large, we cannot trust perturbation theory anymore, since it is a second order contribution, and it becomes of the order of the first order perturbation.

One can do a resummation procedure for resumming the leading logarithms at all orders, and obtain that the effective coupling is

$$J_{\text{eff}} = \frac{J}{1 + J I_2} = \frac{J}{1 + J \rho_0 \ln \frac{T_F}{T}}, \tag{14.34}$$

and thus the interaction diverges at a Kondo temperature

$$T_K = T_F e^{\frac{1}{J\rho_0}}. \tag{14.35}$$

As before, alternatively we can replace T_F with $(D - \epsilon_F)/k_B$, and the resulting formula is a good approximation to experimental values.

14.3 Strong Coupling and the Kondo Problem

As we said, the *Kondo problem* is the fact that we have a diverging resistivity at small temperatures, which is clearly unphysical.

The method of Anderson of "poor man's scaling," based only on a second order perturbation in the effective spin-spin coupling, nevertheless shows how we can avoid this Kondo problem.

The details are somewhat involved, so we will explain only the results. Consider an electronic band $-D < \epsilon_{\vec{k}} < D$, and we want to see how the second order transition amplitudes change when we vary it, $D \to D + \delta D$. The second order analysis in the isotropic limit gives the equation for the coupling J:

$$\frac{dJ}{d \ln D} = J^2 \rho_0. \tag{14.36}$$

The solution of this equation is

$$J(D) = \frac{J(D_0)}{1 - J(D_0)\rho_0 \ln(D/D_0)}. \tag{14.37}$$

Since we are interested in the effective bandwidth explored at finite T, of width $k_B T$ around the Fermi surface, we can set $D/D_0 = k_B T/\epsilon_F = T/T_F$, valid to logarithmic accuracy, as we have explained in the previous section.

To determine the Kondo temperature, note that T_K must be scale invariant, since it is a physical quantity, and independent of our choice of D:

$$D \frac{\partial T_K}{\partial D} = 0. \tag{14.38}$$

Define the dimensionless coupling

$$g = \rho_0 J, \tag{14.39}$$

and its *beta function*, the same one defined in relativistic quantum field theory,

$$\frac{dg}{d \ln D} \equiv \beta(g), \tag{14.40}$$

defining renormalization. Then the scaling equation is simply the statement that, *up to second order in perturbation*, the beta function is

$$\beta(g) = g^2. \tag{14.41}$$

Dimensional analysis tells us that $k_B T_K = D f(g)$, where $f(g)$ is a dimensionless function. From the scale invariance condition, $D \partial T_K / \partial D = 0$, we obtain

$$f(g) + \frac{\partial f}{\partial g} \frac{\partial g}{\partial lnD} = 0, \tag{14.42}$$

with solution

$$f(g) \sim e^{\frac{1}{g}}. \tag{14.43}$$

Finally then, the Kondo temperature is

$$k_B T_K = D e^{\frac{1}{J \rho_0}}, \tag{14.44}$$

as we have argued in the last section.

The complete analysis was due to Wilson, using the numerical renormalization group approach. He found that, as the energy scale goes to zero, the effective coupling diverges, $J \to -\infty$. The divergence in the resistivity is then found to occur at $T = 0$, instead of at T_K, which is an artifact of perturbation theory. Note, however, that experiments show a *finite* resistivity as $T \to 0$ or $D \to \infty$.

In Wilson's analysis, one considers spherical shells in position space and discretizes energy bands. This is the key to avoiding divergences, since the continuum of energies, and the presence of all energy scales, leads to the divergence, which is somewhat similar to what one obtains in QCD at large distances (color confinement). A correct renormalization group procedure gets rid of the divergence.

The discretization of the Kondo Hamiltonian results in a chain Hamiltonian

$$H_K = \sum_{n=0}^{\infty} (f_{n+1}^\dagger f_n + f_n^\dagger f_{n+1}) + H_{int}. \tag{14.45}$$

Renormalization amounts to a discretization only up to an N, followed by introduction of a further point $N + 1$, modifying the Hamiltonian as

$$H_{N+1} = \sqrt{\Lambda} H_N + \sum_\sigma (f^\dagger_{N+1,\sigma} f_{N\sigma} + f^\dagger_{N\sigma} f_{N+1\sigma}). \qquad (14.46)$$

Thus the renormalization group transformation is increasing the number of points, and rescaling the energy by $\sqrt{\Lambda}$: $H_{N+1} = R(H_N)$,

In conclusion, the Kondo problem is a strong coupling problem for which we still need a good resolution.

14.4 Heavy Fermions and the Kondo Lattice

As we already explained, heavy fermion materials are materials in which the effective mass of the electrons is much larger (thousands of times) than the mass of the free electron. One defines the effective mass in correlation to the relaxation time τ,

$$m_* = m \frac{\tau_*}{\tau}, \qquad (14.47)$$

and one has a conductivity

$$\sigma = \frac{ne^2}{m_*} \frac{\tau_*}{1 + \omega^2 \tau_*^2}. \qquad (14.48)$$

Therefore, one can use, at least in some regime, Fermi liquid theory, on which the above picture is based.

A heavy fermion material can be described by the presence of many local impurity moments, forming a lattice called the *Kondo lattice*. Because the lattice preserves translational invariance, elastic scattering conserves momentum, which leads to coherent scattering off the impurities. This in turn leads, in the simplest heavy fermion materials, to a dramatic drop in resistivity below the Kondo temperature T_K. Generically, at high enough temperatures, the heavy fermion materials can be described by a Fermi liquid picture.

Heavy fermion materials have:

- Curie-Weiss susceptibility at high temperature, $\chi \sim 1/(T + \theta)$
- Paramagnetic susceptibility at low temperatures, $\chi \sim$ const.
- Much enhanced linear specific heat at small temperatures, $C_V = \gamma T$
- Quadratic low-temperature resistivity $\rho = \rho_0 + \gamma T^2$.

In Fermi liquid theory, the low-temperature susceptibility and specific heat are given by

$$\chi = \mu_B^2 \frac{\rho_0^*}{1 + F_0^a}$$
$$\gamma = \frac{\pi^2 k_B^2}{3} \rho_0^*, \qquad (14.49)$$

with ratio

$$\frac{\chi}{\gamma} = \left(\frac{\mu_B}{\pi k_B}\right)^2 \frac{3}{1+F_0^a}, \qquad (14.50)$$

and is constant as χ and γ vary by changing ρ_0^*. Here F_0^a is a constant that describes the interaction integrated over angles, as $F_0^a = \int \frac{d\Omega}{4\pi}(-\frac{1}{2})U(2p_F \sin\theta/2)$, and $U(p)$ is the momentum-space interaction. This relation, observed experimentally over a large range of χ and γ, allows us to understand heavy fermions as a lattice version of the Kondo model, with a renormalized ρ_0^*.

A simple Kondo lattice Hamiltonian is

$$H_K = \sum_{\vec{k}} \epsilon_{\vec{k}} c_{\vec{k}\sigma}^\dagger c_{\vec{k}\sigma} + J \sum_j c_{j\alpha}^\dagger (\vec{\sigma})_{\alpha\beta} c_{j\beta} \cdot \vec{S}_j, \qquad (14.51)$$

where $c_{j\alpha} = \frac{1}{\sqrt{N}} \sum_{\vec{k}} c_{\vec{k}\alpha} e^{i\vec{k}\cdot\vec{R}_j}$ is the electron annihilation operator at site j and $\vec{S}_j$ is an impurity spin at site j. It has two scales: the Kondo temperature $T_K \sim De^{-\frac{1}{2J\rho_0}}$ and the RKKY scale (Rudermann, Kittel, Kasuya, Yosida) $E_{RKKY} = J^2 \rho_0$.

At small $J\rho_0$, $E_{RKKY} \gg T_K$, and we have an antiferromagnetic ground state, since the magnetic impurities on the lattice tend to align themselves due to their interaction, whereas the Kondo interaction with the conduction electrons is small and cannot change it. At large $J\rho_0$, $E_{RKKY} \ll T_K$, the Kondo effect is strong, and every site in the lattice coherently scatters the electrons. In this case, we have a Fermi liquid picture. As the temperature is increased above $J^2\rho$, the scale of the second order diverging scattering term in the Kondo effect, both the antiferromagnetic state and the Fermi liquid picture disappear.

Important Concepts to Remember

- The resistivity in some metals with magnetic impurities has a minimum as a function of T, below which it increases logarithmically $\sim -\ln T$ until a Kondo temperature T_K. Its explanation is the Kondo effect.
- The Kondo effect is the interaction of conduction electron spins with the impurity spins, described by the Kondo Hamiltonian. It contains an antiferromagnetic interaction.
- A second order scattering effect of the electrons off impurities leads to a logarithmic divergence $-\ln T$ in the scattering rate, thus in the resistivity. When compared with the leading effects, it leads to a minimum in the resistivity.
- Since this is a perturbative effect comparing leading and subleading terms, it cannot be trusted except at large values. A resumming of leading logs gives an effective coupling that diverges at a Kondo temperature $T_K \sim (T_F \text{ or } D/k_B) e^{-\frac{1}{|J|\rho_0}}$.
- The Kondo problem is how to get rid of this divergence. A second order scaling relation ("poor man's scaling"), based on a quadratic beta function, reproduces T_K.
- Wilson's renormalization group method shows that there is no divergence, but we still need an analytical strong coupling resolution of the Kondo problem.

- Heavy fermion materials have the effective electron mass m_* up to thousands of times larger than the free m_e and obey Fermi liquid theory at least at large temperatures or couplings.
- Most heavy fermion materials are described by a Kondo lattice model, are antiferromagnetic at small coupling, and are a Fermi liquid at large coupling.

Further Reading

For a review of the Kondo model and the Kondo problem, see chapter 7 in [7]. For a review of heavy fermions and the Kondo lattice, see [21].

Where It will be Addressed in String Theory

The Kondo problem will be addressed via AdS/CFT in Chapter 48.

Exercises

(1) Expand the Kondo Hamiltonian (14.1), by writing the fields and spins in components.
(2) Verify the results of the integrals (14.27).
(3) Find the exact relation between $\Gamma^{(0)}$ and $\rho_R^{(0)}$ in the Kondo model analysis.
(4) Consider the 3-loop corrected beta function

$$\beta(g) = g^2 + \alpha g^3, \tag{14.52}$$

where $|\alpha| \ll 1$. Find the modified solution to $f(g)$ in (14.43).
(5) Describe how the RKKY scale $J^2 \rho_0$ appears from the Kondo lattice Hamiltonian (14.51).

   Hint: consider equation (14.36) for the Kondo Hamiltonian (14.1) and the definition of ρ_0.

Hydrodynamics and Transport Properties: From Boltzmann to Navier-Stokes

In this chapter we will learn about transport in classical theory, which is based on the Boltzmann equation, and also about the hydrodynamic expansion, described by the Navier-Stokes equation, and its relativistic generalization.

15.1 Boltzmann Equations

For classical transport, we define a distribution function $f(\vec{r}, \vec{v}, t)$, giving the number of particles per volume of phase space $d^3\vec{r}d^3\vec{v}$ as

$$dN = f(\vec{r}, \vec{v}, t)d^3\vec{r}d^3\vec{v}. \tag{15.1}$$

To describe transport, we follow a volume element along the line of the flow. Then, if there are no collisions, the distribution should be conserved:

$$f(t + dt, \vec{r} + d\vec{r}, \vec{v} + d\vec{v}) = f(t, \vec{r}, \vec{v}). \tag{15.2}$$

Otherwise, we have an explicit time dependence due to the collisions that modifies the probability distribution,

$$dt\frac{\partial f}{\partial t} + d\vec{r} \cdot \vec{\nabla}_{\vec{r}}f + d\vec{v} \cdot \vec{\nabla}_{\vec{v}}f = dt\left(\frac{\partial f}{\partial t}\right)_{\text{coll}}, \tag{15.3}$$

or, noting that $\vec{a} = d\vec{v}/dt$,

$$\frac{\partial f}{\partial t} + \vec{v} \cdot \vec{\nabla}_{\vec{r}}f + \vec{a} \cdot \vec{\nabla}_{\vec{v}}f = \left(\frac{\partial f}{\partial t}\right)_{\text{coll}}, \tag{15.4}$$

which is known as the *Boltzmann equation*.

As we already saw in the case of electronic transport in metals, we often think of the time variation due to collisions by an average, as described by a *relaxation time* $\tau_c(\vec{r}, \vec{v})$, such that

$$\left(\frac{\partial f}{\partial t}\right)_{\text{coll}} = -\frac{f - f_0}{\tau_c}, \tag{15.5}$$

where f_0 is a distribution at thermal equilibrium. The name for τ_c is appropriate, since considering an $f = f(t)$ only, we get

$$\frac{d(f - f_0)}{dt} = -\frac{f - f_0}{\tau_c} \Rightarrow f - f_0 = (f - f_0)(\tau = 0)e^{-\frac{t}{\tau_c}}, \tag{15.6}$$

where f_0 is time-independent, or an exponential relaxation down to f_0.

Thus the most common form of the Boltzmann equation is

$$\frac{\partial f}{\partial t} + \vec{v} \cdot \vec{\nabla}_{\vec{r}} f + \vec{a} \cdot \vec{\nabla}_{\vec{v}} f = -\frac{f - f_0}{\tau_c}. \tag{15.7}$$

We can moreover find conservation equations for the fluid dynamics, which complement the above equation. Define first the number density as

$$n = \int d^3 \vec{v} f. \tag{15.8}$$

Then the average value of any function A is defined as

$$\langle A \rangle = \frac{1}{n} \int d^3 \vec{v} A f. \tag{15.9}$$

Moreover, consider a function $g = g(\vec{v})$ only, which is a function conserved in collisions (in general, we would write a function of the momentum, $g(\vec{p})$).

Multiplying the Boltzmann equation (15.7) by $g(\vec{v})$ and integrating over velocity, $\int d^3 \vec{v}$, we obtain first

$$\int d^3 \vec{v} \, g \frac{\partial f}{\partial t} + \int d^3 \vec{v} g \vec{v} \cdot \vec{\nabla}_{\vec{r}} f + \int d^3 \vec{v} \, g \vec{a} \cdot \vec{\nabla}_{\vec{v}} f = \int d^3 \vec{v} g \left(\frac{\partial f}{\partial t} \right)_{\text{coll}} = 0, \quad (15.10)$$

where the integral on the right-hand side is zero by the assumption of g being conserved in collisions (which means basically the fact that this integral of g is zero). Then, by identifying the resulting integrals with corresponding averages, and writing

$$g \vec{v} \cdot \vec{\nabla}_{\vec{r}} f = \vec{\nabla}_{\vec{r}} (g \vec{v} f) - f \vec{\nabla}_{\vec{r}} (g \vec{v}) = \vec{\nabla}_{\vec{r}} (g(\vec{v}) \vec{v} f)$$

$$g \vec{a} \cdot \vec{\nabla}_{\vec{v}} f = \vec{\nabla}_{\vec{v}} \cdot (g \vec{a} f) - f \vec{\nabla}_{\vec{v}} \cdot (g \vec{a}), \tag{15.11}$$

and integrating the right-hand side of the second equation, the first term is a boundary term assumed to be zero, so we obtain finally

$$\frac{\partial}{\partial t} (n \langle g \rangle) + \vec{\nabla} \cdot (n \langle g \vec{v} \rangle) - n \vec{a} \cdot \langle \vec{\nabla}_{\vec{v}} g \rangle = 0. \tag{15.12}$$

This is a general form of conservation equations for the fluid dynamics, also part of the Boltzmann theory.

15.2 Applications of the Boltzmann Equation

Application of the Boltzmann Equation (15.7): Particle Diffusion
Consider first the case of particle diffusion in one dimension, along the x direction (still for a 3-dimensional fluid), at constant temperature. Then $\vec{\nabla}_{\vec{r}} \to \frac{d}{dx}$, and for $f = f(x)$ only, we get from (15.7) that

$$v_x \frac{df}{dx} = -\frac{f - f_0}{\tau_c}. \tag{15.13}$$

We solve this equation self-consistently, by assuming the zeroth order function is the equilibrium (time-independent) distribution f_0, and substituting it in df/dx on the left-hand side, to find the first order solution

$$f_1 \simeq f_0 - v_x \tau_c \frac{df_0}{dx}. \tag{15.14}$$

For the equilibrium function being the Maxwell-Boltzmann distribution,

$$f_0 = e^{\frac{\mu-\epsilon}{k_B T}} \Rightarrow \frac{df_0}{dx} = \frac{f_0}{k_B T} \frac{d\mu}{dx}. \tag{15.15}$$

We can define the particle flux density along the x direction,

$$J_n^x = \int v_x dn = \int v_x f \tilde{g}(\epsilon) d\epsilon, \tag{15.16}$$

where $\tilde{g}(\epsilon)$ is the density of states, which we saw in Chapter 3 is (there it was used for fermions, but the derivation is independent of that)

$$\tilde{g}(\epsilon) = \frac{1}{2\pi^2} \left(\frac{2m}{\hbar^2} \right)^{3/2} \epsilon^{1/2}. \tag{15.17}$$

Substituting $f \simeq f_1$ from (15.14) inside J_n^x, and using that

$$\int d\epsilon \, v_x f_0(\epsilon) \tilde{g}(\epsilon) = 0, \tag{15.18}$$

since the $+v_x$ terms cancel against the $-v_x$ terms, and assuming that τ_c is independent of $\vec{v}$, we obtain

$$J_n^x = -\frac{d\mu}{dx} \frac{\tau_c}{k_B T} \int d\epsilon \, v_x^2 f_0 \tilde{g}(\epsilon) = -\frac{d\mu}{dx} \frac{\tau_c}{k_B T} \frac{1}{3} \frac{2}{m} \int d\epsilon \frac{mv^2}{2} f_0 \tilde{g}(\epsilon), \tag{15.19}$$

where in the second equality we have used isotropy to say that $\langle v_x^2 \rangle = 1/3 \langle v^2 \rangle$. We recognize the integral in the last equality as $(3/2)nk_B T$ from kinetic theory (this is the kinetic theory definition of temperature), leading to

$$J_n(x) = -\frac{n\tau_c}{m} \frac{d\mu}{dx}. \tag{15.20}$$

But because of the BE distribution (15.15), which implies

$$\mu = k_B T \ln n + \text{const.} \Rightarrow \frac{d\mu}{dx} = \frac{k_B T}{n} \frac{dn}{dx}, \tag{15.21}$$

we finally obtain

$$J_n^x = -\frac{\tau_c k_B T}{m} \frac{dn}{dx} \equiv -D_n \frac{dn}{dx}, \tag{15.22}$$

which is *Fick's diffusion law*, with an *Einstein relation for the diffusivity* D_n that can be rewritten, since by our definition of averages

$$n \left\langle \frac{mv^2}{2} \right\rangle = \int \frac{mv^2}{2} f_0 \tilde{g}(\epsilon) d\epsilon = \frac{3}{2} nk_B T, \tag{15.23}$$

as

$$D_n = \frac{\tau_c k_B T}{m} = \frac{1}{3}\langle v^2 \rangle \tau_c. \tag{15.24}$$

Application of Boltzmann Equation (15.7) to FD Distribution

One can make a similar calculation in the case of fermions, with distribution

$$f_0 = \frac{1}{e^{\frac{\epsilon - \mu}{k_B T}} + 1}, \tag{15.25}$$

and find again Fick's law, with the diffusivity

$$D_n = \frac{1}{3} v_F^2 \tau_c. \tag{15.26}$$

15.3 Nonviscous Hydrodynamics

We consider the laws of the ideal fluid as an application of the conservation equations for Boltzmann theory (15.12).

$g = m$

We consider first $g = m$ in (15.12), since it is obviously conserved in collisions. Since $\rho = m\,n$, we obtain

$$\frac{\partial \rho}{\partial t} + \vec{\nabla} \cdot (\rho \vec{v}) = 0, \tag{15.27}$$

the continuity equation for the mass density. Note that another way to derive it would be to say that the particle flux through the boundary ∂V of the volume V equals minus the time derivative of the mass inside V,

$$\frac{\partial}{\partial t} \int_V \rho dV = -\oint_{\partial V} \rho \vec{v} \cdot d\vec{A}, \tag{15.28}$$

from which the above follows by the use of Stokes's law to convert the integral over area into an integral over volume.

Next, we can use dynamics, in the form of the $\vec{F} = m\vec{a}$ equation for the particles. The force through the boundary ∂V equals the integral of $d\vec{F} = -pd\vec{A}$, and $dm\vec{a} = \rho d\vec{v}/dt dV$, so that

$$\int_V \rho \frac{d\vec{v}}{dt} dV = -\oint pd\vec{A} = -\int dV \vec{\nabla} p, \tag{15.29}$$

where in the second equality we used again Stokes's law. Note that if there are also external forces acting, we should add them to the volume integral on the right hand side. But since $\vec{v} = \vec{v}(\vec{r}, t)$, we have

$$\frac{d\vec{v}}{dt} = \frac{\partial \vec{v}}{\partial t} + (\vec{v} \cdot \vec{\nabla}_{\vec{r}})\vec{v}. \tag{15.30}$$

Therefore, considering the equation over an infinitesimal volume, and dividing by ρ, we get

$$\left(\frac{d\vec{v}}{dt} =\right) \frac{\partial \vec{v}}{\partial t} + (\vec{v} \cdot \vec{\nabla}_{\vec{r}})\vec{v} = -\frac{\vec{\nabla}_{\vec{r}} p}{\rho}(+\vec{g} + \cdots). \tag{15.31}$$

This is Euler's equation, and it is the equation for a nonviscous fluid. For an incompressible fluid, $\rho = $ constant, we write

$$\frac{\partial \vec{v}}{\partial t} + (\vec{v} \cdot \vec{\nabla}_{\vec{r}})\vec{v} = -\frac{\vec{\nabla}_{\vec{r}} p}{\rho}(+\vec{g} + \cdots). \tag{15.32}$$

This generalizes to the Navier-Stokes equation in the case of the viscous fluid, as we will see.

We should also note that the simplest kind of flow is the potential flow: If we consider an irrotational fluid, i.e. $\vec{\nabla}_{\vec{r}} \times \vec{v} = 0$, then $\vec{v} = \vec{\nabla}_{\vec{r}} \phi$, with ϕ called the potential. Then for the incompressible fluid, the continuity equation $\vec{\nabla}_{\vec{r}} \cdot \vec{v} = 0$ becomes

$$\Delta \phi = 0. \tag{15.33}$$

This is a solvable flow, unlike the general case of the Navier-Stokes equation (or even the general Euler equation), which are nonlinear, and very hard to solve except in special cases.

$g = m\vec{v}$

Applying (15.12) for the case of $g = m\vec{v}$ leads to a new conservation equation. We note that $nmv_i = \rho v_i$ and

$$n\langle gv_j\rangle = \rho\langle v_i v_j\rangle = \rho V_i V_j + \rho\langle(v_i - V_i)(v_j - V_j)\rangle \equiv \Pi_{ij}, \tag{15.34}$$

where $V_i = \langle v_i \rangle$ is the average velocity of the flow. Then

$$\Pi_{ij} \equiv \rho V_i V_j + P_{ij}$$
$$P_{ij} = \rho\langle(v_i - V_i)(v_j - V_j), \tag{15.35}$$

and we note that P_{ij} becomes $p\delta_{ij}$ in the case of an isotropic system, where p is the pressure. Indeed, for $i = j$, $\rho\delta v_i^2$ is a kinetic energy with respect to the average flow, responsible for pressure.

Finally then, (15.12) becomes

$$\frac{\partial}{\partial t}(\rho v_i) + \frac{\partial}{\partial x_j}\Pi_{ij}(-nF_i) = 0, \tag{15.36}$$

where we put in brackets the effect of possible external forces. Alternatively, we can derive the same equation by starting from

$$\frac{\partial}{\partial t}(\rho v_i) = \rho\frac{\partial v_i}{\partial t} + \frac{\partial \rho}{\partial t}v_i, \tag{15.37}$$

and using Euler's equation to write the first term on the right-hand side as $-\rho(\vec{v} \cdot \vec{\nabla}_{\vec{r}})v_i - \partial_i p$ and in the second term, $\partial \rho/\partial t = -\vec{\nabla} \cdot (\rho\vec{v})$ from the continuity equation, to obtain finally

$$\frac{\partial}{\partial t}(\rho v_i) = -\frac{\partial}{\partial x_j}(\rho v_i v_j + p\delta_{ij}) = -\partial_i \Pi_{ij}. \tag{15.38}$$

Note that Π_{ij} are the components of the momentum flux density tensor, i.e. the spatial components of the nonrelativistic limit of the energy-momentum tensor. This already implies that the relativistic generalization of this conservation equation will be

$$\partial_\mu T^{\mu\nu} = 0. \tag{15.39}$$

15.4 Viscous Hydrodynamics and Navier-Stokes

We can generalize the conservation equation (15.38) to the nonrelativistic, but viscous, case, using the information of the relativistic generalization. All we need to do in fact is to add an extra term to the momentum flux density tensor,

$$\Pi_{ij} = p\delta_{ij} + \rho v_i v_j + \sigma_{ij}, \tag{15.40}$$

where σ_{ij} is a new, viscous stress tensor, term. Since this term is related to viscosity, which is a type of friction for fluids, it must depend on the gradient (space derivative) of the velocity field $\vec{v}$.

But here we introduce the physics input that the hydrodynamic theory is an effective theory at large scale, understood as an *expansion in the number of derivatives*. Therefore *in the first approximation*, we can consider only terms depending on the first derivatives (we could imagine in principle functions depending independently on the first, second, third, etc. derivatives), and depending *linearly* on them, i.e. the first term in the multi-variable Taylor expansion. But moreover, we consider that for a fluid in uniform rotation, with $\vec{v} = \vec{\Omega} \times \vec{r}$, we will not have any viscosity, which restricts the dependence to dependence on the symmetric part, $\partial_i v_j + \partial_j v_i$ (which vanishes for the uniform rotation, as easily checked).

We can thus expand σ_{ij} in the irreducible representations, the symmetric traceless part and a trace:

$$\sigma_{ij} = \eta \left(\partial_i v_j + \partial_j v_i - \frac{2}{3}\delta_{ij}\partial_k v_k \right) + \zeta \delta_{ij}\partial_k v_k. \tag{15.41}$$

Here η is called *shear viscosity*, and ζ is called bulk viscosity. In fact, one can show that for thermodynamical stability, we must have $\eta > 0$ and $\zeta > 0$. This is physically clear, since viscosity is friction, which generates heat. The thermodynamical condition amounts to the fact that heat can only be generated (but not absorbed) by this process, and the signs were chosen such that the friction effect generates heat.

Substituting the ansatz for $\Pi_{ij} = P_{ij} + \sigma_{ij}$ in the conservation equation, we have

$$\frac{\partial}{\partial t}(\rho v_i) = -\frac{\partial}{\partial x_j} \left[p\delta_{ij} + \rho v_i v_j + \eta \left(\partial_i v_j + \partial_j v_i - \frac{2}{3}\delta_{ij}\partial_k v_k \right) + \zeta \delta_{ij}\partial_k v_k \right], \tag{15.42}$$

and putting the $-v_j \partial_j(\rho v_i)$ term on the left-hand side, we write it in the form of the Euler's equation plus corrections:

$$\rho \left[\frac{\partial v_i}{\partial t} + (\vec{v} \cdot \vec{\nabla}) v_i \right] = -\partial_i p + \partial_j \left[\eta \left(\partial_j v_i + \partial_i v_j - \frac{2}{3} \delta_{ij} \partial_k v_k \right) + \zeta \, \partial_k v_k \right]. \quad (15.43)$$

In general, η and ζ are functions of T and p, but if they are approximately constant, we can rewrite the above as

$$\rho \left[\frac{\partial \vec{v}}{\partial t} + (\vec{v} \cdot \vec{\nabla}) \vec{v} \right] = -\vec{\nabla}_{\vec{r}} p + \eta \Delta \vec{v} + \left(\zeta + \frac{\eta}{3} \right) \vec{\nabla}_{\vec{r}} (\vec{\nabla}_{\vec{r}} \cdot \vec{v}). \quad (15.44)$$

This is the *Navier-Stokes equation*.

15.5 Relativistic Generalization

As we already said, the relativistic generalization of the fluid hydrodynamics equations is the conservation equations for the energy-momentum tensor $T^{\mu\nu}$, the generalization of Π^{ij}, and the charge currents J_I^μ:

$$\nabla_\mu T^{\mu\nu} = 0 \quad \text{and} \quad \nabla_\mu J_I^\mu = 0. \quad (15.45)$$

Here for completeness I used the general relativistic *covariant derivative* ∇_μ for a curved space, though it is not needed in special relativity. The covariant derivative will be described in Part II.

Ideal Fluid
For an ideal fluid with a comoving particle 4-velocity $u^\mu = dx^\mu / d\tau$ (the relativistic frame will be specified better shortly, when introducing the next corrections to this $T_{\mu\nu}$), energy density ρ and pressure p, the energy-momentum tensor is

$$T^{\mu\nu} = \rho u^\mu u^\nu + p(g^{\mu\nu} + u^\mu u^\nu). \quad (15.46)$$

Here

$$P^{\mu\nu} \equiv g^{\mu\nu} + u^\mu u^\nu \quad (15.47)$$

is a projector onto states normal to u^μ, since

$$P^{\mu\nu} u_\mu = 0 \quad \text{and} \quad P^{\mu\rho} P_{\rho\nu} = P^\mu{}_\nu \equiv P^{\mu\rho} g_{\rho\nu}. \quad (15.48)$$

Also, we have the charge currents

$$J_I^\mu = q u^\mu. \quad (15.49)$$

In the nonrelativistic case, $u^\mu \simeq (1, \vec{v})$, so one finds

$$T_{00} = \rho; \quad T_{ij} = \rho v_i v_j + p \delta_{ij}; \quad J_I^\mu = (q_I, q_I \vec{v}). \quad (15.50)$$

We have also an entropy current; however, in the general viscous case, as we said, the entropy current is not conserved, i.e. entropy (thus heat) is produced, due to the friction. In the ideal case, we have as for the charge currents,

$$J^{\mu}_{s,\text{ideal}} = s\, u^{\mu},$$

(15.51)

where s is the entropy density, and thus

$$\nabla_{\mu}(J^{\mu}_{s})_{\text{ideal}} = 0.$$

(15.52)

Dissipative Fluid

As in the nonrelativistic case, the viscous generalization includes extra terms in the energy-momentum tensor and the charge currents:

$$(T_{\mu\nu})_{\text{dissipative}} = \rho u^{\mu}u^{\nu} + p P^{\mu\nu} + \Pi^{\mu\nu}_{(1)}$$
$$(J^{\mu}_{I})_{\text{dissipative}} = q_{I}u^{\mu} + \Upsilon^{\mu}.$$

(15.53)

As in the nonrelativistic case, since the hydrodynamic expansion is an expansion in derivatives, $\Pi^{\mu\nu}_{(1)}$ is again linear in ∂u. In general, we can consider the expansion order by order in derivatives up to any order we want, but here we will restrict ourselves to the first order, to find the relativistic generalization of the Navier-Stokes equation.

Moreover, the energy-momentum tensor, which always has an ambiguity in its definition in flat space, can be defined uniquely by imposing the condition of the *Landau frame*, by demanding that

$$\Pi^{\mu\nu}_{(1)}u_{\mu} = 0; \quad \Upsilon^{\mu}u_{\mu} = 0.$$

(15.54)

We can decompose $\nabla^{\nu}u^{\mu}$ into a part parallel to u^{μ}, and a part transverse to u^{μ}, which can be further decomposed in a trace, the divergence θ, and a traceless part, which contains a symmetric traceless part, the shear $\sigma^{\mu\nu}$, and an antisymmetric part, the vorticity $\omega^{\mu\nu}$, so

$$\nabla^{\nu}u^{\mu} = -a^{\mu}u^{\nu} + \sigma^{\mu\nu} + \omega^{\mu\nu} + \frac{1}{d-1}\theta P^{\mu\nu},$$

(15.55)

where a^{μ} in the part parallel to u^{μ} is the acceleration, and we have

$$a^{\mu} = u^{\nu}\nabla_{\nu}u^{\mu}$$
$$\theta = \nabla_{\mu}u^{\mu} = P^{\mu\nu}\nabla_{\mu}u_{\nu}$$
$$\sigma^{\mu\nu} = \nabla^{(\mu}u^{\nu)} + u^{(\mu}a^{\nu)} - \frac{1}{d-1}\theta P^{\mu\nu}$$
$$\omega^{\mu\nu} = \nabla^{[\mu}u^{\nu]} + u^{[\mu}a^{\nu]}.$$

(15.56)

Here the antisymmetrization and symmetrization is with strength one (we can remove the (anti)symmetrization symbol if we multiply by a tensor that already has the symmetry). But since $\Pi^{\mu\nu}_{(1)}$ is symmetric, and built from ∂u and in the Landau frame, the unique form for it contains the irreducible components $\sigma^{\mu\nu}$ and $\theta P^{\mu\nu}$, with coefficients the shear and bulk viscosities:

$$\Pi^{\mu\nu}_{(1)} = -2\eta\sigma^{\mu\nu} - \zeta\theta P^{\mu\nu}.$$

(15.57)

Then the equation of motion for the fluid hydrodynamics, the relativistic version of the Navier-Stokes equations, comes from substituting the above $T^{\mu\nu}$ in the conservation equation:

$$\nabla_\mu T^{\mu\nu} = 0. \tag{15.58}$$

By a similar reasoning, one finds also the first order charge current:

$$\Upsilon_{I(1)}^\mu = -\tilde{\kappa}_{IJ} P^{\mu\nu} \nabla_\nu \left(\frac{\mu_J}{T} \right) - \mathcal{U}_I l^\mu - \gamma_I P^{\mu\nu} \nabla_\nu T. \tag{15.59}$$

Here T is the temperature, $\tilde{\kappa}_{IJ}, \mathcal{U}_I$, and γ_I are coefficients (like η and ζ), μ_J are the chemical potentials for the charges q_J, and

$$l^\mu = \epsilon_{\nu\rho\sigma}{}^\mu u^\nu \nabla^\rho u^\sigma. \tag{15.60}$$

The fact that this ansatz solves the conservation equations at first order is left as an exercise.

Important Concepts to Remember

- The Boltzmann equation is a transport equation, for the distribution function $f(\vec{r}, \vec{v}, t)$, driven by a collision term.
- One can find also conservation equations for quantities $g(\vec{v})$ conserved in the collisions
- One can apply the Boltzmann equation to find Fick's diffusion law $J_n^x = -D_n dn/dx$.
- One can apply the conservation equations for mass and momentum to find the Euler equation, and the conservation of the momentum flux density Π_{ij} in the nonviscous fluid case.
- The equation of motion in the viscous fluid case is found by adding terms to Π_{ij}, a viscous stress tensor σ_{ij}. Hydrodynamics is an expansion in derivatives, so to first order (Taylor expansion) we have linear dependence on $\partial_i v_j$, more specifically a symmetric traceless part with coefficient η (shear viscosity) and a trace part with coefficient ζ (bulk viscosity).
- Substituting the ansatz in the conservation equation, we obtain the Navier-Stokes equation.
- The relativistic generalization of the equations of motion is $\nabla_\mu T^{\mu\nu} =$ and $\nabla_\mu J_I^\mu = 0$, and in the Landau frame $\Pi^{\mu\nu} u_\mu = 0 = \Upsilon_I^\mu u_\mu$, we have the symmetric traceless part $\sigma^{\mu\nu}$ with coefficient 2η and the trace part $\theta P^{\mu\nu}$ with coefficient $-\zeta$.

Further Reading

Appendix G of [6] for the Boltzmann equation; chapters I and II in [22] for the nonrelativistic hydrodynamics (and chapter XV for the relativistic part); and [23] and [24] for the relativistic part.

Where It will be Addressed in String Theory

Relativistic fluids will be addressed in the fluid-gravity correspondence in Chapter 38, and the nonrelativistic case of the Navier-Stokes equation will be addressed in Chapter 39.

Exercises

(1) For the Boltzmann equation, consider a stationary solution,

$$\frac{\partial f}{\partial t}(\vec{r}, \vec{v}, t) = 0, \tag{15.61}$$

giving an equilibrium distribution

$$f_0(\vec{r}, \vec{v}) = \rho(\vec{r}) \left[\frac{m}{2\pi k_B T(r)} \right]^{1/2} \exp\left[-\frac{m|\vec{v} - \vec{v}_0|^2}{2k_B T(r)} \right]. \tag{15.62}$$

In the case of no external forces, $\rho(\vec{r}) = \rho_0$, and for $T(r) = T_0$ and $\vec{v}_0 = 0$, we have

$$f_0(\vec{v}) = \rho_0 \left[\frac{m}{2\pi k_B T_0} \right]^{1/2} e^{-\frac{mv^2}{2k_B T_0}}. \tag{15.63}$$

Show that it is a solution to the Boltzmann equation.

(2) Consider particle diffusion as an application of the Boltzmann equation for the Bose-Einstein distribution:

$$f_{BE} = \frac{1}{e^{\frac{\epsilon - \mu}{k_B T}} - 1}. \tag{15.64}$$

(3) Apply the conservation equation (15.12) in the text for the case of the kinetic energy of the particle, $g = 1/2 m v_i^2$.

(4) Calculate the energy dissipation (dE_{kin}/dt) in an incompressible fluid, using the Navier-Stokes equations.

(5) Prove that the relativistic dissipative energy-momentum tensor and entropy current to first order satisfy the conservation equations

$$\nabla_\mu T^{\mu\nu} = 0 \quad \text{and} \quad \nabla_\mu J_I^\mu = 0, \tag{15.65}$$

where

$$T^{\mu\nu} = \rho u^\mu u^\nu + p P^{\mu\nu} + \Pi_{(1)}^{\mu\nu}$$
$$J_I^\mu = q_I u^\mu + \Upsilon_{(1)}^\mu. \tag{15.66}$$

PART II

ELEMENTS OF GENERAL RELATIVITY AND STRING THEORY

16 The Einstein Equation and the Schwarzschild Solution

We start Part II with a quick introduction to general relativity, specifically to the Einstein's equation and the Schwarzschild solution. I will assume that people have seen Einstein's equation before, but will not assume that they have taken a full course on general relativity; hence I will describe the general ideas here.

16.1 Special Relativity

To understand what general relativity is, we must first frame special relativity in a certain way, so that we can generalize it. Special relativity starts with the experimental observation that the speed of light is constant in all *inertial* reference frames, hence as theorists we can put it equal to 1, $c = 1$. That translates into the statement that the invariant distance (distance element) is not $dl^2 = d\vec{x}^2$ as in Galilean mechanics, but rather

$$ds^2 = -dt^2 + d\vec{x}^2 = \eta_{ij}dx^i dx^j, \tag{16.1}$$

where $\eta_{ij} = \mathrm{diag}(-1, +1, +1, +1)$, such that $ds^2 = 0$ (for light) in one inertial reference frame is the same in another inertial reference frame.

But that also means that the symmetry group of special relativity is the group that leaves invariant ds^2, that is $SO(1, 3)$, or in general $SO(1, d - 1)$, the Lorentz group. That generalizes the $SO(3)$ of rotations for the invariance of $dl^2 = d\vec{x}^2$, in that the transformations are still linear,

$$x'^i = \Lambda^i{}_j x^j, \tag{16.2}$$

and they leave ds^2 invariant, $ds'^2 = ds^2$.

Special relativity is then the statement that physics is Lorentz invariant.

16.2 The Geometry of General Relativity

As the name suggests, general relativity is then the statement that physics is invariant under a general coordinate transformation $x'^i = x'^i(\{x^j\})$. To obtain that, we must consider a general spacetime, curved instead of flat, defined by the distance between two points, or *line element*

$$ds^2 = g_{ij}(x)dx^i dx^j. \tag{16.3}$$

Here $g_{ij}(x)$ are arbitrary functions that replace the constant η_{ij}, collectively called the *metric*. Sometimes by an abuse of notation, ds^2 (the line element) is called the metric. We note from its definition above that g_{ij} is a symmetric matrix, since it multiplies a symmetric quantity.

Example: the Sphere S^2 To understand the concepts, we start with the simplest example we can consider, the usual 2-sphere. In our usual experience, we define it by embedding in the (spatial flat) Euclidean 3-dimensional space with metric

$$ds^2 = dx_1^2 + dx_2^2 + dx_3^2, \tag{16.4}$$

via the embedding constraint that the radius of the sphere is constant:

$$(x_1)^2 + (x_2)^2 + (x_3)^2 = R^2. \tag{16.5}$$

By taking the derivative of the constraint, we find the relation between the differentials:

$$2(x_1 dx_1 + x_2 dx_2 + x_3 dx_3) = 0 \Rightarrow$$
$$dx_3 = -\frac{x_1 dx_1 + x_2 dx_2}{x_3} = -\frac{x_1}{\sqrt{R^2 - x_1^2 - x_2^2}} dx_1 - \frac{x_2}{\sqrt{R^2 - x_1^2 - x_2^2}} dx_2. \tag{16.6}$$

Substituting x_3 from the constraint and the above dx_3 into the embedding metric, we find the line element:

$$ds^2 = dx_1^2 \left(1 + \frac{x_1^2}{R^2 - x_1^2 - x_2^2}\right) + dx_2^2 \left(1 + \frac{x_2^2}{R^2 - x_1^2 - x_2^2}\right) + 2dx_1 dx_2 \frac{x_1 x_2}{R^2 - x_1^2 - x_2^2}$$
$$\equiv g_{ij}(x)dx^i dx^j, \quad i = 1, 2. \tag{16.7}$$

Thus by solving the embedding constraint and replacing in the Euclidean metric of the embedding space, we find the form of the metric on the curved space S^2. Note, however, that the metric above is valid only on patches, since the points with $x_1^2 + x_2^2 = R^2$ (i.e., $x_3 = 0$) are excluded, because they are singular.

Instead, we can also solve the embedding constraint using some intrinsic coordinates, a parametrization in terms of θ and ϕ, instead of using the Euclidean coordinates themselves. Thus solving the constraint by

$$x_1 = R\cos\theta; \quad x_2 = R\sin\theta\cos\phi; \quad x_3 = R\sin\theta\sin\phi, \tag{16.8}$$

we find the line element

$$ds^2 = R^2(d\theta^2 + \sin^2\theta d\phi^2) \equiv g_{ij}(\theta, \phi)dx^i dx^j, \quad \text{for} \quad x^i = (\theta, \phi). \tag{16.9}$$

The first form of the metric (16.7) was obtained by embedding the 2-dimensional sphere in flat 3-dimensional space with Euclidean signature $(+ + +)$. We might ask whether this is a general feature, and we can represent curved spaces as embeddings in some flat $(d + 1)-$ dimensional space with Euclidean or Minkowski metric.

That is not true, since first the symmetric matrix g_{ij} has $d(d + 1)/2$ components in a d-dimensional space, and we can change it by using d arbitrary functions, the coordinate transformations $x'^i = x'^i(x)$, which means that there are $d(d - 1)/2$ independent functions

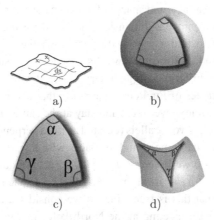

Fig. 16.1 (a) Curved space. The functional form of the distance between two points depends on local coordinates. (b) A triangle on a sphere, made from two meridian lines and a segment of the equator has two angles of 90° (π/2). (c) The same triangle, drawn for a general curved space of positive curvature, emphasizing that the sum of the angles of the triangle exceeds 180° (π). (d) In a space of negative curvature, the sum of the angles of the triangle is below 180° (π).

that we need to specify. Therefore we can embed only the general space in $D = d(d + 1)/2$ dimensional flat space, by specifying the d coordinate transformations $x'^i = x'^i(x)$ and $d(d − 1)/2$ embedding functions, i.e. extra coordinates $X'^a = X'^a(x)$. Note that $d(d − 1)/2 = 1$ only in $d = 3$, allowing for a single extra coordinate for the embedding space.

But more importantly, the signature of the embedding space would not be fixed, since for instance, while some 2-dimensional spaces with $\det(g_{ij}) > 0$ (i.e., with Euclidean signature ++, or locally Euclidean metric), like S^2, can be embedded in 3-dimensional Euclidean space, others, also with $\det(g_{ij}) > 0$, thus with Euclidean signature, or locally Euclidean metric, can be embedded only in flat space with Minkowski signature. The quintessential example in two dimensions (with constant curvature, like the sphere) is *Lobachevski space*, defined by the embedding in three Minkowski dimensions, with metric

$$ds^2 = dx^2 + dy^2 − dz^2, \tag{16.10}$$

by the embedding constraint

$$x^2 + y^2 − x^2 = −R^2. \tag{16.11}$$

Thus in this case, despite its Euclidean signature, the space can be embedded only in a flat space of Minkowski signature. So while we could fix the number of necessary dimensions of the embedding flat space to $d(d + 1)/2$, we could not fix its signature to accomodate all the curved metrics. Then the concept of defining curved spaces by embedding in flat spaces is not very useful, and we must think instead of spaces as being *intrinsically curved*; see Figure 16.1(a). In the example of S^2 above, we must consider the metric in terms of intrinsic coordinates on the sphere, like the usual parametrization in (16.9), without making any reference to an embedding.

Then we can define spaces of positive curvature $R > 0$ (we will explain shortly what that means precisely) like S^2 (which is in fact the unique 2-dimensional space of *constant* positive curvature), and spaces of negative curvature $R < 0$, like Lobachevski space (which is in fact the unique 2-dimensional space of *constant negative curvature*).

One has to find a way to determine whether $R > 0$ or $R < 0$ without using any embedding, and that intrinsic way is based on measuring the sum of angles of a triangle, $\alpha + \beta + \gamma$. In Euclidean geometry, $\alpha + \beta + \gamma = \pi$, which one proves based on the Euclid's postulates of Euclidean geometry, that two parallels (defined as lines perpendicular to the same line) never meet and stay at the same distance from each other. In spaces with $R > 0$ instead, two parallels (defined as lines perpendicular to the same geodesic) converge, so they eventually meet, and in spaces with $R < 0$, the two parallels diverge. The result is that for $R > 0$ we have $\alpha + \beta + \gamma > \pi$, as in Figure 16.1(c), whereas for $R < 0$ we have $\alpha + \beta + \gamma < \pi$, as in Figure 16.1(d). For a space with $R > 0$ like S^2, we can convince ourselves that this is true, since if we consider a triangle made by two meridians, as in Figure 16.1(b), meeting at the North Pole, and an equator (intersecting the meridians at $\pi/2$), the sum of angles is $\pi/2 + \pi/2 + \gamma > \pi$.

16.3 Einstein's Theory: Definition

Now we are ready to formulate Einstein's general relativity theory, based on two physical assumptions:

- Gravity is geometry: The idea is that matter follows geodesics in curved space, and the resulting motion appears to us as due to the effect of gravity. To picture that using an analogy, imagine a 2-dimensional rubber sheet, with a mass deforming the sheet around it. Then a light probe sent close to the mass will move on a geodesic and get deflected, which to a 2-dimensional observer on the sheet would look like the effect of 2-dimensional gravity.
- Matter sources gravity: The opposite is also true, namely the source of the gravity, or geometry, is matter itself: Matter curves the spacetime. In the above picture, the deformation of the rubber sheet was due to a heavy mass.

These two assumptions can be translated into two physical principles, which can be described mathematically, plus an equation for the dynamics of gravity, which will be Einstein's equation.

(1) The first principle is the one we were trying to generalize from the case of special relativity, that *physics is invariant under general coordinate transformations*, not just Lorentz transformations that preserve inertial frames, but the most general transformations $x'^i = x'^i(\{x^j\})$. Indeed, by analogy with the special relativity case, we see that the line element is invariant, since

$$ds^2 = g_{ij}(x)dx^i dx^j = g'_{ij}(x')dx'^i dx'^j. \tag{16.12}$$

Note that this is somewhat different than in the special relativity case, when the Lorentz transformation would leave the Minkowski metric η_{ij} itself invariant. Now the metric changes (if the metric is the same, we say we have an *isometry* of the metric), but the point is that the line element has the same general form, in terms of some metric.

(2) The second principle is the *equivalence principle*, which can be stated as the fact that *there is no difference between acceleration and gravity*, or that "if you are in a freely falling elevator you cannot distinguish it from the case of being weightless (without gravity)," (or "if you are in an elevator maintaining position in a gravitational field, you cannot distinguish this from the case of being uniformly accelerated"). Einstein proposed this elevator Gedanken (thought) experiment to emphasize the physical meaning of the equivalence principle. Note that this is a *local* statement; globally we can make a distinction between uniform acceleration and gravity, since the gravitational field of, say, the Earth has tidal forces, due to the fact that gravity acts slightly differently at different positions. Therefore by measuring those tidal forces we could determine the presence of the gravitational field.

This second principle is mathematically expressed as follows. In the first physics class one takes, one is taught about the fact that there is a Newton's law for dynamics, $\vec{F} = m_i \vec{a}$, and a Newton's law for gravity, $\vec{F}_g = m_g \vec{g}$, and one usually takes for granted that the constant of proportionality in the two is the same, which we call mass m. But logically, these two constants of proportionality are a priori different, and the fact that the *inertial mass* m_i and the *gravitational mass* m_g are the same is a physical principle, the equivalence principle:

$$m_i = m_g. \tag{16.13}$$

If the two would be different, we could decide experimentally our situation inside the elevator, detecting whether we are under the influence of gravity or acceleration.

16.4 Kinematics of Gravity

The dynamics equation that we will write will be the Einstein equation. But in terms of what can we write it? The metric g_{ij} is not a good candidate, since it changes under general coordinate transformations, and has at most $d(d + 1)/2 - d = d(d - 1)/2$ independent degrees of freedom, which would characterize the space, as we saw. Thus we need to define invariant objects in terms of which to write Einstein's equation, and the basic object will be the Riemann tensor.

To define it, first we must define the inverse metric $g^{\mu\nu} \equiv (g_{\mu\nu})^{-1}$,

$$g^{\mu\nu} g_{\nu\rho} = \delta^{\mu}_{\rho}, \tag{16.14}$$

and then define the *Christoffel symbol* $\Gamma^{\mu}_{\nu\rho}$, that plays the role of "gauge field of gravity," by

$$\Gamma^{\mu}_{\nu\rho} = \frac{1}{2} g^{\mu\sigma} (\partial_\rho g_{\nu\sigma} + \partial_\nu g_{\rho\sigma} - \partial_\sigma g_{\nu\rho}). \tag{16.15}$$

Then the Riemann tensor is defined as the "field strength of gravity," in the sense that it mimics the formula for the field strength of a gauge field in an $SO(p, q)$ gauge theory, with indices a, b in the fundamental and (ab) (antisymmetric) for the adjoint, in which case the field strength is written as

$$F_{\mu\nu}^{ab} = \partial_\mu A_\nu^{ab} - \partial_\nu A_\mu^{ab} + A_\mu^{ac} A_\nu^{cb} - A_\nu^{ac} A_\mu^{cb}. \tag{16.16}$$

If we artificially separate the indices in $\Gamma_{\nu\rho}^\mu$ as "gauge" μ, ν and "spacetime" ρ as $(\Gamma^\mu{}_\nu)_\rho$, we can write

$$(R^\mu{}_\nu)_{\rho\sigma} = \partial_\rho(\Gamma^\mu{}_\nu)_\sigma - \partial_\sigma(\Gamma^\mu{}_\nu)_\rho + (\Gamma^\mu{}_\lambda)_\rho(\Gamma^\lambda{}_\nu)_\sigma - (\Gamma^\mu{}_\lambda)_\sigma(\Gamma^\lambda{}_\nu)_\rho. \tag{16.17}$$

Here it is understood that the brackets are artificial, and the "gauge" and "spacetime" indices are the same, and the difference is just to emphasize the similarity with the gauge case.

From the Riemann tensor we can define the *Ricci tensor*

$$R_{\mu\nu} = R^\lambda{}_{\mu\lambda\nu} \tag{16.18}$$

and the *Ricci scalar*

$$R = R_{\mu\nu} g^{\mu\nu} \tag{16.19}$$

as its contractions. Now the Ricci scalar is truly a coordinate invariant (a scalar), so it is a true invariant measure of the curvature of space (though not the only one). Before, when we said $R > 0$ or $R < 0$, we were actually referring to the Ricci scalar.

Tensors

The Riemann tensor and the Ricci tensor are examples of general relativity tensors. As in the case of special relativity tensors, we can define a *contravariant tensor* A^μ as an object that transforms as dx^μ under a general coordinate transformation, i.e. as

$$A'^\mu(x') = \frac{\partial x'^\mu}{\partial x^\nu} A^\nu(x), \tag{16.20}$$

and a *covariant tensor* B_μ as an object that transforms as ∂_μ, or more precisely like the inverse of the above, since actually ∂_μ is not a tensor unless it acts on a scalar, i.e. as

$$B'_\mu(x') = \frac{\partial x^\nu}{\partial x'^\mu} B_\nu(x). \tag{16.21}$$

A general tensor with both covariant and contravariant indices, $A^{\nu_1...\nu_n}_{\mu_1...\mu_m}$ transforms as usual, as the product of the transformations of its indices.

Note that ∂_μ is a tensor only if it acts on a scalar, otherwise it is not, and one must define a *covariant derivative* D_μ for an action on a general tensor. We write its action on a tensor with a covariant and a contravariant index, T_ρ^ν, and it is easy to generalize to the action on a tensor with an arbitrary number of covariant and contravariant indices. Its action is

$$D_\mu T_\rho^\nu = \partial_\mu T_\rho^\nu + \Gamma_{\mu\sigma}^\nu T_\rho^\sigma - \Gamma_{\mu\rho}^\sigma T_\sigma^\nu. \tag{16.22}$$

Note also that the Christoffel symbol $\Gamma_{\nu\rho}^\mu$ itself is not a tensor. This can be understood as follows. Any curved space looks flat on small scales, which means that on a sufficiently small patch we can bring the metric to the Minkowski form, plus quadratic corrections, $g_{\mu\nu} = \eta_{\mu\nu} + \mathcal{O}(\delta x^2)$. But $\Gamma_{\nu\rho}^\mu \sim g^{-1}\partial g$ contains only first derivatives of the metric, which vanish in this reference frame. Thus $\Gamma^\mu_{\nu\rho}$ can be made equal to zero in a small neighborhood, whereas a tensor that is nonzero in a reference frame will be nonzero in any other, since the transformation multiplies it only by factors of $\partial x'/\partial x$.

So to put a special relativity theory in curved space, we need to replace derivatives ∂_μ with covariant derivatives, $\eta_{\mu\mu}$ with $g_{\mu\nu}$, special relativity tensors with general relativity tensors, and the volume of space $d^d x$ with the *invariant volume in curved space*:

$$d^d x \sqrt{-g} \equiv d^d x \sqrt{-\det g_{\mu\nu}}. \tag{16.23}$$

Note that the minus inside the square root is because $\det g_{\mu\nu} < 0$ in Minkowski signature. Invariance of the above volume element means that $d^d x' \sqrt{-g'(x')} = d^d x \sqrt{-g(x)}$.

16.5 Dynamics of Gravity: The Einstein Equation

We are now finally ready to write the dynamics of gravity, in the form of an action for it. The action must be invariant, thus a scalar. It is written as an integral of a Lagrangean density $\mathcal{L}_{\text{gravity}}$ over the invariant volume element, so $\mathcal{L}_{\text{gravity}}$ must be invariant, i.e. a scalar. The simplest one we could think is R; it is the invariant of lowest dimension (dimension 2, since $[\partial_\mu] = 1$ and R has two derivatives, and various dimensionless metrics), and it happens to be compatible with experiments. Therefore the *Einstein-Hilbert (EH) action for gravity* is

$$S_{\text{gravity}} = \frac{1}{16\pi G} \int d^d x \sqrt{-g} R. \tag{16.24}$$

It is important to stress that ultimately, this action for gravity happens to agree well with experiments (including the weak field approximation), and that is the reason it is chosen, but a priori the only constraint was that $\mathcal{L}$ be a scalar, so we could have chosen a function $f(R)$, or other invariant combinations, like R^2, $R^2_{\mu\nu}$, $R^2_{\mu\nu\rho\sigma}$ (all the invariants of dimension 4), or R^3 or $(D_\mu R)^2$ contractions (dimension 6), etc. In fact, in string theory (and more generally in quantum gravity) such terms do appear as small quantum corrections to the classical EH term. Being the term of lowest mass dimension, we could guess that it would be a good starting point for perturbation theory, adding the others at the quantum level, but ultimately it is an issue decided by experiments as to what the action must be.

To vary the gravitational action, we write

$$\delta R = (\delta R_{\mu\nu})g^{\mu\nu} + R_{\mu\nu}\delta g^{\mu\nu}. \tag{16.25}$$

One can show (though I will not do it here) that $\delta R_{\mu\nu}$ generates a total derivative term that vanishes under integration, and therefore does not contribute to the equations of motion.

We can evaluate $\delta\sqrt{-g}$ by writing first

$$\delta\sqrt{-g} = \frac{1}{2}\sqrt{-g}\frac{\delta \det g_{\mu\nu}}{\det g_{\mu\nu}}, \tag{16.26}$$

and then, using the matrix formula

$$\det M = e^{\mathrm{Tr}\log M} \Rightarrow \delta \det M = e^{\mathrm{Tr}\log M} \, \mathrm{Tr}[M^{-1}\delta M] = -\det M \, \mathrm{Tr}[M\delta M^{-1}], \quad (16.27)$$

applied for $M = g_{\mu\nu}$, we obtain

$$\delta\sqrt{-g} = -\frac{1}{2}\sqrt{-g}g_{\mu\nu}\delta g^{\mu\nu}. \tag{16.28}$$

Finally then, the equation of motion is

$$\frac{\delta S_{\mathrm{grav}}}{\delta g^{\mu\nu}} = 0 \Rightarrow R_{\mu\nu} - \frac{1}{2}g_{\mu\nu}R = 0. \tag{16.29}$$

This is Einstein's equation in a vacuum.

Adding Matter

We now consider adding matter to the system. The simplest example would be a scalar field, with Minkowski space action

$$S_\phi^M = -\frac{1}{2}\int d^4x (\partial_\mu\phi)(\partial_\nu\phi)\eta^{\mu\nu}. \tag{16.30}$$

According to the rules explained above, the curved space action is

$$S_\phi = -\frac{1}{2}\int d^4x\sqrt{-g}(D_\mu\phi)(D_\nu\phi)g^{\mu\nu}. \tag{16.31}$$

Just that, *only for scalar fields*, $D_\mu\phi = \partial_\mu\phi$, so the action is

$$S_\phi = -\frac{1}{2}\int d^4x\sqrt{-g}(\partial_\mu\phi)(\partial_\nu\phi)g^{\mu\nu}. \tag{16.32}$$

We can define the energy-momentum tensor (known from electromagnetism, for instance) in a unique way (in Minkowski space, we have an ambiguity in its definition), by the variation of the matter action with respect to the metric, specifically

$$T_{\mu\nu} = -\frac{2}{\sqrt{-g}}\frac{\delta S_{\mathrm{matter}}}{\delta g^{\mu\nu}}. \tag{16.33}$$

Then, for a general action,

$$S = \frac{1}{16\pi G}\int d^dx\sqrt{-g}R + S_{\mathrm{matter}}, \tag{16.34}$$

the variation is written in terms of the gravity variation plus the energy momentum tensor contribution,

$$\delta S = \frac{1}{16\pi G}\int d^dx\sqrt{-g}\delta g^{\mu\nu}\left(R_{\mu\nu} - \frac{1}{2}g_{\mu\nu}R - 8\pi G T_{\mu\nu}\right), \tag{16.35}$$

and therefore the equation of motion is

$$R_{\mu\nu} - \frac{1}{2}g_{\mu\nu}R = 8\pi G T_{\mu\nu}, \tag{16.36}$$

which is Einstein's equation with matter.

As an example, in the case of the above free massless scalar field action, the energy-momentum tensor is (doing the variation with respect to the metric)

$$T^{\phi}_{\mu\nu} = \partial_\mu\phi\partial_\nu\phi - \frac{1}{2}g_{\mu\nu}(\partial\phi)^2. \tag{16.37}$$

Another relevant example is the case of a homogeneous and isotropic system without friction, i.e. a perfect fluid, for which the energy-momentum tensor is

$$T_{\mu\nu} = \text{diag}(\rho, p, p, p). \tag{16.38}$$

16.6 The Newtonian Limit

We now want to consider the Newtonian limit to make contact with Newtonian gravity. Therefore we make a small fluctuation expansion,

$$g_{\mu\nu} = \eta_{\mu\nu} + \kappa_N h_{\mu\nu}, \tag{16.39}$$

where $\kappa_N^2 = 16\pi G$ and $\kappa_N h_{\mu\nu} \ll 1$. Then the Einstein-Hilbert action for gravity turns (to leading order) into the *Fierz-Pauli action for gravity*,

$$S = \int d^4x \mathcal{L}_{FP}; \quad \mathcal{L}_{FP} = \frac{1}{2}h_{\mu\nu,\rho}^2 + h_\mu^2 - h^\mu h_{,\mu} + \frac{1}{2}h_{,\mu}^2, \tag{16.40}$$

where

$$h_{\mu\nu,\rho} \equiv \partial_\rho h_{\mu\nu}; \quad h_\mu = \partial^\nu h_{\nu\mu}; \quad h \equiv h^\mu{}_\mu. \tag{16.41}$$

But since the general coordinate transformation becomes a sort of gauge invariance for small fluctuations (as we leave to show as an exercise), the Fierz-Pauli action has a kind of gauge redundancy in the description, which can be fixed by choosing a gauge.

The usual choice is the *de Donder gauge*, the gauge for gravity that is analogous to the Lorenz (covariant) gauge for gauge fields:

$$\partial^\nu \bar{h}_{\mu\nu} = 0; \quad \bar{h}_{\mu\nu} \equiv h_{\mu\nu} - \eta_{\mu\nu}\frac{h}{2}. \tag{16.42}$$

In this gauge, the Fierz-Pauli action becomes

$$\mathcal{L}_{g.fix} = \frac{1}{2}(\partial_\rho \bar{h}_{\mu\nu})^2, \tag{16.43}$$

so the equation of motion becomes

$$\Box\bar{h}_{\mu\nu} = 0. \tag{16.44}$$

As we see, like in the Lorenz gauge case, in the de Donder gauge the equation of motion becomes simply Klein-Gordon.

Consider now matter in the form of a static pressureless fluid ("dust"), with energy-momentum tensor

$$T_{\mu\nu} = \text{diag}(\rho_m, 0, 0, 0). \tag{16.45}$$

Then, since the Einstein-Hilbert action becomes the gauge-fixed Pauli-Fierz action, we can replace $\Delta(\kappa_N h_{00})$ on the left-hand side of the Einstein equation, and with the above $T_{\mu\nu}$ on the right-hand side, we get

$$\vec{\nabla}^2(\kappa_N h_{00}) = -8\pi G \rho_m. \tag{16.46}$$

But this matches the form of the local Gauss law for gravity,

$$\vec{\nabla}^2 U_{\text{Newton}} = 4\pi G \rho_m \tag{16.47}$$

(the global form, in terms of flux, can be reduced to this local form, just like the global form of the electrostatic Gauss law reduces to the local form $\vec{\nabla}^2 \phi_{\text{electric}} = -\rho_{\text{electric}}/\epsilon_0$). Therefore we can identify

$$\kappa_N h_{00} = -2U_{\text{Newton}}. \tag{16.48}$$

We see that indeed, as advertised, the Einstein-Hilbert action is the correct one, as it reduces to the usual Newtonian gravity in the weak field, nonrelativistic ($v \ll 1$) limit. Moreover, with this identification, now one can prove that in this Newtonian limit, we can always put the metric in the form

$$\begin{aligned}ds^2 &\simeq -(1 + 2U_{\text{Newton}})dt^2 + (1 - 2U_{\text{Newton}})d\vec{x}^2 \\ &= -(1 + 2U_{\text{Newton}})dt^2 + (1 - 2U_{\text{Newton}})(dr^2 + r^2 d\Omega^2).\end{aligned} \tag{16.49}$$

16.7 The Schwarzschild Solution

The simplest solution of general relativity is the Schwarzschild solution, found in 1919 by Schwarzschild (while he was fighting in World War I, in the trenches). It is a solution to the Einstein equations in vacuum ($T_{\mu\nu} = 0$), or more precisely for a point-like source $\rho_m = M\delta^3(\vec{x})$*. Then for weak fields, the equation of motion (16.46) becomes

$$\Delta \kappa_N h_{00} = -8\pi GM\delta^3(\vec{x}), \tag{16.50}$$

with the solution given by the Newton potential of a point mass:

$$\kappa_N h_{00} = -2U = +2\frac{MG}{r}. \tag{16.51}$$

The full solution is actually

$$ds^= -\left(1 - \frac{2MG}{r}\right)dt^2 + \frac{dr^2}{1 - \frac{2MG}{r}} + r^2 d\Omega^2. \tag{16.52}$$

To check how this is consistent with the general weak field form (16.49), note that $1/(1 + 2U) \simeq 1 - 2U$, and redefining $(1 - 2U(r))r^2 = \bar{r}^2$ (which does not affect the coefficient

---

* We will see that the source is shielded by a horizon; but the linearized equations definitely have a source at $r = 0$.

of the dr^2 term to leading order, since the correction generates only a subleading term), we get the form (16.49).

Then, Birkhoff's theorem, proven in 1923, says that for any static and spherical matter distribution, outside it, the Schwarzschild solution is the most general one, i.e. we can always put it in this form via coordinate transformations. Therefore this is indeed a solution in vacuum.

If the solution is valid *all the way down to* $r_H = 2MG > 0$, i.e. if the matter distribution is *inside* r_H ($r_{\text{matter}} < r_H$), we call the solution a *Schwarzschild black hole*. But at $r = r_H$, we have an apparent singularity, which would make no sense: How it is then that we need to put a singularity (mass) at $r = 0$, if we already have a singularity at r_H? In fact, it is not really a singularity, since we can calculate the true invariant measure of curvature, the Ricci scalar, at the r_H, and find[†]

$$R \sim \frac{1}{r_H^2} = \frac{1}{(2MG)^2} = \text{finite}. \tag{16.53}$$

Therefore it is only an issue of using wrong coordinates, and in fact there are good coordinates, the "Kruskal coordinates," where there is nothing special at r_H (no apparent singularity). However, there is a physical significance to r_H, related to its name, an *event horizon*: Light cannot escape from it, so we cannot enter in causal contact with (we have a horizon with respect to events in) the interior $r < r_H$. Indeed, consider the trajectory of light, on the null geodesic $ds^2 = 0$, at constant θ, ϕ, ($d\Omega_2 = 0$), i.e. radial motion. Then we obtain, near $r \simeq r_H$,

$$dt = \frac{dr}{1 - \frac{2MG}{r}} \simeq 2MG\frac{d(r - 2MG)}{r - 2MG} \Rightarrow$$
$$t \simeq 2MG\ln(r - 2MG) \to \infty. \tag{16.54}$$

Therefore it takes an infinite amount of time t (the time measured by the observer at infinity, $r \to \infty$) for light to get to the horizon, or get out of the horizon, leading to the loss of causal contact with the interior of the event horizon.

Important Concepts to Remember

- General relativity is a generalization of special relativity for invariance under general coordinate transformations.
- Curved space is considered as intrinsically curved, defined by the metric, not by an embedding in flat space.
- In Einstein's theory, gravity is geometry, i.e. matter follows geodesics in curved space, and to us it appears as the effect of gravity. Also, matter sources gravity, namely it curves spacetime.
- Physics is general coordinate invariant, and it satisfies the equivalence principle, which can be stated as $m_i = m_g$.

[†] More precisely, all possible invariants are finite, including $R_{\mu\nu\rho\sigma}R^{\mu\nu\rho\sigma}$, so there is no curvature singularity.

- The Christoffel symbol is like the gauge field of gravity, and the Riemann tensor is like its field strength.
- The Einstein-Hilbert action is the integral of the Ricci scalar, but it is just the simplest action we can write, and it agrees with experiments.
- For weak fields, the EH action reduces to the Fierz-Pauli action, and in the de Donder gauge, the equation of motion is simply Klein-Gordon.
- The Schwarzschild solution is the most general solution to the Einstein equations in vacuum, in the case of a static spherically symmetric matter distribution. If it is valid all the way to $r_H = 2MG$, we have a Schwarzschild black hole.
- The event horizon is only an apparent singularity, not a real one, but light cannot escape from it.

Further Reading

A simple introduction to general relativity is found in Part II (chapters 8–11) of [25]. An advanced text is Wald [26].

Exercises

(1) Prove that the general coordinate transformation on $g_{\mu\nu}$,

$$g'_{\mu\nu}(x') = g_{\rho\sigma}(x)\frac{\partial x^\rho}{\partial x'^\mu}\frac{\partial x^\sigma}{\partial x'^\nu}, \qquad (16.55)$$

reduces for infinitesimal transformations to

$$\delta_\xi g_{\mu\nu}(x) = (\xi^\rho\partial_\rho)g_{\mu\nu} + (\partial_\mu\xi^\rho)g_{\rho\nu} + (\partial_\nu\xi^\rho)g_{\rho\nu}. \qquad (16.56)$$

(2) Prove that the commutator of two covariant derivatives, when acting on a covariant vector, gives the action of the Riemann tensor on it:

$$(D_\mu D_\nu - D_\nu D_\mu)A_\rho = -R^\sigma{}_{\rho\mu\nu}A_\sigma. \qquad (16.57)$$

(3) Check explicitly that the metric $g_{\mu\nu}$ is covariantly constant, $D_\mu g_{\nu\rho} = 0$, by substituting the expression for the Christoffel symbols.

(4) Calculate the energy-momentum tensor for a 2-index antisymmetric tensor field $B_{\mu\nu}$ with field strength

$$H_{\mu\nu\rho} = 3\partial_{[\mu}B_{\nu\rho]} \equiv \partial_\mu B_{\nu\rho} + \partial_\nu B_{\mu\rho} + \partial_\rho B_{\mu\nu} \qquad (16.58)$$

and Minkowski space action

$$S = \int d^4x \frac{H^2_{\mu\nu\rho}}{6}. \qquad (16.59)$$

(Write first the action in curved space.)

(5) Calculate the 4-dimensional Newton potential for a spherically symmetric body of constant density ρ and find g_{00} inside it, at least for weak fields (in the Newtonian limit).

(6) Considering that the gauge-fixed Fierz-Pauli action is a good approximation on-shell for the Einstein action at weak fields, calculate the Ricci scalar at the horizon of a very large black hole, and verify that $R \sim 1/r_H^2$.

(7) Assuming that $T_{\mu\nu}$ is traceless, calculate $R_{\mu\nu}$ as a function of $T_{\mu\nu}$ only, in an arbitrary dimension, from the Einstein equation.

17 The Reissner-Nordstrom and Kerr-Newman Solutions and Thermodynamic Properties of Black Holes

In this chapter we will study gravitational solutions with charge and angular momentum that generalize the Schwarzschild solution, and we will formulate and check the thermodynamic properties of black holes.

17.1 The Reissner-Nordstrom Solution: Solution with Electric Charge

We want to construct a gravitational solution that is electrically charged under a $U(1)$ field, with $F_{\mu\nu} = \partial_\mu A_\nu - \partial_\nu A_\mu$, having an electric field

$$F_{rt} = \frac{Q}{4\pi\epsilon_0 r^2}, \tag{17.1}$$

which implies a gauge field

$$A_t = -\frac{Q}{4\pi\epsilon_0 r}, \tag{17.2}$$

with the other components zero.

The energy-momentum tensor for electromagnetism, whose action is $\int d^dx \sqrt{-g}[-F_{\mu\nu}F_{\rho\sigma}\, g^{\mu\rho}g^{\nu\sigma}/4]$ is found to be

$$T_{\mu\nu} = -\frac{2}{\sqrt{-g}}\frac{\delta S}{\delta g^{\mu\nu}} = F_{\mu\rho}F_\nu{}^\rho - \frac{1}{4}g_{\mu\nu}F_{\rho\sigma}F^{\rho\sigma}. \tag{17.3}$$

The energy density is thus

$$T_{00} = +(F_{rt})^2 g^{rr} - \frac{1}{2}g_{00}(F_{rt})^2 g^{rr}g^{tt}$$

$$\simeq +\frac{1}{2}\frac{Q^2}{(4\pi\epsilon_0)^2 r^4}, \tag{17.4}$$

where in the last equality we substituted $g_{\mu\nu} \simeq \eta_{\mu\nu}$, since we are interested only in small fluctuations, and in that case the energy-momentum tensor is already of first order in the fluctuations.

The equation of motion comes from adding the above energy-momentum tensor to the point particle one, $T_{00} = M\delta^4(\vec{r})$, which gives in the Newtonian approximation

$$\vec{\nabla}^2 U_N = 4\pi G\left(M\delta^3(\vec{r}) + \frac{Q^2}{2(4\pi\epsilon_0)^2 r^4}\right). \tag{17.5}$$

This equation is solved by using

$$\vec{\nabla}^2 \frac{1}{r} = -4\pi\delta^3(\vec{r}); \quad \vec{\nabla}^2 \frac{1}{r^2} = \frac{2}{r^4}, \tag{17.6}$$

and we find

$$U_N = -\frac{MG}{r} + \frac{Q^2 G}{16\pi\epsilon_0^2 r^2}. \tag{17.7}$$

From the general form (16.49), with the same observation as in the Schwarzschild case, that redefining $r^2(1 - 2U(r))$ as $\tilde{r}^2$ affects only the subleading terms of g_{rr}, we can embed the metric as $-(1 + 2U(r))dt^2 + dr^2/(1 + 2U(r)) + r^2 d\Omega_2^2$. From it we can guess the full solution, which is in fact

$$ds^2 = -\left(1 - \frac{2MG}{r} + \frac{Q^2 G}{8\pi\epsilon_0^2 r^2}\right)dt^2 + \frac{dr^2}{1 - \frac{2MG}{r} + \frac{Q^2 G}{8\pi\epsilon_0^2 r^2}} + r^2 d\Omega_2^2, \tag{17.8}$$

together with the electric field (17.1). This is the *Reissner-Nordstrom solution*.

17.2 Horizons, BPS Bound, and Extremality

We can check that the Reissner-Nordstrom solution has two *horizons*, i.e. places where the metric becomes singular, which are found from putting g_{00} to zero,

$$1 - \frac{2MG}{r} + \frac{Q^2 G}{8\pi\epsilon_0^2 r^2} = 0, \tag{17.9}$$

solved by (putting $G = 1$)

$$r = r_\pm = M \pm \sqrt{M^2 - \frac{Q^2}{8\pi\epsilon_0^2}}. \tag{17.10}$$

From now on we will write

$$\tilde{Q} \equiv \frac{Q}{\epsilon_0\sqrt{8\pi}}, \tag{17.11}$$

and we put $G = 1$, such that there is no dimensionful quantity anymore. Dimensions can be reobtained from dimensional analysis, with appropriate factors of $G, \hbar$, and c. Then note that the metric can be rewritten as

$$ds^2 = -\Delta dt^2 + \frac{dr^2}{\Delta} + r^2 d\Omega_2^2$$
$$\Delta = \left(1 - \frac{r_+}{r}\right)\left(1 - \frac{r_-}{r}\right), \tag{17.12}$$

and the parameters $\tilde{Q}$ and M are written in terms of the event horizons as

$$M = \frac{r_+ + r_-}{2}; \quad \tilde{Q} = \sqrt{r_+ r_-}. \tag{17.13}$$

The condition that $r_\pm$ be real amounts to the *Bogomolnyi-Prasad-Sommerfeld (BPS) bound*:

$$M \geq \tilde{Q}. \tag{17.14}$$

If the solutions $r_\pm$ would not be real, it would mean that there are no *event horizons* (which appear at real values of $r_\pm$), i.e. that the singularity $r = 0$ is not shielded from causal contact by any horizon. But under very general conditions, based on so-called *energy conditions* (that amount to the fact that the energy density is positive in some frame, believed to hold in any quantum theory), it was proven by Hawking and Penrose in the 1970s (in the so-called *singularity theorems*), that there can be no "*naked singularities,*" i.e. singularities not shielded from causal contact by event horizons.

Therefore the BPS bound must hold. But the BPS bound, discovered in the context of topological solitons (where it is an exact statement), was found later to appear from the (super)symmetry algebra if we embed the solitons in a supersymmetric theory. In turn, if the black holes presented here are also embedded in a locally supersymmetric theory (supergravity), then the same supersymmetry algebra implies the BPS bound, hence the use of the same name for it.

The case of the saturation of the BPS bound, $M = \tilde{Q} = Q/\epsilon_0\sqrt{8\pi}$, so $r_+ = r_- = M$, is called the *extremal* case and is very special.

In that case, the metric becomes

$$ds^2 = -\left(1 - \frac{M}{r}\right)^2 dt^2 + \left(\frac{dr}{1 - \frac{M}{r}}\right)^2 + r^2 d\Omega_2^2, \tag{17.15}$$

and defining $\bar{r} = r - M$, we find

$$ds^2 = -\frac{dt^2}{\left(1 + \frac{M}{\bar{r}}\right)^2} + \left(1 + \frac{M}{\bar{r}}\right)^2 (d\bar{r}^2 + \bar{r}^2 d\Omega_2^2). \tag{17.16}$$

Note that then the metric is written in terms of the function

$$H(\bar{r}) = 1 + \frac{M}{\bar{r}}, \tag{17.17}$$

which is a *harmonic function*, it satisfies the Poisson equation

$$\Delta_{(3)} H(\bar{r}) \propto M\delta^3(\bar{r}). \tag{17.18}$$

Not surprisingly, we find that in these coordinates the exact Einstein's equation reduces to the above Poisson equation, that now $\bar{r} = 0$ is the singularity. This is an odd situation, since in the original coordinates $\bar{r} + M = r = 0$ was the singularity. Moreover, as before, $\bar{r} + M = r = M$, i.e. $\bar{r} = 0$ is also an event horizon. Therefore this is truly an extremal case, since the singularity and the horizon coincide. Add mass to the extremal case, and the horizon sits outside the singularity. Add charge to it, and there is no horizon, and we have a (forbidden) naked singularity.

The metric in terms of $H(r)$ is of a type that applies to many extremal solutions, as we will see later.

We can make an observation, based on the fact that the electromagnetic (Maxwell) theory has electromagnetic duality for the interchange of electric and magnetic charge. That means that we can write the solution with both $\tilde{Q}_e$ and Q_m, giving an electromagnetic field

$$F_{rt} = \frac{\tilde{Q}}{\sqrt{2\pi}\, r}; \quad F_{\theta\phi} = Q_m \sin\theta, \tag{17.19}$$

by simply replacing $\tilde{Q}_e^2$ in the metric with

$$Q^2 = \tilde{Q}_e^2 + Q_m^2. \tag{17.20}$$

17.3 General Stationary Black Hole: The Kerr-Newman Solution

The only other parameter that we can add to the black hole that can be measured at infinity is angular momentum J. Any parameter other than M, Q, and J will be called "hair," and there are so-called "*no-hair theorems*," proven by Hawking and Penrose, which say that under very general conditions only M, Q, and J are allowed as parameters that can be observed at infinity. Of course, we could have "short range hair," like for instance, a massive scalar charge Q_s. But the potential for such a charge would be the Yukawa potential,

$$V(r) = \frac{Q_s e^{-mr}}{r}, \tag{17.21}$$

which therefore cannot be observed at infinity (it decays exponentially). The theorems apply only to long-range interactions.

The black hole with M and J is called the *Kerr solution*, and the most general stationary solution, with M, Q, and J nonzero, is called the *Kerr-Newman solution*. It is given by

$$ds^2 = -\frac{\Delta}{\Sigma}(dt - a\sin^2\theta d\phi)^2 + \frac{\sin^2\theta}{\Sigma}[(r^2+a^2)d\phi - adt]^2 + \Sigma\left(\frac{dr^2}{\Delta} + d\theta^2\right)$$

$$\Sigma \equiv \rho^2 \equiv r^2 + a^2\cos^2\theta$$

$$\Delta = r^2 + a^2 + \tilde{Q}^2 - 2Mr$$

$$A = -\frac{\tilde{Q}r}{\Sigma}[dt - a\sin^2\theta\, d\phi]$$

$$a \equiv \frac{J}{M}. \tag{17.22}$$

The solution with $Q = 0$ is the Kerr solution. As before, we can find a horizon where the metric becomes singular, specifically now when $g_{rr} \to \infty$, or $\Delta = 0$, giving

$$r_\pm = M \pm \sqrt{M^2 - \tilde{Q}^2 - a^2}. \tag{17.23}$$

A *Killing vector* is a vector in the direction of an isometry of the metric. It can be defined more formally, but we simply state that if the metric has an isometry in some direction, we can write a vector that translates in that direction.

In the case of the Kerr-Newman metric, we have the Killing vectors

$$\xi^a \equiv \left(\frac{\partial}{\partial t}\right)^a ; \quad \psi^a \equiv \left(\frac{\partial}{\partial \phi}\right)^a, \tag{17.24}$$

where the notation means that $\xi^a \partial_a = \partial/\partial t$ and $\psi^a \partial_a = \partial/\partial \phi$.

17.4 Quantum Field Theory at Finite Temperature and Hawking Radiation

Black Hole Thermodynamics

We saw that classically, black holes don't emit light, in fact the event horizon is causally disconnected from an observer at infinity. But in a revolutionary paper in 1975, Stephen Hawking proved that quantum mechanically black holes emit what became known as *Hawking radiation*, an (almost) blackbody radiation at a given temperature. The derivation of Hawking is complicated and general, and was proven to be correct, despite many attempts at disproving it. However, now, knowing it is valid, we can use a shortcut that lacks some of the rigor of Hawking's initial proof, yet is good enough for most purposes. The derivation is based on a Wick rotation to Euclidean signature.

Quantum Field Theory at Finite Temperature

The derivation is based on the idea of quantum field theory at finite temperature, which, even though it should be taught in a usual Quantum Field Theory class, sometimes isn't, so I will very quickly review it.

If we want to put a Quantum Field Theory at finite temperature, for the purpose of calculating static quantities (time-independent), we can Wick rotate to Euclidean space, i.e. consider imaginary time, and consider periodic trajectory in the imaginary time, with period $\beta = 1/T$, where T is the temperature. For time-dependent quantities, the issue is much more complicated, and we can calculate only specific quantities, there is no general formalism.

So one obtains that the Euclidean partition function at a finite temperature T, with $\beta = 1/T$, can be written as a path integral over periodic trajectories in the Euclidean time, with period β:

$$Z_E[\beta] = \int_{\phi(\vec{x},t_E+\beta)=\phi(\vec{x},t_E)} \mathcal{D}\phi \, e^{-S_E[\phi]} = \text{Tr}[e^{-\beta \hat{H}}]. \tag{17.25}$$

This is generalized from a derivation in the case of quantum mechanics, as follows. The transition amplitude between the state of position q at time t and the state of position q' at time t' is written as

$$\langle q',t|q,t\rangle = \langle q'|e^{-i\hat{H}(t'-t)}|q\rangle = \sum_{n,m}\langle q'|n\rangle\langle n|e^{-i\hat{H}(t'-t)}|m\rangle\langle m|q\rangle = \sum_n \psi_n(q')\psi_n^*(q)e^{-iE_n(t'-t)}, \tag{17.26}$$

where in the first equality we have used the time evolution operator $U = e^{-i\hat{H}t}$, in the second we have introduced a complete set of eigenstates of the Hamiltonian, satisfying $\sum_n |n\rangle\langle n| = 1$, on both sides of the evolution operator, and in the third we have used the fact that $|n\rangle$ is an eigenstate of the Hamiltonian and $\langle q|n\rangle = \psi_n(q)$. Then consider periodic paths, $q' = q$, for periodicity $t' - t$ Wick rotated to Euclidean space to be $= -i\beta$, and Wick rotate also the action, to give $iS \to -S_E$. Since the transition amplitude can also be written as a path integral

$$\langle q', t|q, t\rangle = \int \mathcal{D}q(t)e^{iS[q(t)]}, \tag{17.27}$$

we obtain (17.25).

Hawking Radiation

Consider therefore the Wick-rotated Schwarzschild black hole solution,

$$ds^2 = +\left(1 - \frac{2MG}{r}\right)d\tau^2 + \frac{dr^2}{1 - \frac{2MG}{r}} + r^2 d\Omega_2^2, \tag{17.28}$$

where $t = -i\tau$. Whereas for the Schwarzschild solution in Minkowski signature we didn't have a problem to go inside the horizon at $r_H = 2MG$, since inside the role of space and time was simply interchanged, as dt^2 has a positive coefficient and dr^2 a negative one, for the Schwarzschild solution in a Euclidean signature that is a problem, since now both the coefficients of the $d\tau^2$ and the dr^2 terms turn from positive to negative, and certainly changing the signature of spacetime is not an option (and besides, we want to work in Euclidean signature anyway).

So it seems that we need to stop at the horizon, a place that was not that remarkable in Minkowski signature, since there was no singularity for it.

But now we obtain something singular in general. Indeed, consider the metric near $r = 2MG$, defining $r - 2MG = \tilde{r}$, such that the metric is (for $G = 1$) approximately

$$ds^2 \simeq \frac{\tilde{r}}{2M}d\tau^2 + 2M\frac{d\tilde{r}^2}{\tilde{r}} + (2M)^2 d\Omega_2^2. \tag{17.29}$$

Defining $\rho = \sqrt{\tilde{r}}$, we find

$$ds^2 \simeq 8M\left[d\rho^2 + \rho^2\left(\frac{d\tau}{4M}\right)^2\right] + (2M)^2 d\Omega_2^2. \tag{17.30}$$

We see that we find a metric of the type $ds^2 \simeq d\rho^2 + \rho^2 d\theta^2$, which is the metric of a flat space. If θ has a general periodicity δ, then the metric is of a *cone*, defined as a slice of angular width δ of flat space centered at $\rho = 0$, with the edges identified. Therefore, unless $\delta = 2\pi$, we have a cone, with a conical singularity at $\rho = 0$ (even though it is flat everywhere else). If instead $\delta = 2\pi$, we have simply smooth flat space.

So, to avoid the unphysical conical singularity of the space at $\rho = 0$, at $r = 2MG$ (at the Euclidean version of the horizon), we need to have periodicity $\delta = 2\pi$ in $\tau/(4M)$, such that the metric near the Euclidean horizon looks like flat space in polar coordinates. That translates into periodicity $\beta = 8\pi M$ for τ, the Euclidean (imaginary) time. According to our general analysis before, that would correspond to putting the Quantum Field Theory in Minkowski space at finite temperature:

$$T_{BH} = \frac{1}{\beta} = \frac{1}{8\pi M}. \tag{17.31}$$

Except there is a catch: Inverting the above relation, to write $M = 1/(8\pi T)$ means that the black hole, viewed as a thermodynamic object, cannot be stable, i.e. in thermal equilibrium with its environment (flat Minkowski space), since

$$C = \frac{\partial M}{\partial T} = -\frac{1}{8\pi T^2} < 0, \tag{17.32}$$

and negative specific heat is an indicator of thermodynamical instability. That means that, really, it is not the Quantum Field Theory on the whole of space that is put at finite temperature, like the general formalism would seem to imply, but rather it is the black hole itself that radiates thermally, with a black hole *Hawking temperature* of

$$T_{\text{Hawking}} = \frac{1}{8\pi M}. \tag{17.33}$$

What then happens is that the black hole radiates heat into the environment, which carries away energy, thus mass, lowering M. This in turn makes T higher, so the black hole radiates away its energy faster, etc. Because of this caveat of $C < 0$, it is not exactly clear that the derivation above for QFT at finite temperature applies, so one would need to use Hawking's original derivation instead for rigor. We will see, however, that in the case of black holes in AdS space we have $C > 0$ instead, so in that case the black hole is at thermal equilibrium with its environment, and we can safely use the above derivation.

But in any case, we would need to use Hawking's version to find the general formula, which is not obvious from the Euclidean analytical continuation procedure (which is obvious how to do only case by case, for a specific metric).

Surface Gravity and General Case

To describe the general formula, we need to define the concept of *surface gravity*. For a general stationary black hole, but in particular for the Kerr-Newman one (which in flat space is the most general stationary black hole with spherical symmetry), we can define the Killing vector

$$\chi^a \equiv \left(\frac{\partial}{\partial t}\right)^a + \Omega_H \left(\frac{\partial}{\partial \phi}\right)^a = \xi^a + \Omega_H \psi^a, \tag{17.34}$$

where Ω_H is the angular velocity of inertial observers at the horizon. This vector is null at the horizon ($\chi_a \chi^a = 0$) and is a generator of the horizon (everywhere parallel to it). Then one can prove that at the horizon we have

$$\nabla^a(\chi^b \chi_b) = -2\kappa \chi^a|_{\text{horizon}}$$
$$\chi^b \nabla_b \chi^a = \kappa \chi^a|_{\text{horizon}}. \tag{17.35}$$

The constant of proportionality κ is called the *surface gravity* of the horizon. From the above, we can define it as a limit on the horizon:

$$\kappa^2 = \lim_{\text{horizon}} \frac{1}{4} \frac{||\nabla^a(\chi^b\chi_b)||^2}{\chi^a\chi_a} = \frac{||\chi^b\nabla_b\chi^a||^2}{\chi^a\chi_a}. \tag{17.36}$$

Note that $\chi^a\chi_a = 0$ at the horizon (χ^a is a null generator of the horizon), and so also the numerators are zero on the horizon, hence the need to define it as a limit, approaching the horizon.

Finally then, in terms of κ, the Hawking temperature of a general black hole is

$$T_{BH} = \frac{\kappa}{2\pi}. \tag{17.37}$$

17.5 The Four Laws of Black Hole Thermodynamics

The four laws of black hole thermodynamics were formulated by Bardeen, Carter, and Hawking in 1973.

To facilitate comparison, let us first write the *four laws of (regular) thermodynamics:*

0th law: There is a quantity called temperature T that is constant over the system at thermodynamic equilibrium.

1st law: The variation of the energy of the system is given by

$$\delta E = T\delta S - p\delta V + \Omega\delta J + \mu\delta N + \cdots \tag{17.38}$$

2nd law: The entropy always increases, $\Delta S \geq 0$, in any process.

3rd law: It is impossible, by any (even if ideal) procedure to reduce the temperature T to zero by a finite sequence of operations.

Then the *four laws of black hole thermodynamics* are as follows:

0th law: The surface gravity κ of a stationary black hole is constant over the event horizon. We see therefore that κ is the analogue of temperature, and stationarity corresponds to thermal equilibrium. The constancy of κ over the horizon of a stationary black hole is quite technical to prove, so we will just state it here.

1st law: For a system in the presence of a black hole, the variation in mass (or energy) is

$$\delta M = \frac{\kappa}{2\pi}\frac{\delta A}{4} + \Omega_H\delta J_H + \Phi\delta Q \left(+ \int \Omega\delta dJ + \int \bar{\mu}\delta dN + \int \bar{\theta}\delta dS\right), \tag{17.39}$$

where the terms in brackets correspond to the presence of matter other than the black hole, J is angular momentum, Φ is the electric potential, Q the charge. We should note here that the four laws were written before Hawking's paper on black hole radiation, so the authors took great care to phrase everything as an analogy, to say that the area acts as "an analogue of entropy," etc. But after the paper of Hawking, when one now knows that $\kappa/(2\pi)$ is the temperature of the black hole, it allows us to identify

$$S_{BH} = \frac{A}{4} \tag{17.40}$$

(in Planck units) as the *entropy of the black hole*. The BH here could stand for black hole, but actually it stands for *Bekenstein-Hawking entropy*, since Bekenstein was actually the first one to suggest that the black hole has an entropy, and to argue that it is proportional to the area. His simple argument obtained a proportionality constant of order one, though not the 1/4.

There is a general, quite complicated, proof in the paper of Bardeen, Carter, and Hawking for this statement, but we will not attempt to reproduce it here. However, since the most general single (spherically symmetric) stationary black hole is the Kerr-Newman black hole, for our purposes it will suffice to check it for this case, which we will do shortly.

2nd law: In any process in the presence of a black hole, $\Delta(S + S_{BH}) \geq 0$. In particular, corrolaries of this statement are the fact that the area of a black hole *classically* never decreases, and with time it could increase. In particular, in the case of a black hole merger, the area of the formed black hole is bigger or equal to the area of the two initial black holes. Of course, in practice the statements that were proven first were these two "corrolaries" (that the area of a black hole never decreases classically, and the area in a black hole merger increases), out of which the general principle was inferred. These proofs were highly technical, so will not be commented upon here.

3rd law: It is impossible by any (no matter how idealized) procedure to reduce the surface gravity κ of a black hole to zero by any finite sequence of operations. This would be technically hard to prove, but since the temperature of a black hole in flat spacetime (in a curved spacetime background things could be more complicated) shows negative specific heat, and thus decreases with increased M, this is very clear, since we could never reach infinite mass through a finite sequence of operations. (Note that quantum mechanically, a black hole would eventually evaporate by radiating away all its energy until a final burst of very high-temperature radiation, leaving behind a state at $T = 0$, but we are talking here only about the classical limit of black hole physics).

Check for Kerr-Newman

We can now check the first law of black hole thermodynamics, as promised, for the Kerr-Newman case (the most general stationary state). We first calculate the quantities involved in the first law.

1. The area of the horizon. First, we note that the horizon must be the outer one, r_+, since that already takes an infinite amount of time to reach. The region even slightly outside the inner horizon is still causally disconnected from the observers at infinity. The area of the horizon is written in general as

$$A_H = \int_{r=r_+} \sqrt{-g_{\theta\theta}g_{\phi\phi}}d\theta d\phi. \tag{17.41}$$

Substituting the metric components from (17.22), we get

$$A_H = \int_{r=r+} \sqrt{\frac{(r^2+a^2)^2 - \Delta a^2 \sin^2\theta}{\Sigma}} \Sigma \sin^2\theta d\theta d\phi = (r_+^2 + a^2)\int_{-1}^{+1} d(\cos\theta)\int_0^{2\pi} d\phi$$
$$= 4\pi(r_+^2 + a^2), \tag{17.42}$$

where we have used that at the horizon, $r = r_+$, we have $\Delta(r_+) = 0$.

2. The angular velocity of inertial observers, Ω_H. For an inertial observer, we have

$$ds^2 = g_{tt}dt^2, \tag{17.43}$$

but moreover we want it to be at θ =constant, $r =$ constant, so $d\phi(g_{\phi\phi}d\phi + g_{t\phi}dt) = 0$, leading to

$$\Omega \equiv \frac{d\phi}{dt} = -\frac{g_{t\phi}}{g_{\phi\phi}} = \frac{a(r^2 + a^2 - \Delta)}{(r^2 + a^2)^2 - \Delta a^2 \sin^2\theta}. \tag{17.44}$$

At the horizon, it tends to

$$\Omega_H = \frac{a}{r_+^2 + a^2} = \frac{4\pi a}{A_H}. \tag{17.45}$$

3. The electric potential of the event horizon is found from the expression for A in (17.22) at $\theta = 0$:

$$\Phi = \frac{\tilde{Q}r_+}{r_+^2 + a^2} = \frac{4\pi\tilde{Q}r_+}{A_H}. \tag{17.46}$$

4. The surface gravity is calculated from the quantity

$$\chi^a\chi_a = g_{tt} + g_{\phi\phi}\Omega_H^2 + 2g_{t\phi}\Omega_H, \tag{17.47}$$

where we have used the definition of χ^a and the fact that $\xi^t = 1$ and $\psi^\phi = 1$ and the rest of the components are 0. Then we have

$$4\kappa^2 = \lim_{\text{horizon}} \frac{[\partial/\partial r(\chi^a\chi_a)]^2 g^{rr} + [\partial/\partial\theta(\chi^a\chi_a)]^2 g^{\theta\theta}}{\chi^a\chi_a}. \tag{17.48}$$

After a calculation that is left as an exercise, one finds

$$\kappa = \frac{\sqrt{M^2 - a^2 - \tilde{Q}^2}}{2M\left(M + \sqrt{M^2 - a^2 - \tilde{Q}^2}\right) - \tilde{Q}^2} = \frac{4\pi(r_+ - M)}{A_H}. \tag{17.49}$$

Then the variation in the first law is

$$\frac{\kappa}{8\pi}\delta A_H = \frac{r_+\delta r_+ + a\delta a}{r_+^2 + a^2}(r_+ - M). \tag{17.50}$$

The other terms in the first law are

$$\Omega_H\delta J_H = \frac{a}{r_+^2 + a^2}\delta(aM) = \frac{a}{r_+^2 + a^2}(a\delta M + M\delta a)$$

$$\Phi\delta\tilde{Q} = \frac{r_+\tilde{Q}\delta\tilde{Q}}{r_+^2 + a^2}. \tag{17.51}$$

Thus the sum of the above three terms gives

$$\frac{\kappa\delta A_H}{8\pi} + \Omega_H\delta J_H + \Phi\delta\tilde{Q} = \frac{1}{r_+^2 + a^2}[a^2\delta M + r_+\{\tilde{Q}\delta\tilde{Q} + (r_+ - M)\delta r_+ + a\delta a\}]. \tag{17.52}$$

Substituting also

$$r_+ - M = \sqrt{M^2 - \tilde{Q}^2 - a^2} \Rightarrow \delta r_+ = \delta M + \frac{M\delta M - \tilde{Q}\delta\tilde{Q} - a\delta a}{\sqrt{M^2 - \tilde{Q}^2 - a^2}}, \qquad (17.53)$$

we finally find

$$\frac{\kappa\delta A_H}{8\pi} + \Omega_H\delta J_H + \Phi\delta\tilde{Q} = \frac{1}{r_+^2 + a^2}[a^2\delta M + r_+(r_+ - M)\delta M + r_+M\delta M] = \delta M,$$

$$(17.54)$$

as expected.

Important Concepts to Remember

- The black hole solution with electric charge (Reissner-Nordstrom) has in general two horizons, r_+ and r_-.
- It obeys the BPS bound $M \geq \tilde{Q}$, and when saturated, the horizons coincide, $r_+ = r_-$, and moreover coincide with the singularity, $\bar{r} = 0$. A $M < \tilde{Q}$ would correspond to a naked singularity, which is excluded by various theorems.
- That extremal solution is written in terms of a harmonic function $H(\bar{r})$ as $ds^2 = -H^{-2}dt^2 + H^2d\vec{x}_\perp^2$.
- The most general stationary black hole is the Kerr-Newman solution, with M, Q, and angular momentum J. Any other charge observed at infinity would be called "hair," and there are "no-hair theorems" forbidding it.
- Killing vectors are isometry directions for the metric.
- Quantum Field Theory at finite temperature T for static quantities can be described by Wick rotation to Euclidean space, and periodicity of this (imaginary) time with period $\beta = 1/T$.
- Black holes radiate quantum mechanically with temperature $T = \kappa/(2\pi)$, becoming $T = 1/(8\pi M)$ for the Schwarzschild case in flat space. These have $C = \partial M/\partial T < 0$, so radiate away their energy.
- The classical black holes obey the same four laws of thermodynamics as usual thermodynamical systems, with Bekenstein-Hawking entropy $S_{BH} = A/4$, where A is the area of the event horizon, and with the horizon as the thermodynamical system.
- The first law of black hole thermodynamics, $\delta M = T\delta S + \Omega_H\delta J_H + \Phi_H\delta\tilde{Q}$, can be checked for the Kerr-Newman solution.

Further Reading

For black holes and their thermodynamics, see chapters 31 and 33 in [27]. Advanced texts are Wald [26] and Hawking and Ellis [28]. The original articles of Hawking [29] on black hole radiation and of Bardeen, Carter, and Hawking [30] on the four laws of black hole thermodynamics are worth reading.

Exercises

(1) Using arguments similar to the ones for the black hole with mass M and charge Q, calculate g_{00} for the case of M, Q and also a "cosmological constant," with $T_{\mu\nu} = \Lambda g_{\mu\nu}$. Suggest a form for the correct complete ds^2.

(2) Check whether the two horizons of the Reissner-Nordstrom solution are event horizons, i.e. whether light cannot escape from them in finite time.

(3) Calculate the norms of the Killing vectors $\xi^a \equiv (\partial/\partial t)^a$ and $\psi^a \equiv (\partial/\partial\phi)^a$ for the Reissner-Nordstrom solution at the horizon, for both nonextremal and extremal cases. Is $(\partial/\partial\theta)^a$ a Killing vector? What about for the Kerr-Newman solution?

(4) Calculate the temperature of the nonextremal Reissner-Nordstrom solution, paralleling what we did in the chapter for the Schwarzschild solution, and check that $T \to 0$ in the extremal limit.

(5) Consider the Kerr-Newman black hole and the limit of zero temperature for it. Does the entropy go to zero in general? How do you understand this result?

(6) Fill in the steps omitted in the calculation of the surface gravity κ of the Kerr-Newman black hole, to obtain

$$\kappa = \frac{4\pi(r_+ - M)}{A_H}. \tag{17.55}$$

18 Extra Dimensions and Kaluza-Klein

In this chapter we will study the concept of extra dimensions and the Kaluza-Klein solution to dealing with them. This is necessary, since, for instance, in superstring theory we are forced by the quantum consistency of the theory to have $D = 10$ total number of dimensions. Also in supergravity, while we can certainly consider the theory in four dimensions by itself, considering the theory in $D = 11$ dimensions allows a certain minimality and unification, since there is a unique 11-dimensional supergravity, and from it, we can obtain various four-dimensional supergravities. For these reasons and others, we are led to consider what we could do with extra dimensions, other than the observed 3+1 ones.

18.1 Kaluza-Klein Theory

The idea of extra dimensions is an old one, by now almost a century old. It was introduced by Theodor Kaluza in 1921 and Oskar Klein in 1926. They considered the possibility that there are extra dimensions, which are, however, very small, thus unobservable. According to Klein, who considered this general plan, we can consider a D-dimensional space that is a product, $K_D = M_4 \times K_n$, of our 4-dimensional space M_4, times an n-dimensional *compact space*, like an n-sphere S^n or an n-torus $T^n = S^1 \times \cdots \times S^1$. The idea is useful, since it can be used for unification of fields, as proposed by Kaluza. The reason we don't see the extra dimensions is that the compact space is very small, so we cannot penetrate it, either with light (so we can actually see it), or with high-energy particles (so we can detect it). The above paradigm became known as Kaluza-Klein theory.

This theory did not receive much interest for a long time, except from Einstein who, after being the referee for Kaluza's paper and keeping it for more than a year trying to find a flaw in it, become interested in the idea of unification through extra dimensions. But then, around the 1970s, it become popular again, after first supergravity became popular (in the late 1970s and early 1980s), and as we said, it naturally considered extra dimensions, and then finally after it became clear (in the early 1980s) that superstring theory is a quantum gravity theory, and it needs 10 dimensions.

But how small can the extra dimensions be? There are two answers we can give:

(1) The experimental one is: *Assuming that there is nothing special about the theory in D dimensions*, they have to be smaller than about $(1 \text{ TeV})^{-1}$, since we reach energies up to about a few TeV in accelerator experiments, and we don't see anything yet. Of course, the assumption can be broken, and there are special theories, like so-called large extra

dimensional theories (à la Arkani-Hamed, Dimopoulos, and Dvali), or Randall-Sundrum theories, where the dimensions can be considerably larger or even infinite.

(2) The theoretical one is: The presence of extra dimensions would modify geometry, thus (according to our analysis of the previous chapters) gravity, so it is possible that these extra dimensions stayed small, rather than become infinite, due to quantum mechanical reasons. That would mean that the natural scale associated with these dimensions would be the quantum gravity scale, the Planck scale $l_P \sim G_N^{1/2}$, the only scale that can be made from G (gravity), $\hbar$ (quantum), and c (relativistic). It is about $(10^{19} \text{ GeV})^{-1}$, thus ridiculously small.

So how do we understand the presence of extra dimensions? We said that the space is a product $M_4 \times K_n$, but it is actually more useful to consider the case when K_n varies slightly as we move in our (3+1)–dimensional spacetime, $K_n = K_n(\vec{x}, t)$. That is a realistic situation, since the volume or shape of the extra dimensions acts as fields that can vary over the 3+1–dimensional spacetime. But a more useful concept is then to consider instead the total spacetime parametrized by $(\vec{x}, t, \vec{y})$, where $\vec{y} \in K_n$. Mathematically, that means that we can consider, instead of fields in just our spacetime, $\phi(\vec{x}, t)$, fields in the total spacetime $\phi(\vec{x}, t, \vec{y})$. But we don't see the extra dimensional space, so we would like a description that doesn't rely on it, i.e. a purely 4-dimensional description.

To understand better the concepts, let us consider first the simplest case of a compact space, the circle S^1. In that case, we know that there is always exists a Fourier expansion in terms of regular fields, times modes on the circle:

$$\phi(\vec{x}, t, y) = \sum_{n \geq 0} e^{\frac{iny}{R}} \phi_n(\vec{x}, t). \tag{18.1}$$

In this case, there is nothing assumed to write this formula; this formula is a general theorem, namely the Fourier theorem. We will call this expansion in general the "KK expansion." Then from the point of view of 3+1 dimensions, we have simply an infinite set of usual (3+1)–dimensional fields $\phi_n(\vec{x}, t)$ that replace the (4+1)–dimensional field.

18.2 The Kaluza-Klein Metrics

There are in the literature three metrics that are sometimes called KK metrics, so we need to differentiate between them.

1. The KK Background Metric

We said that the background space is of the type $M_4 \times K_n$, but that means that this space must be a solution of the equations of motion (including Einstein's equations) of direct product type. That means that, first, the metric is block diagonal and, second, that the nontrivial blocks depend only on the coordinates corresponding to them,

$$g_{\Lambda\Sigma}^{(0)}(\vec{x}, t; \vec{y}) = \begin{pmatrix} g_{\mu\nu}^{(0)}(\vec{x}, t) & 0 \\ 0 & g_{mn}^{(0)}(\vec{y}) \end{pmatrix}, \tag{18.2}$$

where $\vec{y}$ are coordinates on K_n, which corresponds to the second block on the diagonal.

Note that it can also be the case that we expand around a background that is not a solution of the equations of motion (this case has been considered in the literature), but in that case we must be careful with the consequences we derive from it, since the starting point is not very well defined.

But a subtlety must be mentioned here. The metric itself is a field of the theory, i.e. a variable, so when we write that we have a theory on $M_4 \times K_n$, we mean that this is the background, not the full fluctuating metric. If we consider only small fluctuations, the above statement could be considered as a pedantic point, but we will see, especially in Chapter 19, that we can consider instead large deviations for the metric, and then it really matters.

2. The KK Expansion

Once we have a KK background, we can consider the KK expansion around it. This is an *exact* decomposition, the generalization of the Fourier expansion on a circle described before.

As we said, the Fourier theorem states that we can always expand functions on a circle, perhaps also being functions on another space, as

$$\phi(\vec{x}, t; y) = \sum_{n \geq 0} \phi_n(\vec{x}, t) e^{\frac{iny}{R}}. \tag{18.3}$$

The next simplest similar case that should be familiar is the expansion on a 2-sphere S^2, known from electromagnetism and/or quantum mechanics (in that case without the extra dependence on $(\vec{x}, t)$), also a theorem, namely that we can always expand

$$\phi(\vec{x}, t; \theta, \phi) = \sum_{l,m} \phi_{lm}(\vec{x}, t) Y_{l,m}(\theta, \phi). \tag{18.4}$$

Here the $Y_{lm}(\theta, \phi)$ are called the spherical harmonics on S^2, and in both cases, the functions onto which we expand the fields are eigenfunctions of the Laplacean on the extra dimensions:

$$\partial_y^2 e^{\frac{iny}{R}} = -\left(\frac{n}{R}\right)^2 e^{\frac{iny}{R}}$$
$$\Delta_2 Y_{lm}(\theta, \phi) = -\frac{l(l+1)}{R^2} Y_{lm}(\theta, \phi). \tag{18.5}$$

Similarly then, in the general case of an arbitrary extra-dimensional (product) space K_n, (it is a theorem that) we can always write

$$\phi(\vec{x}, t; \vec{y}) = \sum_{q, I_q} \phi_q^{I_q}(\vec{x}, t) Y_q^{I_q}(\vec{y}). \tag{18.6}$$

Here q is an index that measures the eigenvalue of the Laplacean in n dimensions, $\Delta_{(n)}$, like n for the S^1 and l for the S^2, $Y_q^{I_q}(\vec{y})$ is also called a *spherical harmonic*, like in the 2-sphere case, and I_q is an index in some representation (that depends on q) of the symmetry group of K_n. Indeed, consider the case of S^2, which as a manifold is the same as the "coset manifold" $SO(3)/SO(2) = SU(2)/U(1)$ (in general, an n-sphere $S^n = SO(n+1)/SO(n)$), and has the symmetry $SO(3)$ (in general, the symmetry is the group in the numerator of the coset). The $SU(2)$ has representations of given spin j, half integer or integer, but for the bosonic case only the integer representations, labeled by l, are allowed. So the l in the Y_{lm} has the dual role of measuring the eigenvalue of the Laplacean and being an index for the representations of the symmetry group, the analog of q in the general case. In the representation of given l, the index $m = -l, \ldots, 0, \ldots, l$ is the index for the given representation, the analog of I_q in the general case.

Then also $Y_q^{I_q}$ are eigenfunctions of the Laplacean on K_n, with eigenvalues labeled by q:

$$\Delta_{(n)} Y_q^{I_q}(\vec{y}) = -m_q^2 Y_q^{I_q}(\vec{y}). \tag{18.7}$$

We have called the eigenvalue $-m_q^2$, since indeed m_q acts as a mass from the 4-dimensional point of view. Indeed, consider a single field in the KK expansion:

$$\phi(\vec{x}, t; \vec{y}) = \phi_q^{I_q}(\vec{x}, t) Y_q^{I_q}(\vec{y}). \tag{18.8}$$

Then, since the D-dimensional D'Alembertian (wave operator) splits into a 4-dimensional D'Alembertian plus an n-dimensional Laplacean, $\Box_{(D)} = \Box_{(4)} + \Delta_{(n)}$, and $\Box_{(4)}$ only acts on ϕ_q, whereas $\Delta_{(n)}$ only acts on Y_q, we get

$$\Box_{(D)} \Phi(\vec{x}, t; \vec{y}) = (\Box_{(4)} + \Delta_{(n)}) \phi(\vec{x}, t; \vec{y}) = (\Box_{(4)} - m_q^2) \phi(\vec{x}, t; \vec{y}). \tag{18.9}$$

That means that for a field that is D-dimensional massless, i.e. $\Box_{(D)} \Phi(\vec{x}, t; \vec{y}) = 0$, we obtain

$$\left[(\Box_{(4)} - m_q^2) \Phi_q^{I_q}(\vec{x}, t) \right] Y_q^{I_q}(\vec{y}) = 0, \tag{18.10}$$

i.e. the 4-dimensional field looks like a massive field of mass m_q. That incidentally also shows that naively considering a *temporal* compact extra dimension, i.e. a D'Alembertian $\Box_{(4)} - \partial_0^2$, would result in $\Box_{(4)} + m_q^2$, i.e. a tachyon (an object of negative mass-squared), that signals an instability of the system.

So now we can also answer the question: If the space K_n is too small to directly observe, how would we detect it indirectly? From the above description in terms of effective $(3 + 1)$–dimensional fields, we see that to excite a nontrivial field, and thus to access structure on K_n at the level of its $Y_q^{I_q}$ spherical harmonic, we need at least an energy of m_q, to produce the field $\phi_q^{I_q}$ at least at rest, if not in motion.

3. The KK Reduction Ansatz

Finally, we should consider a case when we don't see the extra dimensions, since K_n is very small. Since the KK expansion is in terms of an infinite number of fields, this expansion is clearly not very useful. Instead, we can consider the case that we have an effective description only in terms of the "$n = 0$" fields, or "y-independent" ("$y = 0$") fields. This is called the *KK dimensional reduction ansatz*. Note that in general, we would keep the fields of lowest allowed eigenvalue of the Laplacean $\Delta_{(n)}$, which in general would be a multiplet of fields. In general, this would not be $m_q = 0$, would also not necessarily be $n = 0$ (or $q = 0$), and would not be y-independent, but rather of a given, simplest dependence, $Y_0(\vec{y})$.

So generally, we would have

$$\phi(\vec{x}, t; \vec{y}) = \phi_0(\vec{x}, t) Y_0(\vec{y}) \tag{18.11}$$

for the dimensional reduction ansatz. Also note that this is an *ansatz*, which is a name for an educated guess (in German), meaning that it is not guaranteed to work. The point is that putting $\phi_n = 0$ for $n \neq 0$ is not necessarily a solution to the equations of motion, though for the circle (Fourier) case it is. If it isn't, we say that we have an *inconsistent truncation*, and we will discuss more about this situation (including how one could fix it) in the next chapter.

In conclusion, the KK background metric is a solution to the equations of motion, the KK expansion is a parametrization (always valid), and the KK reduction is an ansatz (not guaranteed to work).

18.3 Fields with Spin

Until now we have kind of assumed that the D-dimensional fields were scalars, we didn't have any Lorentz indices on them. So now we want to understand what happens in the case of fields with (Lorentz) spin. We will see that various components act as different fields (in different 4-dimensional Lorentz spin representations) in four dimensions, and thus we obtain the advertised idea of *unification of different fields into a higher dimension*.

For instance, the simplest case we can consider is a D-dimensional vector, A_M, i.e. D-dimensional electromagnetism. We split the D–dimensional indices into 4-dimensional and n-dimensional, $M = (\mu, m)$, with $\mu = 0, 1, 2, 3$ and $m = 4, \ldots, 4 + n$. Then the vector splits as

$$A_M(\vec{x}, t; \vec{y}) = (A_\mu(\vec{x}, t; \vec{y}), A_m(\vec{x}, t; \vec{y})). \tag{18.12}$$

In this case, A_μ behaves as a 4-dimensional vector (gauge field): If we consider a $(3+1)$–dimensional Lorentz transformation $\Lambda^\mu{}_\nu \in SO(1, 3)$, then A_μ transforms as

$$A'_\mu(\vec{x}', t'; \vec{y}) = \Lambda_\mu{}^\nu A_\nu(\vec{x}, t; \vec{y}). \tag{18.13}$$

On the other hand, for the same $(3+1)$–dimensional Lorentz transformation, A_m act as scalars, i.e. are unchanged,

$$A'_m(\vec{x}', t'; \vec{y}) = A_m(\vec{x}, t; \vec{y}). \tag{18.14}$$

We thus see that the 4-dimensional Lorentz vector A_μ and the n 4-dimensional Lorentz scalars A_m are unified in the same D-dimensional Lorentz vector field, which is the idea of unification mentioned before.

The original D-dimensional theory being (by assumption) fully Lorentz invariant under $SO(1, 3 + n)$, the KK background of a product type (which therefore does not mix μ and m indices) breaks that invariance down to $SO(1, 3) \times SO(n)$. Therefore now the $SO(n)$ acts *only* on $A_m(\vec{x}, t; \vec{y})$ as an *internal* symmetry, transforming the coordinates $\vec{y}$, not $\vec{x}, t$. Specifically,

$$A'_m(\vec{x}, t; \vec{y}') = \Lambda_m{}^n A_n(\vec{x}, t; \vec{y}). \tag{18.15}$$

That means that now $SO(n)$ appears from a 4-dimensional perspective as an internal symmetry like the gauge $SU(3) \times SU(2) \times U(1)$, and not a spacetime symmetry (that would act on the spacetime indices). That means that in the KK theory we have managed a *geometrization of the internal symmetry*.

We see then exemplified the two important reasons to like KK theory:

1. The unification of various different fields into a higher dimension and
2. The geometrization of the internal symmetry.

In this case of the vector A_M, the KK expansion is

$$A_\mu(\vec{x}, t; \vec{y}) = \sum_{q, I_q} A_\mu^{q, I_q}(\vec{x}, t) Y_q^{I_q}(\vec{y})$$

$$A_m(\vec{x}, t; \vec{y}) = \sum_{q, I_q} A^{q, I_q}(\vec{x}, t) Y_m^{q, I_q}(\vec{y}). \tag{18.16}$$

Note here that in the first case the index μ sits on the 4-dimensional fields, whereas in the second the index m sits on the spherical harmonic, since the A_m are scalars. That is as it should be, since it is the spherical harmonic that depends on $\vec{y}$, which is acted upon by the Lorentz transformation. Also note that I_q is an index in the *isometry group of the space*, whereas m is an index in the (local) Lorentz group (the group that acts on the flat space tangent to the manifold at any point). For instance, in the case of the n-sphere, $K_n = S^n = SO(n+1)/SO(n)$, the isometry group (with I_q indices) is $SO(n+1)$, but the Lorentz group is $SO(n)$. In this case it happens that $SO(n) \subset SO(n+1)$, so the transformation under the Lorentz group can be embedded into the isometry transformation, but that is not true in general. In fact, it is the isometry group, and not the Lorentz group, that plays the role of internal symmetry, since it is it that is represented on the 4-dimensional fields. A very particular case is when the two groups coincide, the case of the torus T^n, which is just a slice of flat space with some identifications, so the isometry group is $SO(n)$. Then, as we will shortly see, the spherical harmonics contain something like "$\delta_m^{I_q}$," that allows us to identify the I_q index on the fields as a Lorentz index, acted upon by $\Lambda_m{}^n$.

Note that as far as representation theory goes, for the breaking of the Lorentz symmetry from $SO(1, 3+n)$ to $SO(1,3) \times SO(n)$, the split of the vector representation as $M = (\mu, m)$ can be formally described as

$$R_V = r_V \otimes \mathbb{1} \oplus \mathbb{1} \otimes r_V^{(n)}, \tag{18.17}$$

where R or r refers to representation, and V to the vector representation, and the two factors refer to the two groups. Another way to write it, that emphasizes the dimensions of the representations, would be as

$$\underline{(4+n)}_V = \underline{4}_V \otimes 1 \oplus 1 \otimes \underline{n}_V. \tag{18.18}$$

For gravity, we have a similar situation, with

$$g_{MN}(\vec{x}, t; \vec{y}) = \begin{pmatrix} g_{\mu\nu}(\vec{x}, t; \vec{y}) & g_{m\mu}(\vec{x}, t; \vec{y}) \\ g_{\mu m}(\vec{x}, t; \vec{y}) = g_{m\mu} & g_{mn}(\vec{x}, t; \vec{y}) \end{pmatrix}. \tag{18.19}$$

Since the original g_{MN} is a symmetric tensor, i.e. a metric (graviton), after the breaking of $SO(1, 3 + n) \to SO(1, 3) \times SO(n)$ we have that $g_{\mu\nu}$ is also a symmetric tensor, but now for the $SO(1, 3)$ group, so it is a 4-dimensional metric (the graviton), whereas $g_{\mu m} = g_{m\mu}$ is a $SO(1, 3)$ vector, i.e. transforms as

$$g'_{\mu m}(\vec{x}', t'; \vec{y}) = \Lambda_\mu{}^\nu g_{\nu m}(\vec{x}, t; \vec{y}). \tag{18.20}$$

Then also $g_{mn}(\vec{x}, t; \vec{y})$ are $SO(1, 3)$ scalars, and as before $SO(n)$ acts only on $\vec{y}$, and on the m, n indices.

As a representation, under the split $SO(1, 3 + n) \to SO(1, 3) \times SO(n)$, the split of the symmetric representation above can be described as

$$R_S = r_S \otimes \mathbb{1} \oplus r_V \otimes r_V^{(n)} \oplus \mathbb{1} \otimes r_S^{(n)}, \tag{18.21}$$

where the index S refers to the symmetric representation, or emphasizing the dimensions of representations, as

$$\underline{(4 + n)(5 + n)/2_S} = \underline{10_S} \otimes 1 \oplus \underline{4_V} \otimes \underline{n_V} \oplus 1 \otimes \underline{n(n + 1)/2_S}. \tag{18.22}$$

We have described bosons, we should describe also fermions, i.e. spinor representations. For a spinor ψ_Ω, with Ω in a spinor representation in D dimensions, the index splits into a *product* of a spinor representation for $SO(1, 3)$, with index α, and a spinor representation for $SO(n)$, with index z, i.e. $\Omega = (\alpha z)$, and the spinor is thus a set of Lorentz spinors for $SO(1, 3)$, with a label that is a spinor for the internal $SO(n)$ representation:

$$\psi_\Omega(\vec{x}, t; \vec{y}) = \psi_{\alpha z}(\vec{x}, t; \vec{y}). \tag{18.23}$$

As for the KK expansion, for the spinor it is

$$\psi_{\alpha z}(\vec{x}, t; \vec{y}) = \sum_{q, I_q} \psi_\alpha^{q, I_q}(\vec{x}, t) \eta_z^{q, I_q}(\vec{y}). \tag{18.24}$$

Here η_z are $SO(n)$ spinors called *Killing spinors*. We will not try to give a general definition for them. We will just notice that since we want ψ_Ω to be an anticommuting spinor, and also (especially) the 4-dimensional spinor η_α to be also anticommuting, symmetry properties dictate that the Killing spinors η_z must be *commuting*.

In terms of representations, we can describe the above decomposition of the spinor under the breaking $SO(1, 3 + n) \to SO(1, 3) \times SO(n)$ as

$$R_{sp} = r_{sp} \otimes r_{sp}^{(n)}. \tag{18.25}$$

18.4 The Original Kaluza-Klein Theory

As an example of the formalism, let us consider the case of the original theory proposed by Kaluza and Klein. Their idea was to unify gravity, described by the metric $g_{\mu\nu}$, and electromagnetism, described by the vector B_μ, in a 5-dimensional metric $g_{\Lambda\Sigma}$.

Then:

(1) The KK background metric is a solution of 5-dimensional gravity of $Mink_4 \times S^1$ form,

$$g_{\Lambda\Sigma}^{(0)}(\vec{x}, t; y) = \begin{pmatrix} \eta_{\mu\nu} & \vec{0} \\ \vec{0} & g_{55} = 1 \end{pmatrix}. \tag{18.26}$$

(2) The KK expansion is a parametrization, namely the usual Fourier expansion on the circle,

$$g_{\Lambda\Sigma}(\vec{x}, t; y) = \begin{pmatrix} \eta_{\mu\nu} + \sum_n h_{\mu\nu}^{(n)}(\vec{x}, t) e^{\frac{iny}{R}} & g_{5\mu} = g_{\mu 5} \\ g_{\mu 5}(\vec{x}, t; y) = \sum_n h_{\mu 5}^{(n)}(\vec{x}, t) e^{\frac{iny}{R}} & g_{55} = 1 + \sum_n \phi^{(n)}(\vec{x}, t) e^{\frac{iny}{R}} \end{pmatrix}. \tag{18.27}$$

(3) The KK reduction amounts to putting all the $n \neq 0$ modes to zero,

$$g_{\Lambda\Sigma}(\vec{x}, t; y) = \begin{pmatrix} \eta_{\mu\nu} + h_{\mu\nu}^{(0)}(\vec{x}, t) \equiv g_{\mu\nu}(\vec{x}, t) & g_{5\mu} = g_{\mu 5} \\ g_{\mu 5}(\vec{x}, t; y) = h_{\mu 5}^{(0)}(\vec{x}, t) \equiv B_\mu & g_{55} = 1 + \phi^{(0)}(\vec{x}, t) \equiv \varphi(\vec{x}, t) \end{pmatrix}. \tag{18.28}$$

Note that then $g_{\mu\nu}(\vec{x}, t)$ is the 4-dimensional metric and $B_\mu(\vec{x}, t)$ is the 4-dimensional vector of electromagnetism. This seems fine, except for one thing: It predicts that, besides massless (since the 5-dimensional metric was massless) gravity and electromagnetism, we also would have a massless scalar φ. But this certainly contradicts experiments, since a massless scalar would create a long-range scalar fifth force that would affect the motion of the planets, and it would have been observed since Kepler and Newton, so that is ruled out.

A possible solution would be to put also $\varphi = 1$, i.e. $\phi^{(0)} = 0$, but that turns out (unlike the dimensional reduction itself) to be an inconsistent truncation, that is, it is not a solution of the equations of motion. So we would need to keep this scalar in the theory, and that contradicts experiments, hence the original Kaluza-Klein theory does not work.

Generalization: n-Torus $T^n = (S^1)^n$

The torus is obtained from periodic identifications of a slice (product of intervals) of $\mathbb{R}^n$ (for instance, for T^2 we take a rhomboidal slice generated by two arbitrary vectors, and we identify the opposite sides). But that means that the metric on the torus is flat, $g_{mn}^{(0)} = \delta_{mn}$. Therefore the KK background metric is

$$g_{\Lambda\Sigma}^{(0)}(\vec{x}, t; \vec{y}) = \begin{pmatrix} \eta_{\mu\nu} & \vec{0} \\ \vec{0} & g_{mn} = \delta_{mn} \end{pmatrix}. \tag{18.29}$$

The KK expansion is then simply the product of the Fourier expansions on the circles in T^n:

$$g_{\Lambda\Sigma}(\vec{x}, t; \vec{y}) = \begin{pmatrix} \eta_{\mu\nu} + \sum_{\{n_i\}} h_{\mu\nu}^{\{n_i\}}(\vec{x}, t) e^{\sum_i \frac{in_i y^i}{R_i}} & g_{m\mu} = g_{\mu m} \\ g_{\mu m}(\vec{x}, t; \vec{y}) = \sum_{\{n_i\}} B_\mu^{m\{n_i\}}(\vec{x}, t) e^{\sum_i \frac{in_i y^i}{R_i}} & g_{mn} = \delta_{mn} + \sum_{\{n_i\}} h_{mn}^{\{n_i\}}(\vec{x}, t) e^{\sum_i \frac{in_i y^i}{R_i}} \end{pmatrix}. \tag{18.30}$$

We see therefore that the scalar spherical harmonics are simply the products of the Fourier modes:

$$Y_{\{n_i\}}(\vec{y}) = \prod_i e^{\frac{in_i y^i}{R_i}}.$$ (18.31)

Moreover, for the spherical harmonics Y_{mn}, these reduce to the same scalar spherical harmonics, with some delta functions, since as we advertised, in the torus case the isometry group is simply the Lorentz group $SO(n)$, so we can identify the two, leading to

$$Y^{pr}_{mn\{n_i\}} = \delta^p_m \delta^r_n Y_{\{n_i\}},$$ (18.32)

where the set $(pr\{n_i\})$ is what we mean by (qI_q) in the general case. Then doing the delta functions when multiplying with $h^{pr\{n_i\}}$, we obtain $h^{\{n_i\}}_{mn}$.

Important Concepts to Remember

- We can have a small extra-dimensional compact space K_n, i.e. $M_4 \times K_n$, used for unification of fields in higher dimensions by Kaluza and Klein.
- We can view the KK theory as defined on a D-dimensional space, or as an infinite set of 4-dimensional fields, defined by the KK expansion.
- The KK background metric is a solution of the equations of motion around which we expand the theory, though one can formally consider expansion around a background that is not a solution.
- The KK expansion is a parametrization in terms of an infinite set of 4-dimensional fields, a generalization of the Fourier expansion that can always be performed.
- It is written in terms of spherical harmonics $Y^{qI_q}(\vec{y})$ which are eigenfunctions of the Laplacean organized by q in terms of the eigenfunction m_q, acting like a 4-dimensional mass for the fields, and the index I_q in a representation of the isometry group of the space.
- The KK reduction ansatz is an ansatz for putting all $n \neq 0$ fields to zero, and keeping only the multiplet with the lowest allowed value for the Laplacean, and the simplest $\vec{y}$ dependence, $Y_0(\vec{y})$, so $\phi(\vec{x}, t; \vec{y}) = \phi_0(\vec{x}, t)Y_0(\vec{y})$. It can in principle be an inconsistent truncation, i.e. not a solution of the equations of motion to put $\phi_{n\neq 0} = 0$.
- Fields with spin show unification in the higher dimension, since the higher dimensional Lorentz indices split under the breaking of the Lorentz symmetry from $SO(1, 3 + n)$ to $SO(1, 3) \times SO(n)$, due to the background metric being of block diagonal form. For instance, $M = (\mu, m)$ for a vector index (sum) and $\Omega = (\alpha z)$ for a spin index (product). The spherical harmonics carry all the indices for $SO(n)$ (compact) Lorentz spin, besides the indices for the isometry group representation.
- In the original Kaluza-Klein theory, the KK reduction results in massless fields $(g_{\mu\nu}, B_\mu, \varphi)$, i.e. gravity, electromagnetism, and scalar. But the scalar contradicts experiments, and putting it to zero is not a consistent truncation.
- For the case of the torus $T^n = (S^1)^n$, the scalar spherical harmonics are products of Fourier modes on each circle, and, since the Lorentz and isometry groups are the same,

the spherical harmonics with Lorentz spin indices reduce to the scalar spherical harmonics, with delta functions.

Further Reading

For a more detailed introduction to the KK program in supergravity, see the review [31]. For a specific example and more references, see [32].

Exercises

(1) Write the KK background metric and KK expansion for a compactification of the 8-dimensional metric on $M_4 \times S^2 \times S^2$.

(2) Write the KK reduction of a free 8 dimensional gauge field on $M_4 \times S^2 \times T^2$.

(3) Dimensionally reduce an 11-dimensional metric on $S^2 \times S^2 \times T^3$, down to four dimensions.

(4) Consider the nonlinear version of the original KK metric reduction ansatz:

$$g_{MN} = \phi^{-1/3} \begin{pmatrix} g_{\mu\nu} & B_\mu \phi \\ B_\mu \phi & \phi \end{pmatrix}. \tag{18.33}$$

Prove that the quadratic action for B_μ doesn't contain factors of ϕ, at least in the case that ϕ is a constant.

(5) Prove that for the KK ansatz

$$g_{\mu m}(\vec{x}, \vec{y}) = B_\mu^{AB}(\vec{x}) V_m^{AB}(\vec{y}), \tag{18.34}$$

where (AB) is in the adjoint of a gauge group and $V^{AB} = V^{mAB}\partial_m$ satisfies its nonabelian algebra, then, choosing a general coordinate transformation with parameter

$$\xi_m(\vec{x}, \vec{y}) = \lambda^{AB}(\vec{x}) V_m^{AB}(\vec{y}) \tag{18.35}$$

(which corresponds to an isometry of the metric, V_m^{AB} is called the Killing vector for the isometry), the transformation with parameter $\lambda^{AB}(\vec{x})$ is the nonabelian gauge transformation of $B_\mu^{AB}(\vec{x})$.

(6) Consider Y^A, $A = 1, \ldots, 5$ as Cartesian coordinates for the 4-sphere S^4. Check that

$$Y^{AB} = Y^A Y^B - \frac{1}{5}\delta^{AB}Y^2 \tag{18.36}$$

is a symmetric traceless spherical harmonic. Check that, as a polynomial in five dimensions, we have $\Box_{(5)}Y^{AB} = 0$, and then, defining $Y^A Y^A \equiv r^2$, check that

$$\Box_{(5)}Y^{AB} = -\frac{k(k+3)}{R^2}Y^{AB}. \tag{18.37}$$

19 Electromagnetism and Gravity in Various Dimensions, Consistent Truncations

In this chapter we will continue the study of higher dimensions, defining the generalization of electromagnetism and of gravity in higher dimensions, and learning how to do consistent truncations and nonlinear ansätze.

19.1 Electromagnetism in Higher Dimensions

Electromagnetism is easily generalized to higher dimensions. The Maxwell's equations, the equation of motion

$$\partial^\mu F_{\mu\nu} = j_\nu \tag{19.1}$$

and the Bianchi identity

$$\partial_{[\mu} F_{\nu\rho]} = 0, \tag{19.2}$$

are both trivially generalized (are the same) in higher dimensions. Also, the solution of the Bianchi identity is still

$$F_{\mu\nu} = \partial_\mu A_\nu - \partial_\nu A_\mu. \tag{19.3}$$

In three spatial dimensions, we also write $A_\mu = (-\Phi, \vec{A})$ and define the electric field as

$$E^i = F^{0i} = -F_{0i} \Rightarrow \vec{E} = -\partial_0 \vec{A} - \vec{\nabla}\Phi. \tag{19.4}$$

These clearly also trivially generalize to higher dimensions. However, the magnetic field, defined as

$$B^i = \frac{1}{2}\epsilon^{ijk} F_{jk}, \tag{19.5}$$

doesn't obviously generalize, unless we consider the ϵ tensor in d spatial dimensions, resulting in a magnetic field that is a $(d-2)$−antisymmetric tensor, or $(d-2)$−form.

The local form of Gauss's law,

$$\vec{\nabla}^2 \Phi = -\frac{\rho_e}{\epsilon_0}, \tag{19.6}$$

or

$$\vec{\nabla} \cdot \vec{E} = \frac{\rho_e}{\epsilon_0} \tag{19.7}$$

can be also trivially generalized to higher dimensions.

19.2 Gravity in Higher Dimensions

The Einstein equation was written in a form valid in any dimension,

$$R_{\mu\nu} - \frac{1}{2}g_{\mu\nu}R = 8\pi G_N^{(D)} T_{\mu\nu}, \tag{19.8}$$

the only difference being the use of the D-dimensional Newton constant $G_N^{(D)}$. As before, we still have the Riemann tensor

$$R^{\mu}{}_{\nu\rho\sigma} = \partial_\rho \Gamma^{\mu}{}_{\nu\sigma} - \partial_\sigma \Gamma^{\mu}{}_{\nu\rho} + \Gamma^{\mu}{}_{\lambda\rho}\Gamma^{\lambda}{}_{\nu\sigma} - \Gamma^{\mu}{}_{\lambda\sigma}\Gamma^{\lambda}{}_{\nu\rho}, \tag{19.9}$$

and the usual definition of the Christoffel symbols. From the fact that the Riemann tensor is written as a field strength for the Christoffel symbol, we have the gravitational Bianchi identity,

$$\partial_{[\lambda}R^{\mu}{}_{|\nu|\rho\sigma]} = 0, \tag{19.10}$$

as usual. The local form of the Gauss's law for gravity,

$$\vec{\nabla}^2 U_{\text{Newton}} = 4\pi G_N^{(D)} \rho_m, \tag{19.11}$$

is still the same, since it is found as the weak field limit of the Einstein equation, with the identification

$$\kappa_N h_{00} = -2U_{\text{Newton}}. \tag{19.12}$$

19.3 Spheres in Higher Dimensions

Since we want to generalize the integrated Gauss's law to higher dimensions, I will do a mathematical interlude in order to remember how to calculate the volume of spheres in higher dimensions.

I define the sphere by a constraint in d-dimensional (spatial) Euclidean space,

$$S^{d-1}(R): \quad \sum_{i=1}^{d}(x_i)^2 = R^2, \tag{19.13}$$

and the ball in the same space is

$$B^d(R): \quad \sum_{i=1}^{d}(x_i)^2 \le R^2. \tag{19.14}$$

Then we know that, for instance,

$$vol(S^1(R)) = 2\pi R; \quad vol(S^2(R)) = 4\pi R^2. \tag{19.15}$$

In general, we write as a function of the volume of the *unit sphere* S^{d-1},

$$vol(S^{d-1}(R)) = R^{d-1}vol(S^{d-1}). \tag{19.16}$$

We find this by evaluating in two different ways the same d-dimensional Gaussian integral I_d: On one hand, since $r^2 = \sum_{i=1}^{d}(x_i)^2$,

$$I_d = \int_{\mathbb{R}^d} dx_1 \cdots dx_d e^{-r^2} = \prod_{i=1}^{d} \int_{-\infty}^{+\infty} dx_i e^{-(x_i)^2} = (\sqrt{\pi})^d = \pi^{\frac{d}{2}}, \qquad (19.17)$$

where we have used the value for the 1-dimensional Gaussian integral. On the other hand, using spherical coordinates, and since the integrand is independent on angles, doing trivially the angular integral to obtain S^{d-1}, we obtain

$$I_d = vol(S^{d-1}) \int_0^\infty dr\, r^{d-1} e^{-r^2} = vol(S^{d-1}) \int_0^\infty \frac{dt}{2} t^{\frac{d}{2}-1} e^{-t} = \frac{vol(S^{d-1})\Gamma(d/2)}{2}, \qquad (19.18)$$

where we have used the integral definition of the Gamma function.

Equating the two ways to calculate I_d, we obtain

$$vol(S^{d-1}) = \frac{2\pi^{d/2}}{\Gamma(d/2)}. \qquad (19.19)$$

We can check that indeed we obtain $vol(S^1) = 2\pi$ and $vol(S^2) = 4\pi$, using $\Gamma(3/2) = 1/2\Gamma(1/2)$ and $\Gamma(1/2) = \sqrt{\pi}$ (the 1-dimensional Gaussian integral). We also obtain $vol(S^3) = 2\pi^2$ and $vol(S^4) = 8\pi^3$.

19.4 The Electric Field of a Point Charge in Higher Dimensions

The Electric Field of a Point Charge from the Integrated Gauss's Law

The local form of Gauss's law (19.7) can be integrated in three dimensions on the ball B^3 to give

$$\int_{B^3} dV \vec{\nabla} \cdot \vec{E} = \int_{B^3} dV \frac{\rho_e}{\epsilon_0} = \frac{Q}{\epsilon_0}. \qquad (19.20)$$

On the other hand, from Stokes's theorem, the left-hand side is equal to

$$\int_{S^2(R)} d\vec{S} \cdot \vec{E}. \qquad (19.21)$$

For a point charge, with spherical symmetry, the above integral on the sphere gives

$$vol(S^2) R^2 E(R). \qquad (19.22)$$

Equating with the previous result, we get

$$E(R) = \frac{Q}{vol(S^2)\epsilon_0 R^2} = \frac{Q}{4\pi\epsilon_0 R^2}. \qquad (19.23)$$

We can easily generalize this to d spatial dimensions. Integrating the same (19.7) in d dimensions over the ball B^d, we get

$$\int_{B^d} dV \vec{\nabla} \cdot \vec{E} = \int_{B^d} dV \frac{\rho_e}{\epsilon_0} = \frac{Q}{\epsilon_0}. \tag{19.24}$$

On the other hand, from Stokes's theorem, the left-hand side equals

$$\int_{S^{d-1}(R)=\partial B^d(R)} d\vec{S} \cdot \vec{E}. \tag{19.25}$$

For a point charge, with spherical symmetry, this gives

$$vol(S^{d-1})R^{d-1}E(R). \tag{19.26}$$

Equating with the previous result, we get

$$E(R) = \frac{Q}{R^{d-1}} \frac{1}{vol(S^{d-1})\epsilon_0} = \frac{Q}{R^{d-1}} \frac{\Gamma(d/2)}{2\pi^{d/2}\epsilon_0}. \tag{19.27}$$

19.5 The Gravitational Field of a Point Mass in Higher Dimensions

Integrated Gauss Law for the Gravitational Field of a Point Mass

We repeat the same calculation for the gravitational case. Consider the local form of Gauss's law for gravity in D spacetime dimensions ($d = D - 1$ space dimensions) (19.11). There $G_N^{(D)}$ is defined as in 3+1 dimensions from the coefficient of the Einstein-Hilbert action,

$$\frac{1}{16\pi G_N^{(D)}} \int d^D x \sqrt{-g} R. \tag{19.28}$$

Integrating Gauss's law (19.11) in three spatial dimensions, over the ball B^3, we obtain

$$\int_{B^3} dV \vec{\nabla}^2 U_N = 4\pi G_N^{(D)} \int_{B^3} dV \rho_m = 4\pi G_N^{(D)} M. \tag{19.29}$$

On the other hand, the left-hand side is, using Stokes's theorem,

$$\int_{S^2(R)} d\vec{S} \cdot \vec{\nabla} U_N. \tag{19.30}$$

We also note that $\vec{\nabla} U_N = -\vec{g}_N(R)$, the gravitational acceleration (gravitational force per unit test mass). For a point mass (with spherical symmetry), the integral equals

$$vol(S^2)R^2 |\vec{g}_N(R)|. \tag{19.31}$$

Equating the two ways to calculate the gravitational acceleration, we obtain

$$|\vec{g}(R)| = \frac{4\pi G_N^{(D)} M}{vol(S^2)R^2} = \frac{G_N^{(D)} M}{R^2}. \tag{19.32}$$

Planck Scale

We now make some observations on the Planck scale. The Planck scale is defined in 3+1 dimensions as the unique scale made up of G, $\hbar$, and c, specifically

$$m_{\mathrm{Pl}} = \sqrt{\frac{\hbar c}{G_N}}; \quad l_{\mathrm{Pl}} = \sqrt{\frac{G_N \hbar}{c}}. \tag{19.33}$$

In the theorist's units, with $\hbar = c = 1$, we have $m_{\mathrm{Pl}} = G_N^{-1/2}$ and $l_{\mathrm{Pl}} = G_N^{1/2}$. This is the quantity most used in particle physics. However, for gravity and cosmology, one usually uses the *reduced Planck scale*

$$M_{\mathrm{Pl}} = \frac{m_{\mathrm{Pl}}}{\sqrt{8\pi}} = \frac{1}{\sqrt{8\pi \, G_N}}, \tag{19.34}$$

such that the Einstein-Hilbert action has the coefficient $M_{\mathrm{Pl}}^2/2$:

$$S_{EH} = \frac{M_{\mathrm{Pl}}^2}{2} \int d^4x \sqrt{-g^{(4)}} R^{(4)}. \tag{19.35}$$

Under KK-dimensional reduction on $M_4 \times K_n$, the EH action in $D = 4 + n$ spacetime dimensions (19.28) reduces as follows. We first note that for a diagonal background metric

$$g_{MN}^{(0)} = \begin{pmatrix} g_{\mu\nu}^{(0)} & 0 \\ 0 & g_{mn}^{(0)} \end{pmatrix}, \tag{19.36}$$

the invariant volume element splits up,

$$\sqrt{-g_{(0)}^{(D)}} \, d^D x = \sqrt{-g_{(0)}^{(4)}} \, d^4 x \sqrt{g_{(0)}^{(n)}} \, d^n x, \tag{19.37}$$

and as a result, when considering the kinetic term for gravity (thus considering only the quadratic part, coming from the Ricci tensor, whereas the volume element is considered to be in the background), we have the split

$$\frac{1}{16\pi G_N^{(D)}} \int d^D x \sqrt{-g^{(D)}} R^{(D)} \rightarrow \frac{1}{16\pi G_N^{(D)}} V^{(n)} \int d^4 x \sqrt{-g_{(0)}^{(4)}} R^{(4)} + \cdots \tag{19.38}$$

and by identifying this with the EH action in four dimensions, with coefficient $1/(16\pi G_N^{(4)})$, we get that

$$G_N^{(D)} = G_N^{(4)} V^{(n)}. \tag{19.39}$$

We now want to define it in terms of the (reduced) Planck mass in D dimensions. For that, we must consider the dimensionality. Since $\int d^D x \sqrt{-g} R$ has dimension $-D + 2$ (2 from R and $-D$ from the volume element), the same has to be true for $G_N^{(D)}$, and thus we can define

$$\frac{1}{8\pi G_N^{(D)}} \equiv [M_{\mathrm{Pl}}^{(D)}]^{D-2} \tag{19.40}$$

in general.

In this case, the Einstein-Hilbert action becomes

$$\frac{M_{\mathrm{Pl}}^{D-2}}{2} \int d^D x \sqrt{-g^{(D)}} R^{(D)}. \tag{19.41}$$

We are now finally ready to integrate Gauss's law (19.11) in d space dimensions, obtaining

$$\int_{B^d} dV\, \vec{\nabla}^2 U_N = 4\pi\, G_N^{(D)} \int_{B^d} \rho_m dV = 4\pi\, G_N^{(D)} M. \qquad (19.42)$$

Using Stokes' theorem, the left-hand side equals

$$\int_{S^{d-1}(R)=\partial B^d} d\vec{S}\cdot\vec{\nabla} U_N. \qquad (19.43)$$

For a point mass M (with spherical symmetry), we get

$$vol(S^{d-1})R^{d-1}|\vec{g}_N(R)|. \qquad (19.44)$$

Equating the two calculations, we get

$$|\vec{g}_N(R)| = \frac{4\pi\, G_N^{(D)}\Gamma\left(\frac{D-1}{2}\right)}{2\pi^{\frac{D-1}{2}}}\frac{M}{R^{D-2}}. \qquad (19.45)$$

Since $\vec{g}_N(R) = -\vec{\nabla} U_N$, we obtain the Newtonian potential as

$$U_N(R) = -\frac{4\pi\, G_N^{(D)}}{(D-3)vol(S^{D-2})}\frac{M}{R^{D-3}} = -\frac{2\pi^{\frac{3-D}{2}}\Gamma\left(\frac{D-1}{2}\right)}{D-3}\frac{M G_N^{(D)}}{R^{D-3}} = -\frac{C^{(D)}G_N^{(D)}M}{R^{D-3}}. \qquad (19.46)$$

Of course, since we live in 3+1 dimensions, we must consider doing something with the extra dimensions. One choice is the KK compactification. As an example, we will consider the simplest case of KK compactification, on $T_R^n = (S_R^1)^n$, in which case we have

$$G_N^{(D)} = (2\pi R)^n G_N^{(4)}. \qquad (19.47)$$

We could do a calculation of the gravitational acceleration valid for any distance, but since in any case the most relevant regimes are the extreme cases $r \ll R$ and $r \gg R$, we describe only these. In the $r \ll R$ regime, we have an approximately D-dimensional space, so the law is the one derived above in (19.45):

$$|\vec{g}_N(r)| = \frac{4\pi\, G_N^{(D)}\Gamma\left(\frac{D-1}{2}\right)}{2\pi^{\frac{D-1}{2}}}\frac{M}{r^{D-2}}. \qquad (19.48)$$

By contrast, for $r \gg R$, we have an approximately 4-dimensional potential, and expressing $G_N^{(4)}$ in terms of the D-dimensional Newton's constant, we get

$$|\vec{g}_N(r)| = \frac{G_N^{(4)}M}{r^2} = \frac{G_N^{(D)}M}{(2\pi)^{D-4}R^{D-4}r^2}. \qquad (19.49)$$

So, except for some factors of 2 and π, the result changes only the behavior with r^{D-2} with one with $R^{D-4}r^2$.

19.6 Consistent Truncations

As we mentioned already, when we do a KK-dimensional reduction (truncation of the KK expansion), an issue that appears is whether the truncation is consistent, i.e. whether it satisfies the equations of motion.

What could go wrong? For $\phi_n = 0$ not to solve the equation of motion means that we have something like

$$(\Box - m_n^2)\phi_n = c_n(\phi_0)^2 + \mathcal{O}(\phi_n), \tag{19.50}$$

so putting $\phi_n = 0$ while keeping $\phi_0 \neq 0$ would not be consistent. But such a case arises if we have terms in the action that are linear in the ϕ_n's, so that in the equation of motion we get terms with no ϕ_n, only ϕ_0.

For instance, consider a $\int d^D x \sqrt{-g^{(D)}} \lambda \phi^3$ interaction in $D = d + n$ dimensions ($d = 4$ in the physical case), and write the KK expansion. Since we have $\phi_q^{I_q}(\vec{x}) Y_q^{I_q}(\vec{y})$ terms in the expansion of ϕ, we will also get the term (putting all metric factors to the background, and considering only fluctuating scalars in this contribution)

$$\lambda \int d^n x \sqrt{\det g_{mn}^{(0)}} Y_q^{I_q}(\vec{y}) Y_0^{I_0}(\vec{y}) Y_0^{J_0}(\vec{y}). \tag{19.51}$$

Denoting the integral over $\vec{y}$ as $C_q^{I_q I_0 J_0}$, if $C_q^{I_q I_0 J_0} \neq 0$, we get

$$(\Box - m_q^2)\phi_q^{I_q}(\vec{x}) = \lambda C_q^{I_q I_0 J_0} \phi_0^{I_0}(\vec{x}) \phi_0^{J_0}(\vec{x}) + \mathcal{O}(\phi_{q'}), \tag{19.52}$$

so the truncation will be inconsistent. However, if

$$C_q^{I_q I_0 J_0} = \int d^n x \sqrt{\det g_{mn}^{(0)}} Y_q^{I_q}(\vec{y}) Y_0^{I_0}(\vec{y}) Y_0^{J_0}(\vec{y}) = 0 \tag{19.53}$$

the truncation is consistent. This is what happens, for instance, in the case of the torus T^n, since $Y_0(\vec{y}) = 1$. For S^1, we we would then get

$$C_1 = \int dy Y_n^{I_n} = \int dy e^{\frac{iny}{R}} = 0, \tag{19.54}$$

and for T^n, $C_{T^n} = \prod_n C_1 = 0$. That means that for a torus, the truncation is always consistent, *if we keep all the $n = 0$ modes*.

This has a simple generalization to a space K_n invariant under a group G. If we keep *all* the singlets under G, i.e. Y_0 are all singlets and Y_n are all nonsinglets, then by group invariance $\int Y_n Y_0 Y_0 = 0$, since only the integral of a nonsinglet with another nonsinglet could give a nonzero result.

Considering now the original KK ansatz, note that the original truncation to all the singlets, i.e. $g_{\mu\nu}, B_\mu$, and $\phi = \Phi - 1$ is consistent. However, if we put also $\phi = 0$, the truncation becomes inconsistent.

But can one fix the inconsistency of the truncation? Sometimes one can, though there is no general rule for when one can do so. Certainly it would seem like we should be able to do so, since the theories that we obtain after dimensional reduction are useful theories that can be considered on their own, so it would be strange if we could never decouple them from an infinite series of massive modes. Indeed, at least sometimes one can make a nonlinear redefinition of fields, something of the type

$$
\begin{aligned}
\phi_q' &= \phi_q + a\phi_0^2 \\
\phi_0' &= \phi_0 + \sum_{p,q(incl.0)} c_{pq}\phi_p\phi_q,
\end{aligned} \tag{19.55}
$$

such that after this redefinition there is no inconsistent term in the equation of motion,

$$
(\Box - m_q^2)\phi_q'^{\,I_q} = \mathcal{O}(\phi_{q'}'^{\,I_q'}), \tag{19.56}
$$

so *putting to zero the nonlinearly redefined fields* ϕ_q' is consistent. Alternatively, we could write from the beginning a *nonlinear KK ansatz for the dimensionally reduced fields* ϕ_0 that would solve the equations of motion. Note that there is no a priori reason to write a nonlinear ansatz for the full KK expansion, since the linear KK expansion always works (it is a theorem, a generalization of the Fourier theorem), but instead we need a nonlinear ansatz for the dimensional reduction fields.

This is, for example, what happens for the reductions of 10-dimensional IIB supergravity on $AdS_5 \times S^5$ (though there the full nonlinear consistent truncation is not known yet, it is only partially known), the reduction of 11-dimensional supergravity on $AdS_4 \times S^7$ and $AdS_7 \times S^4$.

Example: Nonlinear Ansatz for Gravity The simplest example is the nonlinear ansatz needed to get the correct d-dimensional Einstein action from the reduction of the D-dimensional Einstein action. We find that the correct nonlinear ansatz is actually

$$
g_{\mu\nu}(\vec{x}, t; \vec{y}) = g_{\mu\nu}(\vec{x}, t) \left[\frac{\det g_{mn}(\vec{x}, t; \vec{y})}{\det g_{mn}^{(0)}(\vec{y})} \right]^{-\frac{1}{d-2}}. \tag{19.57}
$$

To prove that this ansatz gives the full nonlinear Einstein action in d dimensions is complicated, but we can do a simple check.

Indeed, since the Christoffel symbol is of the type $\Gamma \sim g^{-1}\partial g$ and the Ricci tensor of the type $\partial\Gamma + \Gamma\Gamma$, under constant scale transformations, $g_{\mu\nu} \to \lambda g_{\mu\nu}$, $R_{\mu\nu}$ is invariant. But in the Einstein action, we have $\int \sqrt{-g^{(D)}} R^{(D)}$, and from $R^{(D)} = g^{\Lambda\Sigma} R_{\Lambda\Sigma}^{(D)}$, we consider only the term $\tilde{R} = R_{\mu\nu}^{(D)} g^{\mu\nu}$, which contains (among other terms) the Einstein Lagrangean in d dimensions, $R_{\mu\nu}^{(d)} g^{\mu\nu}$ (note that the terms not in $\tilde{R}$ contain the kinetic terms for the scalars and the vectors). Then, under a constant scale transformation, and considering that for the Einstein action we need only to keep the background metric for g_{mn} and $g_{\mu n}$, since do not want the scalars and the vectors *in this* S_{EH}, using the rescaled metric we have

$$
\sqrt{-g^{(D)}}\tilde{R} = \sqrt{-g^{(D)}} R_{\mu\nu}^{(D)} g^{\mu\nu} \to \sqrt{-g^{(d)}} \sqrt{\det g_{mn}} \lambda^{\frac{d}{2}-1} [R_{\mu\nu}^{(d)} g^{\mu\nu} + \cdots], \tag{19.58}
$$

and we want to rewrite this as

$$
\sqrt{-g^{(d)}} \sqrt{\det g_{mn}^{(0)}} [R_{\mu\nu}^{(d)} g^{\mu\nu}], \tag{19.59}
$$

which can be integrated to give the d-dimensional Einstein action. This fixes indeed

$$\lambda = \left[\frac{\det g_{mn}}{\det g_{mn}^{(0)}} \right]^{-\frac{1}{d-2}}. \tag{19.60}$$

Note that, of course, λ was considered constant, whereas the right-hand side isn't, but if we actually consider this $\vec{x}$-dependent λ, the nonlinear reduction works.

As an application of this, consider the original KK reduction ansatz. Then $d = 4$ and $n = 1$, so we get

$$g_{\mu\nu}^{(5)} = g_{\mu\nu}^{(4)} g_{55}^{-1/2}. \tag{19.61}$$

Define then

$$g_{55} \equiv \Phi^{2/3} \tag{19.62}$$

to obtain

$$g_{\mu\nu}^{(5)} = \Phi^{-1/3} g_{\mu\nu}^{(4)}. \tag{19.63}$$

To also obtain the Maxwell action coming from $g_{\mu 5}$, we need to redefine it as well, as

$$g_{\mu 5} = B_{\mu}(\vec{x}, t) g_{55}. \tag{19.64}$$

All in all, the original nonlinear KK ansatz is

$$g_{\Lambda\Sigma} = \Phi^{-1/3}(\vec{x}, t) \begin{pmatrix} g_{\mu\nu}(\vec{x}, t) & B_{\mu}(\vec{x}, t)\Phi(\vec{x}, t) \\ B_{\mu}(\vec{x}, t)\Phi(\vec{x}, t) & \Phi(\vec{x}, t) \end{pmatrix}. \tag{19.65}$$

Important Concepts to Remember

- Electromagnetism and gravity can be extended to higher dimensions.
- Gauss's law leads to an electric or gravitational field proportional to $1/vol(S^{d-1})R^{d-1}$.
- The Newton constant under KK reduction is rescaled as $G_N^{(D)} = G_N^{(4)} V^{(n)}$, and $1/(8\pi G_N^{(D)}) \equiv (M_{\mathrm{Pl}}^{(D)})^{D-2}$.
- At distances much greater than the KK radius, the potentials look 4-dimensional, and at distances much smaller than it, they look D-dimensional.
- A KK reduction can be inconsistent, if $\phi_n = 0$ while $\phi_0 \neq 0$ doesn't solve the equations of motion, or there are terms linear in ϕ_n's in the action. If we keep all the $n = 0$ modes on a torus, or all the singlets under G for a K_n invariant under G, then the truncation is consistent.
- Sometimes one can make a nonlinear redefinition of fields, or write from the beginning a nonlinear KK ansatz for the dimensionally reduced fields, such that the truncation obtained by putting to zero the new ϕ_n's is consistent.
- In the case of gravity, the nonlinear redefinition is known.

Further Reading

For an example of consistent truncation and nonlinear KK ansatz, in the case of $AdS_7 \times S^4$ compactification, see [32].

Exercises

(1) Calculate the d-dimensional Newton potential for a spherically symmetric body of constant density ρ and find g_{00}, at least for weak fields (in the Newtonian limit).

(2) Calculate the electric field for a sphere of radius R and uniformly distributed charge Q, and a grounded conducting infinite plane, on that infinite plane.

(3) Calculate the gravitational acceleration for a KK compactification from 10 dimensions onto $S_R^4 \times S_{R'}^2$, when $R \gg R'$, in the various distance regimes available.

(4) Consider a scalar χ in five dimensions with a canonical kinetic term $(-1/2 \int (\partial_\mu \chi)^2)$ and KK reduce to four dimensions using the original KK ansatz for the metric. Write the nonlinear redefinition for the scalar χ that leads to a canonical kinetic term for the reduced theory.

(5) Check that putting $\Phi(\vec{x}, t) = 1$ in the original KK ansatz leads to an inconsistent truncation.

(6) Consider a nondynamical 5-dimensional gravity (with no kinetic term for it) coupled to a dynamical scalar χ with canonical kinetic term, and consider the original KK ansatz for the metric. Is $\Phi = 1$ still an inconsistent truncation? Why?

20 Gravity Plus Matter: Black Holes and p-Branes in Various Dimensions

In this chapter we will learn how to write solutions describing black holes and their generalizations, p-branes, in various dimensions.

20.1 Schwarzschild Solution in Higher Dimensions

We have seen in the last chapter that in Newtonian gravity we have the same local form of Gauss's law in higher dimensions as in 3+1 dimensions,

$$\vec{\nabla}^2 U_N = 4\pi\, G_N^{(D)} \rho_m, \tag{20.1}$$

and that from it we can extract the Newton potential for a point source,

$$U_N(r) = -\frac{2\pi^{\frac{3-D}{2}}\, \Gamma\left(\frac{D-1}{2}\right) G_N^{(D)} M}{r^{D-3}} \equiv -\frac{C G_N^{(D)} M}{r^{D-3}}. \tag{20.2}$$

We also saw that we still have $\kappa_N^{(D)} h_{00} = -2U_N$ in the general case, where $\kappa_N^{(D)}$ is defined so that the coefficient of the EH action is $1/2\kappa_N^2$:

$$\kappa_N^{(D)} = \sqrt{8\pi\, G_N^{(D)}}. \tag{20.3}$$

In 3+1 dimensions, we have argued for the form of the Schwarzschild solution based on a general form of the small field metric in terms of the Newton potential $U_N(r)$. With the above generalization of $U_N(r)$ to higher dimensions, we can now write the Schwarzschild solution in higher dimensions as

$$\begin{aligned}
ds^2 &= -(1 + 2U_N)dt^2 + \frac{dr^2}{1 + 2U_N} + r^2 d\Omega_{D-2}^2 \\
&= -\left(1 - \frac{2C G_N^{(D)} M}{r^{D-3}}\right)dt^2 + \frac{dr^2}{1 - \frac{2C G_N^{(D)} M}{r^{D-3}}} + r^2 d\Omega_{D-2}^2.
\end{aligned} \tag{20.4}$$

20.2 Reissner-Nordstrom Solution in Higher Dimensions

We have also seen in the last chapter that the electric field of a point charge in D spacetime dimensions is

$$F_{0r} = E(r) = \frac{Q}{\Omega_{D-2}\epsilon_0 r^{D-2}}, \qquad (20.5)$$

where we have denoted $\Omega_{D-2} = vol(S^{D-2})$.

When coupling to gravity, we replace this electric field in the energy-momentum tensor, which has the usual form, since in any dimension D we have

$$T_{\mu\nu} = -\frac{2}{\sqrt{-g}}\frac{\delta S_{\text{matter}}}{\delta g^{\mu\nu}} = F_{\mu\rho}F_{\nu}{}^{\rho} - \frac{1}{4}g_{\mu\nu}(F_{\rho\sigma})^2, \qquad (20.6)$$

so in approximately flat space $g_{\mu\nu} \simeq \eta_{\mu\nu}$ (for use in the linearized Einstein's equation for U_N),

$$T_{00} = (F_{0r})^2 g^{rr} - \frac{1}{4}g_{00}2(F_{0r})^2 g^{rr}g^{00} \simeq \frac{1}{2}(F_{0r})^2 = \frac{(Q/\Omega_{D-2}\epsilon_0)^2}{2r^{2(D-2)}}. \qquad (20.7)$$

We add this T_{00} to the point source, of the form $M\delta^{D-1}(x)$, that generates the Schwarzschild solution, and appears on the right-hand side of the local Gauss's law for gravity:

$$\Delta_{(D-1)}U_N = 4\pi G_N^{(D)}\left[M\delta^{D-1}(x) + \frac{(Q/\Omega_{D-2}\epsilon_0)^2}{2r^{2(D-2)}}\right]. \qquad (20.8)$$

We can now use the fact that

$$\vec{\nabla}^2 \frac{1}{r^{2(D-3)}} = \left(\frac{d^2}{dr^2} + \frac{D-2}{r}\frac{d}{dr}\right)\frac{1}{r^{2(D-3)}} = \frac{2(D-3)^2}{r^{2(D-2)}} \qquad (20.9)$$

and

$$\vec{\nabla}^2 \frac{1}{r^{D-3}} = -\Omega_{D-2}(D-3)\delta^{D-1}(x). \qquad (20.10)$$

The last equation is derived by first noting that at $r \neq 0$ we obtain 0, and then integrating the equation on a small ball $B^{D-1}(r)$ for small r. By the Stokes theorem, the left-hand side becomes

$$\int_{S^{D-2}(r)=\partial B^{D-1}(r)} d\vec{S} \cdot \vec{\nabla}\frac{1}{r^{D-3}} = \Omega_{D-2}r^{D-2}\frac{d}{dr}\frac{1}{r^{D-3}}, \qquad (20.11)$$

and the right-hand side gives $-\Omega_{D-2}(D-3)$, matching it.

From the two above relations, we deduce that

$$U_N(r) = -\frac{MG_N^{(D)}4\pi}{(D-3)\Omega_{D-2}r^{D-3}} + \frac{4\pi G_N^{(D)}Q^2}{4(D-3)^2\Omega_{D-2}^2\epsilon_0^2 r^{2(D-3)}} \equiv -\frac{CMG_N^{(D)}}{r^{D-3}} + \frac{C'Q^2 G_N^{(D)}}{r^{2(D-3)}}. \qquad (20.12)$$

Now we can again use the same general weak field approximation for the metric, with

$$-g_{00} = 1 + 2U_N \equiv F(r), \qquad (20.13)$$

to argue for the form of the of full metric of the charged (Reissner-Nordstrom) black hole, which is

$$ds^2 = -F(r)dt^2 + \frac{dr^2}{F(r)} + r^2 d\Omega_{D-2}^2, \qquad (20.14)$$

exactly as in the Schwarzschild case, or the 3+1–dimensional case.

The metric has two *horizons*, as in the 3+1–dimensional case, which are found from the equation $F(r) = 0$,

$$1 - \frac{2CMG_N^{(D)}}{r^{D-3}} + \frac{2C'Q^2 G_N^{(D)}}{r^{2(D-3)}} = 0, \qquad (20.15)$$

which has the solutions

$$r_\pm^{D-3} = CMG_N^{(D)} \pm \sqrt{(CMG_N^{(D)})^2 - 2C'Q^2 G_N^{(D)}}. \qquad (20.16)$$

That again signifies that we have a *BPS bound*,

$$M^2 \frac{G_N^{(D)}C^2}{2C'} = M^2 8\pi G_N^{(D)} \epsilon_0^2 = \left(\epsilon_0 \kappa_N^{(D)} M\right)^2 \geq Q^2, \qquad (20.17)$$

which arises because of the need to have $r_\pm \in \mathbb{R}$, i.e. to have horizons. Indeed, as we have already said, the presence of singularities unshielded by horizons, i.e. of "naked singularities," is forbidden by various theorems.

In the case of equality (saturation of the BPS bound), i.e for $C^2 M^2 G_N^{(D)} = 2C'Q^2$, we have an "extremal black hole solution," with

$$r_+ = r_- \equiv r_H = (CMG_N^{(D)})^{\frac{1}{D-3}}, \qquad (20.18)$$

and then the function $F(r)$ can be rewritten as

$$F(r) = 1 - 2\left(\frac{r_H}{r}\right)^{D-3} + \left(\frac{r_H}{r}\right)^{2(D-3)} = \left[1 - \left(\frac{r_H}{r}\right)^{D-3}\right]^2. \qquad (20.19)$$

We would next want to make a coordinate transformation such that the spatial part of the metric can be rewritten as

$$\frac{dr^2}{F(r)} + r^2 d\Omega_2^2 = H(\bar{r})^2[d\bar{r}^2 + \bar{r}^2 d\Omega_{D-2}^2]. \qquad (20.20)$$

This implies the two equations for the two unknowns $\bar{r}(r)$ and $H(\bar{r})$:

$$\frac{dr}{1 - \left(\frac{r_H}{r}\right)^{D-3}} = H(\bar{r})d\bar{r}$$

$$r = H(\bar{r})\bar{r} \Rightarrow dr = d\bar{r}(H(\bar{r}) + \bar{r}H'(\bar{r})). \qquad (20.21)$$

Equating the two forms for $dr/d\bar{r}$, we obtain

$$\bar{r}H'(\bar{r}) = -\frac{r_H^{D-3}}{r^{D-3}}H(\bar{r}). \qquad (20.22)$$

At this point it is easier to try the ansatz (guess)

$$H(\bar{r}) = \left[1 + \left(\frac{r_H}{\bar{r}}\right)^{D-3}\right]^{\alpha},$$

(20.23)

for which we have

$$\frac{\bar{r}H'(\bar{r})}{H(\bar{r})} = -\alpha(D-3)\frac{r_H^{D-3}}{\bar{r}^{D-3} + r_H^{D-3}},$$

(20.24)

and from matching with (20.22), we obtain

$$r = \left[\frac{\bar{r}^{D-3} + r_H^{D-3}}{\alpha(D-3)}\right]^{\frac{1}{D-3}}.$$

(20.25)

On the other hand, the ansatz for $H(\bar{r})$ in the second equation in (20.21) gives

$$r = \bar{r}\left[1 + \left(\frac{r_H}{r}\right)^{D-3}\right]^{\alpha}.$$

(20.26)

We see that we have an equality between the two expressions if $\alpha = 1/(D-3)$, and then

$$r = \left[\bar{r}^{D-3} + r_H^{D-3}\right]^{\frac{1}{D-3}},$$

(20.27)

and one can rewrite

$$\sqrt{F(r)} = 1 - \left(\frac{r_H}{r}\right)^{D-3} = \frac{1}{1 + \left(\frac{r_H}{\bar{r}}\right)^{D-3}} = \frac{1}{f(\bar{r})}.$$

(20.28)

Finally then, the metric for the extremal black hole can be rewritten as

$$ds^2 = -[f(\bar{r})]^{-2}dt^2 + [f(\bar{r})]^{\frac{2}{D-3}}(d\bar{r}^2 + \bar{r}^2 d\Omega_{D-2}^2)$$

(20.29)

in terms of the *harmonic function*

$$f(\bar{r}) = 1 + \left(\frac{r_H}{\bar{r}}\right)^{D-3},$$

(20.30)

i.e. a function that satisfies the Poisson equation

$$\Delta_{\bar{r}}^{(D-1)} f(\bar{r}) = -8\pi G_N^{(D)} \delta^{D-1}(\bar{r}).$$

(20.31)

As in the 3+1–dimensional case, there is now not only a horizon at $\bar{r} = 0$ ($r = r_H$ in the original coordinates), but also a singularity at the same place. Then indeed, this is an extremal solution, at the separation point between the solutions with singularities clothed in horizons and solutions with naked singularities.

We could now go back to the general (nonextremal) Reissner-Nordstrom case, and write it as a function of the above harmonic function $f(\bar{r})$ and a Schwarzschild harmonic function for the nonextremality $\mu = M - Q/\kappa_N^{(D)}$; however, we will not do so here.

20.3 P-Branes

Besides black holes, in higher dimensions we could have other generalizations as well. We could have asked about three spatial dimensions as well: why is it that we considered only the point-like black hole, which is described by a harmonic function in the three dimensions of its codimension in spacetime? Why not objects with infinite spatial extension in some directions, like a 1-dimensional object (string), or a 2-dimensional object (wall)? In fact, *for finite spatial extension*, we can consider in our universe such objects, called cosmic strings and domain walls in cosmology, which could have appeared in the beginning of the universe and remained, to be found in the sky. The point is that the harmonic function for a codimension 2-object (a string) is $H \propto \log z$, and for a codimension 1-object (wall) is $H \propto |x|$, and in both cases *for such an object with infinite spatial extension* we would obtain a metric that increases away from the object. Since in our 3+1–dimensions the spacetime is approximately flat, there can be no such objects.

But we don't have this problem if we go to higher dimensions, since, for instance, in D spacetime dimensions, a black hole (point) is codimension $D - 1$, so we could have black holes (points), as well as strings, walls, etc. of codimensions ≤ 3, thus with harmonic functions that decrease toward infinity. We can then safely consider objects with infinite extension in p spatial dimensions, called *p-branes*. The word comes from an extension of the word mem-brane, which now would be a 2-brane. If we consider objects with horizons, like the black holes, we call them black p-branes.

The simplest generalization when we go to higher dimensions is, in the absence of electric charges, the black Schwarzschild p-branes, where we simply add p flat directions of space to the Schwarzschild solution in d dimensions. If we consider a KK reduction on a torus T^p of a D-dimensional theory ($D = d + p$), we just add flat space with periodic identifications to the Schwarzschild solution:

$$ds^2 = -\left(1 - \frac{2C_{D-p}G_N^{(D-p)}M}{r^{D-3-p}}\right)dt^2 + d\vec{x}_p^2 + \frac{dr^2}{1 - \frac{2C_{D-p}G_N^{(D-p)}M}{r^{D-3-p}}} + r^2 d\Omega_{D-2-p}^2. \quad (20.32)$$

However, there are more interesting generalizations, for *charged and extremal p-branes*. But charged here doesn't mean charged under the usual abelian electromagnetic field A_μ. Instead consider a totally antisymmetric tensor field, a $(p + 1)$−form $A_{\mu_1...\mu_{p+1}}$, that is a generalization of the Maxwell field, in the sense that it also has an abelian gauge invariance. Note that there are theorems saying that we cannot have a (usual) nonabelian generalization of the field $A_{\mu_1...\mu_{p+1}}$ (the first that I am aware of is by Teitelboim and Nepomechie [33, 34, 35]), though for reasons that I will not explain, people now try to find some kind of nonabelian generalization for it. One considers the field strength of the antisymmetric tensor field,

$$F_{\mu_1...\mu_{p+2}} = (p + 2)\partial_{[\mu_1}A_{\mu_2...\mu_{p+2}]}, \quad (20.33)$$

and in terms of it, the action

$$-\frac{1}{2(p + 2)!}\int d^Dx\sqrt{-g}F_{(p+2)}^2. \quad (20.34)$$

Since it is written in terms of the field strength, the action has invariance under the gauge transformations

$$\delta A_{\mu_1...\mu_{p+1}} = \partial_{[\mu_1}\Lambda_{\mu_2...\mu_{p+1}]}, \tag{20.35}$$

which leaves $F_{(p+2)}$ invariant. Note that in form language, $F_{(p+2)} = dA_{(p+1)}$ and $\delta A_{(p+1)} = d\Lambda_{(p)}$, and the gauge invariance is a consequence of $d^2 = 0$.

A charged p-brane will therefore be an object (electrically) charged under the field $A_{(p+1)}$. To gain some intuition about this case, we will first understand the Maxwell case better. In electromagnetism, an electron is a solution to the action for the kinetic term of the gauge fields, plus a source term,

$$S = \int d^4x \left[-\frac{1}{4}F_{\mu\nu}^2 + j^\mu A_\mu \right], \tag{20.36}$$

where j^μ is a delta-function source, i.e. in the rest frame it is $j^0 = e\delta^3(x)$ and $j^i = 0$. The action is then the Maxwell action plus a source term

$$S_{\text{source}} = e\int d\tau \frac{\partial t}{\partial\tau}A_0. \tag{20.37}$$

In turn, when coupling this action to gravity, we saw that it leads to the electrically charged Reissner-Nordstrom black hole solution, which besides having the electric field of a point source also has the RN form for the metric.

Another useful observation about electromagnetism is that it has electric-magnetic duality, in the presence of both electric and magnetic charge (or none of either). Then, if j_μ^e is the electric current and j_m^μ is the magnetic current, we have the Maxwell equations

$$\partial^\mu F_{\mu\nu} = j_\nu^e$$
$$\partial_{[\mu}F_{\nu\rho]} = \frac{1}{3!}\epsilon_{\mu\nu\rho\sigma}j_m^\sigma. \tag{20.38}$$

Multiplying the second equation by $\epsilon^{\mu\nu\rho\lambda}$ and considering that $\epsilon_{\mu\nu\rho\sigma}\epsilon^{\mu\nu\rho\lambda} = 3!\delta_\sigma^\lambda$, and defining

$$\epsilon^{\mu\lambda\nu\rho}F_{\nu\rho} = (*F)^{\nu\lambda}, \tag{20.39}$$

we can rewrite the second Maxwell equation as

$$j_m^\lambda = \partial_\mu(\epsilon^{\mu\lambda\nu\rho}F_{\nu\rho}) = \partial_\mu\left((*F)^{\mu\lambda}\right). \tag{20.40}$$

That means that the Maxwell equations are simply interchanged under the *duality transformation*

$$F_{\mu\nu} \leftrightarrow (*F)_{\mu\nu}; \quad j_\mu^e \leftrightarrow j_\mu^m. \tag{20.41}$$

The duality means that the existence of the electric solution (electron) immediately implies also the existence of the magnetic solution (monopole), which can be written in terms of $(*F)_{\mu\nu}$ instead of $F_{\mu\nu}$. When coupling to gravity, that leads to the fact that there is also a Reissner-Nordstrom solution with magnetic charge Q_m.

20.4 P-Brane Solutions

To describe an electric p-brane solution in D dimensions, by analogy with the above electromagnetic case, we consider a p-brane carrying electric charge Q_p with respect to a $(p+1)$–form fields $A_{\mu_1...\mu_{p+1}}$, which must mean that there is a source coupling

$$\int d^D x j^{\mu_1...\mu_{p+1}} A_{\mu_1...\mu_{p+1}}, \tag{20.42}$$

for which the only nonzero components are

$$\int d^D x j^{01...p} A_{01...p}, \tag{20.43}$$

specifically because the source of $A_{01...p}$ is, for a p-brane in its rest frame:

$$j^{01...p} = Q_p \delta^{D-1-p}(x). \tag{20.44}$$

This is an object (infinitely) extended in the p spatial dimensions $x^1, \ldots, x^p$ and in time x^0, and the delta function is transverse to these directions.

Thus the p-brane electric field is the solution to the kinetic term for the gauge field, plus the above source, i.e. to the action

$$S = \int d^D x \left(-\frac{1}{2(p+2)!} F^2_{\mu_1...\mu_{p+2}} + j^{\mu_1...\mu_{p+1}} A_{\mu_1...\mu_{p+1}} \right), \tag{20.45}$$

giving a solution

$$A_{01...p} = \frac{Q_p C_p}{r^{D-3-p}}. \tag{20.46}$$

Here C_p is a constant, chosen such that the above is a harmonic function in the transverse dimensions:

$$\Delta_{(D-p-1)} \frac{Q_p C_p}{r^{D-3-p}} = -Q_p \delta^{(D-1-p)}(x). \tag{20.47}$$

When coupling to gravity, we get solutions that can be described by two harmonic functions, like we argued also in the general RN case. Nevertheless, we find it useful to first start describing the extremal solution, written in terms of a single harmonic function.

Moreover, we are interested in p-brane solutions in *supergravity* and *string theory*. Supergravity is a (locally) supersymmetric theory of gravity, and therefore can be thought of as a field theory of gravity plus matter, including fermions. String theory is a further generalization that can be thought of as (as we will see in later chapters) supergravity coupled to an an infinite set of massive fields.

In these cases then, besides the Einstein-Hilbert kinetic term for gravity and the antisymmetric tensor field action, we also add a coupling to a scalar field called the *dilaton*. Therefore, the generic action we will consider for the supergravity and string theory case is

$$S_D(p) = \frac{1}{2\kappa_N^{(D)}} \int d^D x \sqrt{-g} \left[R - \frac{1}{2(p+2)!} e^{-a(p+1)\phi} F^2_{(p+2)} - \frac{1}{2}(\partial_\mu \phi)^2 \right]. \tag{20.48}$$

The source coupling, in the presence of gravity, could be described as before, as

$$\int d^D x \sqrt{-g} j^{\mu_1 \ldots \mu_{p+1}} A_{\mu_1 \ldots \mu_{p+1}}. \tag{20.49}$$

However, as we saw in the electromagnetism case, by substituting the form of $j^{01 \ldots p}$, we should obtain something like (20.37).

In fact, as we will describe in more detail in the next chapter for the particular case of strings ($p = 1$), one must also consider a general surface spanned in spacetime by this electric p-brane solution, called a *worldvolume*, parametrized by intrinsic coordinates ξ^i, $i = 0, 1, \ldots, p$, and an action for this surface. All in all, the source term is

$$S_{p+1} = T_{p+1} \int d^d \xi \left[-\frac{1}{2} \sqrt{-\gamma} \gamma^{ij} \partial_i X^M \partial_j X^N g_{MN} e^{\frac{a(p+1)\phi}{p+1}} + \frac{p-1}{2} \sqrt{-\gamma} \right.$$
$$\left. - \frac{1}{(p+1)!} \epsilon^{i_1 \ldots i_{p+1}} \partial_{i_1} X^{M_1} \ldots \partial_{i_{p+1}} X^{M_{p+1}} A_{M_1 \ldots M_{p+1}} \right]. \tag{20.50}$$

Here γ^{ij} is the intrinsic metric on the p-brane. The first two terms (on the first line) define the action for the p-brane worldvolume surface defined by the embedding $X^M(\xi^i)$, and the last term (on the second line) is the generalization of the source term in the form (20.37). Note that the spacetime field $A_{M_1 \ldots M_{p+1}}$ ($M_1, \ldots, M_{p+1} = 0, 1, \ldots, D-1$) is "pulled back" onto the worldvolume ξ^i, $i = 0, 1, \ldots, p$ by the $\partial_{i_j} X^{M_j}$.

Then we can describe first the extremal solution to $S_D(p+1) + S_{p+1}$, which corresponds now to having charge density equaling the mass density, or *tension* of the p-brane:

$$\frac{M}{V_p} = T_p = K_p Q_p. \tag{20.51}$$

As before, we have an equality called BPS bound, $T_p \geq K_p Q_p$, and at saturation of the bound, we have an extremal solution, with the horizon on top of the singularity.

For the string theory solution in $D = 10$, the general p-brane solution is

$$ds^2 = e^{-\frac{\phi}{2}} \left[H_p^{-1/2} (-dt^2 + d\vec{x}_p^2) + H_p^{1/2} (dr^2 + r^2 d\Omega_{8-p}^2) \right]$$
$$e^{-2\phi} = H_p^{\frac{p-3}{2}}$$
$$A_{01 \ldots p} = -\frac{1}{2} (H_p^{-1} - 1). \tag{20.52}$$

Note that the metric is expressed in the usual "Einstein frame," meaning that the gravitational action is the usual Einstein-Hilbert term. In string theory, metrics are often quoted in a "string frame," related to the usual metric by the factor $e^{\phi/2}$. We will come back to this issue later.

The function H_p is a harmonic function, that is, for a single p-brane:

$$\Delta_{(9-p)} H_p \propto Q_p \delta^{(9-p)}(x) \Rightarrow H_p(r) = 1 + \frac{C_p Q_p}{r^{7-p}}. \tag{20.53}$$

But we can write *superpositions of parallel p-branes*, simply by adding similar terms in the harmonic function, with different centers:

$$H_p = 1 + \sum_i \frac{C_p Q_{p,i}}{|\vec{x} - \vec{x}_i|^{7-p}}. \tag{20.54}$$

We will not show this here, but in a supersymmetric theory, this configuration is stable, i.e. the configuration is supersymmetric and there is no force between the *p*-branes.

This is an example of a more general story: in a supersymmetric configuration there is no force between the components.

Nonextremal Solutions

Nonextremal solutions are found easily by introducing a nonextremality parameter

$$\mu = T_p - K_p Q_p \tag{20.55}$$

and using it to construct a Schwarzschild harmonic function

$$f(r) = 1 - \frac{\tilde{C}_p \mu_p}{r^{7-p}}. \tag{20.56}$$

Then the metric for the nonextremal solution is obtained by "blackening" the solution, by inserting $f(r)$ as if the solution was Schwarzschild,

$$ds^2 = e^{-\frac{\phi}{2}} \left[H_p^{-1/2}(-f(r)dt^2 + d\vec{x}_p^2) + H_p^{1/2} \left(\frac{dr^2}{f(r)} + r^2 d\Omega_{8-p}^2 \right) \right], \tag{20.57}$$

and leaving the dilaton and $(p + 1)$–form field unmodified.

One can also consider magnetic solutions, the analog of the magnetic monopole in four dimensions, with

$$F_{ij} \sim \frac{\epsilon_{ijk} x^k Q_m}{r^3} = \frac{Q_m}{r^2} \epsilon_{ijk} \frac{x^i}{r} \propto \frac{Q_m}{r^2} \epsilon_{(2)}. \tag{20.58}$$

Here $\epsilon_{(2)}$ is a 2-form that integrated gives the volume of the transverse 2-sphere Ω_2. The magnetically charged solution will have then a solution with a field proportional to the volume form $\epsilon_{(p+2)}$, so we have a magnetic field

$$F_{(p+2)} \propto \epsilon_{(p+2)}. \tag{20.59}$$

The difference is that now this is not a *p*-brane, but rather a $\tilde{p} = (D - 4 - p)$–brane, as we can easily check, since the $(p + 2)$-sphere is transverse to the $\tilde{p}$-brane. The magnetic *p*-brane solution can be thought of as a soliton, like the monopole. In that respect, it would seem like it is the analog of a (singular) Dirac monopole, which needs a source term in the action. However, when solving the equations of motion for the magnetic *p*-branes, one finds that it doesn't really matter if you put the source term or not, one still obtains the solution. In that respect, it is more like a 't Hooft monopole, i.e. a topological field configuration that doesn't need a source term to obtain the solution.

Important Concepts to Remember

- The Schwarzschild solution in D spacetime dimensions is written in terms of a harmonic function in $D - 1$ dimensions.
- The Reissner-Nordstrom solution in D spacetime dimensions has two horizons and obeys a BPS bound $M \kappa_N^{(D)} \epsilon_0 \geq |Q|$.
- The extremal solution has a single horizon, coinciding with the singularity.
- p-branes are solutions with (infinite) extension in p spatial dimensions.
- The black Schwarzschild p-brane in D dimensions is obtained from the Schwarzschild solution in $d = D - p$ dimensions, with a torus for the p dimensions.
- Charged p-brane solutions are charged with respect to a $(p + 1)$–form field $A_{(p+1)}$, a generalization of the Maxwell field with field strength $F_{(p+2)} = dA_{(p+1)}$.
- The electric p-brane solution is found as a solution to the supergravity action, with kinetic term for gravity, the $F_{(p+2)}$ field, and maybe a dilaton scalar, plus a source term involving the coupling of charge to $A_{(p+1)}$, and an action for the p-brane worldvolume.
- The p-brane worldvolume is the hypersurface spanned by the p-brane as it moves in spacetime.
- There is a BPS bound for the tension $T_p = M/V_p$ vs. the charge density Q_p, $T_p \geq K_p Q_p$.
- The extremal solution in string theory is written in terms of a harmonic function H_p in the $9 - p$ transverse dimensions.
- The nonextremal solution is written in terms of two harmonic functions, H_p and a Schwarzschild black hole harmonic function for $\mu_p = T_p - K_p Q_p$, obtained by blackening like for the Schwarzschild solution.
- The magnetic solution has $F_{(p+2)}$ proportional to the volume form in the transverse dimensions, $\epsilon_{(p+2)}$, giving a $\tilde{p} = (D - 4 - p)$–brane solution. It doesn't need a source term.

Further Reading

For p-branes, see the review [36]. See also the "harmonic function rule" for composing p-branes, developed in [37]. See also how to make an extremal solution nonextremal in [38].

Exercises

(1) Using arguments similar to the ones in the text, write g_{00} for the case of the 4-dimensional black hole with mass, electric charge Q_e and magnetic charge Q_m. Suggest a form for ds^2. What changes for ds^2 in higher dimensions?

(2) Write the laws of thermodynamics for a Reissner-Nordstrom black hole in higher dimensions and check that the first law is satisfied, like we did in the four dimensions for the Kerr-Newman case.

(3) Consider the black Schwarzschild *p*-brane in higher dimensions on the torus T^p. What is the mass and temperature of the solution?

(4) Write the equations of motion satisfied by the electric *p*-brane solution in D dimensions, coming from the supergravity action plus source terms.

(5) Describe the duality relation between electric and magnetic *p*-branes that parallels electric-magnetic duality ($Q_e \leftrightarrow Q_m$, $F \leftrightarrow *F$) for the 4-dimensional electromagnetic fields.

(6) Write the solution for a set of parallel extremal *p*-branes situated at different positions (and with the charges of the same sign).

21

Weak/Strong Coupling Dualities in $1+1, 2+1,$ $3+1,$ and $d+1$ Dimensions

In this chapter we will study weak/strong dualities in various dimensions. We have seen in Chapter 5 the first example of such an (exact) duality, the duality between the fermionic massive Thirring model and the bosonic sine-Gordon model, and we have also seen an example in the condensed matter context in Chapter 6, the Kramers-Wannier duality that inverts the temperature of the Ising model in $1+1$ dimensions.

We want now to formalize a bit the idea of duality in the field theory context, by seeing in a few examples how it can be defined as a transformation of the path integral of the theory. Of course, in most examples, it cannot be written down as explicitly, but rather the transformation is defined only on the fields and couplings of the model, and we will see how that happens as well.

To start, we put the transformation defined in Chapter 5, the bosonization transformation that took us between the massive Thirring model and the sine-Gordon model, in the form of such a duality transformation, using the original formulation of Burgess and Quevedo. This will exemplify the general steps: we introduce the path integral in a first order form, with a constraint introduced via Lagrange multipliers, for a *master action*. If instead of solving for the Lagrange multipliers we solve for the original fields, we obtain a dual theory in terms of the Lagrange multipliers.

Consider a Dirac fermion in $1+1$ dimensions interacting with external fields a_i via operators $\mathcal{O}_i$. It has the Minkowski space path integral

$$Z_F[J] = \int \mathcal{D}\psi \exp\left[i \int d^2x \left(-\bar{\psi}\slashed{\partial}\psi + \sum_i a_i \mathcal{O}_i(\psi)\right)\right], \qquad (21.1)$$

where J stands for the set of external sources $\{a_i\}$. Concretely, consider the operators being the usual vector and axial vector currents for the fermions (particles):

$$\sum_i a_i \mathcal{O}_i(\psi) = a_\mu i\bar{\psi}\gamma^\mu\psi + b_\mu i\bar{\psi}\gamma^\mu\gamma_3\psi. \qquad (21.2)$$

We next gauge the global $U(1)$ symmetry ($\psi \to e^{i\alpha}\psi$) by introducing a gauge field A_μ. We also introduce the constraint $F_{\mu\nu} = 0$ (which in $1+1$ dimensions is simply $F_{01} = 0$) with a Lagrange multiplier Λ for $\epsilon^{\mu\nu}F_{\mu\nu}$, which means that in fact the gauge field is trivial, and we can go back to the ungauged theory.

The result is a path integral involving a *master action* for the duality:

$$Z_G = \int \mathcal{D}\psi \mathcal{D}A_\mu \mathcal{D}\Lambda \exp\left(i\int d^2x[\mathcal{L}_F(\psi, a_i) + i\bar{\psi}\gamma^\mu\psi A_\mu + \frac{1}{2}\Lambda\epsilon^{\mu\nu}F_{\mu\nu}]\right)\Delta[\partial \cdot A].$$
$$(21.3)$$

Here $\Delta[\partial \cdot A]$ is a functional delta function that imposes the Lorenz gauge condition $\partial \cdot A = 0$. Indeed, we see that $Z_F = Z_G \equiv Z$, since solving the constraint by varying the Lagrange multiplier means $F_{\mu\nu} = 0$, allowing us to put $A_\mu = 0$ locally, which returns us to the original action. If, instead of that, we integrate over A_μ and ψ, we obtain the *bosonized action as a dual action* (again J stands for the set of external sources, now a_μ and b_μ):

$$Z[J] = \int \mathcal{D}\Lambda \, \exp(iS_B[\Lambda, J]). \tag{21.4}$$

Using the identity (which will not be proven here)

$$\int \mathcal{D}\psi \, \exp\left(i \int d^2x (-\bar{\psi}\partial\!\!\!/\psi + i\bar{\psi}\gamma^\mu\psi A_\mu)\right) = \exp\left(\frac{i}{4\pi} \int d^2x F^{\mu\nu}\frac{1}{\Box}F_{\mu\nu}\right), \tag{21.5}$$

valid in $1+1$ dimensions, and doing also the integral over A_μ via the change of variables $A_\mu = \epsilon_{\mu\nu}\partial^\nu\rho$, we obtain the bosonized Lagrangean in terms of the canonically normalized field $\phi = \sqrt{\pi}\Lambda$-,

$$\mathcal{L}_B(\phi, a, b) = -\frac{1}{2}\partial^\mu\phi\partial_\mu\phi + \frac{1}{\sqrt{\pi}}\partial^\mu\phi b_\mu + \frac{1}{\sqrt{\pi}}\epsilon^{\mu\nu}\partial_\mu\phi a_\nu. \tag{21.6}$$

Comparing with the $a_i \mathcal{O}_i$ term in the Lagrangean in the fermionic picture, we obtain the mapping of the original (fermionic) particle current to the bosonic topological, i.e. soliton, current, as well as the fermionic topological current to bosonic particle current:

$$i\bar{\psi}\gamma^\mu\psi \leftrightarrow -\frac{1}{\sqrt{\pi}}\epsilon^{\mu\nu}\partial_\nu\phi; \quad i\bar{\psi}\gamma^\mu\gamma_3\psi \leftrightarrow \frac{1}{\sqrt{\pi}}\partial^\mu\phi. \tag{21.7}$$

Note that the first bosonic current and the second fermionic one are topological, since they are conserved ($\partial^\mu j_\mu^{\text{top}} = 0$) by topology, i.e. without using any dynamical properties. We see then that we have obtained the same mapping of currents between particle and soliton that was obtained in the massive Thirring vs. sine-Gordon models, even though here there were no explicit self-interactions. The relation was simply a result of bosonization.

21.1 $1+1$ Dimensions: T-Duality in the Path Integral

The previous example, however, is somewhat more convoluted, and we showed it first only because it related to the previously described bosonization. The simplest example of duality is still in $1+1$ dimensions, and it is called T-duality, understood as a symmetry along the worldsheet of the string. We have not yet described what strings are, to say nothing of T-duality, but for the present purposes it suffices to consider it as a duality transformation of a simple Euclidean theory of $1+1$–dimensional bosons.

Consider therefore a Euclidean version of a $1+1$–dimensional "sigma model" for D bosons $X^I, I = 1, \ldots, D$, with action

$$S = \int d^2\sigma \sqrt{\gamma}\gamma^{\mu\nu}g_{IJ}(X)\partial_\mu X^I \partial_\nu X^J. \tag{21.8}$$

It can be split by isolating one of them, call it X^0, which has the property that translation in it is a symmetry of the "metric" g_{IJ}, as

$$S = \int d^2\sigma \sqrt{\gamma}\gamma^{\mu\nu}[g_{00}\partial_\mu X^0 \partial_\nu X^0 + 2g_{0i}\partial_\mu X^0 \partial_\nu X^i + g_{ij}\partial_\mu X^i \partial_\nu X^j]. \quad (21.9)$$

Note that one could add also a term with $\epsilon^{\mu\nu}$ and an antisymmetric B-field $B_{IJ}(X)$,

$$S = \int d^2\sigma \epsilon^{\mu\nu}\partial_\mu X^I \partial_\nu X^J B_{IJ}(X), \quad (21.10)$$

but for simplicity we will consider $B_{IJ} = 0$.

The first step is to write the action in a first order form, with a Lagrange multiplier for a constraint, i.e. to write the master action. As before, we introduce a gauge field V_μ and enforce the vanishing of its curvature by a Lagrange multiplier for $\epsilon^{\mu\nu}\partial_\mu V_\nu$, obtaining the master action

$$S_{\text{master}} = \int d^2\sigma \left\{ \sqrt{\gamma}\gamma^{\mu\nu}[g_{00}V_\mu V_\nu + 2g_{0i}V_\mu \partial_\nu X^i + g_{ij}\partial_\mu X^i \partial_\nu X^j] + 2\epsilon^{\mu\nu}\hat{X}^0 \partial_\mu V_\nu \right\}. \quad (21.11)$$

We see that varying with respect to the Lagrange multiplier $\hat{X}^0$ imposes the flatness of the connection V_μ, i.e. $\partial_{[\mu}V_{\nu]} = 0$, so this allows us to write $V_\mu = \partial_\mu X^0$, which takes us back to the original action.

If instead we vary with respect to V_μ, we obtain

$$V_\mu = \frac{\hat{g}_{00}}{\sqrt{\gamma}}\epsilon^{\nu\rho}\gamma_{\rho\mu}\partial_\nu\hat{X}^0 - g_{0i}\partial_\mu\hat{X}^i, \quad (21.12)$$

where $\hat{X}^i = X^i$, $\hat{g}_{00} = 1/g_{00}$. We then replace this back into the master action and obtain a sigma model in terms of $\hat{X}^I = (\hat{X}^0, \hat{X}^i)$ and $\hat{g}_{IJ}$. Note that the Lagrange multiplier becomes one of the new bosons. Moreover, we now need to include a B-field term (a B-field is generated), so that finally

$$\tilde{S} = \int d^2\sigma \left[\sqrt{\gamma}\gamma^{\mu\nu}\hat{g}_{IJ}\partial^\mu\hat{X}^I \partial_\nu\hat{X}^J + \epsilon^{\mu\nu}\hat{B}_{IJ}\partial_\mu\hat{X}^I \partial_\nu\hat{X}^J \right], \quad (21.13)$$

where the "T-dual variables" are

$$\hat{g}_{00} = \frac{1}{g_{00}}; \quad \hat{g}_{ij} = g_{ij} - \frac{g_{0i}g_{0j}}{g_{00}}; \quad \hat{B}_{0i} = \frac{g_{0i}}{g_{00}}. \quad (21.14)$$

We can equate the V_μ in the dual formulation (21.12) with the one in the original formulation, $V_\mu = \partial_\mu X^0$, to obtain the duality transformation of the fields:

$$\partial_\mu X^0 = \frac{\hat{g}_{00}}{\sqrt{\gamma}}\epsilon^{\nu\rho}\gamma_{\rho\mu}\partial_\nu\hat{X}^0 - g_{0i}\partial_\mu\hat{X}^i = \frac{\hat{g}_{00}}{\sqrt{\gamma}}\epsilon_{\rho\mu}\partial^\rho\hat{X}^0 - g_{0i}\partial_\mu\hat{X}^i. \quad (21.15)$$

Since the "metric" g_{00} appears in front of the kinetic term for X^0, $\partial_\mu X^0 \partial^\mu X^0$, it has the interpretation of a coupling constant. We then see that $\hat{g}_{00} = 1/g_{00}$ implies that the coupling constant is inverted under the duality, i.e. we have a strong/weak duality, that exchanges a strongly coupled theory with a weakly coupled one, *from the point of view of the 1+1 dimensional "worldsheet."*

21.2 $2+1$ Dimensions: Particle-Vortex Duality

We next go one dimension higher, in $2+1$ dimensions, and consider a theory that has a soliton solution, specifically of the vortex type. We will see that now, besides obtaining the features of strong/weak duality from the $1+1$–dimensional case, we also have a duality that interchanges particles and vortices, and their respective currents.

We start therefore with an Abelian-Higgs model, for a $U(1)$ gauge field coupled to a charged complex scalar, with action

$$S = \int d^3x \left[-\frac{1}{2}|D_\mu \Phi|^2 - V(|\Phi|) - \frac{1}{4}F_{\mu\nu}^2 \right]. \tag{21.16}$$

As usual, we have $F_{\mu\nu} = \partial_\mu a_\nu - \partial_\nu a_\mu$ and $D_\mu \Phi = \partial_\mu \Phi - iea_\mu \Phi$. We choose a potential that spontaneously breaks (in the vacuum) the $U(1)$ symmetry and admits vortex solutions. A vortex solution for (Φ, A_μ) is a solution where for $\Phi = \Phi(r)e^{i\alpha(\theta)}$, with (r, θ) polar coordinates for the two spatial directions, one has $\Phi(r = 0) = 0$ and $\alpha(\theta) = N\theta$, such that we have zeroes of Φ with nontrivial monodromy around them. Note that for a vortex, we have $\epsilon^{ij}\partial_i\partial_j\alpha \propto \delta(r)$, so it is a singular solution.

In the presence of such vortices, we split the field as $\Phi = \Phi_0 e^{i\alpha}$, with $\alpha = \alpha_{\text{smooth}} + \alpha_{\text{vortex}}$, where α_{vortex} contains all the nontrivial monodromies of the vortices, i.e. the singular parts, and $\epsilon^{ij}\partial_i\partial_j\alpha_{\text{smooth}} = 0$. Under such a split, the action becomes

$$S = -\frac{1}{2}\int d^3x \left[(\partial_\mu \Phi_0)^2 + (\partial_\mu \alpha_{\text{smooth}} + \partial_\mu \alpha_{\text{vortex}} + ea_\mu)^2 \Phi_0^2 \right]$$
$$- \int d^3x \left[V(\Phi_0) + \frac{1}{4}F_{\mu\nu}^2 \right]. \tag{21.17}$$

We now repeat the procedure from $1+1$ dimensions, writing a first order action by replacing $\partial_\mu \alpha$ with an independent λ_μ and imposing the flatness of its curvature, $\epsilon^{\mu\nu\rho}\partial_\nu\lambda_\rho = 0$, with Lagrange multipliers b_μ. The master action is therefore

$$S_{\text{master}} = \int d^3x \left[-\frac{1}{2}(\partial_\mu \Phi_0)^2 - \frac{1}{2}(\lambda_{\mu,\text{smooth}} + \lambda_{\mu,\text{vortex}} + ea_\mu)^2 \Phi_0^2 + \epsilon^{\mu\nu\rho}b_\mu\partial_\nu\lambda_{\rho,\text{smooth}} \right.$$
$$\left. -V(\Phi_0) - \frac{1}{4}F_{\mu\nu}^2 \right]. \tag{21.18}$$

Varying with respect to b_μ, the constraint is solved by $\lambda_\mu = \partial_\mu \alpha$, which takes us back to the original action. But varying with respect to $\lambda_{\mu,\text{smooth}}$ instead gives

$$(\lambda_\mu + ea_\mu)\Phi_0^2 = \epsilon^{\mu\nu\rho}\partial_\nu b_\rho. \tag{21.19}$$

Substituting in the master action, we obtain the dual action as a function of the Lagrange multipliers b_μ:

$$S_{\text{dual}} = \int d^3x \left[-\frac{(f_{\mu\nu}^b)^2}{4\Phi_0^2} - \frac{1}{2}(\partial_\mu \Phi_0)^2 - e\epsilon^{\mu\nu\rho}b_\mu\partial_\nu a_\rho - \frac{2\pi}{e}b_\mu j_{\text{vortex}}^\mu - V(\Phi_0) - \frac{1}{4}F_{\mu\nu}^2 \right]. \tag{21.20}$$

The original theory has a global $U(1)$ invariance, with current associated with the scalar field Φ,

$$j_\mu = -ie(\Phi^\dagger \partial_\mu \Phi - \Phi \partial_\mu \Phi^\dagger) = e\Phi_0^2 \partial_\mu \alpha, \tag{21.21}$$

whereas in the dual theory we have instead a *vortex current* (also appearing in the dual action), which for constant Φ_0 is

$$j^\mu_{\text{vortex}} = \frac{e}{2\pi}\epsilon^{\mu\nu\rho}\partial_\nu \partial_\rho \alpha = \frac{1}{2\pi \Phi_0^2}\epsilon^{\mu\nu\rho}\partial_\nu j_\rho. \tag{21.22}$$

We see then that the particle and vortex currents are dual to each other, as are the corresponding fields, α and b_μ, via

$$\partial_\mu \alpha + ea_\mu = \frac{1}{\Phi_0^2}\epsilon^{\mu\nu\rho}\partial_\nu b_\rho. \tag{21.23}$$

The gauge field a_μ is unaffected by the duality. We can therefore call this duality a *particle-vortex duality*.

Finally, we observe that again this is a strong/weak duality, since in front of the kinetic term for α, $(\partial_\mu \alpha)^2$, in the original theory, we have Φ_0^2, playing the role of coupling factor $1/g^2$, whereas in the dual theory, in front of the Maxwell kinetic term for b_μ we have $1/\Phi_0^2$, so again we have inverted the coupling, $\tilde{g} = 1/g$. A strongly coupled theory is mapped to a weakly coupled one.

21.3 3+1 Dimensions: Maxwell Duality

Finally, we come to the original duality that historically served as a blueprint for the other examples, the electric-magnetic duality of the Maxwell equations, or Maxwell duality.

The Maxwell action for the electromagnetic field can be written in formal language as

$$S = -\frac{1}{4}\int d^4x F_{\mu\nu}^2 = -\frac{1}{4}\int F \wedge *F. \tag{21.24}$$

In the presence of only electrons, i.e. particles, the Maxwell equations are composed of the equations of motion for the above action with sources,

$$d*F = *J \to \partial_{[M} * F_{NP]} = *J_{MNP} \to \partial_N F^{NP} = J^P, \tag{21.25}$$

and the Bianchi identity, which is postulated (and is equivalent with the absence of magnetic sources) as

$$dF = 0 \to F = dA \to \partial_{[M} F_{NP]} = 0 \to \partial_N * F^{NP} = 0. \tag{21.26}$$

But one can consider as well the existence of magnetic sources, called magnetic monopoles, specifically of the Dirac, or singular, type. Even though we have never seen one in experiments, the simple existence of a single one in the universe has the far-reaching consequence of quantizing charge (as shown by Dirac), which we know to be an experimental fact, so it is an assumption we will make.

These sources are added as sources to the Bianchi identity, i.e. changing it to

$$dF = X; \quad X = d\omega \rightarrow F = dA + \omega;$$
$$\partial_{[M}F_{NP]} = X_{MNP} \rightarrow \partial_M * F^{MN} = *X^P. \tag{21.27}$$

The electron (particle) has the current source in the form of a delta function, specifically

$$(*J)_{123} = e\delta^3(y), \tag{21.28}$$

which implies that the electric charge of a field configuration is given via Gauss's law, by

$$e = \int_{S^2} *F = \int_{M^3} *J. \tag{21.29}$$

Similarly, a single Dirac (singular) monopole will have the source

$$(X)_{123} = g\delta^3(y), \tag{21.30}$$

which implies that the magnetic charge of a field configuration is given via Gauss's law, by

$$g = \int_{S^2} F = \int_{M^3} X. \tag{21.31}$$

On the other hand, the electric and magnetic charges are related by the *Dirac quantization condition*

$$eg = 2\pi n, \quad n \in \mathbb{Z}, \tag{21.32}$$

which will, however, not be proven here. This condition means that if there is a single magnetic monopole of charge g in the universe, it means that electric charge *must* be quantized, namely $e = (2\pi/g)n$.

We can now notice the symmetry of the Maxwell's equations with electric *and* magnetic sources,

$$F \leftrightarrow *F; \quad *J \leftrightarrow X, \tag{21.33}$$

which means exchanging electric and magnetic fields and also the corresponding charges:

$$\vec{E} \rightarrow \vec{B}; \quad \vec{B} \rightarrow -\vec{E}; \quad e \leftrightarrow g. \tag{21.34}$$

Maxwell duality can also be extended to a duality at the level of the path integral, like in the case of $(1+1)$–dimensional T-duality and $(2+1)$–dimensional particle-vortex duality, but we will not do it here.

21.4 $d+1$ Dimensions: Poincaré Duality

Finally, we generalize to dimensions higher than $3 + 1$, to what is generically called Poincaré duality. It is similar to the Maxwell case in the previous subsection, just that instead of gauge fields A_μ we must consider n-form (totally antisymmetric) tensor fields

$A_{\mu_1...\mu_n}$, with field strength

$$F_{\mu_1...\mu_{n+1}} = (n+1)\partial_{[\mu_1}A_{\mu_2...\mu_{n+1}]}, \tag{21.35}$$

and action

$$S = -\int \frac{F_{\mu_1...\mu_{n+1}}^2}{2(n+1)!} = -\int \frac{F_{(n+1)} \wedge *F_{(n+1)}}{2(n+1)!}, \tag{21.36}$$

where the last expression is written in form language. Note that we have a gauge invariance of the action

$$\delta A_{\mu_1...\mu_n} = \partial_{[\mu_1}\Lambda_{\mu_2...\mu_{n+1}]}. \tag{21.37}$$

Considering also a source term, analogous to the electrons in the Maxwell case, just that in this case sourcing a "$(n-1)$-brane," a singularity extended in $n-1$ space directions, the equations of motion are, in form language,

$$d(*F)_{D-n-1} = (*J)_{D-n}. \tag{21.38}$$

The Bianchi identities, postulated to hold so that $F = dA$, are also modified by the presence of "magnetic branes" (generalizations of the magnetic monopoles), with source X, and read

$$dF_{n+1} = X_{n+2} \equiv d\omega_{n+1} \rightarrow F_{n+1} = dA_n + \omega_{n+1}. \tag{21.39}$$

The analog of the electron, the electrically charged singular $(n-1)$−brane has a current source

$$(*J)_{1...D-n} = e_n\delta^{D-n}(y), \tag{21.40}$$

where e_n is the electric charge (or rather, charge density per unit volume of the source), whereas the analog of the monopole, the magnetically charged singular $D-n-3$ brane, has current source

$$X_{1...n+2} = g_{D-n-2}\delta^{n+2}(y), \tag{21.41}$$

where f_{D-n-2} is the magnetic charge density.

The electric-type charges are defined by the Gauss law

$$e_n = \int_{S^{D-n-1}} (*F)_{D-n-1} = \int_{M^{D-n}} *J_{D-n}, \tag{21.42}$$

and the magnetic-type charges are defined by the magnetic Gauss law

$$g_{D-n-2} = \int_{S^{n+1}} F_{n+1} = \int_{M^{n+2}} X_{n+2}. \tag{21.43}$$

These charges obey a generalized Dirac quantization condition

$$e_n g_{D-n-2} = 4\pi \frac{k}{2}. \tag{21.44}$$

In this case, we observe the *Poincaré duality symmetry* of the equations of motion and Bianchi identities in the presence of electric and magnetic charges, which takes the same form as in the Maxwell case:

$$F_{(n+1)} \leftrightarrow (*F)_{(D-n-1)}; \quad (*J)_{(D-n)} \leftrightarrow X_{(n+2)} \Rightarrow e_n \leftrightarrow g_{D-n-2}. \qquad (21.45)$$

Important Concepts to Remember

- Bosonization can be written as a duality of the path integral.
- A duality in the path integral amounts to gauging a symmetry direction, and introducing a first order form with a constraint imposed via Lagrange multipliers, and then integrating out the original fields, to write the action in terms of the Lagrange multipliers.
- T-duality in $1+1$ dimensions is an example of such a duality in the path integral, and it inverts the sigma model metric, meaning it is a strong/weak duality on the string *worldsheet*.
- Particle-vortex duality in $2+1$ dimensions is also an example, and it is also a strong/weak duality, exchanging particles and vortices, both at the level of fields and at the level of currents.
- Maxwell duality exchanges electric and magnetic fields and charges, $F \leftrightarrow *F, *J \leftrightarrow X$, $e \leftrightarrow g, \vec{E} \to \vec{B}, \vec{B} \to -\vec{E}$.
- Poincaré duality generalizes Maxwell duality to higher dimensions, with "p-brane" charges, again exchanging $F \leftrightarrow *F, *J \leftrightarrow X, e \leftrightarrow g$.

Further Reading

Bosonization as duality in $1+1$ dimensions was described in [39]. A review of T-duality in $1+1$ dimension and the $2+1$–dimensional case can be found in [40]. Maxwell duality is described in the review [41]. For Poincaré duality in the context of p-branes, see [36].

Exercises

(1) Derive the general transformation rules for the spacetime fields under T-duality (also called *Buscher's rules*) for the case of nonzero B_{IJ}.

(2) Consider a $U(N)$ version of the Abelian-Higgs action in (21.16), as a trace over the same action, with A_μ and Φ being in the adjoint of $U(N)$. Find a simple way to embed the abelian action and its particle-vortex duality into the $U(N)$ action.

(3) Consider the $3+1$–dimensional action

$$S = \int d^4 x \left[-\frac{W(\phi)}{4} F_{\mu\nu}^2 + \theta \epsilon^{\mu\nu\rho\sigma} F_{\mu\nu} F_{\rho\sigma} - \frac{1}{2}(\partial_\mu \phi)^2 \right]. \qquad (21.46)$$

Is it Maxwell (electric-magnetic) self-dual?

(4) Find a duality relation between the 2+1–dimensional actions

$$S_1 = \int d^3x \left[-\frac{1}{2} m^2 A_\mu A_\mu + \frac{1}{2} m \epsilon^{\mu\nu\rho} A_\mu \partial_\nu A_\rho \right] \qquad (21.47)$$

and

$$S_2 = \int d^3x \left[-\frac{1}{4} F_{\mu\nu} F^{\mu\nu} - \frac{1}{2} m \epsilon^{\mu\nu\rho} A_\mu \partial_\nu A_\rho \right], \qquad (21.48)$$

via the introduction of a master action for the duality.

(5) Find the action Poincaré dual to the action $S_D(d)$ in (20.48) plus the source action (20.50).

22

The Relativistic Point Particle and the Relativistic String

We are now ready to start with string theory proper. In this chapter, we will describe the action, symmetries, and equations of motion of the relativistic string. But these are (with a bit of thought) straightforward generalizations of the particle case, provided we understand it well enough, hence we will start with a rigorous analysis of the relativistic point particle.

22.1 The Relativistic Point Particle

The action for a nonrelativistic point particle is given by its kinetic energy, integrated over time:

$$S = \int dt\, L = \int dt\, \frac{m\dot{\vec{x}}^2}{2}. \tag{22.1}$$

For a relativistic point particle, we must write an action that reduces to this one. A good guess is the rest energy mc^2 times the integral of the total length of the worldline of the particle, i.e. the integral of the invariant element $d\tau$, with

$$d\tau^2 = -\eta_{\mu\nu}dx^\mu dx^\nu. \tag{22.2}$$

Putting $c = 1$ like always, the guess for the action is

$$S = -m \int d\tau = -m \int d\tau \sqrt{-\dot{X}^\mu \dot{X}^\nu \eta_{\mu\nu}}. \tag{22.3}$$

Here the trajectory of the particle is described parametrically in terms of its proper time τ as $X^\mu(\tau)$, leading to

$$\dot{X}^\mu \equiv \frac{dX^\mu}{d\tau}. \tag{22.4}$$

Then, in the nonrelativistic limit, $v/c \ll 1$, *on the worldline trajectory of the particle*,

$$d\tau^2 = dt^2 - \frac{d\vec{x}^2}{c^2} = dt^2 \left(1 - \frac{v^2}{c^2}\right), \tag{22.5}$$

so we obtain

$$S = -mc^2 \int dt \sqrt{1 - \frac{v^2}{c^2}} \simeq \int dt \left[-mc^2 + \frac{mv^2}{2}\right], \tag{22.6}$$

i.e. the nonrelativistic particle action plus a constant (minus the rest energy of the particle integrated over time).

The action we have described is an action on the one-dimensional "worldline" of the particle, in terms of the fields $X^\mu(\tau)$.

Symmetries

The particle action is then

$$S_1 = -m \int d\tau \sqrt{-\eta_{\mu\nu} \frac{dX^\mu}{d\tau} \frac{dX^\nu}{d\tau}}. \tag{22.7}$$

It has manifest spacetime (X^μ) Poincaré symmetry, since the invariance of $ds^2 = -\eta_{\mu\nu}dX^\mu dX^\nu$ defines the Poincaré group. It also has reparametrization invariance, or general coordinate (diffeomorphisms) invariance in one dimension, for arbitrary $\tau' = \tau'(\tau)$. In this case, this is just a parametrization of the trajectory, but the trajectory (path) itself stays the same, so $X'^\mu(\tau'(\tau)) = X^\mu(\tau)$. Then also

$$\frac{dX^\mu}{d\tau} = \frac{dX^\mu}{d\tau'} \frac{d\tau'}{d\tau}, \tag{22.8}$$

and

$$S_1 = -m \int \sqrt{-\eta_{\mu\nu} \frac{dX^\mu}{d\tau'} \frac{dX^\nu}{d\tau'} \frac{d\tau'}{d\tau}} d\tau = -m \int \sqrt{-\eta_{\mu\nu} \frac{dX^\mu}{d\tau'} \frac{dX^\nu}{d\tau'}} d\tau' \tag{22.9}$$

is invariant.

Equations of Motion

The variable in the action is the trajectory $X^\mu(\tau)$, so varying the action we obtain

$$\delta S_1 = -m \int d\tau \, \delta \left(\sqrt{-\dot{X}^\mu \dot{X}_\mu} \right) = -m \int d\tau \left[-\frac{\eta_{\mu\nu} \dot{X}^\mu \delta \dot{X}^\nu}{\sqrt{-\dot{X}^\mu \dot{X}_\mu}} \right]$$

$$= -m \int d\tau \frac{d}{d\tau} \left[\frac{\eta_{\mu\nu} \dot{X}^\nu}{\sqrt{-\dot{X}^\mu \dot{X}_\mu}} \right] \delta X^\mu + \delta X^\mu m u_\mu \big|_{\tau_i}^{\tau_f}. \tag{22.10}$$

In the second line we have partially integrated the derivative on δX^ν. Also noting that $\sqrt{-\dot{X}^\mu \dot{X}_\nu} = 1$ and $\dot{X}^\mu = dX^\mu/d\tau = u^\mu$, and using the 4-momentum of the particle,

$$p_\mu = mc \frac{dX^\mu}{d\tau}, \tag{22.11}$$

from the vanishing of the bulk term we obtain the equation of motion

$$\frac{dp^\mu}{d\tau} = mc \frac{d^2 X^\mu}{d\tau} = 0, \tag{22.12}$$

whereas the boundary term is satisfied if we fix X^μ at the time boundaries.

The equation of motion looks somewhat trivial, the motion of a free particle in flat space. But since we do have a free particle in flat space, it is correct. We can actually improve the situation, and consider instead the particle moving in a gravitational field, replacing $\eta_{\mu\nu} \to g_{\mu\nu}$ in the action. Then it is left as an exercise to show that we obtain the free motion of a particle in a gravitational field, i.e. the geodesic motion. That, however, is nontrivial: We constructed general relativity based on the assumption that the motion of a particle in curved spacetime appears to us as the effect of gravity, so the geodesic motion describes the interaction of the particle with the gravitational field.

We can moreover consider the coupling of the particle action to a background gauge field A_μ with charge q for the particle. We have seen how to do that when talking about p-branes. The general coupling of A_μ with a source term j^μ is

$$\int d^4x A_\mu(x) j^\mu(x). \tag{22.13}$$

For a particle of charge q in the rest frame, $j^\mu(x) = q\delta^3(x)\delta^{\mu 0}$, which when boosted into a general frame becomes

$$j^\mu(X(t)) = q\frac{dX^\mu}{dt}\delta^3(X^i(t)), \tag{22.14}$$

so integrated in the source term gives

$$q\int d^4x \delta^3(X^i(t)) A_\mu(X^\mu(t))\frac{dX^\mu}{dt} = \int d\tau\left(q\frac{dX^\mu}{d\tau}\right) A_\mu(X^\nu(\tau)). \tag{22.15}$$

As we already mentioned, by solving the equations of motion with this source term, we get the electromagnetic field of the point particle, i.e. the interaction of the particle with the electromagnetic field.

We have seen therefore that we can describe the interaction of the particle with $g_{\mu\nu}$ and A_μ in this worldline formalism.

22.2 First Order Action for the Point Particle

We can now put the action of the particle in another form. That is, we write another action for it, which will be equivalent to (22.9) at least classically (on-shell). The idea is to write a first order form for the action, such that it becomes quadratic in the fields $X^\mu(\tau)$. There is a general procedure for this situation, namely we need to introduce an auxiliary field, whose equation of motion is algebraic.

The sought-for auxiliary field is an independent "worldline metric" (in one spacetime dimension) $\gamma_{\tau\tau}(\tau)$. It will be auxiliary since 1-dimensional gravity is not dynamical, we cannot form a nontrivial Einstein action for it (kinetic term). But it is actually better (simpler) to work with the "einbein" field

$$e(\tau) = \sqrt{-\gamma_{\tau\tau}(\tau)}. \tag{22.16}$$

Note that in general, we can replace the metric $g_{\mu\nu}$ in favor of the so-called "vielbein" e^a_μ (from "viel" = many and "bein" = leg in German), through the relation $g_{\mu\nu} = e^a_\mu e^b_\nu \eta_{ab}$. The field was initially introduced in 4 dimensions under the name "vierbein" ("vier" = four in German), and in other dimensions one can use "ein, zwei, drei, etc. -bein" ("ein, zwei, drei" = one, two, three), or in general "vielbein." But in one dimension things are simple, and we use the term einbein.

The general relativistic invariance of the metric, $ds^2 = g_{\mu\nu}dX^\mu dX^\nu = g'_{\mu\nu}dX'^\mu dX'^\nu$ translates into worldline reparametrization invariance of the einbein,

$$e'(\tau')d\tau' = e(\tau)d\tau, \qquad (22.17)$$

which can be rewritten as

$$\frac{[e'(\tau')]^{-1}}{d\tau'} = \frac{[e(\tau)]^{-1}}{d\tau}. \qquad (22.18)$$

We can guess the form of the equivalent worldline action. We want a quadratic action in $\dot{X}^\mu$, so a term will be $\dot{X}^\mu\dot{X}^\nu\eta_{\mu\nu}d\tau$, with a coefficient that depends on $e(\tau)$. But since we want to have the above reparametrization invariance, it really should be $e^{-1}(\tau)$ times a constant. We can also write a constant term, which by dimensional analysis should be proportional to m^2, and by reparametrization invariance should be em^2 times a constant. We can fix the constants by imposing that we get an action equivalent with S_1, leading to

$$S_P = \frac{1}{2}\int d\tau \left[e^{-1}\dot{X}^\mu\dot{X}^\nu\eta_{\mu\nu} - em^2\right]. \qquad (22.19)$$

Symmetries

The action is reparametrization invariant by construction. It is also manifestly spacetime Poincaré invariant, since it depends only on the combination $dX^\mu dX^\nu \eta_{\mu\nu}$.

Equations of Motion

We can solve for the equation of motion of the einbein $e(\tau)$ of S_P, obtaining

$$-e^{-2}\dot{X}^\mu\dot{X}^\nu\eta_{\mu\nu} - m^2 = 0, \qquad (22.20)$$

solved by

$$e = \frac{1}{m}\sqrt{-\eta_{\mu\nu}\dot{X}^\mu\dot{X}^\nu}. \qquad (22.21)$$

Substituting it back in S_P, we obtain

$$S_P \rightarrow S_1 = -m\int d\tau\sqrt{-\eta_{\mu\nu}\dot{X}^\mu\dot{X}^\nu}. \qquad (22.22)$$

Therefore at least on-shell (classically), the two action S_1 and S_P are equivalent.

But S_P is quadratic in $X^\mu(\tau)$, so it is easier to work with. Moreover, unlike S_1, for which the $m \rightarrow 0$ limit is singular, for S_P the limit is easy, and we obtain

$$S_P = \frac{1}{2}\int d\tau e^{-1}\dot{X}^\mu\dot{X}^\nu\eta_{\mu\nu}. \qquad (22.23)$$

Gauge Fixing and Constraints

Even more, we have now one auxiliary field, $e(\tau)$, but also one "gauge invariance," the reparametrization invariance described by the arbitrary function $\tau'(\tau)$, which allows us to fix the einbein to anything we want by fixing a gauge. The most convenient is $e'(\tau') = 1$, which always can be chosen, since now

$$d\tau' = d\tau e(\tau). \qquad (22.24)$$

With this gauge choice, and having thus fixed reparametrization invariance (obtaining only an action with spacetime Poincaré invariance), we obtain

$$S_P = \frac{1}{2} \int d\tau \, \dot{X}^\mu \dot{X}^\nu \eta_{\mu\nu}. \tag{22.25}$$

But if we fix this gauge, we still have to impose the equation of motion of the gauge fixed field, $e(\tau)$, as a constraint on the model, the same way as in electromagnetism fixing the gauge $A_0 = 0$ leads to the Gauss constraint $\partial^i E_i = 0$ (the would-be equation of motion for A_0). So we must impose as a constraint $\delta S/\delta e|_{e=1} = 0$, giving

$$\eta_{\mu\nu} \frac{dX^\mu}{d\tau} \frac{dX^\nu}{d\tau} = 0. \tag{22.26}$$

Note, however, that in S_P, the $e^{-1} = ee^{-2}$ factor can be understood as $-\sqrt{-\det(\gamma_{\tau\tau})}\gamma^{\tau\tau}$ in one dimension. Then, using the general definition of the energy-momentum tensor of $T_{\mu\nu} = -2/\sqrt{-g}\delta S/\delta g^{\mu\nu}$, we see that the above is nothing but the energy-momentum tensor in one dimension:

$$\eta_{\mu\nu} \frac{dX^\mu}{d\tau} \frac{dX^\nu}{d\tau} = T_{\tau\tau} \equiv T = 0. \tag{22.27}$$

Thus fixing the gauge $e = 1$ for reparametrization invariance leads to the constraint $T = 0$.

22.3 Relativistic Strings

The generalization to relativistic strings is straightforward, with a bit of thought. If the particle action S_1 was the length of the worldline, with a coefficient = minus the rest mass, such that it minimizes the time-length between the initial and final points, the action for a string that spans a 1+1–dimensional *worldsheet* as it moves in time (see Figure 22.1) should be the area of this worldsheet, with a coefficient of dimension 2, called the *string tension T*, so

$$S = -T \int dA \equiv -\frac{1}{2\pi\alpha'} \int dA. \tag{22.28}$$

Here we have defined $T = 1/(2\pi\alpha')$ for historical reasons: String theory started as an attempt to describe hadronic physics (what we now know is QCD). In particular, it reproduced a linear plot for M^2 as a function of spin of hadronic excitations, with a coefficient (slope) that was matched against an experimental one, denoted by α'.

The invariant area of the 2-dimensional worldsheet is, as we saw in Chapter 16, $d^2\xi\sqrt{-\det(\gamma_{ab})}$, where $(\xi^a) = (\tau, \sigma)$ for $a = 1, 2$.

Then the *Nambu-Goto action* (that started string theory) is

$$S_{NG} = -\frac{1}{2\pi\alpha'} \int d\sigma \, d\tau \sqrt{-\det(\gamma_{ab})}, \tag{22.29}$$

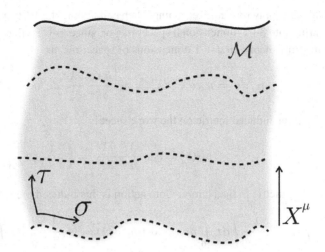

Fig. 22.1   String moving in spacetime parametrized by X^μ spans a worldsheet $\mathcal{M}$ parametrized by σ (coordinate along the string) and τ (worldsheet time).

or even more explicitly

$$S_{NG} = -\frac{1}{2\pi\alpha'} \int_{\tau_i}^{\tau_f} d\tau \int_0^l d\sigma \sqrt{-\gamma_{11}\gamma_{22} + (\gamma_{12})^2}. \tag{22.30}$$

Yoichiro Nambu received the Nobel Prize mostly for the discovery of the Nambu-Goldstone boson and related work, but the development of string theory was also cited. This is to date the only Nobel Prize related to string theory.

Induced Metric

Note that in the above we have not specified whether the metric γ_{ab} is intrinsic (independent) or induced from spacetime by the embedding of the worldsheet in spacetime. In fact it is the latter. To understand this, consider first the familiar case of a 2-dimensional spatial surface embedded in 3-dimensional Euclidean space, with metric $ds^2 = d\vec{X} \cdot d\vec{X}$. If the embedding that describes the surface is $\vec{X}(\xi^i)$, then for the embedding

$$d\vec{X} = \frac{\partial \vec{X}}{\partial \xi^i} d\xi^i, \tag{22.31}$$

which means that *on the surface*, we can substitute the above in the Euclidean metric and obtain

$$ds^2 = \left(\frac{d\vec{X}}{d\xi^i} \cdot \frac{\partial \vec{X}}{\partial \xi^j} \right) d\xi^i d\xi^j \equiv g_{ij}(\xi) d\xi^i d\xi^j. \tag{22.32}$$

That means that the *induced metric* on the surface is

$$g_{ij}(\xi) = \frac{d\vec{X}}{\partial \xi^i} \cdot \frac{\partial \vec{X}}{\partial \xi^j}. \tag{22.33}$$

We can apply the same reasoning when embedding the 1+1–dimensional worldsheet of the string into 3+1–dimensional spacetime, or, since as we will see, we need extra dimensions in string theory, in $d + 1$ dimensions of spacetime, as

$$ds^2 = \eta_{\mu\nu}dX^\mu dX^\nu = \eta_{\mu\nu}\frac{\partial X^\mu}{\partial \xi^a}\frac{\partial X^\nu}{\partial \xi^b}d\xi^a d\xi^b \equiv h_{ab}(\xi)d\xi^a d\xi^b. \qquad (22.34)$$

Then the induced metric on the worldsheet is

$$h_{ab}(\xi) = \eta_{\mu\nu}\frac{\partial X^\mu}{\partial \xi^a}\frac{\partial X^\nu}{\partial \xi^b} = \begin{pmatrix} \dot{X}^2 & \dot{X}\cdot X' \\ \dot{X}\cdot X' & X'^2 \end{pmatrix}_{ab}. \qquad (22.35)$$

The metric in the Nambu-Goto action is this induced metric h_{ab}, so we have

$$S_{NG} = -\frac{1}{2\pi\alpha'}\int d\tau \int_0^l d\sigma \sqrt{-\det(h_{ab}(X))} = -\frac{1}{2\pi\alpha'}\int d\tau \int_0^l d\sigma \sqrt{-\dot{X}^2 X'^2 + (\dot{X}\cdot X')^2}. \qquad (22.36)$$

Symmetries

1. The action is manifestly spacetime Poincaré invariant, since it is written in terms of $dX^\mu dX^\nu \eta_{\mu\nu}$.

2. It is also manifestly reparametrization invariant, under $\xi^a = \xi^a(\xi')$, since it is written in terms of the invariant volume of the worldsheet, $d^2\xi\sqrt{-\det(h_{ab})}$, and we saw in Chapter 16 that this is invariant under general coordinate transformations. But we can also see explicitly that it is still true for the induced metric as well, since

$$h_{ab}(\xi) = \frac{\partial X}{\partial \xi'^c}\cdot\frac{\partial X}{\partial \xi'^d}\frac{\partial \xi'^c}{\partial \xi^a}\frac{\partial \xi^d}{\partial \xi^b} = h_{cd}(\xi')\frac{\partial \xi'^c}{\partial \xi^a}\frac{\partial \xi^d}{\partial \xi^b}, \qquad (22.37)$$

leading to the required transformation law

$$d^2\xi\sqrt{-\det(h_{ab})} = \sqrt{-\det(h'_{ab})}\left|\frac{\partial \xi'}{\partial \xi}\right|d^2\xi = \sqrt{-\det(h'_{ab})}d^2\xi'. \qquad (22.38)$$

22.4 First Order Action for the Relativistic String

We can write an action for the relativistic string that is equivalent on-shell (classically) with S_{NG}, in a similar way to the action for the massless particle, by introducing an independent (auxiliary) metric γ_{ab} on the string worldsheet. We can guess it by simply generalizing the case of S_P for the particle, and find

$$S_P[X, \gamma] = -\frac{1}{4\pi\alpha'}\int_M d\tau\, d\sigma\sqrt{-\det(\gamma_{ab})}\gamma^{ab}\partial_a X^\mu \partial_b X^\nu \eta_{\mu\nu}. \qquad (22.39)$$

This is the *Polyakov action*, called this despite being actually found by Brink, DiVecchia, Howe, Deser, and Zumino.

We should comment that there can be no kinetic term for the metric (Einstein-Hilbert term), since in two dimensions the Einstein action is a topological invariant,

$$\int d^2x\sqrt{-\gamma}R^{(2)} = \chi,\tag{22.40}$$

the Euler number, depending only on the topology of the manifold. Then even if added, it would not contribute to the equations of motion.

Now we can check that we get the Nambu-Goto action after solving for γ_{ab}: If we vary the intrinsic (independent) metric on the worldsheet, γ_{ab}, the variation of the Polyakov action is

$$\delta S_P = -\frac{1}{4\pi\alpha'}\int_M d\tau\, d\sigma\sqrt{-\gamma}\delta\gamma^{ab}\left[\partial_a X^\mu \partial_b X^\nu \eta_{\mu\nu} - \frac{1}{2}\gamma_{ab}(\gamma^{cd}\partial_c X^\mu \partial_d X^\nu \eta_{\mu\nu})\right]$$

$$= -\frac{1}{4\pi\alpha'}\int_M d\tau\, d\sigma\sqrt{-\gamma}\delta\gamma^{ab}\left[h_{ab} - \frac{1}{2}\gamma_{ab}(\gamma^{cd}h_{cd})\right].\tag{22.41}$$

Therefore the equation of motion for γ_{ab}, $\delta S/\delta\gamma^{ab} = 0$ gives, after multiplying the resulting square bracket by h^{ab} to find $2^2 = (\gamma_{ab}h^{ab})^2$, and substituting the result back in the equation of motion,

$$\gamma_{ab} = h_{ab},\tag{22.42}$$

i.e., the intrinsic metric is the induced metric. Replacing this in the Polyakov action gives back the Nambu-Goto action, $S_P \to S_{NG}$, as we can easily check, since $S_P = -1/(4\pi\alpha')\int\sqrt{-\gamma}\,\gamma^{ab}h_{ab}$ and $\gamma^{ab}h_{ab} = 2$.

Symmetries

1. The Polyakov action is again manifestly spacetime Poincaré invariant, since it is written in terms of $dX^\mu dX^\nu \eta_{\mu\nu}$.

2. It is also worldsheet reparametrization (general coordinate transformations or diffeomorphism) invariant, since the fields X^μ act as scalars on the worldsheet, i.e. $X'^\mu(\tau', \sigma') = X^\mu(\tau, \sigma)$, whereas γ_{ab} is a metric, i.e. it transforms in the usual way:

$$\gamma_{ab}(\tau, \sigma) = \gamma_{cd}(\tau', \sigma')\frac{\partial\xi'^c}{\partial\xi^a}\frac{\partial\xi'^d}{\partial\xi^b}.\tag{22.43}$$

3. In addition, we have a new type of symmetry, called Weyl invariance, which amounts to multiplying the metric by a function (local parameter), called a "conformal factor,"

$$X'^\mu(\tau, \sigma) = X^\mu(\tau, \sigma)$$

$$\gamma'_{ab}(\tau, \sigma) = e^{2\omega(\tau, \sigma)}\gamma_{ab}(\tau, \sigma).\tag{22.44}$$

We see that then, in two dimensions, $\sqrt{-\det(\gamma_{ab})} \to e^{2\omega}\sqrt{-\det(\gamma_{ab})}$, compensated by $\gamma^{ab}h_{ab} \to \gamma^{ab}h_{ab}$.

Note that now we can offer an deeper explanation as to why there was no cosmological constant term in the Polyakov action, a term given by simply the invariant volume of the independent metric, like the m^2 term in the particle case, or the $(d - 2)T_p$ term in the p-brane case described in Chapter 21. Of course, we have the explanation that, for the analogy with the particle, we have $m = 0$ now (relativistic string), and on the other hand, if we add a cosmological constant term, we don't have the equivalence with the Nambu-Goto action. But now we have a deeper explanation: Such a term would not be Weyl invariant, and as we will see, Weyl invariance is central to the analysis of the string, because it will lead to 2-dimensional conformal invariance, responsible for most string properties.

Equations of Motion

As we saw in previous chapters, the energy-momentum tensor is given by the variation of the matter action with respect to the inverse metric. In the case of the string action, there is no gravitational action (since it would lead to a topological number, i.e. an integer), so the variation is of the full action. Using the normalization traditional in string theory (since there is no gravitational action to compare against, the normalization of the energy-momentum tensor is irrelevant for most things), we have

$$T_{ab}(\tau, \sigma) \equiv -4\pi \frac{1}{\sqrt{-\gamma}} \frac{\delta S_P}{\delta \gamma^{ab}} = \frac{1}{\alpha'} \left(\partial_a X^\mu \partial_b X_\mu - \frac{1}{2} \gamma_{ab} \partial_c X^\mu \partial^c X_\mu \right). \quad (22.45)$$

Then the equation of motion for γ^{ab} is simply

$$T_{ab} = 0. \quad (22.46)$$

The energy-momentum tensor is covariantly conserved, as usual, $\nabla_a T^{ab} = 0$. But since we are in a Weyl invariant theory, i.e. a theory invariant under the multiplication of the metric by a common function (as opposed to simply invariant under variations of the metric), the Weyl invariance can be written as

$$\gamma^{ab} \frac{\delta S}{\delta \gamma^{ab}} = 0, \quad (22.47)$$

which means that the energy-momentum tensor is *traceless off-shell* (since on-shell the whole T_{ab} is zero):

$$T^a{}_a = 0. \quad (22.48)$$

The X^μ equation of motion is found by varying the action with respect to it and partially integrating the derivative on X:

$$\delta_X S_P = \frac{1}{2\pi\alpha'} \int d\tau \int_0^l d\sigma \sqrt{-\gamma} \delta X^\mu \nabla^2 X_\mu - \frac{1}{2\pi\alpha'} \int d\tau \sqrt{-\gamma} \delta X^\mu \partial^\sigma X_\mu|_{\sigma=0}^{\sigma=l}. \quad (22.49)$$

Note that there would be a boundary term in time as well, but we assume that is zero, like for the particle. The equation of motion for X^μ is obtained by putting to zero the bulk term, obtaining

$$\nabla^2 X^\mu = 0, \quad (22.50)$$

i.e. the wave equation in two dimensions.

Then by putting to zero also the boundary term, we obtain boundary conditions. We can have the *Neumann boundary condition*

$$\partial^\sigma X^\mu(\tau, 0) = \partial^\sigma X^\mu(\tau, l) \quad (22.51)$$

(Neumann since the derivative of the field X^μ is fixed at the boundary) or the *Dirichlet boundary condition*

$$\delta X^\mu(\tau, 0) = \delta X^\mu(\tau, l). \tag{22.52}$$

(Dirichlet since the field itself is fixed: $\delta X^\mu = 0$ means $X^\mu = $ constant.)

But note now that we actually have two options: The above conditions are only if the string is *open*, i.e. its endpoints are not identified. But we can also have a *closed string*, i.e. the boundary condition is

$$X^\mu(\tau, 0) = X^\mu(\tau, l), \tag{22.53}$$

so the variations δX^μ at the two endpoints are also identical.

22.5 Gauge Fixing and Constraints

The metric γ_{ab} on the worldsheet has three independent components, γ_{11}, γ_{12} and γ_{22}. For any general coordinate invariant theory in two dimensions, we have invariance under two arbitrary functions $\xi^a(\xi')$, so we can use them to fix two of the metric components but are left with another one unfixed. However, for a theory that is also Weyl invariant, like string theory, we have three arbitrary functions, $\xi^a(\xi')$ and $\omega(\xi)$, which we can use to fix the whole metric. In fact, we can fix it to η_{ab}, though we will not prove it here (one should be able to reach the gauge using the arbitrary functions, which is not obvious, but it can be done). This gauge is usually called *conformal gauge* (though strictly speaking, conformal gauge would be when we fix the metric to a conformal factor times η_{ab}, using only general coordinate invariance, and the η_{ab} gauge would be called a unit gauge, but that distinction is not often made).

In the conformal gauge, the Polyakov action becomes

$$S_P = -\frac{1}{4\pi\alpha'} \int d^2\sigma\, \eta_{\mu\nu} \partial_a X^\mu \partial_b X^\nu \eta^{ab}, \tag{22.54}$$

and its equation of motion is

$$\left(\frac{\partial^2}{\partial\tau^2} - \frac{\partial^2}{\partial\sigma^2} \right) X^\mu = 0. \tag{22.55}$$

But like in the case of the particle, fixing the gauge for general coordinate invariance and Weyl invariance means that we need to impose the would-be equation for γ_{ab} (in the $\gamma_{ab} = \eta_{ab}$ gauge) as a constraint (analog of the Gauss constraint for electromagnetism). As we saw, the equation of motion for γ_{ab} was

$$T_{ab} = 0, \tag{22.56}$$

so this is imposed as a constraint in the conformal gauge.

Nambu-Goto in the Static Gauge

We can instead consider a *static gauge* for the general coordinate invariance, fixing $X^0 = \tau$ (by using ξ^τ). Then also $\vec{X} = \vec{X}(\tau, \sigma)$, and the Nambu-Goto action becomes

$$S_{NG} = -\frac{1}{2\pi\alpha'} \int d\sigma \; d\tau \sqrt{(\dot{X} \cdot X')^2 - \dot{X}^2 X'^2}$$

$$= -\frac{1}{2\pi\alpha'} \int_0^l d\sigma \int d\tau \sqrt{|\vec{X}'|^2}, \tag{22.57}$$

where the second line is for *static* strings, $\vec{X} = \vec{X}(\sigma)$, so that $\dot{X} \cdot X' = 0$ and $\dot{X}^2 = -1$, $X'^2 = \vec{X}'^2$. We finally obtain

$$S = -\Delta t T \int d\sigma \frac{|d\vec{X}|}{d\sigma} = -\Delta t |\delta \vec{X}| T. \tag{22.58}$$

But for a static solution, the action is minus the energy times the time interval, so we obtain the energy as

$$E = |\Delta \vec{X}| T, \tag{22.59}$$

justifying the name of string tension for T, since it is energy per unit length of the static string.

String Endpoints

We now want to understand the implications of the boundary conditions. For Dirichlet boundary conditions, we will obtain D-branes, which will be discussed later. Here we will concentrate on the Neumann boundary conditions, and we will show that it implies that the endpoints move at the speed of light. This is to be expected, since we are in a relativistic theory, so it is the only logical possibility.

To prove the statement, note first that the energy-momentum tensor in the conformal gauge $g_{ab} = \eta_{ab}$ is

$$T_{ab} = \frac{1}{\alpha'} \left(\partial_a X^\mu \partial_b X_\mu - \frac{1}{2} \eta_{ab} \partial_c X^\mu \partial^c X_\mu \right). \tag{22.60}$$

In components, we obtain

$$\alpha' T_{00} = \dot{X}^2 + \frac{1}{2}(-\dot{X}^2 + X'^2) = \frac{1}{2}(\dot{X}^2 + X'^2)$$

$$\alpha' T_{11} = X'^2 - \frac{1}{2}(-\dot{X}^2 + X'^2) = \frac{1}{2}(\dot{X}^2 + X'^2)$$

$$\alpha' T_{01} = \dot{X} \cdot X'. \tag{22.61}$$

Then the equations of motion $T_{00} = T_{11} = 0$ are $\dot{X}^2 + X'^2 = 0$. Since at the string endpoints $X'(\tau, \sigma) = 0$ by the Neumann boundary condition, the equations of motion imply that on the endpoints

$$\dot{X}^2 = 0 \Rightarrow \frac{\partial X^\mu}{d\tau} \frac{\partial X_\mu}{d\tau} = 0, \tag{22.62}$$

i.e. they move at the speed of light (on $ds^2 = 0$).

Important Concepts to Remember

- The relativistic particle action is $S_1 = -m \int d\tau$, and is reparametrization invariant and spacetime Poincaré invariant, and has as equation of motion $dp^\mu/d\tau = 0$.
- Coupling it to gravity $g_{\mu\nu}$ or a background gauge field A_μ on the worldline, we obtain the interaction of the particle with the spacetime fields.
- The equivalent quadratic particle action S_P is also spacetime Poincaré invariant and reparametrization invariant, but is written in terms of an auxiliary einbein on the worldline.
- Fixing the gauge $e = 1$ for reparametrization in the $m \to 0$ case, we must impose the equation of motion of e, $T = 0$.
- The Nambu-Goto string action is minus the string tension times the area of the string worldsheet, using the induced metric on the worldsheet, and is spacetime Poincaré and reparametrization invariant.
- The classically equivalent Polyakov action is written in terms of an independent worldsheet metric γ_{ab} and is quadratic in $X^\mu(\xi)$.
- S_P is spacetime Poincaré invariant, reparametrization invariant, and Weyl invariant, allowing us to fix a conformal gauge $\gamma_{ab} = \eta_{ab}$.
- The equation of motion for X^μ is the wave equation $\partial^2 X^\mu = 0$, and we must impose the equation of motion for γ_{ab}, $T_{ab} = 0$ as a constraint.
- In the static gauge, the energy of the string is the tension times the length of the string.
- The endpoints of the Neumann open string move at the speed of light.

Further Reading

Chapters 1, 2 in [42], vol. I, chapters 5, 6, 7 in [43], chapter 1 in [11], chapter 7 in [8].

Exercises

(1) Calculate the Hamiltonian for the massive relativistic particle action, both for S_1 and for S_P.

(2) Calculate the equation of motion of a free particle in a gravitational field (the geodesic equation) for the particle action S_1, with $\eta_{\mu\nu} \to g_{\mu\nu}$, and specialize to the Newtonian limit to recover the motion in Newtonian gravity.

(3) Calculate the X^μ equation of motion for the string actions S_P and S_{NG} in the static gauge and show that the one for S_P reduces to the one for S_{NG}.

(4) Instead of a 1+1–dimensional worldsheet for a string, consider a 3+1–dimensional "worldvolume" for a "3-brane." Write the equivalent of the NG action for it and its symmetries.

(5) Consider an open string moving in D flat dimensions, with $p + 1$ dimensions having Neumann boundary conditions and $D - p - 1$ having Dirichlet boundary conditions. Write the most general solution for the equations of motion for X^μ.

(6) Calculate the Hamiltonian and the equations of motion for the Nambu-Goto string in static gauge.

(7) Consider a Nambu-Goto string in static gauge moving on the space $\mathbb{R}_t \times S^2$, with metric

$$ds^2 = -dt^2 + \sin^2\theta d\phi^2 + d\theta^2, \tag{22.63}$$

and an open string configuration where $\theta =$ independent of the time t. Write the Nambu-Goto action on this ansatz and solve it to find the string solution.

Lightcone Strings and Quantization

In this chapter we continue the description of strings, with their quantization, and the analysis of the important light-cone gauge for the string.

23.1 Conformal Invariance

We have seen in the last chapter that we can use the three invariances of the theory, the two diffeomorphisms and the one Weyl invariance, to put the worldsheet metric in the conformal gauge, $\gamma_{ab} = \eta_{ab}$, and the Polyakov action became

$$S_P = -\frac{1}{4\pi\alpha'} \int_{-\infty}^{+\infty} \int_0^l d\sigma \, \eta^{ab} \partial_a X^\mu \partial_b X^\nu \eta_{\mu\nu}. \tag{23.1}$$

The equations of motion of this action (in conformal gauge) were

$$\left(\frac{\partial^2}{\partial\tau^2} - \frac{\partial^2}{\partial\sigma^2}\right) X^\mu = 0. \tag{23.2}$$

But often when we fix a gauge, we are left with a residual gauge invariance: Even if the number of functions fixed equals the number of invariances, one can still have invariance under transformations with parameters of restricted (not general) dependence. A standard example is the electromagnetism in the Lorenz gauge $\partial^\mu A_\mu = 0$, which still allows us to fix the radiation gauge by $A_0 = 0$.

The same thing happens now. Defining the variables

$$\sigma^\pm = \tau \pm \sigma, \tag{23.3}$$

the transformation

$$\sigma^+ \to \tilde{\sigma}^+ = f(\sigma^+); \quad \sigma^- \to \tilde{\sigma}^- = g(\sigma^-) \tag{23.4}$$

is a residual gauge invariance, which is identified with the same *conformal invariance* we have identified in the context of condensed matter theory. Note, first, that the above is a general coordinate transformation (diffeomorphism) of a particular type, since the arguments of f and g are restricted to σ^+ or σ^-, instead of a general (σ, τ). Second, note that now the flat worldsheet metric is changed as follows:

$$ds^2 = -d\sigma^+ d\sigma^- = \frac{d\tilde{\sigma}^+}{f'} \frac{d\tilde{\sigma}^-}{g'} = (f'(\tilde{\sigma}^+)g'(\tilde{\sigma}^-))^{-1}[-d\tilde{\sigma}^+ d\tilde{\sigma}^-]. \tag{23.5}$$

This is of the general type

$$ds^2 = d\vec{x}^2 = e^{2\omega(\vec{x}')}d\vec{x}'^2 \tag{23.6}$$

which is the definition of conformal invariance – invariance under a transformation of flat space that multiplies the metric by a conformal factor. Moreover, we see that compared with the definition of conformal invariance in the condensed matter case (when we were referring to invariance of a spatial theory, i.e. on a Euclidean space), we have now a Minkowski space definition, so via Wick rotation, instead of z and $\bar{z}$, we have σ^+ and σ^-.

Finally, note that after the transformation to $\tilde{\sigma}^\pm$, which was a particular general coordinate transformation, we can use the Weyl invariance of the theory to make a particular type of compensating Weyl transformation, with

$$\gamma_{ab} \rightarrow (f'(\tilde{\sigma}^+)g'(\tilde{\sigma}^-))\gamma_{ab}, \tag{23.7}$$

to return back to the conformal gauge metric η_{ab}. Therefore, indeed, a combination of a special kind of diffeomorphism, and a compensating special kind of Weyl transformation, is a residual gauge invariance, which leaves the conformal gauge metric unchanged.

Moreover, after such a transformation, the redefined time coordinate $\tilde{\tau}$, equaling

$$\tilde{\tau} = \frac{1}{2}(\tilde{\sigma}^+(\sigma^+) + \tilde{\sigma}^-(\sigma^-)), \tag{23.8}$$

satisfies the wave equation, since

$$\left(\frac{\partial^2}{d\sigma^2} - \frac{\partial^2}{\partial\tau^2}\right)\tilde{\tau} = \left(\frac{\partial}{\partial\sigma} - \frac{\partial}{\partial\tau}\right)\left(\frac{\partial}{\partial\sigma} + \frac{\partial}{\partial\tau}\right)\tilde{\tau} = -4\frac{\partial}{\partial\sigma^+}\frac{\partial}{\partial\sigma^-}\tilde{\tau} = 0, \tag{23.9}$$

where we have used $\sigma^\pm = \tau \pm \sigma$, which means

$$\partial_\pm = \frac{1}{2}(\partial_\tau \pm \partial_\sigma). \tag{23.10}$$

23.2 Light-Cone Gauge and Mode Expansions

23.2.1 Light-Cone Gauge

But that means that the τ coordinate after the conformal transformation, $\tilde{\tau}$, satisfies the same wave equation as X^μ, so we can choose a gauge (for the conformal invariance! we are still in the conformal gauge) where τ is equal to one of the X^μ's.

It is useful to introduce *light-cone coordinates* in spacetime, defined as

$$X^+ = \frac{X^0 + X^{D-1}}{\sqrt{2}}; \quad X^- = \frac{X^0 - X^{D-1}}{\sqrt{2}}, \tag{23.11}$$

in which case the product of two vectors is written as

$$V \cdot W = -V^-W^+ - V^+W^- + V^iW^i. \tag{23.12}$$

Then the *light-cone gauge* is the choice $\tilde{\tau} = X^+/(2\alpha' p^+) + $ constant, or more precisely

$$X^+(\sigma, \tau) = x^+ + (2\alpha')p^+\tau. \tag{23.13}$$

The reason is that we want to think of X^+ as lightcone time, as is usual in lightcone quantization.

23.2.2 Mode Expansions

The most general solution of the wave equation for X^μ, written as

$$\frac{\partial}{\partial\sigma^+}\frac{\partial}{\partial\sigma^-}X^\mu = 0 \tag{23.14}$$

is

$$X^\mu(\sigma, \tau) = X_R^\mu(\sigma^-) + X_L^\mu(\sigma^+), \tag{23.15}$$

where X_L^μ and X_R^μ are arbitrary functions called the left-moving and the right-moving modes, respectively.

But we still need to impose the correct boundary conditions on these solutions.

Closed String

In this case, we have periodicity,

$$X^\mu(\tau, l) = X^\mu(\tau, 0), \tag{23.16}$$

for $l = 2\pi$, which gives the general periodic solution

$$X_R^\mu(\tau - \sigma) = \frac{1}{2}x^\mu + \frac{\alpha'}{2}p^\mu(\tau - \sigma) + \frac{i\sqrt{2\alpha'}}{2}\sum_{n\neq 0}\frac{1}{n}\alpha_n^\mu e^{-in(\tau-\sigma)}$$

$$X_L^\mu(\tau + \sigma) = \frac{1}{2}x^\mu + \frac{\alpha'}{2}p^\mu(\tau + \sigma) + \frac{i\sqrt{2\alpha'}}{2}\sum_{n\neq 0}\frac{1}{n}\tilde{\alpha}_n^\mu e^{-in(\tau+\sigma)}. \tag{23.17}$$

Here we have denoted the constant in both terms by by $x^\mu/2$ since we want to obtain x_μ as the zero mode of X^μ, we have called the linear term $\alpha' p^\mu/2(\tau \mp \sigma)$ since we want to obtain $\alpha' p^\mu\tau$ as the linear mode of X^μ, and the normalization of the oscillating exponentials was chosen for reasons that will become clear soon. We can also define

$$\alpha_0^\mu = \sqrt{\frac{\alpha'}{2}}p^\mu; \quad \tilde{\alpha}_0^\mu = \sqrt{\frac{\alpha'}{2}}p^\mu, \tag{23.18}$$

which allows us to include the momentum term among the α_n^μ terms. Then the general solution for X^μ is

$$X^\mu(\tau, \sigma) = x^\mu + \alpha' p^\mu\tau + \frac{i\sqrt{2\alpha'}}{2}\sum_n\frac{1}{n}\left[\alpha_n^\mu e^{-in(\tau-\sigma)} + \tilde{\alpha}_n^\mu e^{+in(\tau+\sigma)}\right]. \tag{23.19}$$

Since X^μ are coordinates, we need X_L^μ and X_R^μ to be real. Since complex conjugation is seen to be equivalent to changing the sign of n, for reality we must have

$$\alpha_{-n}^\mu = (\alpha_n^\mu)^\dagger; \quad \tilde{\alpha}_{-n}^\mu = (\tilde{\alpha}_n^\mu)^\dagger. \tag{23.20}$$

Open String

For the open string, the boundary conditions can be either Neumann or Dirichlet. But the Dirichlet case will be treated separately, as it relates to D-branes, to be studied in Chapter 24. We will therefore consider here only the case of Neumann boundary conditions, i.e. free boundary conditions for the endpoints of the string, which move at the speed of light, as we saw in Chapter 22. The boundary condition for an open string of length $l = \pi$ is

$$X'^{\mu}|_{\sigma=0,\pi} = 0. \tag{23.21}$$

For X', this would be the Dirichlet boundary condition relevant for a vibrating violin string, so X'^{μ} is expanded in $\sin(n\sigma)$ modes, i.e. X^{μ} is expanded in $\frac{1}{n}\cos n\sigma$ modes, plus a linear term. But we still need to satisfy the condition for a general solution of the type $X^{\mu} = X_R^{\mu}(\sigma^-) + X_L^{\mu}(\sigma^+)$, which leads uniquely to an expansion in terms of $\frac{1}{n}\cos n\sigma e^{-in\tau} = \frac{1}{2n}(e^{-in(\tau-\sigma)} + e^{-in(\tau+\sigma)})$. All in all, the most general solution that satisfies the Neumann boundary conditions is

$$X^{\mu}(\sigma,\tau) = x^{\mu} + (2\alpha')p^{\mu}\tau + i\sqrt{2\alpha'}\sum_{n\neq0}\frac{1}{n}\alpha_n^{\mu}e^{-in\tau}\cos n\sigma. \tag{23.22}$$

We note then that this is the same expansion as for the closed string, if we identify $\tilde{\alpha}_n^{\mu} = \alpha_n^{\mu}$ and

$$\alpha_0^{\mu} \to \alpha_0^{\mu} + \tilde{\alpha}_0^{\mu} = \sqrt{2\alpha'}p^{\mu}. \tag{23.23}$$

23.3 Constraints and Hamiltonian

We saw that the constraints coming from the γ_{ab} equation, $T_{ab} = 0$, took the form

$$T_{00} = T_{11} = \frac{1}{2\alpha'}(\dot{X}^2 + X'^2) = 0; \quad T_{01} = T_{10} = \frac{1}{\alpha'}\dot{X} \cdot X' = 0. \tag{23.24}$$

In light-cone coordinates,

$$T_{++} = \frac{1}{2}(T_{00} + T_{01}) = \frac{1}{\alpha'}\partial_+X \cdot \partial_+X = \frac{1}{4\alpha'}(\dot{X} + X')^2$$

$$T_{--} = \frac{1}{2}(T_{00} - T_{01}) = \frac{1}{\alpha'}\partial_-X \cdot \partial_-X = \frac{1}{4\alpha'}(\dot{X} - X')^2. \tag{23.25}$$

On-shell (for the general solution of the wave equation), we have

$$T_{++} = \frac{1}{\alpha'}\dot{X}_L^2$$

$$T_{--} = \frac{1}{\alpha'}\dot{X}_R^2. \tag{23.26}$$

Since the Polyakov action in conformal gauge is

$$S_P = \frac{1}{4\pi\alpha'}\int d\tau \int d\sigma\,(\dot{X}^2 - X'^2), \tag{23.27}$$

the canonical momentum conjugate to X^μ is

$$P_\tau^\mu = \frac{\delta S}{\delta \dot{X}_\mu} = \frac{1}{2\pi\alpha'}\dot{X}^\mu, \tag{23.28}$$

so the Hamiltonian is

$$H = \int_0^l d\sigma\,(\dot{X}_\mu P_\tau^\mu - L) = \frac{1}{4\pi\alpha'}\int_0^l d\sigma\,(\dot{X}^2 + X'^2) = \frac{1}{2\pi}\int_0^l d\sigma\,T_{00}. \tag{23.29}$$

For an *on-shell open string*, with $l = \pi$, using the orthonormality relations for sines and cosines,

$$\int_0^\pi \cos n\sigma \cos m\sigma = \delta_{n+m} \tag{23.30}$$

and a similar one with sines, one can calculate that

$$H = \frac{1}{2}\sum_{n=-\infty}^{+\infty} \alpha_{-n}^\mu \alpha_n^\mu. \tag{23.31}$$

For an *on-shell closed string*, with $l = 2\pi$, using the same orthonormality relations, we get

$$H = \frac{1}{2}\sum_{n=-\infty}^{+\infty} (\alpha_{-n}^\mu \alpha_n^\mu + \tilde{\alpha}_{-n}^\mu \tilde{\alpha}_n^\mu). \tag{23.32}$$

Note that in both cases we included the $n = 0$ mode in the sum.

We now calculate the (Fourier) modes of the constraints. For the *on-shell closed string*, we find for the modes of the two constraints

$$L_m \equiv \frac{1}{2\pi}\int_0^{2\pi} d\sigma\,e^{-im\sigma}\,T_{--} = \frac{1}{2\pi\alpha'}\int_0^{2\pi} d\sigma\,e^{-im\sigma}\,\dot{X}_R^2 = \frac{1}{2}\sum_{n=-\infty}^{+\infty}\alpha_{m-n}^\mu \alpha_n^\mu$$

$$\tilde{L}_m \equiv \frac{1}{2\pi}\int_0^{2\pi} d\sigma\,e^{-im\sigma}\,T_{++} = \frac{1}{2\pi\alpha'}\int_0^{2\pi} d\sigma\,e^{-im\sigma}\,\dot{X}_L^2 = \frac{1}{2}\sum_{n=-\infty}^{+\infty}\tilde{\alpha}_{m-n}^\mu \tilde{\alpha}_n^\mu, \tag{23.33}$$

whereas for the on-shell *open string*, the α_n^μ and $\tilde{\alpha}_n^\mu$ modes are identified, so we find a single set of constraint modes, defined as

$$L_m = \frac{1}{2\pi}\int_0^\pi d\sigma\,\left(e^{im\sigma}\,T_{++} + e^{-im\sigma}\,T_{--}\right) = \frac{1}{2\pi\alpha'}\int_{-\pi}^\pi d\sigma\,e^{im\sigma}\,T_{++}$$

$$= \frac{1}{8\pi\alpha'}\int_{-\pi}^\pi d\sigma\,e^{im\sigma}\,(\dot{X} + X')^2 = \frac{1}{2}\sum_{n=-\infty}^{+\infty}\alpha_{m-n}^\mu \alpha_n^\mu. \tag{23.34}$$

We see that the Hamiltonian equals the constant mode of the constraint:

$$H = L_0 \quad \text{open}$$
$$= L_0 + \tilde{L}_0 \quad \text{closed}. \tag{23.35}$$

Then in the open string case, the $H = L_0 = 0$ constraint becomes, separating the α_0^μ part,

$$M^2 \equiv -p_\mu p^\mu = -\frac{\alpha_0^2}{2\alpha'} = \frac{1}{\alpha'} \sum_{n \geq 1} \alpha_{-n}^\mu \alpha_n^\mu, \qquad (23.36)$$

where in the last equality, we have used the fact that $\alpha_{-n}^\mu = (\alpha_n^\mu)^\dagger$ to equate the sum over $n > 0$ with the sum over $n < 0$, leading to $2 \sum_{n \geq 1}$.

Similarly, in the closed string case, the constraint $H = L_0 + \tilde{L}_0 = 0$ becomes, after separating the α_0^μ part

$$M^2 \equiv -p_\mu p^\mu = -\frac{\alpha_0^2 + \tilde{\alpha}_0^2}{\alpha'} = \frac{2}{\alpha'} \sum_{n \geq 1} (\alpha_{-n}^\mu \alpha_n^\mu + \tilde{\alpha}_{-n}^\mu \tilde{\alpha}_n^\mu). \qquad (23.37)$$

Constraints in the Light-Cone Gauge for the Open String

Until now we have not used the light-cone gauge condition, but now we explore its consequences. Since $X^+(\sigma, \tau) = x^+ + (2\alpha')p^+\tau$, by comparison with the general X^μ expansion, we have

$$\alpha_n^+ = 0, \quad \forall n \neq 0. \qquad (23.38)$$

From the lightcone gauge condition, we also obtain

$$\dot{X}^+ \pm X'^+ = (2\alpha')p^+. \qquad (23.39)$$

The constraints $T_{++} = T_{--} = 0$, called the *Virasoro constraints*, become in lightcone coordinates

$$(\dot{X} \pm X')^2 = 0 \Rightarrow \dot{X}^- \pm X'^- = \frac{(\dot{X}^i \pm X'^i)^2}{2(2\alpha')p^+}. \qquad (23.40)$$

Substituting the expansions for X^i and X^-, we obtain

$$\alpha_n^- = \frac{\sqrt{2\alpha'}}{2p^+} \sum_{m \in \mathbb{Z}} \alpha_{n-m}^i \alpha_m^i, \qquad (23.41)$$

and from the $n = 0$ mode we obtain

$$M^2 = 2p^+ p^- - p^i p^i = 2 \sum_{n \geq 1} \alpha_{-n}^i \alpha_n^i. \qquad (23.42)$$

For the closed string, we have a similar relation.

In conclusion, we see that X^+ is fixed by the lightcone gauge, and then X^- is fixed by the Virasoro constraints. That leaves only X^i as physical degrees of freedom.

23.4 Quantization of the String

In order to quantize the theory, we must impose equal-time commutation relations between the canonically conjugate variables. The canonical momentum conjugate to X^μ is, as we saw, $P_\mu = \dot{X}^\mu / (2\pi\alpha')$.

We first calculate the equal-time Poisson brackets in the classical theory:

$$[X^{\mu}(\sigma, \tau), X^{\nu}(\sigma', \tau)]_{P.B.} = [P^{\mu}(\sigma, \tau), P^{\nu}(\sigma', \tau)]_{P.B.} = 0$$
$$[P^{\mu}(\sigma, \tau), X^{\nu}(\sigma', \tau)]_{P.B.} = \delta(\sigma - \sigma')\eta^{\mu\nu}. \tag{23.43}$$

Substituting the on-shell mode expansion for X^{μ} and $P^{\mu} = \dot{X}^{\mu}/(2\pi\alpha')$, like in the case of usual quantum field theory, we find the commutation relations for the modes:

$$[\alpha_m^{\mu}, \alpha_n^{\nu}]_{P.B.} = [\tilde{\alpha}_m^{\mu}, \tilde{\alpha}_n^{\nu}]_{P.B.} = im\delta_{m+n}\eta_{\mu\nu}$$
$$[\alpha_m^{\mu}, \tilde{\alpha}_n^{\nu}]_{P.B.} = 0$$
$$[p^{\mu}, x^{\nu}]_{P.B.} = \eta^{\mu\nu}$$
$$[p^{\mu}, p^{\nu}]_{P.B.} = [x^{\mu}, x^{\nu}]_{P.B.} = 0. \tag{23.44}$$

But we also need to impose the Virasoro constraints $(\dot{X} \pm X')^2 = 0$ $(T_{++} = T_{--} = 0)$, or equivalently, their Fourier components $L_m = 0$ and $\tilde{L}_m = 0$.

Quantization then proceeds as usual by replacing the Poisson brackets $[,]_{P.B.}$ by $-i$ times the commutator, $-i[,]$, and replacing functions of phase space with operators.

The basic commutation relations become

$$[X^{\mu}(\sigma, \tau), X^{\nu}(\sigma', \tau)] = [P^{\mu}(\sigma, \tau), P^{\nu}(\sigma', \tau)] = 0$$
$$[P^{\mu}(\sigma, \tau), X^{\nu}(\sigma', \tau)] = -i\delta(\sigma - \sigma')\eta^{\mu\nu}, \tag{23.45}$$

and from them, on-shell we obtain the commutation relations for the modes:

$$[\alpha_m^{\mu}, \alpha_n^{\nu}] = [\tilde{\alpha}_m^{\mu}, \tilde{\alpha}_n^{\nu}] = m\delta_{m+n}\eta_{\mu\nu}$$
$$[\alpha_m^{\mu}, \tilde{\alpha}_n^{\nu}] = 0$$
$$[p^{\mu}, x^{\nu}] = -i\eta^{\mu\nu}$$
$$[p^{\mu}, p^{\nu}] = [x^{\mu}, x^{\nu}] = 0. \tag{23.46}$$

We note that these commutation relations are like the ones for the creation and annihilation operators of the harmonic oscillator ($a_n^{\dagger}$ and a_n), just with a rescaling, since $\alpha_{-n}^{\mu} \propto (a_n^{\mu})^{\dagger}$. The required rescaling is

$$\alpha_m^{\mu} = \sqrt{m}a_m^{\mu}; \quad \alpha_{-m}^{\mu} = \sqrt{m}a_m^{\dagger\mu}; \quad m > 0. \tag{23.47}$$

But we see that we have unphysical modes in the above procedure, as seen from the $\eta^{\mu\nu}$ on the right-hand side of the commutator of α_m^{μ} with $\alpha_n^{\dagger\nu}$. The solution is to do a familiar kind of (old) covariant quantization in the presence of constraints, of the Gupta-Bleuler type used in electromagnetism. That is, we impose the constraints $L_m = \tilde{L}_m = 0$ as operatorial conditions on physical states. There is one subtlety, namely that the constraints $L_0 = 0$ and $\tilde{L}_0 = 0$ contain quantum ordering ambiguities, so will get modified by a constant.

We will not continue with the old covariant quantization of the string, though we could.

23.5 Light-Cone Gauge Quantization

Instead, we will use the physical light-cone gauge (similar to a unitary gauge in quantum field theory), to quantize just the physical, transverse modes X^i.

At the quantum level, there are ordering ambiguities since the modes α_n^μ, $\tilde{\alpha}_n^\mu$ become operators, which means that in light-cone gauge the condition (23.41) for the *open string* becomes modified to

$$\alpha_n^- = \frac{\sqrt{2\alpha'}}{p^+} \left[\frac{1}{2} \sum_{i=1}^{D-2} \sum_{m \in \mathbb{Z}} : \alpha_{n-m}^i \alpha_m^i : -a\delta_{n,0} \right]. \tag{23.48}$$

Note that only the $n = 0$ mode is ambiguous, since then $\alpha_{-n}^i \alpha_n^i = (\alpha_n^i)^\dagger \alpha_n^i$.

Then X^+ is eliminated by the lightcone gauge condition, and X^- is eliminated by the above relation, including the p^- relation (for $n = 0$). There are no constraints left, and so this is a physical gauge, where the only independent oscillators are α_n^i, and only they are quantized.

The $n = 0$ mode in the case of the *open string* for α_n^- gives the mass constraint,

$$M^2 = 2p^+ p^- - p^i p^i = \frac{1}{\alpha'}(N - a), \tag{23.49}$$

where N is a kind of number operator,

$$N = \sum_{n \geq 1} \alpha_{-n}^i \alpha_n^i = \sum_{n \geq 1} n a_n^{\dagger i} a_n^i, \tag{23.50}$$

and n is called the *level* of the string mode.

String Modes and Phonons

As we see, this is analogous to the case where we studied in the condensed matter part, of the phonon modes in a material, with energies $\hbar\omega_n \sim \hbar n$, just that the formula is for M^2 instead of E. Like in that case where we had an infinite number of different types of phonons, each mode with an arbitrary occupation number, now we have an infinite number of different kinds of particles, each with its own M^2, and for each type we can have an arbitrary occupation number. These different kinds of particles are "composed" of some basic building blocks, with $M_n = n/\alpha'$ and occupation number $N_n = a_n^{\dagger i} a_n^i$, so therefore the mass is quantized in units of $1/\alpha'$.

Lightcone Hamiltonian

The formula for M^2 implies also that

$$H = p^- = \frac{p^i p^i}{2p^+} + \frac{1}{2\alpha' p^+}(N - a). \tag{23.51}$$

Since $p^- = p_+$, it is the conjugate of X^+, which is our lightcone time, therefore it has the interpretation of energy. This is then the *lightcone Hamiltonian*.

We have not yet fixed the ordering constant a. We can fix it in the following way. The simplest mode that we can have has a single oscillator, in the lowest n (i.e. lowest mass) level, a state of the type $a_1^{\dagger i}|0\rangle$, where $|0\rangle$ is a vacuum to be defined shortly. But we see that such a mode is something of the type $|i\rangle$, where $i = 1, \ldots, D - 2$, i.e. it looks like a transverse vector in D dimensions. But that makes sense only if the state is a massless vector, i.e. a gauge field, since if it were massive, it would have $D - 1$ physical components, not $D - 2$. But since

$$N(a_1^{\dagger i})|0\rangle = (a_1^{\dagger i}a_1^i)(a_1^{\dagger i})|0\rangle + \sum_{n \neq 1} nN_n(a_1^{\dagger i})|0\rangle = a_1^{\dagger i}|0\rangle, \tag{23.52}$$

such a state would have mass $1 - a$, which implies we need $a = 1$.

But on the other hand, we can compute it directly. Like for any sum of harmonic oscillator Hamiltonians (for example, in the case of a quantum field), the classical $\sum_{n,i} \hbar\omega_n a_n^{\dagger i}a_n^i$ is replaced at the quantum level (when we carefully replace the operators in H by their formulas in terms of $a_n^{\dagger i}$ and a_n^i) by the symmetrically ordered

$$\sum_{n,i} \hbar\omega_n \frac{a_n^{\dagger i}a_n^i + a_n^i a_n^{\dagger i}}{2} = \sum_{n,i} \hbar\omega_n \left(a_n^{\dagger i}a_n^i + \frac{1}{2} \right). \tag{23.53}$$

Therefore in our case, with $\hbar\omega_n = n$, we have the zero point energy

$$\sum_n \sum_{i=1}^{D-2} \frac{n}{2} = \frac{D-2}{2} \sum_{n \geq 1} n. \tag{23.54}$$

For the $\sum_{n \geq 1} n$, we can use the formula of zeta function regularization. Riemann defined the zeta function as

$$\zeta(s) = \sum_{n \geq 1} \frac{1}{n^s}. \tag{23.55}$$

The function is defined over the complex plane and is analytic everywhere except at negative integers, $s = -p$. But as it happens, there is a unique, finite, analytical continuation (in the complex plane) at the negative integers, so we can define a zeta function fully analytic over the complex plane. At $s = -1$, the value is $-1/12$, so

$$\zeta(-1) = \sum_{n \geq 1} n = -\frac{1}{12}. \tag{23.56}$$

Therefore normal ordering constant a, i.e. the zero point energy of the string, is

$$-a = \frac{D-2}{2} \sum_{n \geq 1} n = -\frac{D-2}{24}. \tag{23.57}$$

But since we have established that we need $a = 1$ for consistency, it means we also need $D = 26$, i.e. that *the bosonic string lives in 26 dimensions.*

We have obtained this result in one simple way, but there are others, all related to the quantum consistency of the theory: the absence of quantum anomalies (quantum breaking of symmetries valid classically) in symmetries that cannot have them. For instance, the absence of anomalies in Lorentz invariance and conformal (Weyl) invariance lead to the same $D = 26$.

23.6 String Spectrum

In order to construct the spectrum of the bosonic string, we need first to define the vacuum. We have seen that the quantum operators that obey commutation relations include not only the oscillators α_n^i, but also the zero modes p^μ, since the string has an overall (zero mode) momentum. That means that the vacuum, unlike the case of quantum field theory, has also a given momentum, i.e. the vacuum is a state $|0; k\rangle$, and it must satisfy

$$\alpha_m^i |0; k\rangle = 0, \forall m > 0$$
$$p^+ |0; k\rangle = k^+ |0; k\rangle; \quad p^i |0; k\rangle = k^i |0; k\rangle. \tag{23.58}$$

The first line is the usual one, saying that annihilation operators kill the vacuum, but the second says that the vacuum has momentum k.

Open String Spectrum

A general state of the *open string* is found by acting with a number N_{in} of creation operators in each *in* mode:

$$|N, k\rangle = \left[\prod_{i=1}^{D-2} \prod_{n \geq 1} \frac{(a_n^{\dagger i})^{N_{in}}}{\sqrt{N_{in}!}} \right] |0; k\rangle = \left[\prod_{i=1}^{D-2} \prod_{n \geq 1} \frac{(\alpha_n^{\dagger i})^{N_{in}}}{\sqrt{n^{N_{in}} N_{in}!}} \right] |0; k\rangle. \tag{23.59}$$

Now, we note that the vacuum is a scalar ϕ (has no spacetime index) and has $M^2 = -1/\alpha'$, i.e. it is a tachyon. Of course, a (scalar) tachyon in the spectrum means only that the potential for it has a negative curvature $M^2 = V''(\phi)$ at $\phi = 0$, like in the case of symmetry breaking (the Higgs potential). What is unclear is whether the potential has a true minimum at some nonzero ϕ (in a nonperturbative region), like the Higgs potential does, or whether it goes all the way to $-\infty$, signaling an unfixable instability. In the first case, the instability would be fixed, and an expansion around the correct, nonperturbative, vacuum would lead only to $M^2 > 0$. As of yet, it is not known in which of these cases the bosonic string actually belongs. For other tachyons, like tachyons on D-branes, it is known that there is a true nonperturbative vacuum, so this leads to the hope that the same could be true for the bosonic string.

One observation is that the spectrum of the bosonic string has an infinite number of fields (infinite number of particles of increasing mass), all interacting, so an instability in one mode implies an instability for the whole string theory. This is the reason that instead of the bosonic string, for concrete applications one uses the *superstring*, which has a condition that eliminates the tachyon from the spectrum, making the theory stable.

At the next level, in terms of increasing mass, we have the state

$$a_n^{\dagger i} |0\rangle = \alpha_n^{\dagger i} |0\rangle, \tag{23.60}$$

which we have used to fix $a = 1$ by imposing it to be massless, $M = 0$. This is a massless vector state, as we saw.

Closed String Spectrum

For the closed string, we can make a similar analysis. The difference is that now we have two $n = 0$ constraints, $L_0 - a = 0$ and $\tilde{L}_0 - a = 0$. They give two equivalent formulas,

$$M^2 = \frac{4}{\alpha'} \sum_{n \geq 1} (\alpha^i_{-n} \alpha^i_n - 1)$$

$$= \frac{4}{\alpha'} \sum_{n \geq 1} (\tilde{\alpha}^i_{-n} \tilde{\alpha}^i_n - 1) \tag{23.61}$$

or another way of writing them is

$$M^2 = \frac{2}{\alpha'} \sum_{n \geq 1} (\alpha^i_{-n} \alpha^i_n + \tilde{\alpha}^i_{-n} \tilde{\alpha}^i_n - 2) = \frac{2}{\alpha'} (N + \tilde{N} - 2)$$

$$N = \tilde{N} \tag{23.62}$$

In this way of writing them, the first is $H = L_0 + \tilde{L}_0 - 2a = 0$, and the second is $P = L_0 - \tilde{L}_0 = 0$ (no worldsheet momentum on the string).

We could again fix a in the same way as in the open string, or simply assume that it is the same as above, which turns out to be true.

The vacuum of the closed string is again a scalar tachyon with momentum, $|0; k\rangle$ for the same reason. The only difference is that now the mass of tachyon is $M^2 = -4/\alpha'$. We would think that the next state would be at $M^2 = -2/\alpha'$, but we must actually have $N = \tilde{N}$ on the state, therefore the next possible state has $N = \tilde{N} = 1$, which gives $M^2 = 0$. The state is a tensor

$$a^{\dagger i}_1 \tilde{a}^{\dagger j}_1 |0; k\rangle = \alpha^{\dagger i}_1 \alpha^{\dagger j}_1 |0\rangle \equiv |ij\rangle. \tag{23.63}$$

This state can be decomposed in irreducible representations of $SO(D - 2)$ as the following:

- a symmetric traceless tensor $|((ij))\rangle$, which can be identified with the *graviton* g_{ij}
- an antisymmetric tensor $|[ij]\rangle$, called the B-field or *Kalb-Ramond field* B_{ij}
- a scalar trace, $|ii\rangle$, called the *dilaton* ϕ.

We now note that if we didn't put $a = 1$, we would have a massive symmetric traceless tensor field with transverse components, which would not make sense since a massive symmetric traceless tensor would have $D - 1$ values for the indices, not $D - 2$.

We also note that since we have a graviton in the spectrum of a quantum theory, string theory is a theory of quantum gravity. It now seems obvious, but it took a long time before string theory was thought of as such, since initially it was invented as a theory to explain hadrons, which were better described by QCD, yet people kept trying to use it for the same.

Important Concepts to Remember

- String theory has conformal invariance, which appears as a residual gauge invariance in the conformal gauge, under a combined diffeomorphism and compensating Weyl transformation.

- We can still fix the lightcone gauge condition $X^+ = x^+ + p^+\tau$.
- The expansion of the string is in terms of x^μ, p^μ and α_n^μ, $\tilde{\alpha}_n^\mu$ for the closed string, and the open string has $\alpha_n^\mu = \tilde{\alpha}_n^\mu$.
- The Fourier modes of the Virasoro constraints $(\dot{X} \pm X')^2$ are L_m and $\tilde{L}_m$, and need to be imposed on states. But for $m = 0$, $H = L_0$ for the open string and $H = L_0 + \tilde{L}_0$ for the closed string.
- In the lightcone gauge, X^+ is fixed by the gauge, and X^- is fixed by the constraints.
- When quantizing, we obtain harmonic oscillators $\alpha_{-m}^\mu = \sqrt{m}a_\mu^{\dagger\mu}$ and we must impose the constraints, with the $m = 0$ constraints modified to $L_0 - a$ and $\tilde{L}_0 - a$. If we impose them on states, we obtain old covariant quantization of Gupta-Bleuler type.
- For lightcone quantization, we quantize just the physical modes α_n^i.
- The string spectrum is made of an infinite number of different kinds of particles of increasing mass, via $M^2 = (N - 1)/\alpha'$ for the open string and $M^2 = 4(N + \tilde{N} - 2)/\alpha'$ for the closed string, with $N = \tilde{N}$, and $N = \sum_n nN_n$.
- The vacuum of the bosonic string (in both closed and open versions) is a scalar tachyon, signaling an instability. It will be removed in the superstring.
- The open string has a massless vector, and the closed string has a massless graviton, B_{ij} and ϕ, indicating it is a theory of quantum gravity.
- The bosonic string theory lives in $D = 26$ dimensions, for quantum consistency, including the absence of Lorentz and conformal anomalies.

Further Reading

Chapters 1 and 2 in [42], vol. I, chapters 9, 12, and 13 in [43], chapter 1 in [11], chapter 7 in [8].

Exercises

(1) Consider the same conformal invariance in 3+1 dimensions, with infinitesimal transformation $x'_\mu = x_\mu + v_\mu(x)$. Show that the most general solution is

$$v_\mu(x) = a_\mu + \omega_{\mu\nu}x^\nu + \lambda x_\mu + b_\mu x^2 - 2x_\mu b \cdot x, \qquad (23.64)$$

where $\omega_{\mu\nu} = -\omega_{\nu\mu}$ is antisymmetric.

(2) Consider the modes L_m and $\tilde{L}_m$ of the closed string constraints. Show that classically (using the Poisson brackets) they satisfy two copies of the *Virasoro algebra*:

$$[L_m, L_n]_{P.B.} = (m - n)L_{m+n}. \qquad (23.65)$$

(3) Calculate the worldsheet momentum P_σ of the bosonic open string.

(4) Consider the particle action in the light-cone gauge $X^+(\tau) = \tau$. Write the first order action and the corresponding Hamiltonian.

(5) Write the light-cone gauge string action on $\mathbb{R}_t \times S^2$ and write its equations of motion.

(6) Write the most general solution for the open string $X^\mu(\tau, \sigma)$ with one Neumann boundary condition and one Dirichlet (fixed).

(7) Write the states of the first massive levels of the bosonic open string and closed string.

24 — D-Branes and Gauge Fields

In this chapter we will describe a new kind of object, which is different than the particles and strings that we can easily visualize. It turns out that string theory has some interesting nonperturbative objects called *D-branes*, which have spatial extension in more than one direction. Of course, in our 3+1–dimensional world we could visualize only some walls, or membranes, with spatial extension in two dimensions, but in higher dimensions (we just saw in Chapter 23 that in bosonic string theory we have 26 dimensions) we can have more choices that are more difficult to visualize. It will turn out that the D-branes carry gauge fields on them, so we will obtain gauge theories on their "*worldvolume.*"

24.1 D-Branes

In the last chapters we have considered only Neumann (free) boundary conditions for the open string. In this chapter we will explore the consequences of having also Dirichlet (fixed) boundary conditions. The conditions are

$$\delta X(\tau, \sigma = 0) = 0; \quad \delta X(\tau, \sigma = \pi) = 0. \tag{24.1}$$

Note first that we can have different boundary conditions for different X's, and moreover we could also have different boundary conditions at each end, i.e. a string with Neumann boundary condition at one end and Dirichlet boundary condition at the other (ND string). However, we will not analyze this case, and will consider only NN and DD strings.

Consider then the case that $p + 1$ directions (including time) have Neumann (free) boundary conditions, and $D - p - 1$ directions have Dirichlet (fixed) boundary conditions, so the endpoints of the string are free to move in a $(p + 1)$–dimensional "wall" in D-dimensional spacetime. Such a "wall" is called a Dp-brane; see Figure 24.1. The word p-brane comes from an extension of the word mem-brane, as we already mentioned, and the D stands for Dirichlet. The two endpoints of the string could be on the same brane, or on different branes.

So it would seem like Dp-branes are simply some boundary conditions for the string. They were in fact considered in some papers even from the early 1970s, but people didn't take them very seriously, since these boundary conditions seem to break the translational invariance of the theory, and that wasn't supposed to happen. But in 1989, Dai, Leigh, and Polchinski proved that the wall is actually a dynamical object (called a D-brane). That increased interest, since certainly a dynamical object can break translational invariance (a particle or a string certainly do), though it was only in 1995, with Polchinski's seminal

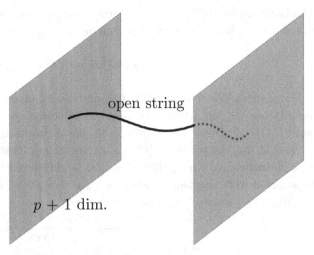

open string

$p + 1$ dim.

Fig. 24.1 Open string between two D-p-branes ($p + 1$–dimensional "walls").

paper (to be described better a bit later in the book) that proved that the mass and charges of the D-branes are identical to those of string extremal p-brane solutions, and thus that D-branes are some nonperturbative solutions of string theory, somewhat like solitons, that D-branes entered the mainstream, and the "second superstring revolution" was started. The interest was due to the fact that now we could use duality arguments, symmetries, etc. to say something about nonperturbative physics. D-branes were a major ingredient in the nonperturbative analyses. A D-brane is an unusual object: A soliton would have a mass $\propto 1/g^2$, whereas a fundamental particle would have a mass that is independent of the coupling, $M \sim 1$, but a D-brane will be some intermediate object, with $M \propto 1/g$.

In this chapter we will use lightcone coordinates, which will split as $\{X^+, X^-, X^a\}$, forming the NN directions, and $\{X^i\}$, forming the DD directions.

24.2 The D-Brane Action: DBI

The dynamical D-branes should have an action minimizing its worldvolume, for the same reason as the action for the string minimizes its worldsheet. Therefore we expect the action, at least as far as gravity is concerned, to be

$$S_p = -T_p \int d^{p+1}\xi \sqrt{-\det(h_{ab})}. \tag{24.2}$$

But again for the same reasons as in the case of the string, the metric h_{ab} cannot be an independent metric on the worldvolume, but rather must be the induced metric on the worldvolume, from spacetime:

$$h_{ab}(\xi) = \frac{\partial X^\mu}{\partial \xi^a} \frac{\partial X^\nu}{\partial \xi^b} g_{\mu\nu}(X). \tag{24.3}$$

This is derived, like in the case of the string, from equating the metric on the worldvolume with the metric in spacetime, restricted on the worldvolume, i.e. from

$$ds^2 = h_{ab}d\xi^a d\xi^b = dX^\mu dX^\nu g_{\mu\nu}. \tag{24.4}$$

Moreover, we saw at the end of last chapter that the massless modes of the closed string are $|IJ\rangle = \alpha^I_{-1}\tilde{\alpha}^J_{-1}|0\rangle$, with I, J transverse, which decomposes in the symmetric traceless part $|((IJ))\rangle$, identified with the metric g_{IJ}, the antisymmetric part $|[IJ]\rangle$, identified with the B-field B_{IJ}, and the trace, $|II\rangle$, called the dilaton. That means that we always have the combination $g_{\mu\nu} + \alpha' B_{\mu\nu}$ (the α' is there for dimensional reasons, as $B_{\mu\nu}$ is defined to be an object with dimension 2, whereas $g_{\mu\nu}$ is dimensionless), so the D-brane action for coupling to $g_{\mu\nu}$ and $B_{\mu\nu}$ must be

$$S_p = -T_p \int d^{p+1}\xi \sqrt{-\det\left(\frac{\partial X^\mu}{\partial \xi^a}\frac{\partial X^\nu}{\partial \xi^b}(g_{\mu\nu} + \alpha' B_{\mu\nu})\right)}. \tag{24.5}$$

We can fix the diffeomorphism invariance of the p-brane (defined by $p+1$ functions $\xi'^a = \xi'^a(\xi)$) by choosing a *static gauge*:

$$X^a = \xi^a; \quad a = 0, 1, \ldots, p. \tag{24.6}$$

Then the transverse coordinates

$$X^i(\xi^a) = \frac{\phi^i(\xi^a)}{\sqrt{T_p}} \tag{24.7}$$

are fields on the worldvolume. We expand the spacetime metric as

$$g_{\mu\nu}(X) = \eta_{\mu\nu} + 2\kappa_N h_{\mu\nu}(X). \tag{24.8}$$

The action for the scalars ϕ^i, at $h_{\mu\nu} = B_{\mu\nu} = 0$ is then

$$S_\phi = -T_p \int d^{p+1}\xi \sqrt{-\det\left(\eta_{ab} + \frac{\partial_a\phi^i \partial_b\phi^i}{T_p}\right)}$$

$$= -T_p \int d^{p+1}\xi \sqrt{\frac{1}{(p+1)!}\epsilon^{a_1\ldots a_{p+1}}\epsilon^{b_1\ldots b_{p+1}}\left(\eta_{a_1b_1} + \frac{\partial_{a_1}\phi^i \partial_{b_1}\phi^i}{T_p}\right)\cdots}$$

$$\overline{\cdots\left(\eta_{a_{p+1}b_{p+1}} + \frac{\partial_{a_{p+1}}\phi^i \partial_{b_{p+1}}\phi^i}{T_p}\right)} \tag{24.9}$$

If we have a single $\phi^i = \phi$, i.e. if $D - p - 1 = 1$, then if we have at least two factors of $\partial\phi\partial\phi$, we have

$$\partial_{[a_1}\phi^i \partial^{[b_1}\phi^i \partial_{a_2]}\phi^j \partial^{b_2]}\phi^j = 0, \tag{24.10}$$

but as we can see, if we have several ϕ^i, it is not true. Then, exactly for a single ϕ, and approximately (as an expansion in $\partial\phi/\sqrt{T_p}$) for several, we have

$$\det\left(\eta_{ab} + \frac{\partial_a\phi^i\partial_b\phi^i}{T_p}\right) = \frac{1}{(p+1)!}\left(\epsilon^{a_1\dots a_{p+1}}\epsilon_{a_1\dots a_{p+1}}\right.$$

$$\left. + n\epsilon^{a_1\dots a_p a_{p+1}}\epsilon_{a_1\dots a_p}{}^{b_{p+1}}\frac{\partial_{a_{p+1}}\phi^i\partial_{b_{p+1}}\phi^i}{T_p} + \cdots\right)$$

$$= \simeq 1 + \frac{\partial_a\phi^i\partial^a\phi^i}{T_p}, \tag{24.11}$$

leading to the action

$$S_\phi \simeq -T_p \int d^{p+1}\xi \sqrt{1 - \frac{\partial_a\phi^i\partial^a\phi^i}{T_p}}$$

$$\simeq -\int d^{p+1}\xi\left[T_p + \frac{1}{2}\partial^a\phi^i\partial_a\phi^i + \cdots\right]. \tag{24.12}$$

We can also calculate the action for $h_{\mu\nu}$ to first order in $h_{\mu\nu}$ and first order in ϕ^i. Using the fact that $\eta_{ai} = 0$, but $h_{ai} \neq 0$, we obtain

$$S_p = -T_p \int d^{p+1}\xi \sqrt{-\det(\eta_{ab} + 2\kappa_N h_{ab} + 4\kappa_N h_{ai}\partial_b\phi^i + \cdots)}$$

$$= -T_p \int d^{p+1}\xi \sqrt{\frac{1}{(p+1)!}\left((p+1)! + (p+1)4\kappa_N h_{ai}\frac{\partial_b\phi^i}{\sqrt{T_p}}p!\delta_a^b\right)}$$

$$\simeq -T_p \int d^{p+1}\xi \sqrt{1 + 4\kappa_N h_{ai}\frac{\partial_a\phi^i}{\sqrt{T_p}} + \cdots}. \tag{24.13}$$

Together with the scalar kinetic term, we have

$$S_p = -\int d^{p+1}\xi\left[T_p + \frac{1}{2}\partial_a\phi^i\partial^a\phi^i + 2\kappa_N\sqrt{T_p}h_{ai}\partial^a\phi^i + \cdots\right]. \tag{24.14}$$

Therefore the vertex for coupling of the canonically normalized fields ϕ^j and h_{ai} is (see Figure 24.2a)

$$+2\kappa_N\sqrt{T_p}ip^a\delta_i^j. \tag{24.15}$$

But note that g_{ai} is a closed string mode and lives in the whole of spacetime, whereas ϕ^i is defined only on the worldvolume and is the position of the D-brane, where the open string ends; therefore it is an open string mode. It is in fact the mode of an open string that lives on the Dp-brane, i.e. that has both ends on the same D-brane.

The interpretation of the above vertex is therefore of a closed string (with mode h_{ai}) coming from the bulk of spacetime toward the D-brane, and colliding with it, and making it oscillate, i.e. exciting the open string mode ϕ^i; see Figure 24.2(b) and Figure 24.3(a).

From the point of view of string theory, the closed string opens up as it collides with the D-brane, and turns into an open string that lives on the D-brane. There is a string calculation

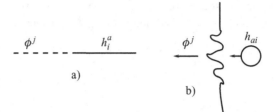

Fig. 24.2 (a) The scalar-graviton coupling in field theory (between ϕ^j and h_i^a). (b) The scalar-graviton coupling in string theory (between ϕ^j and h_{ai}).

that we can do that corresponds to this process. It is an infinitely thin closed string, opening up as it nears the D-brane, and joining into a disk on the D-brane; see Figure 24.3(b). (In string theory, the calculation is for a disk inside the D-brane, with a "vertex operator" that creates a closed string at the center.) Comparing the result of the string calculation with the above vertex in the D-brane field theory, one can fix the parameter T_p (D-brane tension) in the D-brane action. The result is

$$T_p = \frac{1}{(2\pi\alpha')^2 g_{p+1}^2},\tag{24.16}$$

where g_{p+1} is the coupling of the field theory in $p+1$ dimensions, given by

$$g_{p+1}^2 = (2\pi)^{p-2} g_s \alpha'^{\frac{p-3}{2}}.\tag{24.17}$$

We still need to introduce the dependence of the D-brane action on the dilaton ϕ, the trace $\alpha_{-1}^i \alpha_{-1}^i |0; k\rangle$. The dilaton has the property that its VEV gives the closed string coupling constant via the relation

$$g_s = e^{\langle\phi\rangle}.\tag{24.18}$$

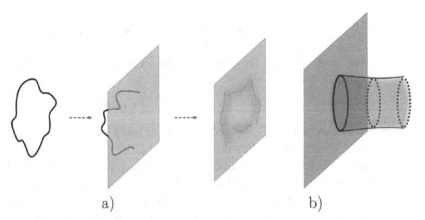

Fig. 24.3 (a) Closed string colliding with a D-brane, exciting an open string mode and making it vibrate. (b) String worldsheet corresponding to it, with a closed string tube coming from infinity and ending on the D-brane as an open string boundary.

Fig. 24.4 Two open string splitting interactions can be glued on the edges to give a closed string interaction ("pair of pants"), therefore $g_{YM}^2 = g_s$.

Because of this fact, string theory has no free parameters, since even the coupling is given simply by a vacuum solution of string theory, and the same applies for any other parameters. In front of the closed string action, we have by definition $1/g_s^2$, but on the other hand

$$g_s = g_o^2, \qquad (24.19)$$

where g_o is the open string coupling. The reason is that a closed string is made of two open strings joined together, and the same applies for a string interaction: the basic interaction of closed strings, the "pair of pants," for a closed string to split into two (which therefore comes with a factor of g_s), can be "cut open" into two basic interactions of open strings to split into two (each of which therefore comes with a factor of g_0), glued together; see Figure 24.4. Then the D-brane action, which is an action for open string modes, comes with a factor of

$$\frac{1}{g_o^2} = \frac{1}{g_s} = e^{-\langle\phi\rangle} \qquad (24.20)$$

in front. Then by consistency, we must have $e^{-\phi}$ in the action, leading to

$$S_p = -T_p \int d^{p+1}\xi \, e^{-\phi} \sqrt{-\det\left(\frac{\partial X^\mu}{\partial \xi^a}\frac{\partial X^\nu}{\partial \xi^b}(g_{\mu\nu} + \alpha' B_{\mu\nu})\right)}. \qquad (24.21)$$

But there are other fields living on the worldvolume. We have seen in the last chapter that the open string with Neumann boundary conditions has a massless vector A_a, interpreted now as a gauge field living on the worldvolume of the D-brane. Its action must reduce in the weak field limit to the Maxwell action, $-(F_{\mu\nu})^2/4$. To find the nonlinear action, one can try various possibilities and obtain that the simplest guess is actually correct, namely that F_{ab} appears inside the determinant,

$$S_p = -T_p \int d^{p+1}\xi \, e^{-\phi}\sqrt{-\det(g_{ab} + \alpha' B_{ab} + 2\pi\alpha' F_{ab})}, \qquad (24.22)$$

just that the h_{ab} and B_{ab} are *induced* on the worldvolume,

$$h_{ab} = \frac{\partial X^\mu}{\partial \xi^a}\frac{\partial X^\nu}{\partial \xi^b}g_{\mu\nu}; \quad B_{ab} = \frac{\partial X^\mu}{\partial \xi^a}\frac{\partial X^\nu}{\partial \xi^b}B_{\mu\nu}, \qquad (24.23)$$

whereas F_{ab} is defined *intrinsically* on the worldvolume.

The above action, called *the Dirac-Born-Infeld (DBI) action* is actually the full (bosonic) action for the D-brane in the bosonic string theory. The action for just F_{ab} (at $h_{\mu\nu} = B_{\mu\nu} = 0$) was found by Born and Infeld in 1934, in an effort to avoid all divergences and singularities in electromagnetism. Indeed, the action has only nonsingular "electron" solutions, with finite energy and energy density at the origin (at the source). Moreover, the action is the unique nonlinear completion of the Maxwell action, without extra derivatives and written only in terms of $F_{\mu\nu}$, that is both causal and has a single characteristic surface (wavefront surface for the propagation of the perturbations), as opposed to two. This result was proved by Plebanski. The action was studied further, and generalized to include scalars, by Dirac, hence Dirac's name is associated also with it.

24.3 The WZ Term

Superstring Theory and Its Light Modes

As we said in previous chapters, bosonic string theory is replaced by superstring theory in order to get rid of instabilities. In *superstring theory*, the fields $g_{\mu\nu}$, $B_{\mu\nu}$, and ϕ are the bosonic fields of the so-called NS-NS (NS stands for Neveu-Schwarz) sector, but there are also bosonic fields in the Ramond-Ramond (RR) sector, which are p-form (totally antisymmetric tensor) fields.

WZ Term

The D-brane action has therefore an extra term, the *Wess-Zumino (WZ) term*, that describes the coupling to the p-form fields. In a supersymmetric background, the result of the WZ term is to cancel the $-T_p \int d^{p+1}\xi$ constant term in the action. Part of the WZ term is the source coupling to an $A_{(p+1)}$ form field, of the type already described for the extremal p-branes:

$$\mu_p \int d^D x\, j^{\mu_1 \ldots \mu_{p+1}} A_{\mu_1 \ldots \mu_{p+1}} = \mu_p \int d^{p+1}\xi\, A_{01\ldots p}. \tag{24.24}$$

The constant μ_p is the charge density and is related to the tension (energy density) T_p by

$$\mu_p = T_p e^{\phi - \langle\phi\rangle}. \tag{24.25}$$

In fact, there is a coupling to other p'-forms, all packaged in the formal expression for the WZ term:

$$S_{\text{WZ}} = \mu_p \int_{p+1} e^{\wedge(\alpha' B_{ab} + 2\pi\alpha' F_{ab})} \wedge \sum_n A_{(n)}. \tag{24.26}$$

Here the exponential is understood in the wedge sense (wedge product instead of usual product), and out of the resulting total form, we keep only the $(p+1)$–form that can be integrated over a $(p+1)$–dimensional worldvolume.

Chan-Paton Factors

Until now we have considered a single D-brane and as a consequence, the endpoints of the string were identical.

But it was realized soon (in fact, much before D-branes were defined properly, in the context of Neumann strings, i.e. for "$D25$-branes" in the bosonic string) that one can add labels to each string endpoint. So consider the label $|i\rangle$ for one endpoint, and $|j\rangle$ for the

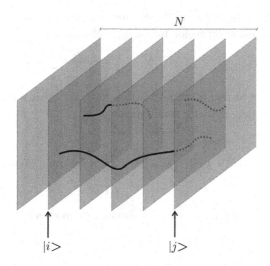

Fig. 24.5 The endpoints of the open string are labeled by the D-brane they end on (out of N D-branes), here $|i\rangle$ and $|j\rangle$.

other, where $i, j = 1, \ldots, N$. This means that the string is an $N \times N$ matrix, matching with the adjoint $(\underline{N} \otimes \underline{\bar{N}})$ of $U(N)$. One considers then the states with wavefunctions

$$|k; a\rangle = \sum_{i,j=1}^{N} |k; ij\rangle \lambda_{ij}^{a}, \tag{24.27}$$

where λ_{ij}^{a} are a basis of adjoint matrices for $U(N)$. Then the states above, for $|k; ij\rangle$ having a vector index as well, are adjoint vectors, which can be identified with $U(N)$ gauge fields. So naturally in this way we can construct nonabelian gauge theories. Moreover, it is easy now to understand that $|i\rangle$ is an index for the D-brane on which the endpoint of the string ends (see Figure 24.5), so in the presence of N D-branes we obtain an $U(N)$ gauge theory.

The full nonabelian D-brane action is not known (though some comments will be made in the next chapter about that), but at least at the quadratic level, we expect the nonabelian action to be the usual

$$S = \int d^{p+1}\xi (-2) \operatorname{Tr} \left[-\frac{1}{2} D_a \phi^i D^a \phi^i - \frac{1}{4} F_{ab} F^{ab} \right]. \tag{24.28}$$

24.4 Quantization of Open Strings on *Dp*-Branes

Consider open strings between two *parallel Dp*-branes, situated at x_1^i and x_2^i. To quantize the open strings, we must first review the quantization of the NN (free) directions, then move to the new DD (fixed) directions. We saw that we had in general (denoting the lightcone transverse coordinates by $X^I = (x^a, x^i)$)

$$\dot{X}^- \pm X'^- = \frac{1}{2\alpha'} \frac{1}{2p^+} (\dot{X}^I \pm X''^I)^2 = \frac{1}{2\alpha'} \frac{1}{2p^+} \left[(\dot{X}^i \pm X'^i)^2 + (\dot{X}^a \pm X'^a)^2 \right]. \tag{24.29}$$

For NN coordinates, in the previous chapter we have found that we have also (from the explicit mode expansion for the on-shell string)

$$\dot{X}^a \pm X'^a = \sqrt{2\alpha'} \sum_{n \in \mathbb{Z}} \alpha_n^a e^{-in(\tau \pm \sigma)}. \tag{24.30}$$

We want to find a similar relation for X^i, so that we can substitute in (24.29). All the coordinates satisfy the wave equation

$$\partial_+ \partial_- X^\mu = 0, \tag{24.31}$$

in particular X^i, so we can write the general solution of the equation of motion for X^i as

$$X^i = f^i(\tau + \sigma) + g^i(\tau - \sigma). \tag{24.32}$$

But the X^i must satisfy the DD boundary conditions

$$X^i(\tau, \sigma = 0) = x_1^i; \quad X^i(\tau, \sigma = \pi) = x_2^i. \tag{24.33}$$

Imposing the $\sigma = 0$ boundary condition on the general solution for X^i, we find

$$g(\tau) = x_1^i - f(\tau), \tag{24.34}$$

and substituting back in X^i we find

$$X^i(\sigma, \tau) = x_1^i + f^i(\tau + \sigma) - f^i(\tau - \sigma). \tag{24.35}$$

Imposing now the $\sigma = \pi$ boundary condition, we get

$$f^i(\tau + \pi) - f^i(\tau - \pi) = (x_2^i - x_1^i) \Rightarrow f^i(x + 2\pi) = f^i(x) + x_2^i - x_1^i. \tag{24.36}$$

Therefore the function f^i can be expanded in sines and cosines, modulo a linear piece, as

$$f^i(x) = \frac{1}{2\pi}(x_2^i - x_1^i)x + \sqrt{\frac{\alpha'}{2}} \sum_{n \geq 1} (\tilde{f}_n^i \cos nx + f_n^i \sin nx). \tag{24.37}$$

Substituting in $X^i(\sigma, \tau)$ and using the fact that $\cos(a + b) - \cos(a - b) = -2 \sin a \sin b$, $\sin(a + b) - \sin(a - b) = 2 \cos a \sin b$, we obtain

$$X^i(\sigma, \tau) = x_1^i + \frac{\sigma}{\pi}(x_2^i - x_1^i) + \sqrt{2\alpha'} \sum_{n \geq 1} (f_n^i \cos n\tau - \tilde{f}_n^i \sin n\tau) \sin n\sigma$$

$$= x_1^i + \frac{\sigma}{\pi}(x_2^i - x_1^i) + \sqrt{2\alpha'} \sum_{n \neq 0} \frac{1}{n} \alpha_n^i e^{-in\tau} \sin n\sigma. \tag{24.38}$$

Here we have rewritten the sum over $n \geq 1$ modes with f_n^i and $\tilde{f}_n^i$ as a sum over $n \neq 0$ with α_n^i modes, identifying $\alpha_n^i + \alpha_{-n}^i = 2nf_n^i$ and $\alpha_n^i - \alpha_{-n}^i = 2in\tilde{f}_n^i$.

We can also include the $n = 0$ mode in the sum, by defining

$$\alpha_0^i = \frac{1}{\pi\sqrt{2\alpha'}}(x_2^i - x_1^i), \tag{24.39}$$

to have finally

$$X^i(\sigma, \tau) = x_1^i + \sqrt{2\alpha'} \sum_{n \in \mathbb{Z}} \frac{1}{n} \alpha_n^i e^{-in\tau} \sin n\sigma. \tag{24.40}$$

Then we can calculate that we have

$$\dot{X}^i \pm X'^i = \sqrt{2\alpha'} \sum_{n \in \mathbb{Z}} \alpha_n^i e^{-in(\tau \pm \sigma)}. \tag{24.41}$$

This is the same as (24.30), except for an overall $\pm$, and the fact that α_0^i is now different.

At this point, we can quantize the strings, as before: Write, first, Poisson brackets for $X^i(\sigma, \tau)$, then turn them to commutators, and calculate the commutators for the oscillator modes. One finds as before

$$[\alpha_m^i, \alpha_n^j] = m \delta^{ij} \delta_{m+n,0}, \tag{24.42}$$

for $m, n \neq 0$. Substituting both relations (24.30) and (24.41) in (24.29), we obtain

$$2p^+ p^- = \frac{1}{\alpha'} \left[\alpha' p^a p^a + \frac{1}{2} \alpha_0^i \alpha_0^i + \sum_{n \geq 1} (\alpha_{-n}^a \alpha_n^a + \alpha_{-n}^i \alpha_n^i) - 1 \right]. \tag{24.43}$$

With respect to the pure NN case, now the only difference is that α_0^i is not related to p^i, so we don't form the combination $p^a p^a + p^i p^i = p^I p^I$. The point is that since the string is fixed in the X^i directions, it cannot have a momentum in them.

Instead, now we can calculate the $(p+1)$–dimensional mass squared (the mass squared on the worldvolume),

$$M^2 \equiv 2p^+ p^- - p^a p^a = \frac{1}{2\alpha'} \alpha_0^i \alpha_0^i + \frac{1}{\alpha'} \left[\sum_{n \geq 1} (\alpha_{-n}^a \alpha_n^a + \alpha_{-n}^i \alpha_n^i) - 1 \right]$$

$$= \left(\frac{x_2^i - x_1^i}{2\pi\alpha'} \right)^2 + \frac{1}{\alpha'} (N^\perp - 1), \tag{24.44}$$

where

$$N^\perp \equiv \sum_{n \geq 1} \sum_a n a_n^{\dagger a} a_n^a + \sum_{n \geq 1} \sum_i n a_n^{\dagger i} a_n^i. \tag{24.45}$$

D-Brane States

We can now construct the ground state of the D-brane theory, as a state with momentum p^+, p^a, but also with Chan-Paton factors $\tilde{i}$, $\tilde{j}$. If we have only two D-branes, then $\tilde{i}$, $\tilde{j} = 1, 2$. The ground state is thus

$$|0; p^+, p^a; [\tilde{i}, \tilde{j}]\rangle. \tag{24.46}$$

Its mass is

$$M^2 = -\frac{1}{\alpha'} + \left(\frac{x_2^i - x_1^i}{2\pi\alpha'} \right)^2. \tag{24.47}$$

So it used to be a scalar tachyon (since we are in the bosonic string), but now it is modified by the string tension $1/(2\pi\alpha')$ times the length between the D-branes, i.e. the classical energy of a string stretched between the D-branes. In particular, the scalar ground state can be massless, if

$$|x_2^i - x_1^i| = 2\pi\sqrt{\alpha'}. \tag{24.48}$$

A general state is obtained by acting on the vacuum with an arbitrary number of creation operators, so

$$|\psi\rangle = \left[\prod_a \prod_{n \geq 1} \frac{(a_n^{\dagger a})^{N_{na}}}{\sqrt{N_{na}!}} \right] \left[\prod_i \prod_{m \geq 1} \frac{(a_n^{\dagger i})^{N_{ni}}}{\sqrt{N_{ni}!}} \right] |0; p^+, p^a[\tilde{i}\tilde{j}]\rangle. \qquad (24.49)$$

In particular, we have the states

$$a_1^{\dagger i}|0; p^+, p^a[\tilde{i}\tilde{j}]\rangle \equiv |\phi^{i[\tilde{i}\tilde{j}]}\rangle$$
$$a_1^{\dagger 0}|0; p^+, p^a[\tilde{i}\tilde{j}]\rangle \equiv |A_a^{[\tilde{i}\tilde{j}]}\rangle \qquad (24.50)$$

with masses

$$M^2 = \left(\frac{x_2^i - x_1^i}{2\pi\alpha'} \right)^2. \qquad (24.51)$$

If $x_2^i \neq x_1^i$, the states are massive, and then the $A_a^{[\tilde{i}\tilde{j}]}$ states are vectors, which must then eat a scalar to become massive via the Higgs mechanism.

But for *coincident* (yet different) D-branes, the states are massless and $A_a^{[\tilde{i}\tilde{j}]}$ are $U(N)$ Yang-Mills fields (massless vectors), and $\phi^{i[\tilde{i}\tilde{j}]}$ are Yang-Mills-charged $U(N)$ scalars, the modes that we have anticipated from the general analysis of the D-brane action.

Note then an unusual fact. The ϕ^i's in the abelian case corresponded to positions of the D-branes in the transverse space, but now in the nonabelian case we have N D-branes, yet a matrix of $N \times N$ ϕ^i's! That is one of the examples of the fact that strings see a different, *quantum* geometry than point particles, a fuzzier version of classical geometry. In particular, coordinates of N objects are not scalars anymore, but $N \times N$ matrices. The diagonal elements still have the interpretation of usual positions: indeed, in a Higgs phase, when we separate the D-branes, the off-diagonal gauge fields eat off-diagonal scalars to become massive, leaving as massless fields only the diagonal ($U(1)$) gauge fields on each D-brane and the diagonal ($U(1)$) scalars, corresponding to the positions of the D-branes. But when the D-branes are on top of each other, the off-diagonal components of the scalars are new, *quantum* degrees of freedom, that have no interpretation in classical geometry.

Important Concepts to Remember

- The bosonic D-brane action in the bosonic string is the DBI action: the area of the world-volume, with the induced metric, with $g_{\mu\nu} \to g_{\mu\nu} + \alpha' B_{\mu\nu}$, and $2\pi F_{ab}$ for the intrinsic gauge field added to the metric h_{ab} as well, and an overall $e^{-\phi}$ in the action.
- In a static gauge, $X^a = \xi^a$, and the transverse scalars X^i become fields on the worldvolume, like A_a and the spacetime fields h_{ab} and B_{ab} and ϕ.
- The graviton, and in general all closed string modes living in the bulk, couple to the scalars, and in general all the open string modes living on the D-brane.
- By comparing the field theory computation in the D-brane effective action with a string computation, we can fix the constants in the D-brane action, like T_p.
- The (closed) string coupling is given by the VEV of the dilaton, $g_s = e^{\langle\phi\rangle}$, and the closed string is the open string squared, so $g_s = g_0^2$.

- In superstring theory, the bosonic closed string modes $g_{\mu\nu}$, $B_{\mu\nu}$, ϕ form the NS-NS sector, but besides extra fermions, we also have an RR sector, formed of p-form (totally antisymmetric tensor) fields.
- The coupling of the D-brane to the RR fields takes the form of a source coupling, $\int_{p+1} A_{(p+1)}$, and some interactions, together forming the WZ term $\mu_p \int_{p+1} e^{\wedge(\alpha'B+2\pi\alpha'F)} \wedge \sum_n A_{(n)}$.
- We can add Chan-Paton factors to the string endpoints, leading to a nonabelian gauge theory on multiple coincident D-branes.
- The mass of open strings in between D-branes is modified by the tension of the string times the distance between the D-branes, and the states that are massless for coincident branes are ϕ^i and A_a.
- In the case of coincident branes, we have a $U(N)$ gauge theory, which means that the scalars ϕ^i, denoting transverse positions of the D-branes, are actually $N \times N$ matrices for N D-branes, leading to a quantum geometry.

Further Reading

Chapter 15 in [43], chapter 5 in [44].

Exercises

(1) Show that the D-brane action is reparametrization and Poincaré invariant, and find the condition to have also Weyl invariance.

(2) For a D3-brane in static gauge in four and five dimensions, at $B_{\mu\nu} = F_{\mu\nu} = 0$, calculate the determinant to write the explicit form of the D-brane action.

(3) Find the vertex coupling between a metric h_{ab} and two gauge fields A_c and A_d, in the D-brane action.

(4) Consider open strings near an *orbifold point* in X^{25}, where $X^{25}(\sigma, \tau)$ is identified with $-X^{25}(\sigma, \tau)$, and a D-brane at a small distance L from it. Quantize nontrivial strings that start and end on the D-brane (but don't live completely on it).

(5) Consider a D4-brane in 10 dimensions sliding in the sixth dimension on a metric

$$ds^2 = H_4^{-1/2}(r)d\vec{x}_{4+1}^2 + H_4^{+1/2}(r)(dr^2 + r^2 d\Omega_4^2), \tag{24.52}$$

where

$$H_4(r) = 1 + \frac{Q}{r^3} \tag{24.53}$$

is a harmonic function in the transverse space. Calculate the potential $V(r)$ and equations of motion for motion in r for a D4-brane and an anti-D4-brane (the only difference between them is in the sign of the WZ term).

(6) Explicitly calculate the DBI action expanded to order $(B_{ab})^2$, then write it as a function of the matrices

$$G = g - (2\pi\alpha')^2 Bg^{-1}B$$
$$\theta = -(2\pi\alpha')^2 g^{-1}Bg + \mathcal{O}(B^3), \qquad (24.54)$$

and the coupling

$$G_s = g_s[1 - (\pi\alpha')^2 \,\mathrm{Tr}[(g^{-1}B)^2] + \mathcal{O}(B^4)]. \qquad (24.55)$$

(7) Consider a D3-brane, extending in x_0, x_1, x_2 and $x_3 \equiv z$, ending on a D5-brane, extending in x_0, x_1, x_2 and x_4, x_5, x_6, where spherical coordinates for x_4, x_5, x_6 are (r, θ, ϕ). Consider solutions to the D5-brane action with magnetic flux

$$F = (2\pi\alpha')n \sin\theta \, d\theta d\phi. \qquad (24.56)$$

Calculate the reduced action $S = S[z(r)]$, and from it the solution with "n units of charge" for $z(r)$.

25 Electromagnetic Fields on D-Branes: Supersymmetry and $\mathcal{N} = 4$ SYM, T-Duality of Closed Strings

In this chapter we continue the discussion of the field theory on D-branes, in particular in the supersymmetric case. After defining supersymmetry, we consider a deeper analysis of the nonabelian and nonlinear version of the DBI action and its supersymmetric case that leads to the important theory of $\mathcal{N} = 4$ SYM, after which we start to address an important symmetry of string theory, T-duality.

25.1 Low-Energy D-Brane Action

We saw in the last chapter that among the states of the open strings stretching between D-branes, the massless states of the bosonic string are the scalars ϕ^i and the gauge fields A_a. Their linearized action is

$$S = \int d^{p+1}\xi(-2)\,\mathrm{Tr}\left[-\frac{1}{2}D_a\phi^i D^a\phi^i - \frac{1}{4}F_{ab}F^{ab}\right]. \tag{25.1}$$

However, as we mentioned already, the bosonic string is unstable because of the tachyon in its spectrum, so instead, one must consider a supersymmetric theory, the superstring. Supersymmetry is a symmetry between bosons and fermions, so the superstring has fermions as well. In the supersymmetric theory, fermions will be present as well on the D-brane, so to (25.1) we should add a fermionic kinetic term also.

In the superstring theory:

1. The tachyon instability is cured, because of a projection ("GSO projection") on the set of states imposed by supersymmetry that removes the tachyon.
2. One obtains extra states in the spectrum. At the massless level, besides the bosonic string states $g_{\mu\nu}$, $B_{\mu\nu}$, and ϕ, now called NS-NS states, we also have some p-form states $\{A_{(p)}\}$ forming the RR sector, and fermions forming the NS-R and R-NS sectors. Together, they form the *supergravity* theory, which is a low-energy limit of string theory.

Critical Dimension for the Superstring

Moreover, the argument about cancellation of quantum anomalies, or vacuum energy, for the string, which led to $D = 26$ dimensions for the bosonic string, can now be modified. We can understand that the fermions will give new contributions to the zero-point energy sum calculation, and now consistency (the absence of anomalies) leads to $D = 10$. We will, however, not show this here.

25.2 Supersymmetry

Supersymmetry is a symmetry between bosons and fermions. To understand what it is, we consider the simplest possible example, which occurs in 1+1 dimensions. A fermion in 1+1 dimensions has two complex components, but if we impose a Weyl condition (chiral fermion, i.e. with only half of the components) or Majorana condition (reality condition for the components), we obtain either one complex, or two real components, respectively.

On-Shell Supersymmetry

Consider the minimal system made up of a real scalar ϕ and a fermion ψ. The free action for such a system is

$$S = -\frac{1}{2} \int d^2x [\bar{\psi} \slashed{\partial} \psi + (\partial_\mu \phi)^2]. \tag{25.2}$$

Since we want to have a symmetry between the boson and the fermion, the boson ϕ must vary into the fermion ψ times the parameter ϵ. But for Lorentz invariance, ϵ must be a spinor, and we must have the combination $\bar{\epsilon}\psi$, so

$$\delta\phi = \bar{\epsilon}\psi = \bar{\epsilon}_\alpha \psi^\alpha. \tag{25.3}$$

But since we can easily see that the mass dimension of ϕ is zero, $[\phi] = 0$ (since $[d^2x] = -2$ and $[\partial^2] = +2$ and the action must have dimension 0), whereas the fermion has dimension 1/2, $[\psi] = 1/2$ (since $[d^2x] = -2$ and $[\slashed{\partial}] = 1$), it follows that the parameter ϵ has mass dimension $-1/2$, $[\epsilon] = -1/2$. The variation of ψ should now also be proportional to ϕ and ϵ, just that if that is all, the dimensions don't match, as $[\phi] = 1/2$, but $[\phi\epsilon] = -1/2$. That means that we must add something (not a field) of dimension 1 without vector Lorentz indices, which uniquely determines it to be $\slashed{\partial} = \partial_\mu \gamma^\mu$, so

$$\delta\psi = \slashed{\partial}\phi\epsilon. \tag{25.4}$$

In principle one could have a coefficient in front of $\delta\psi$, but it is actually 1. Indeed, we can check that the above transformation rules leave the action invariant *off-shell*, i.e. without using the equations of motion $\slashed{\partial}\psi = 0$ and $\Box\phi = 0$. It is left as an exercise to prove it. But the supersymmetry algebra is realized only *on-shell* on the fields.

It was in fact expected that the symmetry would be valid only on-shell since, even though we don't usually think of it this way for bosonic symmetries, a symmetry relates a degree of freedom with another (think of an internal group symmetry on a bosonic field, $\delta\phi^I = \Lambda^I{}_J\phi^J$). But off-shell, in our case the fermion has two real degrees of freedom, whereas the real scalar has one. On the other hand, on-shell the fermion has only one real degree of freedom, since the Dirac equation is a matrix equation that defines one component in terms of the other, whereas the Klein-Gordon equation restricts only the functional form of the field, not the number of fields, so the boson also has one real degree of freedom. Therefore on-shell we have the same number of degrees of freedom, hence we could have a supersymmetry, and in fact we do.

We could also find interactions that preserve this supersymmetry, leading to interacting supersymmetric models.

Off-Shell Supersymmetry

We can also find an off-shell extension of supersymmetry. It is clear what we should do by counting of degrees of freedom. We need to add to the action a bosonic auxiliary field, with one off-shell degree of freedom, but none on-shell. The simplest choice is correct, just an auxiliary real scalar F with the action $+ \int d^2x F^2/2$. One could guess the modifications to the supersymmetry transformation rules. Since the F equation of motion is $F = 0$, δF should also be an equation of motion, which can only be the fermionic one, i.e. we need $\delta F = \bar{\epsilon} \not{\partial} \psi$. It could in principle have a coefficient in front, but it is actually 1. We could also add a term with the equation of motion $F = 0$ in $\delta \psi$, i.e. $\delta_{\text{extra}} \psi = F \epsilon$, perhaps with a coefficient, but the coefficient is actually 1. All in all then,

$$\delta \phi = \bar{\epsilon} \psi; \quad \delta \psi = \not{\partial} \phi \epsilon + F \epsilon; \quad \delta F = \bar{\epsilon} \not{\partial} \psi. \tag{25.5}$$

This is in fact an off-shell supersymmetry of the free action.

One can also have more than one supersymmetries, with parameters ϵ^I, $I = 1, \ldots, \mathcal{N}$. For the supersymmetries to be different, they must take us to different fields from the same one, for instance, $\delta \phi = \bar{\epsilon}_I \psi^I$.

In four dimensions, if we want to have at most spin one, i.e. gauge fields, fermions, and scalars, the maximum number of supersymmetries is $\mathcal{N} = 4$. This is found from an analysis of the possible helicity states. One finds that a combination of the supersymmetries lowers the helicity, and having more than four supersymmetries would lower the helicity from the maximal $+1$ to one less than -1, corresponding to a field of spin higher than one.

25.3 $\mathcal{N} = 4$ Super Yang-Mills in $3+1$ Dimensions

Definition

Consider the theory on the $3+1$–dimensional worldvolume of D3-branes in superstring theory. Superstring theory means $D = 10$, which means that we have $D - p - 1 = 10 - 3 - 1 = 6$ scalars corresponding to the six transverse directions to the D-brane (the D-brane scalars are fluctuations of the D-brane in these directions). We are interested in the D3-brane theory since we live in $3+1$ dimensions, and we could find useful for various applications (perhaps as a toy model) the D-brane worldvolume theory. We also have gauge fields, and fermions, such that we form a supersymmetric theory. In fact, it must have the maximal amount of supersymmetry, $\mathcal{N} = 4$, as we can easily see. Indeed, in this case there must be four fermions, ψ^I with $I = 1, \ldots, 4$, since each supersymmetry takes us from the gauge field to a different fermion. But then the degree of freedom counting works on-shell, since we have eight bosonic degrees of freedom (six real scalars and one gauge field with two degrees of freedom), which in four dimensions matches the number of degrees of freedom of four fermions (an off-shell Weyl fermion in four dimensions has two complex components, reduced on-shell to two real, or one complex degree of freedom). We see then that the counting would not match with another number of supersymmetries.

Symmetries

The four fermions ψ^I live in the fundamental representation of $SU(4)$, and the six real scalars ϕ^i, $i = 1, \ldots, 6$ can be written in the antisymmetric representation of

$SU(4) = SO(6)$, as $\phi^{[IJ]}$, and A_μ is a singlet of $SU(4)$. All the fields are in the adjoint of the $U(N)$ gauge group, with index a. Then we can guess part of the supersymmetry transformation rules, since we can match $\delta\phi$ by putting the correct indices as

$$\delta\phi^{a[IJ]} = \frac{i}{2}\epsilon^{[I}\psi^{J]a}, \tag{25.6}$$

where we have put the coefficient $i/2$ here, so that it doesn't appear anywhere else. For the variation of the gauge field, it still varies in $\bar{\epsilon}$ times the fermion ψ^{aI}, but to match the Lorentz index μ on A_μ we must introduce the constant vector matrix γ_μ, and since A is a $SU(4)$ singlet, the I index is summed:

$$\delta A_\mu^a = \bar{\epsilon}_I \gamma_\mu \psi^{aI}. \tag{25.7}$$

The variation $\delta\psi^{aI}$ is complicated and will not be written here.

Action

Then the action for the interacting $\mathcal{N} = 4$ supersymmetric theory, called $\mathcal{N} = 4$ Supersymmetric Yang-Mills (SYM), is

$$S_{\mathcal{N}=4\,SYM} = \int d^4x (-2)\,\mathrm{Tr}\left[-\frac{1}{4}F_{\mu\nu}^2 - \frac{1}{2}\bar{\psi}_I \slashed{D}\psi^I - \frac{1}{2}D_\mu\Phi_{IJ}D^\mu\Phi^{IJ} \right.$$
$$\left. -g\bar{\psi}^I[\Phi_{IJ},\psi^J] - \frac{g^2}{4}[\Phi_{IJ},\Phi_{KL}][\Phi^{IJ},\Phi^{KL}] \right]. \tag{25.8}$$

The terms on the first line are just the linearized (quadratic) action, coupled minimally to the gauge field. Introducing the minimal coupling in the free theory uniquely selects the interaction terms needed for $\mathcal{N} = 4$ supersymmetry (in other words, the $\mathcal{N} = 4$ SYM theory is unique) as being the terms on the second line, the most general renormalizable interaction terms. The first one is a Yukawa interaction, where imposing the matching of the indices, together with the antisymmetry in $[IJ]$, selects the $\bar{\psi}[\Phi, \psi]$ form. The second is a ϕ^4 interaction, which for adjoint fields must be of the type $[\phi, \phi]^2$ for invariance reasons, and where the IJ indices are contracted in the natural order. Again the coefficients of these two interaction terms are fixed by imposing invariance under supersymmetry.

25.4 Nonlinear Born-Infeld Action

We now describe a few things about the nonlinear BI part in 3+1 dimensions, i.e. the nonlinear action at $B_{\mu\nu} = \Phi^i = 0$, $e^\phi = e^{\langle\phi\rangle} = g_s$:

$$S = -T_3 \int d^{3+1}x\left[\sqrt{-\det(\eta_{ab} + 2\pi\alpha'F_{ab})} - 1 \right]. \tag{25.9}$$

Note that the -1 comes from the WZ term in a supersymmetric background. Consider $2\pi\alpha' = 1$ temporarily; we will reintroduce it at the end by dimensional analysis. Then, defining as usual the magnetic field $B_i = \frac{1}{2}\epsilon_{ijk}F_{jk}$ and the electric field $E_i = F_{0i}$, we can calculate the matrix

$$M_{ab} = \eta_{ab} + F_{ab} = \begin{pmatrix} -1 & -E_1 & -E_2 & -E_3 \\ E_1 & +1 & B_3 & -B_2 \\ E_2 & -B_3 & +1 & B_1 \\ E_3 & B_2 & -B_1 & +1 \end{pmatrix}. \tag{25.10}$$

The determinant inside the square root of the BI action is then (the proof is a simple exercise)

$$-\det M_{ab} = 1 - (\vec{E}^2 - \vec{B}^2) - (\vec{E} \cdot \vec{B})^2. \tag{25.11}$$

But on the other hand we have

$$\frac{1}{2}(\vec{E}^2 - \vec{B}^2) = -\frac{1}{4}F_{\mu\nu}F^{\mu\nu}$$

$$\vec{E} \cdot \vec{B} = -\frac{1}{8}\epsilon^{\mu\nu\rho\sigma}F_{\mu\nu}F_{\rho\sigma} \equiv -\frac{1}{4}\tilde{F}^{\mu\nu}F_{\mu\nu}, \tag{25.12}$$

where we have defined the dual field $\tilde{F}^{\mu\nu} = \frac{1}{2}\epsilon^{\mu\nu\rho\sigma}F_{\rho\sigma}$, so we can rewrite the action as (reintroducing the factors of $(2\pi\alpha')$ by dimensional analysis)

$$S_3 = -T_3 \int d^{3+1}x \left[\sqrt{1 + (2\pi\alpha')^2 \frac{1}{2}F_{\mu\nu}F^{\mu\nu} - (2\pi\alpha')^4 \left(\frac{1}{4}\tilde{F}^{\mu\nu}F_{\mu\nu}\right)^2} - 1 \right]. \tag{25.13}$$

Note that this formula is valid only in 3+1 dimensions, i.e. for D3-branes, since we have $\epsilon^{\mu\nu\rho\sigma}F_{\mu\nu}F_{\rho\sigma}$ in it, which is defined only in 3+1 dimensions.

We see that for consistency with the quadratic action (25.1) (written in the form with $1/g_{3+1}^2$ in front and $D_\mu = \partial_\mu - ieA_\mu$), we must have

$$T_3 = \frac{1}{(2\pi\alpha')^2 g_{3+1}^2}. \tag{25.14}$$

This in fact agrees with the general formula (24.16).

The action in terms of the electric and magnetic fields is finally

$$S_3 = -\frac{1}{(2\pi\alpha')^2 g_{3+1}^2} \int d^{3+1}x \left[\sqrt{1 - (2\pi\alpha')^2(\vec{E}^2 - \vec{B}^2) - (2\pi\alpha')^4(\vec{E} \cdot \vec{B})^2} - 1 \right]. \tag{25.15}$$

At zero magnetic field ($\vec{B} = 0$), this gives

$$S_3 = -\frac{1}{(2\pi\alpha')^2 g_{3+1}^2} \int d^{3+1}x \left[\sqrt{1 - (2\pi\alpha')^2\vec{E}^2} - 1 \right]. \tag{25.16}$$

That means that we have a bound on the electric field,

$$|\vec{E}| \leq \frac{1}{2\pi\alpha'} = E_{\text{crit}}, \tag{25.17}$$

since the action must be real, and moreover, it cannot change from real to imaginary along the solution. In fact, the maximal electric field is reached at the core of the "electron" solution, thus avoiding the *classical* singularity at $r = 0$ for electromagnetism. This is in fact the reason that Born and Infeld introduced their theory, as a theory that did not need quantum effects to get rid of unphysical singularities and divergent energy densities at the position of the electron.

25.5 Closed Strings on Compact Spaces

T-Duality of Closed Strings

We now move to a different topic: a duality symmetry of string theory on compact spaces called *T-duality*. We consider therefore bosonic strings on the simplest compact manifold, a circle S^1 of radius R, which means we identify

$$X^{25} \sim X^{25} + 2\pi R. \tag{25.18}$$

But since we have a closed string, it can wind m times around the compact direction, since the coordinate X^{25} is identified modulo $2\pi R$:

$$X^{25}(\tau, \sigma = 2\pi) - X^{25}(\tau, \sigma = 0) = 2\pi m R \sim 0. \tag{25.19}$$

We define the *winding* w by

$$w = \frac{mR}{\alpha'}. \tag{25.20}$$

The boundary condition on the closed string is now instead of simply periodic, periodic modulo the winding:

$$X^{25}(\tau, \sigma + 2\pi) = X^{25}(\tau, \sigma) + 2\pi \alpha' w. \tag{25.21}$$

But otherwise, the closed string still satisfies the wave equation, so it still has the general solution

$$X^{25}(\tau, \sigma) = X_L^{25}(\tau + \sigma) + X_R^{25}(\tau - \sigma). \tag{25.22}$$

Denote $u = \tau + \sigma$ and $v = \tau - \sigma$. Imposing now the boundary condition on this solution, we obtain

$$X_L^{25}(u + 2\pi) + X_R^{25}(v - 2\pi) = X_L^{25}(u) + X_R^{25}(v) + 2\pi \alpha' w \Rightarrow$$
$$X_L^{25}(u + 2\pi) - X_L^{25}(u) = X_R^{25}(v) - X_R^{25}(v - 2\pi) + 2\pi \alpha' w. \tag{25.23}$$

Note that on the left we have a function of u, and on the right a function of v, so each side must be independently equal to zero, i.e. periodic, except for a linear piece in the expansion of X_L and X_R. Therefore the general expansion for the periodic closed string is still valid, with just a modification in the linear term:

$$X_L^{25}(u) = x_{0L}^{25} + \sqrt{\frac{\alpha'}{2}} \tilde{\alpha}_0^{25} u + i \sqrt{\frac{\alpha'}{2}} \sum_{n \neq 0} \frac{\tilde{\alpha}_n^{25}}{n} e^{-inu}$$

$$X_R^{25}(v) = x_{0R}^{25} + \sqrt{\frac{\alpha'}{2}} \alpha_0^{25} v + i \sqrt{\frac{\alpha'}{2}} \sum_{n \neq 0} \frac{\alpha_n^{25}}{n} e^{-inv}. \tag{25.24}$$

The difference is that now the boundary condition, applied to the above X_L^{25} and X_R^{25}, gives

$$\frac{1}{\sqrt{2\alpha'}}(\tilde{\alpha}_0^{25} - \alpha_0^{25}) = w. \tag{25.25}$$

We have seen before that the momentum canonically conjugate to X^{25} is given by the variation with respect to $\dot{X}^{25}$, giving

$$p = \frac{1}{2\pi\alpha'} \int_0^{2\pi} d\sigma \dot{X}^{25} = \frac{1}{2\pi\alpha'} \int_0^{2\pi} d\sigma (\dot{X}_L^{25} + \dot{X}_R^{25}). \tag{25.26}$$

Substituting the expansion for the on-shell bosonic string with winding, we obtain

$$p = \frac{1}{\sqrt{2\alpha'}}(\alpha_0^{25} + \tilde{\alpha}_0^{25}). \tag{25.27}$$

Solving the relations for p and w, we obtain

$$\alpha_0^{25} = \sqrt{\frac{\alpha'}{2}}(p - w); \quad \tilde{\alpha}_0^{25} = \sqrt{\frac{\alpha'}{2}}(p + w). \tag{25.28}$$

Substituting these values in the expansion for X_L^{25} and X_R^{25} and adding them, we obtain

$$
\begin{aligned}
X^{25}(\tau, \sigma) &= X_L^{25}(\tau + \sigma) + X_R^{25}(\tau - \sigma) \\
&= x_0^{25} + \alpha' p \tau + \alpha' w \sigma + i\sqrt{\frac{\alpha'}{2}} \sum_{n \neq 0} \frac{e^{-in\tau}}{n}(\alpha_n^{25} e^{in\sigma} + \tilde{\alpha}_n^{25} e^{-in\sigma}),
\end{aligned} \tag{25.29}
$$

where

$$x_0^{25} = x_{0L}^{25} + x_{0R}^{25}. \tag{25.30}$$

In this expression, the winding w is discrete, but also the momentum p is discrete, since on a compact space, the translation by $2\pi R$ must be identified with the identity, i.e. $e^{i2\pi Rp} = 1$, so

$$p = \frac{n}{R}. \tag{25.31}$$

Constraints and Spectrum

Among the constraints, the nonzero modes of the Virasoro constraints, L_n and $\tilde{L}_n$, are unmodified. The only ones that are modified are the zero modes of the Virasoro constraints, L_0 and $\tilde{L}_0$ ($L_0 - 1$ and $\tilde{L}_0 - 1$ at the quantum level), which give the mass formula and the worldsheet momentum constraint.

For them, the only modification is in the different form of α_0^{25} and $\tilde{\alpha}_0^{25}$, but as a function of α_n^μ and $\tilde{\alpha}_n^\mu$, they are the same. Then, separating the 25th coordinate, and calling the rest I, split in the lightcone gauge into $I = (+, -, i)$, we have

$$
\begin{aligned}
L_0 &= \frac{\alpha_0^I \alpha_0^I + \alpha_0^{25} \alpha_0^{25}}{2} + N^\perp = \frac{\alpha'}{4}(-2p^+ p^- + p^i p^i) + \frac{\alpha_0^{25} \alpha_0^{25}}{2} + N^\perp \\
\tilde{L}_0 &= \frac{\tilde{\alpha}_0^I \tilde{\alpha}_0^I + \tilde{\alpha}_0^{25} \tilde{\alpha}_0^{25}}{2} + \tilde{N}^\perp = \frac{\alpha'}{4}(-2p^+ p^- + p^i p^i) + \frac{\tilde{\alpha}_0^{25} \tilde{\alpha}_0^{25}}{2} + \tilde{N}^\perp.
\end{aligned} \tag{25.32}
$$

(1) The worldsheet momentum constraint, $L_0 - \tilde{L}_0 = 0$, becomes now

$$
\begin{aligned}
L_0 - \tilde{L}_0 &= \frac{\alpha_0^{25}\alpha_0^{25} - \tilde{\alpha}_0^{25}\tilde{\alpha}_0^{25}}{2} + N^\perp - \tilde{N}^\perp \\
&= -\alpha' pw + N^\perp - \tilde{N}^\perp = 0,
\end{aligned}
\tag{25.33}
$$

leading to the *modified level matching condition*

$$
N^\perp - \tilde{N}^\perp = \alpha' pw = nm.
\tag{25.34}
$$

(2) On the other hand, the worldsheet Hamiltonian condition, coming from the sum of the constraints, $L_0 + \tilde{L}_0 - 2 = 0$, gives the mass spectrum for the compactified theory (dimensionally reduced to 25 dimensions),

$$
\begin{aligned}
M_{\text{compact}}^2 &= -p^I p_I = 2p^+ p^- - p^i p^i = \frac{2}{\alpha'}(L_0 + \tilde{L}_0 - 2) + 2p^+ p^- - p^i p^i \\
&= \frac{\alpha_0^{25}\alpha_0^{25} + \tilde{\alpha}_0^{25}\tilde{\alpha}_0^{25}}{\alpha'} + \frac{2}{\alpha'}(N^\perp + \tilde{N}^\perp - 2),
\end{aligned}
\tag{25.35}
$$

so that finally

$$
M_{\text{compact}}^2 = p^2 + w^2 + \frac{2}{\alpha'}(N^\perp + \tilde{N}^\perp - 2).
\tag{25.36}
$$

We note that the term with w^2 appears since it is the classical energy of the string wound m times around the circle:

$$
E = TL = \frac{1}{2\pi\alpha'}2\pi|m|R = \frac{|m|R}{\alpha} = |w|.
\tag{25.37}
$$

25.6 T-Duality of the Spectrum and of the Background

T-Duality

Writing explicitly the spectrum, we have

$$
M_{\text{compact}}^2 = \left(\frac{n}{R}\right)^2 + \left(\frac{mR}{\alpha'}\right) + \frac{2}{\alpha'}(N^\perp + \tilde{N}^\perp - 2).
\tag{25.38}
$$

In this form, we notice a coincidence of the spectrum, that we call *T-duality*. If we define the dual radius

$$
\tilde{R} = \frac{\alpha'}{R},
\tag{25.39}
$$

then if we exchange R with $\tilde{R}$ at the same time as we exchange n with m, we obtain the same spectrum:

$$
M^2(R; n, m) = M^2(\tilde{R}; m, n).
\tag{25.40}
$$

But note that R is an adjustable parameter characterizing the vacuum, i.e. a *modulus*. In other words, one considers string theory in a background, corresponding to a vacuum of the theory, and in this case, the vacuum is a compactified theory with a certain R. Perhaps in fact, in the full interacting theory, it is not a real vacuum, and there is some potential that favors an R over another. But at this level in the analysis, we consider that all R's are possible vacua. In this case, T-duality tells us that string theory in the vacuum with a small circle R ($\ll \sqrt{\alpha'}$) is equivalent with string theory in a vacuum with a large circle $\tilde{R}$ ($\gg \sqrt{\alpha'}$).

We also note that this is a duality in string theory, due to the fact that strings can wind around a compact dimension, but it is not a duality symmetry in field theory.

Exchanging p with w amounts, in terms of the oscillators, with the exchange

$$\tilde{\alpha}_0^{25} \leftrightarrow \tilde{\alpha}_0^{25}; \quad \alpha_0^{25} \leftrightarrow -\alpha_0^{25}. \tag{25.41}$$

We then note that we can also exchange the nonzero modes in a similar fashion,

$$\tilde{\alpha}_n^{25} \leftrightarrow \tilde{\alpha}_n^{25}; \quad \alpha_n^{25} \leftrightarrow -\alpha_n^{25}, \tag{25.42}$$

and the spectrum is still invariant, since under this exchange, $N^\perp$ and $\tilde{N}^\perp$ are invariant.

Then, adding also the exchange of the zero modes

$$x_0^{25} = x_{0L}^{25} + x_{0R}^{25} \leftrightarrow q_0^{25} = x_{0L}^{25} - x_{0R}^{25}, \tag{25.43}$$

we find that under this T-duality we exchange the $X^{25}(\tau, \sigma) = X_L^{25}(\tau + \sigma) + X_R^{25}(\tau - \sigma)$ with

$$X'^{25}(\tau, \sigma) = X_L(\tau + \sigma) - X_R(\tau - \sigma)$$
$$= q_0^{25} + \alpha' w \tau + \alpha' p \sigma + i\sqrt{\frac{\alpha'}{2}} \sum_{n \neq 0} \frac{e^{-in\tau}}{n} (\tilde{\alpha}_n^{25} e^{-in\sigma} - \alpha_n^{25} e^{in\sigma}). \tag{25.44}$$

So far, we have seen that T-duality is a symmetry of free string theory, i.e. of the mass spectrum. But in fact, we can show that T-duality respects also the interactions of the theory, so it is extended to a full quantum symmetry. It is a duality symmetry, that exchanges one background with another, so is not a symmetry in the usual sense, but rather one that exchanges one background (vacuum) around which we consider the theory, with another.

Since as we already said, the (interacting) massless modes of the superstring form the low energy supergravity theory, the duality symmetry also acts on the supergravity fields, giving the so-called *Buscher rules*. For the NS-NS sector fields $g_{\mu\nu}$, $B_{\mu\nu}$, and ϕ, they are

$$\tilde{g}_{00} = \frac{1}{g_{00}}$$
$$\tilde{g}_{0i} = \frac{B_{0i}}{g_{00}}; \quad \tilde{g}_{ij} = g_{ij} - \frac{g_{0i}g_{0j} - B_{0i}B_{0j}}{g_{00}}$$
$$\tilde{B}_{0i} = \frac{g_{0i}}{g_{00}}; \quad \tilde{B}_{ij} = B_{ij} + \frac{g_{0i}B_{0j} - B_{0i}g_{0j}}{g_{00}}$$
$$\tilde{\phi} = \phi - \frac{1}{2}\log(g_{00}). \tag{25.45}$$

We will not prove them here, but they can be obtained from a transformation of the string path integral. We note that the first relation is equivalent to the relation $R/\sqrt{\alpha'} = \sqrt{\alpha'}/\tilde{R}$.

The Buscher rules define a dual supergravity background, but the corresponding T-duality is not a symmetry in the usual sense, it just maps one background (vacuum solution) into another, yet the supergravity theory around one doesn't look like the supergravity theory around the other. We need the full string theory for that to be true.

We can generalize this T-duality in several ways. One obvious way is to make T-duality on several circles, i.e. on a torus $T^n = (S^1)^n$. One generalization is called "*nonabelian T-duality*," though it is not as well understood as the usual, abelian, T-duality, and there are many open questions. Another one is a mathematical transformation called "*mirror symmetry*," that exchanges some topological numbers characterizing a general compact space, as well as generally exchanging continuous parameters known as *Kähler moduli* with other continuous parameters known as *complex structure moduli*. Mirror symmetry was understood to be a symmetry of string theory, in the sense that string theory in a background vacuum with a compact space K is equivalent with the theory in the background with the mirror symmetric compact space K'. In the case of the circle, mirror symmetry reduces to T-duality.

Important Concepts to Remember

- Superstring theory is a supersymmetric version of string theory, with fermions, with the tachyon projected out, thus stable, and with extra states. It lives in $D = 10$ spacetime dimensions.
- The low-energy limit of string theory is a supergravity theory, with fields in the NS-NS sector, the R-R sector, the NS-R sector, and the R-NS sector.
- Supersymmetry is a symmetry between bosons and fermions, of the general type $\delta\phi = \bar{\epsilon}\psi$ and $\delta\psi = \not{\partial}\phi\epsilon$.
- We can have several supersymmetries, each taking a boson into a different fermion, like $\delta\phi \sim \bar{\epsilon}_i\psi^i$.
- The low-energy theory on N D3-branes is $U(N)$ $\mathcal{N}=4$ Supersymmetric Yang-Mills (SYM).
- The Born-Infeld theory on D3-branes has a form with a maximal electric field, $E_{\text{crit}} = 1/(2\pi\alpha')$.
- For closed strings on compact spaces, the momentum is quantized as usual, $p = n/R$, but strings can also wind around the compact dimension, with winding $w = mR/\alpha'$.
- The string spectrum is symmetric under the exchange of $R \leftrightarrow \tilde{R} = \alpha'/R$ and the simultaneous exchange of momentum and winding, i.e. n with m, called T-duality.
- At the level of the compact X, considering that T-duality changes the oscillators with a sign, it changes $X^{25} = X_L^{25} + X_R^{25}$ with $X'^{25} = X_L^{25} - X_R^{25}$.
- T-duality is true also for interactions, being a full duality symmetry of string theory. On the supergravity modes (the interacting massless string modes), T-duality is described by the Buscher rules.

Further Reading

Chapters 17 and 19 in [43], chapters 4 and 5 in [44], chapter 8 in [11].

Exercises

(1) For the original 3+1–dimensional BI action, in terms of $F_{\mu\nu}$ in flat space, calculate the equations of motion (the "Maxwell equations") as a function of $\vec{E}$ and $\vec{D} = \partial\mathcal{L}/\partial\vec{E}$ and solve them for the "electron," the solution of unit electric charge. Show that it has a maximum electric field, reached at $r = 0$.

(2) Consider a D3-brane moving in the AdS_5 space with metric

$$ds^2 = \frac{r^2}{R^2}(-dt^2 + d\vec{x}^2) + \frac{R^2}{r^2}dr^2 \tag{25.46}$$

and 4-form

$$C_{0123} = \frac{r^4}{R^4}. \tag{25.47}$$

Calculate explicitly the action in the static gauge, and expand it to order $(F_{\mu\nu})^4$.

(3) Consider the action for a D3-brane moving in flat five dimensions at $B_{\mu\nu} = 0$ and $\vec{B} = 0$. Write the action as a function of $\vec{E}$ and X, the fifth dimension. Then after writing canonical momenta for the gauge field and X, write the Hamiltonian (energy) for the field. Is it bounded?

(4) Consider T-duality of closed strings on T^2. Write the mass formula, and find the symmetry group at the "self-dual point," $R_1 = R_2 = \sqrt{\alpha'}$.

(5) Calculate the T-dual solution to (on X_3)

$$ds^2 = dx_3^2 + dx_1^2 + (dx_2 - mx_1 dx_3)^2$$
$$e^\phi = e^{\phi_0}; \quad B = 0. \tag{25.48}$$

using the Buscher rules.

Dualities and M Theory

In this chapter we treat systematically dualities, and show how under them all string theories are interconnected and a manifestation of a single underlying theory.

26.1 T-Duality of Open Strings

We have seen that we can define a symmetry called T-duality as acting on the radius, by $R \to \tilde{R} = \alpha'/R$, together with the exchange of momentum p with winding w, and that it can be extended to a duality on the coordinates $X(\tau, \sigma)$, from $X = X_L + X_R$ to $X' = X_L - X_R$.

We now understand how to generalize this to open strings. We consider open strings with Neumann boundary conditions (NN strings) in a compact direction X^{25}. Then the solution is, as we saw,

$$X^{25}(\tau, \sigma) = x_0^{25} + \sqrt{2\alpha'}\alpha_0^{25}\tau + i\sqrt{2\alpha'}\sum_{n \neq 0}\frac{1}{n}\alpha_n^{25}\cos n\sigma\, e^{-in\tau}, \qquad (26.1)$$

where $\alpha_0^{25} = \sqrt{2\alpha'}p = \sqrt{2\alpha'}n/R$. As we have shown, we can write it in a form like for the closed string, just with $\alpha_n^{25} = \tilde{\alpha}_n^{25}$, i.e. as

$$X^{25}(\tau, \sigma) = X_L^{25}(\tau + \sigma) + X_R^{25}(\tau - \sigma)$$

$$X_L^{25} = \frac{x_0^{25} + q_0^{25}}{2} + \sqrt{\frac{\alpha'}{2}}\alpha_0^{25}(\tau + \sigma) + \frac{i}{2}\sqrt{2\alpha'}\sum_{n \neq 0}\frac{1}{n}\alpha_n^{25}e^{-in\tau}e^{-in\sigma}$$

$$X_R^{25} = \frac{x_0^{25} - q_0^{25}}{2} + \sqrt{\frac{\alpha'}{2}}\alpha_0^{25}(\tau - \sigma) + \frac{i}{2}\sqrt{2\alpha'}\sum_{n \neq 0}\frac{1}{n}\alpha_n^{25}e^{-in\tau}e^{+in\sigma}. \qquad (26.2)$$

Then T-duality acts as for the closed string, by exchanging $X^{25}(\tau, \sigma)$ with

$$X'^{25}(\tau, \sigma) = X_L^{25}(\tau + \sigma) - X_R^{25}(\tau - \sigma)$$

$$= q_0^{25} + \sqrt{2\alpha'}\alpha_0^{25}\sigma + \sqrt{2\alpha'}\sum_{n \neq 0}\frac{1}{n}\alpha_n^{25}e^{-in\tau}\sin n\sigma. \qquad (26.3)$$

We notice that this is the same expansion as the expansion for a string between two D-branes (24.38), just that with

$$\alpha_0^{25} = \frac{1}{\sqrt{2\alpha'}}\frac{x_2^{25} - x_1^{25}}{\pi}. \qquad (26.4)$$

But substituting in our case $\alpha_0^{25} = \sqrt{2\alpha'}n/R$, we obtain

$$X'^{25}(\tau, \sigma = \pi) - X'^{25}(\tau, \sigma = 0) = 2\alpha' p\pi = 2\pi\alpha'\frac{n}{R} = 2\pi\tilde{R}n = x_2'^{25} - x_1'^{25}, \quad (26.5)$$

as well as

$$X'^{25}(\tau, 0) = q_0^{25}, \quad (26.6)$$

identified with $x_1'^{25}$.

The interpretation is then that there is a D-brane at $x_1'^{25} = q_0^{25}$, and strings with one end on it wind n times around the T-dual circle, after which they return, to end on the same D-brane.

Moreover, we have

$$\partial_\sigma X^{25}(\tau, \sigma) = \frac{dX_L}{du}(u = \tau + \sigma) - \frac{dX_R}{dv}(v = \tau - \sigma) = \partial_\tau X'^{25}(\tau, \sigma) \quad (26.7)$$

and

$$\partial_\tau X^{25}(\tau, \sigma) = \frac{dX_L}{du}(u = \tau + \sigma) + \frac{dX_R}{dv}(v = \tau - \sigma) = \partial_\sigma X'^{25}(\tau, \sigma). \quad (26.8)$$

That means that the Neumann boundary condition $\partial_\sigma X^{25} = 0$ becomes the Dirichlet boundary condition $\partial_\tau X'^{25} = 0$ (so that $\delta X'^{25} = 0$), and vice versa, the Dirichlet boundary condition $\partial_\tau X^{25} = 0$ becomes the Neumann boundary condition $\partial_\sigma X'^{25} = 0$.

A formal interpretation of the initial Neumann string is as a string ending on *spacetime-filling* D25-branes. Since a D-brane has Neumann boundary conditions on the parallel directions, and Dirichlet on the transverse directions, it follows that the T-duality, exchanging Neumann with Dirichlet, exchanges the dimensionality of the D-brane. In the case above, the D25-brane turns into a D24-brane with transverse direction X'^{25}. Repeating the T-duality, we see that any Dp-brane turns into either a D$(p + 1)$-brane or a D$(p - 1)$-brane, depending on whether the T-duality direction is transverse or parallel, respectively.

26.2 T-Duality and Chan-Paton Factors

The case treated above led to a single D-brane after T-duality. But we know we can attach Chan-Paton factors to the ends of Neumann open strings, leading to states $|ij\rangle$ in the adjoint of $U(N)$, thus ending on several D25-branes. We now consider T-duality in this nonabelian case.

Wilson Line and Loop

For that, we need to introduce the concept of a Wilson line and Wilson loop. Consider the object

$$W \equiv e^{iq\oint dxA_x} \equiv e^{iw} \quad (26.9)$$

in the abelian case first. The integral is taken either over an open path, in which case we call it a Wilson line, or a closed path C, in which case we call it a Wilson loop. We will be interested in the case of closed paths, winding over the compact direction, so we will have *Wilson loops*. We can see that a Wilson loop is gauge-invariant. Indeed, under a gauge transformation that changes the gauge field as

$$A_x \rightarrow A_x + \partial_x \lambda, \tag{26.10}$$

and with group element

$$U = e^{i\lambda} \tag{26.11}$$

that must be single valued under a full translation around a compact circle in X^{25}, $X^{25} \rightarrow X^{25} + 2\pi R$, it follows that the gauge transformation must also be periodic:

$$q\lambda(x + 2\pi R) = q\lambda(x) + 2\pi m. \tag{26.12}$$

Then under a gauge transformation, w changes by

$$\delta \oint dx A_x = \oint dx \partial_x \lambda = \lambda(2\pi R) - \lambda(0) = 2\pi m, \tag{26.13}$$

so W changes as

$$W \rightarrow e^{2\pi i m} W = W, \tag{26.14}$$

i.e. it is gauge invariant.

Then the quantity $w = q \oint A_x dx$ is an angle θ between 0 and 2π. We can realize the Wilson line by a constant gauge field, related to this θ by

$$q A_x = \frac{\theta}{2\pi R}. \tag{26.15}$$

When we add a coupling to a gauge field A_μ in a particle action, in the Hamiltonian one replaces p_μ with $p_\mu - q A_\mu$, where p_μ is the canonical momentum, so it is quantized on the compact space, for the same reason as before, namely that the wave functions e^{ipx} must be single-valued. Then one has the replacement $p_{25} \rightarrow p_{25} - q A_{25}$ as a gauge-invariant momentum, which means replacing n/R with

$$\frac{n}{R} - \frac{\theta}{2\pi R}. \tag{26.16}$$

We now consider a string with $U(N)$ Chan-Paton factors, i.e. with a general constant gauge field $A_{25} \in U(N)$. Using a gauge transformation, we can always diagonalize A_{25}, to obtain

$$A_{25} = \frac{1}{2\pi R} (\theta_1, \ldots, \theta_N). \tag{26.17}$$

This will give a nonabelian Wilson loop. Such a gauge field breaks the gauge symmetry from $U(N)$ to $U(1)^N$, which means that the remaining gauge invariance after the gauge transformation that diagonalized A_{25} is the abelian $U(1)^N$.

The Chan-Paton state $|ij\rangle$ has charge $+1$ under $U(1)_i$ and -1 under $U(1)_j$, which means that its gauge-invariant momentum is

$$p = \frac{2\pi n + \theta_i - \theta_j}{2\pi R}. \tag{26.18}$$

Substituting this in the formula for M^2, we obtain the spectrum for the open strings as

$$M^2 = \left(\frac{2\pi n - (\theta_j - \theta_i)}{2\pi R}\right)^2 + \frac{1}{\alpha'}(N^\perp - 1). \qquad (26.19)$$

But replacing $R = \alpha'/\tilde{R}$, the resulting spectrum has $\tilde{R}\theta_i$ appearing as positions of D24-branes in X^{25}, so now after T-duality, the open strings wind n times around the dual circle, but then end up on a different brane, situated at a distance $\tilde{R}\Delta\theta$ from the first one.

Moreover, we can check that the distance between the endpoints of the string is really the one above by integrating:

$$X'^{25}(\tau, \pi) - X'^{25}(\tau, 0) = \int_0^\pi d\sigma\, \partial_\sigma X'^{25} = \int_0^\pi d\sigma\, \partial_\tau X^{25} = 2\alpha' p(\pi) = 2\pi\alpha'\left(\frac{n}{R} + \frac{\Delta\theta}{2\pi R}\right)$$

$$= \tilde{R}(2\pi n + \Delta\theta). \qquad (26.20)$$

Here we have used the explicit oscillator form of X^{25} and took ∂_τ and integrated over σ to obtain the result.

26.3 Supergravity Actions

We have seen that the supersymmetric string lives in $D = 10$ dimensions and has supersymmetry. But actually there are $\mathcal{N} = 2$ supersymmetries in 10 dimensions. We can understand this as follows: the action has the maximal amount of supersymmetry possible for a theory with gravity in it. But in four dimensions, that corresponds to $\mathcal{N} = 8$ supersymmetries (the argument goes as follows: one supersymmetry lowers helicity by one half, so by acting with eight supersymmetries we start from the maximal $+2$ helicity and end up in the minimal -2 helicity; if we had more than $\mathcal{N} = 8$, it would lead to helicities $|h| > 2$, i.e. spins higher than 2). Since in four dimensions a Majorana spinor has four real components (a Weyl spinor has two complex components), this gives 32 supercharges. In terms of the minimal 10-dimensional spinor, with 16 real components, this gives $\mathcal{N} = 2$ supersymmetries. But the two spinors can be chosen to have the same, or opposite chirality (chirality means eigenvalue of Γ_{11}, which is the analog of γ_5 in four dimensions: the product of all the gamma matrices).

If the two chiralities are the same,

$$\Gamma_{11}\psi_i = \psi_i, \quad i = 1, 2, \qquad (26.21)$$

we are in *type IIB* theory. If the chiralities are opposite,

$$\Gamma_{11}\psi_i = (-1)^i \psi_i, \qquad (26.22)$$

then we are in the *type IIA* theory.

The massless states of the closed superstring include the graviton $g_{\mu\nu}$, so the low-energy limit (for energies $\ll 1/\sqrt{\alpha'}$) of superstring theory gives a supersymmetric theory of gravity, or *supergravity*, with the field content of the massless modes of the closed superstring.

Type IIA superstring theory leads to type IIA supergravity, and type IIB superstring theory leads to type IIB supergravity.

The NS-NS sector contains bosonic fields that are common to the IIA and IIB superstring, as well as the bosonic string theory: the graviton $g_{\mu\nu}$, the B-field $B_{\mu\nu}$, and the dilaton Φ. The R-R sector contains p-form fields (totally antisymmetric tensor fields) that differ between the IIA and IIB theories. The fermionic sectors are the NS-R and the R-NS sectors, which will not be described here.

Type IIA contains odd p-forms, specifically a 1-form $A_{(1)} = A_\mu$, with field strength $F_{\mu\nu} = \partial_\mu A_\nu - \partial_\nu A_\mu$ ($F_{(2)}$), and a 3-form $A_{(3)} = A_{\mu\nu\rho}$, with field strength $F_{\mu\nu\rho\sigma} = 4\partial_{[\mu}A_{\nu\rho\sigma]}$ ($F_{(4)}$). Since $A_{(p+1)}$ couples to an electric Dp-brane source, the electric branes of type IIA are a D0-brane and a D2-brane. But for a $F_{(p+2)} = *F_{(D-p-2)}$, we have a magnetic brane (coupling electrically to the dual field), with $\tilde{p} = D - p - 4$. In the case of the $D = 10$-dimensional superstring theory this gives $\tilde{p} = 6 - p$, therefore we have also magnetic D4-branes and D6-branes, coupling to $F_{(4)}$ and $F_{(2)}$, respectively.

Type IIB contains even forms, specifically a 0-form (scalar) $A_{(0)} = a$ (axion), with field strength $F_\mu = \partial_\mu a$ ($F_{(1)}$), a 2-form $A_{(2)} = A_{\mu\nu}$ with field strength $F_{\mu\nu\rho} = 3\partial_{[\mu}A_{\nu\rho]}$ ($F_{(3)}$), and a 4-form $A_{(4)}^+ = A_{\mu_1...\mu_4}^+$, with field strength $F_{(5)}^+ = F_{\mu_1...\mu_5}^+$ that is *self-dual*:

$$F_{\mu_1...\mu_5}^+ = \frac{1}{5!}\epsilon_{\mu_1...\mu_5}{}^{\mu_6...\mu_{10}}F_{\mu_1...\mu_6}^+. \tag{26.23}$$

According to the same rules as above, a couples with an electric D(-1)-brane source, $A_{(2)}$ with an electric D1-brane source, and $A_{(4)}^+$ with an electric D3-brane source, and we also have magnetic D7-brane, D5-brane, (and D3-brane) coupling to the same, respectively. Note that the fact that $F_{(5)}^+$ is self-dual can be found from the representation theory of the Lorentz group, the irreducible representation being self-dual. In turn, that could be understood since the self-dual condition *can* be imposed on the field, thus reducing the representation.

We finally write the bosonic parts of the supergravity actions. They are naturally written in the "string frame," i.e. in terms of a metric related to the usual "Einstein frame" metric $G_{\mu\nu}^{(E)}$, by

$$G_{\mu\nu} = e^{\frac{\Phi}{2}}G_{\mu\nu}^{(E)}. \tag{26.24}$$

That is because the action then comes with a factor $e^{-2\Phi}$ in front of the NS-NS sector action, which means it has also a $1/g_s^2 = e^{-2\langle\Phi\rangle}$ factor, as expected.

The IIA bosonic action is

$$S_{IIA} = \frac{1}{2\kappa_{10}^2}\int d^{10}x\left\{\sqrt{-G}\left[e^{-2\Phi}\left(R + 4\partial_\mu\Phi\partial^\mu\Phi - \frac{1}{2}|H_3|^2\right) - \frac{1}{2}|F_2|^2 - \frac{1}{2}|\tilde{F}_4|^2\right]\right.$$
$$\left. - \frac{1}{2}B_2 \wedge F_4 \wedge F_4\right\}, \tag{26.25}$$

where

$$\tilde{F}_4 = dA_3 - A_1 \wedge H_3. \tag{26.26}$$

Here we used the notation

$$\int d^D x \sqrt{-G} |F_p|^2 \equiv \int D^D x \sqrt{-G} \frac{1}{p!} G^{M_1 N_1} \dots G^{M_p N_p} F_{M_1 \dots M_p} F_{N_1 \dots N_p}, \tag{26.27}$$

$F_{(p+2)} = dA_{(p+1)}$ means that $F_{\mu_1 \dots \mu_{p+2}} = (p+2) \partial_{[\mu_1} A_{\mu_2 \dots \mu_{p+2}]}$, and we have $F_2 = dA_1, F_4 = dA_3, H_3 = dB_2$.

The first line in the IIA action is the NS-NS sector action, which comes with the factor of $1/g_s^2$, the second line is the R-R sector action, without the coupling, and the third line is a Chern-Simons–type term (written without the metric, only with $\epsilon^{\mu_1 \dots \mu_D}$, which takes values ± 1, and of the type gauge field wedge field strength, wedge field strength).

For the type IIB bosonic action, there is no covariant form, since there is no covariant action for a self-dual field. There are usually three ways to deal with a self-dual field: write a Lorentz noncovariant form (choose a special direction), impose the self-duality as a constraint on the equations of motion, and introduce auxiliary fields. Usually for type IIB one presents the second form, imposing the self-duality of F_5^+ as a constraint.

Then the IIB bosonic action is

$$S_{IIB} = \frac{1}{2\kappa_{10}^2} \int d^{10}x \left\{ \sqrt{-G} \left[e^{-2\Phi} \left(R + 4 \partial_\mu \Phi \partial^\mu \Phi - \frac{1}{2} |H_3|^2 \right) \right. \right.$$
$$\left. - \frac{1}{2} |F_1|^2 - \frac{1}{2} |\tilde{F}_3|^2 - \frac{1}{4} |\tilde{F}_5|^2 \right]$$
$$\left. - \frac{1}{2} A_4 \wedge H_3 \wedge F_3 \right\}. \tag{26.28}$$

We used the same notation as for the IIA action, with $F_3 = dA_2, F_5 = dA_4, H_3 = dB_2$, and

$$\tilde{F}_3 = F_3 - A_0 \wedge H_3$$
$$\tilde{F}_5 = F_5 - \frac{1}{2} A_2 \wedge H_3 + \frac{1}{2} B_2 \wedge F_3. \tag{26.29}$$

We then impose $\tilde{F}_5 = *\tilde{F}_5$ as a constraint, on the equations of motion. It is because of this constraint that we have the 1/4 factor instead of 1/2 in front of $|\tilde{F}_5|^2$, since we have not yet imposed it, so we have twice as many components.

26.4 S-Duality of Type IIB Theory

The type IIB supergravity action has a certain symmetry. To make it manifest, we first construct the complex field

$$\tau \equiv a + i e^{-\Phi}, \tag{26.30}$$

and from it the matrix

$$\mathcal{M}_{ij} \equiv \frac{1}{\mathrm{Im}\tau} \begin{pmatrix} |\tau|^2 & -\mathrm{Re}\tau \\ -\mathrm{Re}\tau & 1 \end{pmatrix}. \tag{26.31}$$

We also construct out of the NS-NS and the R-R 3-forms the vector

$$F_3^i = \begin{pmatrix} H_3 \\ F_3 \end{pmatrix}. \tag{26.32}$$

In terms of these objects, and the Einstein metric $G_{\mu\nu}^{(E)}$, the type IIB supergravity action becomes

$$S_{IIB} = \frac{1}{2\kappa_{10}^2} \int d^{10}x \sqrt{-G_E} \left[R_E - \frac{\partial_\mu \bar{\tau} \partial^\mu \tau}{2(\mathrm{Im}\tau)^2} - \frac{1}{2}\mathcal{M}_{ij}F_3^i F_3^j - \frac{1}{4}|\tilde{F}_5|^2 \right]$$
$$- \frac{\epsilon_{ij}}{8\kappa_{10}^2} \int d^{10}x A_4 \wedge F_3^i \wedge F_3^j. \tag{26.33}$$

We now can check that it is invariant under the $Sl(2,\mathbb{R})$ symmetry, with matrix element

$$\begin{pmatrix} a & b \\ c & d \end{pmatrix}, \tag{26.34}$$

with $ad - bc = 1$, acting on τ as

$$\tau' = \frac{a\tau + b}{c\tau + d}, \tag{26.35}$$

and on the vector F_3^i as

$$F_3'^i = \Lambda^i{}_j F_3^j, \tag{26.36}$$

where

$$\Lambda^i{}_j = \begin{pmatrix} d & c \\ b & a \end{pmatrix}. \tag{26.37}$$

In terms of Λ, the matrix $\mathcal{M}_{ij}$ transforms as

$$\mathcal{M}' = (\Lambda^{-1})^T \mathcal{M} \Lambda^{-1}. \tag{26.38}$$

The fields $\tilde{F}_5$ and $G_{\mu\nu}^{(E)}$ are invariant under $Sl(2,\mathbb{R})$.

However, while $Sl(2,\mathbb{R})$ is an invariance of the IIB supergravity action (on the self-duality constraint), it is not an invariance of the full string theory. Only a subgroup $Sl(2,\mathbb{Z})$ survives at the quantum level as an invariance of the full quantum string theory. From that point of view, the extended invariance of the special subset of massless fields is a coincidence.

Included in the $Sl(2,\mathbb{Z})$ invariance of the full string theory is the transformation with matrix

$$\begin{pmatrix} a & b \\ c & d \end{pmatrix} = \begin{pmatrix} 0 & -1 \\ 1 & 0 \end{pmatrix}. \tag{26.39}$$

Indeed, it has integer coefficients, and $ad - bc = 1$. But for these values, the transformation on τ is

$$\tau' = -\frac{1}{\tau}. \tag{26.40}$$

If we consider the case of zero axion, $a = 0$, i.e. $\tau = ie^{-\Phi}$, then the above transformation gives

$$e^{-\Phi'} = e^{+\Phi} \Rightarrow \Phi' = -\Phi. \tag{26.41}$$

That, in turn, means that

$$g'_s = \frac{1}{g_s}, \tag{26.42}$$

since $g_s = e^{\langle \Phi \rangle}$. Therefore this is a *nonperturbative duality* (weak/strong). The same transformation on the 3-form vector F_3^i gives

$$F_3'^i = \begin{pmatrix} 0 & 1 \\ -1 & 0 \end{pmatrix} \Rightarrow H_3' = F_3; \quad F_3' = -H_3, \tag{26.43}$$

so we exchange the NS-NS with the R-R 3-form field strengths, and the 5-form field strength is invariant, $\tilde{F}_5' = \tilde{F}_5$. The term *S-duality* refers to the strong/weak duality $\tau' = -1/\tau$, though by extension, it also refers to the full $Sl(2; \mathbb{Z})$ duality group.

26.5 M-Theory

Type IIA string theory at (very) strong coupling, $g_s = e^{\langle \Phi \rangle} \to \infty$, is called M-theory. It was defined in a very influential 1995 paper by Edward Witten ("String theory in various dimensions"), where it was also shown that the various string theories are connected to each other through a chain of dualities. The reason that IIA string theory at strong coupling deserves the new name of M-theory is that in this limit, the theory *becomes 11-dimensional*. One does not yet have a good (nonperturbative) definition of M-theory, only definitions in various corners of parameter ("moduli") space and in various limits. What we can say is that the low-energy limit of M-theory, like the low-energy limit of weak coupling IIA string theory, is a supergravity.

In this case, it is an 11-dimensional supergravity. Since the maximal, $\mathcal{N} = 2$ supersymmetry in 10 dimensions (with 32 supercharges) organizes into a single supersymmetry in 11 dimensions (as the minimal 11-dimensional spinor has 32 components), there is a unique, $\mathcal{N} = 1$, supergravity in 11 dimensions. It contains the metric g_{MN}, the 3-form field $A_{(3)} = A_{MNP}$, with field strength

$$F_{(4)} : F_{M_1 \ldots M_4} = 4\partial_{[M_1} A_{M_2 M_3 M_4]}. \tag{26.44}$$

There is also a fermion (the gravitino) $\psi_{M\alpha}$, but we will not discuss it. The 11-dimensional supergravity bosonic action is

$$S_{11} = \frac{1}{2\kappa_{11}^2} \int d^{11}x \sqrt{-G} \left(R - \frac{1}{2}|F_4|^2 \right) - \frac{1}{6\kappa_{11}^2} \int A_3 \wedge F_4 \wedge F_4. \quad (26.45)$$

The KK dimensional reduction ansatz from 11 dimensions to 10-dimensional string theory is

$$ds^2 = G_{MN}^{(11)} dx^M dx^N = e^{-\frac{2\Phi}{3}} G_{\mu\nu}^{S,(10)} dx^\mu dx^\nu + e^{\frac{4\Phi}{3}} (dx^{10} + A_\mu dx^\mu)^2, \quad (26.46)$$

together with the ansatz for the 3-form:

$$A_{MNP} : (A_{\mu\nu 11} \equiv B_{\mu\nu}^{(IIA)}; \quad A_{\mu\nu\rho} \equiv A_{\mu\nu\rho}^{(IIA)}). \quad (26.47)$$

The metric ansatz is a full reduction ansatz, since Φ parametrizes $G_{10,10}$ through $G_{10,10} = e^{\frac{4\Phi}{3}}$, then A_μ parametrizes $G_{\mu 10}$ through $G_{\mu 10} = e^{\frac{4\Phi}{3}} A_\mu$, and $G_{\mu\nu}^{S,(10)}$ parametrizes $G_{\mu\nu}$ through $G_{\mu\nu} = e^{-\frac{2\Phi}{3}} G_{\mu\nu}^{S,(10)} + e^{\frac{4\Phi}{3}} A_\mu A_\nu$.

One can then check that S_{11} reduces to S_{IIA}. The fact that under the KK reduction we obtain the IIA supergravity is due to the fact that the single 11-dimensional spinor splits into two 10-dimensional spinors of opposite chiralities.

Relation among Parameters in 10 Dimensions and 11 Dimensions

The 11-dimensional supergravity compactified on a circle of radius R has two parameters: κ_{11} (or l_P, the 11-dimensional Planck length) and R, to be matched against the two parameters of 10-dimensional string theory, α' and g_s.

To derive that, we first consider that the metric $g_{10,10} = e^{\frac{4\Phi}{3}}$ is understood, at least in the vacuum, as $(R/l_P)^2$ (the metric should have an explicit factor of R^2, but if we want x^{10} to be dimensional, then we need to make R^2 dimensionless as $(R/l_P)^2$). That means that

$$\left(\frac{R}{l_P} \right)^2 = e^{\frac{4\langle\Phi\rangle}{3}} = g_s^{4/3} \Rightarrow R = l_P g_s^{2/3}. \quad (26.48)$$

It follows that we can define the first 10-dimensional parameter as

$$g_s = (R/l_P)^{3/2}. \quad (26.49)$$

The compactified gravitational action will give

$$\frac{1}{2\kappa_{11}^2} \int d^{11}x \sqrt{-G_{E,11}} R_{E,11} \rightarrow \frac{2\pi R}{2\kappa_{11}^2} \int d^{10}x \sqrt{-G_{S,10}} e^{-2\Phi} R_{S,10}, \quad (26.50)$$

to be identified with

$$\frac{1}{2\kappa_{10}^2} \int d^{10}x \sqrt{-G_{S,10}} e^{-2(\Phi-\langle\Phi\rangle)} R_{S,10}. \quad (26.51)$$

But on the other hand by definition

$$2\kappa_{11}^2 \equiv (2\pi)^8 l_P^9, \quad (26.52)$$

and a similar relation in 10 dimensions (in a general dimension, $2\kappa_D^2 = (2\pi)^{D-3} l_{P,D}^{D-2}$, and in string theory $l_P = \sqrt{\alpha'} g_s^{1/4}$):

$$2\kappa_{10}^2 \equiv g_s^2 (2\pi)^7 (\sqrt{\alpha'})^8 \quad (= (2\pi)^7 l_{P,10}^8). \tag{26.53}$$

That leads to

$$l_P^9 = R g_s^2 (\sqrt{\alpha'})^8 = l_P g_s^{8/3} (\sqrt{\alpha'})^8 \Rightarrow l_P = g_s^{1/3} \sqrt{\alpha'}. \tag{26.54}$$

Combining with the first relation, we obtain

$$R = l_P g_s^{2/3} = g_s \sqrt{\alpha'}, \tag{26.55}$$

i.e. *R is g_s in string units*. Thus indeed, as promised, the string coupling acts as an 11th dimension.

26.6 The String Duality Web

We have defined type IIA and type IIB superstring theories. In fact, they are T-dual to each other. In particular, that means that under T-duality, the Dp-branes turn into D$(p+1)$–branes or D$(p-1)$–branes, and since Dp-branes are sources for $A_{(p+1)}$, it means that T-duality changes even for odd p-forms. We have also seen that type IIA superstring theory at strong coupling is M-theory.

But besides the type IIA and type IIB string theories, there are also the *type I and heterotic superstrings*, both of which have $\mathcal{N} = 1$ supersymmetry in 10 dimensions, and have as low-energy a $\mathcal{N} = 1$ supergravity coupled to vector superfields, i.e. gauge fields and their superpartners, for the gauge groups $SO(32)$ and $E_8 \times E_8$. It will not be explained here what type I and heterotic theories mean, but we will relate them to each other and to M-theory.

The choice of gauge groups is important. In a seminal 1987 paper, Michael Green and John Schwarz proved that there are only a few choices of gauge groups that eliminate anomalies. Quantum anomalies represent quantum breaking of symmetries that are respected classically. If the symmetry is global, it is fine, but if the symmetry is local, it affects the number of degrees of freedom (gauge modes are redundant), so quantum breaking of a local symmetry would amount to different numbers of degrees of freedom at the classical and quantum level, which is clearly absurd. On the other hand, theories with chiral fields have potential anomalies, so cancelling the anomalies of local symmetries is an important consistency condition on field theories.

The *Green-Schwarz anomaly cancellation mechanism* proved that for a few discrete choices of gauge group, which excluding very large abelian groups (with $U(1)^p$ factors, p large) restricts to $SO(32)$ and $E_8 \times E_8$, anomalies cancel. The $SO(32)$ group appears in both possibilities, type I and heterotic string, whereas $E_8 \times E_8$ is only heterotic.

It was realized, however, that all the apparently different string theories are related by dualities. First, the two $SO(32)$ theories are related by S duality. That means that the type I and heterotic dilatons are related by $\phi_I = -\phi_h$, or $g_s^I = 1/g_s^h$, as well as $\tilde{F}_3^I = \tilde{H}_3^h$. Then the $SO(32)$ heterotic string, compactified on S^1, is T-dual to the $E_8 \times E_8$ heterotic string, also

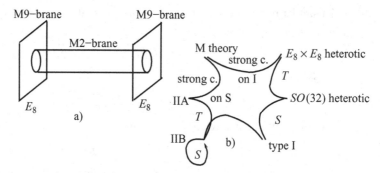

Fig. 26.1   (a) M2-brane between two M9-branes in M theory. (b) The string duality web.

compactified on S^1. But the T-dual of the type I $SO(32)$ string, compactified on S^1, is T-dual to the type IIA string compactified on the interval $S^1/\mathbb{Z}_2$. Note that $S^1/\mathbb{Z}_2$ means that we identify the circle with $\theta \in [-\pi, \pi]$ under $\theta \sim -\theta$, leading to the interval $\theta \in [0, \pi]$. Since type IIA is M-theory on S^1, type I $SO(32)$ string theory on S^1, the strong coupling (S-dual) of the heterotic $SO(32)$ string is T-dual to M-theory on $S^1 \times (S^1/\mathbb{Z}_2)$. And since the heterotic $E_8 \times E_8$ is the T-dual of the heterotic $SO(32)$ string, we arrive at the picture of the heterotic $E_8 \times E_8$ theory being the strong coupling limit of M-theory on the interval $S^1/\mathbb{Z}_2$, with "M9-branes" at the end of each interval. M9-branes are defined analogously to Dp-branes, as endpoints of strongly coupled strings, or "M2-branes," and are 10-dimensional objects in 11 dimensions, with the interval as the transverse coordinate. The M2-branes, or membranes, are the fundamental objects of M-theory, reducing under KK reduction to 10 dimensions to the (fundamental) strings of string theory. Each M9-brane carries an E_8 gauge group, for a total of $E_8 \times E_8$, see Figure 26.1(a). Besides M2-branes and M9-branes, M-theory also has M5-branes.

The resulting chain of dualities linking the various theories together is known as the duality web, and has a pictorial representation in terms of various corners of the same figure; see Figure 26.1(b).

Important Concepts to Remember

- T-duality on open strings exchanges Neumann with Dirichlet boundary conditions, thus leading to D-branes. It changes a Dp-brane into a D$(p + 1)$–brane or a D$(p - 1)$–brane, depending on whether the T-duality is transverse or parallel to the Dp-brane.
- T-duality on open strings with Chan-Paton factors (multiple D25-branes), with Wilson lines characterized by diagonal parameters θ_i, leads to multiple D-branes with positions $\tilde{R}\theta_i$ on the compact circle.
- Superstring theory lives in $D = 10$ and has two supersymmetries. For supersymmetries with same chirality we get type IIB theory, and for opposite chirality type IIA theory.
- The low energy of the string is a supersymmetric theory of the superstring mass-less modes, i.e. supergravity, with bosonic theory composed of the NS-NS sector $g_{\mu\nu}, B_{\mu\nu}, \Phi$, the R-R-sector with p-forms, and a CS term.

- The RR sector of type IIA has odd forms $A_{(1)}$ and $A_{(3)}$ with even p brane sources, and the RR sector of type IIB has even forms $A_{(0)}$, $A_{(2)}$, and $A_{(4)}^+$, with odd p brane sources.
- The type IIB supergravity has $Sl(2, \mathbb{R})$ S-duality, but at the quantum level, i.e. in the full string theory, only a $Sl(2, \mathbb{Z})$ subgroup is an invariance. It contains an element that acts as $\Phi' = -\Phi$, or $g'_s = 1/g_s$.
- Type IIA string theory at strong coupling is 11-dimensional M-theory, which at low energy gives the unique ($\mathcal{N} = 1$) 11-dimensional supergravity. The relation between parameters is $g_s = (R/l_P)^{3/2}$ and $R = g_s \sqrt{\alpha'}$.
- All string theories are related by dualities, into the string duality web: IIA is T-dual to IIB; IIA is M-theory on S^1; $SO(32)$ type I is S-dual to $SO(32)$ heterotic, which is T-dual to $E_8 \times E_8$ heterotic; type I $SO(32)$ on S^1 is T-dual to M theory on $S^1 \times (S^1/\mathbb{Z}_2; E_8 \times E_8$ heterotic is the strong coupling limit of M-theory on the interval $S^1/\mathbb{Z}_2)$.

Further Reading

Chapters 12 in [11], vol. II, chapter 8 in [45].

Exercises

(1) *(Buscher T-duality)*
   Consider the string action at $\phi = \phi_0$, $B_{\mu\nu} = 0$,

$$S = \int d^2\sigma \sqrt{\gamma}\gamma^{ab}g_{\mu\nu}\partial_a X^\mu \partial_b X^\nu, \qquad (26.56)$$

   and a metric $g_{\mu\nu} = (g_{ij}, g_{00}, g_{0i})$. Prove that

$$S = \int d^2\sigma \left[\sqrt{\gamma}\gamma^{ab}(g_{00}V_a V_b + 2g_{0i}V_a\partial_b X^i + g_{ij}\partial_a X^i\partial_b X^j) + 2\epsilon^{ab}\hat{X}^0\partial_a V_b\right] \qquad (26.57)$$

   is a first order form for (26.56), and that by solving for V_a, we get the T-dual string action (in T-dual background), with the $X^0 \leftrightarrow \hat{X}^0$ relation being the same as the $X \leftrightarrow X'$ relation for T-duality of the open string.
(2) Prove that the nonabelian Wilson loop,

$$W[C] = \text{Tr}\, P\{e^{iq \oint_C A \cdot dx}\} \qquad (26.58)$$

   is gauge invariant.
(3) Consider a square torus T^2 of radii $R_1 = R_2 = 1$ and D-branes situated at positions $(0, 0)$, $(0, \pi)$, $(\pi, 0)$, and (π, π) on the torus. Describe the possible strings (ending on D-branes) and their masses, using a condensed matter analogy.
(4) Check that the A_3 action in 11-dimensional supergravity leads to the correct p-form action in type IIA supergravity, under dimensional reduction.
(5) Verify the invariance of the type IIB action under $Sl(2, \mathbb{R})$.

(6) Prove that "D0-branes" ("D-particles") in type IIA theory, with masses $M_n = (|n|/\sqrt{\alpha'})/g_s$, where n is an integer, can be understood as momentum modes in the 11-dimensional M-theory.

(7) Check that the type IIA and type IIB supergravities, as a limit of their string theories, are T-dual to each other upon compactification on a circle, by checking that the field content is consistent with the T-duality rules.

27

The AdS/CFT Correspondence: Definition and Motivation

We finally come to the most important tool we have for using string theory and its low-energy gravitational theory, for strongly coupled field theories and condensed matter systems, the AdS/CFT correspondence. We first need to define better a conformal field theory (CFT) in a general dimension, and AdS space of a general dimensionality, after which we will describe the duality relation between them.

27.1 Conformal Invariance and Conformal Field Theories in $D > 2$ Dimensions

We start by reviewing some ideas about conformal invariance that we described in Part I (in Chapter 8), and then used a bit for string theory. We consider a flat space (either Euclidean or Minkowski in terms of its signature) and a quantum field theory on it. Then conformal transformations are generalizations of scale transformations, for which $x'^{\mu} = \lambda x^{\mu}$ implies

$$ds^2 = d\vec{x}'^2 = \lambda^2 d\vec{x}^2. \tag{27.1}$$

A conformal transformation is a general coordinate transformation $x^{\mu} \rightarrow x'^{\mu}(x)$, such that the metric is instead multiplied by a *local* ("conformal") factor:

$$ds^2 = dx'^{\mu} dx'_{\mu} = [\Omega(x)]^{-2} dx^{\mu} dx_{\mu}. \tag{27.2}$$

So conformal invariance is a kind of local scale invariance. In fact, most (if not all) theories with scale invariance (at the quantum level) have also conformal invariance. It is actually an issue of current research whether there exist scale-invariant theories that are not conformal invariance, and the answer seems to be no.

We emphasize that, even if we have formulated the conformal transformation as a general coordinate transformation, conformal invariance is an invariance *of flat space*, i.e. we check that after this transformation the flat space action still is a flat space action (is independent of the conformal factor, so it is still written with the flat metric).

In previous chapters we have focused on conformal theories in $D = 2$ spacetime dimensions, where we have seen that conformal transformations are all the holomorphic transformations, thus making an infinite algebra, the Virasoro algebra, with generators L_n.

Here, however, we are interested in the case $D \neq 2$, in particular $D = 4$, so the conformal group is finite dimensional. Consider an infinitesimal conformal transformation,

$x'^{\mu} = x^{\mu} + v^{\mu}(x)$, leading to a $\Omega(x) = 1 - \sigma_v(x)$. Replacing the transformation in the metric, we calculate that

$$\partial_{\mu} v_{\nu} + \partial_{\nu} v_{\mu} = 2\sigma_v \delta_{\mu\nu}. \tag{27.3}$$

Taking the trace, we obtain $\sigma_v(x) = \partial \cdot v/d$. The most general solution for this equation is

$$v_{\mu}(x) = a_{\mu} + \omega_{\mu\nu} x_{\nu} + \lambda x_{\mu} + b_{\mu} x^2 - 2x_{\mu} b \cdot x, \tag{27.4}$$

where $\omega_{\mu\nu} = -\omega_{\nu\mu}$ and $\sigma_v(x) = \lambda - 2b \cdot x$.

We see then that λ is a scale transformation, corresponding to a generator D, a_{μ} is a translation, corresponding to P_{μ}, $\omega_{\mu\nu}$ is a Lorentz rotation, corresponding to $J_{\mu\nu}$, and b_{μ} is a transformation called a "special conformal transformation," with generator K_{μ}. Together, there are $1 + d + d(d-1)/2 + d = (d+1)(d+2)/2$, the same number of degrees of freedom as an antisymmetric matrix Ω_{MN} with $d+2$ components. In fact, we see that the generators organize into the generators of $SO(2,d)$, as follows:

$$\bar{J}_{MN} = \begin{pmatrix} J_{\mu\nu} & \bar{J}_{\mu,d+1} & \bar{J}_{\mu,d+2} \\ -\bar{J}_{\mu,d+1} & 0 & D \\ -\bar{J}_{\mu,d+2} & -D & 0 \end{pmatrix}. \tag{27.5}$$

We see first that we have $\bar{J}_{d+1,d+2} = D$, but the other generators organize in a slightly different way than what we would expect:

$$\bar{J}_{\mu,d+1} = \frac{K_{\mu} - P_{\mu}}{2}, \quad \bar{J}_{\mu,d+2} = \frac{K_{\mu} + P_{\mu}}{2}. \tag{27.6}$$

The conformal field theory that we are mostly interested in will be $\mathcal{N} = 4$ SYM in $3+1$ dimensions, which was indeed found to be conformal, though we will not prove it here.

27.2 AdS Space

We have defined what CFT in AdS/CFT means; next we define AdS, which stands for Anti–de Sitter space, a certain special curved space.

Consider curved spaces in d dimensions. For Euclidean signature, the spaces of maximal symmetry are the sphere S^d, which can be embedded in $d + 1$–dimensional Euclidean space, and the Lobachevski space, which can be embedded in $d + 1$–dimensional Minkowski space. For Minkowski signature, the spaces of maximal symmetry are the de Sitter space dS_d, which can be embedded in a Minkowski space in $d + 1$ dimensions, and Anti–de Sitter space AdS_d, which can be embedded in a $d + 1$–dimensional space with signature $(2, d-1)$.

Sphere

The sphere is defined by embedding in Euclidean $d + 1$–dimensional space, with metric

$$ds^2 = dx_1^2 + \cdots + dx_d^2 + dx_{d+1}^2, \tag{27.7}$$

by imposing the constraint

$$(x_1)^2 + \cdots + (x_d)^2 + (x_{d+1})^2 = R^2, \tag{27.8}$$

which respects the same symmetry, the rotation symmetry $SO(d+1)$ of the $d+1$–dimensional embedding space. Therefore S^d is invariant under $SO(d+1)$.

Lobachevski Space

It is defined by embedding in $d+1$–dimensional Minkowski space, with metric

$$ds^2 = dx_1^2 + \cdots + dx_d^2 - dx_{d+1}^2, \tag{27.9}$$

by imposing the constraint

$$(x_1)^2 + \cdots + (x_d)^2 - (x_{d+1})^2 = -R^2, \tag{27.10}$$

which respects the same symmetry, the Lorentz symmetry $SO(1,d)$ of the embedding space. Therefore Lobachevski space is invariant under $SO(1,d)$.

de Sitter Space

It is the Minkowski signature analog of the sphere, i.e. it is defined from it by a Wick rotation, by embedding in $d+1$–dimensional Minkowski space, with metric

$$ds^2 = -dx_0^2 + dx_1^2 + \cdots + dx_{d-1}^2 + dx_{d+1}^2 = R^2, \tag{27.11}$$

by imposing the constraint

$$-(x_0)^2 + (x_1)^2 + \cdots + (x_{d-1})^2 + (x_{d+1})^2 = R^2, \tag{27.12}$$

which respects the same symmetry, the Lorentz symmetry $SO(1,d)$ of the embedding space. Therefore de Sitter space dS_d is invariant under $SO(1,d)$, and its Euclidean version is the sphere, which is therefore denoted sometimes by EdS_d.

Anti–de Sitter Space

It is the Minkowski signature analog of Lobachevski space, i.e. it is defined by a Wick rotation from it, by embedding into $d+1$–dimensional space with two negative signs, with metric

$$ds^2 = -dx_0^2 + dx_1^2 + \cdots + dx_{d-1}^2 - dx_{d+1}^2, \tag{27.13}$$

by imposing the constraint

$$-(x_0)^2 + (x_1)^2 + \cdots + (x_{d-1})^2 - (x_{d+1})^2 = -R^2, \tag{27.14}$$

which respects the same symmetry, $SO(2,d-1)$, of the embedding space. Therefore Anti–de Sitter space AdS_d is invariant under $SO(2,d-1)$, and its Euclidean version is Lobachevski space, which is therefore denoted sometimes by $EAdS_d$.

We note then that the explicit symmetry of AdS_{d+1} is the same as the conformal symmetry of $Minkowski_d$, namely $SO(2,d)$. That is the first hint that there should be a relation between the two, and indeed there is, the AdS/CFT correspondence. It is interesting that the relation was known for a long time before Juan Maldacena's paper, yet nobody knew what it really meant. There was just some speculation that the smallest representation of $SO(2, d)$, the "singleton," was related to theories in Minkowski space. It is now known that it is a representation that lives at the boundary of AdS_{d+1} and is identified with the sources for the conformal field theory in $Minkowski_d$.

Other than the implicit form of the metric of AdS_d obtained by solving the constraint of embedding, there are explicit forms for the metric. One of them is the *Poincaré metric*:

$$ds^2 = -\frac{R^2}{x_0^2}\left(-dt^2 + \sum_{i=1}^{d-2} dx_i^2 + dx_0^2\right). \tag{27.15}$$

Here R is the radius of AdS, the scale similar to the radius of the sphere. Even though the coordinates have the maximum possible extent, $-\infty < t, x^i < +\infty$ and $0 < x_0 < +\infty$, these coordinates do not cover the whole of AdS space. This is an example of the fact that in general relativity sometimes fully extended coordinates cover only part of the space, like we saw first in the black hole case. The Poincaré coordinates cover only what is known as the *Poincaré patch* of AdS space.

In fact, there are coordinates that cover the whole of AdS space, known as *global coordinates*. They give the metric for AdS_d:

$$ds^2 = R^2(-\cosh^2 \rho\, d\tau^2 + d\rho^2 + \sinh^2 \rho\, d\vec{\Omega}_{d-2}^2). \tag{27.16}$$

Note that this is the analytical continuation (Wick rotation) of the metric of a sphere S^d, which can be written in terms of the metric $d\Omega_{d-2}^2$ of the unit S^{d-2} sphere as

$$ds^2 = R^2(\cos^2 \psi\, d\theta^2 + d\psi^2 + \sin^2 \psi\, d\vec{\Omega}_{d-2}^2). \tag{27.17}$$

We can say something about the boundary of AdS space as follows. One can use *Penrose diagrams* to describe the causal and topological properties of gravitational spaces. These are diagrams that change scales, bringing infinity to a finite distance so they can be drawn, while keeping causal relations and topology unchanged. One makes coordinate transformations and drops conformal factors in the metric. The easiest to understand is for 2-dimensional Minkowski space, with metric

$$ds^2 = -dt^2 + dx^2, \tag{27.18}$$

where $-\infty < t, x < +\infty$. We transform first to lightcone coordinates

$$u_\pm = t \pm x \Rightarrow ds^2 = -du_+ du_-, \tag{27.19}$$

$-\infty < u_\pm < +\infty$ and then to new, finite extent lightcone coordinates $\tilde{u}_\pm$ defined by

$$u_\pm = \tan \tilde{u}_\pm \tag{27.20}$$

followed by going back to timelike and spacelike coordinates,

$$\tilde{u}_\pm = \frac{\tau \pm \theta}{2}. \tag{27.21}$$

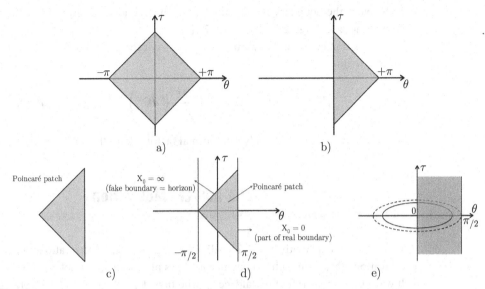

Penrose diagrams. (a) Penrose diagram of 2-dimensional Minkowski space. (b) Penrose diagram of 3-dimensional Minkowski space. (c) Penrose diagram of the Poincaré patch of Anti–de Sitter space. (d) Penrose diagram of global AdS_2 (2-dimensional Anti–de Sitter), with the Poincaré patch emphasized; $x_0 = 0$ is part of the boundary, but $x_0 = \infty$ is a fake boundary (horizon). (e) Penrose diagram of global AdS_d for $d \geq 2$. It is half the Penrose diagram of AdS_2 rotated around the $\theta = 0$ axis.

Finally, we obtain the metric

$$ds^2 = \frac{1}{4 \cos^2 \tilde{u}_+ \cos^2 \tilde{u}_-}(-d\tau^2 + d\theta^2), \qquad (27.22)$$

which after dropping the conformal factor becomes again a flat space metric, just with $|\tilde{u}_\pm| \leq \pi/2$, so $|\tau \pm \theta| \leq \pi$. This is a diamond region, as in Figure 27.1(a). For a *4-dimensional* Minkowski space, the initial metric is the same (after dropping the $r^2 d\Omega^2$ term), $ds^2 = -dt^2 + dr^2$, but with $r \geq 0$. This translates into $\theta \geq 0$ in the final form, leading to a triangle region instead of a diamond, as in Figure 27.1(b). That is the Penrose diagram of 4-dimensional flat space.

The Penrose diagram of the Poincaré patch of AdS_d space can be found in a similar manner, and one finds the same triangle region (times the sphere), since the Poincaré metric is conformal to flat space, as seen from (27.15), as in Figure 27.1(c). There are a priori two (non-lightlike) potential boundaries, $x_0 = 0$ and $x_0 = +\infty$. The second is a fake boundary, as one can continue the AdS space past it, but $x_0 = 0$ is part of the real boundary. As we see from (27.15), the metric at this boundary is (a diverging constant times) the metric of $d-1$–dimensional Minkowski space.

The Penrose diagram of the whole AdS space can be found from the metric in global coordinates, and one finds a cylinder, i.e. a strip times a sphere (rotating the strip around one of its sides), and the Poincaré patch is a triangle embedded into the strip. For this full

AdS space, the boundary is the boundary of the cylinder, i.e. $\mathbb{R}_t \times S^{d-2}$, the vertical line being time t; see Figure 27.1(d) and 27.1(e).

The relation between the two boundaries is a conformal transformation of the boundary. Indeed, writing

$$ds^2 = dr^2 + r^2 d\Omega = r^2 \left(\frac{dr^2}{r^2} + d\Omega^2 \right) = e^{-2\tau}(d\tau^2 + d\Omega^2), \qquad (27.23)$$

where $e^{-\tau} = r$, we see that only a conformal factor relates the $\mathbb{R}^{d-1}$ and $\mathbb{R}_t \times S^{d-2}$ metrics.

27.3 AdS/CFT Motivation

We now turn to a "motivation" for AdS/CFT. It is not quite a derivation, rather a "heuristic derivation" that is enough to convince us of its plausibility, but not enough to be rigorous. It was proposed in 1997 by Maldacena, who thus defined the AdS/CFT correspondence.

The motivation rests on the very important observation, referenced before, made by Polchinski in 1995, that D-branes are the same as the extremal ($M = Q$) p-brane solutions of supergravity, an observation that started the "second superstring revolution." This equality, between the D-branes, defined abstractly as the endpoints of open strings, and extremal p-branes, gravitational (curved) space solutions of string theory, was found by comparing their masses and charges, which were found to match.

Thus D-branes curve space, and give a solution, in the case of D3-branes (solution of type IIB supergravity):

$$ds^2 = H^{-1/2}(r)d\vec{x}_{||}^2 + H^{1/2}(r)(dr^2 + r^2 d\Omega_5^2)$$
$$F_5 = (1 + *)(dt \wedge dx_1 \wedge dx_2 \wedge dx_3 \wedge (d\, H^{-1}))$$
$$H(r) = 1 + \frac{R^4}{r^4}$$
$$R = 4\pi g_s N \alpha'^2; \quad Q = g_s N. \qquad (27.24)$$

This solution has a horizon at $r = r_h = 0$, thus has no temperature. If one modifies the extremal ($M = Q$) solution by adding a small mass δM, one creates a small horizon at $r = r_h > 0$, thus giving temperature and Hawking radiation.

But we can understand this Hawking radiation process also from the point of view of the D3-branes (endpoints of open strings). The process corresponding to it is one of two open strings living on the D3-brane colliding and forming a closed string. But the closed string is not restricted to live on the D3-brane, and instead can move off into the bulk of the space, as Hawking radiation; see Figure 27.2.

This process then means that there must be a relation between the theory on the open strings on the D3-branes, which is $\mathcal{N} = 4$ SYM plus corrections, and the theory of the closed strings, i.e. supergravity plus corrections, in the space curved by the D3-brane. By matching the D3-brane and extremal p-brane points of view, we make this relation precise.

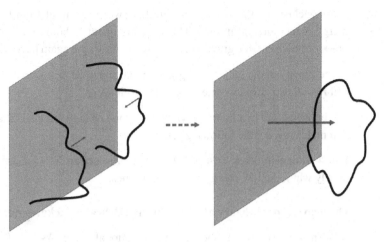

Fig. 27.2   Two open strings living on a D-brane collide and form a closed string that can then peel off and go away from the brane.

From the D3-brane point of view, the action for the interacting system can be written as

$$S = S_{\text{bulk}} + S_{\text{brane}} + S_{\text{interaction}}, \tag{27.25}$$

where the interaction part $S_{\text{interaction}}$ is proportional to the 10-dimensional Newton constant κ_N, which itself is proportional to g_s (since gravity is a closed string interaction, with coupling g_s), and also with α'^2 for dimensional reasons, since $2\kappa_N^2$ has dimension -8 in 10 spacetime dimensions. The result is that $S_{\text{interaction}} \propto \alpha'^2 \to 0$ as $\alpha' \to 0$. In the same $\alpha' \to 0$ limit, the bulk string action S_{bulk} becomes the supergravity action S_{sugra} and, since $\kappa_N \to 0$ also, it becomes *free* (super)gravity. On the other hand, as we saw, in this $\alpha' \to 0$ limit, the D3-brane becomes $\mathcal{N} = 4$ SYM. We note that $\alpha' \to 0$ is a low-energy limit: since α' has dimensions, one must really construct a dimensionless object that can go to zero, and this is $\alpha' E^2$.

All in all, in the D3-brane point of view, we have two low-energy systems in the $\alpha' \to 0$ limit:

- free gravity in the bulk of the gravitational space
- 4-dimensional $\mathcal{N} = 4$ SYM on the D3-brane.

Moving on to the extremal p-brane point of view, the energy E_p at a point p with radius r is related to the energy E as measured by an observer at infinity (which measures the time t) by

$$E_p \sim \frac{d}{d\tau} = \frac{1}{\sqrt{g_{00}}}\frac{d}{dt} \sim \frac{E}{\sqrt{g_{00}}} \Rightarrow$$
$$E \sim H^{-1/4}E_p \sim rE_p \to 0 \quad \text{for} \quad r \to 0. \tag{27.26}$$

Here we have used that as $r \to 0$, $H(r) \sim R^4/r^4 \propto r^{-4}$, and we kept the energy E_p, measured in the theory in the bulk, fixed.

In conclusion, at $r \to 0$, we are in the low-energy limit from the point of view of the energy E measured at infinity. The same argument as before gives that at large distances, $\delta r \to \infty$, we have free gravity in the bulk, therefore we again have two low-energy systems:

- free gravity in the bulk of the gravitational space (at $\delta r \to \infty$)
- at $r \to 0$, we have low-energy gravity excitations.

Identifying the two points of view, since one of the systems is the same, the other must also match, giving the following relation:

The 4-dimensional $\mathcal{N} = 4$ SYM with gauge group $SU(N)$, at large N (we will see shortly why), the low energy on the D3-branes,
=

The gravitational theory at $r \to 0$ in the D3-brane background, for $\alpha' \to 0$.

We now want to define better the gravitational space. As $r \to 0$, $H(r) \sim R^4/r^4$, so the metric is

$$ds^2 \simeq \frac{r^2}{R^2}(-dt^2 + d\vec{x}_3^2) + \frac{R^2}{r^2}dr^2 + R^2 d\Omega_5^2. \tag{27.27}$$

Making the change of variables $r/R = R/x_0$, we obtain the metric

$$ds^2 = R^2 \frac{-dt^2 + d\vec{x}_3^2 + dx_0^2}{x_0^2} + R^2 d\Omega_5^2, \tag{27.28}$$

which is the metric of $AdS_5^R \times S_R^5$ in Poincaré coordinates.

Since the metric of $AdS_5 \times S^5$ was derived in the $r \to 0$ limit of the D3-brane metric, to "undo" the limit, we must take $r \to \infty$ ($x_0 \to 0$) in $AdS_5 \times S^5$, which leads to the boundary of the space, as we saw. This is where the gauge theory lives, and for the Poincaré coordinates used here, the boundary metric is $\mathbb{R}^4$ (flat Minkowski space).

The relation is therefore a type of *holography*, where a $d + 1$-dimensional theory can be represented on a d-dimensional surface (the boundary), without losing information. From usual holography, we know that the only way this can happen is if the $d + 1$–dimensional information is encoded nonlocally onto d dimensions.

We can make the gravitational space even more precisely defined. The harmonic function behaves at $r \to 0$ as $H(r) \sim R^4/r^4 \sim \alpha'^2/r^4$, so

$$E = E_p H^{-1/4} \propto \frac{E_p r}{\sqrt{\alpha'}} = \frac{E_p \sqrt{\alpha'} r}{\alpha'}. \tag{27.29}$$

But we want to compare the gravity (string) theory with the gauge theory, so we must keep fixed the energy in both. That means that we keep the gauge theory energy E, as well as the string theory in string units, $E\sqrt{\alpha'}$, fixed. In turn that means that

$$U \equiv \frac{r}{\alpha'} \tag{27.30}$$

is fixed. Substituting $r = U\alpha'$ and $R^4 = \alpha'^2 4\pi g_s N$ in the $AdS_5 \times S^5$ metric, we obtain

$$ds^2 = \alpha' \left[\frac{U^2}{\sqrt{4\pi g_s N}}(-dt^2 + d\vec{x}_3^2) + \sqrt{4\pi g_s N} \left(\frac{dU^2}{U^2} + d\Omega_5^2 \right) \right]. \tag{27.31}$$

We see that, except for an overall α' factor, the rest of the metric is written in terms of finite quantities.

We can also relate g_s with g_{YM}^2. We have seen that the fact that the closed string interaction looks like an open string interaction squared means that $g_s = g_o^2$, and g_o is proportional to the coupling of the D3-brane theory, i.e. the YM coupling g_{YM}. One can find the exact relation as

$$4\pi g_s = g_{YM}^2. \tag{27.32}$$

27.4 AdS/CFT Definition and Limits

To be able to use the heuristic derivation above, we must be able to use the gravitational solution, which means that we should be able to neglect string corrections.

But in string theory there are two types of corrections. There are α' corrections, which are quantum worldsheet string corrections, and g_s corrections, which are quantum space-time string corrections.

The α' corrections are small if the curvatures are large with respect to $\sqrt{\alpha'}$, i.e. the corrections come in units of α'/R^2. Therefore the condition for small $\alpha'\mathcal{R}$ corrections is

$$R = \sqrt{\alpha'}(4\pi g_s N)^{1/4} \gg \sqrt{\alpha'} \Rightarrow g_s N \gg 1. \tag{27.33}$$

From the point of view of the gauge theory, that gives the fact that the *'t Hooft coupling* is large:

$$\lambda \equiv g_{YM}^2 N \gg 1. \tag{27.34}$$

As 't Hooft showed, in a $SU(N)$ gauge theory, the effective loop counting parameter, which defines perturbation theory, is not g_{YM}^2, but rather $g_{YM}^2 N$. Therefore perturbation theory would be $\lambda \ll 1$, but here we are in the *nonperturbative regime* $\lambda \gg 1$.

There are also quantum string corrections, which are proportional to g_s, so for these to be small, we need to have $g_s \ll 1$.

But together with $g_s N \ll 1$, it means that we need $N \to \infty$. The limit is then the one defined by 't Hooft, of $N \to \infty$, $g_{YM} \to 0$, with $\lambda = g_{YM}^2 N$ fixed, just that now it is fixed to a very large ($\gg 1$) number.

Then the map that we have heuristically derived is *between $\mathcal{N} = 4$ SYM in this 't Hooft limit at $\lambda \gg 1$, and string theory in the supergravity limit, in $AdS_5 \times S^5$*. But this map is a *duality*, like the S-duality from the last chapter, since when $\mathcal{N} = 4$ SYM is strongly coupled, string theory is weakly coupled, and vice versa. Indeed, we have seen that the α' corrections in the gravity (string) side are really

$$\frac{\alpha'}{R^2} = \frac{1}{\sqrt{g_{YM}^2 N}} = \frac{1}{\sqrt{\lambda}} \tag{27.35}$$

corrections, whereas the $\mathcal{N} = 4$ SYM corrections are corrections in λ. This allows us to use classical supergravity to say something about nonperturbative $\mathcal{N} = 4$ SYM, when normal methods cannot help us.

One could in principle imagine three scenarios:

- That only in the above limit we have matching between the two sides
- That α' corrections also match, but g_s corrections don't and
- That both α' and g_s corrections match.

It is in fact this last, strongest form of duality which that was found, through innumerable examples, to be correct. For any $g_{YM}^2 = 4\pi g_s$ and for any N, we expect the two sides to agree, based on the fact that all tests so far have been found to be correct.

27.5 State Map

We now want to define the AdS/CFT correspondence map explicitly, and we start with the map between states of the two theories. We consider gauge-invariant operators $\mathcal{O}$ in $\mathcal{N} = 4$ SYM, which are necessarily composite (since there are no gauge-invariant fields in the theory), and belong to some representation I_n of the global symmetry group $SO(6) = SU(4)$, which rotates the six real scalar among each other (and the four complex spinors). The operators must be gauge invariant, since there is no gauge group on the AdS side, hence we cannot match objects that have gauge indices.

To find the corresponding object in $AdS_5 \times S^5$, we consider the KK expansion of fields on S^5, of the general type

$$\phi(x, y) = \sum_n \sum_{I_n} \phi_{(n)}^{I_n}(x) Y_{(n)}^{I_n}(y), \tag{27.36}$$

where $Y_{(n)}^{I_n}$ are spherical harmonics on the S^5, in the representation I_n of the global symmetry $SO(6)$ of S^5, and n is the level, which measures the eigenvalue of $\square$ on S^5.

Then the relation is between a field $\phi_{(n)}^{I_n}$ in the representation I_n and of mass m, and an operator $\mathcal{O}_n^{I_n}$ in the same representation, with (conformal) dimension

$$\Delta = \frac{d}{2} + \sqrt{\frac{d^2}{4} + m^2 R^2}, \tag{27.37}$$

where d is the dimension of the CFT ($d = 4$ in our case). This precise formula is valid for scalars; for other fields it is modified.

The total bosonic symmetry group (there are also fermionic symmetries) of both theories is $SO(2, 4) \times SO(6)$, where $SO(2, 4)$ is the conformal group on the CFT side and the isometry group of AdS on the AdS side, and $SO(6)$ is the global, R-symmetry group on the CFT side, and the isometry group of S^5 on the AdS side. Note that the six real scalars Φ^i of $\mathcal{N} = 4$ SYM, which are rotated by $SO(6)$, correspond to the six coordinates X^i transverse to the D3-branes (on which we have $\mathcal{N} = 4$ SYM).

Important Examples

Other than scalar fields, two very important examples merit attention, since they will be much used in the following:

- A *global symmetry current* J_μ in the field theory for a symmetry G is mapped to a gauge field A_μ (for the same *local* symmetry G) in the gravity dual. Note that A_μ is massless, and the current J^μ has mass dimension $d - 1$. The gauging of a global symmetry involves adding the coupling $\int d^d x J^\mu A_\mu$ to the action, so this is a consistent choice, as we will see in the next section.

- The *energy momentum tensor* $T_{\mu\nu}$ can be thought of as an example of the above, in the case that the symmetry is translations, with generators P_ν. In the gravity dual, local translations are really general coordinate transformations, whose "gauge field" is the metric (graviton) $g_{\mu\nu}$, which is therefore the gravity dual field for the operator $T_{\mu\nu}$. Again this makes sense, since adding a coupling of matter to gravity is done through the term $\int d^d x \sqrt{-g} T_{\mu\nu} \delta g^{\mu\nu}$.

27.6 Witten (or GKPW) Construction

Next, we show the prescription, found by Witten, based on earlier work in momentum space by Gubser, Klebanov, and Polyakov (thus often called GKPW construction) of how to map observables on both sides, for the states we defined. Consider a massless scalar ϕ. By the above map, it would correspond to an operator with dimension $\Delta = d = 4$. But we can calculate that such a massless field will go to a constant (as a function of x_0; it can still depend on $\vec{x}$) at the boundary of AdS space, $\phi(\vec{x}, x_0) \to \phi_0(\vec{x})$. The natural interpretation for such a ϕ_0, defined on the 4-dimensional Minkowski boundary is as a source for the operator $\mathcal{O}$ corresponding to it.

We consider therefore the (Euclidean signature) partition function for the operator $\mathcal{O}$, with source ϕ_0:

$$Z_{\mathcal{O}}[\phi_0] = \int \mathcal{D}[SYM\ fields] e^{-S_{\mathcal{N}=4\ SYM} + \int d^4 x \mathcal{O}(x)\phi_0(x)}. \qquad (27.38)$$

The use of the partition function is, as usual, that it is the generating functional of correlators of $\mathcal{O}$, which can be found by taking derivatives at zero:

$$\langle \mathcal{O}(x_1) \ldots \mathcal{O}(x_n) \rangle = \frac{\delta^n}{\delta\phi_0(x_1) \ldots \delta\phi_0(x_n)} Z_{\mathcal{O}}[\phi_0] \Big|_{\phi_0 = 0}. \qquad (27.39)$$

The natural guess is then that this partition function can be calculated in the string side by equating with the string theory partition function on the AdS background, with the boundary source ϕ_0, $Z_{\text{String,AdS}}[\phi[\phi_0]]$, and this is the prescription of Witten:

$$Z_{\mathcal{O}}^{CFT}[\phi_0] = Z_{string}^{AdS}[\phi[\phi_0]]. \qquad (27.40)$$

Here ϕ is a string (supergravity) field in the bulk sourced (classically) by the boundary value ϕ_0, hence we write $\phi[\phi_0]$. At this moment, we have taken ϕ to stand for any generic string (supergravity) field, though in the case ϕ is not a scalar, we will have also Lorentz indices that will make the equations more complicated. But if we are in the $\alpha' \to 0, g_s \to 0$ limit, the string theory becomes classical supergravity, and as usual, the partition function becomes just $\mathcal{N}e^{-S_{\text{classical}}}$, since the path integral fluctuations just select the classical solution and give a constant normalization, thus

$$Z_{string}^{AdS}[\phi[\phi_0]] = e^{-S_{\text{supergravity}}[\phi[\phi_0]]}, \tag{27.41}$$

where we find the classical solution for ϕ as a function of its boundary value ϕ_0 and replace in the on-shell action.

In conclusion, Witten's prescription is

$$Z_{\mathcal{O}}^{CFT}[\phi_0] = \int \mathcal{D}[SYM\,fields]e^{-S_{N=4\,SYM}+\int d^4x\mathcal{O}(x)\phi_0(x)} = Z_{\text{class.sugra}}^{AdS}[\phi_0] = e^{-S_{class.sugra}[\phi[\phi_0]]}. \tag{27.42}$$

Note that this prescription is for Euclidean signature, thus it maps the two dual theories in their Wick rotated form. The issue of Wick rotation back to Minkowski space is very subtle. Some things about it will be said in later chapters.

27.7 Generalizations

This initial duality between $\mathcal{N} = 4$ SYM in 3+1 dimensions and $AdS_5 \times S^5$ has been generalized to many other cases. First, there are other maximally supersymmetric cases. Then there are also cases with less, or even no supersymmetry, and/or no conformal invariance. In these more general cases, we speak of a duality between a gauge theory and its *gravity dual*, so in these cases, at least if there is no conformal invariance, the technically correct term would be *gauge/gravity duality*, instead of AdS/CFT correspondence.

There are two important cases with maximal supersymmetry. One is between a 2+1–dimensional theory called "ABJM theory," after its discoverers, Aharony, Bergman, Jafferis, and Maldacena, and type IIA string theory in the $AdS_4 \times \mathbb{CP}^3$ background. Note that $\mathbb{CP}^3 = S^7/Z_k|_{k\to\infty}$, so this duality is also to a space of the type $AdS \times S$. The ABJM model and its dual will be treated in more detail in Chapter 42. The other example is between a peculiar 6-dimensional conformal field theory with (0,2) supersymmetry and M-theory on $AdS_7 \times S^4$ background. Note that the 6-dimensional theory is an interacting theory, for which there is no good perturbative description in terms of a Lagrangean. In fact, for some time it was believed that there were no nontrivial interacting theories in six-dimensions, until this example appeared in string theory. But the theory has string-like degrees of freedom, since the dual gravitational theory, M-theory, is believed to have fundamental membrane degrees of freedom (objects with spatial extension in two directions), and the one dimension corresponds to the holography direction r (or x_0).

General Properties

We now enumerate several general properties of gravity duals:

- The gauge theory lives at the boundary of the gravity dual, if such a place can be defined, or really nowhere, if not. Of course, in general there is also a compact space X_m, on top of the $d + 1$–dimensional holographic space.

- The energy scale is $U = r/\alpha'$. Indeed, it has dimensions of energy, and it can be shown that in the case of AdS_5 with a cut-off, it corresponds to energy. The same applies in a general situation: this holographic $d + 1$'th dimension is always understood as the energy of the field theory. Therefore making explicit the energy dependence of the field theory allows one to describe it holographically in $d + 1$ dimensions.
- The global symmetry of the field theory is the global symmetry of the extra-dimensional space X_m (which after KK reduction on X_m turns into a local – gauge – symmetry of the $d + 1$–dimensional space).
- Supergravity fields in the gravity dual correspond to SYM operators made up from adjoint fields, i.e. "glueball"-like operators.
- Quarks in the gauge theory, i.e. fields in the fundamental representation of the gauge group and in some representation R of the global symmetry group G, forming R-charged composite operators, are mapped, *in the gravity dual*, to SYM fields for the group G (note that supergravity can be coupled to SYM fields, so there is nothing wrong with having SYM fields in the gravity dual).
- The mass spectrum of the tower of glueballs associated with a certain operator $\mathcal{O}$ in the gauge theory is mapped to the mass spectrum of the wave equation for the supergravity field that couples to $\mathcal{O}$ in the gravity dual. This spectrum is discrete if the gauge theory has a mass gap, which can be understood by the fact that light travels a finite time in the gravity dual between the endpoints, i.e. it acts like a quantum mechanical box, with discrete spectrum.
- Wavefunctions $e^{ik\cdot x}$ of the states in the gauge theory are mapped to wavefunctions $\Phi = e^{ik\cdot x}\Psi(U, X_m)$ of the gravity dual, where U is the holographic coordinate, and X_m are coordinates on the compact space, and Ψ is a wavefunction for these extra coordinates.

Important Concepts to Remember

- Conformal invariance in $D \neq 2$ is invariance under transformations of the coordinates of flat space that multiply the flat space metric with a conformal factor, and give the group $SO(2, D)$.
- de Sitter is the Minkowski signature analog of the sphere, and Anti–de Sitter is the Minkowski signature analog of Lobachevski space.
- AdS space has Poincaré coordinates that cover only the Poincaré patch, and global coordinates that cover the whole space.
- The Penrose diagram of AdS space is a cylinder, with boundary $\mathbb{R}_t \times S^{d-2}$, and of the Poincaré patch is a triangle (of revolution) inside the cylinder, with boundary $\mathbb{R}^4$, conformal to $\mathbb{R}_t \times S^{d-2}$.
- AdS/CFT is obtained by comparing the two points of view for D3-branes: as endpoints of open strings, and as extremal p-brane solutions of string theory.
- The usual AdS/CFT is the relation between $3+1$–dimensional $\mathcal{N} = 4$ SYM with gauge group $SU(N)$, at large N, and string theory in $AdS_5 \times S^5$ background, where the gauge theory lives at the boundary of space. It is a holographic duality (strong/weak coupling).

- The heuristic derivation is for the limit $\alpha' \to 0$, $g_s \to 0$, which maps to the 't Hooft limit $N \to \infty$, $g_{YM}^2 \to 0$, with $\lambda = g_{YM}^2 N$ fixed *and large* ($\gg 1$), but the duality is believed to be valid at all g_s and N.
- The fields in the KK expansion on S^5 of 10-dimensional fields, with representation I_n on S^5, are mapped to gauge invariant operators $\mathcal{O}^{I_n}$ of $\mathcal{N} = 4$ SYM, in the same representation, in the gravity dual, with dimension related to the mass of the field.
- The Euclidean partition function for the operator $\mathcal{O}$ in CFT with source ϕ_0, $Z_{\mathcal{O}}^{CFT}[\phi_0]$, equals the string partition function in AdS_5 for fields with boundary value ϕ_0, i.e. in the $\alpha' \to 0$, $g_s \to 0$ limit, $e^{-S_{\mathrm{sugra}}[\phi[\phi_0]]}$.
- General pairs of gauge theories and a corresponding gravity dual in $d + 1$ dimensions, with some compact space, and the gauge theory living at its boundary, give a gauge/gravity duality.
- The energy scale is $U = r/\alpha'$, so holography is the statement that adding the energy in the description, we obtain a $d + 1$–dimensional gravitational space.
- The global symmetry of the gauge theory is the global symmetry of the extra dimensional space.
- Supergravity fields in the gravity dual couple to glueball operators, whereas SYM vector fields for a group G couple to G-charged operators, made up of "quarks" in the fundamental representation of the gauge group and a representation of the global group G ("G-charged").
- Mass spectra of glueballs are mapped to mass spectra of fields in the gravity dual, and wavefunctions in the gauge theory are mapped to wavefunctions (with a wavefunction for the extra dimension) in the gravity dual.

Further Reading

For more details, see chapters 10 and 11 in [8], as well as [46]. A useful very early review is [47]. Maldacena's original paper is [48]. The Witten prescription was described in [49], based on the earlier work in momentum space by Gubser, Klebanov, and Polyakov, in [50].

Exercises

(1) Derive the conformal algebra in terms of P_μ, $J_{\mu\nu}$, K_μ, D from the $SO(d, 2)$ algebra, with the definitions in the text.

(2) Prove that the coordinate transformation

$$X_0 = R \cosh \rho \cos \theta$$
$$X_i = R \sinh \rho \, \Omega_i$$
$$X_{d+1} = R \cosh \rho \sin \tau \qquad (27.43)$$

takes one between the global metric in AdS and the embedding in the flat space of signature $(2, d - 1)$.

(3) Check that the $r \to 0$, $\alpha' \to 0$ limit in the text takes the Dp-brane metric to $AdS_{p+2} \times S^{8-p}$ only for $p = 3$.

(4) Consider the KG equation $(\Box - m^2)\phi = 0$ in the Poincaré coordinates for AdS_{d+1}. Show that near the boundary $x_0 = 0$, the two independent solutions go like $x_0^{2h_\pm}$, where

$$2h_\pm = \frac{d}{2} \pm \sqrt{\frac{d^2}{4} + m^2 R^2}. \tag{27.44}$$

(5) Consider the solution for N extremal M5-branes, with metric

$$ds^2 = f_5^{-1/3}(r)(-dt^2 + d\vec{x}_5^2) + f_5^{2/3}(r)(dr^2 + r^2 d\Omega_4^2), \, f_5(r) = 1 + \frac{\pi N l_P^3}{r^3}, \tag{27.45}$$

and the 4-form

$$F_4 = *(dt \wedge dx^1 \wedge \ldots \wedge dx^5 \wedge d(G^{-1})). \tag{27.46}$$

Show that the limit $r \to 0$, $l_P \to 0$, with $U^2 \equiv r/l_P^3$ fixed leads to the $AdS_7 \times S^4$ metric with a constant and quantized F_4, and calculate the radii R_{S^4} and R_{AdS_7}.

(6) Show that for AdS space in global coordinates, it takes a finite time to reach the boundary from an arbitrary point, but an infinite time to reach the center $\rho = 0$.

(7) Show that, if the classical solution for a scalar in AdS in Poincaré coordinates is written as

$$\phi_i(z_0, \vec{z}) = \int d^d z K_{B,\Delta_i}(z_0, \vec{z}; \vec{x}) \phi_{0i}(\vec{x}), \tag{27.47}$$

as a function of the boundary value $\phi_{0i}(\vec{x})$, using the Witten prescription, the 3-point function of scalars (coming from a supergravity action with interaction term $\mathcal{L}_{\text{int}} = \lambda \phi_1 \phi_2 \phi_3$) can be written as

$$\langle \mathcal{O}(x_1) \mathcal{O}(x_2) \mathcal{O}(x_3) \rangle = -\lambda \int \frac{d^d z dz_0}{z_0^{d+1}} K_{B,\Delta_1}(z_0, \vec{z}; \vec{x}) K_{B,\Delta_2}(z_0, \vec{z}; \vec{x}_2) K_{B,\Delta_3}(z_0, \vec{z}; \vec{x}_3). \tag{27.48}$$

PART III

APPLYING STRING THEORY TO CONDENSED MATTER PROBLEMS

28 The pp Wave Correspondence: String Hamiltonian from $\mathcal{N} = 4$ SYM

The first application of string theory to condensed matter that we consider relates to the appearance of spin chains. In this chapter we show a first step toward that, obtaining the "pp wave correspondence," that leads to a discrete ("chain") Hamiltonian from $\mathcal{N} = 4$ SYM, with a particular gravitational dual, of the "pp wave" type. In the next chapter we will generalize this construction to spin chains.

28.1 PP Waves

PP waves, or "parallel-plane" waves, are plane-fronted gravitational waves. That is, they are solutions to the Einstein equations for perturbations moving at the speed of light, with a plave wave front. They can be defined in a general curved background, but for the purposes of this chapter, it suffices to consider the case of flat background, in which case the pp wave metric is

$$ds^2 = 2dx^+dx^- + (dx^+)^2 H(x^+, x^i) + \sum_i dx_i^2. \tag{28.1}$$

As we can see, it is a perturbation of flat space in lightcone coordinates, moving on the $x^+ = 0$ lightcone, and defined by the function $H(x^+, x^i)$.

PP waves are the only (nontrivial) cases in general relativity when the linearized solution is actually exact, and they obey a superposition principle. Otherwise, general relativity is highly nonlinear, as seen from the Einstein equations written in terms of $g_{\mu\nu}$, and solutions don't obey a superposition principle.

As one might expect, the only nonzero component for the Ricci tensor of the pp wave is R_{++}. It should be proportional to g_{++}, the only nontrivial component of the metric. By dimensional analysis, and since $R_{\mu\nu}$ contains two derivatives, it should be proportional to ∂_i^2, the only possible general coordinate invariant, since $\partial_+\partial_i$ or $\partial_+\partial_+$ are not invariant. Note that the metric inverse in the $(+-)$ space is

$$\begin{pmatrix} 0 & 1 \\ 1 & H \end{pmatrix}^{-1} = \begin{pmatrix} -H & 1 \\ 1 & 0 \end{pmatrix}, \tag{28.2}$$

so $g^{++} = 0$. The proper coefficient is obtained from a calculation, as

$$R_{++} = -\frac{1}{2}\partial_i^2 H(x^+, x^i). \tag{28.3}$$

Since $g^{++} = 0$, it follows that $R = 0$ as well.

11-Dimensional Supergravity pp Waves

In 11-dimensional supergravity, the pp wave metric ansatz is the same as above, just that the solution is completed by a nonzero 4-form field strength:

$$F_{(4)} = dx^+ \wedge \phi_{(3)} : \quad F_{(4)+\mu_1\mu_2\mu_3} = \phi_{(3)\mu_1\mu_2\mu_3}. \tag{28.4}$$

From the "Maxwell" equations for $F_{(4)}$ (the equation $dF_{(4)} = 0$, implied by $F = dA$, and the field equation $d * F_{(4)} = 0$), we have

$$d\phi_{(3)} = 0 : \quad \partial_{[\mu_1}\phi_{(3)\mu_2\mu_3\mu_4]} = 0$$

$$d * \phi_{(3)} = 0 : \quad \partial_{[\mu_1}\epsilon_{\mu_2\dots\mu_8]}{}^{\mu_9\mu_{10}\mu_{11}} = 0 \Leftrightarrow \partial^{\mu_1}\phi_{(3)\mu_1\mu_2\mu_3} = 0. \tag{28.5}$$

On the other hand, from Einstein's equations, now reducing to $R_{++} = 8\pi G_N T_{++}$, we obtain

$$-\frac{1}{2}\partial_i^2 H = \frac{2}{4!}\phi_{\mu\nu\rho}\phi^{\mu\nu\rho} \Rightarrow -\partial_i^2 H = |\phi|^2, \tag{28.6}$$

with the usual notation for the contraction $|\phi|^2$.

One solution to the above equations of motion is in the form of an H expressed as a quadratic form,

$$H = \sum_{ij} A_{ij} x^i x^j, \tag{28.7}$$

where

$$\text{Tr}\, A = -\frac{1}{2}|\phi|^2. \tag{28.8}$$

One can find solutions that preserve (some) supersymmetry as well. In general, since $\delta bose = fermi\,\epsilon$ and $\delta fermi = bose\,\epsilon$, and on a fixed background, the fermions cannot have VEVs, since they would break Lorentz invariance, it follows that we get a trivial relation $\delta bose = 0$ and a nontrivial one, $0 = \delta fermi = bose\,\epsilon$. One can solve this relation to find the bosonic fields as a function of position, as well as (in general) to restrict ϵ in some way, correspondingly to a subset of supersymmetry being preserved.

However, in the case of pp waves any solution preserves at least 1/2 of the supersymmetry, and it was found in 1984 by Kowalski-Glikman that there is even a solution that preserves the full (maximal) supersymmetry, with

$$A_{ij} x^i x^j = -\sum_{i=1,2,3} \frac{\mu^2}{9}x_i^2 - \sum_{i=4}^{9} \frac{\mu^2}{36}x_i^2$$

$$\phi_{(3)} = \mu\, dx^1 \wedge dx^2 \wedge dx^3. \tag{28.9}$$

That is highly nontrivial, since there are very few fully supersymmetric backgrounds. In fact, after the fully supersymmetric pp wave was found, a theorem was proven, that the only maximally supersymmetric solutions of 11-dimensional supergravity are 11-dimensional Minkowski space, $AdS_4 \times S^7$, $AdS_7 \times S^4$, and the maximally supersymmetric pp wave.

Then, in 1990, Horowitz and Steif proved that for pp waves, there are no (on-shell) corrections to the α' corrections to the equations of motion. That is so, since, for instance, $R = g^{++}R_{++} = 0$, and it is true of all possible relevant contractions, of R^{n+1} type, that would appear as α'^n corrections. That in turn means that pp waves are exact string solutions, and therefore are highly special.

10-Dimensional Supergravity pp Waves

For AdS/CFT, the relevant case is 10-dimensional type IIB supergravity, with its $AdS_5 \times S^5$ solution.

In the type IIB supergravity, there exist also pp wave solutions, of general type with harmonic function

$$H = \sum_{ij} A_{ij} x^i x^j, \tag{28.10}$$

constant dilaton $\phi = \phi_0$, and self-dual 5-form field strength

$$F^+_{(5)} = dx^+ \wedge (\omega + *\omega) : \quad F^+_{(5)+\mu_1...\mu_4} = \omega_{\mu_1...\mu_4}, \quad F^+_{(5)+\mu_5...\mu_8} = \omega_{\mu_5...\mu_8}. \tag{28.11}$$

As before, the Maxwell equations for $F^+_{(5)}$, $dF^+_{(5)} = 0 = d * F^+_{(5)}$ imply that

$$d\omega = d * \omega = 0 : \quad \partial_{[\mu_1} \omega_{\mu_2...\mu_5]} = \partial^{\mu_1} \omega_{\mu_1...\mu_4} = 0, \tag{28.12}$$

whereas the Einstein equations imply that

$$-\frac{1}{2} \partial_i^2 H = \frac{1}{48} \omega_{\mu_1...\mu_4} \omega^{\mu_1...\mu_4} \Rightarrow -\partial_i^2 H = |\omega|^2. \tag{28.13}$$

But there exists also a maximally supersymmetric solution, with

$$H = \frac{\mu^2}{64} \sum_i x_i^2, \quad \omega_{(4)} = \frac{\mu}{2} dx^1 \wedge dx^2 \wedge dx^3 \wedge dx^4. \tag{28.14}$$

As in the 11-dimensional case, the only maximally supersymmetric solutions are 10-dimensional Minkowski space, $AdS_5 \times S^5$, and the maximally supersymmetric pp wave. It is then perhaps not surprising that the pp wave is a certain limit of $AdS_5 \times S^5$.

28.2 The Penrose Limit in Gravity

Indeed, the concept of pp wave features prominently in a certain limit of gravitational spacetimes, defined by Penrose. Penrose proved a theorem, saying that: Near a null geodesic (the path of a light ray) in any metric, the spacetime becomes a pp wave.

Formally, we express this as follows. In the neighborhood of a null geodesic defined by $V = Y^i = 0$, $U = \tau$, we can always put the metric in the form (as Penrose showed)

$$ds^2 = dV \left(dU + \alpha dV + \sum_i \beta_i dY^i \right) + \sum_{ij} C_{ij} dY^i dY^j, \tag{28.15}$$

where, as we see, U and V are lightcone coordinates, and Y^i are transverse coordinates. We can then rescale the coordinates as

$$U = u; \quad V = \frac{v}{R^2}; \quad Y^i = \frac{y^i}{R} \tag{28.16}$$

and take the limit $R \to \infty$, which takes us close to the $V = 0$ null geodesic. By taking the limit, we obtain a pp wave metric in the u, v, y^i coordinates. More precisely, the metric is

$$ds^2 = 2dudv + g_{ij}(u)dy^idy^j, \tag{28.17}$$

where $g_{ij}(u)$ is obtained from $C_{ij}(U, V, Y^i)$ by putting $V = Y^i = 0$ and $U = u$, which is the pp wave in *Rosen coordinates*. But one can make a transformation to the standard form in *Brinkmann coordinates*,

$$ds^2 = 2dx^+dx^- + H(x^+, x^a)(dx^+)^2 + d\vec{x}^2, \tag{28.18}$$

where

$$H(x^+, x^a) = A_{ab}(x^+)x^ax^b. \tag{28.19}$$

To obtain this coordinate transformation, we first write the metric $g_{ij}(u)$ in terms of vielbeins e_i^a as

$$g_{ij}(u) = e_i^a(u)e_j^b(u)\delta_{ab}, \tag{28.20}$$

define the inverse vielbein e_a^i, and the vielbeins are put (using the local Lorentz transformations acting on it) into a form satisfying

$$\dot{e}_{ai}(u)e_b^i(u) = \dot{e}_{bi}(u)e_a^i(u). \tag{28.21}$$

Here the dot refers to d/du. Then we find

$$g_{ij}dy^idy^j = \left(dx^a - \dot{e}_i^a e_c^i x^c du\right)\left(dx^b - \dot{e}_j^b e_d^j x^d du\right)\delta_{ab}, \tag{28.22}$$

so the transformation between the Rosen and Brinkmann coordinates is

$$u = x^+$$
$$v = x^- + \frac{1}{2}\dot{e}_{ai}e_b^i x^a x^b$$
$$y^i = e_a^i x^a, \tag{28.23}$$

and we have

$$A_{ab} = \ddot{e}_{ai}e_b^i. \tag{28.24}$$

The interpretation of the Penrose limit procedure is the following. Consider a boost along a direction called x, while taking the overall scale of the metric to infinity. The boost is

$$t' = \cosh\beta\, t + \sinh\beta\, x; \quad x' = \sinh\beta\, t + \cosh\beta\, x, \tag{28.25}$$

leading to

$$x' - t' = e^{-\beta}(x - t); \quad x' + t' = e^{+\beta}(x + t). \tag{28.26}$$

Considering the boost with $e^\beta = R$, where $R \to \infty$ is the overall scale of the metric, and scaling all the coordinates, (t, x, y^i), by $1/R$, we obtain (28.16).

We now prove that the Penrose limit of $AdS^5 \times S^5$ for a geodesic on S^5 is the maximally supersymmetric pp wave. Consider a null geodesic rotating around an equator of S^5 and staying in the center of AdS_5 in global coordinates. The metric of $AdS_5 \times S^5$ is

$$ds^2 = R^2\left(-\cosh^2 \rho d\tau^2 + d\rho^2 + \sinh^2 \rho d\Omega_3^2\right) + R^2\left(\cos^2 \theta d\psi^2 + d\theta^2 + \sin^2 \theta d\Omega_3^{2}\right).$$

$$(28.27)$$

The equator of S^5 is defined by $\theta = 0$ and is parametrized by ψ, and the center of AdS_5 is $\rho = 0$, with time coordinate τ. Then to be near the null geodesic, we consider small θ and ρ, leading to the metric

$$ds^2 \simeq R^2\left[-(1 + \rho^2)d\tau^2 + d\rho^2 + \rho^2 d\Omega_3^2\right] + R^2\left[(1 - \theta^2)d\psi^2 + d\theta^2 + \theta^2 d\Omega_3^{2}\right].$$

$$(28.28)$$

For the Penrose limit, define the lightcone coordinates

$$\tilde{x}^{\pm} = \frac{\tau \pm \psi}{\sqrt{2}},$$

$$(28.29)$$

and rescale

$$\tilde{x}^+ = x^+; \quad \tilde{x}^- = \frac{x^-}{R^2}; \quad \rho = \frac{r}{R}; \quad \theta = \frac{y}{R}.$$

$$(28.30)$$

Then, since we have

$$R^2(d\psi^2 - d\tau^2) = -2R^2 d\tilde{x}^+ d\tilde{x}^- = 2dx^+ dx^-$$

$$R^2 d\tau^2 \simeq \frac{(dx^+)^2}{2}$$

$$R^2 d\psi^2 \simeq \frac{(dx^+)^2}{2},$$

$$(28.31)$$

we obtain the pp wave metric

$$ds^2 = -2dx^+ dx^- - \mu^2(r^2 + y^2)(dx^+)^2 + d\vec{y}^2 + d\vec{r}^2,$$

$$(28.32)$$

after rescaling $x^+ \to \sqrt{2}\mu x^+$, $x^- \to x^-/\mu\sqrt{2}$. This is the maximally supersymmetric pp wave of type IIB string theory.

28.3 The Penrose Limit of AdS/CFT

We have seen that AdS/CFT relates $\mathcal{N} = 4$ SYM to string theory in $AdS_5 \times S^5$, and that the Penrose limit of $AdS_5 \times S^5$ is the maximally supersymmetric pp wave. It is therefore natural to ask what is the corresponding limit on the $\mathcal{N} = 4$ SYM side. This was defined in the paper by David Berenstein, Juan Maldacena, and myself (BMN), in 2002, and it became known as the *BMN limit*.

The energy in AdS_5 is $E = i\partial/\partial\tau$, i.e. the generator of time translations in τ, the global time coordinate. Also, the angular momentum J (Noether charge of rotations) for rotations in the plane of coordinates X^5, X^6 transverse to the D3-branes, with angle ψ between them, is $J = -i\partial_\psi$. In the $AdS_5 \times S^5$ limit, the angle ψ is the angle along the equator of S^5 defined previously.

On the other hand, the AdS/CFT map relates the energy E with the conformal dimension Δ of an operator in $\mathcal{N} = 4$ SYM. We will not explain this map further, but it can be found rigorously. On the other hand, the angular momentum J for rotating X^5 with X^6, corresponding to scalars Φ^5 and Φ^6 in $\mathcal{N} = 4$ SYM, is identified with the $U(1) \subset SO(6) = SU(4)$ charge that rotates these scalars.

Then the momenta $p^\pm$ on the pp wave are related through the coordinate redefinitions as

$$p^- = -p_+ = i\frac{\partial}{\partial x^+} = i\frac{\partial}{\partial \tilde{x}^+} = \frac{i}{\sqrt{2}}(\partial_\tau + \partial_\psi) = \frac{\Delta - J}{\sqrt{2}}$$

$$p^+ = -p_- = i\frac{\partial}{\partial x^-} = -i\frac{\partial}{R^2 \partial \tilde{x}^-} = \frac{i}{\sqrt{2}R^2}(\partial_\tau - \partial_\psi) = \frac{\Delta + J}{\sqrt{2}R^2}. \tag{28.33}$$

Finally, we introduce the same rescaling we introduced on the coordinates, $p^- \to \mu\sqrt{2}p^-$, $p^+ \to p^+/\mu\sqrt{2}$, leading to

$$\frac{p^-}{\mu} = \Delta - J; \quad 2\mu p^+ = \frac{\Delta + J}{R^2}. \tag{28.34}$$

But to define the string theory on the pp wave, we need to keep $p^\pm$ fixed as we take $R \to \infty$, which means to keep $\Delta - J$ fixed and $(\Delta + J)/R^2$ fixed, so $\Delta \simeq J \sim R^2 \to \infty$. Therefore we have obtained that the *Penrose limit is a large R-charge limit in the gauge theory.*

Note that in $\mathcal{N} = 4$ SYM we have a BPS condition (bound) $\Delta \geq |J|$, similar to the BPS condition $M \geq |Q|$ for black holes and black p-branes. We have already mentioned that in supersymmetric theories such a bound arises from the supersymmetry algebra, so it is not surprising that we obtain it again here. In the string theory on the pp wave, this BPS bound translates into the statement that $p^\pm \geq 0$.

To define better the large R charge limit, we note that the R^2, identified with the radius of $AdS_5 \times S^5$ (the overall scale of the metric), is written in gauge theory variables as

$$\frac{R^2}{\alpha'} = \sqrt{4\pi g_s N} = \sqrt{g_{YM}^2 N}, \tag{28.35}$$

so for g_s fixed, we have $J \sim R^2 \sim \sqrt{N}$, and therefore

$$\frac{J}{\sqrt{N}} = \text{fixed} \tag{28.36}$$

defines the large R-charge limit.

28.4 The String Hamiltonian on the pp Wave

Our aim is to identify quantum states and a Hamiltonian acting on them on the two sides of the correspondence. We start with the string side. The bosonic string action in the pp wave is

$$S = -\frac{1}{2\pi\alpha'} \int_0^l d\sigma \int d\tau \frac{1}{2}\sqrt{-\gamma}\gamma^{ab}\left[-2\partial_a x^+\partial_b x^- - \mu^2 x_i^2 \partial_a x^+\partial_b x^+ + \partial_a x^i\partial_b x^i\right],$$

(28.37)

where $x^i = (\vec{r}, \vec{y})$. We use the conformal gauge $\sqrt{-\gamma}\gamma^{ab} = \eta^{ab}$ and the lightcone gauge $x^+ = (x_0^+ +)p^+\tau$. But we want rather to write it as $x^+(\tau, \sigma) = \tau$, which means to rescale $\tau \to \tau/p^+$, under which the wave equation

$$\frac{\partial^2}{\partial\tau^2}x^i = \frac{\partial^2}{\partial\sigma^2}x^i$$

(28.38)

becomes

$$\frac{\partial}{\partial\tau^2}x^i = \frac{1}{(p^+)^2}\frac{\partial^2}{\partial\sigma^2}x^i.$$

(28.39)

But also we want to change from $l = 2\pi$ to an arbitrary l, in units of α', so the wave equation becomes

$$\frac{\partial^2}{\partial\tau^2}x_i = c^2\frac{\partial^2}{\partial\sigma^2}x_i,$$

(28.40)

where

$$c = \frac{l}{2\pi\alpha'p^+}.$$

(28.41)

Since we still want to use $c = 1$, we get

$$l = 2\pi\alpha'p^+.$$

(28.42)

In the new lightcone gauge, $x^+ = \tau$, we have

$$\eta^{ab}\partial_a x^+\partial_b x^- = 0; \quad \eta^{ab}\partial_a x^+\partial_b x^- = -1,$$

(28.43)

so the bosonic string action becomes

$$S = -\frac{1}{2\pi\alpha'}\int d\tau \int_0^{2\pi\alpha'p^+} d\sigma\left[\frac{1}{2}\partial_a x^i\partial^a x^i + \frac{\mu^2}{2}x_i^2\right].$$

(28.44)

The only difference with respect to the action in flat space is the mass of the fields, the transverse coordinates x^i. The equation of motion is now

$$(-\partial_\tau^2 + \partial_\sigma^2)x^i - \mu^2 x^i = 0.$$

(28.45)

A free wave mode

$$x^i \propto e^{-i\omega_n\tau + ik_n\sigma}$$

(28.46)

has momentum k_n defined by the Fourier expansion of the circle (closed string) of length $l = 2\pi\alpha'p^+$ after our rescalings, to be (both on the pp wave and in flat space)

$$k_n = \frac{n}{\alpha'p^+}.$$

(28.47)

The difference is that now imposing the equation of motion leads to

$$\omega_n = \sqrt{\mu^2 + \frac{n^2}{(\alpha' p^+)^2}}, \tag{28.48}$$

instead of $\omega_n = k_n$ as in flat space.

For an open string in flat space, we have already calculated the open string lightcone Hamiltonian:

$$H_{l.c.} = p^- = -p_+ = \frac{p^i p^i - 1/\alpha'}{p^+} + \frac{1}{2} \sum_{n \geq 1} \omega_n N_n. \tag{28.49}$$

But on the pp wave the x^i's are massive, and therefore are confined near $x = 0$ and have no zero modes p^i, but simply modes of $n = 0$ (that now are massive also). We are interested in the closed string, for which we have both left-movers and right-movers, and we put them together in $\sum_{n \in \mathbb{Z}}$, $n > 0$ being left-movers, $n < 0$ right-movers, and $n = 0$ added. The closed string Hamiltonian is therefore

$$H_{l.c.} = \sum_{n \in \mathbb{Z}} \omega_n N_n. \tag{28.50}$$

Translating into SYM variables, with

$$\frac{E}{\mu} = \Delta - J; \quad 2\mu p^+ = \frac{\Delta + J}{R^2} \simeq \frac{2J}{R^2} = \frac{2J}{\alpha' \sqrt{g_{YM}^2 N}}, \tag{28.51}$$

we obtain the relation

$$(\Delta - J)_n = \frac{\omega_n}{\mu} = \sqrt{1 + \frac{g_{YM}^2 N}{J^2} n^2}. \tag{28.52}$$

The limit in which this is valid is $\mu \alpha' p^+ = $ fixed, so

$$\frac{g_{YM}^2 N}{J^2} = \text{fixed}. \tag{28.53}$$

28.5 Bosonic SYM Operators for String Theory Modes

We want to construct bosonic SYM operators dual to states of the string on the pp wave. For that, we analyze the properties of the fields in SYM, to use them to construct the operators.

The R-charge J rotates the X^5, X^6 spacetime coordinates, corresponding to the Φ^5, Φ^6 fields, so we can define an object with $U(1)$ R-charge $J = +1$ (by definition), the complex field

$$Z = \Phi^5 + i\Phi^6, \tag{28.54}$$

that will transform under the $e^{i\alpha}$ $U(1)$ gauge transformation as $Z \to e^{i\alpha} Z$. Then $\bar{Z}$ will have R-charge $J = -1$, and the rest of the scalars, Φ^m, $m = 1, \ldots, 4$ have R-charge $J = 0$.

It is also useful to consider the derivative of Z, but since we are interested in building gauge-invariant operators, we need objects that transform covariantly under the gauge group $SU(N)$, so we consider instead the covariant derivative

$$D_\mu Z = \partial_\mu Z + [A_\mu, Z]. \qquad (28.55)$$

Then, in order to consider operators dual to states of given energy $E = \Delta - J$, we classify the fields according to their $\Delta - J$ values. The minimum value is obtained for Z, which has dimension $\Delta = 1$ and $J = 1$, so $\Delta - J = 0$. It is the unique field with this minimal value. Next, for $\Delta - J = 1$, we have the scalars Φ^m, with $\Delta = 1$ and $J = 0$, and the covariant derivatives $D_\mu Z$, with dimension $\Delta = 2$ and R-charge $J = 1$. There are no other covariant objects with $\Delta - J = 1$ that we can build out of the bosonic fields Z, Φ^m, and A_μ. Finally, $\bar{Z}$ has $\Delta - J = 2$, since $\Delta = 1$ but $J = -1$.

We then note that the objects of unit $\Delta - J$ are the eight fields $(D_\mu Z, \Phi^m)$, which therefore correspond to the eight transverse string oscillators at level zero on the pp wave, $a_0^{\dagger i}$. To construct the operator dual to a vacuum state of momentum p^+, $|0; p^+\rangle$, of $E = \Delta - J = 0$ and J units of p^+, we must consider an object made up of J Z's. Since $\mathcal{N} = 4$ SYM is a theory with only adjoint fields, the only way to make a gauge-invariant object out of them is to take the trace. Normalizing the operator, the vacuum state corresponds to the operator

$$|0; p^+\rangle = \frac{1}{\sqrt{J}N^{J/2}} \mathrm{Tr}[Z^J]. \qquad (28.56)$$

Note that in any conformal field theory, there is a well-defined "operator-state correspondence," a one-to-one map between operators and states, so putting the equality sign between a state and an operator is not a nonsensical statement.

Since we argued that $a_0^{\dagger i}$ correspond to $(D_\mu Z, \Phi^m)$, excited states made up of $a_0^{\dagger i}$ are obtained by inserting the corresponding fields (consider Φ^m, for example) inside the trace. But since there is no reason to favor one position over another, we must sum over all the positions inside the trace. If we have a single excitation, however, that would be redundant because of the cyclicity of the trace, but if we consider two excitations, it becomes relevant. Then the state with two level zero excitations is

$$a_0^{\dagger m} a_0^{\dagger r} |0; p^+\rangle = \frac{1}{\sqrt{J}N^{\frac{J}{2}+1}} \sum_{l=1}^{J} \mathrm{Tr}[\Phi^m Z^l \Phi^r Z^{J-l}]. \qquad (28.57)$$

On the other hand, excitations constructed out of level $n > 0$ oscillators $a_n^{\dagger i}$ need to have momentum $e^{\frac{inx}{L}}$ around the circle of the string (the σ coordinate of the closed string of length L), so we must consider the phase $e^{\frac{2\pi inl}{J}}$ (periodic under $l \to l + J$) associated with the insertion of the operator Φ at site l inside the trace. So the state with one excitation of Φ^4, say, would be

$$a_n^{\dagger 4} |0; p^+\rangle = \frac{1}{\sqrt{J}} \sum_{l=1}^{J} \frac{1}{\sqrt{J}N^{\frac{J+1}{2}}} \mathrm{Tr}[Z^l \Phi^4 Z^{J-l}] e^{\frac{2\pi inl}{J}}. \qquad (28.58)$$

Just because of the cyclicity of the trace, such an operator would actually be zero, being proportional (after putting Φ^4 in the first position by cyclicity for all terms in the sum) to $\sum_{l=1}^{J} e^{\frac{2\pi inl}{J}} = 0$. But that is fine; in fact, this maps to the statement that for a closed string we have the level matching condition $N = \tilde{N}$, so we need to have the same number of excitations going to the left (with positive momentum n) and to the right (with negative momentum, i.e. $-n$). So the first nonzero state is with an excitation of momentum n and one of momentum $-n$, corresponding to an operator where we put one field on the first position (by the cyclicity of the trace), and the other at site l, so, say,

$$a_n^{\dagger 4} a_{-n}^{\dagger 1} |0; p^+\rangle = \frac{1}{\sqrt{J}} \sum_{l=1}^{J} \frac{1}{N^{\frac{J}{2}+1}} \operatorname{Tr}[\Phi^1 Z^l \Phi^4 Z^{J-l}] e^{\frac{2\pi i n l}{J}}. \tag{28.59}$$

These operators have been called BMN operators in the literature, and we note that they are easily defined in the "dilute gas approximation," with only a few "impurities" of the Φ^m type, to use a condensed matter terminology. In Chapter 29 we will describe how to go away from this limit, and construct general operators corresponding to a spin chain of a particular type.

28.6 Discretized String Action from $\mathcal{N} = 4$ SYM

So we have obtained a picture of string states on the pp wave mapped to SYM operators that look like discretized closed strings of a very large number of sites or, in another interpretation to be developed next chapter, a spin chain with J sites. But on such states we can act with Hamiltonians, and we want to find a Hamiltonian acting on the string states (discretized string) that corresponds to a "Hamiltonian" acting on the SYM operators (remember that because of the conformal field theory operator-state correspondence this is not an empty concept). Here we will only sketch the proof of the mapping between these two Hamiltonians.

Via the operator-state correspondence, an operator of the type

$$\mathcal{O} = \frac{1}{\sqrt{J} N^{\frac{J+1}{2}}} \operatorname{Tr}[Z^l \Phi Z^{J-l}] \tag{28.60}$$

is mapped to a state of the type

$$|a_l^\dagger\rangle \equiv \operatorname{Tr}[(b^\dagger)^l a^\dagger (b^\dagger)^{J-l}]|0\rangle, \tag{28.61}$$

where $b^\dagger$ creates a Z and $a^\dagger$ a Φ. Note that the creation operators still have adjoint $SU(N)$ indices, hence we are still writing the trace. We replace this description with an equivalent one, where we define a creation operator $a_l^\dagger$ that creates an $a^\dagger$ at site l among the J $b^\dagger$'s.

The action of the interaction part of the SYM Hamiltonian,

$$H_{\text{int}} = -g_{YM}^2 \operatorname{Tr}\left[\sum_{I>J} [\Phi^I, \Phi^J][\Phi_I, \Phi_J]\right], \tag{28.62}$$

which in terms of Z and Φ^m becomes

$$H_{\text{int}} = -g_{YM}^2 \operatorname{Tr}\left[\sum_m [Z, \Phi^m][\bar{Z}, \Phi^m]\right], \tag{28.63}$$

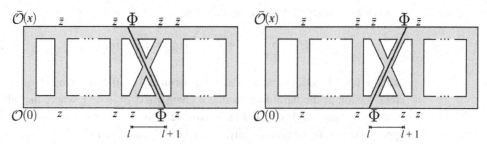

Fig. 28.1 Planar one-loop Feynman diagrams for the 2-point function of $\mathcal{O}$. (a) Planar one-loop Feynman diagram with hopping from $l + 1$ to l. (b) Planar one-loop diagram with hopping from l to $l + 1$.

on the operator-states defined before, is in terms of Feynman diagrams acting on the operators, for the 2-point functions that define their conformal dimension Δ (since $\Delta - J$ is the energy of the string on the pp wave).

The interaction SYM Hamiltonian in terms of the $b, b^\dagger$ and the relativistic field $\phi = (a + a^\dagger)/\sqrt{2}$ is

$$H_{\text{int}} \to 2g_{YM}^2 N \sum_m [b^\dagger, \Phi^m][b, \Phi^m]. \tag{28.64}$$

Drawing Feynman diagrams between two operators with J sites and one impurity Φ^m, a continuous thick band being a $Z\bar{Z}$ line and a band with a black line being a Φ^m line, we have only two planar diagrams constructed from H_{int}, one where we connect the Φ^m at site l with a Φ^m at site $l - 1$, and we have a $Z\Phi^m\bar{Z}\Phi^m$ interaction in the middle, and one where we connect site l with site $l + 1$, as in Figure 28.1(a) and 28.1(b).

The reason that we consider only planar diagrams is that in the large N limit, as 't Hooft proved, the nonplanar diagrams (diagrams that cannot be drawn on a plane, since we cross with a line *over* another) are subleading with respect to planar ones with factors of $1/N^2$. We can see this easily with a 3-loop example, using 't Hooft's double line notation for adjoint fields. Indeed, adjoint fields $A_\mu^a (T_a)_{ij}$ have two fundamental indices i, j, and the fundamental indices are conserved by propagators, $\propto \delta_{ik}\delta_{jl}$ for propagation between (ij) and (kl), and vertices. Then a line loop corresponds to a factor of $\delta_i^i = N$. For the planar diagram with a double-line loop with two propagators dividing if further into three, as in Figure 28.2(a), we get a factor of N^4 from the four single-line loops, whereas for the nonplanar version with the endpoints on one side of the two propagators interchanged, as in Figure 28.2(b), we have N^2 only.

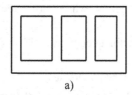
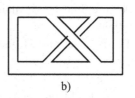

a)   b)

Fig. 28.2 (a) Planar 3-loop diagram. (b) Its nonplanar counterpart.

After a calculation, we find the Hamiltonian

$$H \sim \sum_l a_l^\dagger a_l + \frac{g_{YM}^2 N}{(2\pi)^2}\left[\left(a_l + a_l^\dagger\right) - \left(a_{l+1} + a_{l+1}^\dagger\right)\right]^2. \qquad (28.65)$$

But the first (kinetic) term is a sum of harmonic oscillator terms at each site, the discrete version of the harmonic oscillator chain $\int dx[\dot{\Phi}^2 + \Phi^2]$, where the relativistic field is $\Phi(x, t) \sim (a_l + a_l^\dagger)/\sqrt{2}$, and the second term is the discretized version of Φ'^2, with an appropriate normalization. Finally, we find the Hamiltonian

$$H = \int_0^L d\sigma \frac{1}{2}[\dot{\Phi}^2 + \Phi'^2 + \Phi^2], \qquad (28.66)$$

where the length L of the discrete "chain" is

$$L = \frac{2\pi J}{\mu\sqrt{g_{YM}^2 N}} = 2\pi\alpha' p^+. \qquad (28.67)$$

This in turn implies the action

$$S = \int d\tau \int_0^L d\sigma \frac{1}{2}[\dot{\Phi}^2 - \Phi'^2 - \Phi^2], \qquad (28.68)$$

which is the action of a string on the pp wave.

In conclusion, we have found that the Hamiltonian that acts on the large R-charge BMN operator sector is a discretized version of the string Hamiltonian on the pp wave, and the BMN operators are mapped to states that look like the states of a spin chain of J sites, with magnon impurities. We will develop that picture further in the next chapter.

Important Concepts to Remember

- pp waves are plane-fronted gravitational wave solutions to the Einstein's equations for perturbations moving at the speed of light.
- 11-dimensional supergravity has a unique maximally supersymmetric pp wave solution, which together with 11-dimensional Minkowski and $AdS_4 \times S^7, AdS_7 \times S^4$ make up all the maximally supersymmetric backgrounds.
- 10-dimensional type IIB superstring theory has a unique maximally supersymmetric pp wave solution, which together with 10-dimensional Minkowski and $AdS_5 \times S^5$ make up all the maximally supersymmetric backgrounds.
- pp waves are exact string solutions, having no α' corrections to the equations of motion.
- Near a null geodesic in any metric, the space looks like a pp wave. To obtain it, the Penrose limit boosts along a direction, while taking the overall scale of the metric to infinity.
- The Penrose limit of $AdS_5 \times S^5$ for geodesics in S^5 is the maximally supersymmetric type IIB pp wave.
- The Penrose limit of AdS/CFT on the gauge theory side corresponds to taking a sector of operators of large R-charge, with $\Delta - J \sim \mathcal{O}(1)$ and $\Delta + J \sim R^2 \to \infty$. It corresponds to $J \sim \sqrt{N} \to \infty$, with $g_{YM}^2 N/J^2$ fixed.

- The string Hamiltonian on the pp wave has $\omega_n = \sqrt{\mu^2 + n^2/(\alpha'p^+)^2}$, mapped to $(\Delta - J)_n = \sqrt{1 + g_{YM}^2 N^2 n^2/J^2}$.
- The vacuum of the pp wave string, $|0; p^+\rangle$, is mapped to $\text{Tr}[Z^J]$, with $Z = \Phi^5 + i\Phi^6$ having R-charge $+1$, and the excitations are mapped to insertions of $(D_\mu Z, \Phi^m)$ with momentum factor $e^{2\pi i n l/J}$ at site l, summed over l, leading to the BMN operators.
- The action of the SYM interaction Hamiltonian on the SYM operators, mapped to states, is via Feynman diagrams for the 2-point functions giving the dimension Δ, and corresponds to a discretized Hamiltonian for the string on the pp wave, or a spin chain of J sites.

Further Reading

The pp wave correspondence was started by Berenstein, Maldacena, and myself (BMN) in [51]. For a review of the pp wave correspondence, see, for instance, the lectures of Plefka [52]. The 11-dimensional maximally supersymmetric pp wave as a solution to supergravity was described by Kowalski-Glikman in [53]. The Penrose limit was defined in [54], but its physical meaning was clarified in [51]. The 10-dimensional maximally supersymmetric pp wave in type IIB supergravity was found in [55]. It was shown to be the Penrose limit of $AdS_5 \times S^5$ in [51, 56]. The Aichelburg-Sexl shockwave was found in [57]. It was proven that pp waves are solutions of string theory exact in α' by Horowitz and Steif in [58].

Exercises

(1) Find the pp (Aichelburg-Sexl) shockwave solution moving in flat space, corresponding to a massless "photon" source $T_{++} = p\delta(x^+)\delta(x^i)$, solving for $H(x^+, x^i)$.
(2) Repeat the exercise for a "graviton" with $T_{++} = 0$, but the shockwave moving also on $x^+ = x^i = 0$.
(3) Consider the Penrose limit in $AdS_5 \times S^5$ for a null geodesic moving in AdS_5 (and fixed in S^5). Show that it gives 10-dimensional Minkowski space.
(4) Consider the operator

$$\mathcal{O} = \det Z = \frac{1}{N!}\epsilon^{a_1...a_N}\epsilon^{b_1...b_N}Z_{a_1 b_1}...Z_{a_N b_N}. \tag{28.69}$$

Does it correspond to a state on the pp wave? What kind of Feynman diagrams would contribute to its anomalous dimension?
(5) Write the *fermionic* $\mathcal{N} = 4$ SYM fields with $\Delta - J = 1$, and all fields (including derivatives) with $\Delta - J = 2$.
(6) By considering a Bogoliubov transformation

$$b_n = \alpha_n a_n + \beta_n a_n^\dagger, \tag{28.70}$$

with α_n and β_n c-numbers, diagonalize the Hamiltonian

$$H = \sum_n \left[a_n^\dagger a_n + \frac{\mu^2}{2}(a_n + a_n^\dagger)^2 \right].$$ (28.71)

(7) Consider the Feynman diagram in Figure 28.3.

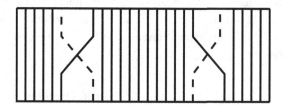

Fig. 28.3 Feynman diagram.

Is its contribution considered by the Hamiltonian H? Why?

(8) Find and prove the correctness of the flat space limit of the pp wave Hamiltonian, its eigenenergies, and its eigenstates.

29

Spin Chains from $\mathcal{N} = 4$ SYM

In Chapter 28, we have seen that we can derive from $\mathcal{N} = 4$ SYM a Hamiltonian acting on states in a certain limit that corresponds to a discretized string living on the pp wave (Penrose limit of $AdS_5 \times S^5$). In this chapter, we will generalize the Hamiltonian to a Hamiltonian for a spin chain, and see that for a subsector of states, and at one-loop only, the spin chain becomes our standard example for spin chains: the $XXX_{1/2}$ Heisenberg spin chain.

More precisely, we saw that in the Penrose limit of AdS/CFT, we can define states $|a_l^+\rangle$ for "impurities" corresponding to ϕ's, inserted at site l along a string of $b^\dagger$'s corresponding to Z's, and we can define a discrete Hamiltonian acting on them. In the continuum limit for the discrete chain, the Hamiltonian becomes the one for the string on the pp wave. We will call the case with a few "impurities" the "dilute gas approximation."

29.1 Cuntz Oscillators, Eigenstates, and Eigenenergies from the pp Wave

In this chapter we will denote the total length of the spin chain by L instead of J, in order not to confuse with the coupling of the Heisenberg spin chain.

We said that one obtains the Hamiltonian

$$H \sim \sum_{l=1}^{L} \left\{ a_l^\dagger a_l + \frac{g_{YM}^2 N}{(2\pi)^2} \left[(a_l + a_l^\dagger) - (a_{l+1} + a_{l+1}^\dagger) \right]^2 \right\}. \tag{29.1}$$

However, the creation operators in the above are not the usual ones satisfying $[a_i, a_j^\dagger] = \delta_{ij}$, but rather a new kind of oscillators called "Cuntz oscillators," satisfying the conditions

$$a_i a_i^\dagger = 1; \quad a_i^\dagger a_i = \mathbb{1} - (|0\rangle\langle 0|)_i. \tag{29.2}$$

To diagonalize the Hamiltonian, we first go to Fourier modes a_n on the discretized circle (spin chain), by

$$a_j = \frac{1}{\sqrt{L}} \sum_{n=1}^{L} e^{\frac{2\pi i j n}{L}} a_n. \tag{29.3}$$

For the Fourier modes a_n, and *only when considering the action of the Cuntz oscillators on the states of the "dilute gas approximation,"* i.e. on states

$$|\psi_{\{n_i\}}\rangle = |0\rangle_1 \dots |n_{i_1}\rangle \dots |n_{i_k}\rangle \dots |0\rangle, \tag{29.4}$$

one obtains the usual algebra of creation and annihilation operators,

$$[a_n, a_m]|\psi_{\{n_i\}}\rangle = \left(\delta_{nm} - \frac{1}{L}\sum_k e^{2\pi i i_k \frac{m-n}{L}}\right)|\psi_{\{n_i\}}\rangle, \tag{29.5}$$

as one can easily check. Then in the dilute gas approximation we can write

$$[a_n, a_m^\dagger] \simeq \delta_{mn} + \mathcal{O}\left(\frac{1}{L}\right). \tag{29.6}$$

Finally making the change from (a_n, a_{J-n}) to $(c_{n,1}, c_{n,2})$ by

$$a_n = \frac{c_{n,1} + c_{n,2}}{\sqrt{2}}; \quad a_{J-n} = \frac{c_{n,1} - c_{n,2}}{\sqrt{2}}, \tag{29.7}$$

motivated by the fact that on the closed string we know we should have left-moving modes and right-moving modes, and a level-matching condition saying the numbers of the two types must match, the Hamiltonian becomes diagonal, with eigenvalues (for the eigenstates created by $c_{n,1}, c_{n,2}$)

$$\omega_n = \sqrt{1 + \frac{g_{YM}^2 N}{\pi^2} \sin^2 \frac{\pi n}{L}}. \tag{29.8}$$

The Fock states corresponding to these eigenenergies are

$$c_{n,1|2}^\dagger|0\rangle = \frac{a_n^\dagger \pm a_{J-n}^\dagger}{\sqrt{2}}|0\rangle = \frac{1}{\sqrt{L}}\sum_j \frac{e^{\frac{2\pi ijn}{L}} \pm e^{-\frac{2\pi ijn}{L}}}{\sqrt{2}} a_j^\dagger|0\rangle, \tag{29.9}$$

which are mapped to the BMN operators (with two states) and factors of $\cos(2\pi jn/L)$ and $i\sin(2\pi ijn/L)$:

$$\frac{1}{\sqrt{L}}\sum_{l=1}^{L}\frac{1}{N^{\frac{L}{2}+1}}\operatorname{Tr}[\Phi^1 Z^l \Phi^1 Z^{L-l}]\left[\cos\left(\frac{2\pi inl}{L}\right) \quad \text{or} \quad i\sin\left(\frac{2\pi inl}{L}\right)\right]. \tag{29.10}$$

We see that for $n \ll L$, the eigenenergies and eigenstates reduce to the one of the Hamiltonian of the string on the pp wave. The $n \sim L$ case is also correct, but it corresponds to a different kind of string in AdS space, not just to a string on the pp wave = Penrose limit of $AdS_5 \times S^5$.

We also note that ω_n is valid to *all orders* in $g_{YM}^2 N$, but it is exact only for the "dilute gas" approximation, with the number of impurities $M \ll L$. In this limit, we obtain a physical picture like the one in Figure 29.1.

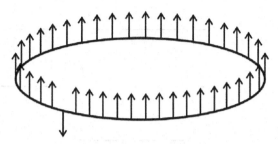

Fig. 29.1   A periodic spin chain of the type that appears in the pp wave string theory. All spins are up, except one excitation has one spin down.

29.2 The $SO(6)$ Spin Chain from $\mathcal{N} = 4$ SYM

But we can generalize to a *spin chain*, with a number of magnon excitations $M \sim L$. In that case, the spin chain in the dilute gas approximation will be described by the Hamiltonian of the string on the pp wave, or alternatively in terms of the Cuntz oscillators above.

The full spin chain with insertions of the scalars Φ^i will be a spin chain for "spins" in a representation R of $SO(6)$, the symmetry group of the scalars. In the full supersymmetric case, the spin chain will have "spins" in the $SU(2, 2|4)$ supergroup, the symmetry of the whole theory. We have noted in Chapter 6 that this is a generalization of the usual spin chain, with spins in the spin j representation of $SU(2)$.

The full spin chain is complicated, with nontrivial spins and a nontrivial all-loop structure, so it is not clear whether it is useful for condensed matter, though it is an interesting integrable mode. But the case that has received most attention is of an $SU(2)$ invariant subsector of the $SO(6)$ spin of $\mathcal{N} = 4$ SYM, which was found to reduce at 1-loop to the Heisenberg Hamiltonian. We will study it after the general case.

$SO(6)$ Spin Chain

We want to consider the 1-loop action of $\mathcal{N} = 4$ SYM Feynman diagrams on operators composed of all the scalars $\Phi^I, I = 1, \ldots, 6$:

$$\mathcal{O}[\psi] = \psi^{I_1 \ldots I_L} \, \text{Tr}[\Phi_{I_1} \ldots \Phi_{I_L}]. \tag{29.11}$$

In the 't Hooft limit $g_{YM} \to 0, N \to \infty, g_{YM}^2 N = $ fixed, only planar diagrams contribute, as we saw in the last chapter. The interacting Hamiltonian is

$$H_{\text{int}} = -g_{YM}^2 \, \text{Tr}\left([\Phi_I, \Phi_J][\Phi^I, \Phi^J]\right). \tag{29.12}$$

Then we consider the fact that $\mathcal{O}$ is mapped to states $|\psi\rangle$ (through the operator-state correspondence of conformal field theory cited before), and we can act on states with a Hamiltonian defined through the Feynman diagrams for the 2-point function. Since we consider planar diagrams only, the action of the Hamiltonian will be local on the spin chain (if we try to connect sites situated far apart, we will have to go over lines (propagators) situated in the way, as we can convince ourselves from the Figure 29.2).

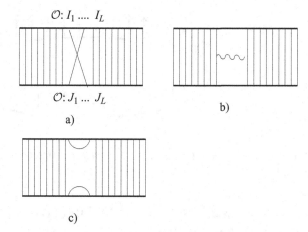

$$\mathcal{O}: I_1 \dots I_L$$

$$\mathcal{O}: J_1 \dots J_L$$

a)   b)

c)

Fig. 29.2 (a) Feynman diagram for the 2-point function of $\mathcal{O}^{I_1 \dots I_L}$ with $\mathcal{O}^{J_1 \dots J_L}$ in the full $SO(6)$ model given by the scalar interaction vertex, contributing to P term. (b) Feynman diagram given by the gauge interaction, contributing to the P term. (c) Feynman diagram contributing to the K term.

Then the action of this Hamiltonian is defined through:

- The identity operator, coming from Feynman diagrams in which two free scalar propagators in between the two operators are connected by gluons
- The trace operator,

$$K^{J_l J_{l+1}}_{I_l I_{l+1}} = \delta_{I_l, I_{l+1}} \delta^{J_l, J_{l+1}}, \tag{29.13}$$

coming from Feynman diagrams that contract two of the fields of the same operator, as in Figure 29.2(c) and

- The permutation operator,

$$P^{J_l, J_{l+1}}_{I_l, I_{l+1}} = \delta^{J_{l+1}}_{I_l} \delta^{J_l}_{I_{l+1}}, \tag{29.14}$$

coming from the same kind of Feynman diagrams as in last chapter: a 4-scalar interaction vertex for two pairs of neighboring scalars, where the same scalar at position l is contracted via the vertex with a scalar at position $l + 1$ or $l - 1$, as in Figure 29.2(a), as well as the gauge interactions, as in Figure 29.2(b).

The action of the Hamiltonian corresponds to the action of a matrix of anomalous dimensions in the gauge theory, since energy E is mapped to anomalous dimension Δ. This *anomalous dimension matrix* is defined as follows. Consider the renormalization of the composite operators $\mathcal{O}$. In quantum field theory, besides the renormalization of the fundamental fields, if we have composite operators, they also require an (independent) renormalization, due to the new divergences appearing when we consider several fields at the same spacetime point. This renormalization is in general a matrix in operator space:

$$\mathcal{O}^A_{\text{ren}} = Z^A{}_B \mathcal{O}^B. \tag{29.15}$$

Then we define the anomalous dimension matrix the way we would define the anomalous dimension of a field, only considering that we have matrices now, i.e. as

$$\Gamma = \frac{dZ}{d \ln \Lambda} \cdot Z^{-1}. \tag{29.16}$$

Then the eigenvectors $\mathcal{O}_n$ of Γ are multiplicatively renormalized, and have a definite anomalous dimension γ_n (eigenvalue of Γ), such that

$$\langle \mathcal{O}_n^{\text{ren}}(x) \mathcal{O}_n^{\text{ren}}(y) \rangle = \langle Z \cdot \mathcal{O} Z \cdot \mathcal{O} \rangle = \frac{\text{const.}}{|x - y|^{2(L+\gamma_n)}}. \tag{29.17}$$

We consider the Feynman diagrams for the above 2-point function. For instance, the Feynman diagram that gives the operator proportional to the identity gives a Z-factor contribution:

$$Z^{...J_l J_{l+1}...}_{...I_l I_{l+1}...} = \mathbb{1} - \frac{g_{YM}^2 N}{16\pi^2} \ln \Lambda \delta_{I_l}^{J_l} \delta_{I_{l+1}}^{J_{l+1}}. \tag{29.18}$$

In total, considering all the Feynman diagrams, we obtain

$$Z^{...J_l J_{l+1}...}_{...I_l I_{l+1}...} = \mathbb{1} - \frac{g_{YM}^2 N}{16\pi^2} \ln \Lambda \left(\delta_{I_l I_{l+1}} \delta^{J_l J_{l+1}} + 2\delta_{I_l}^{J_l} \delta_{I_{l+1}}^{J_{l+1}} - 2\delta_{I_l}^{J_{l+1}} \delta_{I_{l+1}}^{J_l} \right). \tag{29.19}$$

That means that our "Hamiltonian," the anomalous dimension matrix Γ, is given at 1-loop by

$$\text{``}H\text{''} = \Gamma = \frac{g_{YM}^2 N}{16\pi^2} \sum_{l=1}^{L} (K_{l,l+1} + 2 - 2P_{l,l+1}). \tag{29.20}$$

This is then interpreted as a Hamiltonian acting on a spin chain with $SO(6)$-valued spins. At higher loops, the Hamiltonian becomes even more complicated, and corresponds to some unusual spin chain, which is not clear if it has condensed matter applications. In fact, once one includes fermions, the total symmetry group is $SU(2, 2|4)$, so the full spin chain is for "spins" belonging to this supergroup.

29.3 The $SU(2)$ Sector and the Dilatation Operator

But as we mentioned, if we consider simply a subset of states invariant under $SU(2)$, and restrict to 1-loop, we obtain something very well studied: the Heisenberg Hamiltonian.

To do this, we simulate the case of the pp wave spin chain, just that since we want an equal number of the basis Z fields and of the "impurities," invariant under an $SU(2)$, we need to construct the two *complex* scalars

$$Z = \Phi^1 + i\Phi^2; \quad W = \Phi^3 + i\Phi^4. \tag{29.21}$$

From them, we construct the operators

$$\mathcal{O}_\alpha^{J_1, J_2} = \text{Tr}[Z^{J_1} W^{J_2}] + \text{perms.}, \tag{29.22}$$

where the permutations always have a total number J_1 of Z fields and J_2 of W fields.

$$H^{(1)}:$$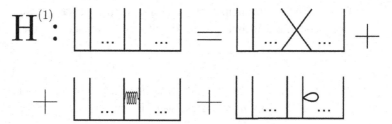

Fig. 29.3 The action of the anomalous dimension matrix ("Hamiltonian") on operators through Feynman diagrams is defined by the renormalization factor. Relevant diagrams are shown.

To formalize the anomalous dimension matrix as a Hamiltonian, we consider a related way to define it, as a "dilatation operator" $\mathcal{D}$ acting on operators $\mathcal{O}$ as a matrix:

$$\mathcal{D} \circ \mathcal{O}_\alpha^{J_1 J_2}(x) = \sum_\beta \mathcal{D}_{\alpha\beta} \mathcal{O}_\beta^{J_1 J_2}(x). \tag{29.23}$$

The dilatation operator as a symmetry operator defines the behavior of operators under dilatations (scalings), i.e. defines the anomalous dimension. Defining it through an expansion in perturbation theory,

$$\mathcal{D} = \sum_{n \geq 0} \mathcal{D}^{(n)}, \tag{29.24}$$

where n is the loop order ($\mathcal{D}^{(n)} \sim \mathcal{O}(g_{YM}^{2n})$), it is obtained by attaching half of the legs of Feynman diagrams to the operator, as in Figure 29.3.

Then diagonalizing $\mathcal{D}$ (the "Hamiltonian"), we find the eigenstates of $\mathcal{D}$:

$$\mathcal{D} \circ \mathcal{O}_A' = \Delta_A \mathcal{O}_A'. \tag{29.25}$$

The diagonalization is found through a matrix $V_{\alpha A}$, so

$$\mathcal{O}_\alpha = V_{\alpha A} \mathcal{O}_A' \Rightarrow \mathcal{D} \circ \mathcal{O}_A = (V_{\alpha A} \Delta_A V_{A\beta}^{-1}) \mathcal{O}_\beta. \tag{29.26}$$

The 2-point function of $\mathcal{O}_\alpha$ is a sum of a tree-level and a 1-loop term:

$$\langle \mathcal{O}_\alpha(x) \bar{\mathcal{O}}_\beta(0) \rangle = \frac{S_{\alpha\beta}}{|x|^{2(J+2)}} + \frac{T_{\alpha\beta}}{|x|^{2(J+2+\Delta_A)}}, \tag{29.27}$$

where $T_{\alpha\beta} = \langle \mathcal{O}_\alpha H \bar{\mathcal{O}}_\beta \rangle_{\text{one-loop}}$. On the other hand, because of the diagonalization procedure, we can write

$$\langle \mathcal{O}_\alpha(x) \bar{\mathcal{O}}_\beta(0) \rangle = V_{\alpha A} V_{\beta B}^* \langle \mathcal{O}_A(x) \bar{\mathcal{O}}_B'(0) \rangle = V_{\alpha A} V_{\beta B}^* \frac{\delta_{AB} C_A}{|x|^{2(J+2+\Delta_A)}}$$

$$= V_{\alpha A} V_{\beta B}^* \frac{\delta_{AB} C_A}{|x|^{2(J+2)}} (1 - 2\Delta_A \log(|x|\Lambda). \tag{29.28}$$

Here we have assumed that the diagonal operators $\mathcal{O}_A'$ are orthogonal, but not normalized, giving a factor of C_A in the 2-point function, and we have introduced the dimensional transmutation scale Λ together with the anomalous dimension Δ_A. Defining further

$C_{AB} \equiv \delta_{AB} C_A$ and $\Delta_{AB} = \delta_{AB} \Delta_A$, we finally have

$$\langle \mathcal{O}_\alpha(x) \bar{\mathcal{O}}_\beta(0) \rangle = \frac{1}{|x|^{2(J+2)}} \left[(VCV^\dagger)_{\alpha\beta} - 2(VC\Delta V^\dagger)_{\alpha\beta} \log(|x|\Lambda) \right]. \quad (29.29)$$

Finally, we can also express $\mathcal{D}$ in terms of

$$\check{Z}_{ij} \equiv \frac{d}{dZ_{ij}}, \quad (29.30)$$

which removes one scalar (with gauge indices ij) from the operator.

At tree level, the dilatation operator is trivial, i.e. the identity operator, measuring the length $J_1 + J_2$ of the chain, so we can express it at

$$\mathcal{D}^{(0)} = \mathrm{Tr}[Z\check{Z} + W\check{W}]. \quad (29.31)$$

On the other hand, at 1-loop, it is the interaction Hamiltonian, with one of the commutators replaced by $[\check{Z}, \check{W}]$ (since this is the action of the Feynman diagram on the operator):

$$\mathcal{D}^{(1)} = -\frac{g_{YM}^2}{8\pi^2} \, \mathrm{Tr}([Z, W][\check{Z}, \check{W}]). \quad (29.32)$$

At the planar level, this equals the restriction of the anomalous dimension matrix Γ for the $SO(6)$ spin chain to the $SU(2)$ sector. In this case, the trace operator $K_{l,l+1}$ doesn't act on the $SU(2)$ sector, since it would mean to contract a Z field with a Z field (or W with W) in the same operator, but K instead contracts Z with $\bar{Z}$ and W with $\bar{W}$. Then we obtain the planar 1-loop dilatation operator in the $SU(2)$ sector as

$$\mathcal{D}^{(1)}_{\mathrm{planar}} = \frac{g_{YM}^2 N}{8\pi^2} \sum_{i=1}^{L} (\mathbb{1}_{i,i+1} - P_{i,i+1}). \quad (29.33)$$

This is exactly the Hamiltonian of the XXX spin 1/2 Heisenberg spin chain, with coupling $J = -\lambda/(16\pi^2)$.

29.4 Bethe Ansatz Results

We now go back to the results of the coordinate Bethe ansatz for the Heisenberg spin chain, obtained in Chapter 6, and see what they can tell us about $\mathcal{N} = 4$ SYM.

The cyclicity of the trace in $\mathcal{N} = 4$ SYM is seen to be mapped to the fact that the total momentum on the spin chain with M magnon excitations is zero:

$$P = \sum_{j=1}^{M} p_j = 0. \quad (29.34)$$

In turn, this corresponds, when mapping via AdS/CFT to a closed string in $AdS_5 \times S^5$, to the vanishing of the worldsheet momentum. Note that from the point of view of the Heisenberg spin chain, this is the condition to have a closed (periodic) spin chain, but moreover choosing the solution of P strictly zero, not equivalent to zero.

That means that the first excitation is a 2-magnon excitation,

$$\psi(p_1, p_2) = \sum_{x_1, x_2} |x_1, x_2\rangle, \tag{29.35}$$

with energy given by the sum of the magnon energies:

$$E = \frac{\lambda}{16\pi^2} 8 \left[\sin^2 \frac{p_1}{2} + \sin^2 \frac{p_2}{2} \right]. \tag{29.36}$$

In this case, since we have $P = p_1 + p_2 = 0$, we just get twice the 1-magnon energy. In general, however, for M magnons, we get

$$E = \frac{\lambda}{16\pi^2} \sum_{j=1}^{L} 8 \sin^2 \frac{p_j}{2} = \frac{\lambda}{8\pi^2} \sum_{j=1}^{L} \frac{1}{u_j^2 + 1/4}, \tag{29.37}$$

where $u = (\cot p/2)/2$, which gives now the anomalous dimension of the SYM operator.

The Bethe equations for 2-magnons, with $p_2 = -p_1$, so $u_2 = -u_1$, reduce to a single one:

$$e^{ip_1 L} = S(p_1, p_2) = \frac{\cot p_1/2 - \cot p_2/2 + 2i}{\cot p_1/2 - \cot p_2/2 - 2i} = \frac{2 \cot p_1/2 + 2i}{2 \cot p_1/2 - 2i} = \frac{e^{ip_1/2}}{e^{-ip_1/2}} = e^{ip_1}. \tag{29.38}$$

The solution is

$$p_1 = \frac{2\pi n}{L - 1}, \tag{29.39}$$

which leads to the anomalous dimension

$$\gamma = \frac{\lambda}{2\pi^2} \sin^2 \frac{\pi n}{L - 1}. \tag{29.40}$$

Substituting $-p_2 = p_1 = 2\pi n/(L - 1)$ in the ansatz for the wavefunction, one obtains, as we saw in (6.74), the 2-magnon state

$$|\psi(n)\rangle = C_n \cos \left(\pi n \frac{2l + 1}{L - 1} \right) |x_2 + l; x_2\rangle, \tag{29.41}$$

mapping to the SYM 2-magnon operator:

$$\mathcal{O}^{L,2} = C_n \sum_{l=1}^{L} \cos \left[\pi n \frac{2l + 1}{L - 1} \right] \text{Tr}[W Z^l W Z^{L-l}]. \tag{29.42}$$

Note that for small n and $L \to \infty$, the operator turns into a 2-magnon BMN operator:

$$\sum_{l=1}^{L} \cos \frac{2\pi n l}{L} \text{Tr}[W Z^l W Z^{L-l}]. \tag{29.43}$$

In general, for a number M of magnons much smaller than L and for $n \ll L$, the spectrum of operators is given by insertions of

$$a_n^\dagger = \frac{1}{\sqrt{L}} \sum_{l=1}^{L} e^{\frac{2\pi i n l}{L}} \sigma_l^-, \tag{29.44}$$

where σ_l^- are Pauli matrices at site l. These operators are BMN operators, showing that the dilute gas approximation for the spin chain gives the spectrum dual to the string on the pp wave, as expected. Moreover, in this limit, we have

$$\gamma = \Delta - L - M = \frac{\lambda}{8\pi^2} \sum_{k=1}^{M} 4\sin^2 \frac{p_k}{2} \simeq \frac{\lambda}{8\pi^2} \sum_{k=1}^{M} p_k^2 = \frac{\lambda}{2L^2} \sum_{k=1}^{} n_k^2. \quad (29.45)$$

Therefore in the dilute gas approximation, the independent magnons correspond to the independent string oscillators on the pp wave. But note that even in the case that $M \sim L$, which could even be small, we still have a Fock spectrum, corresponding to independent string excitations in $AdS_5 \times S^5$.

Important Concepts to Remember

- The oscillators in the Hamiltonian acting on SYM states are Cuntz oscillators.
- In the dilute gas approximation, we obtain a diagonal Hamiltonian that contains an all-loop result, but is exact only in the given approximation, and with eigenstates given by the BMN operators.
- In general, the Hamiltonian acts on a spin chain derived from $\mathcal{N} = 4$ SYM, with "spins" in $SO(6)$ in the bosonic case.
- The $SO(6)$ spin chain has a Hamiltonian derived from an anomalous dimension matrix for the operators with L scalars.
- It can also be defined as a dilatation operator, acting on operators by attaching Feynman diagrams to it.
- If we restrict to the $SU(2)$ sector, made up of operators containing only two complex fields, Z and W, the dilatation operator giving the Hamiltonian at 1-loop becomes the Heisenberg $XXX_{1/2}$ spin chain Hamiltonian.
- Magnon excitations in the dilute gas approximation of the Heisenberg spin chain map to BMN operators, corresponding to independent string oscillators on the pp wave.
- But in general, magnons are mapped to independent string oscillators in $AdS_5 \times S^5$.

Further Reading

The spin chains were defined in $\mathcal{N} = 4$ SYM by Minahan and Zarembo in [59]. Good reviews (though at an advanced level) are [60] and [61].

Exercises

(1) Write the Fock state $c_{n,1}^{\dagger I_1} c_{n,1}^{\dagger I_2} c_{n,1}^{\dagger I_3} c_{n,1}^{\dagger I_4} |0\rangle$ in terms of the Cuntz oscillators, and the corresponding 4-magnon SYM operator.

(2) Consider the explicit action of the $SO(6)$ Hamiltonian Γ on the operator $\mathcal{O}[\Phi_1 \Phi_1 \Phi_3 \Phi_4 \Phi_1]$ and estimate the size of the 1-loop correction to the anomalous dimension. How is this consistent (or not) with the BMN estimate?

(3) Show that the Feynman diagram (4-vertex at y with two lines ending on $\mathcal{O}(x)$ and two lines ending on $\mathcal{O}(0)$), as in Figure 29.4,

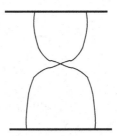

Fig. 29.4   Feynman diagram.

where the lines are scalars, gives a contribution

$$\sim \frac{1}{4\pi^2 |x|^2} \log(|x|\Lambda) + \text{finite}. \tag{29.46}$$

(4) Write explicitly the correct 3-magnon eigenstate with total momentum zero, and deduce its corresponding spin chain operator in $\mathcal{N} = 4$ SYM. Show that for $J \to \infty$ it reduces to the BMN operator.

(5) Consider the Heisenberg $XXX_{1/2}$ spin chain with four sites. Using the coordinate Bethe ansatz, write *all* its eigenstates and eigenenergies.

(6) For the above case (Exercise 5), write the corresponding spin chain operators in $\mathcal{N} = 4$ SYM.

30

The Bethe Ansatz: Bethe Strings from Classical Strings in AdS

In the last chapter we have seen how to obtain the Heisenberg spin chain Hamiltonian from the $\mathcal{N} = 4$ SYM. But that did not really use string theory, we learned only how to relate the spin chain to a quantum field theory. In this chapter, we will use the AdS/CFT correspondence to write a map between spin chains and strings in AdS space.

30.1 Thermodynamic Limit of the Bethe Ansatz for the $SU(2)$ Sector

The AdS/CFT correspondence maps solutions to the Bethe ansatz equations for the spin chain to some classical strings in $AdS_5 \times S^5$. We have seen that magnon excitations in the $SU(2)$ sector of $\mathcal{N} = 4$ SYM, in the "dilute gas approximation," $M/L \to 0$, are mapped to strings on the pp wave, which is a Penrose limit of $AdS_5 \times S^5$, around a geodesic in the center of AdS_5 and rotating on an equator of S^5.

In the case of large M, we expect to have semiclassical strings (large), as opposed to pointlike ones. Moreover, if we look at the $SU(2) = SO(3)$ sector, we expect to move in an $S^3 \subset S^5$. Indeed, an n-sphere $S_n = SO(n+1)/SO(n)$ is in general a coset manifold, but for $n = 3$ we have $SO(4)/SO(3) \simeq SO(3)$, so it is a group manifold as well.

One can match various classical string solutions in AdS space against solutions of the Bethe ansatz equations, but instead we will effectively treat all the solutions together, by considering the equation in the thermodynamic limit, $M, L \to \infty$, with M/L fixed.

Thermodynamic Limit of Bethe Equations

In Chapter 7 we have seen the M Bethe equations for the M rapidities of the Heisenberg $XXX_{1/2}$ spin chain:

$$e^{ip_i L} = \prod_{k=1;\ k \neq i}^{M} S(p_i, p_k), \tag{30.1}$$

where the momenta can be expressed in terms of rapidities $u_k = \frac{1}{2} \cot p_k / 2$. Then the Bethe equations become

$$\left(\frac{u_i - i/2}{u_k + i/2} \right)^L = \prod_{k=1;\ k \neq i}^{M} \left(\frac{u_i - u_k - i}{u_i - u_k + i} \right), \quad k = 1, \ldots, M. \tag{30.2}$$

Taking the log on both sides of the equation leads to

$$L \log\left(\frac{u_i - i/2}{u_i + i/2}\right) = \sum_{\substack{k=1, \\ k \neq i}}^{M} \log\left(\frac{u_i - u_k - i}{u_i - u_k + i}\right) - 2\pi i n_i. \tag{30.3}$$

Note that we have put a $-2\pi i n_i$ term for the branch of the log, and that n_i are arbitrary integers associated with each root u_i, so the set $\{n_i\}$ are quantum numbers for the multiparticle system $\{u_i\}$.

Assuming self-consistently that $p_i \sim \mathcal{O}(1/L)$, which leads to $u_i \sim \mathcal{O}(L)$, and thus defining

$$x_i = \frac{u_i}{L} = \text{finite}, \tag{30.4}$$

the (log of the) Bethe equations becomes

$$\frac{1}{x_i} - 2\pi n_i = \frac{2}{L} \sum_{\substack{k=1; \\ k \neq i}}^{M} \frac{1}{x_i - x_k}. \tag{30.5}$$

The Bethe roots u_i can be anywhere in the complex plane, they do not need to be necessarily real. The only condition is that the total energy of the system,

$$\frac{E}{2J} = \frac{L}{2} - \sum_{j=1}^{M} \frac{1}{u_j^2 + 1/4}, \tag{30.6}$$

is real, which leads to the condition that for each Bethe root u_i in the solution $\{u_i\}_{i=1,\dots,M}$, we also must have its complex conjugate u_i^*. In the thermodynamic limit $L \to \infty$, for two roots having the same real parts $\text{Re}(u_k) = \text{Re}(u_j)$, the fermionic nature of the chain requires that the imaginary parts are different (since otherwise the wavefunctions are zero, as we saw in Chapter 7). Considering the Bethe equations in (30.2), since $L \to \infty$, the left-hand side can only be 0 or ∞, which means the right-hand side must have a zero or a pole. But the right-hand side is a product of terms, so we must have at least one pair (k, j) such that $u_k = u_j \pm i$, which is of the type described before, of equal real part, but differing imaginary part. The interpretation in this case is of two magnons of the same energy (related to the same real parts of the rapidities) that form a bound state by splitting their imaginary parts in the u plane.

30.2 Bethe Strings and their Equations

So for each j we find a k such that $u_k = u_j \pm i$, and we can then do the same for k, etc., until finally we have an array with $u_k = \text{Re}(u) + ik$. But plugging in (30.5), we see that on such an array, n_i must be constant, so $n_i = n_C$ characterizes the array, and then the k's represent the quantum numbers characterizing the solution $\{u_k\}_k$. Since $u_{k+1} - u_k = i$, then $x_{k+1} - x_k = i/L \to 0$, even when $M \to \infty$, so the array becomes a smooth line. But moreover, in this $M \to \infty$ limit, we note that since $u_{k+1} - u_k$ is not exactly i, but differs by small terms of order $1/L$, which can also change the real part, when we consider x_k's with k's differing by order $M \propto L$, the real parts can change too. That results in the fact that the previously vertical lines actually can curve. In summary, in the thermodynamic limit the roots u_i accumulate on *smooth contours* called *Bethe strings*. They are symmetric from the real axis, since for every u_k, we must have also u_k^*.

To match against the equations we will derive from AdS space, we will turn the Bethe equations into integral equations. The procedure is the usual one originally used in continuum electrodynamics, defining an electron density from a sum of Dirac delta functions. We thus define the *Bethe root density*

$$\rho \equiv \frac{1}{L} \sum_{j=1}^{M} \delta(x - x_j), \tag{30.7}$$

such that the normalization is

$$\int_{\mathcal{C}} dx \rho(x) = \frac{M}{L}. \tag{30.8}$$

Here $\mathcal{C}$ is the contour in x space that corresponds to the Bethe string.

Then the thermodynamic limit of the Bethe equations, (30.5), becomes

$$2P \int_{\mathcal{C}} dy \frac{\rho(y)}{y - x} = -\frac{1}{x} + 2\pi n_{\mathcal{C}(u)} \quad \text{for} \quad x \in \mathcal{C}, \tag{30.9}$$

where P denotes the principal part of the integral.

In the thermodynamic limit, $u = xL$ and $\frac{1}{L} \sum_i \to \int dx$, for the energy (30.6), we obtain

$$\frac{E}{2J} = \left(\frac{L}{2} \right) - \frac{1}{L} \int_{\mathcal{C}} dx \frac{\rho(x)}{x^2}. \tag{30.10}$$

Putting $x = 0$ in (30.9), or rather from the total momentum (in the thermodynamic limit $Lx_k = u_k = 1/p_k$), equal to $2\pi m$ from the general quantization condition,

$$\hat{P} = \sum_i p_i \to \sum_i \frac{1}{u_i} = \frac{1}{L} \sum_i \frac{1}{x_i} = 2\pi m, \tag{30.11}$$

we obtain the integral relation

$$\int dx \frac{\rho(x)}{x} = 2\pi m. \tag{30.12}$$

The integral equations (30.7), (30.9), (30.10), and (30.12) can be written independently for each Bethe string (smooth component $\mathcal{C}_n$ of the Bethe root contour), and will correspond to integral equations for macroscopic strings in AdS space.

30.3 Long Strings in AdS

We now construct the string theory dual to the $SU(2)$ sector. As we said, since $SU(2) = S^3$ as a manifold, we consider an $S^3 \subset S^5$. The six $\mathcal{N} = 4$ SYM scalars Φ^m correspond to the six Euclidean coordinates X^m for the S^5. Therefore for the $SU(2)$ sector, with $Z = \Phi^1 + i\Phi^2$ and $W = \Phi^3 + i\Phi^4$, we must consider the Euclidean coordinates X^i for the S^3, i.e. with $X^i X^i = 1$, and form

$$Z = X^1 + iX^2; \quad W = X^3 + iX^4. \tag{30.13}$$

From them, define the $SU(2)$ *group element* (not element of the algebra, which is generated just by the Pauli matrices σ_A, $A = 1, 2, 3$)

$$g = \begin{pmatrix} Z & W \\ -\bar{W} & \bar{Z} \end{pmatrix} = \begin{pmatrix} X^1 + iX^2 & X^3 + iX^4 \\ -X_3 + iX_4 & X_1 - iX_2 \end{pmatrix} = X^1 \mathbb{1} + iX_4 \sigma_1 + iX_3 \sigma_2 + iX_2 \sigma_3 \equiv iX^i \tilde{\sigma}_i. \tag{30.14}$$

Note then that indeed, $\det g = Z\bar{Z} + W\bar{W} = \sum_i (X^i)^2 = 1$, and moreover

$$\mathrm{Tr}[gg^\dagger] = 2 \sum_i (X^i)^2 = 2, \tag{30.15}$$

where $g^{-1} = g^\dagger$ since $g \in SU(2)$. We have defined $\tilde{\sigma}_i = (-i\mathbb{1}, \sigma_3, \sigma_2, \sigma_1)$.

Define also the currents

$$j_a = g^{-1} \partial_a g = \begin{pmatrix} X^1 - iX^2 & -X^3 - iX^4 \\ X^3 - iX^4 & X^1 + iX^2 \end{pmatrix} \partial_a \begin{pmatrix} X^1 + iX^2 & X^3 + iX^4 \\ -X^3 + iX^4 & X^1 - iX^2 \end{pmatrix}. \tag{30.16}$$

Explicitly, that gives

$$\begin{aligned}
j_a = g^{-1}\partial_a g &= (X^1 \mathbb{1} - iX^4\sigma_1 - iX^3\sigma_2 - iX^2\sigma_3)\partial_a(X^1 \mathbb{1} + iX^4\sigma_1 + iX^3\sigma_2 + iX^2\sigma_3) \\
&= \mathbb{1}\left(\sum_i X^i \partial_a X^i \right) + i\sigma_1(X^1\partial_a X^4 - X^4\partial_a X^1 + X^3\partial_a X^2 - X^2\partial_a X^3) \\
&\quad + i\sigma_2(X^1\partial_a X^3 - X^3\partial_a X^1 + X^2\partial_a X^4 - X^4\partial_a X^2) \\
&\quad + i\sigma_3(X^1\partial_a X^2 - X^2\partial_a X^1 + X^4\partial_a X^3 - X^3\partial_a X^4) \\
&\equiv \frac{\sigma_A j_a^A}{2i}.
\end{aligned} \tag{30.17}$$

Note that since $\sum_i X^i \partial_a X^i = \partial_a(X^i X^i)/2 = 0$, there is no component of j_a along the identity, so we have expanded only in the Pauli matrices, thus $\mathrm{Tr}(j_a) = 0$.

More generally, we see that $\mathrm{Tr}[(j_a)^{2n+1}] = 0$ (no sum over a), since to obtain a nonzero result we would have to use at least one ϵ_{ABC} in the expansion of a product of Pauli matrices, $\sigma_A\sigma_B = \delta_{AB} + i\epsilon_{ABC}\sigma_C$, to get a nonzero result, and then we would get $\dots \epsilon_{ABC} j_a^A j_a^B = 0$ by symmetry.

We can now write the string action in conformal gauge on the embedding space (with the constraint $\sum_i (X^i)^2 = 1$ added), as (considering that the string is at a fixed position in AdS_5, and in the part of S^5 transverse to the S^3)

$$\begin{aligned}
S &= \frac{1}{4\pi\alpha'} \int_0^{2\pi} d\sigma \int d\tau \left[-R^2(\partial_a X^0)^2 + R^2(\partial_a X^i)^2 \right] \\
&= \frac{R^2}{4\pi\alpha'} \int_0^{2\pi} d\sigma \int d\tau \left[-(\partial_a X^0)^2 + (\partial_a X^i)^2 \right].
\end{aligned} \tag{30.18}$$

Here we can translate into SYM variables and write $R^2/(4\pi\alpha') = \sqrt{\lambda}/(4\pi)$, and we still have to impose $X^i X^i = 1$. The action can be rewritten in terms of the currents, noting that

$$\mathrm{Tr}[(j_a)^2] = \mathrm{Tr}[g^{-1}(\partial_a g)g^{-1}(\partial^a g)] = -\mathrm{Tr}[(\partial_a g^{-1})(\partial^a g)]$$
$$= -\mathrm{Tr}[(\partial_a X^1 \mathbb{1} - i\partial_a X^4 \sigma_1 - i\partial_a X^3 \sigma_2 - i\partial_a X^2 \sigma_3)$$
$$\times (\partial^a X^1 \mathbb{1} + i\partial^a X^4 \sigma_1 + i\partial^a X^3 \sigma_2 + i\partial^a X^2 \sigma_3)]$$
$$= -2\sum_i [(\partial_a X^i)(\partial^a X^i)]. \tag{30.19}$$

Note that this relation is true without summing over a. Then we can rewrite the string action in conformal gauge as

$$S = -\frac{\sqrt{\lambda}}{4\pi} \int d\tau \int_0^{2\pi} d\sigma \left[\frac{\mathrm{Tr}[(j_a)^2]}{2} + (\partial_a X^0)^2 \right]. \tag{30.20}$$

The equation of motion for X^0 is then the usual one:

$$\partial_+ \partial_- X^0 = 0. \tag{30.21}$$

Next we consider the equation of motion for $g = iX^A \tilde{\sigma}_A$, using the fact that

$$\frac{\delta j_a(\sigma)}{\delta(X^A(\sigma')\tilde{\sigma}^A)} = \frac{g^{-1}}{2} \partial_a \delta(\sigma - \sigma'). \tag{30.22}$$

The equation of motion is then, after multiplying with g and canceling some terms:

$$\eta^{ab} \partial_a j_b = 0. \tag{30.23}$$

Using lightcone coordinates, with $\partial_\pm = \partial_\tau \pm \partial_\sigma$, we obtain

$$\partial_+ j_- + \partial_- j_+ = 0. \tag{30.24}$$

Since we used the action in the conformal gauge, we must impose the constraints that replace the equations of motion of the gauge fixed degrees of freedom, i.e. the Virasoro constraints

$$(\partial_\pm X^i)^2 = 0 \Rightarrow (\partial_\pm X^0)^2 = (\partial_\pm X^i)^2 = -\frac{1}{2} \mathrm{Tr}[(j_\pm)^2]. \tag{30.25}$$

But the equation of motion for X^0 is solved in the usual way by

$$X^0 = \kappa\tau = \kappa \frac{(\tau+\sigma) + (\tau-\sigma)}{2}, \tag{30.26}$$

with κ a constant, which means that the Virasoro constraints reduce to

$$\mathrm{Tr}[(j_\pm)^2] = -2\kappa^2. \tag{30.27}$$

Charges for Operators

The string action has an $SU(2)_L \times SU(2)_R$ global symmetry defined by the action from the left and from the right on g with an element $h \in SU(2)$, i.e. $g \to gh$ and $g \to hg$. The Noether currents for these global symmetries are j_a and

$$l_a = gj_a g^{-1} = \partial_a g \cdot g^{-1} \equiv \frac{\sigma^A}{2i} l_a^A. \tag{30.28}$$

The charges associated with the two currents are

$$Q_L^a = \frac{\sqrt{\lambda}}{4\pi} \int_0^{2\pi} d\sigma \, l_\tau^A; \quad Q_R^a = \frac{\sqrt{\lambda}}{4\pi} \int_0^{2\pi} d\sigma \, j_\tau^A. \tag{30.29}$$

We want to study strings that correspond to particular operators in SYM, operators of the type

$$\mathcal{O} = \mathrm{Tr}[Z^{L-M} W^M] + \text{perms.}, \tag{30.30}$$

with total number $L - M$ of Z's and M of W's. Under $g \to gh$, (ZW) transforms as a doublet,

$$(Z \quad W) \to (Z \quad W) \begin{pmatrix} \cdot & \cdot \\ \cdot & \cdot \end{pmatrix}, \tag{30.31}$$

this being the first row in g, whereas under $g \to hg$,

$$\begin{pmatrix} Z \\ -\bar{W} \end{pmatrix} \to \begin{pmatrix} Z \\ -\bar{W} \end{pmatrix} \begin{pmatrix} \cdot & \cdot \\ \cdot & \cdot \end{pmatrix}, \tag{30.32}$$

this being the first column in g.

The first transformation means that the charges under the third component of the group are

$$Q_R^3(Z) = +1; \; Q_R^3(W) = -1,$$
$$Q_L^3(Z) = +1; \; Q_L^3(-\bar{W}) = -1 \Rightarrow Q_L^3(W) = +1. \tag{30.33}$$

We then find the charges of the operators dual to strings as

$$Q_R^3(\mathcal{O}) = L - 2M; \quad Q_R^3(\mathcal{O}) = L. \tag{30.34}$$

30.4 Classical String Equations and Bethe String Limit

Since $j_a = g^{-1}\partial_a g$, we find explicitly

$$\partial_+ j_- - \partial_- j_+ + [j_+, j_-] = \partial_+(g^{-1}\partial_- g) - \partial_-(g^{-1}\partial_+ g) + g^{-1}(\partial_+ g)g^{-1}\partial_- g - g^{-1}(\partial_- g)g^{-1}\partial_+ g$$
$$= 0. \tag{30.35}$$

We found zero after expanding the terms, and canceling them one by one.

We can put this equation (identity) together with the equation of motion $\partial_+ j_- + \partial_- j_+ = 0$ into a single one, by defining

$$J_\pm = \frac{j_\pm}{1 \mp x}, \tag{30.36}$$

which means that

$$\partial_+ J_- - \partial_- J_+ + [J_+, J_-] = \frac{\partial_+ j_-}{1+x} - \frac{\partial_- j_+}{1-x} + \frac{[j_+, j_-]}{1-x^2}$$
$$= [(\partial_+ j_- - \partial_- j_+ + [j_+, j_-]) - x(\partial_+ j_- + \partial_- j_+)]/(1-x^2)$$
$$= 0 \tag{30.37}$$

for every x.

The advantage of this is that we can define a "monodromy" (object that defines parallel transport around a closed curve) of a "connection" $J_\pm$ by

$$\Omega(x) \equiv P \exp\left[-\int_0^{2\pi} d\sigma J_\sigma\right] = P \exp\left[\int_0^{2\pi} d\sigma \frac{1}{2}\left(\frac{j_+}{x-1} + \frac{j_-}{x+1}\right)\right]. \tag{30.38}$$

"Connection" is a mathematical name for gauge field, and indeed this monodromy is basically a Wilson loop (as defined in Section 26.2) for the gauge field $A_\pm = J_\pm$ around the circle in σ. It defines a monodromy, since that is true of the Wilson loop. Note that we have used $J_\sigma = (J_+ - J_-)/2$.

Finally, from the monodromy we can define the real function $p(x)$ by

$$\mathrm{Tr}[\Omega(x)] \equiv 2\cos p(x) = e^{ip(x)} + e^{-ip(x)}. \tag{30.39}$$

This is a good definition, since as we saw, $\mathrm{Tr}[(j_a)^{2n+1}] = 0$, so we also have $\mathrm{Tr}[(\int d\sigma J_\sigma)^{2n+1}] = 0$, so only even powers survive in $\mathrm{Tr}[\Omega(x)]$, and thus can be written in terms of the even function cos. Moreover, since $\mathrm{Tr}[(j_\pm)^2] = -2\kappa^2 < 0$, one really needs an i in the exponential defining $\mathrm{Tr}[\Omega]$, leading to a real value for the $\cos p(x)$.

From the definition of $p(x)$ as derived from the trace of an object with singularities at $x = \pm 1$, $p(x)$ also has singularities at $x = \pm 1$. Using the above observations, and the fact that $\mathrm{Tr}[(j_\pm)^2] = -2\kappa^2$, so near $x = \pm 1$, $-\int d\sigma J_\sigma \to \pi\kappa/(\sqrt{2}(x \mp 1))$, we obtain

$$p(x) \simeq -\frac{\pi\kappa}{x \mp 1} + \cdots \tag{30.40}$$

near $x \simeq \pm 1$.

At $x \to \infty$, we expand in $1/x$ and obtain

$$\mathrm{Tr}\,\Omega \simeq 2 + \frac{1}{x^2}\left(\frac{1}{2}\int_0^{2\pi} d\sigma_1 \int_0^{2\pi} d\sigma_2\right) \mathrm{Tr}(j_\tau(\sigma_1)j_\tau(\sigma_2)) + \cdots$$
$$= 2 - \frac{4\pi^2}{\lambda}\frac{Q_R^2}{x^2} + \cdots$$
$$= 2 - \frac{4\pi^2}{\lambda x^2}(L - 2M)^2 + \cdots, \tag{30.41}$$

where in the first line we have used $j_+ + j_- = 2j_\tau$ and in the last line we have substituted the value of Q_R for the operators whose string duals we are considering.

That leads to

$$p(x) \simeq -\frac{2\pi(L - 2M)}{\sqrt{\lambda}x} + \cdots \tag{30.42}$$

at $x \to \infty$.

As $x \to 0$, expanding $J_\pm(x) \simeq j_\pm(1 \pm x) + \cdots$, and writing

$$J_\sigma(x) = \frac{J_+ - J_-}{2} \simeq j_\sigma + xj_\tau + \cdots = g^{-1}(\partial_\sigma + xl_\tau + \cdots)g, \tag{30.43}$$

we obtain

$$\text{Tr}[\Omega] \simeq 2 + \frac{x^2}{2} \int_0^{2\pi} d\sigma_1 d\sigma_2 \, \text{Tr}[l_\tau(\sigma_1)l_\tau(\sigma_2)]$$
$$= 2 - \frac{4\pi^2 Q_L^2}{\lambda} x^2 + \cdots$$
$$= 2 - \frac{4\pi^2 L^2}{\lambda^2} x^2 + \cdots, \tag{30.44}$$

where in the last line we have substituted the value of Q_L for the operators whose string dual we are considering.

We can now extract the behavior of $p(x)$ at $x \to 0$ as

$$p(x) \simeq 2\pi m + \frac{4\pi^2 L^2}{\lambda} x^2 + \cdots. \tag{30.45}$$

Here m is an arbitrary integer, introduced because of the ambiguity when taking $\cos^{-1}$, i.e. defining various Riemann sheets for the branch cut starting at $x = 0$ in the complex x plane.

Since $p(x)$ has singularities at $x = \pm 1$, it is useful to subtract them, and define a function that has no poles, only branch cut singularities:

$$G(x) \equiv p(x) + \frac{\pi\kappa}{x+1} + \frac{\pi\kappa}{x-1}. \tag{30.46}$$

A function with only branch cut singularities is determined simply by its discontinuity across the cut,

$$G(x + i0) - G(x - i0) \equiv 2\pi i\rho(x), \tag{30.47}$$

via the dispersion relation

$$G(x) = \int_C dy \frac{\rho(y)}{x - y}, \tag{30.48}$$

where C is a contour passing through the cut. The cut extends from $x = 0$ to $x = \infty$. Then on this contour,

$$p(x + i0) - p(x - i0) = 2\pi n_k, \tag{30.49}$$

for $x \in C_k$, since as we saw, $p(x)$ at $x \to 0$ has Riemann sheets differing by 2π times an integer.

Together with the fact that in $G(x)$ we have the contribution of the subtracted poles, we obtain for $x \in C_k$

$$2P \int_C dy \frac{\rho(y)}{x - y} = \frac{2\pi\kappa}{x-1} + \frac{2\pi\kappa}{x+1} + 2\pi n_k. \tag{30.50}$$

The worldsheet energy E of the string in AdS, $\delta S/\delta \dot{X}^0$, must be identified with the conformal dimension Δ via the AdS/CFT map, so

$$\Delta = \frac{\sqrt{\lambda}}{2\pi} \int_0^{2\pi} d\sigma \, \partial_\tau X^0 = \sqrt{\lambda}\kappa. \tag{30.51}$$

Here we have used $X^0 = \kappa\tau$.

We can obtain three more integral equations for $p(x)$ by considering the expansions at $x = 0$ and $x = \infty$.

Using the behavior at infinity of $p(x)$, we can calculate from the $1/x$ term

$$\begin{aligned}
\int_C dx \rho(x) &= \frac{1}{2\pi i} \int dx \left[p(x + i0) - p(x - i0) + \frac{2\pi\kappa}{x + i0} - \frac{2\pi\kappa}{x - i0} \right] \\
&= \int_C dx \left[\frac{1}{x + i0} - \frac{1}{x - i0} \right] \left[\frac{2\pi}{\sqrt{\lambda}x}(2M - L) + 2\pi\kappa \right] \\
&= \frac{2\pi}{\sqrt{\lambda}}(\Delta + 2M - L).
\end{aligned} \tag{30.52}$$

Using the behavior at $x = 0$ of $p(x)$ and of the pole, we can calculate from the constant term in x (giving the coefficient of $\int_C dx/x$)

$$\begin{aligned}
\int_C dx \frac{\rho(x)}{x} &= \frac{1}{2\pi i} \int_C \frac{dx}{x} \left[p(x + i0) - p(x - i0) + \frac{2\pi\kappa(x + i0)}{x^2 - 1} - \frac{2\pi\kappa(x - i0)}{x^2 - 1} \right] \\
&= 2\pi m.
\end{aligned} \tag{30.53}$$

Using the same behavior, we can calculate from the term linear in x (now giving the coefficient of $\int_C dx/x$):

$$\begin{aligned}
\int_C dx \frac{\rho(x)}{x^2} &= \frac{1}{2\pi i} \int_C \frac{dx}{x^2} \left[p(x + i0) - p(x - i0) + \frac{2\pi\kappa(x + i0)}{x^2 - 1} - \frac{2\pi\kappa(x - i0)}{x^2 - 1} \right] \\
&= \frac{2\pi}{\sqrt{\lambda}}(\Delta - L).
\end{aligned} \tag{30.54}$$

30.5 Massive Relativistic Systems from AdS Space and the TBA?

Finally, we turn to a more speculative approach on how to deal with *massive* relativistic systems described by the thermodynamic Bethe ansatz (TBA) by using AdS space and holography. This section is based on ideas I learned from Pedro Vieira [62].

Consider some massive quantum field theory with eigenstates of the Hamiltonian that start with m_1 (the lightest). Consider the scattering of external states of mass m. Then the 4-point S-matrix of the states of mass m should be, near an m_k pole, of the type

$$S \sim \frac{g_k^2}{s - m_k^2}, \tag{30.55}$$

where $s = (p_1 + p_2)^2$ is the Mandelstam variable and m_k is the mass of the physical state. This then defines the 3-point coupling g_k of two m states and a m_k. In particular, we are interested in the case $m_k = m_1$ and $g_k = g_1$, which is the leading contribution to the S-matrix.

Consider next putting this massive quantum field theory inside AdS space, *without modifying AdS space (no backreaction)*. If $m_j R \gg 1$, then the scattering can be considered to happen on a size much smaller than R, so that space looks approximately flat in it. But let the external states of mass m propagate until the AdS boundary. Since $m_j R \gg 1$, they will be of order $\sim e^{-m_j R} <<< 1$ near the boundary.

Since now we have a quantum field theory in an AdS background, it should be dual to some conformal theory, by the same logic that we described for AdS/CFT. Note that we will not have a conformal *field* theory in this case, since there will be no energy-momentum tensor $T_{\mu\nu}$, which is dual to gravity fluctuations in AdS space, which are frozen.

Consider a 3-point function with boundary values x_1, x_2, x_3 for particles with general masses m_1, m_2, m_3 (correlator) for this QFT in AdS space. Since the interaction is almost pointlike, the 3-point function reduces to three geodesics from a common point P in the interior to the three points, each branch being the propagator, proportional to $e^{-m_i \Delta L_{P-x_i}}$. One needs to optimize the point P to obtain a maximum result for the correlator (there is an integration over P that effectively selects the P with the minimal value for the sum of geodesic lengths, such that $\prod_i e^{-m_i \Delta L_{P-x_i}}$ is maximal).

But from the point of view of the conformal theory on the boundary, we expect to obtain a 3-point function consistent with conformal invariance. This restricts it to be of the form

$$\frac{C_{123}}{|x_1 - x_2|^{\Delta_1 + \Delta_2 - \Delta_3}|x_2 - x_3|^{\Delta_2 + \Delta_3 - \Delta_1}|x_3 - x_1|^{\Delta_3 + \Delta_1 - \Delta_2}}, \tag{30.56}$$

where

$$\Delta_j = m_j R \gg 1. \tag{30.57}$$

The coupling g_{ijk} between m_i, m_j, m_k states in the massive QFT is found from C_{ijk} in the conformal theory on the boundary by a rescaling, $g_{ijk} = C_{ijk}/\sqrt{\Delta_i \Delta_j \Delta_k}$.

In this way, we now have a map between a massive relativistic QFT in d-dimensional *flat space* (since the interacting region in AdS space is very small, almost flat) and a conformal theory in $d - 1$ (also flat) dimensions. If the d-dimensional massive relativistic QFT is integrable, like, for instance, in the case $d = 2$ for the sine-Gordon model, we can write a thermodynamical Bethe ansatz.

But one can use now a kind of reverse holographic duality, and use conformal theory results on the $d - 1$ (one for sine-Gordon) dimensional boundary to learn about the massive quantum field theory and its thermodynamical Bethe ansatz. This has not been done until now, though it seems possible to do.

Important Concepts to Remember

- In the thermodynamic limit of the Bethe ansatz, we obtain integral equations for $x = u/L$, which is finite in the limit.

- In this limit, the Bethe roots, which are normally separated by $u_k = u_j \pm i$, form a smooth continuous contour, called the Bethe string.
- One can write integral equations for a density $\rho(x)$ of Bethe roots valid for each Bethe string.
- The string dual to the $SU(2)$ sector is a string moving in an $S^3 \subset S^5$ and fixed in AdS_5.
- The monodromy of the current J_σ defines a function $p(x)$ that obeys integral equations of the type of the Bethe equations.

Further Reading

For AdS strings see the review [61]. See also the review [60] for $\mathcal{N} = 4$ SYM.

Exercises

(1) Consider a Bethe string solution like in Figure 30.1, situated on a circle (two arcs, bounded by (a,b) and (−b, −a)). If C_1 has $n_{C_1(u)} = -1$, what is $n_{\bar{C}_1}(u)$?

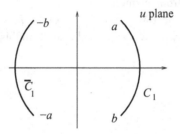

u plane

Fig. 30.1   Bethe strings in the u plane.

(2) Consider the Bethe string given by the interval $(-d, d)$ on the imaginary axis, with density

$$\sigma(x) = 2, \quad -c < x < c$$
$$\tilde{\sigma}(x), \quad c < c < d$$
$$\tilde{\sigma}(-x), \quad -d < x < -c, \tag{30.58}$$

and zero for the rest, where

$$\tilde{\sigma}(x) = \frac{1}{2\pi}\sqrt{(d^2 - x^2)(x^2 - c^2)}\left[-\frac{1}{xcd} + 4\int_{-c}^{c} \frac{dy}{x - y} \frac{1}{\sqrt{(d^2 - y^2)(c^2 - y^2)}} \right] \tag{30.59}$$

and

$$c = \frac{1}{8K(r)}, \quad d = \frac{1}{8\sqrt{r}K(r)}, \quad r = \frac{c^2}{d^2}, \tag{30.60}$$

K and E being the first and second elliptic integrals:

$$K(q) = \int_0^{\pi/2} \frac{d\phi}{\sqrt{1 - q \sin^2 \phi}}; \quad E = \int_0^{\pi/2} d\phi \sqrt{1 - q \sin^2 \phi}. \qquad (30.61)$$

Calculate the energy and the normalization N/L.

(3) Consider the "folded string," whose action on the ansatz (that includes $\dot{t} = \kappa$, $\dot{\phi}_1 = \omega_1$, $\dot{\phi}_2 = \omega_2$) is

$$S = -\frac{\sqrt{\lambda}}{4\pi} \int d\tau \int_0^{2\pi} d\sigma \, [\kappa^2 + \psi'^2 - \cos^2 \psi \, \omega_1^2 - \sin^2 \psi \, \omega_2^2] \qquad (30.62)$$

and

$$q \equiv \sin^2 \psi_0 - \frac{\kappa^2 - \omega_1^2}{\omega_2^2 - \omega_1^2}. \qquad (30.63)$$

Find the equations of motion and calculate

$$E \equiv \frac{\partial L}{\partial \dot{t}}; \quad J_1 \equiv -\frac{\partial L}{\partial \dot{\phi}_1}; \quad J_2 \equiv -\frac{\partial L}{\partial \dot{\phi}_2}. \qquad (30.64)$$

(4) Consider the string in $AdS_3 \times S^3$. Write the full bosonic action in conformal gauge in terms of group elements.

(5) Consider the BMN limit, $M \sim \mathcal{O}(1)$, $L \to \infty$ for the AdS integrability equations. What can you say about the density ρ in this limit?

In this chapter we will describe in more generality integrability at the quantum level, with the particular emphasis on spin chains. We will then see how it manifests itself in string theory, via AdS/CFT, in particular in the case of $\mathcal{N} = 4$ SYM.

31.1 Hints of Integrability

In Chapter 31 we derived the integral equations for classical (long) strings in AdS space (replacing $\kappa = \Delta/\sqrt{\lambda}$):

$$2P \int_C dy \frac{\rho(y)}{x - y} = 2\pi n_k + \frac{2\pi\Delta}{\sqrt{\lambda}} \frac{2x}{x^2 - 1}$$

$$\int_C dx \rho(x) = \frac{2\pi}{\sqrt{\lambda}}(\Delta + 2M - L)$$

$$\int_C dx \frac{\rho(x)}{x} = 2\pi m$$

$$\int_C dx \frac{\rho(x)}{x^2} = \frac{2\pi}{\sqrt{\lambda}}(\Delta - L). \tag{31.1}$$

Note that we have already started the comparison with $\mathcal{N} = 4$ SYM by replacing the charges Q_R and Q_L with their values on the operators whose long string duals we are studying, and by replacing κ with $\Delta/\sqrt{\lambda}$. Further rescaling

$$x \to \frac{4\pi L x}{\sqrt{\lambda}}, \tag{31.2}$$

we obtain the equations

$$2P \int_C dy \frac{\rho(y)}{x - y} = \frac{x}{x^2 - \frac{\lambda}{16\pi^2 L^2}} \frac{\Delta}{L} + 2\pi n_k$$

$$\int_C dx \rho(x) = \frac{M}{L} + \frac{\Delta - L}{2L}$$

$$\int_C dx \frac{\rho(x)}{x} = 2\pi m$$

$$\frac{\lambda}{8\pi^2 L} \int_C dx \frac{\rho(x)}{x^2} = \Delta - L = \frac{\lambda}{8\pi^2} H^{(1-loop)}. \tag{31.3}$$

Then in the limit $\lambda/L^2 \to 0$, we obtain the thermodynamic limit of the Bethe equations from Chapter 30. This is as it should be, since in fact before we obtained the Bethe ansatz only from the 1-loop Hamiltonian, so we are indeed in the limit $\lambda/L^2 \to 0$. At this point, then, it must be puzzling why a perturbative calculation in SYM matches a perturbative one in AdS space (given that AdS/CFT is a strong/weak coupling duality), but we will see later what is supposed to be the complete all-order result.

We can now match individual solutions of the two types of equations. Individual Bethe strings (solutions of the Bethe ansatz equations in the thermodynamic limit) are matched to individual long strings in AdS space. For matching the two sides, one first finds the Bethe curve (string) as a solution to the integral Bethe ansatz equations, then one calculates the resolvent $G(x)$ defined in a similar way to the AdS case, namely by

$$G(x) = \frac{1}{L} \sum_{i=1}^{L} \frac{1}{x - x_i} = \int_C dy \frac{\rho(y)}{x - y}, \tag{31.4}$$

from it the density $\rho(x)$, and from it the energy, which is matched against the energy of the long string in AdS space. One finds agreement for various types of strings: folded string (a string that goes straight, and then returns back folding in on itself), circular string, etc.

As we saw, this matching is related to matching of integrals of motion on the two sides of the correspondence. It is therefore necessary to understand better integrability.

31.2 Integrability and the Yangian

In Part we have described integrability abstractly, as arising from the representations of the Yang-Baxter equations, which generated a transfer matrix $T(u)$ and from it an infinite set of conserved charges (integrals of motion) through the algebraic Bethe ansatz.

But here we want to start in a different way, and we will show later that is in fact equivalent. We describe integrability in terms of the existence of an infinite dimensional algebra called *the Yangian*, defined abstractly as follows.

The *Yangian algebra* $Y(G)$ is an associative Hopf algebra generated by elements J^A and Q^B obeying a regular Lie algebra for J^A,

$$[J^A, J^B] = f^{AB}{}_C J^C, \tag{31.5}$$

and a commutation relation defined by the same structure constants for Q with J,

$$[J^A, Q^B] = f^{AB}{}_C Q_C, \tag{31.6}$$

together with the so-called *Serre relations*:

$$[Q^A, [Q^B, J^C]] + [Q^B, [Q^C, J^A]] + [Q^C, [Q^A, J^B]] = \frac{1}{24} f^{ADK} f^{BEL} f^{CFM} f_{KLM} \{J_D, J_E, J_F\}$$
$$\tag{31.7}$$

and

$$[[Q^A, Q^B], [J^C, Q^D]] + [[Q^C, Q^D], [J^A, Q^B]]$$
$$= \frac{1}{24}(f^{AGL} f^{BEM} f^{KFN} f_{LMN} f^{CD}{}_K + f^{CGL} f^{DEM} f^{KFN} f_{LMN} f_K{}^{AB})\{J_G, J_E, J_F\}. \quad (31.8)$$

Here J^A are the generators of a semisimple Lie algebra for a group G and $\{A, B, C\}$ denotes the symmetric product.

Then the Yangian basis is J_n^A, where $J_0^A = J^A$, $J_1^A = Q^A$ and for $n > 1$, J_n^A is an n-local operator appearing in the $(n-1)$–form commutator (repeated commutator, antisymmetrized) of the Q^A's.

It is left as an exercise to prove that for $SU(2)$, the first Serre relation (31.7) is trivial, and for $SU(N)$ with $N \geq 3$, the first Serre relation (31.7) implies the second one (31.8).

2-Dimensional Models
Specializing to models in 1+1 dimensions, with symmetries defined by the Lie algebra for a group G, $[T_A, T_B] = f_{AB}{}^C T_C$, we obtain conserved currents j_μ^A, i.e. $\partial^\mu j_\mu^A = 0$.

Then constructing the current

$$j_\mu = \sum_A j_\mu^A T_A, \quad (31.9)$$

if it satisfies the condition to be a flat connection (gauge field with zero field strength), i.e. if

$$\partial_\mu j_\nu - \partial_\nu j_\mu + [j_\mu, j_\nu] = 0, \quad (31.10)$$

we can construct a Yangian as follows. First, the conserved charges of the current j_μ^A are as usual,

$$J^A = \int_{-\infty}^{+\infty} dx\, j_0^A(x, t), \quad (31.11)$$

but moreover, we also have the new conserved charges

$$Q^A = f^A{}_{BC} \int_{-\infty}^{+\infty} dx \int_x^{+\infty} dy\, j_0^B(x, t) j_0^C(y, t) - 2 \int_{-\infty}^{+\infty} dx\, j_0^A(x, t), \quad (31.12)$$

which is therefore 2-local (double integral), and J^A, Q^A generate the Yangian algebra.

31.3 Quantum Integrability for Spin Chains

Specializing further to the case of spin chains, i.e. having a discrete spatial coordinate x, defined by a site i at which we have "spins" in the representation $\mathcal{R}$ of the group G, we have a Yangian symmetry defined as follows. The charge J^A is found as usual, the integral over x turning into a sum over i:

$$J^A = \sum_i J_i^A. \quad (31.13)$$

If moreover $G = SU(N)$, then we can also write the charge Q^A as

$$Q^A = f^A{}_{BC} \sum_{i<j} J_i^B J_j^C, \tag{31.14}$$

where with respect to the continuum case, the double integral has turned into a double sum.

In this $SU(N)$ case, the T_A are traceless $N \times N$ matrices (otherwise arbitrary), with the Yangian defined by the algebra

$$[J_b^a, J_d^c] = \delta^c{}_b J_d^a - \delta^a{}_d J_b^c$$
$$[J_b^a, Q_d^c] = \delta^c{}_b Q_d^a - \delta^a{}_d Q_b^c, \tag{31.15}$$

and the Serre relations

$$[J_b^a, [Q_d^c, Q_f^e]] - [Q_b^a, [J_d^c, Q_f^e]]$$
$$= \frac{h^2}{4} \sum_{p,q} \left([J_b^a, [J_p^c J_d^p, J_q^e J_f^q]] - [J_p^a J_b^p, [J_d^c, J_q^e J_f^q]] \right). \tag{31.16}$$

Note that the second Serre relation is implied by the first in the case of $SU(N)$, as we already mentioned.

Now we want to make contact with the other definition of integrability, from Chapter 7. In that case, one had a Lax operator $L_j(u)$ at site j, out of which one constructed the *transfer matrix*, or monodromy matrix:

$$T(u) = L_L(u) \cdots L_2(u) L_1(u). \tag{31.17}$$

The Lax operator and the monodromy were found by solving the Yang-Baxter equation in the algebraic Bethe ansatz.

But the Lax operator $L_j(u)$ acted as a discrete connection (gauge field) for the Lax equation

$$\psi_{j+1} = L_j \psi_j, \tag{31.18}$$

which was the discrete version of the Dirac equation

$$(\partial_x + A_x)\psi(x) = 0, \tag{31.19}$$

leading to a monodromy

$$T(x, u) = P \exp\left[-\int_0^x dy A_x(y, u) \right]. \tag{31.20}$$

The Lax connection is also flat.

The existence of the monodromy matrix $T(u)$ meant that there are $L - 1$ charges Q_k that commute with the Hamiltonian and between themselves,

$$[Q_k, Q_l] = 0; \quad [Q_k, H] = 0, \tag{31.21}$$

obtained as the coefficients of a series expansion around $u = 0$ of $\text{Tr}\, T(u)$. In particular, Q_k involves *at least* k-neighboring interactions (so that Q_1, identified with H, has next to neighbor interactions, $\sum_k (\ldots)_k (\ldots)_{k+1}$).

We can now make the connection with the definition of integrability in terms of the Yangian. The Yangian charges J_n^A arise from the coefficients of the expansion of $T(u)$ (without the trace! since the Yangian charges have gauge indices) *around $u = \infty$*. For the case of $SU(N)$, the expansion is

$$T(u) = 1 + h \sum_{n=0}^{\infty} u^{-n-1} t^{(n)}, \tag{31.22}$$

and the relation of the coefficients with the Yangian charges is, for the two that generate the Yangian,

$$P^A = J_0^A = t^{(0)}; \quad Q^A = J_1^A = t^{(1)} - \frac{h}{2} t^{(0)} \cdot t^{(0)}. \tag{31.23}$$

By contrast, the expansion that gives the Bethe ansatz charges Q_k is around $u = 0$ and is defined as follows. For the trace, one can write

$$\operatorname{Tr} T(u) = 2 \cos P(u), \tag{31.24}$$

since the eigenvalues of $T(u)$ are opposite phases ($e^{\pm ip}$). Then the expansion of $P_0(u)$ contains a pole, and a regular function (Taylor-expandable):

$$P_0(u) = -\frac{sL}{u} + \sum_{n=0}^{\infty} u^n I_n. \tag{31.25}$$

Here I_n are the integrals of motion, and I_0 is the momentum, I_1 is the energy, etc.

Note that here u, also called λ, is a spectral parameter, not the individual Bethe root.

31.4 Classical Integrability for Strings in AdS Space

We can easily match the formalism with the formalism in AdS space. As we saw, $J_\pm = j_\pm/(1 \mp x)$ are the analogs of the Lax connections, also being flat connections:

$$\partial_- J_+ - \partial_+ J_- + [J_+, J_-] = 0. \tag{31.26}$$

In that case, the monodromy of this connection,

$$\Omega(x) = P \exp\left[-\int_0^{2\pi} d\sigma J_\sigma\right], \tag{31.27}$$

gives both the Yangian charges, through its expansion at $x = \infty$, and the integrals of motion, through its expansion at $x = 0$.

We have seen that in the last chapter, since

$$\operatorname{Tr} \Omega(x) = 2 \cos p(x) \tag{31.28}$$

was expanded, and we have found at least the charges Q_R (in the expansion at $x = \infty$), and Q_L (in the expansion at $x = 0$).

The Dirac equation for the connection J_σ is now

$$[\partial_\sigma - J_\sigma]\psi = 0 \Rightarrow \left[\partial_\sigma - \frac{1}{2}\left(\frac{j_+}{x-1} + \frac{j_-}{x+1}\right)\right]\psi = 0, \qquad (31.29)$$

and it implies that $\Omega(x)$ is a monodromy for $\psi(\sigma, x)$:

$$\psi(\sigma + 2\pi; x) = \Omega(x)\psi(\sigma; x). \qquad (31.30)$$

Diagonalizing $\Omega(x)$, the eigenvalues are opposite phases, $e^{\pm ip(x)}$, which means that $\mathrm{Tr}\,\Omega(x) = 2\cos p(x)$.

The definition of the monodromy from the Dirac equation is given by first writing an infinitesimal relation, from

$$(\partial_\sigma + A_\sigma)\psi(\sigma) = 0 \Rightarrow \psi(\sigma + d\sigma) - \psi(\sigma) = d\sigma A_\sigma \psi(\sigma), \qquad (31.31)$$

which integrates to

$$\psi(\sigma_2) = P\exp\left[-\int_{\sigma_1}^{\sigma_2} d\sigma A_\sigma\right]\psi(\sigma_1), \qquad (31.32)$$

defining the monodromy $\Omega(x)$.

Note that we can discretize the infinitesimal relation above to

$$\psi_{j+1} - \psi_j = A_j\psi_j \Rightarrow \psi_{j+1} = (1 + A_j)\psi_j = L_j\psi_j, \qquad (31.33)$$

which is the form in terms of the Lax equation (in terms of the Lax connection).

As we have seen, we have matched the *quantum* integrability of the SYM spin chain with the *classical* integrability of the (long) strings in AdS. Indeed, the strings in AdS were classical, thus long (so that quantum fluctuations are unimportant), and on the SYM side this corresponds to the highly quantum regime. However, while we should have been at large SYM coupling, we actually investigated only the 1-loop Hamiltonian of the spin chain until now.

31.5 Higher Loops in $\mathcal{N} = 4$ SYM and All-Loop Bethe Ansatz

But people have considered all loops before now. First, the spin chain Hamiltonian of SYM was calculated explicitly at the first few loops. Some results for the dilatation operator (Hamiltonian) are the 2-loop result

$$\mathcal{D}_{2-\mathrm{loop}} = \sum_{j=1}^{L}\left(-\vec{\sigma}_j \cdot \vec{\sigma}_{j+2} + 4\vec{\sigma}_j \cdot \vec{\sigma}_{j+1}\right) \qquad (31.34)$$

and the first term of the 3-loop result

$$\mathcal{D}_{3-\mathrm{loop}} = -\sum_{j=1}^{L}\vec{\sigma}_j \cdot \vec{\sigma}_{j+3} + \cdots \qquad (31.35)$$

It was found that the Hamiltonian continues to be *integrable* at higher loops. In fact, there are results proving integrability of the $\mathcal{N} = 4$ SYM at all loops, at least in interesting spin chain sectors, but also in general (with less rigor).

Moreover, integrability plus the correct BMN limit uniquely fixes the Hamiltonian up to 5-loops.

In fact, Beisert, Dippel, and Staudacher [65] have proved that there exists an all-loop Hamiltonian, defined implicitly through its Bethe equations, and it reproduces the 5-loop Hamiltonian fixed as above. There are some particular type of corrections that appear at even higher loops, which we will describe shortly, but until then let us describe the BDS Bethe ansatz. In fact, it turns out that the ansatz is actually just asymptotic (asymptotic Bethe ansatz), i.e. it is strictly speaking valid only at $L \to \infty$.

The K Bethe ansatz equations (for a system of K magnons) have the same functional form as the ones for the Heisenberg Hamiltonian,

$$e^{ip_k L} = \prod_{j=1, j \neq k}^{K} S(p_k, p_j), \tag{31.36}$$

where also the 2-body S-matrix has the same functional form when written as a function of the rapidities u_k,

$$S(p_k, p_j) = \frac{u(p_k) - u(p_j) + i}{u(p_k) - u(p_j) - i}, \tag{31.37}$$

and the only difference is in the relation of the rapidity functions to the momenta p:

$$u(p) = \frac{1}{2} \cot \frac{p}{2} \sqrt{1 + 8g^2 \sin^2 \frac{p}{2}}. \tag{31.38}$$

Then the conserved charges (integrals of motion) are found, as a sum over sites,

$$Q_r = \sum_{k=1}^{N} q_r(p_k), \tag{31.39}$$

where the charge per site is

$$q_r(p) = \frac{2 \sin\left[\frac{r-1}{2} p\right]}{r - 1} \left(\frac{\sqrt{1 + 8g^2 \sin^2 \frac{p}{2}} - 1}{2g^2 \sin \frac{p}{2}} \right)^{r-1}. \tag{31.40}$$

In particular, for $r = 2$ we obtain the Hamiltonian (energy), $H = Q_2$, and the energy per site is

$$h = q_2(p) = \frac{1}{g^2} \left[\sqrt{1 + 8g^2 \sin^2 \frac{p}{2}} - 1 \right]. \tag{31.41}$$

The total dilatation operator is the (rescaled) total energy:

$$\mathcal{D} = g^2 \sum_k h(p_k). \tag{31.42}$$

Another parametrization for the all-loop Bethe ansatz is better for the purposes of generalizing. Define

$$x(u) \equiv \frac{u}{2} + \frac{u}{2}\sqrt{1 - \frac{2g^2}{u^2}}, \qquad (31.43)$$

which inverts to

$$u(x) = x + \frac{g^2}{2x}. \qquad (31.44)$$

Then e^{ip} is rewritten as

$$e^{ip} = \frac{x(u + i/2)}{x(u - i/2)}. \qquad (31.45)$$

Defining $u_k \equiv u(p_k)$ (at site k), and moreover

$$x_k^{\pm} \equiv x(u_k \pm i/2) \Rightarrow p_k = -i \log \frac{x_k^+}{x_k^-}, \qquad (31.46)$$

the conserved charges per site are

$$q_r(x_k) = \frac{i}{r-1}\left(\frac{1}{(x_k^+)^{r-1}} - \frac{1}{(x_k^-)^{r-1}}\right). \qquad (31.47)$$

The Bethe ansatz equations, defining the asymptotic Bethe ansatz, are now rewritten as

$$\left(\frac{x(u_k + i/2)}{x(u_k - i/2)}\right)^L = \prod_{j=1, j \neq k}^{K} \frac{u_k - u_j + i}{u_k - u_j - i} \Rightarrow$$

$$\left(\frac{x_k^+}{x_k^-}\right)^L = \prod_{j=1, j \neq k}^{K} \frac{x_k^+ - x_k^-}{x_k^- - x_j^+} \frac{1 - \frac{g^2}{2x_k^+ x_j^-}}{1 - \frac{g^2}{2x_k^- x_j^+}} \equiv \prod_{j=1, j \neq k}^{K} S_0(p_k, p_j). \qquad (31.48)$$

The full result for the Bethe ansatz equations is obtained by multiplying the S-matrix $S_0(p_k, p_j)$ by a complicated phase $e^{2i\theta(p_k, p_j)}$:

$$S_0(p_k, p_j) \to S_0(p_k, p_j)e^{2i\theta(p_k, p_j)}. \qquad (31.49)$$

We have not mentioned the applicability of the Bethe ansatz. It was originally written for the $SU(2)$ sector. Other sectors were later defined, with $Sl(2)$ and $SU(1|1)$ symmetries. They are all embedded into an $SU(2|3)$ sector, with fields $(Z, \phi^1, \phi^2|\psi^1, \psi^2)$ (ψ^1, ψ^2 are fermionic), and moreover for the whole $\mathcal{N} = 4$ SYM theory one can write an $SU(2, 2|4)$ symmetric spin chain.

It was found by Beisert [66] that S_0 is fixed by symmetry, already in the $SU(2|3)$ sector, and also in the full $SU(2, 2|4)$, whereas the phase $e^{2i\theta}$ is not, so it is a dynamical quantity. However, various arguments, including crossing symmetry and a transcendentality principle, were used to completely fix the function form of the phase as well, in [67]. This final asymptotic Bethe ansatz, with $S_0 e^{2i\theta}$, is believed to hold for the whole theory.

In conclusion, we have a map between integrability in $\mathcal{N} = 4$ SYM and integrability in AdS space that allows one to fix the Bethe ansatz for the asymptotic spin chains, and thus theoretically solve the theory, at least for large operators.

Important Concepts to Remember

- The integral equations for SYM and AdS strings match in the limit $\lambda/L^2 \to 0$ (since we have used the 1-loop ansatz in the SYM case).
- Integrability can be defined by having an infinite dimensional algebra called Yangian, generated by a usual Lie algebra for J^A, together with a new generator Q^A, obeying Serre relations, via the multiple (n-form) commutators of Q^A's.
- In 2-dimensional models, a current $j_\mu = j_\mu^A T_A$ that is a flat connection gives both the J^A charges as usual integrals, and the Q^A charges as 2-local objects (double integrals).
- For $SU(N)$ spin chains, $J^A = \sum_i J_i^A$ and $Q^A = f^A{}_{BC} \sum_{i<j} J_i^B J_j^C$.
- The monodromy matrix $T(u)$ constructed in the algebraic Bethe ansatz by solving the Yang-Baxter equation for a particular representation contains in the expansion at $u = 0$ of $P_0(u)$, with $\mathrm{Tr}\, T(u) = 2 \cos P_0(u)$, all the integrals of motion Q_k (or I_n), whereas in the expansion at $u = \infty$ it contains all the Yangian charges J_n^A.
- In AdS space we have a monodromy matrix $\Omega(x)$ made out of a flat connection $J_\pm$, mapped to the Lax connection of the spin chain, and its Dirac equation is related to the Lax equation.
- The higher loop spin chain Hamiltonian of $\mathcal{N} = 4$ SYM continues to be integrable.
- Integrability plus the BMN limit fixes the Hamiltonian up to five loops.
- The all-loop Hamiltonian is defined implicitly through an asymptotic Bethe ansatz, which in fact fixes all the charges (integrals of motion) $q_k(p)$.
- We can rewrite the Bethe ansatz in the spectral parameter plane, and then there is a 2-body S-matrix S_0 that is completely fixed by the total symmetry of the SYM, and a dynamical phase that is unfixed by it. The whole spin chain, and in particular exact subsectors like $Sl(2)$, $SU(1|1)$, obey the same Bethe ansatz.

Further Reading

The review [61] for integrability in AdS, [63] and [64] for the Yangian, and for higher loops the original papers [65], [66], and [67].

Exercises

(1) Verify that in the BMN limit $M \to \mathcal{O}(1)$, $L \to \infty$, the AdS/SYM Bethe string equations give the expected BMN scaling.
(2) Verify that for $SU(2)$ the first Serre relation is trivial, and that for $SU(3)$ the first Serre relation implies the other.

(3) For the 2-dimensional case with Lie algebra $[T_A, T_B] = f_{AB}{}^C T_C$, calculate the commutator $[J^A, Q^B]$.

(4) Consider the monodromy matrix of the Heisenberg $XXX_{1/2}$ model:

$$T(\lambda) = \frac{1}{2} W(\lambda) \, \mathbb{1} - \frac{i}{2} W^{-1}(\lambda) Q^i \sigma_i$$

$$W(\lambda) = \sqrt{2 + \sqrt{4 - \vec{Q}^2(\lambda)}}$$

$$\vec{Q}^2(\lambda) = 4 \sin^2(2P_0(\lambda)) \tag{31.50}$$

and for an individual magnon k, we have the conserved charges

$$q_r(u_k) = \frac{i}{r-1} \left[\frac{1}{(u_k + i/2)^{r-1}} - \frac{1}{(u_k - i/2)^{r-1}} \right]. \tag{31.51}$$

Write explicitly the transfer matrix $T(\lambda)$ up to (including) two loops, using the general 2-loop expansion.

AdS/CFT Phenomenology: Lifshitz, Galilean, and Schrödinger Symmetries and their Gravity Duals

In this chapter we switch gears from exact statements, for a well-defined (through a heuristic derivation from a system of branes) pair of dual theories, to a more phenomenological approach, designed to deal with certain more familiar condensed matter systems.

The approach we start to examine here is therefore less ambitious, and tries to capture some of the relevant strongly coupled physics in condensed matter in an effective gravitational (geometric) picture, but without the claim to have an exact description, like in previous chapters.

32.1 Lifshitz Symmetry and its Phenomenological Gravity Dual

In some condensed matter systems, near a fixed point (i.e., near a phase transition), one finds a "dynamical scaling," or "Lifshitz scaling," i.e., space and time scale differently at these "Lifshitz points": instead of the relativistic scaling where time scales the same way as space, $\vec{x} \to \lambda\vec{x}; t \to \lambda t$, now we have

$$t \to \lambda^z t; \quad \vec{x} \to \lambda\vec{x}. \tag{32.1}$$

Here z is called the dynamical critical exponent.

A model field theory with Lifshitz scaling is the "Lifshitz field theory" with $z = 2$, which arises near multicritical points of some known materials, with Lagrangean

$$\mathcal{L} = \int d^d x dt [(\partial_t \phi)^2 - k(\vec{\nabla}^2 \phi)^2]. \tag{32.2}$$

It is easy to see that it indeed shows Lifshitz scaling with exponent $z = 2$.

As we said, we will consider a "phenomenological AdS/CFT" approach, in which we try to realize the symmetry group geometrically. The simplest example, in the relativistic case, is of the usual AdS/CFT type: the isometry group of AdS_{d+1} is $SO(2, d)$, the same as the conformal group of d-dimensional Minkowski space. Therefore we can assume that AdS/CFT still holds for any gravitational theory in AdS_{d+1}, even without a derivation from a string theory duality, just that now we don't know what the field theory is, we can only guess. That is, we assume any gravitational theory in an AdS background is holographic. There is by now a lot of evidence for this.

The same principle is now applied to cases without conformal symmetry, with the extra complication that we need to find a "gravity dual" that has the right symmetry, matching the symmetry algebra we want to describe.

In particular, in order to realize geometrically in $d+1$ dimensions the Lifshitz scaling, a simple gravity dual was proposed,

$$ds_{d+1}^2 = L^2 \left(-\frac{dt^2}{u^{2z}} + \frac{d\vec{x}^2}{u^2} + \frac{du^2}{u^2} \right), \tag{32.3}$$

where

$$d\vec{x}^2 = dx_1^2 + \cdots + dx_{d-1}^2 \tag{32.4}$$

and $0 < u < +\infty$.

We immediately see that the metric is invariant under a scaling transformation that generalizes Lifshitz scaling. From the condition that $d\vec{x}^2/u^2$ be invariant we find that the transformation is

$$t \to \lambda^z t; \quad \vec{x} \to \lambda \vec{x}; \quad u \to \lambda u. \tag{32.5}$$

The generator of infinitesimal scaling transformations is then

$$D = -i(zt\partial_t + x^i\partial_i + u\partial_u). \tag{32.6}$$

The one problem with the above gravity dual is that the metric is not geodesically complete, and $u = \infty$ (which corresponds, according to the AdS/CFT dictionary we found, to the IR, or low energy) is a pp curvature singularity: even though the Ricci scalar is

$$R = -\frac{2}{L^2}(z^2 + 2z + 3), \tag{32.7}$$

so it is constant, there is a singularity in the Riemann tensor. This means that we cannot define things exactly at $u = \infty$, we need to regularize a bit away from it.

Other Killing vectors (isometries) of the metric, which generate the isometry algebra, are the usual Poincaré generators: the energy (time translation generator) H, the momentum (translation generator) P_i, and Lorentz generator M_{ij}:

$$H = -i\partial_t; \quad P_i = -i\partial_i; \quad M_{ij} = -i(x^i\partial_j - x^j\partial_i). \tag{32.8}$$

A simple calculation of the algebra finds that the usual Poincaré algebra for H, P_i, M_{ij} is extended to

$$\begin{aligned}
[D, H] &= z\partial_t = izH \\
[D, P_i] &= \partial_i = iP_i \\
[D, M_{ij}] &= 0 \\
[M_{ij}, P_k] &= \delta_i^k\partial_j - \delta_k^j\partial_i = i(\delta_k^i P_j - \delta_k^j P_i) \\
[M_{ij}, M_{kl}] &= i(\delta_{ik}M_{jl} - \delta_{jk}M_{il} + \delta_{il}M_{kj} - \delta_{jl}M_{ki}) \\
[P_i, P_j] &= 0.
\end{aligned} \tag{32.9}$$

This is the Lifshitz algebra, realized geometrically in the $d+1$–dimensional metric.

We have found by trial and error a metric that has the required symmetry, but a natural question arises: Is this metric a solution of some gravitational theory? Indeed, we have examples of actions that generate such a solution for $d = 3$, namely the action found in [68],

$$S = \int d^4x \sqrt{-g}(R - 2\Lambda) - \frac{1}{2}\int \left(\frac{1}{l^2}F_{(2)} \wedge *F_{(2)} + F_{(3)} \wedge *F_{(3)}\right) - c\int B_{(2)} \wedge F_{(2)},$$
(32.10)

where $B_{(2)}$ is the analog of the B-field,

$$F_{(2)} = dA_{(1)}; \quad F_{(3)} = dA_{(2)},$$
(32.11)

and the cosmological constant is related to the parameters L and z of the solution (gravity dual) by

$$\Lambda = -\frac{z^2 + z + 4}{2L^2}.$$
(32.12)

Such an action can appear in string theory.

Other matter can also be added to gravity with the cosmological constant to obtain another example of relativistic gravitational action that has the same solution:

$$S = \frac{1}{2\kappa_N^2}\int dt d^d x dr \sqrt{-g}\left[R - 2\Lambda - \frac{1}{4}F_{\mu\nu}F^{\mu\nu} - \frac{1}{2}m^2 A_\mu A^\mu\right].$$
(32.13)

This action was found in [69].

One can, however, also obtain the metric background dual to a nonrelativistic field theory in a theory of nonrelativistic gravity, specifically the "Horava gravity" (that has an action that treats space and time differently), as was done in [70].

32.2 Galilean and Schrödinger Symmetries and their Phenomenological Gravity Duals

We can realize geometrically via the same "phenomenological AdS/CFT" approach also larger algebras. These are relevant for the study of cold atoms (atoms at super-small temperatures) and fermions at unitarity, i.e. fermions with infinite scattering length and zero range interaction. These still have Lifshitz scaling, with the same generators $\{M_{ij}, P_i, H, D\}$, but moreover have another symmetry, the "Galilean boosts," i.e. the non-relativistic (Galilean) version of boosts,

$$t \to t; \quad x_i \to x_i - v_i t$$
(32.14)

as well as another symmetry we can identify with a conserved rest mass (invariant under Galilean boosts), or equivalently (if we divide by the mass of one particle), the particle number N. Together, these generators form a *conformal Galilean algebra*. We will find the algebra from the gravity dual, so we will not cite it here.

In the case of $z = 2$, there is an extra symmetry generator C, a "special conformal generator," and the resulting extended algebra is called the "Schrödinger algebra." Note that (by an abuse of notation) sometimes one refers to the conformal Galilean algebra at any z as Schrödinger algebra. The Schrödinger algebra is so named because it is the symmetry of the Schrödinger equation of the free particle.

It is presently not known how to write a gravity dual for the conformal Galilean algebra in $d + 1$ dimensions, as is usual, though there are hints that there should exist such a metric. Instead, to realize geometrically the conformal Galilean algebra, we must generalize AdS/CFT to write a $d + 2$–dimensional gravity dual, specifically

$$ds^2 = L^2 \left(-\frac{dt^2}{u^{2z}} + \frac{d\vec{x}^2}{u^2} + \frac{du^2}{u^2} + \frac{2dt\, d\xi}{u^2} \right). \tag{32.15}$$

The extra coordinate ξ appears as a lightcone-type coordinate, which will be related to how we embed in string theory.

Note that this metric is not time reversal invariant, unlike the metric dual to the Lifshitz symmetry, $t \to -t$ changing the sign of $dt\, d\xi$ term. We also note that the metric is now nonsingular (it doesn't have anymore a pp curvature singularity), since it is conformal to a pp wave spacetime, by removing the L^2/u^2 factor.

The scaling symmetry of the metric is now modified with respect to the gravity dual of the Lifshitz symmetry, by the need to scale also ξ. To make $dtd\xi/u^2$ invariant, we must have $\xi \to \lambda^{2-z}\xi$, so the symmetry is

$$t' = \lambda^z t, \quad \vec{x}' = \lambda\vec{x}; \quad u' = \lambda u; \quad \xi' = \lambda^{2-z}\xi. \tag{32.16}$$

The generator of infinitesimal scaling transformations is then

$$D = -i(zt\partial_t + x^i\partial_i + u\partial_u + (2 - z)\xi\partial_\xi), \tag{32.17}$$

and the form of the Poincaré generators H, P_i, M_{ij} is the same.

In the gravity dual, the Galilean boost $\vec{x}' = \vec{x} - \vec{v}t$ must be supplemented by a transformation of ξ to make a symmetry of the metric. We find that we need

$$\xi' = \xi + \frac{1}{2}(2\vec{v} \cdot \vec{x} - v^2 t), \tag{32.18}$$

where the first term is linear in the parameter $\vec{v}$ and the second is a nonlinear term, neglected for infinitesimal transformations. Therefore the generator of Galilean boosts is

$$K_i = -i(x^i\partial_\xi - t\partial_i). \tag{32.19}$$

In addition to this, it is clear that translations in ξ is also a symmetry, which we can identify with the particle number

$$N = -i\partial_\xi. \tag{32.20}$$

The algebra of these symmetry operators is easily found by commuting the operators to be the following:

$$[K_i, P_j] = \delta_{ij}\partial_\xi = i\delta_{ij}N$$
$$[D, K_i] = zt\partial_i - x^i\partial_\xi + (2 - z)x^i\partial_\xi - t\partial_i = (1 - z)iK_i$$
$$[K_i, M_{ij}] = t(\delta_{ij}\partial_j - \delta_{jk}\partial_i) + \delta_{jk}x^i\partial_\xi - \delta_{ik}x^i\partial_\xi = i(\delta_{jk}K_i - \delta_{ik}K_j)$$
$$[K_i, H] = -\partial_i = -iP_i$$
$$[K_i, N] = [H, N] = [P_i, N] = [M_{ij}, N] = 0$$
$$[D, N] = (2 - z)\partial_\xi = (2 - z)iN$$
$$[D, H] = ziH$$
$$[D, P_i] = iP_i. \tag{32.21}$$

This is the conformal Galilean algebra.

In the case of $z = 2$, we also have a special conformal generator C, for the symmetry

$$u \to (1 - at)u$$
$$x^i \to (1 - at)x^i$$
$$t \to (1 - at)t$$
$$\xi \to \xi - \frac{a}{2}(\vec{x}^2 + u^2). \tag{32.22}$$

The generator of infinitesimal transformations is then

$$C = +i\left[t(u\partial_u + x^i\partial_i + t\partial_t) + \frac{\vec{x}^2 + u^2}{2}\partial_\xi\right]. \tag{32.23}$$

We easily find that the extra commutators are given by (for $z = 2$)

$$[D, C] = -2iC$$
$$[H, C] = +iD$$
$$[C, P_i] = iK_i$$
$$[M_{ij}, C] = [K_i, C] = 0. \tag{32.24}$$

This is the Schrödinger algebra.

32.3 String Theory Embeddings of Schrödinger Symmetry

The metric in the case of $z = 2$ dual to the Schrödinger algebra can be realized in string theory, as was done in [71, 72, 73]. As we saw, translations in ξ corresponded to the particle number N, conserved in the field theory. That suggests the fact that one needs to consider a discrete light cone quantization (DLCQ) of some theory, with the compact direction ξ, such that we consider the sector of given momentum in ξ, $p_\xi = N/R$.

Indeed, the procedure that gives the field theory with Schrödinger symmetry and its gravity dual is a DLCQ on the field theory side, killing off one field theory coordinate (ξ). In the gravity dual, the procedure is the so-called null Melvin twist (written explicitly for the relevant case in one of the exercises), a set of steps, each of which keeps us within string theory. The procedure is applied to the $AdS_5 \times S^5$ solution, leading to a gravity dual of a 2+1–dimensional theory.

For generality, we describe the result for the case of nonzero temperature, to be explained in the next chapter. The solution after the null Melvin twist is

$$ds^2 = r^2 \left[-\frac{\beta^2 r^2 f(r)}{k(r)}(dt+dy)^2 - \frac{f(r)}{k(r)}dt^2 + \frac{dy^2}{k(r)} + d\vec{x}^2 \right]$$
$$+ \frac{dr^2}{r^2 f(r)} + \frac{(d\psi + A)^2}{k(r)} + d\Sigma_4^2, \tag{32.25}$$

where

$$f(r) = 1 - \frac{r_+^4}{r^4}$$

$$k(r) = 1 + \beta^2 \frac{r_+^4}{r^4}, \tag{32.26}$$

and this metric has a nonzero temperature given by

$$T = \frac{r_+}{\pi \beta}. \tag{32.27}$$

At $T = 0$, we see that $k(r) = f(r) = 1$. Then, by Kaluza-Klein reducing on ψ and Σ_4, we obtain the $z = 2$ gravity dual to Schrödinger symmetry.

However, the field theory obtained by the null Melvin twist construction is not a usual, local, field theory, but rather a complicated nonlocal theory, characterized by the existence of a dipole, hence being called a "dipole theory."

Use of AdS/CFT

There are many uses for AdS/CFT in the phenomenological context of the Schrödinger symmetry, the simplest being the use of n-point functions from the partition function. As explained in Chapter 27, the basic AdS/CFT relation is the identification of the partition functions on the two sides of the correspondence,

$$Z_{\text{CFT}}[\phi_0] = Z_{\text{sugra}}[\phi[\phi_0]], \tag{32.28}$$

where ϕ_0 in the CFT is a source for the operator coupling to ϕ, and in the supergravity is a boundary source for the field ϕ, considered on-shell.

Then for a scalar, we find the behavior near the boundary as

$$\phi(r) \propto r^\Delta e^{-i\omega t + i\vec{k}\cdot\vec{x} + il\xi}, \tag{32.29}$$

where the power Δ is now slightly modified with respect to the AdS case, as

$$\Delta_\pm = \frac{d}{2} + 1 \pm \sqrt{\left(\frac{d}{2}+1\right)^2 + m^2 L^2 + \delta_{z,2} l^2}. \tag{32.30}$$

Important Concepts to Remember

- In a phenomenological approach to AdS/CFT, one uses symmetries to define the gravity/field theory dual pair.
- Near fixed points (phase transitions) in condensed matter, often one has Lifshitz scaling $t \to \lambda^z t, x^i \to \lambda x^i$.
- The invariance group of Lifshitz models is realized geometrically in a gravity dual with a term $-dt^2/u^{2z}$.
- The Lifshitz gravity dual can be realized as a solution in both relativistic and non-relativistic gravitational theories.
- The conformal Galilean algebra includes Galilean boosts and a particle number, and is realized geometrically in $d+2$ dimensions by adding a term with $2dt\, d\xi/u^2$ to the Lifshitz metric.
- In the case $z = 2$, the conformal Galilean algebra becomes the Schrödinger algebra, the algebra of symmetries of the Schrödinger equation of the free particle, which includes a special conformal generator C.
- The Schrödinger gravity dual can be embedded in string theory, by a null Melvin twist of the $AdS_5 \times S^5$ duality, leading to a nonlocal "dipole" field theory.
- One can use AdS/CFT in the usual way for conformal Galilean theories, with just a simple modification to the $\Delta(m)$ formula.

Further Reading

For mode details, see the review [1]. The gravity dual of Lifshitz symmetry was found in [68], and the gravity dual of conformal Galilean and Schrödinger symmetries was found in [74] and [75]. A relativistic gravity action that has the Lifshitz spacetime as a solution was found in [68] and [69], and a nonrelativistic gravity embedding was found in [70]. The string theory embedding of the gravity dual of Schrödinger symmetry was found in [71, 72, 73].

Exercises

(1) Prove explicitly the invariance of the gravity dual metric for the Lifshitz invariance under $\{P_i, H, M_{ij}, D\}$.

(2) Prove explicitly the invariance of the gravity dual metric for the conformal Galilean invariance under $\{P_i, H, M_{ij}, D, K_i, N\}$ and for $z = 2$ also under C.

(3) Calculate the Ricci scalar for the Lifshitz metric and show that it is given by

$$R = -\frac{2}{L^2}(z^2 + 2z + 3).\tag{32.31}$$

(4) Calculate the Ricci scalar for the conformal Galilean metric.

(5) The null Melvin twist for the $AdS_5 \times S^5$ geometry:

$$ds^2 = r^2(-dt^2 + d\vec{x}_2^2 + dy^2) + \frac{d\vec{r}^2}{r^2} + (d\psi + A)^2 + d\Sigma_{\mathbb{CP}^2}^2$$

$$F_{(5)} = dC_{(4)} = 2(1 + *)d\psi \wedge J \wedge J, \tag{32.32}$$

where $dA + 2J$ and $vol(\mathbb{CP}^2) = J \wedge J/2$, is defined by the sequence:

(a) Boost by amount γ on the y direction

(b) T-dualize on y.

(c) Twist (replace) the one-form $\sigma = d\psi$ as $\sigma \to \sigma + \alpha dy$.

(d) T-dualize again on y.

(e) Boost by $-\gamma$ along y.

(f) Scale the boost and the twist as: $\gamma \to \infty$ and $\alpha \to 0$, where $\beta = \alpha e^\gamma/2 = $ fixed. Show that after this procedure, one obtains

$$ds^2 = r^2(-2dudv - r^2du^2 + d\vec{x}^2) + \frac{dr^2}{r^2} + (d\psi + A)^2 + d\Sigma_{\mathbb{CP}^2}^2$$

$$F_{(5)} = 2(1 + *)d\psi \wedge J \wedge J$$

$$B_{(2)} = r^2du \wedge (d\psi + A), \tag{32.33}$$

where $u = \beta(t + y)$, $v = (t - y)/(2\beta)$.

(6) Calculate the holographic 2-point function of scalars for the $z = 2$ Schrödinger case, where the scalar has the usual kinetic term, $\int d^d x \sqrt{-g}[-(\partial_\mu \phi)^2)/2]$.

33 Finite Temperature and Black Holes

In this chapter we will learn how to introduce finite temperature in AdS/CFT, and we will learn that the correct procedure is to introduce a black hole in the gravity dual.

33.1 Finite Temperature and Hawking Radiation

Finite Temperature in Quantum Field Theory

To do that, we must first quickly remember how to introduce finite temperature in Quantum Field Theory. A bit more detailed discussion was done in Chapter 17.

One starts by considering a quantum mechanical amplitude between two states with positions q and q', at times t and t', written as a sum over states,

$$\langle q', t'|q, t\rangle = \sum_n \psi_n(q')\psi_n^*(q)e^{-iE_n(t'-t)}, \tag{33.1}$$

and also as a path integral,

$$\langle q', t'|q, t\rangle = \int \mathcal{D}q(t)e^{iS[q(t)]}, \tag{33.2}$$

and we Wick rotate to Euclidean time. Considering closed *periodic* Euclidean paths,

$$q' = q(t_E + \beta) = q = q(t_E), \tag{33.3}$$

one obtains two ways to express the amplitude, either as a trace giving the partition function at temperature $T = 1/\beta$,

$$\langle q, t'|q, t\rangle = \text{Tr}[e^{-\beta\hat{H}}] = Z[\beta], \tag{33.4}$$

or as a path integral over the periodic Euclidean paths:

$$\langle q, t'|q, t\rangle = \int_{q(t_E+\beta)=q(t_E)} \mathcal{D}q(t_E)e^{-S_E[q(t_E)]}. \tag{33.5}$$

In this way we find that the statistical mechanical partition function at temperature $T = 1/\beta$ is given by the quantum field theory partition function on closed periodic Euclidean paths of period β.

Black Hole Temperature

In turn, this was used to calculate the Hawking temperature of the Schwarzschild black hole. One Wick rotates the black hole to

$$ds^2 = +\left(1 - \frac{2MG}{r}\right)d\tau^2 + \frac{dr^2}{1 - \frac{2MG}{r}} + r^2 d\Omega_2^2. \tag{33.6}$$

Then, from the condition of absence of conical singularities at the Euclidean horizon (the condition that the Euclidean horizon is actually a flat space, i.e. a plane), one obtains that τ has period $\beta = 8\pi M$. The previous argument about periodic Euclidean time giving finite temperature suggested then that the temperature of the black holes is

$$T = \frac{1}{\beta} = \frac{1}{8\pi M}. \tag{33.7}$$

However, this doesn't mean that we can put a quantum field theory at finite temperature by adding a black hole, since the specific heat of the Schwarzschild black hole is negative,

$$C = -\frac{\partial M}{\partial T} = -8\pi M^2 < 0, \tag{33.8}$$

which means that the black hole is thermodynamically unstable, i.e. not in equilibrium with its environment. It simply means that the black hole has the temperature given above, but not that the whole space is at this temperature.

We will see, however, that adding a black hole in AdS space leads to a positive specific heat, hence in that case we have an equilibrium situation, of thermodynamic stability, and therefore in this case we can deduce that the whole space, including the field theory at its boundary, is put at finite temperature.

Before deriving that however, we prove that adding a black hole gives mass to fermions, but not to bosons, leading to the breaking of supersymmetry, since unbroken supersymmetry requires equal masses for bosons and fermions.

33.2 Fermions and Spin Structures in Black Holes and Supersymmetry Breaking

At $r \to \infty$, the Euclidean (Wick-rotated) Schwarzschild black hole becomes

$$ds^2 \simeq d\tau^2 + dr^2 + r^2 d\Omega_2^2, \tag{33.9}$$

with the periodicity $\tau \sim \tau + \beta$, which means that the space is $S_\tau^1 \times \mathbb{R}^3$, i.e. the KK vacuum (or background) for compactification from four to three flat dimensions. One can expand around the S^1 in Fourier modes (the "spherical harmonics"), and dimensionally reduce to the lowest modes on S^1, as we already described.

However, in a theory with fermions on a compact direction, the fermions can acquire a phase $e^{i\alpha}$ when going around a circle like S_τ^1 at $r = \infty$, i.e. $\psi \to e^{i\alpha}\psi$. Such an α is called a *spin structure*, and $\alpha = 0$ and $\alpha = \pi$ are always fine on a circle, independent of the theory, since Lorentz invariance dictates that the fermions appear in the Lagrangean always pairwise, so $\mathcal{L}$ is invariant under $\psi \to -\psi$. For additional symmetries acting on ψ, other α's could be allowed, i.e. to have periodicity around the circle only modulo multiplication by an element of the symmetry group.

Near the horizon of the Euclidean black hole, $r = 2MG$, the metric becomes

$$ds^2 \simeq \frac{r - 2MG}{2MG} d\tau^2 + 2MG \frac{[d(r - 2MG)]^2}{r - 2MG} + (2M)^2 d\Omega_2^2$$
$$\simeq 8M^2 G^2 \left(d\rho^2 + \rho^2 \frac{d\tau^2}{(4MG)^2} \right) + (2M)^2 d\Omega_2^2, \tag{33.10}$$

where in the second line we have substituted $\rho \equiv \sqrt{r - 2MG}$. We see that the space becomes approximately of $\mathbb{R}^2 \times S^2$ type, with the (ρ, τ) coordinates describing a plane in polar coordinates (note that without the periodicity imposed, it would be a cone with a conical singularity at the euclidean horizon $\rho = 0$). But the space $\mathbb{R}^2 \times S^2$ is simply connected, which means that we can smoothly shrink a loop, situated anywhere on it, to zero. That means that there are no nontrivial fermion phases (spin structures) on it, since otherwise we would get ψ being undefined at the final point, toward where the nontrivial circle would shrink. That means that there is a unique spin structure. But we still have to relate it to what happens at infinity, where we want to define the spin structure of KK compactification.

At $r = \infty$, going around the circle is the transformation $\tau \to \tau + \beta$, which at the horizon corresponds to a rotation in the $\mathbb{R}^2$ plane ($\theta = \tau/4\pi$, and $\theta \to \theta + 2\pi$ is a rotation). But a fermion can be defined as an object that returns to itself only after a 4π rotation in a plane, i.e. it changes by a sign, $\psi \to -\psi$ under a 2π rotation in the spatial plane. This is the unique spin structure at the horizon.

Translating this structure into $r = \infty$, we see that the unique spin structure is *antiperiodic at infinity*, i.e. it is of the type $\psi = \psi(\theta)$. In turn, that means that the fermion has a nontrivial mass under dimensional reduction on the S^1 at infinity. Indeed, a fermion that is massless in four dimensions (before the KK reduction) satisfies the Dirac equation

$$\not{\partial}\psi = 0 \Rightarrow \not{\partial}^2\psi = \Box_{4d}\psi = 0, \tag{33.11}$$

where we have used $\not{\partial}\not{\partial} = \partial_\mu \partial_\nu \frac{\{\gamma^\mu, \gamma^\nu\}}{2} = \partial^2$. But under dimensional reduction to three dimensions,

$$\Box_{4d}\psi = \left(\Box_{3d} + \frac{\partial^2}{\partial\theta^2} \right)\psi = (\Box_{3d} - m^2)\psi, \tag{33.12}$$

since the variation of $\psi(\theta)$ leads to $\partial^2\psi/\partial\theta^2 = -m^2\psi$. In turn, that means that 4-dimensional massless fermions KK reduced to three dimensions become massive.

On the other hand, bosons don't become massive, since we can have bosons that are periodic at infinity, meaning they don't depend on θ, i.e. $\partial^2\phi/\partial\theta^2 = 0$, so

$$\Box_{4d}\phi = \left(\Box_{3d} + \frac{\partial^2}{\partial\theta^2} \right)\phi = \Box_{3d}\phi. \tag{33.13}$$

This leads to the fact that the spin structure imposed by the black hole breaks supersymmetry. This is as it should be, since in general finite temperature breaks supersymmetry.

33.3 Witten Prescription for Finite Temperature in AdS/CFT

We now want to describe the prescription for putting a quantum field theory at finite temperature in AdS/CFT, found by Witten. The prescription is to put a black hole in AdS space, the gravity dual to the (conformal) quantum field theory.

The metric of AdS_{n+1} space in global coordinates can be written as

$$ds^2 = -\left(\frac{r^2}{R^2} + 1\right)dt^2 + \frac{dr^2}{\frac{r^2}{R^2} + 1} + r^2 d\Omega_{n-1}^2, \tag{33.14}$$

where in four dimensions

$$R^2 = -\frac{3}{\Lambda} \tag{33.15}$$

and for AdS space $\Lambda < 0$ is a *negative cosmological constant*.

We note that for $r \ll R$, we can describe the metric in the Newtonian approximation, used extensively in Part II, with the identification

$$2U_N = \frac{r^2}{R^2}. \tag{33.16}$$

Then introducing a black hole in AdS space is done in the same way as we introduced a black hole in flat space, namely by modifying the Newtonian potential by the addition of the term from a point mass,

$$2U_N \to 2U_N - \frac{w_n M}{r^{n-2}}, \tag{33.17}$$

for AdS_{n+1}. The fully nonlinear solution is obtained in the same way as for all black holes we studied, by modifying g_{00} and $1/g_{rr}$ in the same way, as

$$
\begin{aligned}
ds^2 &= -\left(\frac{r^2}{R^2} + 1 - \frac{w_n M}{r^{n-2}}\right)dt^2 + \frac{dr^2}{\frac{r^2}{R^2} + 1 - \frac{w_n M}{r^{n-2}}} + r^2 d\Omega_{n-1}^2 \\
&= -f(r)dt^2 + \frac{dr^2}{f(r)} + r^2 d\Omega_{n-1}^2.
\end{aligned}
\tag{33.18}
$$

The AdS_{n+1} space (at $M = 0$) solves the Einstein equation

$$R_{\mu\nu} = -\frac{n}{R^2} g_{\mu\nu} \equiv \Lambda g_{\mu\nu}, \tag{33.19}$$

i.e. the energy-momentum tensor is a *cosmological constant*, a constant times the metric. Moreover, one finds

$$w_n = \frac{8\pi G_N}{(n-2)\Omega_{n-1}} \tag{33.20}$$

as we have seen in Chapter 19. Indeed, for $n = 3$ (AdS_4), with $\Omega_2 = 4\pi$, one finds $w_3 = 2G_N$, as expected.

We define the outer horizon of the black hole the largest solution of the equation $g_{00} = 0$:

$$\frac{r_+^2}{R^2} + 1 = \frac{w_n M}{r_+^{n-2}}. \tag{33.21}$$

Expanding around the outer horizon, $r = r_+ + \delta r$, $\delta r \ll r_+$, gives the metric

$$ds^2 \simeq -\left(\frac{2r_+}{R^2} + \frac{(n-2)w_n M}{r_+^{n-1}}\right)\delta r dt^2 + \frac{(d\delta r)^2}{\delta r\left(\frac{2r_+}{R^2} + \frac{(n-2)w_n M}{r_+^{n-1}}\right)} + r_+^2 d\Omega_{n-1}^2. \tag{33.22}$$

Calling by the name of $A = f'(r_+)$ the factor multiplying δr, and defining $\rho = \sqrt{\delta r}$, so $d(\delta r)^2/\delta r = 4d\rho^2$, the metric can be written as

$$ds^2 \simeq \frac{4}{A}\left(\delta\rho^2 + \rho^2\frac{A^2\delta t^2}{4}\right) + r_+^2 d\Omega_{n-1}^2, \tag{33.23}$$

so to avoid conical singularities in the metric, the periodicity of $At/2$ must be 2π, i.e. the periodicity of t must be

$$\beta = \frac{4\pi}{A} = \frac{4\pi}{f'(r_+)} = \frac{4\pi}{\frac{nr_+}{R^2} + \frac{n-2}{r_+}}, \tag{33.24}$$

where we have used the defining equation of r_+ to replace $w_n M/r_+^{n-1}$ with $1/r_+ + r_+/R^2$. That finally leads to a temperature of the AdS black hole of

$$T = \frac{1}{\beta} = \frac{f'(r_+)}{4\pi} = \frac{nr_+^2 + (n-2)R^2}{4\pi R^2 r_+}. \tag{33.25}$$

We now turn to the analysis of the $T(M)$ curve. We first note from (33.21) that as $M \to 0$, then $r_+ \to 0$ as well, and then from (33.25), $T \to \infty$. In this regime, since the black hole is very small, it does not feel the size of AdS, and it feels almost in flat space. Then in this regime, the flat space formula is approximately valid: in $n = 3$ (AdS_4), we would have $T = 1/(8\pi M)$. On the other hand, for $M \to \infty$, we obtain from (33.21) that $r_+ \to \infty$, which from (33.25) gives $T \to \infty$ as well. That means that there is a minimum in between, for $M_{\min}$, giving $T_{\min}$.

To find it, we take d/dM of (33.21), multiplied by r_+^{n-2}, which gives

$$\frac{dr_+}{dM}\left[\frac{nr_+^{n-1}}{R^2} + (n-2)r_+^{n-3}\right] = w_n, \tag{33.26}$$

and since the bracket is positive, it means that $dr_+/dM > 0$, so to solve for the minimum of $T(M)$, via

$$0 = \frac{dT}{dM} = \frac{dT}{dr_+}\frac{dr_+}{dM}, \tag{33.27}$$

it suffices to solve $dT/dr_+ = 0$, which we find from (33.25),

$$\frac{dT}{dr_+} = \frac{n}{4\pi R^2} - \frac{n-2}{4\pi r_+^2} = 0, \tag{33.28}$$

resulting in

$$r_+ = R\sqrt{\frac{n-2}{n}} \Rightarrow T_{\min} = \frac{nr_+}{2\pi R} = \frac{\sqrt{n(n-2)}}{2\pi R}. \tag{33.29}$$

Then on the low M branch, $M < M(T_{\min})$, we have negative specific heat, $C = \partial M/\partial T < 0$, which means that the black hole is thermodynamically unstable, like the Schwarzschild black hole (in flat space) that it becomes when $M \to 0$. On the $M > M(T_{\min})$ branch, however, we find positive specific heat, $C = \partial M/\partial T > 0$, which would seem to imply thermodynamic stability. However, $C > 0$ is only a necessary condition, not a sufficient one, for thermodynamic stability. For true stability, we also need that the free energy in the presence of the black hole is less than the free energy in its absence, i.e. that

$$F_{BH-AdS} < F_{AdS}. \tag{33.30}$$

Since the partition function is related to the free energy by

$$Z = e^{-\beta F}, \tag{33.31}$$

where $\beta = 1/T$, and on the other hand in the gravity side of AdS/CFT, the (classical) partition function is approximately

$$Z_{\text{sugra}} \simeq e^{-S_E}, \tag{33.32}$$

where S_E is the Euclidean action, we obtain

$$S_E = \frac{F}{T}, \tag{33.33}$$

so the condition for stability is

$$F_{AdS-BH} - F_{AdS} = T(S_{AdS-BH} - S_{AdS}) < 0. \tag{33.34}$$

The calculation is a bit long, since the gravitational action for AdS space (and the AdS black hole) is divergent, so we need to find a regularization that matches with the asymptotics of the AdS black hole, but the final result is that the temperature obeys

$$T > T_1 \equiv \frac{n-1}{R} > T_{\min}, \tag{33.35}$$

to have thermodynamic stability. So only for $M > M(T_1) > M(T_{\min})$ do we have a thermodynamically stable black hole. Finally, we have the picture in Figure 33.1.

The physical interpretation is as follows. In AdS space in global coordinates, the black hole appears as a smaller cylinder inside the AdS cylinder, and as we have said, light takes a finite time from the horizon, which corresponds to a point inside AdS, until the boundary of AdS. So the Hawking radiation emitted by the black hole in AdS space at temperature T takes a finite time to go from the horizon (emission) to the boundary and back. Therefore the radiation eventually returns and feeds back into the black hole some of the energy lost as radiation. If the black hole is very small, it radiates a lot, and by the time the radiation has returned, it has already shrunk too much, and the returned radiation cannot keep it from shrinking, since as it shrinks, its temperature (and thus the rate of radiation loss) increases further. If the radiation is large enough, however, we can reach an equilibrium, as

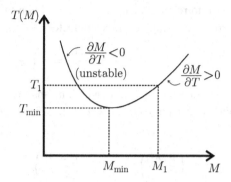

Fig. 33.1 $T(M)$ for the AdS black hole. The lower M branch is unstable, having $\partial M/\partial T < 0$. The higher M branch has $\partial M/\partial T > 0$, and above T_1 it is stable.

the returning radiation can compensate for the rate of energy loss, all the more so since as the black hole would shrink, its temperature (and the rate of energy loss) would decrease. The intermediate case on the high-mass branch, with $T_{\min} < T < T_1$ is still unstable, since the decrease in temperature as the black hole shrinks through radiation is not enough to stem the loss until the radiation returns from the boundary (the temperature is always larger than $T_{\min}$).

33.4 The Witten Metric

For the AdS black hole to put the boundary conformal field theory at finite temperature, we need to consider one more ingredient. The AdS black hole metric at infinity is

$$ds^2_{r \to \infty} \simeq \left(\frac{r}{R}dt\right)^2 + \left(\frac{R}{r}dr\right)^2 + r^2 d\Omega^2_{n-1}, \tag{33.36}$$

which corresponds, at fixed r, to a product of a circle of radius $\frac{r}{R}\frac{1}{T}$ and a sphere of radius r:

$$S^1_{\frac{r}{R} \times \frac{1}{T}} \times S^{n-1}_r. \tag{33.37}$$

But the boundary theory, $\mathcal{N} = 4$ SYM, is conformal invariant, so only relative scales matter, not absolute ones. On the other hand, we want the boundary theory to live in compactified flat space, i.e. $S^1 \times \mathbb{R}^{n-1}$, instead of $S^1 \times S^{n-1}$. This is possible only if the radius of S^{n-1} with respect to the radius of S^1 (since only relative scales matter in a conformal theory) goes to infinity, i.e. if

$$\frac{r}{\frac{r}{R}\frac{1}{T}} = RT \to \infty. \tag{33.38}$$

That means that we must take the temperature of the black hole to infinity, $T \to \infty$, which on the stable high-mass branch of $T(M)$ means that we must take $M \to \infty$ as well. But then we need to rescale the coordinates to obtain finite quantities in the metric at infinity, which corresponds to the CFT. In the factor

$$\frac{r^2}{R^2} - \frac{w_n M}{r^{n-2}} = r^2 \left(\frac{1}{R^2} - \frac{w_n M}{r^n} \right), \tag{33.39}$$

we want the bracket to be finite, which means that $r \propto M^{1/n}$. Since we also want to have $r^2 dt^2$ finite, we need to have $t \sim 1/r \sim M^{-1/n}$. Including finite contributions, we make the rescalings

$$r = \left(\frac{w_n M}{R^{n-2}} \right)^{\frac{1}{n}} \rho; \quad t = \left(\frac{w_n M}{R^{n-2}} \right)^{-\frac{1}{n}} \tau; \quad d\Omega_i = \left(\frac{w_n M}{R^{n-2}} \right)^{-\frac{1}{n}} dx_i \tag{33.40}$$

as $M \to \infty$, and substitute in the AdS black hole metric. Then the metric becomes

$$ds^2 \simeq \left(\frac{\rho^2}{R^2} - \frac{R^{n-2}}{\rho^{n-2}} \right) d\tau^2 + \frac{d\rho^2}{\frac{\rho^2}{R^2} - \frac{R^{n-2}}{\rho^{n-2}}} + \rho^2 \sum_{i=1}^{n-1} dx_i^2. \tag{33.41}$$

This is the gravity dual to the finite temperature $\mathcal{N} = 4$ SYM and was found by Witten, hence I will refer to it as the Witten metric.

Taking into account the rescalings, we can calculate the period of τ as

$$\beta_1 = \frac{4\pi}{R}, \tag{33.42}$$

leading to a temperature in units of R, as felt by conformal field theory, of

$$T = \frac{R}{\beta_1} = \frac{n}{4\pi}. \tag{33.43}$$

In the case of $n = 4$, corresponding to $\mathcal{N} = 4$ SYM, we obtain $T = 1/\pi$.

The Witten metric for $n = 4$ can be written as

$$ds^2 = \frac{\rho^2}{R^2} \left[\left(1 - \frac{R^4}{\rho^4} \right) d\tau^2 + R^2 d\vec{x}^2 \right] + R^2 \frac{d\rho^2}{\rho^2 \left(1 - \frac{R^4}{\rho^4} \right)}. \tag{33.44}$$

Rescaling the coordinates as

$$\frac{\rho}{R} = \frac{U}{U_0}; \quad \tau = t \frac{U_0}{R}; \quad \vec{x} = \vec{y} \frac{U_0}{R^2}, \tag{33.45}$$

and adding back the $R^2 d\Omega_5^2$ factor that was neglected until now (we had worked only with the AdS_{n+1} factor), we obtain the solution

$$ds^2 = \frac{U^2}{R^2} [-f(U) dt^2 + d\vec{y}^2] + R^2 \frac{dU^2}{U^2 f(U)} + R^2 d\Omega_5^2$$

$$f(U) = 1 - \frac{U_0^4}{U^4}. \tag{33.46}$$

This metric is a nonextremal version of the Poincaré patch of AdS, but it can also be obtained as the near-horizon limit of a nonextremal D3-brane.

We note an interesting fact: we took a certain limit of the black hole in global AdS space, and we obtained, after some rescaling of coordinates, a nonextremal version of the Poincaré patch of AdS. Therefore in the case of the nonextremal metric (at $M \to \infty$), there is no difference between global and Poincaré coordinates, there is a single metric describing it.

We can also make the usual redefinition of coordinates,

$$\frac{U}{R} = \frac{R}{z}; \quad \frac{U_0}{R} = \frac{R}{z_0}, \tag{33.47}$$

to put the metric in the form

$$ds^2 = \frac{R^2}{z^2}\left[-f(z)dt^2 + d\bar{y}^2 + \frac{dz^2}{f(z)}\right] + R^2 d\Omega_5^2$$

$$f(z) = 1 - \frac{z^4}{z_0^4}. \tag{33.48}$$

In these coordinates, the temperature is

$$T = \frac{1}{\pi z_0}. \tag{33.49}$$

Finally, in this chapter we have discussed only the case of AdS_{n+1}, but the procedure described here applies to any gravity dual: if we want to put the boundary theory to a finite temperature, we need to add a black hole in the gravity dual (make the gravity dual nonextremal).

We have seen an example of this more general situation at the end of the last chapter, when we described the gravity dual for a Schrödinger invariant theory at finite temperature, arising from string theory. In that case, since the gravity dual arose as a certain limit of $AdS_5 \times S^5$ at finite temperature, we know the limit we took is correct. But otherwise, if we don't derive the metric from a known string theory duality, we would need to check the limit on parameters we need to take, and if in that limit we have thermodynamical stability, as we did here in the case of AdS space.

Important Concepts to Remember

- A quantum field theory is put at finite temperature by considering closed periodic paths in Euclidean time.
- The Hawking temperature of a black hole is found by imposing the absence of conical singularities at the horizon of the Wick-rotated (Euclidean) black hole, and finding the periodicity β, resulting via the above at a temperature $T = 1/\beta$.
- For a Schwarzschild black hole, $C = \partial M/\partial T < 0$, so the black hole is thermodynamically unstable, leading to the absence of equilibrium with the background.
- A space with a black hole has a unique spin structure, corresponding to fermions that are antisymmetric around the circle at infinity.
- Dimensionally reducing on the circle at infinity, we break supersymmetry, since fermions get a mass, but bosons don't.
- To consider a quantum field theory at finite temperature in AdS/CFT, we put a black hole in the gravity dual.

- For the black hole in AdS space, $T(M)$ has a minimum, and the low-M branch has $C < 0$ and becomes the flat space black hole for $M \to 0$, whereas the high-M branch has $C > 0$ and is thermodynamically stable for $T > T_1 > T_{min}$, when $F_{AdS-BH} < F_{AdS}$.
- To obtain a gravity dual to the finite T CFT, we need to send the mass to infinity, $M \to \infty$ and rescale the coordinates, obtaining the Witten metric.
- The Witten metric, obtained from AdS in global coordinates, is only the nonextremal Poincaré metric, obtained as the near-horizon limit of the nonextremal D3-branes, thus there is a unique metric (global or Poincaré) for the finite temperature case.
- For any gravity dual, the finite temperature version is obtained by putting a black hole in it, but in general we need to check the thermodynamic stability and the limit we need to impose.

Further Reading

The Witten prescription for finite temperature AdS/CFT was done in [76].

Exercises

(1) Prove that the temperature of the nonextremal $z = 2$ Schrödinger metric from string theory, from the last chapter, is

$$T = \frac{r_+}{\pi \beta}. \tag{33.50}$$

(2) Consider the AdS-Reissner-Nordstrom solution in AdS_{d+1},

$$ds^2 = \frac{R^2}{z^2}\left(-f(z)dt^2 + d\vec{x}^2 + \frac{dz^2}{f(z)}\right)$$

$$f(z) = 1 - (1 + kz_+^2\mu^2)\left(\frac{z}{z_+}\right)^d + kz_+^2\mu^2\left(\frac{z}{z_+}\right)^{2(d-1)}$$

$$K \equiv \frac{(d-2)\kappa_{N,d+1}^2}{(d-1)g^2R^2}. \tag{33.51}$$

Prove that its temperature is given by

$$T = \frac{1}{4\pi z_+}(d - K(d-2)z_+^2\mu^2). \tag{33.52}$$

(3) Consider the near-extremal D3-brane metric

$$ds^2 = \alpha'\left\{\frac{U^2}{R^2}[-f(U)dt^2 + d\vec{y}^2] + R^2\frac{dU^2}{U^2f(U)} + R^2d\Omega_5^2\right\}$$

$$f(U) = 1 - \frac{U_0^4}{U^4}, \tag{33.53}$$

where $U_0 = \pi T R^2$.

(a) Prove that T is the temperature of this metric and $T \to 0$ gives the near-horizon extremal D3-brane metric, $AdS_5 \times S^5$.

(b) Prove that a light ray takes a finite time to travel between the boundary at $U = \infty$ and the horizon at $U = U_0$.

(4) Check that the rescaling

$$U = \rho \frac{U_0}{R}; \quad t = \tau \frac{R}{U_0}; \quad \vec{y} = \vec{x} \frac{R^2}{U_0} \tag{33.54}$$

on the metric in Exercise 3 gives the Witten finite T metric, if $R = $ AdS radius.

(5) Calculate the entropy of the Witten metric.

(6) Calculate the area of the worldsheet of a string of length L on the boundary that falls straight down from a small distance ϵ from the boundary down to the horizon in the Witten metric.

Hot Plasma Equilibrium Thermodynamics: Entropy, Charge Density, and Chemical Potential of Strongly Coupled Theories

In the previous chapter we have seen how to introduce finite temperature in AdS/CFT, and thus how to describe strongly coupled quantum field theories at finite temperature via a gravitational theory: we must put a black hole in the gravity dual and consider a scaling limit.

The strongly coupled field theory at finite temperature can be described as a hot plasma, i.e. a state of matter at high temperature where we do not form intermediate constituents like atoms and molecules, but we have a "soup" of the fundamental degrees of freedom, albeit in the cases of interest they will be strongly coupled among each other.

The toy model case (the "hydrogen atom" or "harmonic oscillator" of AdS/CFT) is $\mathcal{N} = 4$ SYM, and by putting it at finite temperature we obtain a strongly coupled plasma, which will be studied in the next section. The realistic model for whose sought-for description this toy model was introduced was QCD, and specifically the experimentally observed strongly coupled plasma obtained in high-energy heavy ion collisions at the RHIC experiment in Brookhaven and at the ALICE experiment in the LHC. We will explain why we believe we can use the $\mathcal{N} = 4$ SYM toy model to describe QCD at finite temperature, when the theories are quite different.

But we can also think of more general scenarios. In condensed matter physics, one can also consider strongly interacting hot plasmas. For instance, one could think of heavy fermion materials described by the Kondo lattice at high temperatures, where very strongly correlated electrons at high temperatures could be thought of as a hot plasma. In any case, the AdS/CFT correspondence allows us to describe the properties of strongly interacting high-temperature plasmas, so in this chapter we will study their static (equilibrium) properties, and in Chapter 36 we will move on to dynamic and nonequilibrium properties, after learning about the use of retarded Green's functions defined spectral function in AdS/CFT.

34.1 $\mathcal{N} = 4$ SYM Plasma from AdS/CFT: Entropy, Energy Density, Pressure

To start off, we calculate bulk thermodynamic properties of the $\mathcal{N} = 4$ SYM plasma, described by the AdS_5 black hole in the scaling limit of the last chapter, i.e. the Witten metric.

We start with the entropy, from which we will derive the energy density and pressure as well.

As we explained in Part II when talking about black hole thermodynamics, the entropy of a black hole is given by the Bekenstein-Hawking formula: namely it is the area of the horizon over 4 in Planck units, i.e. for the rescaled AdS_5 black hole we have

$$S = \frac{A_H}{4G_{N,5}}, \tag{34.1}$$

where A_H is the area of the horizon. From the Witten metric (33.48), the area of the horizon, located at $z = z_0$, is

$$A_H = \frac{R^3}{z_0^3} \int dy_1 dy_2 dy_2. \tag{34.2}$$

By the AdS/CFT correspondence, the entropy of a black hole in AdS_5 is equated with the entropy of the finite temperature $\mathcal{N} = 4$ SYM plasma in the strong coupling limit, $\lambda = g_{YM}^2 N \to \infty$, and the volume in 3+1 dimensions that this entropy corresponds to is $\int dy_1 dy_2 dy_3$, the spatial volume of the plasma. Therefore, while the total entropy of the plasma is infinite, the *entropy density* is finite and is

$$s = \frac{S}{\int dy_1 dy_2 dy_3} = \frac{R^3}{4G_{N,5} z_0^3}. \tag{34.3}$$

To finalize the calculation, we need to find the 5-dimensional Newton constant, $G_{N,5}$ as a function of the radius of AdS_5 (and S^5), R. In string theory, one can derive the normalization of the Einstein action and obtain

$$S_g = \frac{1}{(2\pi)^7 g_s^2 \alpha'^4} \int d^{10}x \sqrt{-g} R, \tag{34.4}$$

and we can identify the coefficient as

$$2\kappa_{N,10}^2 = 16\pi G_{N,10} = (2\pi)^7 g_s^2 \alpha'^4. \tag{34.5}$$

On the other hand, for $AdS_5 \times S^5$ we have the radius R defined by

$$R^4 = \alpha'^2 g_{YM}^2 N = \alpha'^2 (4\pi g_s) N, \tag{34.6}$$

so we obtain

$$G_{N,10} = \frac{\pi^4}{2N^2} R^8. \tag{34.7}$$

But we are interested in the *5-dimensional* Newton's constant, obtained by dimensional reduction on the S^5 of radius R and volume $R^5 \Omega_5 = R^5 \pi^3$. In the dimensional reduction, the $1/(16\pi G_{N,10})$ gets multiplied by this volume, coming from $\int d^5x \sqrt{-g_{\Omega_5}}$, so that

$$G_{N,5} = \frac{G_{N,10}}{R^5 \pi^3} = \frac{\pi}{2N^2} R^3. \tag{34.8}$$

Finally, then, we find the entropy density of the strongly coupled $\mathcal{N} = 4$ SYM plasma as

$$s_{\lambda \to \infty} = \frac{\pi^2}{2} N^2 T^3. \tag{34.9}$$

Note that we did the above calculation for the case of a black hole in the $AdS_5 \times S^5$ background, corresponding to $\mathcal{N} = 4$ SYM, but the ideas of the calculation should be general. We can do this calculation also in the case of generic gravity dual backgrounds, for any boundary quantum field theory. In particular, in the "phenomenological" (AdS/CMT) cases presented before, we also have an AdS background, so the entropy density calculation is the same, only the value of $G_{N,5}$ as a function of the parameters of the theory changes.

Now we can calculate also the energy density $\epsilon = E/V$ and the pressure P of the strongly coupled $\mathcal{N} = 4$ SYM plasma, from the thermodynamic relations

$$s = \frac{\partial P}{\partial T}; \quad \epsilon = -P + Ts. \tag{34.10}$$

Then we find

$$P_{\lambda \to \infty} = \frac{\pi^2}{8} N^2 T^4$$

$$\epsilon_{\lambda \to \infty} = \frac{3\pi^2}{8} N^2 T^4. \tag{34.11}$$

We can compare these results with the weak coupling results. The entropy density of a free boson is

$$s = \frac{2\pi^2 T^3}{45}, \tag{34.12}$$

and that of a free fermion is 7/8 of the bosonic result.

In $\mathcal{N} = 4$ SYM we have: two degrees of freedom for the gauge field A_μ, six for the scalars Φ^I, in total eight bosonic degrees of freedom, and a matching eight fermionic degrees of freedom, and all are in the adjoint of $SU(N)$, which has $N^2 - 1$ components. We therefore find for the total weak coupling entropy density of the $\mathcal{N} = 4$ SYM plasma (we approximate $N^2 - 1 \simeq N^2$ due to the large N limit)

$$s_{\lambda \to 0} = \left(8 + 8\frac{7}{8}\right)(N^2 - 1)\frac{2\pi^2 T^3}{45} \simeq \frac{2\pi^2}{3} N^2 T^3. \tag{34.13}$$

Again using the thermodynamic relations, we also find

$$P_{\lambda \to 0} = \frac{\pi^2}{6} N^2 T^4$$

$$\epsilon_{\lambda \to 0} = \frac{\pi^2}{2} N^2 T^4. \tag{34.14}$$

Therefore the ratios of the strong to weak coupling results is

$$\frac{s_{\lambda \to \infty}}{s_{\lambda \to 0}} = \frac{P_{\lambda \to \infty}}{P_{\lambda \to 0}} = \frac{\epsilon_{\lambda \to \infty}}{\epsilon_{\lambda \to 0}} = \frac{3}{4}. \tag{34.15}$$

The ratios of entropy densities can be interpreted as the ratio of effective free degrees of freedom, which reduces by a factor of 3/4 from weak to strong coupling. But in the case of QCD, the ratio is found to be (from lattice calculations) of about 80%, very close to the 75% of $\mathcal{N} = 4$ SYM. This is one of the examples of calculations that convinced people that *at finite temperature* the calculations using $\mathcal{N} = 4$ SYM should be good approximations of the ones for QCD, though it is not yet exactly clear why this is the case. This was taken to be a form of *universality*, i.e. that at finite temperature, the real-world QCD and finite temperature toy models like $\mathcal{N} = 4$ SYM should not differ too much from a qualitative point of view.

In the case of the phenomenological backgrounds used for condensed matter, we don't have the above concern, since in any case we do not have a rigorous way of deriving the dual field theory, so we can simply apply the AdS black hole calculation in the same phenomenological way as for other observables.

34.2 Adding Chemical Potential and Charge Density

Next, we consider processes that involve exchange of particles with the plasma. Therefore, we need to consider how to add chemical potential and charge. We will still use as an example our toy model ("hydrogen atom") of $\mathcal{N} = 4$ SYM, which has only a global $SU(4) = SO(6)$ R-symmetry. Since the charge will be added by sourcing one of the global symmetry generators, adding chemical potential and charge can refer to only a subgroup of the $SU(4)$ R-symmetry. By the AdS/CFT correspondence, the global current for this charge will be sourced by a gauge field for the corresponding *local* symmetry in AdS space.

For a generic phenomenological construction for condensed matter, we will assume that there is a global symmetry in the conformal field theory used for condensed matter, and that its current is sourced by the gauge field of the same, but local, symmetry in AdS space.

In the AdS/CFT correspondence, the coupling of the current J_μ to its source a_μ, identified with the source on the boundary of AdS space for A_μ, is $\int d^d x J^\mu a_\mu$. In particular, the charge density J^0 of the conformal field theory couples to a_0, the boundary source for the A_0 field in AdS space. Then, introducing charge J^0 in the CFT corresponds to having a classical solution for the gauge field of the type

$$ A = A_0(z)dt + \cdots \to a_0 dt, \tag{34.16} $$

as we approach the boundary at $z \to 0$. This is the condition to have an electrically charged classical solution in an AdS background, with the electric charge Q in AdS giving rise to an electric charge density J^0 in the CFT. Moreover, in the effective action for the CFT, the coupling $\int J^0 a_0$ must be identified with the chemical potential term $q\mu$ (where the integral of J^0 gives q), where μ is the chemical potential for the corresponding R-charge. Then the boundary condition for the gauge field in AdS space is

$$ A \to \mu dt \quad \text{as} \quad z \to 0. \tag{34.17} $$

The solution we are after, the Reissner-Nordstrom solution in AdS space (with a negative cosmological constant), is found from the Einstein-Maxwell system with a negative cosmological constant. The solution in AdS background in Poincaré coordinates is found similarly to the solution in flat space: we simply replace $f(z)$ in the Witten metric (33.48) with an expression with electric charge, in a similar way to what happened in (20.12). The solution is found to have metric

$$ds^2 = \frac{R^2}{z^2}\left(-f(z)dt^2 + d\vec{x}^2 + \frac{dz^2}{f(z)}\right),$$

$$f(z) = 1 - \left(1 + Kz_+^2\mu^2\right)\left(\frac{z}{z_+}\right)^d + Kz_+^2\mu^2\left(\frac{z}{z_+}\right)^{2(d-1)},$$

$$K \equiv \frac{(d-2)\kappa_{N,d+1}^2}{(d-1)g^2R^2}, \tag{34.18}$$

where the solution is written in $d + 1$ dimensions for generality, R is the radius of AdS_{d+1}, μ is a charge-type parameter, $z = z_+$ corresponds to the horizon of the black hole solution, and $\kappa_{N,d+1}$ is the Newton constant in $d + 1$ dimensions. The gauge field of the solution is

$$A_0 = \mu\left[1 - \left(\frac{z}{z_+}\right)^{d-2}\right], \tag{34.19}$$

and we see that it does satisfy the condition $A \to \mu dt$ as $z \to 0$, as we wanted. We might think that this is a constant, which could be dropped from the gauge field (since only its derivative, the field strength, is gauge-invariant), but when interacting with gravity, we need to have $A = 0$ at the horizon $z = z_+$, since otherwise we have a singularity for fluctuations at the horizon.

We can calculate the temperature of the AdS Reissner-Nordstrom metric above, in the same way as we calculated already for other black holes, and obtain the result

$$T = \frac{1}{4\pi z_+}\left(d - K(d-2)z_+^2\mu^2\right), \tag{34.20}$$

which we leave as an exercise to prove.

To find the rest of the thermodynamics for the field theory at chemical potential μ, dual to the metric (34.18), we need to calculate the thermodynamic potential. Since we are at fixed temperature (calculated above) and fixed μ, the thermodynamic potential is the grand canonical potential $\Omega = \Omega(\mu, V, T) = U - TS - \mu N$, with differential

$$d\Omega = -SdT - PdV - Nd\mu. \tag{34.21}$$

The thermodynamic potential is the one appearing in the relevant partition function. Since in the conformal field theory we have

$$Z_{\text{CFT}} = e^{-\beta\Omega}, \tag{34.22}$$

and this equals the supergravity partition function

$$Z_{\text{sugra}} = e^{-S_{\text{sugra}}(T,\mu,V)}, \tag{34.23}$$

it follows that $\Omega = T S_{\text{sugra}}(T, \mu, V)$.

The necessary condition for the existence of this Ω is that the on-shell supergravity action $S_{\text{sugra}}(T, V, \mu)$ does not have any boundary terms, i.e. the boundary terms vanish, when we put $a_0 = $ constant. For the Einstein-Maxwell action in AdS space this is satisfied, hence we have

$$\Omega = TS_{\text{sugra}}(T, V, \mu). \tag{34.24}$$

The result of the calculation, left as an exercise, is

$$\Omega(T(z_+), V_{d-1}, \mu) = -\frac{R^{d-1}}{2\kappa_N^2 z_+^d}(1 + Kz_+^2\mu^2)V_{d-1}. \tag{34.25}$$

To find the charge density in the CFT in this case, we consider the 1-point function

$$\rho = \langle J^0 \rangle = \left.\frac{\delta S_{\text{sugra}}}{\delta a_0}\right|_{a_0=0}. \tag{34.26}$$

Until now we have considered constant chemical potential μ, which gave the thermodynamic potential Ω. But it would be useful to consider constant charge density ρ instead, so that the thermodynamic potential is the free energy $F = \Omega + \mu Q$. As we see from this relation, by comparison with the constant μ case, we need a term linear in μ in the supergravity action proportional to the thermodynamic potential. We can obtain this by adding a boundary term to the supergravity action,

$$+\frac{1}{g^2}\int_{z\to 0}d^dx\sqrt{-h}n^aF_{ab}A^b, \tag{34.27}$$

where h_{ab} is the boundary metric, and n^a is a fixed vector. By adding this term to the supergravity action, it means that we fix n^aF_{ab} instead of A_b on the boundary, since it amounts to a Legendre transform of the on-shell action, but since a_0, the boundary value of A_0, is the chemical potential, now we fix the charge density, and conjugate to it in the boundary CFT.

34.3 Phenomenological Plasma Models

Consider a strongly coupled quantum field theory in flat space. One can write a phenomenological model for a gravity dual (gravitational background where the dual gravitational theory lives) that respects the symmetries and the qualitative nonperturbative physics of the field theory. Then based on it, one can try to fit several observables to fix the parameters of the gravity dual, and then predict other observables.

This method has been applied mostly to QCD, usually at high temperature to have some interesting thermodynamics, for which there is a lot of experimental information. In the case of condensed matter systems this program was developed less, but the same general ideas apply. Because of this situation, we will describe the case of QCD, simply as a guideline to what we can achieve.

To define the phenomenological model in an AdS background, we need at least Einstein gravity and a cosmological constant, since we have AdS space and $g_{\mu\nu}$ is dual to the energy-momentum tensor $T_{\mu\nu}$ of the field theory. We also need a gauge field A_μ, with electric charge corresponding to the chemical potential for the corresponding global symmetry of the field theory.

Finally, we need a scalar field (the "dilaton"), which will correspond to a scalar operator in the field theory. We know that in condensed matter applications such a scalar operator could be responsible for condensation and superconductivity, but in the case of QCD a bifundamental scalar $X^{\alpha\beta}$ in the gravity dual is related to the chiral order parameter $\bar{q}_R^\alpha q_L^\beta$, a scalar operator responsible for the symmetry-breaking $SU(N_f)_L \times SU(N_f)_R \to SU(N_F)_V$. More generally, we can say that it is needed to obtain a gravitational background dual (obtained by the coupling of gravity to the dilaton) to a field theory with a running coupling constant.

Consider then the general holographic model for a 5-dimensional Einstein-Maxwell-dilaton system, with a potential $V(\phi)$ and a function $f(\phi)$ in front of the Maxwell term,

$$S = \frac{1}{16\pi G_5} \int_{\mathcal{M}} d^5x \sqrt{-g} \left[R - \frac{1}{2}(\partial_\mu \phi)^2 - V(\phi) - \frac{f(\phi)}{4} F_{\mu\nu}^2 \right] + S_{\text{GHY}} + S_{\text{CT}}, \quad (34.28)$$

where S_{GHY} is the usual Gibbons-Hawking-York boundary term for gravity, a boundary term necessary to be added to the gravitational action when considering spacetimes with boundaries,

$$S_{\text{GHY}} = \frac{1}{8\pi G_5} \int_{\partial \mathcal{M}} d^4x \epsilon \sqrt{h} K, \quad (34.29)$$

with h the determinant of the boundary metric, $\epsilon = +1$ for $\partial\mathcal{M}$ timelike and $= -1$ for $\partial\mathcal{M}$ spacelike, and K the trace of the extrinsic curvature of the boundary $\partial\mathcal{M}$,

$$K_{ij} = -\nabla_i n_j, \quad (34.30)$$

with n_i the unit tangent to geodesics normal to the surface $\partial\mathcal{M}$. S_{CT} is a holographic counterterm, which appears when we treat correctly the regularization of space near the boundary, a procedure known as *holographic renormalization*, which will not be explained here.

This general holographic model is defined by the two functions $V(\phi)$, the potential for the dilaton, and $f(\phi)$, the kinetic function for the Maxwell field. We now turn to a phenomenological determination for them, using particular QCD data to fix them, though other ways of fixing are possible.

We want to have an AdS background, so the potential $V(\phi)$ needs to have a negative cosmological constant among its terms. We also consider a mass term, as the simplest thing we can have at small ϕ, leading to the perturbative potential

$$V(\phi) = -\frac{12}{L^2} + \frac{1}{2}m^2\phi^2 + \mathcal{O}(\phi^4), \quad (34.31)$$

where L is the radius of AdS space. But we must also ensure that the potential has a correct speed of sound, which we will describe in more detail in the next subsection. This is not possible if such a potential is valid at all ϕ. Instead, for the purely exponential potential

$$V = V_0 e^{\gamma\phi}, \quad (34.32)$$

with $V_0 < 0$, the equations of motion can be solved analytically, and one finds a speed of sound

$$c_s^2 = \frac{d\log T}{d\log s} = \frac{1}{3} - \frac{\gamma^2}{2}, \quad (34.33)$$

but on the other hand the space is not asymptotically AdS. One must then instead consider a potential that interpolates between (34.31) at small ϕ and (34.32) at large ϕ, such that we have an asymptotic AdS_5 space, but we also have a temperature-dependent speed of sound $c_s(T)$.

QCD Models

The simplest example of such a potential, with $\phi \leftrightarrow -\phi$ symmetry, is

$$V = \frac{-12\cosh(\gamma\phi) + b\phi^2}{L^2}, \tag{34.34}$$

and works quite well in reproducing the temperature dependence of the speed of sound $c_s(T)$ of hot plasmas obtained in lattice QCD. Even better, with an added bonus that one gets very close to having a second-order phase transition behavior for deconfinement in QCD, is the potential

$$V = \frac{-12\cosh(\gamma\phi) + b_2\phi^2 + b_4\phi^4 + b_6\phi^6}{L^2}. \tag{34.35}$$

Note that it is of the same type as the one above, but with independent ϕ^4 and ϕ^6 coefficients instead. From a numerical fit to $c_s(T)$ of QCD from the one obtained from the black hole solution in the general potential, one finds $\gamma \simeq 0.606$, $b_2 \simeq 1.975$, $b_4 = -0.030$, and $b_6 = -0.0004$.

On the other hand, since the mass m of scalar fields in the asymptotically AdS gravity dual of radius L is related to the dimension Δ of operators in the field theory by the relation

$$m^2 L^2 = \Delta(\Delta - 4), \tag{34.36}$$

valid in $d = 4$ dimensions (see (27.37)), one finds that the operator whose gravity source is ϕ has a dimension of $\Delta \simeq 3.9$.

In this way we have fixed $V(\phi)$ (assuming an ansatz and then fitting the coefficients) using $c_s(T)$, but this observable is quite insensitive to $f(\phi)$, so we need another way to fix it.

One such way is to fit the baryon susceptibility at zero chemical potential, defined as the limit

$$\chi_2(\mu = 0) \equiv \lim_{\mu \to 0} \frac{\rho(\mu)}{\mu}, \tag{34.37}$$

where $\rho = N/V$ is the particle number density for a chemical potential μ. We have seen in the previous section how to calculate these quantities in the gravity dual. To do that, one first considers a black hole ansatz of the type

$$ds^2 = e^{2A(r)}[-h(r)dt^2 + d\vec{x}^2] + \frac{e^{2B(r)}dr^2}{h(r)}; \quad A = A_\mu dx^\mu = \Phi(r)dt; \quad \phi = \phi(r). \tag{34.38}$$

One should really do a numerical fit for $f(\phi)$, though it seems that a simple guess gives a very good fit:

$$f(\phi) = \frac{\operatorname{sech}\left[\frac{6}{5}(\phi - 2)\right]}{\operatorname{sech}\frac{12}{5}}. \tag{34.39}$$

To check it, one first finds a formula for $\chi_2(\mu = 0; T)$ in the general black hole ansatz that will not be deduced here,

$$\frac{\chi_2(\mu = 0; T)}{T^2} = \frac{L}{16\pi^2} \frac{s}{T^3} \frac{1}{\int_{r_H}^{\infty} dr e^{-2A} f(\phi)^{-1}},\qquad(34.40)$$

and verifies that the resulting $\chi_2(\mu = 0; T)$ function fits well the lattice result for QCD.

Now that we have completely fixed the action, and the corresponding black hole solutions, we can derive any other thermodynamic or transport quantity and find *predictions* for QCD that can be tested.

The general procedure shown here can be applied to other examples: fix the parameters or functions in the action by fitting certain observables, then calculate other observables by using the derived action.

34.4 Speed of Sound

We now turn to some details of the calculation of the speed of sound.

For the calculation of the speed of sound, it is unnecessary to consider a chemical potential or particle density, hence we can consider $\Phi(r) = 0$, i.e. a trivial gauge field. Therefore consider the black hole ansatz (34.38) at $\Phi(r) = 0$. The horizon of this black hole is the largest (closest to the boundary) value of r, called r_H, where $h(r_H) = 0$, so is the surface ($r = r_H, t = $ const.). The area of this surface is

$$A_H = e^{3A(r_H)} \int dx_1 dx_2 dx_3,\qquad(34.41)$$

and thus the entropy density of the dual field theory equals the entropy of the black hole, i.e. its area A_H over 4 in Planck units, divided by the 3-dimensional volume:

$$s = \frac{S}{\int dx_1 dx_2 dx_3} = \frac{A_H}{4G_5 \int dx_1 dx_2 dx_3} = \frac{e^{3A(r_H)}}{4G_5}.\qquad(34.42)$$

On the other hand, the temperature of the black hole is found in the usual way, by analytical continuation to Euclidean space, and imposing that the horizon $r = r_H$ becomes the origin of flat space in polar coordinates, instead of a generic conical point. The metric near the horizon is approximated by $h(r) \simeq h'(r_H)(r - r_H), A(r) \simeq A(r_H), B(r) \simeq B(r_H)$, for an induced Euclidean metric (t_H is the Euclidean time)

$$ds^2 \simeq e^{2A(r_H)} h'(r_H)(r - r_H) dt_E^2 + \frac{e^{2B(r_H)}}{h'(r_H)(r - r_H)} dr^2$$
$$= \frac{4e^{2B(r_H)}}{h'(r_H)} \left(d\rho^2 + \frac{h'(r_H)^2 e^{2(A(r_H)-B(r_H))}}{4} \rho^2 dt_E^2 \right),\qquad(34.43)$$

where $\rho = \sqrt{r - r_H}$. From the condition that $h'(r_H) e^{A(r_H)-B(r_H)}/2 \times t_E$ has periodicity 2π, and that the temperature is the inverse of the periodicity β of the Euclidean time t_E, we obtain

$$T = \frac{1}{\beta} = \frac{|h'(r_H)| e^{A(r_H) - B(r_H)}}{4\pi}. \tag{34.44}$$

The speed of sound in the plasma is defined by

$$c_s^2 = \frac{d \log T}{d \log s}, \tag{34.45}$$

so that $T \sim s^{c_s^2}$, and can thus be calculated from

$$c_s^2 = \frac{d[A(r_H) - B(r_H) + \log |h'(r_H)|]}{3 dA(r_H)}. \tag{34.46}$$

As mentioned in the previous section, for an exponential potential (34.32), one finds that the speed of sound is constant and given by (34.33).

One can calculate other static quantities as well, like various susceptibilities, and make predictions once we have fixed $V(\phi)$ and $f(\phi)$, but we will not do that here.

Important Concepts to Remember

- Static plasma observables are obtained by putting a black hole the gravity dual and taking a scaling limit. One can use the "hydrogen atom" (toy model) of $\mathcal{N} = 4$ SYM for strongly coupled field theories at finite temperature.
- The energy density ϵ, pressure P, and entropy density s at strong coupling in $\mathcal{N} = 4$ SYM are found to be 3/4 of the perturbative results, meaning a reduction by 3/4 in effective number of degrees of freedom. This is close to the 80% reduction observed in lattice QCD.
- Chemical potential (for an R-charge) is added in the gravity dual by making charged black hole solutions, with respect to a charge dual to the R-charge, such that near the boundary $A \to \mu dt$.
- To fix the charge density ρ instead, one needs to add a boundary term to the supergravity action in the gravity dual, thus effectuating a Legendre transform.
- Phenomenological models are found by writing an AdS action for an Einstein-Maxwell-dilaton system, with unknown potential $V(\phi)$ and kinetic function $f(\phi)$, and fixing those by matching with experimental or lattice data for some observables.
- In QCD, we can fix $V(\phi)$ from the speed of sound $c_s(T)$ and $f(\phi)$ from the quark susceptibility $\chi_2(\mu = 0; T)$.
- The speed of sound is found by calculating s, T, and then $c_s^2 = d \log T / d \log s$.

Further Reading

A review of plasmas in AdS/CFT can be found in [77]. On how to add chemical potential to AdS/CFT, see the original paper [78] and the review [1]. The calculation that fixes the phenomenological potential $V(\phi)$ via the speed of sound was done in [79], whereas the calculation of $f(\phi)$ via the quark susceptibility $\chi_2(\mu = 0, T)$ was done in [80]. The calculation of the speed of sound $c_s(T)$ for the purely exponential potential was done in [81].

Exercises

(1) Calculate the entropy density, energy density and pressure for field theory corresponding to the gravity dual in (34.18).
(2) Show that the temperature of the AdS-Reissner-Nordstrom solution (34.18) is given by

$$T = \frac{1}{4\pi z_+} \left(d - K(d-2)z_+^2 \mu^2 \right). \tag{34.47}$$

(3) Calculate the grand canonical potential for the AdS-Reissner-Nordstrom solution, by calculating the regularized on-shell action and subtracting the contribution of AdS space, to find

$$\Omega(T(z_+), V_{d-1}, \mu) = -\frac{R^{d-1}}{2\kappa_N^2 z_+^d}(1 + K z_+^2 \mu^2)V_{d-1}. \tag{34.48}$$

(4) Calculate the form of the action (34.28) on the ansatz (34.38).
(5) Find the equations of motion for $A(r), B(r), h(r), \Phi(r), \phi(r)$ in the ansatz (34.38), using the results from the previous exercise.

35 Spectral Functions and Transport Properties

In this chapter we will describe spectral functions, useful quantities that can be used to calculate transport properties, and show how they can be calculated using the AdS/CFT correspondence.

35.1 Retarded Green's Functions and Susceptibilities

The spectral functions are based on *retarded Green's functions*, defined for observables $\mathcal{O}_A$ and $\mathcal{O}_B$ as

$$G^R_{\mathcal{O}_A \mathcal{O}_B}(\omega, \vec{k}) = -i \int d^{d-1}x \, dt \, e^{i\omega t - i\vec{k}\cdot\vec{x}} \theta(t) \langle [\mathcal{O}_A(t, \vec{x}), \mathcal{O}_B(0, 0)] \rangle. \tag{35.1}$$

Note the step (Heaviside) function $\theta(t)$, which means that the function is indeed *retarded*, i.e. causal. The retarded Green's function describes the evolution of small perturbations about equilibrium in linear response theory, specifically the way operator $\mathcal{O}_A$ responds (linearly) under a perturbation by a source coupling to $\mathcal{O}_B$.

Indeed, we will prove that for small perturbations (neglecting nonlinear terms), we have

$$\delta \langle \mathcal{O}_A \rangle(\omega, \vec{k}) = G^R_{\mathcal{O}_A \mathcal{O}_B} \delta\phi_{B(0)}(\omega, \vec{k}). \tag{35.2}$$

To prove it, we start with the time-dependent perturbation to the Hamiltonian defined by the source $\delta\phi_{B(0)}$, coupling to the operator $\mathcal{O}_B$:

$$\delta H(t) = \int d^{d-1}x \, \delta\phi_{B(0)}(t, \vec{x}) \mathcal{O}_B(t, \vec{x}). \tag{35.3}$$

We are after the variation in the VEV of $\mathcal{O}_A$, defined in a quantum theory at finite temperature as

$$\langle \mathcal{O}_A \rangle(t, \vec{x}) = \text{Tr}[\rho(t) \mathcal{O}_A(\vec{x})], \tag{35.4}$$

where $\rho(t)$ is the density matrix of the full Hamiltonian, thus time-dependent, and satisfying the evolution equation

$$i\partial_t \rho = [H_0 + \delta H(t), \rho]. \tag{35.5}$$

Going to the interaction picture, so that the time dependence due to H_0 is absorbed into the time dependence of $\mathcal{O}_A$, and we are left only with the time dependence coming from

$\delta H(t)$, we obtain

$$\langle \mathcal{O}_A \rangle (t, \vec{x}) = \text{Tr}[\rho_0 U^{-1}(t) \mathcal{O}(\vec{x}) U(t)]. \tag{35.6}$$

Here now the density matrix at temperature T is only for H_0,

$$\rho_0 = e^{-\frac{H_0}{T}}, \tag{35.7}$$

and the time evolution operator is (as usual in the interaction picture) only for $\delta H(t)$:

$$U(t) = \text{T} e^{-i \int_0^t dt' \, \delta H(t)}. \tag{35.8}$$

Then the variation of the operator VEV is found (substituting the various terms) to be

$$\delta \langle \mathcal{O}_A \rangle (t, \vec{x}) = -i \, \text{Tr} \left[\rho_0 \int_0^t dt' [\mathcal{O}_A(t, \vec{x}), \delta H(t')] \right]$$

$$= -i \int_0^t dt' d^{d-1} x' \langle [\mathcal{O}_A(t, \vec{x}), \mathcal{O}_B(t', \vec{x}')] \rangle \delta \phi_{B(0)}(t', \vec{x}'), \tag{35.9}$$

where in the second line we have substituted δH in terms of $\delta \phi_{B(0)}$. Doing a Fourier transform over t and $\vec{x}$, we obtain

$$\delta \langle \mathcal{O}_A(\omega, \vec{k}) \rangle = G^R_{\mathcal{O}_A \mathcal{O}_B}(\omega, \vec{k}) \delta \phi_{B(0)}(\omega, \vec{k}). \tag{35.10}$$

q.e.d.

The function we are interested in is the retarded Green's function in full Fourier space, but to find its properties, consider Fourier transforming back over ω:

$$G^R_{\mathcal{O}_A \mathcal{O}_B}(t, \vec{k}) = \int \frac{d\omega}{2\pi} e^{-i\omega t} G^R_{\mathcal{O}_A \mathcal{O}_B}(\omega, \vec{k}). \tag{35.11}$$

Then we can use the fact that the retarded Green's function is supposed to be causal, so it must vanish for $t < 0$ (as we saw, in t space it was defined with a $\theta(t)$). But in turn, when thinking of it as an integral over ω, we would like to calculate it using complex function theory, i.e. as a contour integral in ω space. To be able to close the contour in the upper half plane (Im $\omega > 0$), we need the following two properties:

(1) $G^R_{\mathcal{O}_A \mathcal{O}_B}(\omega, \vec{k})$ analytic in ω in the upper-half plane Im $\omega > 0$.

(2) $G^R_{\mathcal{O}_A \mathcal{O}_B}(\omega, \vec{k}) \to 0$ as $|\omega| \to \infty$.

Indeed, then, for $t < 0$, closing the contour by an infinite semicircle in the upper-half plane, to make the contour Γ in Figure 35.1, the factor $e^{-i\omega t}$ (becoming $e^{-|t| Im(\omega)}$), together with property (2), guarantees that the integral over the semicircle is zero. But the full integral over the contour Γ is also zero, since there are no poles in the upper-half plane, so the integral over the real line (defining $G^R(t, \vec{k})$) is also zero.

Moreover, the properties (1) and (2) imply that we can also write the representation as a Γ contour integral over the complex ω plane, denoted by z or ζ:

$$G^R(z) = \oint_\Gamma \frac{d\zeta}{2\pi i} \frac{G^R}{\zeta - z}. \tag{35.12}$$

This follows since then again the integral over the semicircle at infinity vanishes because of property (2), and the full integral over Γ is given simply by the pole at $z = \zeta$, which results in the left-hand side, as property (1) guarantees there are no other poles.

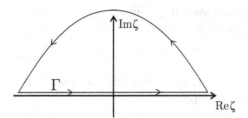

Contour of integration for G^R.

Splitting this relation into its real and imaginary parts, we obtain the *Kramers-Kronig relations*, which are in fact valid for any functions that satisfy properties (1) and (2):

$$\text{Re}G^R(\omega) = P \int_{-\infty}^{+\infty} \frac{d\omega'}{2\pi} \frac{\text{Im}G^R(\omega')}{\omega' - \omega}$$

$$\text{Im}G^R(\omega) = -P \int_{-\infty}^{+\infty} \frac{d\omega'}{2\pi} \frac{\text{Re}G^R(\omega')}{\omega' - \omega}. \tag{35.13}$$

By considering $\omega \to 0$ in the first relation above, we obtain

$$\chi \equiv \lim_{\omega \to 0} \text{Re}\, G^R_{\mathcal{O}_A \mathcal{O}_B}(\omega, \vec{x}) = \int_{-\infty}^{+\infty} \frac{d\omega'}{2\pi} \frac{\text{Im}G^R_{\mathcal{O}_A \mathcal{O}_B}(\omega', \vec{x})}{\omega'}. \tag{35.14}$$

This is a thermodynamic "sum rule," since G^R appears both outside and inside the integral sign in the above, and the quantity χ it defines is a static thermodynamic susceptibility,

$$\chi_{AB} = \frac{\partial \langle \mathcal{O}_A \rangle}{\partial \phi_{B(0)}}, \tag{35.15}$$

in the same way as, say, the electric susceptibility is $\chi_{E,E} = \partial D / \partial E$, for a perturbation of the Hamiltonian by $\int \vec{D} \cdot \delta \vec{E}$.

35.2 Spectral Functions

One can write a spectral representation for G^R, which follows from the definition of G^R, by inserting a complete set $\mathbb{1} = \sum_n |n\rangle \langle n|$ between $\mathcal{O}_A$ and $\mathcal{O}_B$. After some algebra, we find

$$G^R_{\mathcal{O}_A \mathcal{O}_B}(\omega, \vec{k}) = \sum_{mn} e^{-\frac{E_n}{T}} \left(\frac{A_{nm} B_{mn} \delta^{(d)}(k_{nm} - k)}{E_n - E_m + \omega + i\epsilon} - \frac{A_{mn} B_{nm} \delta^{(d)}(k_{mn} - k)}{E_m - E_n + \omega + i\epsilon} \right), \tag{35.16}$$

where

$$H_0 |m\rangle = E_m |m\rangle$$
$$k_{nm} = k_n - k_m$$
$$A_{mn} = \langle m | \mathcal{O}_A(0,0) | n \rangle$$
$$B_{mn} = \langle m | \mathcal{O}_B(0,0) | n \rangle. \tag{35.17}$$

If we are considering a single operator $\mathcal{O}_A$, then from the sum rule (35.14), we see that for χ we have the *spectral function*

$$\chi_A = -\mathrm{Im}\, G^R_{\mathcal{O}_A \mathcal{O}_B}(\omega, \vec{k}), \qquad (35.18)$$

since χ is given by a spectral integral (sum) similar to the above spectral representation, a function that satisfies

$$\omega \chi_A(\omega, \vec{k}) \geq 0. \qquad (35.19)$$

We can define more relevant correlation functions. First, we can define the *advanced Green's function*

$$G^A_{\mathcal{O}_A \mathcal{O}_B}(\omega, \vec{k}) = +i \int d^{d-1}x \int dt\, e^{i\omega t - i\vec{k}\cdot\vec{x}} \theta(-t) \langle [\mathcal{O}_A(t, \vec{x}), \mathcal{O}_B(0, 0)] \rangle, \qquad (35.20)$$

which indeed is zero for $t > 0$. Note the $+$ sign in front of the definition of G^A. In position space,

$$G^A_{\mathcal{O}_A \mathcal{O}_B}(t, \vec{x}) = +i\theta(-t) \langle [\mathcal{O}_A(t, \vec{x}), \mathcal{O}_B(0, 0)] \rangle$$
$$G^R_{\mathcal{O}_A \mathcal{O}_B}(t, \vec{x}) = -i\theta(t) \langle [\mathcal{O}_A(t, \vec{x}), \mathcal{O}_B(0, 0)] \rangle. \qquad (35.21)$$

Since $\theta(t) + \theta(-t) = 1$, we can also define in position space

$$\rho_{\mathcal{O}_A \mathcal{O}_B}(t, \vec{x}) = \langle [\mathcal{O}_A(t, \vec{x}), \mathcal{O}_B(0, 0)] \rangle = i(G^R_{\mathcal{O}_A \mathcal{O}_B}(t, \vec{x}) - G^A_{\mathcal{O}_A \mathcal{O}_B}(t, \vec{x})), \qquad (35.22)$$

and in momentum (Fourier) space

$$\rho_{\mathcal{O}_A \mathcal{O}_B}(\omega, \vec{k}) = \int d^{d-1}x \int dt\, e^{i\omega t - i\vec{k}\cdot\vec{x}} \langle [\mathcal{O}_A(t, \vec{x}), \mathcal{O}_B(0, 0)] \rangle$$
$$= i(G^R_{\mathcal{O}_A \mathcal{O}_B}(t, \vec{x}) - G^A_{\mathcal{O}_A \mathcal{O}_B}(t, \vec{x})). \qquad (35.23)$$

Then $\rho_{\mathcal{O}_A \mathcal{O}_B}(\omega, \vec{k})$ is a *spectral function for* $G^{R,A}_{\mathcal{O}_A \mathcal{O}_B}(\omega, \vec{k})$, since from its definition, we can write a spectral integral representation,

$$G^{R,A}(\omega, \vec{k}) = \int \frac{d\omega'}{2\pi} \frac{\rho(\omega', \vec{k})}{\omega - \omega' \pm i\epsilon}, \qquad (35.24)$$

since by substituting the definition of ρ on the right-hand side, we obtain

$$\int \frac{d\omega'}{2\pi} \int d^{d-1}x\, dt\, e^{i\omega' t - i\vec{k}\cdot\vec{x}} \frac{\langle [\mathcal{O}_A(t, \vec{x}), \mathcal{O}_B(0, 0)] \rangle}{\omega - \omega' \pm i\epsilon}. \qquad (35.25)$$

Then we use the fact that

$$\mp\theta(\pm t)e^{i\omega t} = \int_{-\infty}^{+\infty} \frac{d\omega'}{2\pi i} \frac{e^{i\omega' t}}{\omega - \omega' \pm i\epsilon}, \qquad (35.26)$$

proved as follows. For the upper sign, we close the contour in the upper-half plane, and since $e^{i\omega' t}$ vanishes for $t > 0$ on the infinite semicircle in the upper-half plane (for $\mathrm{Im}\, \omega' > 0$), the integral is equal to the integral over Γ, given by the pole at $\omega' = \omega + i\epsilon$, where the result is $e^{i\omega}$. For $t < 0$, we must close the contour in the lower-half plane to obtain a vanishing integral over the infinite semicircle in the lower-half plane, and then inside the contour there are no poles, giving zero. A similar analysis can be done for the lower sign, giving $\theta(-t)$. Using the above integral relation, we obtain indeed (35.24).

If $\mathcal{O}_A$ and $\mathcal{O}_B$ are Hermitian, then ρ is Hermitian too, which means it has real diagonal elements, then using the relation

$$\frac{1}{x \pm i\epsilon} = P\frac{1}{x} \mp i\pi \delta(x), \tag{35.27}$$

for the integrand, where P is the principal part, we have

$$\frac{\rho}{\omega - \omega' \pm i\epsilon} = \frac{\rho}{\omega - \omega'} \mp i\pi \rho\delta(\omega - \omega'), \tag{35.28}$$

where the first term is purely real and the second is purely imaginary. Then taking the real and imaginary parts of (35.24), we obtain

$$\mathrm{Re}G^R(\omega, \vec{k}) = \mathrm{Re}G^A(\omega, \vec{k}) = P\int \frac{d\omega'}{2\pi} \frac{\rho(\omega', \vec{k})}{\omega - \omega'}, \tag{35.29}$$

as well as

$$\mathrm{Im}G^R(\omega, \vec{k}) = -\mathrm{Im}G^A(\omega, \vec{k}) = -\frac{1}{2}\rho(\omega, \vec{k}), \tag{35.30}$$

where the last equality is obtained remembering that $\rho = i(G^R - G^A)$. Then the spectral function is

$$\rho = -2\mathrm{Im}G^R(\omega, \vec{k}), \tag{35.31}$$

making contact with the previous definition of a spectral function χ_A for χ in (35.18).

35.3 Kubo Formulas

We now present applications of spectral functions to calculate various observables.

1. Electric Conductivity
The spectral functions can be used for many things, but in particular we are interested in transport properties. The first such example is the electric conductivity, which is the response of current to a perturbation in the electric field, or rather, in the gauge potential generating it. Thus consider a perturbation δA_i (in the gauge $A_0 = 0$), such that

$$E_j \equiv F_{0j} = \partial_t \delta A_j, \tag{35.32}$$

and go to Fourier (ω) space, so that

$$E_j = -i\omega\delta A_{j(0)}. \tag{35.33}$$

Then the linear response in ω space of the (VEV of the) current to it is

$$\langle J_x \rangle = \sigma E_x = -i\omega\delta A_{x(0)}, \tag{35.34}$$

but on the other hand it is, by the general theory of retarded Green's functions and linear response, equal to

$$G^R_{J_x J_x}(\omega, \vec{k})\delta A_{x(0)}, \tag{35.35}$$

which leads to the relation

$$\sigma(\omega, \vec{k}) = \frac{iG^R_{J_x J_x}(\omega, \vec{k})}{\omega}, \tag{35.36}$$

known as *Kubo's formula for the electric conductivity.*

More precisely, usually we are interested in a real conductivity only, in which case we have

$$\sigma(\omega, \vec{k}) = -\frac{\text{Im}G^R_{J_x J_x}(\omega, \vec{k})}{\omega}, \tag{35.37}$$

and specializing for the DC conductivity, we get

$$\sigma(0, \vec{k}) = -\lim_{\omega \to 0} \frac{\text{Im}G^R_{J_x J_x}(\omega, \vec{k})}{\omega}, \tag{35.38}$$

more usually called the Kubo formula.

2. Shear Viscosity

The next example is the shear viscosity, i.e. the normal viscosity that creates a gradient in the x direction for the velocity in the transverse y direction (shear). We have seen in Part I of the course that the shear viscosity arises, in a relativistic model, from the expansion in derivatives (on the 4-velocity) of the energy-momentum tensor. At order zero, we get the ideal (nonviscous) fluid, and at first order we get the viscosity terms. More precisely, we have

$$\begin{aligned} T_{\mu\nu} = {} & \rho u_\mu u_\nu + P(g_{\mu\nu} + u_\mu u_\nu) \\ & + 2\eta \left[\frac{\nabla_\mu u_\nu + \nabla_\nu u_\mu}{2} - \frac{1}{d-1}(\nabla_\lambda u^\lambda)(g_{\mu\nu} + u_\mu u_\nu) \right] \\ & + \zeta(\nabla_\lambda u^\lambda)(g_{\mu\nu} + u_\mu u_\nu). \end{aligned} \tag{35.39}$$

We consider the case of a perturbation of a zeroth order fluid at rest, with $u^\mu = (1, 0, 0, 0)$, by adding a small metric perturbation in directions (xy), i.e. h_{xy}, and expanding to linear order. Ignoring the bulk viscosity ζ term, we get

$$T_{xy} = P\,h_{xy} + \eta\partial_t h_{xy} + \mathcal{O}(h^2_{xy}) + \mathcal{O}(\partial^2 h_{xy}). \tag{35.40}$$

Going to Fourier space, and considering that the Fourier transform of the constant first term is a delta function $\delta(\omega)$ that can be ignored as long as we deal only with $\omega > 0$, we obtain

$$T_{xy}(\omega) = -\eta i\omega h_{xy} + \mathcal{O}(h^2_{xy}) + \mathcal{O}(\partial^2 h_{xy}), \tag{35.41}$$

but on the other hand from the retarded Green's function analysis in linear response, we know it should equal

$$G^R_{T_{xy} T_{xy}} h_{xy}, \tag{35.42}$$

hence we get the formula

$$\eta(\omega, \vec{k}) = \frac{iG^R_{T_{xy}T_{xy}}(\omega, \vec{k})}{\omega},$$
(35.43)

which is the *Kubo formula for the shear viscosity.*

As in the case of the electric conductivity, one usually is interested in a real viscosity, in which case we get

$$\eta(\omega, \vec{k}) = -\frac{\mathrm{Im}G^R_{T_{xy}T_{xy}}(\omega, \vec{k})}{\omega},$$
(35.44)

and even more specifically, for the static shear viscosity

$$\eta(0, \vec{0}) = -\lim_{\omega \to 0} \frac{\mathrm{Im}G^R_{T_{xy}T_{xy}}(\omega, \vec{0})}{\omega},$$
(35.45)

more usually called the Kubo formula for the shear viscosity (than the form above).

35.4 AdS/CFT in Minkowski Space at Finite Temperature

The proposal to extend AdS/CFT to Minkowski space (remember that we have really defined AdS/CFT in Euclidean space, and the analytical continuation is more subtle than one could think), and, moreover at finite temperature, was done by Son and Starinets in 2002. As we saw in the last chapter, finite temperature is obtained by introducing a black hole in the gravity dual, which leads to having two boundaries, both in the holographic (($d + 1$)th) coordinate z, one the usual one at "infinity," and the other at the horizon of the black hole, where we also have singularities that require a definition of the boundary conditions.

The Son and Starinets finite temperature prescription is an algorithmic one. In Euclidean space, correlation functions are obtained from the on-shell supergravity action, written as a function of the boundary values ϕ_0. Assume that now we can also write the on-shell action in momentum space for the d coordinates of the boundary, and as a boundary term for the holographic coordinate z, namely as

$$S_{\text{on-shell}} = \int \frac{d^d k}{(2\pi)^d} \phi_0(-\vec{k}) \mathcal{F}(\vec{k}, z) \phi_0(\vec{k}) \Big|^{z=z_H}_{z=z_B},$$
(35.46)

where z_H is the horizon location and z_B is the boundary location at infinity. Then the prescription gives the retarded Green's function as

$$G^R(\vec{k}) = -2\mathcal{F}(\vec{k}, z)|_{z=z_B}.$$
(35.47)

If one applies the AdS/CFT prescription for the shear viscosity of a gravity dual at finite temperature (with a black hole), one finds that under very general conditions (including isotropy and Einstein gravity, which if violated, could lead to other results, as was found in explicit examples) the ratio of shear viscosity to entropy density is

$$\frac{\eta}{s} = \frac{1}{4\pi}.$$

(35.48)

We will not describe this calculation here (we will do it later, in Chapter 38), but it is by now a famous one.

There is another way to think about the retarded Green's function, which leads to a different way to calculate it. We know that the retarded Green's function is

$$G^R_{\mathcal{O}_A \mathcal{O}_B} = \frac{\delta \langle \mathcal{O}_A \rangle}{\delta \phi_{B(0)}} \Bigg|_{\delta\phi=0}.$$

(35.49)

But we saw that the basis of the AdS/CFT correspondence is the equality of the CFT and bulk partition functions, depending on the source $\phi_{(0)}$, coupling to $\mathcal{O}$ and giving the boundary value for the gravity dual field ϕ:

$$Z_{\text{CFT}}[\phi_{(0)}] = Z_{\text{bulk}}[\phi_{(0)}].$$

(35.50)

In Minkowski space, putting in the correct factors of i, we have

$$Z_{\text{CFT}} = \int \mathcal{D}\text{fields} e^{iS + i \int \phi_{(0)} \mathcal{O}},$$

(35.51)

and on the gravity dual side,

$$Z_{\text{bulk}} = e^{iS_{\text{on-shell}}[\phi(\phi_{(0)})]}.$$

(35.52)

Then taking a derivative with respect to $\phi_{(0)}$ on both sides of the equality, we obtain the one-point function as

$$\langle \mathcal{O} \rangle = -i \frac{\delta Z_{\text{bulk}}[\phi_{(0)}]}{\delta \phi_{(0)}} = \frac{\delta S_{\text{on-shell}}[\phi_{(0)}]}{\delta \phi_{(0)}},$$

(35.53)

as a function of $\phi_{(0)} \neq 0$.

On the other hand, in the AdS metric

$$ds^2 = \frac{R^2}{z^2} (\gamma_{\mu\nu} dx^\mu dx^\nu + dz^2),$$

(35.54)

where the boundary is at $z = 0$, the expansion of the field ϕ near the boundary is

$$\phi(z) = \left(\frac{z}{R}\right)^{d-\Delta} \phi_{(0)} + \left(\frac{z}{R}\right)^{\Delta} \phi_{(1)} + \cdots,$$

(35.55)

where $\phi_{(0)}$ is the source on the boundary, giving a *non-normalizable term* (when integrating over z close to 0, since $\Delta > d/2$ by the BF bound (37.2) and $\sqrt{-g} \propto (R/z)^{d+1}$), and $\phi_{(1)}$ gives a normalizable term.

Then we can write

$$\langle \mathcal{O} \rangle = \frac{1}{\sqrt{-g}|_{\text{bd}}} \frac{\delta S_{\text{on-shell}}}{\delta \phi_{(0)}(z)} \Bigg|_{\phi_{(0)}=0} = \left(\frac{R}{z}\right)^{\Delta} \frac{1}{\sqrt{-\gamma}} \frac{\delta S_{\text{on-shell}}}{\delta \phi(z)} \Bigg|_{\phi_{(0)}=0},$$

(35.56)

where $\phi_{(0)}(z)$ is obtained from $\phi(z)$ by taking out a $(z/R)^{d-\Delta}$ factor. Since in general the action receives, besides the kinetic term, containing derivatives, also a boundary term, that can be fixed through a procedure called holographic renormalization (that will not be explained here), we write

$$S_{\text{on-shell}} = S[\partial_\mu \phi] + S_{\text{boundary}}, \tag{35.57}$$

where $S[\partial_\mu \phi]$ is the action *before* the on-shell procedure, leading to

$$\langle \mathcal{O} \rangle = \left(\frac{R}{z} \right)^\Delta \frac{1}{\sqrt{\gamma}} \left(-\frac{\delta S[\partial_\mu \phi(\phi_{(0)})]}{\delta \partial_z \phi(z)} - \frac{\delta S_{\text{boundary}}}{\delta \phi(z)} \right) \Bigg|_{\phi_{(0)}=0}. \tag{35.58}$$

Through holographic renormalization, one finds for a canonical scalar a boundary action

$$S_{\text{boundary}} = \frac{\Delta - d}{2R} \int d^d x \sqrt{-\gamma} \phi^2(z), \tag{35.59}$$

which leads to

$$\langle \mathcal{O} \rangle = -\lim_{z \to 0} \left(\frac{R}{z} \right)^\Delta \left[\frac{z}{R} \partial_z \phi|_{\phi_{(0)}=0} + \frac{\Delta - d}{2R} 2\phi|_{\phi_{(0)}=0} \right]$$
$$= -\frac{2\Delta - d}{R} \phi_{(1)}. \tag{35.60}$$

Finally then, the retarded Green's function is obtained as

$$G^R_{\mathcal{O}_A \mathcal{O}_B} = \frac{\delta \langle \mathcal{O}_A \rangle}{\delta \phi_{B(0)}} \Bigg|_{\delta\phi=0} = \frac{2\Delta_A - d}{R} \frac{\delta \phi_{A(1)}}{\delta \phi_{B(0)}}, \tag{35.61}$$

i.e., as the variation of the normalizable mode with respect to the non-normalizable mode.

This formulation is in fact equivalent with the formulation of Son and Starinets, and which one is applied depends on the ease of calculation.

35.5 Other Transport Properties

We have seen in Part I that the transport properties with respect to different coordinates mix (so that, for instance, the electric conductivity, or the shear viscosity, are matrices), but in fact the transport properties of different operators also mix in general.

If we introduce a nonzero chemical potential in the theory $\mu \neq 0$, thus in general having a finite charge density, then the electric current and heat (energy) current (flux) mix; i.e., we have a thermoelectric (Nernst) effect, that a temperature gradient generates an electric current, $J_x = -\alpha \nabla_x T$, and the usual fact that an electric field generates heat flux (through the resistance), $Q_x = \alpha T E_x$. The diagonal elements are $J_x = \sigma E_x$ and $Q_x = -\bar{\kappa} \nabla_x T$, leading to the matrix

$$\begin{pmatrix} \langle J_x \rangle \\ \langle Q_x \rangle \end{pmatrix} = \begin{pmatrix} \sigma & \alpha T \\ \alpha T & \bar{\kappa} T \end{pmatrix} \begin{pmatrix} E_x \\ -\frac{\nabla_x T}{T} \end{pmatrix}. \tag{35.62}$$

Then one finds, through a similar analysis as for σ and η, the Kubo formulas

$$\alpha(\omega)T = +\frac{iG^R_{Q_xJ_x}(\omega)}{\omega}$$

$$\bar{\kappa}(\omega)T = +\frac{iG^R_{Q_xQ_x}(\omega)}{\omega}. \tag{35.63}$$

Also as before, usually we want real transport coefficients, so we obtain

$$\alpha(\omega)T = -\frac{\mathrm{Im}G^R_{Q_xJ_x}(\omega)}{\omega}$$

$$\bar{\kappa}(\omega)T = -\frac{\mathrm{Im}G^R_{Q_xQ_x}(\omega)}{\omega}, \tag{35.64}$$

or moreover for the static quantities:

$$\alpha(0)T = -\lim_{\omega\to 0}\frac{\mathrm{Im}G^R_{Q_xJ_x}(\omega)}{\omega}$$

$$\bar{\kappa}(0)T = -\lim_{\omega\to 0}\frac{\mathrm{Im}G^R_{Q_xQ_x}(\omega)}{\omega}. \tag{35.65}$$

Important Concepts to Remember

- Retarded Green's functions give the linear response of a VEV of an operator from a perturbation of a source of another operator, $G^R_{\mathcal{O}_A\mathcal{O}_B} = \delta\langle\mathcal{O}_A\rangle/\delta\phi_{(0)B}$.
- The real part of G^R is a thermodynamic susceptibility, obtained as a spectral integral representation from the imaginary part of G^R, which is therefore a spectral function.
- The VEV of the commutator of the operators is $\rho = i(G^R - G^A)$, which acts as a spectral function for $G^{R,A}$. In the case of Hermitian operators, we obtain the same spectral function as above, since then $\rho = -2\mathrm{Im}G^R$.
- The electric conductivity is found from a Kubo formula, $\sigma = iG^R_{J_xJ_x}(\omega)/\omega$, and the shear viscosity is found from a similar Kubo formula, $\eta = iG^R_{T_{xy}T_{xy}}/\omega$.
- The AdS/CFT prescription in Minkowski space at finite temperature is given in one way by writing the on-shell action $S = \int \phi_0\mathcal{F}\phi_0|_{\mathrm{bd.}}$, as $G^R = -2\mathcal{F}$.
- In another way, we find $G^R_{\mathcal{O}_A\mathcal{O}_B} = (2\Delta_A - d)/R\delta\phi_{A(1)}/\delta\phi_{B(0)}$, i.e. by varying the normalizable mode in terms of the non-normalizable mode.
- Heat and electric conductivities mix, and all the elements of the matrix also have Kubo formulas.

Further Reading

For more details, see the review by Hartnoll [1]. The prescription of Son and Starinets was introduced in [82].

Exercises

(1) Using the spectral representation for G^R, show that for any vector v_A

$$i\omega v_A [G^R_{\mathcal{O}_A \mathcal{O}_B}(\omega, x - x') - G^R_{\mathcal{O}_B \mathcal{O}_A}(\omega, x - x')] v_B \geq 0. \qquad (35.66)$$

(2) Using the property in Exercise 1, show that the time average over one cycle of the dissipation power (rate of work) for the system is positive.

(3) Calculate holographically the spectral function for a scalar operator with $\Delta = 2$ in $d = 3$.

(4) Find the Kubo formula for the bulk viscosity ζ, paralleling what was done for σ and η.

(5) Using the definition of G^R and of the time-reversal operator, prove that for a time-reversal invariant system,

$$G^R_{\mathcal{O}_A \mathcal{O}_B} = \pm G^R_{\mathcal{O}_B \mathcal{O}_A}, \qquad (35.67)$$

and the sign is "+" if the $\mathcal{O}_A$ and $\mathcal{O}_B$ have the same time-reversal symmetry; also that from this follows the symmetry of the matrices of transport properties.

36 Dynamic and Nonequilibrium Properties of Plasmas: Electric Transport, Langevin Diffusion, and Thermalization via Black Hole Quasi-Normal Modes

In this chapter, we continue the analysis of strongly coupled hot plasmas started in Chapter 34 with dynamic and nonequilibrium properties. Electric charge transport will be related to Green's functions in the gravity dual, and by applying the membrane paradigm, which will be described more formally in Chapter 38, we will be able to calculate the transport at the horizon of a black hole.

During transport, the momentum distribution broadens due to Langevin diffusion, from Brownian motion, as the charge carriers collide with the plasma. We will see how to understand it from the motion of strings in the gravity dual. Finally, an initial departure from equilibrium results in thermalization via thermal collisions. In the gravity dual, these will be shown to be described by the quasi-normal modes of a black hole.

36.1 Electric Charge Transport in the Plasma and the Membrane Paradigm

In the previous chapter, we have calculated the Kubo formula for electric charge transport, i.e. the conductivity σ. For the general AC conductivity we have the formula

$$\sigma(\omega, \vec{k}) = \frac{iG^R_{J_x J_x}(\omega, \vec{k})}{\omega}, \tag{36.1}$$

relating it to the retarded correlator, which can be calculated using AdS/CFT.

Consider an AdS Schwarzschild black hole in 3+1 dimensions, inside which we consider the gauge field perturbation that gives rise to the retarded Green's function and the conductivity, with metric

$$ds^2 = \frac{r_H^2}{L^2 u^2}\left(-f(u)dt^2 + dx^2 + dy^2\right) + \frac{L^2 du^2}{u^2 f(u)}, \tag{36.2}$$

where

$$f(u) = 1 - u^3,$$
$$r_H = \frac{4\pi T L^2}{3}. \tag{36.3}$$

Note that we have defined $u = r_H/r$, where r is the usual radial coordinate coming from the D-brane action. Then the boundary is at $u = 0$ ($r = \infty$), and the horizon is at $u = 1$ ($r = r_H$).

However, there is a simpler way to calculate the *DC* conductivity from a reinterpretation of a term in the action. Consider first a simple action in $d+1$ dimensions,

$$S_{em} = -\int_{\Sigma} d^{d+1}x \sqrt{-g} \frac{1}{4g_{d+1}^2(r)} F_{MN} F^{MN},$$ (36.4)

where $g_{d+1}(r)$ is an r-dependent "coupling," that can arise due to background fields that couple to the field strengths. From the bulk action, when varying to obtain the equations of motion, we obtain a boundary term that must be canceled, so we need to add

$$S_{\text{"boundary"}} = \int_{\Sigma} d^d x \sqrt{-\gamma} \left(\frac{j^{\mu}}{\sqrt{-\gamma}}\right) A_{\mu},$$ (36.5)

where j^{μ} is a conjugate momentum to A_{μ}, obtained by varying the action with respect to $\partial_r A_{\mu}$,

$$j^{\mu} = -\frac{1}{g_{d+1}^2(r)} \sqrt{-g} F^{r\mu},$$ (36.6)

and $\gamma_{\mu\nu}$ is the induced metric on the "stretched horizon" (a surface close to the black hole event horizon) Σ, thought of as a "boundary." Then we can identify a horizon ("membrane") current by varying the boundary term in the action with respect to A_{μ}:

$$J_{\text{horizon}}^{\nu} = \frac{\delta S_{\text{"boundary"}}}{\delta A_{\mu}} = \frac{j^{\nu}}{\sqrt{-\gamma}} = -\frac{1}{g_{d+1}^2(r)} \sqrt{g_{rr}} F^{r\nu}(r_H).$$ (36.7)

One has to impose "infalling" boundary conditions for the gauge field at the horizon $r = r_H$ ($u = 1$), meaning that it can depend only on the combination $dv = dt + \sqrt{g_{rr}/g_{tt}}dr$ (more on that in Section 3 of this chapter), which means that at the horizon $r \to r_H$, in the gauge $A_r = 0$, we obtain

$$\partial_r A_i = \sqrt{\frac{g_{rr}}{g_{tt}}} \partial_t A_i \Rightarrow F_{ri} = \sqrt{\frac{g_{rr}}{g_{tt}}} F_{ti},$$ (36.8)

so that we have a linear relation between the current and the electric field at the horizon

$$J_{\text{horizon}}^i = -\frac{1}{g_{d+1}^2(r_H)} \sqrt{g_{tt}} F_t^i = \frac{1}{g_{d+1}^2(r_H)} \hat{E}^i,$$ (36.9)

where $\hat{E}^i$ is an electric field as measured at infinity, leading to a conductivity *as viewed from infinity*, i.e. in the dual field theory, but calculated *at the horizon*:

$$\sigma_{\text{field theory}} = \sigma_{\text{horizon}} = \frac{1}{g_{d+1}^2(r_H)}.$$ (36.10)

One can generalize the calculation for an Maxwell action with a "tensor" coupling constant, for simplicity in 3+1 dimensions,

$$S = \int d^4x \sqrt{-g} \left[-\frac{1}{8g_4^2} F_{\mu\nu} X^{\mu\nu\rho\sigma} F_{\rho\sigma} \right],$$ (36.11)

and then in a similar way one obtains the DC conductivity of the field theory from the horizon as

$$\sigma_{\text{field theory}} = \sigma_{\text{horizon}} \equiv \sigma(\omega = 0, k = 0) = \frac{1}{g_4^2}\sqrt{-g}\sqrt{-X^{xtxt}X^{xrxr}}\bigg|_{r=r_H}. \qquad (36.12)$$

We see that we have obtained a rather unusual situation: we want to calculate the conductivity of the field theory, which lives at the boundary of the gravity dual, but the result is expressed in terms of a formula at the horizon, which is a different "boundary" surface.

We can understand this as follows: AdS/CFT is an example of holography, where the physics in the bulk is equal to the physics on the boundary. On the other hand, in Chapter 38 we will see that there is another important "holographic" relation in gravitational physics, the "membrane paradigm," that states that under very specific conditions, the gravitational physics away from a black hole (i.e., in the "bulk") can be described in terms of the physics of a very thin "membrane" situated just outside the event horizon of the black hole. Here we will describe a very specific form of the general membrane paradigm of Chapter 38, and give some speculations based on the general form. Putting together the two relations, AdS/CFT and the membrane paradigm, we can describe the physics on the boundary at infinity in terms of the physics at the horizon.

We now try to understand better the calculation of the DC conductivity of the field theory from the horizon. For concreteness, consider the specific case when

$$X_{\mu\nu}{}^{\rho\sigma} = 2\delta_{[\mu\nu]}{}^{[\rho\sigma]} - 8\gamma L^2 C_{\mu\nu}{}^{\rho\sigma}, \qquad (36.13)$$

corresponding to a usual Maxwell term, plus a coupling of the field strength to the Weyl tensor, giving a total vector action

$$S_{\text{vector}} = \frac{1}{g_4^2}\int d^4x\sqrt{-g}\Big[-\frac{1}{4}F_{\mu\nu}F^{\mu\nu} + \gamma L^2 C_{\mu\nu\rho\sigma}F^{\mu\nu}F^{\rho\sigma}\Big]. \qquad (36.14)$$

The *Weyl tensor* is defined in terms of the *Schouten tensor*, which in 3+1 dimensions is

$$S_{\alpha\beta} = \frac{1}{2}\left(R_{\alpha\beta} - \frac{R}{6}g_{\alpha\beta}\right), \qquad (36.15)$$

as follows:

$$C_{\alpha\beta}{}^{\gamma\delta} = R_{\alpha\beta}{}^{\gamma\delta} - 4S_{[\alpha}{}^{[\gamma}\delta_{\beta]}^{\delta]}. \qquad (36.16)$$

Consider a Fourier transform of the gauge field A_μ over the 2+1–dimensional boundary coordinates, x, y, t,

$$A_\mu(t, x, y, u) = \int \frac{d^3q}{(2\pi)^3}e^{i\vec{q}\cdot\vec{x}}A_\mu(\vec{q}, u), \qquad (36.17)$$

where $\vec{q}\cdot\vec{x} = -t\omega + q^x x + q^y y$ and $q^\mu = (\omega, q, 0)$. The boundary condition at the horizon $u = 1$ is that the gauge field has a zero there, i.e. that near $u = 1$ we can write

$$A_y(\vec{q}, u) = (1 - u)^b F(\vec{q}, u), \qquad (36.18)$$

with $F(\vec{q}, u)$ regular there.

The AdS/CFT prescription is given by the Kubo formula (36.1), for which we need to know the retarded correlator. But by the general prescription for correlators that we have defined in Part II, we need to calculate the on-shell supergravity action as a function of the boundary values for A_μ, and differentiate twice with respect to A_μ at $A_\mu = 0$. The A_μ part of the supergravity action is given in (36.14), so substituting the Fourier transformed A_μ in it, we get a term defined on the total boundary of the space:

$$S_{yy} = -\frac{1}{2g_4^2} \int d^3x \left[\sqrt{-g} g^{uu} g^{yy} (1 - 8\gamma L^2 C_{uy}{}^{uy}) A_y(u, \vec{x}) \partial_u A_y(u, \vec{x}) \right]\Big|_{\text{boundary}}. \quad (36.19)$$

Evaluating on the AdS-Schwarzschild black hole solution in (36.2), we get

$$S_{yy} = -\frac{2\pi T}{3g_4^2} \int d^3x \left[(1 - u^3)(1 + 4\gamma u^2) A_y(u, \vec{x}) \partial_u A_y(u, \vec{x}) \right]\Big|_{\text{boundary}}. \quad (36.20)$$

If the boundary of the space is considered to be entirely at $u = 0$ (at "infinity"), then we get

$$S_{yy} = -\frac{2\pi T}{3g_4^2} \int d^3x \left[A_y(u, \vec{x}) \partial_u A_y(u, \vec{x}) \right]\Big|_{u=0}$$

$$\equiv \int \frac{d^2\vec{q}}{(2\pi)^3} \frac{1}{2} A_y(-\vec{q}) G_{yy}(\vec{q}) A_y(\vec{q}), \quad (36.21)$$

where we have defined the 2-point retarded correlator for A_y:

$$G_{yy}(\omega, q = 0) = -\frac{4\pi T}{3g_4^2} \frac{\partial_u A_y(u, \omega)}{A_y(u, \omega)}\Big|_{u\to 0}. \quad (36.22)$$

We can now substitute in the Kubo formula (36.1) and obtain the conductivity of the field theory

$$\sigma_{\text{field theory}} = \frac{1}{3g_4^2} \text{Im} \frac{\partial_u A_y}{w A_y}\Big|_{u\to 0}, \quad (36.23)$$

where $w \equiv \omega/4\pi T$. Note that this expression is entirely at the boundary, unlike the previous expression (36.12), which was at the horizon. The problem is that we now need to solve for A_y on the whole space, since the boundary condition (36.18) is at the horizon, whereas we need the corresponding solution at the boundary at infinity.

But we are interested only in the DC conductivity, which means we can consider only the solution at small ω, i.e. small w, which is found to be

$$A_y(u) \simeq (1 - u)^{-iw} \left[F_1(u) + w F_2(u) \right], \quad (36.24)$$

where $F_1(u) = C$ (constant) and $F_2'(0) = iC(2 + 12\gamma)$. Knowing also that $F_1(u)$ and $F_2(u)$ are well-behaved at the horizon $u = 1$ allows us to calculate the DC conductivity at infinity by substituting A_y in (36.23), to obtain

$$\sigma(\omega \to 0) = \frac{1}{3g_4^2} \text{Im}\left[i + \frac{F_2'(0)}{F_1(0)} \right] = \frac{1}{3g_4^2}[1 + (2 + 12\gamma)] = \frac{1 + 4\gamma}{g_4^2}. \quad (36.25)$$

This is the same result as was obtained from the calculation at the horizon (36.12) by substituting the form of the tensor X in (36.13).

We can now reinterpret the calculation (36.12) at the horizon as follows. In the first way of calculating the conductivity, we have added a boundary term at the horizon to the supergravity action to cancel it on the equation of motion (on-shell), and this boundary term was interpreted as defining a current, which was found to be proportional to the electric field at infinity. But we can instead consider the on-shell action, the generator of the retarded correlator via the Kubo formula, reduced to a boundary term in (36.20), as a boundary term *at the horizon*, and use it to calculate a "retarded correlator" for the Kubo formula now at the horizon. Note that in so doing, we must ignore the boundary at infinity ($u = 0$), since otherwise we will get a diverging contribution (much larger than the one at the horizon), so one must look for a reinterpretation of the Kubo formula.

The boundary term (36.20), only at the horizon $u = 1$, gives

$$S_{yy} = -\frac{2\pi T}{3g_4^2}(1 + 4\gamma) \int d^3x \left[3(1 - u)A_y(u, \vec{x})\partial_u A_y(u, \vec{x})\right]\Big|_{u \to 1}, \qquad (36.26)$$

which defines the "retarded correlator"

$$G_{yy}(\omega, q = 0) = -\frac{4\pi T}{3g_4^2}(1 + 4\gamma) \left[3(1 - u)\frac{\partial_u A_y(u, \omega)}{A_y(u, \omega)}\right]\Big|_{u \to 1}. \qquad (36.27)$$

Considering also that at the horizon we have the boundary condition (36.18), which gives $\partial_u A_y/w \simeq i/(1 - u)$, we obtain for the DC conductivity

$$\sigma(\omega \to 0) = \frac{1}{3g_4^2}(1 + 4\gamma)\text{Im}\left[3(1 - u)\frac{i}{1 - u}\right] = \frac{1 + 4\gamma}{g_4^2}. \qquad (36.28)$$

Notice that the factors of $1 + 4\gamma$ and -3 enter in a completely different way from the normal Kubo formula calculation above, so it is highly nontrivial that one obtains the same result. This calculation is a new type of "membrane paradigm," where we take advantage of an AdS/CFT type calculation, but defined at the horizon instead of at the boundary.

So how can we understand the fact that the usual Kubo calculation at the boundary and the "membrane paradigm" calculations at the horizon give the same result? In the particular case of the DC conductivity, for diagonal metric backgrounds

$$ds^2 = -g_{tt}dt^2 + g_{rr}dr^2 + g_{ij}dx^i dx^j, \qquad (36.29)$$

with the isotropy condition $g_{ij} = g_{zz}\delta_{ij}$, the equations of motion in the zero momentum limit can be shown to give

$$\partial_r j^i = 0 + \mathcal{O}(\omega F_{it}); \quad \partial_r F_{it} = 0 + \mathcal{O}(\omega j^i). \qquad (36.30)$$

The proof is left as an exercise.

The relation (36.9), written in terms of j^i, the momentum conjugate to A_μ, instead of in terms of J_{horizon}^i, becomes

$$j^i(r_H) = \frac{1}{g_{d+1}^2}\sqrt{\frac{-g}{g_{rr}g_{tt}}}g^{zz}\Big|_{r_H} F_{it}(r_H). \qquad (36.31)$$

But the equations of motion (36.30) mean that this relation between j^i and F_{it} at zero momentum, i.e. the DC conductivity, continues to be valid at all r, until the boundary at infinity. That means that the DC conductivity of the field theory, calculated at infinity using the usual AdS/CFT, should give the same result as the calculation at the horizon:

$$\sigma_{\text{field theory}}(\omega = 0) = \sigma_{\text{horizon}}(\omega = 0) = \frac{1}{g_{d+1}^2}\sqrt{\frac{-g}{g_{rr}g_{tt}}}g^{zz}\Bigg|_{r_H}. \tag{36.32}$$

We could also calculate a general transport coefficient χ in the case of a quantity associated with a scalar ϕ, with kinetic action

$$S_{\text{kin}} = -\frac{1}{2}\int d^{d+1}x\sqrt{-g}F(r)g^{\mu\nu}\partial_\mu\phi\partial_\nu\phi. \tag{36.33}$$

The result for the real transport coefficient is given again by a Kubo formula *at the horizon*:

$$\chi = -\lim_{\omega\to 0}\lim_{\vec{k}\to 0}\frac{\text{Im}G^R(\omega,\vec{k})}{\omega} = \lim_{\omega\to 0}\lim_{\vec{k}\to 0}F(r_H)\sqrt{\frac{-g}{g_{rr}g_{tt}}}\Bigg|_{r_H}. \tag{36.34}$$

The proof is left as an exercise.

Note that this very concrete and specific proof of the membrane paradigm, equating the DC conductivity at the horizon with the one at the boundary, in the field theory, is valid only for the DC case, i.e. for $\omega = 0$, and only for diagonal gravity dual metrics. But we will see in Chapter 38 that the membrane paradigm is a much more general statement, so even though we have no explicit proof, there should be a similar equivalence between boundary calculations and horizon calculations for more general transport coefficients.

36.2 Momentum Broadening, Langevin Diffusion, and its Gravity Dual

As a charged particle moves through a plasma, it experiences a broadening of its momentum, due to the effect of random collisions, leading to Langevin diffusion.

Momentum Broadening and Langevin Diffusion

The force acting on the particle has a "drag" component, a frictional term proportional to the momentum, with the proportionality coefficient being the drag coefficient η_D, and a random component F_i, with a part transverse to the momentum, and a part longitudinal to it,

$$\frac{dp_i}{dt} = -\eta_D(p)p_i + F_i^L + F_i^T, \tag{36.35}$$

such that the correlators are defined by coefficients κ_L and κ_T via

$$\langle F_i^L(t_1)F_j^L(t_2)\rangle = \kappa_L(p)\hat{p}_i\hat{p}_j\delta(t_1 - t_2)$$
$$\langle F_i^T(t_1)F_j^T(t_2)\rangle = \kappa_T(p)(\delta_{ij} - \hat{p}_i\hat{p}_j)\delta(t_1 - t_2), \tag{36.36}$$

where $\hat{p}_i$ is a unit vector parallel to the momentum. An Einstein relation relates κ_L with η_D by

$$\eta_D = \frac{\kappa_L}{2TE}. \tag{36.37}$$

In the case of a relativistic particle, $E = \gamma m = m/\sqrt{1 - v^2}$, whereas for a nonrelativistic one we have simply $E = m$. For nonrelativistic particles, $\kappa_L = \kappa_T$ by isotropy, and another Einstein relation relates them to the diffusion coefficient D by

$$\kappa_L = \kappa_T = \frac{2T^2}{D}. \tag{36.38}$$

Because of these relations, it is actually enough to know a single coefficient, for instance the drag coefficient. In the $\mathcal{N} = 4$ SYM plasma at strong coupling, through the use of the black hole in $AdS_5 \times S^5$ gravity dual, one can calculate

$$\eta_D = \frac{\pi}{2}\sqrt{g_{YM}^2 N}\frac{T^2}{m}, \tag{36.39}$$

though we will not describe all the details of the calculation.

One can write down a worldline action for the particle moving through the plasma, which includes the random component, through the coupling $\int d\tau X_i(\tau)F^i(\tau)$ (and a driving component $\int d\tau X_\mu(\tau)\mathcal{F}^\mu(\tau)$). In general, the variation of the action with respect to X_i ($i = 1, 2, 3$) is given by two terms, with one containing a Heaviside function, that is basically the response to a driving perturbation (a friction term), and the random component,

$$\frac{\delta S}{\delta X_i(t)} = \int_{-\infty}^{+\infty} d\tau \theta(\tau)C^{ij}(\tau)X_j(t - \tau) + F^i(t), \tag{36.40}$$

where C^{ij} is called the *memory kernel*,

$$\theta(\tau)C^{ij}(\tau) = G_R^{ij}(\tau) \tag{36.41}$$

is the *retarded Green's function for* $\mathcal{F}^i(t)$ (conjugate to $X_i(t)$, from the action), as defined in the last chapter, and in general the Gaussian random variable F^i has a correlation function

$$\langle F^i(t)F^j(t')\rangle = A^{ij}(t - t'). \tag{36.42}$$

The quantities C^{ij} and A^{ij} are the antisymmetric and symmetric parts of the correlator, in the operator formalism being

$$C^{ij}(t) = G_{a-sym}^{ij}(t) = \langle[\mathcal{F}^i(t), \mathcal{F}^j(0)]\rangle; \quad A^{ij}(t) = G^{ij,sym}(t) = \langle\{\mathcal{F}^i(t), \mathcal{F}^j(0)\}\rangle. \tag{36.43}$$

We have seen in the last chapter that $\theta(t)$ times the antisymmetric term C^{ij} is the retarded Green's function. We have also seen that C^{ij} is the Fourier transform of the spectral function $\rho^{ij}(\omega)$,

$$C^{ij}(t) = \int_{-\infty}^{+\infty} d\omega e^{-i\omega t}\rho^{ij}(\omega), \tag{36.44}$$

whereas the retarded Green's function is written in terms of it as

$$G_R^{ij} = \frac{1}{2\pi}\int_{-\infty}^{+\infty} d\omega'\frac{\rho^{ij}(\omega')}{\omega - \omega' + i\epsilon}. \tag{36.45}$$

If the correlation functions A^{ij} vanish for $t - t' \gg \tau_c$, with τ_c a *correlation time*, we obtain the local Langevin equation written at the beginning of the section. Indeed, then the noise correlator becomes a white noise,

$$A^{ij}(t - t') \simeq \kappa^{ij}\delta(t - t'); \quad \text{for} \quad t - t' \gg \tau_c, \tag{36.46}$$

and in fact the diffusion constants κ^{ij} reduce to κ_L and κ_T defined before. One can define γ^{ij} as the integral of C^{ij}:

$$C^{ij}(t) = \frac{d}{dt}\gamma^{ij}(t). \tag{36.47}$$

Then the friction term involving the retarded propagator $\theta(\tau)C^{ij}(\tau)$ reduces to the usual drag force, since it becomes (by a partial integration) in the large time limit $\tau \gg \tau_c$:

$$\int_0^{+\infty} d\tau C^{ij}(\tau)X_j(t - \tau) \simeq \left(\int_0^{+\infty} d\tau\gamma^{ij}(\tau)\right)\dot{X}_j(t). \tag{36.48}$$

Moreover, for a free relativistic particle,

$$\frac{\delta S}{\delta X_i(t)} = \frac{dp^i}{dt}, \tag{36.49}$$

so all in all we obtain the usual Langevin equations that we started off the section with:

$$\frac{dp^i}{dt} = -\eta^{ij}\dot{X}_j(t) + F^i(t)$$
$$\langle F^i(t)F^j(t')\rangle = \kappa^{ij}\delta(t - t'). \tag{36.50}$$

Now the friction coefficients η^{ij} are the integrals of γ^{ij}, and they are given by Kubo formulas of their own. Indeed, $C^{ij} = d\gamma^{ij}/dt$ are spectral functions for F^i, conjugate to X_i, and the η^{ij} are transport coefficients in the linear response relation between an applied force F^i and a resulting equilibrium velocity $\dot{X}_j$ such that $p^i = $ constant, in the same way as the electric conductivity σ was a transport coefficient for the linear response relation between applied $\vec{E}$ and resulting current $\vec{j}$, with the spectral function being for J_μ, conjugate to A_μ. One finds therefore

$$\eta^{ij} = \int_0^{+\infty} d\tau\gamma^{ij}(\tau) = -\lim_{\omega \to 0}\frac{\mathrm{Im}G_R^{ij}(\omega)}{\omega}, \tag{36.51}$$

but we will leave the details of the proof as an exercise. On the other hand, the diffusion coefficients are given, as before, by the symmetric part of the Green's function:

$$\kappa^{ij} = \lim_{\omega \to 0} G_{\mathrm{sym}}^{ij}(\omega). \tag{36.52}$$

The relation between G_R^{ij} and G_{sym}^{ij} for a canonical ensemble at temperature T is found to be

$$G_{\mathrm{sym}}^{ij}(\omega) = -\coth\frac{\omega}{2T}\mathrm{Im}G_R^{ij}(\omega), \tag{36.53}$$

so it is sufficient to calculate the retarded Green's function G_R to find all the Langevin dynamics.

For short enough times, shorter than a relaxation time τ_D, but larger than the correlation time τ_c, we obtain a momentum broadening. We consider the small broadening around $\vec{p}_0$, defined by $\vec{p} \simeq p_0 \vec{v}_0/v_0 + \delta\vec{p}$. To solve for it, we can treat the velocity as approximately constant, and solve the Langevin equations (36.50), by first linearizing them,

$$\frac{d\delta p_i^T}{dt} = -\eta_{D,0}^T \delta p_i^T + F_i^T$$

$$\frac{d\delta p_i^L}{dt} = -\eta_{D,0}^L p_i^0 + \left[\eta_D^L + p\frac{\partial \eta_D^L}{\partial p} \right]_{p_0} \delta p_i^L + F_i^L, \qquad (36.54)$$

and then calculating the average broadening in the transverse and longitudinal momenta one finds

$$\langle (p_T)^2 \rangle = 2\kappa_T t; \quad \langle (p - p_0)^L \rangle = \kappa_L t. \qquad (36.55)$$

The transverse momentum broadening can be calculated as follows. Neglect the drag force $\eta_{D,0}^T$ term (in the correct calculation it will not play any role), since we are at short times. Then

$$\delta p_i^T(t) = \int_0^t dt_1 F_i^T(t_1), \qquad (36.56)$$

and calculating the average momentum broadening we get

$$\langle (p_T(t))^2 \rangle = \sum_{i,j=1,2} \delta^{ij} \langle \delta p_i^T(t) \delta p_j^T(t) \rangle = \sum_{i,j=1,2} \delta^{ij} \int_0^t dt_1 \int_0^t dt_2 \langle F_i^T(t_1) F_j^T(t_2) \rangle$$

$$= \sum_{i,j=1,2} \delta^{ij} \int_0^t dt_1 \int_0^t dt_2 A_{ij}(t_1 - t_2) \simeq 2\kappa_T t. \qquad (36.57)$$

The longitudinal broadening can be calculated similarly. This increase of the broadening with time t is a general characteristic of Brownian motion, so we see that the momentum fluctuations have a Brownian-like diffusion.

Gravity Dual

We will focus on our standard toy model, $\mathcal{N} = 4$ SYM, since things are easier to write, but anything we say here can be applied to more general field theories/gravity duals. As we saw, to find the Langevin dynamics we need to find the retarded Green's function G^R.

The gravity dual to the hot $\mathcal{N} = 4$ SYM plasma is the AdS_5-Schwarzschild black hole, or Witten metric, written as (similarly to (36.2) in 3+1 dimensions)

$$ds^2 = \frac{L^2}{z_H u} \left(-f(u)dt^2 + d\vec{x}^2 + \frac{z_H^2}{f(u)} du^2 \right), \qquad (36.58)$$

where

$$f(u) = 1 - u^4; \quad z_H = \frac{1}{\pi T}. \qquad (36.59)$$

The drag coefficient of the plasma (36.39) is found from a moving, trailing string, that moves in the x^1 direction of the $u = 0$ boundary with velocity v, and drops all the way to the $u = 1$ horizon, with classical configuration

$$X^1 = X_0^1(t, u) = vt + \frac{vz_H}{2}\left(\arctan(u) + \log\sqrt{\frac{1-u}{1+u}}\right). \qquad (36.60)$$

It corresponds to a charged particle (quark in the case of QCD) moving through the plasma at constant velocity, and feeling a drag force (friction).

But consider now fluctuations around this string configuration, $X^i = X_0^i + \delta X^i(t, u)$. The fluctuations of the string *at the boundary*, where the string endpoint corresponds to the position of the particle (quark) moving through the plasma, are considered through the above analysis as the ones coupling to the forces, whose correlators (36.43) we need to calculate the transport coefficients.

But we know how to calculate the retarded correlators from AdS/CFT: using the Son and Starinets prescription from last chapter. First, write the on-shell action in the form (35.46), and then the Green's function is given by (35.47). One expands the Nambu-Goto action in fluctuations around the background solution (36.60), and obtains the quadratic action (dropping the constant and linear parts)

$$
\begin{aligned}
\delta S_{NG} &= -\frac{1}{2\pi\alpha'}\delta\int d^2\sigma\sqrt{-\det G_{\mu\nu}\partial_\alpha X^\mu\partial_\beta X^\nu} \\
&= -\frac{1}{2}\int dt\, du\, \mathcal{G}^{\alpha\beta}\partial_\alpha\delta X^i\partial_\beta\delta X^i \\
&= -\frac{1}{2}\int dt\, du\left(\mathcal{G}_L^{\alpha\beta}\partial_\alpha\delta X^1\partial_\beta\delta X^1 + \mathcal{G}_T^{\alpha\beta}\sum_{i=2,3}\partial_\alpha X^i\partial_\beta X^i\right). \qquad (36.61)
\end{aligned}
$$

There we parametrized the worldsheet by the static gauge choice $\sigma^\alpha = (t, u)$ that fixes worldsheet reparametrizations. But we are interested in the *on-shell* action generated by some source $\delta X^i(u \to 0)$ on the boundary at infinity. Therefore we partially integrate this action, and use the resulting equations of motion for the fluctuations,

$$\partial_\alpha(\mathcal{G}^{\alpha\beta}\partial_\beta\delta X^i(t, u)) = 0, \qquad (36.62)$$

to cancel the resulting bulk term in the Nambu-Goto action for the fluctuations, and obtain

$$
\begin{aligned}
\delta S_{NG}(\text{on-shell}) &= -\frac{1}{2}\int dt\, du\, \partial^\beta(\mathcal{G}^{\alpha\beta}(\partial_\alpha\delta X^i)\delta X^i) \\
&= -\frac{1}{2}\int dt\left[\delta X^i\mathcal{G}^{\alpha u}\partial_\alpha\delta X^i\right]_{u=1}^{u=0}, \qquad (36.63)
\end{aligned}
$$

which is finally in the form (35.46).

Moreover, we need to solve the equations of motion for δX^i, which are identical for all $i = 1, 2, 3$, in ω space, to find the Green's function in ω space, as we need, so we expand the solutions of the equations of motion in functions $\phi_0(\omega)$ as

$$\delta X^i = \int_{-\infty}^{+\infty}\frac{d\omega}{2\pi}\phi_0(\omega)\Psi(\omega, t, u), \qquad (36.64)$$

where $\phi_0(\omega)^* = \phi_0(-\omega)$, and solve the equations of motion for a Ψ *normalized to unity at the boundary* and with infalling boundary conditions at the horizon, i.e. solutions that near the horizon at $u = 1$ depend only on the infalling coordinate $u_- = t$. Note that this is really a kind of bulk-to-boundary propagator, since we can multiply this Ψ solution with an arbitrary source $\delta\phi(\omega, x^i)$, and therefore in using Ψ's we already factorize these sources, called $\phi_0(\vec{k})$ in (35.47). The retarded solution needed for this propagator are of the type

$$\Psi_R(\omega, t, u) = e^{-i\omega t}\psi_R(\omega, u), \qquad (36.65)$$

satisfy $\Psi_R(\omega, t, 0) = e^{-i\omega t}$ and $\Psi_R(\omega, t, u)^* = \Psi_R(-\omega, t, u)$, and one can solve for them numerically, though I will not show how to do that explicitly here. Finally, the Son and Starinets prescription using Ψ_R's gives simply

$$G^R(\omega) = -\left.\Psi_R^*(\omega, t, u)\mathcal{G}^{u\beta}\partial_\beta\Psi_R(\omega, t, u)\right|_{u=0}. \qquad (36.66)$$

With this Green's function, one can calculate all the Langevin dynamics, as we said before.

36.3 Thermalization from Quasi-Normal Modes of the Dual Black Hole

Finally, the last issue we want to study in this chapter is the decay toward equilibrium of a perturbation of the thermal state by some operator. Since the thermal state corresponds to a black hole, this decay should correspond to the decay of a perturbation in the field dual to the operator. Therefore we need to study the decay modes of fields in the vicinity of an AdS black hole, what are known as *quasi-normal modes*. In the following we will explain what they are and how they are calculated.

We are interested in Schwarzschild black holes in AdS space, for an arbitrary mass M, corresponding to an arbitrary temperature T, in units of the AdS radius R, and not for $M \to \infty$, like in the case of the Witten metric, usually used to describe finite temperature plasmas at a fixed temperature taken to be $T = 1/\pi$ (one can consider also the $M \to \infty$ limit, but we will not do it here). The metric in d dimensions was, as we saw,

$$ds^2 = -f(r)dt^2 + \frac{dr^2}{f(r)} + r^2 d\Omega_{d-2}^2, \qquad (36.67)$$

where

$$f(r) = \frac{r^2}{R^2} + 1 - \left(\frac{r_0}{r}\right)^{d-3}$$
$$M = \frac{(d-2)\Omega_{d-2}r_0^{d-3}}{16\pi G_{N,d}}, \qquad (36.68)$$

so the horizon $r = r_+, t$ fixed has spherical symmetry, as opposed to planar for the Witten metric. The temperature of the metric was found as (see (33.25))

$$T = \frac{f'(r_+)}{4\pi}. \qquad (36.69)$$

The simplest perturbation to consider is a scalar field perturbation, corresponding to the decay of perturbations defined by scalar operators in the plasma. One must therefore consider a scalar perturbation obeying the Klein-Gordon equation in the AdS black hole background:

$$\Box \phi = 0. \tag{36.70}$$

Considering modes an expansion in spherical harmonics $Y(\vec{\Omega})$ and in $e^{-i\omega t}$ as

$$\phi(t, r, \vec{\Omega}) = r^{\frac{d-2}{2}} \psi_l(r) Y_l(\vec{\Omega}) e^{-i\omega t}, \tag{36.71}$$

and redefining the radius by $dr_* = dr/f(r)$, the wave equation reduces to a flat space case with a potential:

$$\left[\frac{\partial^2}{\partial r_*^2} + \omega^2 - \tilde{V}(r_*) \right] \psi_l(r_*) = 0. \tag{36.72}$$

The potential $\tilde{V}$ one obtains is positive, vanishes at the horizon $r = r_+$ ($r_* = -\infty$), and diverges at $r = +\infty$ (a finite r_*).

The quasi-normal modes of the black hole in flat space are defined as follows. The potential $\tilde{V}$ in this case vanishes at $r = +\infty$ since the space is flat there, so we can always define waves coming from infinity (infinite potential), scattering off the potential and being partly reflected and partly absorbed by a black hole. Quasi-normal modes are modes that are purely outgoing at infinity, $\phi \sim e^{-i\omega(t-r_*)}$, and purely ingoing (or "infalling") at the horizon, $\phi \sim e^{+i\omega(t+r_*)}$ ($u = t - r_*$, $v = t + r_*$).

For a black hole in asymptotically AdS space where the potential at infinity is divergent, at a generic ω there are solutions that are ingoing at the future horizon and outgoing at the past horizon. But for a discrete set of complex ω's, the quasi-normal modes, the solutions are purely ingoing (infalling) near the horizon, i.e. $e^{-i\omega(t+r_*)}$. Consider the black hole metric in $(r, v = t + r_*)$ coordinates:

$$ds^2 = -f(r)dv^2 + 2dv\, dr + r^2 d\Omega_{d-2}^2. \tag{36.73}$$

Consider the modification of the expansion in (36.71) by replacing $e^{-i\omega t}$ with the ingoing factor, $e^{-i\omega v}$:

$$\phi(v, r, \vec{\Omega}) = r^{\frac{d-2}{2}} \psi_l(r) Y_l(\vec{\Omega}) e^{-i\omega v}. \tag{36.74}$$

As we said, generically near the horizon $r = r_+$ there will be outgoing modes as well, i.e. $e^{-i\omega(t-r_*)}$. Since near the horizon we have

$$r_* = \int^{r_*} \frac{dr}{f(r)} \simeq \frac{1}{f'(r_+)} \ln(r - r_+), \tag{36.75}$$

these modes can be rewritten as

$$e^{-i\omega(t-r_*)} = e^{-i\omega(t+r_*)} e^{2i\omega r_*} \simeq e^{-i\omega v} (r - r_+)^{\frac{2i\omega}{f'(r_+)}} = e^{-i\omega v} (r - r_+)^{\frac{i\omega}{2\pi T}}. \tag{36.76}$$

That means that the outgoing modes are smooth at the horizon $r = r_+$ only if $2i\omega/f'(r_+)$ is an integer, otherwise we have a singularity. But that selects a discrete set of ω's, the quasi-normal modes ω_n, which must be found numerically.

One can show that the quasi-normal modes ω_n are complex (cannot be purely real or purely imaginary), and moreover *for the decaying (thermalizing) case* $\text{Im}(\omega_n) < 0$ (so that

$e^{i\omega_n t}$ decays), though I will not explain more here. Therefore one writes

$$\omega_n = \omega_{n,r} - i\omega_{n,i}, \tag{36.77}$$

where $\omega_{n,r}, \omega_{n,i} \in \mathbb{R}_+$.

The quasi-normal frequencies determine the decay of the field at late times, which in AdS space is always exponential, i.e. specifically, substituting into the ansatz (36.74):

$$\phi \sim e^{-\omega_{n,i}(t+r_*)}. \tag{36.78}$$

Of course, that means that the leading decay is due to the lowest quasi-normal frequency "ω_1." One might be tempted to think that there would be a regime of power law decay when the mode decayed enough (like there would be in asymptotically flat space), but in AdS space there isn't.

As we explained at the beginning of the section, this exponential decay due to the quasi-normal modes corresponds in the field theory to the decay of a perturbation of the thermal state generated by an operator $\mathcal{O}$ dual to ϕ, i.e. to the "thermalization" of the perturbation.

Important Concepts to Remember

- Electric charge transport in a plasma can be determined by the Kubo formula calculated in the gravity dual, but also by a "membrane paradigm" calculation near the horizon of the black hole, because we can define a "membrane current" there.
- The membrane paradigm in general says that the bulk physics near a black hole can be described holographically by a membrane close to the horizon of a black hole. Added to the AdS/CFT holography, the physics on the boundary should be described by physics on the membrane near the horizon of the gravity dual black hole.
- For the DC conductivity we can show that the membrane paradigm calculation equals the boundary calculation, and the reason is that then the equations of motion make σ be constant on slices from the horizon to the boundary.
- We can define a kind of new AdS/CFT by considering the on-shell gravity action only at the horizon instead of at the boundary at infinity, i.e. as a function of horizon sources only, not boundary sources.
- Momentum broadening in the plasma due to Langevin diffusion is defined by three coefficients, $\eta_D, \kappa_L, \kappa_T$, related by Einstein relations, so only one is independent.
- One can relate the diffusion coefficients to the symmetric and antisymmetric parts of a Green's function for positions of the particle, conjugate to driving forces, which themselves are related to the retarded Green's function through relations coming from complex analysis.
- The broadening of momentum on intermediate scales $\tau_c \ll \tau \ll \tau_D$ is of Brownian motion type, i.e. $\langle (p_T)^2 \rangle = 2\kappa_T t$ and $\langle (p_L - p_0)^2 \rangle = \kappa_L t$.
- In the gravity dual, for $\mathcal{N} = 4$ SYM the AdS_5-Schwarzschild black hole or Witten metric, we must consider a trailing classical string, with one end moving at constant speed on the boundary, and the other end trailing all the way down to the horizon, and consider fluctuations δX^i of the string around the classical configuration, dual to the fluctuations on the particle moving through the plasma.

- The necessary retarded Green's function that defines the Langevin dynamics is found through the Son and Starinets prescription from the eigenfunctions for fluctuations $\delta X^i = \int \phi_0 \Psi$, in ω space, around the classical string, by $G^R = -\Psi_R^* \mathcal{G}^{u\beta} \partial_\beta \Psi_R|_{u=0}$.
- Thermalization, i.e. the exponential decay, of a perturbation of the plasma by an operator $\mathcal{O}$, corresponds to the exponential decay due to quasi-normal modes for the field ϕ dual to $\mathcal{O}$ of black holes in the gravity dual.
- Quasi-normal modes in AdS are modes that are purely ingoing (infalling) near the horizon $r = r_+$, $\sim e^{-i\omega(t+r_*)}$. They are a discrete set $\omega_n = \omega_{n,r} - i\omega_{n,i}$, with $\omega_{n,r}, \omega_{ni}$ a set of positive real numbers, found numerically from the condition that there are no outgoing modes.
- The decay of the mode ϕ is determined as $\sim e^{-\omega_{n,i}(t+r_*)}$, and the leading decay is due to the lowest quasi-normal mode ω_1.

Further Reading

Electric charge transport from the membrane paradigm was calculated in [83]. For a reinterpretation and generalization of the result see [84]. The broadening of momentum through Langevin processes (Brownian motion) in the plasma as done here was first described in [85], based on the earlier calculation in [86]. A complete account in a general background is described in [87]. Thermalization through the use of black hole quasi-normal modes in the gravity dual was first described in [88].

Exercises

(1) Calculate the Weyl tensor for the metric (36.2).
(2) Show that the zero momentum limit of the Maxwell equation in a general diagonal background gives the relations (36.30).
(3) Show that the transport coefficient for a scalar field ϕ is given by (36.34).
(4) Prove the Einstein relation

$$\eta_D = \frac{\kappa_L}{2TE}. \tag{36.79}$$

(5) Prove the Kubo formula

$$\eta^{ij} = \int_0^{+\infty} d\tau \gamma^{ij}(\tau) = -\lim_{\omega \to 0} \frac{\operatorname{Im} G_R^{ij}(\omega)}{\omega}. \tag{36.80}$$

(6) Show that the moving, trailing string in the Witten metric (36.58) has the form (36.60), by using the equations of motion coming from the Nambu-Goto action for the string in the background (36.58).
(7) Calculate the equation of motion for $\psi(r)$ in the expansion (36.74) in the AdS Schwarzschild background.

37

The Holographic Superconductor

The holographic superconductor was defined over a long period, with the contribution of many people, but the most important contributions were done in 2008, first by Gubser, and then the model was refined by Hartnoll, Herzog, and Horowitz.

The holographic superconductor is another example of the "phenomenological AdS/CFT" that we have seen first in the description of nonrelativistic systems with Lifshitz and conformal Galilean symmetry. That is, we don't start with a well-defined string theory duality, but rather we assume that there is an AdS/CFT duality that would describe the system of interest, and introduce ingredients necessary to define it, without deriving them from a fundamental description.

37.1 Holographic Superconductor Ingredients

In order to define a holographic superconductor, we need to have the following ingredients:

- We need an AdS background, since we want to describe an approximately conformal field theory (conformal at zero temperature). Moreover, the superconductors that are not well understood are high T_c ones, which are non-Fermi liquids (cannot be described by the standard Fermi liquid theory), that are at strong coupling, and so can admit a weak coupling gravitational description in the dual. Also, many of them are layered (for instance, cuprates and organic superconductors), and so can be approximated as being 2+1–dimensional. All in all, it means that we need an *AdS₄ background*.
- We want to describe charge transport, so we need to describe a conserved $U(1)$ current J_μ in the field theory that couples in the gravity dual to a *gauge field A_μ* (A_μ is a source for the operator J_μ, through a coupling $\int J^\mu(x)A_\mu(x)$).
- We want to describe a superconductor, therefore to have symmetry breaking, i.e. to have some composite operators condensing. We will consider at least one, $\mathcal{O}$, that must condense, i.e. to have a VEV $\langle \mathcal{O} \rangle \neq 0$ in the superconducting phase. This operator will couple to a field ϕ in the bulk. For a usual, *s*-wave superconductor (the notation is the one from orbitals, with *s*-wave meaning with angular momentum $l = 0$), we therefore need a field ϕ with spin $s = 0$, which couples to the gauge field, to influence the charge transport, so it will be a *charged scalar ψ*. If we want to describe a *p*-wave superconductor (*p*-wave denotes angular momentum $l = 1$), it would mean a field of spin $s = 1$, i.e. a vector V_μ, and if we would want to describe a *d*-wave superconductor (*d*-wave denotes angular momentum $l = 2$), it would mean a field of spin $s = 2$, i.e. a tensor $B_{\mu\nu}$.

- Finally, we want to describe a theory at finite temperature, and, as we saw in the last Chapters 33 through 36, that means that we must introduce a *black hole* in the gravity dual.

With these ingredients, it follows that the Lagrangean for the gravity dual will be

$$\mathcal{L} = \frac{1}{2\kappa^2}\left(R + \frac{d(d-1)}{L^2}\right) - \frac{1}{4g^2}F_{\mu\nu}^2 - |(\partial_\mu - iqA_\mu)\psi|^2 - m^2|\psi|^2 - V(|\psi|). \quad (37.1)$$

Here d is the dimension of the boundary field theory, $d = 3$ in the case of AdS_4, $d(d-1)/L^2 \equiv -\Lambda$ is a *cosmological constant*: the AdS background satisfies $R_{\mu\mu} \propto \Lambda g_{\mu\nu}$ (an energy-momentum tensor that is a constant times the metric), corresponding to a constant term in the gravitational action, a cosmological constant. Also $V(|\psi|)$ is a potential.

We will consider the case when $V = 0$, and m^2 corresponding to a field ψ that is stable at infinity. In Minkowski space, stability is simple: a field with $m^2 \geq 0$ is stable, since that means that the curvature of the potential is positive, $V''(\phi) \geq 0$, so the field cannot slip away to infinity, whereas $m^2 < 0$ is unstable, since it means that $V''(\phi) < 0$, and the field can slip away to infinity, lowering its energy. However, in an AdS background, Breitenloher and Freedman found that there is a modification of this rule due to the extra terms from curved background in the Klein-Gordon equation, and we can have stability even if $m^2 < 0$, as long as it is not too negative, specifically satisfying the *Breitenloher-Freedman (BF) bound*. For scalars, the bound is

$$m^2 \geq -\frac{d^2}{4L^2} \quad (37.2)$$

for AdS_{d+1}. For AdS_4, that gives $m^2 L^2 \geq -9/4$.

Since we want to describe a superconducting system, it means that for $T < T_c$, we must have a condensing operator, $\langle \mathcal{O} \rangle \neq 0$. But since $\mathcal{O}$ couples to ψ through the coupling $\int \mathcal{O}\psi$, it means that $\langle \mathcal{O} \rangle \neq 0$ can happen only if $\psi \neq 0$ (consider the equations of motion of the action with this coupling to see this). That means that in the gravity dual, the $\psi = 0$ solution is thermodynamically unstable toward a solution with $\psi \neq 0$. That is, if we start with the $\psi = 0$ solution, then a small perturbation will lead it to a solution with $\psi \neq 0$, including near the horizon of the black hole, where the temperature of the black hole originates. Moreover, this has to happen only for $T < T_c$, whereas for $T > T_c$, $\psi = 0$ must be the thermodynamically stable solution.

But by dimensional reasons, to have a $T_c \neq 0$, we must introduce a scale that can generate it. It can only be done through either the introduction of a chemical potential μ or a nonzero charge density ρ. The AdS/CFT dictionary relates μ to the VEV of A_0: indeed, the coupling $\int J^\mu A_\mu$ means that for $\mu = 0$ and for constant A_0 (near the boundary of AdS_4) we obtain $A_0 \int J^0 \equiv A_0 Q$ (the charge is $Q = \int J^0$), which is indeed the usual chemical potential coupling μN or μQ. Moreover, we saw in Chapter 35 that the leading term in the near-boundary expansion of a field in AdS, which for a massless field is just a constant, is a non-normalizable mode. We also have another independent mode, a normalizable one, that appears as a subleading mode in the near-boundary expansion of the field. Finally, the higher order terms in the near-boundary expansion are dependent on the normalizable and non-normalizable modes. In the case of A_0, the other independent mode defines the charge density ρ, so all in all at $r \to \infty$ (near the boundary), we have the expansion

$$A_0(r) \to \mu - \frac{\rho}{r} + \cdots \tag{37.3}$$

So either way, if we keep $\mu \neq 0$ or $\rho \neq 0$, the gauge field A_0 is turned on.

37.2 Superconducting Black Holes

Ansatz

The needed ansatz for the holographic dual to the superconductor is then

$$ds^2 = g_{tt}(r)dt^2 + g_{rr}(r)dr^2 + ds_2^2(r)$$
$$A_\mu dx^\mu = \Phi(r)dt; \quad \psi = \psi(r). \tag{37.4}$$

On this ansatz, the terms involving ψ in the Lagrangean become

$$-|(\partial_\mu - iqA_\mu)\psi|^2 - m^2|\psi|^2 \to -g^{tt}q^2\Phi^2|\psi|^2 - g^{rr}|\partial_r\psi|^2 - m^2|\psi|^2$$
$$= -g^{rr}|\partial_r\psi|^2 - (m^2 + g^{tt}q^2\Phi^2)|\psi|^2. \tag{37.5}$$

Since $g^{tt} < 0$, we note that the "effective mass squared" m_{eff}^2 that appears in front of $|\psi|^2$ is smaller than the mass squared:

$$m_{\text{eff}}^2 = m^2 + g^{tt}q^2\Phi^2 < m^2. \tag{37.6}$$

Moreover, since $g^{tt} < 0$ outside the horizon and $-g^{tt} \to \infty$, for finiteness of m_{eff}^2, we need to have $\Phi = 0$ at the horizon. Also, if Φ drops to zero slower or in the same way as $|g^{tt}|$ rises, it means that we could have m_{eff}^2 negative and below the BF bound just outside the horizon, i.e. it could become unstable there. As we saw, this means that the $\psi = 0$ solution is unstable toward perturbations, and we can create a black hole with nonzero scalar at the horizon, which means a condensate of $\langle \mathcal{O} \rangle$ in the dual. But if the scalar is unstable at the horizon (where the scalar is still zero) and stable at infinity, it will become nonzero throughout the solution, leading to a *hairy black hole*. We have mentioned in Part II that there are "no-hair theorems," i.e. theorems that say that a black hole is characterized at infinity only by mass M, angular momentum J, and electric and magnetic charges Q_e, Q_m, but not scalar charges, but those theorems assumed a Minkowski background. In AdS background one can have "scalar hair," i.e. we can have a nonzero scalar that survives near the boundary.

We have seen that in Minkowski space, the one-point function (operator VEV) is found by differentiation of the partition function $Z_{\text{CFT}} = Z_{\text{bulk}} = e^{iS_{\text{bulk}}}$,

$$\langle \mathcal{O} \rangle = -i\frac{\delta Z_{\text{bulk}}[\phi_{(0)}]}{\delta \phi_{(0)}} = \frac{\delta S_{\text{bulk}}[\phi_{(0)}]}{\delta \phi_{(0)}}, \tag{37.7}$$

and that in AdS_{d+1}, near the boundary at $z = 0$, a scalar field behaves as

$$\phi(z) = \left(\frac{z}{R}\right)^{d-\Delta} \phi_{(0)} + \left(\frac{z}{R}\right)^{\Delta} \phi_{(1)} + \cdots \tag{37.8}$$

with the first term being non-normalizable, and the second normalizable. The mass m of the scalar field ϕ is related to the dimension Δ of the operator dual to it by $m^2 R^2 = \Delta(\Delta - 3)$, and then the 1-point function is

$$\langle \mathcal{O} \rangle = \frac{2\Delta - d}{R} \phi_{(1)}. \tag{37.9}$$

That means that to get a nonzero operator VEV (condensate in the superconducting state), we need to have a perturbation by a scalar (dual to the operator) normalizable mode $\phi_{(1)}$, and this needs to happen for $T < T_c$ only.

Background

We saw that we need to introduce a nonzero scalar ψ (for a nonzero VEV) and a nonzero electric potential Φ (to lower m_{eff}^2). In principle, we would need to consider these fields together with gravity, and solve together the coupled system. But this can be done only numerically, and it would be hard to learn any useful physics this way. Instead, one considers a *background*, and in it we consider the fields ψ and Φ in a probe approximation without backreaction, i.e. we consider that they don't change the background.

In *fundamental* AdS/CFT (derived from a brane system in string theory), the probe approximation is usually to consider one brane or $N_f \ll N$ branes as probes in a background created by other N branes, and in that case one can consider it as an approximation in N_f/N. But in the case under analysis, we are using a "probe approximation" for ψ and Φ, without having an obvious small parameter (like N_f/N before), so it is not clear why one can use it, other than that it still gives the correct physics. We have to remember, however, that the starting point is phenomenological (it is an ansatz, it is not derived), so the probe approximation for it, even though it is not clear if it is a good approximation, can be thought of as simply changing the phenomenological AdS/CFT description (being from a phenomenological standpoint as good a starting point as the full one).

As a background we can consider either the AdS_4-Reissner-Nordstrom black hole background, as Gubser did, or the neutral black hole, as Hartnoll, Herzog, and Horowitz did in their first paper.

The AdS-RN background is written like all black holes in terms of a function appearing in g_{tt} and in g^{rr}, as

$$ds^2 = -f(r)dt^2 + \frac{dr^2}{f(r)} + r^2 d\Omega_{2(k)}^2$$

$$f(r) = k - \frac{2M}{r} + \frac{Q^2}{4r^2} + \frac{r^2}{R^2}$$

$$\Phi(r) = \frac{Q}{r} - \frac{Q}{r_H}; \quad \psi = 0$$

$$d\Omega_{k=1}^2 = d\theta^2 + \sin^2\theta d\phi^2$$

$$d\Omega_{k=0}^2 = dx^2 + dy^2. \tag{37.10}$$

We note here that the $k = 1$ is the usual black hole, which in the *extremal* case has horizon of type $AdS_2 \times S^2$, but we also have the $k = 0$ black hole, with horizon of the type $AdS_2 \times \mathbb{R}^2$, which is what we will be interested in. We also note that we have added a constant term to Φ, besides the usual electric potential, to have $\Phi(r = r_H) = 0$, which we argued that we need.

The neutral black hole background is obtained by putting $Q = 0$, but also considering $k = 0$, so $d\Omega_2^2 = dx^2 + dy^2$, and

$$f(r) = \frac{r^2}{R^2} - \frac{2M}{r}. \tag{37.11}$$

Probe Approximation

The equations of motion in the probe approximation are the following:

- The scalar ψ equation on the ansatz,

$$\frac{1}{\sqrt{-g}} \partial_r \sqrt{-g} g^{rr} \partial_r \psi + m_{\text{eff}}^2 \psi = 0, \tag{37.12}$$

which explicitly becomes

$$\psi'' + \left(\frac{f'}{f} + \frac{2}{r} \right) \psi' + \frac{\Phi^2}{f^2} \psi - \frac{m^2}{f} \psi = 0. \tag{37.13}$$

Note that it is written in a form in terms of $f(r)$ and m_{eff}^2, which is therefore valid in both cases.

- The equation of motion for Φ in the background is

$$\Phi'' + \frac{2}{r} \Phi' - \frac{2\psi^2}{f} \Phi = 0, \tag{37.14}$$

which is the same equation that one would get for a massive vector field (in the absence of a vector potential $\vec{A}$). For the perturbation Φ, we again must impose the boundary condition that $\Phi = 0$ for normalizability.

We will be interested in the case of $m^2 = -2/R^2$, which is obtained for either $\Delta = 1$ or $\Delta = 2$. Then, at infinity, we have

$$\psi(r) \simeq \frac{\psi^{(1)}}{r} + \frac{\psi^{(2)}}{r^2} + \cdots$$
$$\Phi(r) \simeq \mu - \frac{\rho}{r} + \cdots. \tag{37.15}$$

Unlike other cases, now it happens that *both* $\psi^{(1)}$ and $\psi^{(2)}$ modes are normalizable, so according to our previous analysis, correspond to scalar VEVs for operators (condensates),

$$\langle \mathcal{O}_i \rangle = \sqrt{2} \psi^{(i)}, \quad i = 1, 2. \tag{37.16}$$

We can choose either one of the scalar modes to vanish, and calculate the VEV that is generated by the other. Of course, in the above near-boundary behavior, we have only two independent parameters, not four. Putting $\mu = 0$ and considering only $\rho \neq 0$, we can calculate T_c, $\psi^{(i)}$ and thus $\langle \mathcal{O}_i \rangle$.

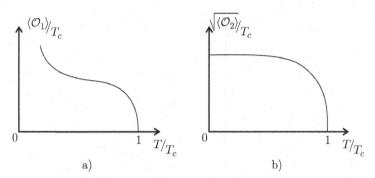

Fig. 37.1   (a) Condensate 1 as a function of temperature. (b) Condensate 2 as a function of temperature.

Numerically, one finds the results sketched in Figure 37.1. Moreover, near $T = T_c$, one finds a square root vanishing, specifically with

$$\langle \mathcal{O}_1 \rangle \simeq 9.3 T_c \left(1 - \frac{T}{T_c} \right)^{1/2}$$

$$\langle \mathcal{O}_2 \rangle \simeq 144 T_c \left(1 - \frac{T}{T_c} \right)^{1/2}.$$

(37.17)

One also finds numerically the relation between the critical temperature T_c and the input, the charge density ρ:

$$T_c(\mu = 0) \simeq 0.118 \sqrt{\rho}.$$

(37.18)

Note that if we consider instead of $\mu = 0$ and nonzero ρ, a nonzero chemical potential $\mu \neq 0$ and $\rho = 0$, we obtain a critical temperature that is proportional to μ instead, as it should be for dimensional reasons:

$$T_c(\rho = 0) \simeq 0.0588 \mu \propto \mu.$$

(37.19)

37.3 Transport and Mass Gap

We can calculate the instability of the AdS_4-RN black hole background under a perturbation $\psi = \psi(r) e^{-i\omega t}$. Indeed, as expected, if we take this background and consider $T < T_c$, the fact that the black hole is expected to superconduct means that there must exist a solution with ingoing (infalling) boundary conditions at the horizon (the wave going into the black hole) and with Im $\omega > 0$, i.e. the mode is actually exponentially growing. We have seen in Chapter 36 that these are black hole *quasi-normal modes*, and the superconducting conditions is therefore that they move into the upper-half plane (Im $\omega_n > 0$) at $T < T_c$. That is what we expect: the $\psi = 0$ mode is thermodynamically unstable, and numerically one finds a growing mode that ends only on the numerically backreacted full charged hairy black hole (with scalar "hair").

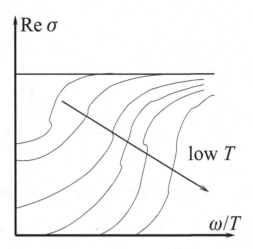

Fig. 37.2 Conductivity vs. rescaled frequency for various temperatures.

Conductivity

One can now calculate various properties of the superconductor from the holographic dual. For condensed matter physics, transport is important, especially charge transport. Therefore we want to calculate the conductivity.

Consider therefore a perturbation in the Maxwell field, for the x component,

$$\delta A_x = \delta A_x(r)e^{i\omega t}, \tag{37.20}$$

for which one finds the equation of motion:

$$A_x'' + \frac{f'}{f}A_x' + \left(\frac{\omega^2}{f^2} - 2\frac{\psi^2}{f^2}\right)A_x = 0. \tag{37.21}$$

We have seen the expansion at infinity of $A_0 = \Phi$, and for A_x it is similar: instead of μ we have a source $A_x^{(0)}$ for J_x (the coupling in the action is $\int J_x A_x$), and instead of $\rho = \langle J_0 \rangle$ we have $\langle J_x \rangle \equiv A_x^{(1)}$ (the normalizable mode of A_x), so

$$A_x = A_x^{(0)} + \frac{\langle J_x \rangle}{r} + \cdots \tag{37.22}$$

Then, as we already saw, the conductivity is given by the linear response of the current to the electric field $E_x = -\partial_t \delta A_x \to i\omega\delta A_x$:

$$\sigma(\omega) = \frac{\langle J_x \rangle}{E_x} = \frac{\langle J_x \rangle}{i\omega A_x} = \frac{-iA_x^{(1)}}{\omega A_x^{(0)}}. \tag{37.23}$$

One obtains numerically the result for Re $\sigma(\omega/T)$ sketched in Figure 37.2. We see that as $T \to T_c$, we conductivity goes very fast to a constant, but at low temperature, it goes slower to the constant, and it starts off at Re $\sigma = 0$, i.e. there is a *frequency (energy) gap*, below which there is no conductivity, $\sigma = 0$ for $\omega < \omega_g$. Moreover, one finds that

$$\omega_g \simeq (q\langle \mathcal{O} \rangle)^{\frac{1}{\Delta}}, \tag{37.24}$$

which becomes exact in the probe approximation.

One also finds that one has

$$\frac{\hbar\omega_g}{k_B T_c} \simeq 8.4, \tag{37.25}$$

and in any case is close to 8. But as we saw in Part I, for low T_c superconductors, which are weakly coupled and described well by BCS theory, where we have the well-known result $\hbar\omega_g/(k_B T_c) \simeq 3.54$, we generically measure $\hbar\omega_g/(k_B T_c) \simeq 3.5$. On the other hand, for high T_c cuprate superconductors, which are strongly coupled, we have experimentally also $\hbar\omega_g/(k_B T_c) \simeq 8$, so the holographic superconductor describes this very well.

However, there are things that are hard to understand as well. For a weakly coupled superconductor (BCS theory), we have

$$\omega \simeq 2E_g, \tag{37.26}$$

where E_g is the energy gap in the charged spectrum, i.e. the energy of an "electron." Indeed, in this case, Cooper pairs, the pairs of electrons of opposite momentum and spin that form the effective scalar that condenses, have negligible binding energy, so the energy of the Cooper pair is twice the energy of the electron.

At strong coupling, however, we need not have this equality, and in fact if we do, it is a puzzle, since we expect the strong interactions to modify it. For a superconductor, we can measure E_g independently from the Boltzmann suppression factor of the real part of the conductivity at small temperatures:

$$\mathrm{Re}\,\sigma(T \to 0) \simeq e^{-\frac{E_g}{k_B T}} \quad \text{for} \quad E_g/T \gg 1. \tag{37.27}$$

Then one finds for our holographic superconductor that $E_g \neq \hbar\omega_g/2$ in general, *except* for the case $\Delta = 1$ or 2 (corresponding to $m^2 R^2 = -2$), which we were describing before. In that case, we actually have $E_g = \hbar\omega_g/2$, and it becomes hard to explain why it happens for a strongly coupled system, when it should be only for a weakly coupled system. It might be a signal of some symmetry, or something else, but it is a puzzle.

37.4 Breitenlohner-Freedman Bound and Superconductor

We have seen how we get a superconducting gravity dual, but we must understand why it happened. One can calculate m_{eff}^2 at the horizon in the *extremal* AdS_4-RN black hole, and obtain

$$m_{\mathrm{eff}}^2 = m^2 - \frac{\gamma^2 q^2}{2R^2} < m^2, \tag{37.28}$$

where q is the charge and

$$\gamma^2 = \frac{g^2 2R^2}{\kappa_{N,4}^2}. \tag{37.29}$$

On the other hand, to obtain an instability at the horizon, the effective mass square needs to violate the BF bound *at the horizon*. This will be for a *near-extremal* solution, for which we have an AdS_2 factor, but the explanation is generalized. We note here that the BF bound is valid only in AdS space, not in a general space. At infinity, we have AdS_4, so we have the bound $m^2 \geq -9/(4R^2)$, but on the other hand at the horizon, we have $AdS_2 \times S^2$ or $AdS_2 \times \mathbb{R}^2$, so we have a bound for AdS_2, $m^2 \geq -1/(4R_2^2)$, where R_2 is the radius of AdS_2. But R_2 is related to the radius R of AdS_4 by $R_2^2 = R^2/6$, as we have left it to prove in an exercise. Then there is an instability at the horizon if we have

$$-\frac{1}{4} \geq m_{\text{eff}}^2 R_2^2 = \frac{R^2}{6}\left(m^2 - \frac{\gamma^2 q^2}{2R^2}\right). \tag{37.30}$$

Substituting $m^2 R^2 = \Delta(\Delta - 3)$, we obtain a bound on the dimension Δ of operators that can condense:

$$q^2 \gamma^2 \geq 3 + 2\Delta(\Delta - 3). \tag{37.31}$$

We note that the bound on the effective mass at the horizon is weaker than at infinity,

$$m_{\text{eff}}^2 R^2 \geq -\frac{6}{4} \quad \text{vs.} \quad m^2 R^2 \geq -\frac{9}{4}, \tag{37.32}$$

and moreover $m_{\text{eff}}^2 < m^2$, which means that we can have fields that are stable at infinity and unstable at the horizon.

So the ingredients necessary for holographic superconductivity of the gravity dual are:

- We need to have a BF found that is weaker at the horizon than at infinity.
- Because of the coupling of the scalar to A_0, the effective mass squared is lowered at the horizon, $m_{\text{eff}}^2 < m^2$ at the horizon.

On the other hand, as noted before, an "experimental" way to gauge whether we have a superconducting black hole is to see whether there exists a T_c such that for $T < T_c$, the black hole quasi-normal modes move in the upper-half plane.

37.5 Holographic Josephson Junctions

We have seen how to construct a holographic superconductor, but there is an important effect that uses superconductivity as well as a normal state, the Josephson effect, in a superconductor-normal-superconductor (SNS) junction, superconductor-insulator-superconductor (SIS), or generally Josephson junction. We have seen in Part I that for a thin normal/insulator state, we obtain an oscillating (with respect to the gauge-invariant quantum phase γ) current

$$J = J_{\text{max}} \sin \gamma, \tag{37.33}$$

even in the absence of an applied voltage. This is the DC Josephson effect.

To simulate it, we can consider the same action (37.1) that was used for the holographic description of the superconductor. We have seen that one description of the superconductor was using a (planar) AdS-Schwarzschild black hole. If one considers a nonzero chemical potential to generate the superconductivity (a constant in the asymptotics of the gauge field at infinity), then we saw in (37.19) that we obtain a critical temperature that is proportional to the chemical potential.

But that offers a simple way to model an SNS Josephson junction. Consider a space-varying chemical potential $\mu(\vec{x})$ at fixed temperature (given by the black hole), such that in most of the space T is below the T_c given by μ, thus the system in a superconducting state, but in a narrow gap is above it, thus the system is in a normal state.

The ansatz is for the scalar ψ to have a quantum phase φ, and the boundary gauge field to have all components turned on in general, i.e. for the simplest case of $d = 3$:

$$\psi = |\psi|e^{i\varphi}; \quad A = A_t dt + A_r dr + A_x dx. \tag{37.34}$$

All fields are real functions of r and x, but not of the time t. Define also the gauge-invariant field

$$M = A - d\varphi. \tag{37.35}$$

For concreteness, one could consider $m^2 R^2 = -2$, which is above the Breitenlohner-Freedman bound at infinity.

For boundary conditions at infinity, we still have for the scalar

$$|\psi| = \frac{|\psi^{(1)}|(x)}{r} + \frac{|\psi^{(2)}|(x)}{r^2} + \mathcal{O}\left(\frac{1}{r^3}\right), \tag{37.36}$$

corresponding to the two possible operators being turned on, and for concreteness one can choose $|\psi^{(1)}|(x) = 0$, in which case we have $\langle \mathcal{O}_2 \rangle = \sqrt{2}|\psi^{(2)}|(x)$ turned on.

The advantage of this choice is that this does not change the asymptotics of the gauge-invariant field M to the level we are interested. Its asymptotics at the boundary at infinity $r \to +\infty$ of M is as for A:

$$M_t = \mu(x) - \frac{\rho(x)}{r} + \mathcal{O}\left(\frac{1}{r^2}\right)$$

$$M_r = \mathcal{O}\left(\frac{1}{r^3}\right)$$

$$M_x = \nu(x) + \frac{J}{r} + \mathcal{O}\left(\frac{1}{r^2}\right). \tag{37.37}$$

As usual, μ is the chemical potential, ρ is the charge density, ν is superfluid velocity and J is its current (through the obvious relativistic invariance, we can make the generalization of $M_t \to M_x$ giving $\mu \to \nu$ and $-\rho \to J$). From the equations of motion near $r = +\infty$, we find that J is a constant of x, so one can give $\mu(x)$ and J as boundary conditions for the fields.

Since $M_x = \partial_x \varphi - A_x$, $\int_{\text{gap}} M_x = -\int_{\text{gap}} dx \partial_x \varphi + \int_{\text{gap}} dx A_x$, the gauge-invariant phase difference across the gap γ is given by

$$\gamma \equiv -\int_{\text{gap}} dx M_x = \Delta\varphi - \int_{\text{gap}} dx \, A_x, \tag{37.38}$$

which means that from the boundary conditions at infinity, we have

$$\gamma = -\int_{-\infty}^{+\infty} dx[v(x) - v(\infty)]. \tag{37.39}$$

At the horizon $r = r_H$, we need to impose regularity, which implies $M_t = 0$. Finally, since we want an SNS junction, we need to impose a profile for $\mu(x)$ such that $\mu(x = +\infty) = \mu(x = -\infty)$, and near $x = \pm\infty$, we approach an x-independent solution.

To choose the functional form of the chemical potential profile $\mu(x)$, consider that by (37.19) we must have $T_c \simeq 0.0588\mu(\infty)$ for the superconductor, and the temperature T of the black hole must be smaller, $T < T_c$, but we want μ to be approximately constant in a small region $x = (-l/2, +l/2)$ such that the locally defined critical temperature $T_{c,0} \simeq 0.0588\mu(0)$ is smaller than T. Then we have a superconductor for most of x, only in the thin region we have a normal state.

One example of a profile, defined by a width σ, that gives this dependence is

$$\mu(x) = \mu(\infty)\left\{1 - \frac{1-\epsilon}{2\tanh(l/2\sigma)}\left[\tanh\left(\frac{x+l/2}{\sigma}\right) - \tanh\left(\frac{x-l/2}{\sigma}\right)\right]\right\}, \tag{37.40}$$

but others are also possible. For this profile, $T_{c,0} = \epsilon T_c$.

One can then input this $\mu(x)$ profile and a constant J into the numerics, and calculate the resulting γ from (37.39) and check that it satisfies (37.33). One also finds the relation

$$\frac{J_{\max}}{T_c^2} = A_0 e^{-\frac{l}{\xi}}, \tag{37.41}$$

for $\xi \ll l$, with $A_0 \simeq 17.96$ and the coherence length $\xi \simeq 1.17$, as well as

$$\frac{\langle \mathcal{O}(x=0, J=0)\rangle}{T_c^2} = A_1 e^{-\frac{l}{2\xi'}}, \tag{37.42}$$

for $\xi' \ll l$, with $A_1 \simeq 33.52$ and $\xi' \simeq 1.26$, though it should be the same as ξ. Otherwise these relations are as predicted from condensed matter.

Alternative Description and AC Josephson Effect

An alternative construction for the SNS Josephson junction contains more physical effects, in particular it includes the AC Josephson effect, as well as having a *nonabelian junction*, but it does not easily generalize to nonzero temperature. The nonabelian junction will be discussed in the context of the holographic Berry phase later.

As the holographic superconductor/superfluid one considers the $p+1$–dimensional $SU(N_c)$ gauge theory on the worldvolume of N_c coincident Dp–branes at large N_c, and adds a small number of D$(p+2)$–branes, for concreteness considering only one. One also separates one of the Dp-branes from the rest in direction X^1 transverse to them by a quantity Δx^1, which has the effect of Higgsing the gauge group, i.e. breaking $SU(N_c) \to SU(N_c - 1) \times U(1)$ through the presence of the scalar condensate

$$\langle\Phi^1\rangle = \text{diag}(\Delta x^1, 0, \dots, 0). \tag{37.43}$$

As we have described before, the diagonal elements of the scalars in the Dp-brane worldvolume correspond to positions in the transverse directions, and in the above, the first diagonal element (for the unbroken $U(1)$) corresponds to the (relative) position of the separated brane. But the presence of a condensate is interpreted, from the point of the initial group $SU(N_c)$, as evidence of superconducting behavior (remember that in the usual $U(1)$ superconductor, the condensate breaks the gauge group $U(1)$ through the Higgs effect; in the nonabelian case, the gauge group can be only partially broken by the condensate through the Higgs effect).

For an idealized holographic Josephson junction, we want to have a surface that connects two regions of space in which the condensate has the same modulus, but a different phase. Indeed, the scalars corresponding to transverse coordinates naturally form complex variables, so we want them to differ in the two regions of space only by a phase, i.e. angle between coordinates.

For a large N_c, one can replace the Dp-branes with their supergravity background (for $p \leq 4$) for the metric, dilaton, and $(p+1)$-form C_{p+1}, since they will strongly curve the spacetime,

$$ds^2 = \left(\frac{r}{R}\right)^{\frac{7-p}{2}} \eta_{\mu\nu} dx^\mu dx^\nu + \left(\frac{R}{r}\right)^{\frac{7-p}{2}} dr^2 + R^2 \left(\frac{r}{R}\right)^{\frac{p-3}{2}} d\Omega_{8-p}^2$$

$$e^\phi = e^{\phi_0} \left(\frac{R}{r}\right)^{\frac{(7-p)(3-p)}{4}}$$

$$C_{p+1} = \frac{1}{g_s} \left(\frac{r}{R}\right)^{7-p} dr \wedge dx^0 \wedge \ldots \wedge dx^p, \tag{37.44}$$

where the $(8-p)$−sphere is decomposed in terms of $(6-p)$−spheres and 2-spheres as

$$d\Omega_{8-p}^2 = d\theta^2 + \cos^2\theta \; d\Omega_{6-p}^2 + \sin^2\theta d\phi^2. \tag{37.45}$$

Note that at fixed Ω_{6-p}, i.e. at $d\Omega_{6-p} = 0$, we have a 2-sphere, $d\theta^2 + \sin^2\theta d\phi^2$.

The single D$(p+2)$−brane, however, can be thought of as a probe in this geometry, since it will negligibly curve the background. Consider it in the directions t (time), $p-1$ of the space directions x^i (less x^p), the radial direction r and a 2-sphere inside Ω_{8-p}, the one parametrized by θ and ϕ. Thus the nontrivial directions for the Dp-brane are $x^0 = Rt$ and $x^p = Rz$, and on the D$(p+2)$ are (time and) (ϕ, θ, r).

Now consider the Dp-brane that was separated from the stack, and gives a condensate generating superconductivity. It also becomes a probe brane, but it must be at fixed ϕ, θ, r. Yet for the Josephson junction, we want to have two regions of space that differ by their value for the phase of a complex scalar for the transverse position. Such a phase becomes an angle on the sphere in the gravity dual, as we have argued when defining AdS/CFT. Therefore the picture that arises is as follows. At fixed r, we consider an S^2 surface spanned by the D$(p+2)$−brane. Then consider the coordinate z, transverse to it, but along the Dp-brane. This Dp-brane is split in two halfs by the D$(p+2)$−brane situated at $z = 0$, each half ending at a different position (ϕ, θ) on it, as in Figure 37.3. If we want to consider only the usual, Abelian, phase of the condensate, we consider a single angle, for instance, ϕ. Then for each Dp-brane half one writes the fluctuation ansatz ($r = Ru$):

$$X^0 = t; \quad X^\theta = \frac{\pi}{2}; \quad X^\phi = \phi(t, z); \quad X^u = u + \delta u(t, z). \tag{37.46}$$

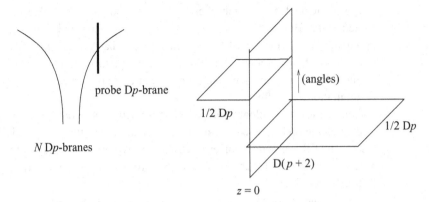

Fig. 37.3 Brane system: the probe Dp-brane in the background of N Dp-branes, and the D($p + 2$)–brane breaking the Dp-brane in two halves, at different angular positions on it.

The endpoints of the two Dp-brane halfs, giving the phases of the condensate on the two sides of the junction, are taken to be at $\phi = 0$ and $\phi = \pi$.

These endpoints, from the point of view of the D($p + 2$)–brane, look like magnetic monopoles, for the magnetic flux $(p + 3)$–dimensional dual to $F_{\theta u} = \partial_{[\theta} A_{u]}$, i.e. for $\mathcal{F}_{p+1} = *_{p+3} F_2 = d\mathcal{A}_p$. However, in a supersymmetric configuration, we have no force on them, and to generate such a force, that could create a Josephson current, we need a magnetic field that couples to the monopole charge. But $\mathcal{A}_p$ couples naturally to the p-dimensional intersection of the Dp-brane with the D($p + 2$)-brane, i.e. to the "magnetic monopoles," via

$$S_p = \int_{z=0} d^p x \mathcal{A}_{M_1 \cdots M_p}(X) \frac{\partial X^{M_1}}{\partial t} \frac{\partial X^{M_2}}{\partial x^1} \cdots \frac{\partial X^{M_p}}{\partial x^{p-1}}. \tag{37.47}$$

On the $(p + 1)$–dimensional worldvolume of the Dp-brane, *in the linearized approximation*, the kinetic term (coming from the volume term of the DBI action) is $T_p \int d^{p+1} x \sqrt{-g} \partial^\mu X^M \partial_\mu X_M$, and when varying it, we generate a p-dimensional boundary term that pairs up with the variation of the above S_p to give

$$T_p \sqrt{-g} g_{MN} g^{zz} \partial_z X^N = \mathcal{F}_{M M_1 \cdots M_p} \frac{\partial X^{M_1}}{\partial t} \frac{\partial X^{M_2}}{\partial x^1} \cdots \frac{\partial X^{M_p}}{\partial x^{p-1}}. \tag{37.48}$$

If we choose the magnetic potential $\mathcal{A}_{\phi 12 \ldots p}(X) = B \cos(\Omega X^0)$, then from the above equation and the background (37.44) we obtain the boundary condition for ϕ:

$$T_p u_0^2 R^{p+1} \partial_z \phi(z = 0) = B \sin(\Omega t). \tag{37.49}$$

On the other hand, we can define the superfluid current J_a^μ as the conjugate of the phase ϕ_a of the condensate,

$$J_a^\mu = \frac{\delta S}{\delta(\partial_\mu \phi_a)} \simeq -T_p u_0^2 R^{p+1} \partial^\mu \phi, \tag{37.50}$$

where we have used the free action $\int \partial X^M \partial X_M$, valid in the linearized approximation. For $a = \phi$ and $\mu = 0$, we have the charge at the endpoint of the Dp-brane, so its time variation, giving the current through the junction, is

$$\frac{d}{dt}Q = \frac{\partial}{\partial t}J_\phi^0(z=0) = -B\Omega\sin(\Omega t). \tag{37.51}$$

By charge conservation, $dQ/dt = -dJ^z/dz$, so

$$\frac{dJ^z}{dz} = B\Omega\sin(\Omega t), \tag{37.52}$$

which can be integrated over a width l of the junction to

$$J^z = B\Omega l\sin(\Omega t). \tag{37.53}$$

This is the *AC Josephson effect*.

We can also obtain an approximate DC Josephson effect as a certain limit of the above. Consider a cosine magnetic flux $B\sin(\Omega t)$ instead of the sinusoidal one, so that

$$\frac{dJ^z}{dz} = -B\Omega\cos(\Omega t), \tag{37.54}$$

and consider the limit $\Omega \to 0, B \to \infty$, with $B\Omega =$ fixed. Then integrating the above over l, we obtain

$$J^z \simeq -(B\Omega)l. \tag{37.55}$$

Thus in this way we can obtain the AC Josephson effect, but the fact that the junction is normal (so above a T_c) is obscured, since we don't have a temperature, like in the previous setup. Introducing temperature turns out to be challenging.

Important Concepts to Remember

- For a holographic high T_c layered superconductor, we need an AdS_4 background, with a bulk gauge field, bulk charged scalar, and black hole.
- For the scalar to be superconducting, we need that $\psi = 0$ to be thermodynamically unstable toward a "hairy" solution with $\psi \neq 0$, for $T < T_c$, where the scalar is stable at infinity, but unstable at the horizon. Then $m_{\text{eff}}^2 < m^2$ because of the coupling to the gauge field A_0.
- The background in which we can put the scalar and the electric potential is an AdS_4-Reissner-Nordstrom black hole, or a neutral AdS_4 black hole with planar horizon.
- The existence of a T_c requires nonzero chemical potential μ or charge density ρ, found in the asymptotic $A_0 = \mu - \rho/r + \cdots$
- For $\Delta = 1$ or 2, we obtain $\langle\mathcal{O}_i\rangle \propto (1 - T/T_c)^{1/2}$ as $T \to T_c$, and $T_c \propto \sqrt{\rho}$.
- The conductivity shows an energy gap, with $\hbar\omega_g/k_B T_c \simeq 8.4$, close to the result for the strong-coupling high T_c cuprate superconductors, but unlike those for $\Delta = 1$ or 2, we have $\hbar\omega_g = 2E_g$.
- There is a bound on the dimensions of operators that superconduct, $3 + 2\Delta(\Delta - 3) \leq q^2\gamma^2$.

- To superconduct, we need a BF bound at the horizon that is weaker than at infinity, allowing for a field stable at infinity, but unstable at the horizon, and the coupling of ψ with A_0, that means a lowering of m_{eff}^2 at the horizon with respect to infinity.
- When superconducting, the black hole quasi-normal modes move into the upper-half plane.

Further Reading

See the review by Hartnoll [1]. The original papers for the holographic superconductor are [89] and [90]. For the holographic Josephson junction, the original paper is [91], and the alternative viewpoint, that obtains also an AC Josephson effect, was presented in [92].

Exercises

(1) Write the equations of motion for a radial ansatz in AdS_4 with scalar ψ, electric potential Φ, and magnetic field $B = B(r)$, which corresponds to magnetic field in the boundary theory (why?).

(2) Calculate the entropy and the temperature of the AdS_4-RN black hole, and check the laws of black hole thermodynamics in this case.

(3) Write the equations of motion for small fluctuations in the scalar ψ, electric potential Φ, and vector potential $\vec{A}$ around the AdS_4-RN black hole.

(4) Prove that the AdS_4-RN background reduces to $AdS_2 \times S^2$ at the horizon, with $R_2^2 = R^2/6$. Calculate the BF bound for Φ and for the vector potential $\vec{A}$ at the horizon.

(5) (Holographic p-wave superconductor) Consider the AdS_4-Schwarzschild black hole and the ansatz for an $SU(2)$-gauge field inside it,

$$A = \Phi(r)\tau^3 dt + w(r)\tau^1 dx, \tag{37.56}$$

where τ_1, τ_2, τ_3 are the Pauli matrices divided by 2. Write down the YM equations on the ansatz, and calculate the effective mass for the "W bosons" $A^\pm$, corresponding to $\tau^\pm$ in the presence of an "electromagnetic" background $\Phi(r)$.

(6) (Holographic isotropic p-wave superconductor) Show that if we add a term $w(r)\tau^2 dx$ to the A in Exercise 5, we obtain an "isotropic holographic p-wave superconductor," i.e. that the condensate obeys a combined gauge+rotation invariance.

The Fluid-Gravity Correspondence: Conformal Relativistic Fluids from Black Hole Horizons

In this chapter we will learn about a map between fluids, governed by a relativistic version of the Navier-Stokes equation, and gravity, governed by the Einstein equation. In particular, conformal relativistic fluids will be mapped to black hole horizons.

38.1 Relativistic Fluids

We start by reviewing some concepts that we learned in Part I. Nonrelativistic ideal fluids are governed by the continuity equation

$$\frac{\partial \rho}{\partial t} + \vec{\nabla} \cdot (\rho \vec{v}) = 0 \tag{38.1}$$

and the Euler equation

$$\frac{\partial \vec{v}}{\partial t} + (\vec{v} \cdot \vec{\nabla})\vec{v} = -\frac{\vec{\nabla} P}{\rho} + \vec{g} + \cdots, \tag{38.2}$$

which is derived from the conservation equation for the stress tensor,

$$\frac{\partial}{\partial t}(\rho v_i) = -\frac{\partial}{\partial x_j}\pi_{ij}. \tag{38.3}$$

The fluid equations are a derivative expansion, specifically derivatives of the velocity, for any (strongly coupled, so that we have a good perturbation) quantum system. Then the expansion up to second order is written in terms of an ideal fluid, plus a viscous term,

$$\pi_{ij} = P\delta_{ij} + \rho v_i v_j + \sigma_{ij}, \tag{38.4}$$

where the viscous stress tensor σ_{ij} is to first order

$$\sigma_{ij} = \eta\left(\frac{\partial v_i}{\partial x_j} + \frac{\partial v_j}{\partial x_i} - \frac{2}{3}\delta_{ij}\frac{\partial v_k}{\partial x_k}\right) + \zeta \delta_{ij}\frac{\partial v_k}{\partial x_k}. \tag{38.5}$$

Here η is the shear viscosity, which describes how viscosity in a tube slows down the fluid through the interaction with the boundary, and ζ is the bulk viscosity, and both must be positive, $\eta \geq 0, \zeta \geq 0$.

Replacing this stress tensor in the conservation equation leads to the Navier-Stokes equation:

$$\rho\left[\frac{\partial \vec{v}}{\partial t} + (\vec{v} \cdot \vec{\nabla})\vec{v}\right] = -\vec{\nabla} P + \eta \Delta \vec{v} + \left(\zeta + \frac{\eta}{3}\right)\vec{\nabla}(\vec{\nabla} \cdot \vec{v}). \tag{38.6}$$

Note that for an incompressible fluid, $\rho = $ constant, so the continuity equation gives $\vec{\nabla} \cdot \vec{v} = 0$, which means that the ζ term in the Navier-Stokes equation vanishes.

The relativistic generalization of the Navier-Stokes equation is obtained from the relativistic generalization of the conservation equation, i.e. the conservation of the energy-momentum tensor:

$$\nabla_\mu T^{\mu\nu} = 0. \tag{38.7}$$

Like in the nonrelativistic case, we expand the energy-momentum tensor in a perfect fluid term and a viscous term,

$$T^{\mu\nu} = \rho u^\mu u^\nu + P P^{\mu\nu} + \pi^{\mu\nu}, \tag{38.8}$$

where

$$P^{\mu\nu} = g^{\mu\nu} + u^\mu u^\nu \tag{38.9}$$

is a projector that we easily see that satisfies

$$u_\mu P^{\mu\nu} = 0, \quad P^{\mu\nu} g_{\mu\nu} = d - 1, \tag{38.10}$$

since $u^\mu u_\mu = -1$. The viscous term, to the leading order in the expansion in derivatives, is

$$\pi^{\mu\nu}_{(1)} = -2\eta \sigma^{\mu\nu} - \zeta \theta P^{\mu\nu}$$
$$\sigma^{\mu\nu} = \nabla^{(\mu} u^{\nu)} + a^{(\nu} u^{\mu)} - \frac{1}{d-1} \theta P^{\mu\nu}$$
$$\theta = \nabla_\mu u^\mu$$
$$a^\mu = u^\nu \nabla_\nu u^\mu. \tag{38.11}$$

Here a_μ is the acceleration.

We have written the expansion of $T_{\mu\nu}$, which is the current of the momentum, but we can write a similar relation for any conserved current, which we call J_I^μ for charge I. We also have an ideal fluid term, which is just $q_I u^\mu$, where q_I is the charge indexed by I, and a viscous term Υ_I^μ:

$$J_I^\mu = q_I u^\mu + \Upsilon_I^\mu. \tag{38.12}$$

The viscous term is expanded to the first order in derivatives as

$$\Upsilon^\mu_{(1)I} = -\tilde{\chi}_{IJ} P^{\mu\nu} \nabla_\nu \left(\frac{\mu_J}{T} \right) - \mathcal{V}_I l^\mu - \gamma_I P^{\mu\nu} \nabla_\nu T, \tag{38.13}$$

where

$$l^\mu = \epsilon_{\alpha\beta\gamma}{}^\mu u^\alpha \nabla^\beta u^\gamma, \tag{38.14}$$

$\tilde{\chi}_{IJ}, \mathcal{V}_I, \gamma_I$ are coefficients and μ_I are the chemical potentials associated with the charges q_I.

The relativistic version of the Navier-Stokes is obtained by substituting the energy-momentum tensor in the conservation equation $\nabla_\mu T^{\mu\nu} = 0$.

In this chapter we want to study fluids under the AdS/CFT correspondence. Since the fluid is naturally strongly coupled, if we make it also conformal, we can put it as a theory on the boundary of a space with AdS asymptotics. It has to be only asymptotically AdS, since we want to have at least temperature in the CFT on the boundary, therefore we need a black hole in the gravity dual, which means the space can at most be asymptotically AdS.

The goal is to obtain the relativistic Navier-Stokes equations in the boundary CFT from the Einstein equations for the black hole in the bulk, defining a *fluid-gravity correspondence*.

38.2 Conformal Fluids

Since we want to study conformal field theories, we need to understand the properties of a conformal fluid. Consider a background metric $g_{\mu\nu}$ (naturally $\eta_{\mu\nu}$, but we can be more general). Then a Weyl transformation of the metric,

$$g_{\mu\nu} = e^{2\phi}\tilde{g}_{\mu\nu}, \quad g^{\mu\nu} = e^{-2\phi}\tilde{g}^{\mu\nu}, \tag{38.15}$$

implies for the 4-velocity u^{μ}, since $u_{\mu}u^{\mu} = u^{\mu}u^{\nu}g_{\mu\nu} = -1$:

$$u^{\mu} = e^{-\phi}\tilde{u}^{\mu}. \tag{38.16}$$

More generally, we say we have a conformal tensor $Q^{\nu_1...\nu_n}_{\mu_1...\mu_m}$ if under the Weyl transformation it transforms as

$$Q^{...}_{...} = e^{-w\phi}\tilde{Q}^{...}_{...}, \tag{38.17}$$

where w is called the conformal weight of the tensor. Note the similarity with the notion of primary field (tensor field) for conformal field theories in two dimensions, which we described previously.

The energy-momentum tensor $T_{\mu\nu}$, defined as

$$T^{\mu\nu} = -\frac{2}{\sqrt{-g}}\frac{\delta S}{\delta g_{\mu\nu}}, \tag{38.18}$$

strictly speaking the one called the *Belinfante tensor*, to be a conformal tensor, must transform with weight $w_T = d + 2$,

$$T^{\mu\nu} = e^{-(d+2)\phi}\tilde{T}^{\mu\nu}, \tag{38.19}$$

since the action S has weight zero, and $\sqrt{g}g_{\mu\nu}$ has weight $-(d+2)$.

Moreover, for a Weyl invariant action S, and given that the Weyl transformation is defined by $g^{\mu\nu}\delta/\delta g^{\mu\nu}$, we see that we obtain a traceless energy-momentum tensor:

$$T^{\mu}_{\ \mu} = 0. \tag{38.20}$$

For an ideal fluid, this gives (since as we saw $g^{\mu\nu}P_{\mu\nu} = d - 1$ and $g^{\mu\nu}u_{\mu}u_{\nu} = -1$),

$$P = \frac{\rho}{d-1}, \tag{38.21}$$

which implies that the *sound velocity* c_s is

$$c_s^2 \equiv \frac{\partial P}{\partial \rho} = \frac{1}{d-1} \Rightarrow c_s = \frac{1}{\sqrt{d-1}}. \tag{38.22}$$

But there is a subtlety. In two dimensions, Weyl invariance, together with diffeomorphism invariance, is needed for conformal invariance, so a conformal invariant Belinfante $T_{\mu\nu}$ is automatically Weyl invariant, thus traceless. But in four dimensions, Weyl invariance is not needed for conformal invariance, and in fact there are many examples of conformal fields theories with Belinfante tensor (the energy-momentum tensor defined via $g_{\mu\nu}$) that is not traceless. Instead, we must remember that the energy-momentum tensor suffers from the Noether ambiguity: we can add a total divergence of a tensor $J_{\rho\mu\nu} = -J_{\mu\rho\nu} = -J_{\nu\mu\rho}$,

$$T_{\mu\nu} \to T_{\mu\nu} + \partial^\rho J_{\rho\mu\nu}, \tag{38.23}$$

since this doesn't change its conservation equation $\nabla^\mu T_{\mu\nu} = 0$. Under this ambiguity, we can find an improved energy-momentum tensor that is traceless.

It is left as an exercise to show that the conserved current is a tensor of weight d, and the temperature T ($E = k_B T$) is a tensor of weight one:

$$J^\mu = e^{-d\phi} \tilde{J}^\mu, \quad T = e^{-\phi} \tilde{T}. \tag{38.24}$$

Since $T^{\mu\nu} = \rho u^\mu u^\nu + \cdots$, it follows that ρ is a tensor of weight d ($u^\mu u^\nu$ has weight 2), and T has weight one, so

$$\rho \sim T^d. \tag{38.25}$$

Then, since $P = \rho/(d-1)$, so is P. Then from the Gibbs-Duhem relation described in Chapter 1, which after dividing with the volume V is

$$P + \rho = sT + \mu_I q_I + \cdots, \tag{38.26}$$

we obtain for the entropy density

$$s \sim T^{d-1}. \tag{38.27}$$

Finally then, we can write the energy-momentum tensor of an ideal *conformal* fluid

$$(T^{\mu\nu})_{\text{ideal,conformal}} = \alpha T^d (g^{\mu\nu} + d u^\mu u^\nu), \tag{38.28}$$

where α is a constant.

For $\theta = \nabla_\mu u^\mu$, we first notice that at most it could have weight one, since

$$\nabla_\mu u^\nu = \partial_\mu u^\nu + \cdots = e^{-\phi} \tilde{\nabla}_\mu \tilde{u}^\nu + \cdots, \tag{38.29}$$

but in fact it doesn't transform covariantly under Weyl transformations, but rather as

$$\theta = \nabla_\mu u^\mu = e^{-\phi} (\tilde{\theta} + (d-1)\tilde{u}^\sigma \partial_\sigma \phi). \tag{38.30}$$

That means that in $T^{\mu\nu}$, its coefficient, the bulk viscosity ζ, must vanish for a conformal theory. On the other hand, one finds that $\sigma^{\mu\nu}$, the traceless velocity derivative, does transform covariantly, with weight 3, since $P^{\mu\nu}$ has weight 2 and u^μ has weight 1, so

$$\sigma^{\mu\nu} = P^{\lambda(\mu}\nabla_\lambda u^{\nu)} - \frac{1}{d-1}P^{\mu\nu}\nabla_\lambda u^\lambda = e^{-3\phi}\tilde{\sigma}^{\mu\nu}. \tag{38.31}$$

Finally then, the energy-momentum tensor of a conformal fluid, in the expansion up to second order, is

$$T^{\mu\nu}_{\text{conf.fluid}} = \alpha T^d(g^{\mu\nu} + du^\mu u^\nu) - 2\eta\sigma^{\mu\nu}. \tag{38.32}$$

Moreover, one can find the scaling of η with temperature:

$$\eta = T^{d-1}\hat{\eta}(\mu_I/T). \tag{38.33}$$

Similarly to the above, one finds that for a conformal fluid, $\gamma_I = 0$, which is left as an exercise to prove. Then the current for a conformal fluid is

$$J^\mu_{I,\text{conf.fluid}} = q_I u^\mu - \tilde{\chi}_{IJ}P^{\mu\nu}\nabla_\nu\left(\frac{\mu_J}{T}\right) - \mathcal{V}_I l^\mu. \tag{38.34}$$

38.3 Fluid Equations from Black Holes

Description of the Conformal Fluid in Asymptotically AdS Space
To describe the conformal fluid via AdS/CFT, we consider the (planar) Schwarzschild black hole in AdS_{d+1}, dual to a finite temperature conformal field theory on the d-dimensional boundary,

$$ds^2 = -r^2 f\left(\frac{r}{r_+}\right)dt^2 + \frac{dr^2}{r^2 f\left(\frac{r}{r_+}\right)} + r^2\delta_{ij}dx^i dx^j, \tag{38.35}$$

where

$$f(r) = 1 - \frac{1}{r^d} \tag{38.36}$$

and the temperature is related to the r_+ parameter by

$$T = \frac{d}{4\pi}r_+. \tag{38.37}$$

Boosting the solution along x^i, we obtain the relativistic solution,

$$ds^2 = \frac{dr^2}{r^2 f\left(\frac{r}{r_+}\right)} + r^2\left[-f\left(\frac{r}{r_+}\right)u_\mu u_\nu + P_{\mu\nu}\right]dx^\mu dx^\nu, \tag{38.38}$$

where the 4-velocity is as usual

$$u^0 = \frac{1}{\sqrt{1-\beta^2}}; \quad u^i = \frac{\beta^i}{\sqrt{1-\beta^2}}. \tag{38.39}$$

We can find the energy-momentum tensor $T_{\mu\nu}$ on the boundary from the above metric, with the prescription that will be presented shortly, and we find that we have an ideal conformal fluid as in (38.32), with the parameter

$$\alpha = \frac{\pi^d}{16\pi\, G_N^{(d+1)}}. \tag{38.40}$$

Until now we have described a uniform system, corresponding to the black hole. But from the point of view of fluids, this is not a very interesting system: we want to have some spatial dependence for u^μ and T, so we promote the parameters to fields, by making $r_+ = r_+(x^\alpha)$ (thus $T = T(x^\alpha)$) and $\beta^i = \beta^i(x^\alpha)$ (so $u^\mu = u^\mu(x^\alpha)$). However, if we keep the above black hole coordinates while making this replacement, we obtain that the coordinates are not regular on the future horizon. Instead, we must work with "Eddington-Finkelstein coordinates" for the boosted black hole:

$$ds^2 = -2u_\mu dx^\mu dr - r^2\left(\frac{r}{r_+}\right)u_\mu u_\nu dx^\mu dx^\nu + r^2 P_{\mu\nu} dx^\mu dx^\nu. \tag{38.41}$$

Promoting r_+ (or $T = r_+ d/(4\pi)$) to a field $r_+(x^\alpha)$ and $\beta^i = \beta^i(x^\alpha)$, we get the metric

$$ds^2 = -2u_\mu(x^\alpha)dx^\mu dr - r^2\left(\frac{r}{r_+}\right)u_\mu(x^\alpha)u_\nu(x^\rho)dx^\mu dx^\nu + r^2 P_{\mu\nu}(x^\alpha)dx^\mu dx^\nu$$
$$\equiv g_{\mu\nu}^{(0)} dx^\mu dx^\nu. \tag{38.42}$$

Of course, while the boosted black hole was a solution of the Einstein equations, after changing the parameters into fields, it is not a solution anymore, so we must correct this (zeroth order) metric order by order in perturbation theory in the number of field theory derivatives, where we consider

$$\frac{\partial u}{T}, \quad \frac{\partial \log T}{T} \sim \mathcal{O}(\epsilon) \ll 1. \tag{38.43}$$

Then, splitting $M = (\mu, r)$, we find as an expansion in ϵ:

$$g_{MN} = g_{MN}^{(0)}(T(\epsilon, x^\rho), u^\alpha(\epsilon, x^\rho)) + \epsilon g_{MN}^{(1)}(T(\epsilon, x^\rho), u^\alpha(\epsilon, x^\rho))$$
$$+ \epsilon^2 g_{MN}^{(2)}(T(\epsilon, x^\rho), u^\alpha(\epsilon, x^\rho))$$
$$T = T^{(0)} + \epsilon T^{(1)} + \mathcal{O}(\epsilon^2)$$
$$u^\alpha = u_{(0)}^\alpha + \epsilon u_{(1)}^\alpha + \mathcal{O}(\epsilon^2). \tag{38.44}$$

One can then solve Einstein's equation order by order, and one finds a very complicated solution, which will not be reproduced here.

38.4 Viscosity over Entropy Density from Black Holes

It remains to give the prescription for finding the CFT energy-momentum tensor from the bulk gravity. The Belinfante energy-momentum tensor is

$$T_{\mu\nu} = -\frac{2}{\sqrt{-g}}\frac{\delta S}{\delta g^{\mu\nu}}, \tag{38.45}$$

but on the other hand in AdS/CFT one defines the partition function as

$$Z_{CFT} = \int \mathcal{D}(...)e^{-S+\int \mathcal{O}\phi_0} = Z_{sugra} = e^{-S_{sugra}[\phi(\phi_0)]}, \qquad (38.46)$$

which leads to the 1-point functions

$$\langle \mathcal{O} \rangle = \left. \frac{\delta S}{\delta \phi_0} \right|_{\phi_0=0}. \qquad (38.47)$$

Considering now $\mathcal{O}$ to be $T_{\mu\nu}$ in the CFT, which means that ϕ_0 is the metric tensor on the boundary, $g_{bd.}^{\mu\nu}$, we obtain naively

$$\langle T_{\mu\nu} \rangle_{CFT} = \frac{\delta S_{sugra}}{\delta g_{bd.}^{\mu\nu}}. \qquad (38.48)$$

The true story is more complicated than that, however. We need to regulate the asymptotically AdS_{d+1} spacetime at $r = r_\infty$, and write the metric there as the boundary metric times r_∞ factors. Moreover, we need to introduce some gravitational boundary terms, and use the Einstein equation (since the supergravity action is on-shell). In the end, introducing n^μ as the outward normal to the $r = r_\infty$ surface, and introducing the "extrinsic curvature" $K_{\mu\nu} = g_{\mu\rho} \nabla^\rho n_\nu$, the correct prescription is

$$T^{\mu\nu} = \lim_{r_\infty \to \infty} \frac{r_\infty^{d-2}}{16\pi G_N^{(d+1)}} \left[K^{\mu\nu} - Kg^{\mu\nu} - (d-1)g^{\mu\nu} - \frac{1}{d-2}\left(R^{\mu\nu} - \frac{1}{2}g^{\mu\nu}R \right) \right]. \qquad (38.49)$$

After a long calculation, one finds that the energy-momentum tensor is that of a fluid with the value of α described above for the ideal part, and the viscous part

$$\pi_{(1)}^{\mu\nu} = 2\eta\sigma^{\mu\nu}, \qquad (38.50)$$

where the shear viscosity is

$$\eta = \frac{N^2}{8\pi^2}(\pi T)^3, \qquad (38.51)$$

in the case of $d = 4$, for $\mathcal{N} = 4$ SYM. In the same theory, at strong coupling, the entropy density is the same as the entropy density of an order $\mathcal{O}(N^2)$ free degrees of freedom, differing from the number of weak coupling (perturbative) degrees of freedom by a factor of 3/4,

$$s = \frac{\pi^2}{2}N^2 T^3, \qquad (38.52)$$

compared with the weak coupling result

$$s = \frac{2\pi^2}{3}N^2 T^3. \qquad (38.53)$$

Finally then, we obtain

$$\frac{\eta}{s} = \frac{1}{4\pi}. \qquad (38.54)$$

In a general dimension, we obtain

$$\eta = \frac{1}{16\pi\, G_N^{(d+1)}} \left(\frac{4\pi}{d}T\right)^{d-1} \tag{38.55}$$

and

$$\frac{1}{16\pi\, G_N^{(d+1)}} \Rightarrow \frac{N^2}{8\pi^2}, \tag{38.56}$$

so we obtain as before $\eta/s = 1/(4\pi)$.

38.5 Membrane Paradigm: Relativistic Fluid Equations from Black Hole Horizons

Heretofore until now, we have described the correspondence between a fluid at the boundary of AdS space (with a black hole) and the gravity in the bulk. But there is another development of interest, which is independent of AdS/CFT, but can be combined with it: we can also describe the gravity in the bulk with a black hole as a *fluid at the horizon* of the black hole, described by what was called the *membrane paradigm*. Comparing with AdS/CFT, we see that the fluid at the horizon of the black hole can be identified with the fluid at the boundary.

The idea of the membrane paradigm is an old one, that was started by the Ph.D. thesis of Thibault Damour in 1979, and it was developed further by Kip Thorne, Prince, and MacDonald, who wrote a book about it in 1986 [93].

The idea, originally defined for astrophysical black holes, i.e. black holes which we could in principle detect through telescope observations, is that from the point of view of the observer at infinity (that uses the telescopes), the black hole is equivalent with a thin fluid membrane that lives just outside the event horizon of the black hole. Thus we can forget about the black hole per se, and consider how we would detect the fluid membrane that replaces the black hole.

Already in Damour's thesis it was shown that the Einstein equations near the horizon of the black hole reduce to the Navier-Stokes equations for the fluid, but it was found that for the astrophysical black holes (in flat four dimensions), one gets unphysical parameters in the NS equations, ones that violate thermodynamical constraints.

However, the new development that was found in the context of the fluid-gravity correspondence (AdS/CFT) in, among other papers, [94] and [95], is that in the case of an AdS background we obtain physical parameters, thus having a perfectly well-defined map from the Einstein equations to the Navier-Stokes equations.

Consider coordinates $x^A = (r, x^\mu)$ for the black hole, with r being transverse to the horizon, and the horizon condition being be $r = r_{II}$, or more generally $F(x^A) = 0$. Consider also the vector

$$l^A = g^{AB}\partial_B r, \tag{38.57}$$

or in general $l^A = g^{AB}\partial_B F$, which is both normal and tangent at the horizon (possible due to the singularity of the metric at the horizon), and thus tangent to the null generators of the horizon (which can be taken to be $\partial_B r$). It is a null vector, since

$$g_{AB}l^A l^B = g^{AB}\partial_A r\partial_B r = 0, \tag{38.58}$$

because $g^{rr} = 0$ and $\partial_\mu r = 0$ at the horizon. Moreover, $l^A = (0, l^\mu)$, since $l^r = g^{rr} = 0$.

Considering a basis e_μ^A for the horizon, define the horizon metric

$$\gamma_{\mu\nu} = g_{AB}e_\mu^A e_\nu^B, \tag{38.59}$$

or using a null vector (can be taken to be $m^A = (1, \vec{0})$) everywhere transverse to the horizon and normalized by $l_A m^A = 1$, then

$$\gamma_{AB} = g_{AB} - l_A m_B - l_B m_A, \tag{38.60}$$

satisfying

$$\gamma_{AB}l^B = g_{AB}l^B - l_A - 0 = 0 \Rightarrow \gamma_{\mu\nu}l^\nu = 0. \tag{38.61}$$

Define the *horizon expansion*

$$\theta = \nabla_A l^A \tag{38.62}$$

and the traceless *horizon shear*

$$\sigma_{AB}^{(H)} = \gamma_A{}^C \gamma_B{}^D \nabla_{(C}l_{D)} - \frac{\theta}{d}\gamma_{AB}, \tag{38.63}$$

not coincidentally using the same notation as for the fluid quantities θ and $\sigma_{\mu\nu}$. Then from $\nabla_\mu(l_A l^A = 0)$, we get

$$l_A \nabla_\mu l^A = 0, \tag{38.64}$$

so $\nabla_\mu l^A$ is tangent to the horizon (being transverse to l_A, which is null), therefore it can be expanded in the horizon basis e_μ^A:

$$\nabla_\mu l^A = \Theta_\mu{}^\nu e_\nu^A. \tag{38.65}$$

Note that in the $l^A = (0, l^\mu)$ basis we have

$$\Theta_\mu{}^\nu = \nabla_\mu l^\nu. \tag{38.66}$$

Then $\Theta_\mu{}^\nu$ is a "*Weingarten map*" from the space tangent to the horizon to itself. This Weingarten map decomposes in components as

$$\gamma^A{}_D \gamma^C{}_B \Theta_C{}^D = \sigma^{(H)}{}_B{}^A + \frac{\theta}{d}\gamma_B{}^A$$
$$\Theta_A{}^B l^A = \kappa(x)l^B$$
$$\Theta_A{}^B m_B \gamma^A{}_C \equiv \Omega_C, \tag{38.67}$$

where Ω_c is the horizon "*momentum*" and $\kappa(x)$ is the surface gravity, defined equivalently by

$$l^B \nabla_B l^A = \kappa(x)l^A, \tag{38.68}$$

in the same way that it was defined from the Killing vector χ^A in Part II. Consider any matrix of the form

$$G_{\mu\nu} = \lambda \gamma_{\mu\nu} - b_\mu b_\nu, \tag{38.69}$$

where $b_\mu c^\mu \neq 0$ and its inverse, then we can write

$$\Theta_\mu{}^\nu = \lambda \theta_{\mu\rho}(G^{-1})^{\rho\nu} + c_\mu l^\nu, \tag{38.70}$$

where

$$\theta_{\mu\nu} = \sigma^{(H)}_{\mu\nu} + \frac{\theta}{d}\gamma_{\mu\nu} \tag{38.71}$$

and $c_\mu l^\mu = \kappa$, the surface gravity.

Consider the Einstein equations with a cosmological constant (since we want to be in AdS background)

$$R_{AB} - \frac{1}{2}g_{AB}R + \Lambda g_{AB} = 8\pi G_N T^{\text{matter}}_{AB}. \tag{38.72}$$

Projecting them on a "boundary" surface, in particular the horizon, reduces them to the "*Gauss-Codazzi equations*" for the embedding, which give

$$R_{AB}l^A e^B_\nu = \bar{D}_\mu(\Theta_\mu{}^\nu - \Theta_\rho{}^\rho \delta^\nu_\mu), \tag{38.73}$$

where the covariant derivative with bar is transverse to the horizon:

$$\bar{D}_{\bar{e}_\mu} e^A_\nu = \Gamma^\sigma_{\mu\nu} e^A_\sigma. \tag{38.74}$$

Considering the contraction of the Einstein equation on $l^A l^B$, we obtain the "*focusing equation*":

$$-l^\mu \nabla_\mu \theta + \kappa(x)\theta - \frac{\theta}{d} - \sigma^{(H)}_{\mu\nu}\sigma^{\mu\nu}_{(H)} - 8\pi G_N T^{\text{matter}}_{AB}l^A l^B = 0. \tag{38.75}$$

This is left as an exercise to prove.

Otherwise, using the modified tensor

$$\tilde{T}_{(H)\mu}{}^\nu = v(\Theta_\mu{}^\nu - \kappa \delta^\nu_\mu), \tag{38.76}$$

where the scalar density v is the area of the horizon, the Gauss-Codazzi equations (projected Eintein's equations) become

$$R_{\mu\nu}S^\nu = D^{(G)}_\nu(\lambda v \theta_{\mu\rho}(G^{-1})^{\rho\nu}) + v\theta\partial_\mu \ln\sqrt{\lambda} + c_\mu \partial_\nu S^\nu + 2S^\nu \partial_{[\nu}c_{\mu]} - v\partial_\mu\theta, \tag{38.77}$$

where

$$S^\mu \equiv vl^\mu \tag{38.78}$$

is an entropy current defined by the horizon area.

Application to Black Hole (Brane)

Now we apply the formalism to the black hole (brane) solution,

$$ds^2_{(0)} = -2u_\mu dx^\mu dr - r^2 f(r)u_\mu u_\nu dx^\mu dx^\nu + r^2 P_{\mu\nu}dx^\mu dx^\nu, \tag{38.79}$$

where

$$f(r) = 1 - \left(\frac{r_H}{r}\right)^d; \quad T = \frac{d}{4\pi}r_H; \quad u^\mu = (\gamma, \gamma v^i), \tag{38.80}$$

so in $d = 4$ we obtain

$$ds^2 = -2u_\mu dx^\mu dr + \frac{\pi^4 T^4}{r^2}u_\mu u_\nu dx^\mu dx^\nu + r^2 \eta_{\mu\nu}dx^\mu dx^\nu. \tag{38.81}$$

The horizon is at $r = r_H$, and the normal vector is then $l^A = (0, u^\mu)$, as expected.

As before, we change the parameters into fields, $u^\mu(x^\rho)$, $T(x^\rho)$, so it is no longer a solution. But one finds to zeroth order

$$\begin{aligned}
\kappa(x) &= 2\pi T(x) \\
\Theta^{(0)\nu}{}_\mu &= -\kappa(x)u_\mu u^\nu = -2\pi T(x)l^\mu l^\nu \\
C^{(0)}_\mu &= -2\pi T(x) = -\kappa(x)u_\mu \\
S^{(0)} &= 4su^\mu = 4sl^\mu,
\end{aligned} \tag{38.82}$$

and the Gauss-Codazzi equation becomes to zeroth order

$$c^{(0)}_\mu \partial_\nu S^{(0)\nu} + 2S^{(0)\nu}\partial_{[\nu}c^{(0)}_{\mu]} = \mathcal{O}(\epsilon^2), \tag{38.83}$$

and after substituting the values for these objects, and using the thermodynamic relations $\epsilon + P = Ts$ and $dp = sdT$, we obtain the conservation equation for the ideal conformal fluid (with $\epsilon = (d - 1)P$):

$$\partial_\nu T^{(0)\nu}_\mu = \partial_\nu[(\epsilon + P)u_\mu u^\nu + P\delta^\nu_\mu] = 0. \tag{38.84}$$

At the first viscous order (second order in the fluid expansion), we obtain the relativistic Navier-Stokes equation with $\eta = s/(4\pi)$:

$$\partial_\nu(T^{(0)\nu}_\mu + T^{(1)\nu}_\mu) = \partial_\nu\left[(\epsilon + P)u_\mu u^\nu + P\delta^\nu_\mu - \frac{s}{2\pi}\pi^{(1)\nu}_\mu\right] = \mathcal{O}(\epsilon^3). \tag{38.85}$$

In conclusion, the membrane paradigm was the fact that we have obtained the Navier-Stokes equations from the Einstein equations on the horizon (the Gauss-Codazzi equations). Note, however, that the fact that we were in AdS background (with a cosmological constant) was crucial. Indeed, without it, the entropy current equation would get an unphysical contribution from the second time derivative of the entropy, and from a negative bulk viscosity. But the source of this can be understood: hydrodynamics as an effective description is valid only as long as the Knudsen number is much less than one,

$$Kn \equiv \frac{l_{\text{corr}}}{L_s} \ll 1, \tag{38.86}$$

where l_{corr} is the correlation length and L_s is the minimal scale for variation of the fields. For a black hole in flat spacetime, $l_{\text{corr}} \sim T^{-1} \sim r_H$, and also $L_s \sim r_H$, so $Kn \sim 1$, and the fluid approximation is not valid. Note that L cannot be larger than r_H, else a perturbation would modify the solution. But in an AdS background we can have $L_s \gg r_H$, hence $Kn \ll 1$.

In the context of hydrodynamics, described by the Navier-Stokes equation, one case of interest is turbulent flow, i.e. turbulence. It is defined in terms of the *Reynolds number*,

$$Re = \frac{Lv}{\nu}, \tag{38.87}$$

where v is the velocity, ν is the viscosity, and L is the characteristic length of the velocity gradient, usually defined by the forcing of the perturbation. For turbulent flows, $Re \gg 100$.

Finally, taking the membrane paradigm in conjunction with the AdS/CFT correspondence, we see that the fluid at the boundary of the AdS background can be identified with the fluid at the horizon of the black hole in AdS space.

Important Concepts to Remember

- The relativistic Navier-Stokes equation comes from the conservation of the energy-momentum tensor, $\nabla_\mu T^{\mu\nu}$, expanded in derivatives (on the velocity field).
- The fluid expansion is an expansion in derivatives, valid for strongly coupled systems.
- In AdS/CFT, we consider the conformal field theory as a fluid, so we get a "fluid-gravity correspondence."
- For a conformal fluid, the energy-momentum tensor is traceless and of weight $d + 2$, the speed of sound is $c_s = 1/\sqrt{d-1}$, with $\eta = \gamma_I = 0$, and $\epsilon \sim P \sim T^d$.
- More precisely, for a Weyl-invariant theory the Belinfante $T_{\mu\nu}$ is automatically traceless, otherwise in a conformal theory the Belinfante tensor is not necessarily traceless, though we can define an improved tensor that is.
- To describe conformal fluids via AdS/CFT, we consider a boosted black hole (brane) in AdS background (and in Eddington-Finkelstein coordinates), and we replace $u^\mu \to u^\mu(x)$, $T \to T(x)$, and solve the Einstein equation order by order in derivatives.
- The boundary energy-momentum tensor, $\langle T_{\mu\nu} \rangle$, is defined by a regularized version of the variation of the supergravity action with respect to the boundary metric.
- For such a black hole, one finds that the ratio of the shear viscosity and entropy density is $\eta/s = 1/(4\pi)$.
- The membrane paradigm says that we can replace a black hole by a thin fluid membrane just outside the horizon of the black hole. Then the Einstein equations projected onto the horizon (the Gauss-Codazzi equations) become the Navier-Stokes equations.
- In flat background, the membrane paradigm doesn't quite work, since we obtain $\zeta < 0$ and nonconservation of the entropy current (extra terms in it), because the Knudsen number of the fluid is of order one, but in AdS space $Kn \ll 1$ and there are no problems.

Further Reading

The reviews [23] and [24] give a good introduction to the fluid-gravity correspondence, though without the membrane paradigm. For that, read the original papers [94] and [95], as well as the reviews [96] and [97]. For the original membrane paradigm, see the book by Thorne, Pierce, and Macdonald [93].

Exercises

(1) Prove that for a conformal fluid we have, under a Weyl transformation $g_{\mu\nu} = e^{2\phi}\tilde{g}_{\mu\nu}$, the current scaling $J^{\mu} = e^{-d\phi}\tilde{J}^{\mu}$ and the temperature scaling $T = e^{-\phi}\tilde{T}$.

(2) Prove that for a conformal fluid we have $\gamma_I = 0$.

(3) Show that the boosted AdS black hole can be written as

$$ds^2 = -2u_{\mu}dx^{\mu}dr - r^2 f(r/r_+)u_{\mu}u_{\nu}dx^{\mu}dx^{\nu} + r^2 P_{\mu\nu}dx^{\mu}dx^{\nu}. \qquad (38.88)$$

(4) Prove that for the boosted black hole (brane) solution, with $u(x)$, $T(x)$, we have the surface gravity $\kappa(x) = 2\pi T(x)$ at the leading order.

(5) Prove that the null geodesic focusing equation,

$$ -l^{\mu}\nabla_{\mu}\theta + \kappa(x)\theta - \frac{\theta^2}{d} - \sigma^{(H)}_{\mu\nu}\sigma^{\mu\nu}_{(H)} - 8\pi T^{\text{matter}}_{AB}l^A l^B = 0, \qquad (38.89)$$

is obtained from the Einstein equation.

39

Nonrelativistic Fluids: From Einstein to Navier-Stokes and Back

In this chapter, we will first describe how to obtain the nonrelativistic limit of the case from the previous chapter, with the real Navier-Stokes equation obtained from the Einstein equation. Then we will show how to describe the boost-invariant conformal Bjorken flow from the Einstein equation, and then finally how to take a general fluid (solution to the Navier-Stokes equation) and find a solution to the Einstein equation.

39.1 The Navier-Stokes Scaling Limit

The relativistic version of the Navier-Stokes equation, described at the end of the last chapter, was obtained from $\nabla_\mu T^{\mu\nu} = 0$ as

$$0 = \nabla_\mu(\rho u^\mu u^\nu + P P^{\mu\nu} + 2\eta\sigma^{\mu\nu} - \zeta\theta P^{\mu\nu}). \tag{39.1}$$

But now we want to find a nonrelativistic scaling limit that reproduces the Navier-Stokes equation. To scale nonrelativistically, we certainly need that

$$v^i \sim \partial_i \sim \epsilon \to 0, \tag{39.2}$$

but by thinking of the limit on the energy ($E - E_0 = mv^2/2$), we derive that we need also

$$\partial_t \sim \epsilon^2 \sim \delta T(x) = T(x) - T_0. \tag{39.3}$$

Then by taking $\epsilon \to 0$, we obtain the nonrelativistic limit, namely the Navier-Stokes equation

$$\rho[\partial_t \vec{v} + (\vec{v} \cdot \vec{\nabla})\vec{v}] + \vec{\nabla}P = \eta\Delta\vec{v}. \tag{39.4}$$

The proof is left as an exercise.

Some observations are in order:

(1) The fluid is nonrelativistic, since $v \ll 1$.

(2) Since for a conformal fluid the speed of sound is $c_s = 1/\sqrt{d-1} \sim \mathcal{O}(1)$, we have $v \ll c_s$, and since the sound wave is a compression wave, it means that the fluid is effectively incompressible.

(3) Then the $\mu = 0$ component of the relativistic equation becomes $\vec{\nabla} \cdot \vec{v} = 0$, which is the continuity equation for the incompressible fluid, with $\rho = $ constant.

(4) The $\mu = i$ component of the relativistic equation becomes the Navier-Stokes equation with

$$\eta = \frac{1}{2\pi T_0}.$$ (39.5)

Once we have defined this nonrelativistic limit, we can take it also on the gravity dual side.

In the gravity dual, we consider the boosted black hole (black brane) solution from Chapter 38:

$$ds^2 = -2u_\mu dx^\mu dr - r^2 \left(\frac{r}{r_+}\right) u_\mu u_\nu dx^\mu dx^\nu + r^2 P_{\mu\nu} dx^\mu dx^\nu.$$ (39.6)

We write $u^\mu \simeq (1, v^i)$ and keeping only the terms of order one and of order v, but not v^2, we obtain the nonrelativistically boosted solution,

$$ds^2 = -r^2 f\, dt^2 + 2dtdr + r^2 \sum_i dx_i dx_i - 2\pi^4 \frac{T^4}{r^4} v_i dx^i dt - 2v_i dx^i dt,$$ (39.7)

as we can easily check. This can be obtain also by a Galilean boost of the black hole.

We then turn the parameters into fields, as before, and get

$$ds^2 = -r^2 f(t, x^i)\, dt^2 + 2dtdr + r^2 \sum_i dx_i dx_i - 2\pi^4 \frac{T(t, x^i)^4}{r^4} v_i(t, x^i) dx^i dt$$

$$- 2v_i(t, x^i) dx^i dt,$$ (39.8)

as well as

$$T = T_0(1 + \epsilon^2 P(t, x^i)).$$ (39.9)

In the temperature we keep the order ϵ^2 term, since we will see later that it is required for consistency, to get the Navier-Stokes equation. Then $P(t, x^i)$ is identified with the pressure. The induced metric at the horizon $r = r_H$, substituting the horizon value $r_H = \pi T \simeq \pi T_0$, is found to be

$$ds_H^2 = h_{ij}(dx^i - v^i dt)(dx^j - v^j dt),$$ (39.10)

where

$$h_{ij}^{(0)} = (\pi T_0)^2 \delta_{ij}.$$ (39.11)

Note that here we have introduced for free the $\mathcal{O}(v^2)$ term, since we are in the nonrelativistic limit, and we have substituted $r = r_H$ in the metric.

From the metric at the horizon, we find also the horizon expansion

$$\theta = \partial_i v^i,$$ (39.12)

which is left as an exercise to prove, as well as the horizon shear and momentum, and the surface gravity

$$\sigma_{ij} = \frac{1}{2}(\pi T_0)^2 \left(\partial_i v_j + \partial_j v_i - \frac{2(\vec{\nabla} \cdot \vec{v})}{d} \delta_{ij} \right)$$

$$\Omega_i = 2\pi T_0 v_i$$

$$\kappa(x) = 2\pi T_0 (1 + \epsilon^2 P(x)). \tag{39.13}$$

The Einstein equation projected onto the null surface of the horizon, by contracting with $l^A l^B$, gives the focusing equation:

$$-\partial_t \theta - v^i \partial_i \theta + \kappa(x)\theta - \frac{\theta^2}{d} - \sigma_{ij}\sigma^{ij} = 0. \tag{39.14}$$

This is the nonrelativistic version of the null focusing equation presented in Chapter 38. Moreover, we can compute the scaling of each term with ϵ, considering that $\partial_i \sim v_i \sim \epsilon$ and $\partial_t \sim \epsilon^2$, thus in the focusing equation all terms are of order ϵ^4, except the $\kappa(x)\theta$ term, which contains an ϵ^2 piece and an ϵ^4 piece (from the κ expansion). Thus the equation at order ϵ^2 gives $\theta = 0$,

$$\partial_i v^i = 0, \tag{39.15}$$

which is the incompressible continuity equation.

On the other hand, the contraction of the Einstein equation with $l^A e^B_\nu$ gives the Gauss-Codazzi equation:

$$(\partial_t + \theta)\Omega_i + v^j \partial_j \Omega_i + \Omega_j \partial_i v^j = -\partial_i \kappa(x) + \partial_j \sigma^j{}_i - \frac{1}{d} \partial_i \theta. \tag{39.16}$$

Here the indices are raised and lowered with $h^{(0)}_{ij} = (\pi T_0)^2 \delta_{ij}$, so, for instance, $\sigma^j{}_i$ has no factor of $(\pi T_0)^2$ in front. Also using the continuity equation $\theta = \partial_i v^i = 0$, we finally obtain the Navier-Stokes equation

$$\partial_t \vec{v} + (\vec{v} \cdot \vec{\nabla})\vec{v} + \vec{\nabla}P = \eta \Delta \vec{v}, \tag{39.17}$$

where

$$\eta = \frac{1}{2\pi T_0}, \tag{39.18}$$

as we advertised.

39.2 Relativistic Hydrodynamics and Bjorken Flow

We now move to a more specialized topic. We can find the gravity dual of a relativistic "*Bjorken flow*" for a boost-invariant expanding fluid, in particular plasma, useful for heavy ion collisions as found at the RHIC (Brookhaven) and ALICE (LHC at CERN) experiments. This is the first example of a way to go from a solution of the Navier-Stokes equation to a solution of the Einstein equations. We will later see how we can generalize to other fluid cases.

Kinematics

To describe the gravity dual of Bjorken flow, we start with a definition of the kinematics and of the Bjorken scaling.

For a relativistic motion in a direction x^1, a useful variable, used in high-energy physics, is the *spacetime rapidity* (i.e., in position space):

$$\eta = \frac{1}{2} \log \frac{x^0 + x^1}{x^0 - x^1} = \frac{1}{2} \log \frac{x^+}{x^-}. \tag{39.19}$$

We also define the lightcone coordinates $x^\pm = x^0 \pm x^1$, which implies $\partial_\pm = \frac{1}{2}(\partial_0 \pm \partial_1)$. Note that a boost by a parameter β simply shifts the rapidity η by β, since the boost acts on $x^\pm$ by multiplication with $e^{\pm\beta}$. Defining also the proper time

$$\tau = \sqrt{(x^0)^2 - (x^1)^2} = \sqrt{x^+ x^-}, \tag{39.20}$$

we have the parametrization

$$x^\pm = \tau e^{\pm\eta}. \tag{39.21}$$

We define also the (momentum-space) *rapidity*

$$y = \frac{1}{2} \log \frac{E + p}{E - p}. \tag{39.22}$$

Applying this to the case of a fluid, with velocity field u^μ, thus 4-momentum $p^\mu = mu^\mu$, and lightlike components $u^\pm = u^0 \pm u^1$, the rapidity becomes

$$y = \frac{1}{2} \log \frac{u^+}{u^-}. \tag{39.23}$$

But since $-1 = u^\mu u_\mu \simeq -u^+ u^-$ (since the transverse components of the fluid velocity are assumed to be negligible), it follows that there is no analog of the τ variable in velocity space, and we have the parametrization

$$u^\pm = e^{\pm y}. \tag{39.24}$$

Again, a boost by β amounts to a shift of y.

We specialize to the case of a perfect fluid, with energy-momentum tensor

$$T^{\mu\nu} = (\epsilon + P)u^\mu u^\nu + P\eta^{\mu\nu}. \tag{39.25}$$

Consider also an equation of state

$$\epsilon = \tilde{w}P. \tag{39.26}$$

Note that the tilde is because it is more common is to define $P = w\epsilon$ instead. The conformal case then corresponds to $\tilde{w} = 3$ ($w = 1/3$). The conservation equation $\partial_\mu T^{\mu\nu} = 0$ in lightcone coordinates becomes

$$\partial_\pm T^{01} + \frac{1}{2}\partial_+(T^{11} \pm T^{00}) - \frac{1}{2}\partial_-(T^{11} \mp T^{00}) = 0. \tag{39.27}$$

Substituting the perfect fluid energy-momentum tensor, we obtain the equations

$$\tilde{w}\partial_+ \log P = -\frac{(1+\tilde{w})^2}{2}\partial_+ y - \frac{\tilde{w}^2 - 1}{2}e^{-2y}\partial_- y$$

$$\tilde{w}\partial_- \log P = +\frac{(1+\tilde{w})^2}{2}\partial_- y + \frac{\tilde{w}^2 - 1}{2}e^{2y}\partial_+ y, \tag{39.28}$$

which is left as an exercise to prove.

Further, using the thermodynamics relations $d\epsilon = Tds + \cdots$ and (the Gibbs-Duhem relation) $\epsilon + P = Ts$, together with the equation of state $\epsilon = \tilde{w}P$ and the ansatz for the entropy density $s = s_0 T^\alpha$, we easily find

$$\epsilon = \tilde{w}P = \epsilon_0 T^{\tilde{w}+1}; \quad s = s_0 T^{\tilde{w}}, \tag{39.29}$$

which implies the entropy density scaling with the energy density

$$s \sim \epsilon^{\frac{\tilde{w}}{\tilde{w}+1}}. \tag{39.30}$$

In the conformal case we obtain

$$\epsilon = \epsilon_0 T^4; \quad P = P_0 T^4; \quad s = s_0 T^3. \tag{39.31}$$

We are now ready to define the behavior of the boost-invariant expanding fluid (plasma), relevant for the QGP. Bjorken made the ansatz for the expanding fluid to be boost-invariant. The condition is that the thermodynamic quantities are independent of η or y (which are, up to a constant, the boost), but depend only on τ.

The ansatz is that the fluid behaves in the same way as the coordinates under the boost, i.e. that the origin of y and η is the same, or that

$$y = \eta. \tag{39.32}$$

Replacing $2y = 2\eta = \log(x^+/x^-)$ in (39.28), we obtain

$$\tilde{w}\partial_+ \log P = -\frac{1+\tilde{w}}{2x^+}$$

$$\tilde{w}\partial_- \log P = -\frac{1+\tilde{w}}{2x^-}, \tag{39.33}$$

whose solution is

$$P = \frac{\epsilon}{\tilde{w}} = P_0(x^+ x^-)^{-\frac{\tilde{w}+1}{2\tilde{w}}} = P_0 \tau^{-\frac{\tilde{w}+1}{\tilde{w}}}. \tag{39.34}$$

In the conformal case, $\tilde{w} = 3$, we obtain

$$P = \frac{\epsilon}{3} = P_0(x^+ x^-)^{-2/3} = P_0 \tau^{-4/3} \propto T^4, \tag{39.35}$$

so

$$\epsilon = \epsilon_0 \tau^{-4/3}; \quad T = T_0 \tau^{-1/3}. \tag{39.36}$$

This fluid is indeed boost invariant, since P, ϵ, s, T are independent of η or y, but depend only on τ.

AdS/CFT Ansatz

Now that we have defined the kinematics and the Bjorken flow, we are ready to define its gravity dual.

Since the asymptotic metric is part of the ansatz for the field theory, therefore is varying, it is important to choose a system of coordinates, which is equivalent to imposing some constraints on the form of the metric.

We consider the "*Fefferman-Graham*" system of coordinates for the asymptotically AdS metric, written as

$$ds^2 = \frac{g_{\mu\nu}(x^\rho, z)dx^\mu dx^\nu + dz^2}{z^2}, \tag{39.37}$$

where near $z = 0$ (the boundary),

$$g_{\mu\nu}(x^\rho, z) = \eta_{\mu\nu} + z^4 g_{\mu\nu}^{(4)}(x^\rho) + \cdots \tag{39.38}$$

Note that generally the dependence on z of the metric would start at order z^2, but in these coordinates it starts at order z^4 instead.

With this choice of coordinates, one can show (though I will not prove it here) that the formalism sketched in the last chapter gives the VEV of the CFT energy-momentum tensor as

$$\langle T_{\mu\nu}(x^\rho)\rangle = \frac{N_c^2}{2\pi^2}g_{\mu\nu}^{(4)}(x^\rho). \tag{39.39}$$

For a static, uniform plasma of constant density ρ, the dual geometry is the planar (brane) AdS-Schwarzschild black hole, the same from the last chapter (which was boosted to obtain the dual to the fluid), though with redefined coordinate $r/r_0 = z_0/z$, thus with metric

$$ds^2 = -\frac{1 - \tilde{z}^4/\tilde{z}_0^4}{\tilde{z}^2}dt^2 + \frac{d\vec{x}^2}{\tilde{z}^2} + \frac{d\tilde{z}^2}{\tilde{z}^2\left(1 - \frac{\tilde{z}^4}{\tilde{z}_0^4}\right)}. \tag{39.40}$$

Making a further change of coordinates

$$\tilde{z} = \frac{z}{\sqrt{1 + \frac{z^4}{z_0^4}}} \Rightarrow \tilde{z}_0 = \frac{z_0}{\sqrt{2}}, \tag{39.41}$$

we obtain the metric

$$ds^2 = -\frac{(1 - z^4/z_0^4)^2}{(1 + z^4/z_0^4)z^2}dt^2 + (1 + z^4/z_0^4)\frac{d\vec{x}^2}{z^2} + \frac{dz^2}{z^2}. \tag{39.42}$$

The temperature of the fluid is identified with the Hawking temperature of this black hole, which is ($z_0 = 1/r_0$ from the last chapter)

$$T = T_{BH} = \frac{1}{\pi\tilde{z}_0} = \frac{\sqrt{2}}{\pi z_0}. \tag{39.43}$$

The entropy is found by considering the area of the horizon A_h and dividing it by $4G_N^{(5)}$. In AdS/CFT, the $1/G_N^{(5)}$ is found, by writing the on-shell supergravity action in $\mathcal{N} = 4$ SYM variables, to be proportional to N_c^2, more precisely, as we said in the last chapter, $(16\pi G_N^{(d+1)})^{-1} = N_c^2/(8\pi^2)$. Since the area of the horizon of the black hole (at $t = $ constant and $\tilde{z} = \tilde{z}_0$) is $(\int dx^1 dx^2 dx^3)/\tilde{z}_0^3$, the entropy density, i.e. the entropy divided by the volume in the CFT, $\int dx^1 dx^2 dx^3$, is

$$s = \frac{N_c^2}{2\pi\tilde{z}_0^3} = \frac{\pi^2}{2}N_c^2 T^3, \tag{39.44}$$

where we have substituted the value of the black hole temperature as a function of $\tilde{z}_0$.

39.3 Boost-Invariant Conformal Flows from Black Hole Horizon

One can show that imposing the conservation equation $\partial_\mu T^{\mu\nu} = 0$ for a boost-invariant ideal expanding fluid, i.e. with $\epsilon = \epsilon(\tau)$, and $P_i = P_i(\tau)$, as well as the tracelessness condition for a conformal fluid, $T^\mu{}_\mu = 0$, we obtain

$$T_{\mu\nu} = \text{diag}\left(\epsilon(\tau), -\tau^3\frac{d}{d\tau}\epsilon(\tau) - \tau^2\epsilon(\tau), \epsilon(\tau) + \frac{\tau}{2}\frac{d}{d\tau}\epsilon(\tau), \epsilon(\tau) + \frac{\tau}{2}\frac{d}{d\tau}\epsilon(\tau)\right). \tag{39.45}$$

Note that the fluid is relativistic in the direction x^1, which is why it is preferred in the above energy-momentum tensor.

Since the fluid is boost invariant, it is only a function of τ (not of η or y), and since it is conformal, the large τ behavior (late time asymptotics) can only be a power law (since for an exponential behavior we need a mass scale, in a conformal theory we can have only power laws):

$$\epsilon(\tau) \sim \frac{1}{\tau^s} + \cdots \quad \text{at} \quad \tau \to \infty. \tag{39.46}$$

From energy positivity we need $s \leq 4$, and $s \geq 0$ since the expanding plasma must lower its energy in time, leading to

$$0 \leq s \leq 4. \tag{39.47}$$

Of course, as we saw before, Bjorken flow is $s = 4/3$, obeying this constraint.

The most general ansatz for the gravity dual of this flow that is consistent with the symmetries is

$$ds^2 = \frac{-e^{a(\tau,z)}d\tau^2 + \tau^2 e^{b(\tau,z)}dy^2 + e^{c(\tau,z)}dx_\perp^2 + dz^2}{z^2}. \tag{39.48}$$

One then solves the Einstein equation as a power series in z (the distance from the boundary), introducing the scaling variable $v = z/\tau^{s/4}$. One finds that for generic s the solution is singular, and the only nonsingular solution is obtained for $s = 4/3$, i.e. for Bjorken flow. Moreover, we are showing here the result only for the leading (ideal fluid) part of the expanding fluid flow, but the same analysis can be made for the higher orders in the fluid expansion (in derivatives of the velocity), giving the viscosity term at the next order, etc. The condition of avoiding singular solutions always selects the physical case.

The nonsingular metric is then

$$ds^2 = \frac{1}{z^2}\left[-\frac{\left(1 - \frac{e_0}{3}\frac{z^4}{\tau^{4/3}}\right)^2}{1 + \frac{e_0}{3}\frac{z^4}{\tau^{4/3}}}d\tau^2 + \left(1 + \frac{e_0}{3}\frac{z^4}{\tau^{4/3}}\right)(\tau^2 dy^2 + dx_\perp^2)\right] + \frac{dz^2}{z^2}. \quad (39.49)$$

The energy density is

$$\epsilon = \frac{e_0}{\tau^{4/3}}. \quad (39.50)$$

Note that this is like an AdS-Schwarzschild black hole with an "*effective horizon*" that is time-dependent, with

$$z_0 = \left(\frac{3}{e_0}\right)^{1/4}\tau^{1/3}, \quad (39.51)$$

which leads to

$$T = \frac{\sqrt{2}}{\pi z_0} = \frac{\sqrt{2}}{\pi}\left(\frac{e_0}{3}\right)^{1/4}\tau^{-1/3}$$

$$s \propto \frac{\tau}{z_0^3} = \text{const.} \quad (39.52)$$

Of course, the metric is not a real black hole, since a real black hole has no time dependence, and so the effective horizon as well is not a real horizon. But because the time variation is not too fast, we can approximate with this adiabatic picture.

39.4 Einstein's Equations Solutions from Solutions of Navier-Stokes Equation

The same logic from here can be applied to find a solution of the Einstein equation from a general solution to the Navier-Stokes equation, that is, in a *nonconformal, nonrelativistic case.*

As we saw, for the nonrelativistic limit, we need a scaling of the type

$$v_i(x^i, \tau) = \epsilon v_i(\epsilon x^i, \epsilon^2 \tau)$$
$$P(x^i, \tau) = \epsilon^2 P(\epsilon x^i, \epsilon^2 \tau). \quad (39.53)$$

Then higher order corrections are relativistic ones, and corrections in derivatives ∂_i also give higher orders in the hydrodynamics expansion.

As in the particular case of Bjorken flow above, for gravity duals of physical systems, we impose that the perturbations of the horizon of the black hole are nonsingular, which selects the physical case. Therefore a general T_{ab} cannot be found in the gravity dual, unless it is physical.

Cauchy Data

To solve the Einstein equations, we need to impose Cauchy data on a surface.

We consider the past null infinity $\mathcal{I}^-$, composed of the regions $x^+ \to -\infty$ and $x^- \to -\infty$, for the asymptotic (field theory) metric

$$ds^2_{p+2} = -dx^+ dx^- + dx_i dx_i. \tag{39.54}$$

One also considers the regularized "boundary" Σ_c, defined by

$$x^+ x^- = -4r_c \tag{39.55}$$

and $x^+ > 0$. Then the (regulated) Cauchy data are given on the surface $\Sigma_c \cup \mathcal{I}^-$, where the metric on Σ_c is $\gamma_{ab} = \eta_{ab}$ and the "extrinsic curvature" is K_{ab} (see the last chapter), leading to an energy-momentum tensor

$$T_{ab} = 2(\gamma_{ab} K - K_{ab}). \tag{39.56}$$

Considering so-called *Rindler coordinates* (associated with accelerated observers), with

$$ds^2_{p+2} = -r\, d\tau^2 + 2d\tau dt + dx_i dx_i, \tag{39.57}$$

then the surface Σ_c is defined by $r = r_c$.

One then finds at the linearized level

$$r_c^{3/2} T^{\tau i} = v^i$$
$$r_c^{3/2} T^{ij} = -\eta \partial^{(i} v^{j)}, \tag{39.58}$$

with the viscosity $\eta = r_c$ and v^i satisfying at the linearized level the Navier-Stokes equation. The analysis can be extended to the nonlinear level as well, though it is more complicated.

Important Concepts to Remember

- The nonrelativistic scaling limit for the conservation equation to give the Navier-Stokes equation is $v^i \sim \partial_i \sim \epsilon$ and $\partial_t \sim \epsilon^2 \sim \delta T(x) = T(x) - T_0$.
- The gravity dual to the Navier-Stokes flow is obtained from the nonrelativistic limit of the gravity dual (boosted planar AdS black hole, or brane), and one obtains $\eta = 1/(2\pi T_0)$.
- Bjorken flow (scaling) is $y = \eta$, i.e. the spacetime rapidity equals the rapidity of the velocity flow, and it leads to a boost invariant fluid, with dependence of thermodynamic objects only on the proper time τ, but not on rapidities (which are changed by boosts).
- Imposing a boost-invariant flow, with a general $\epsilon(\tau)$ asymptotically $\sim \tau^{-s}$ (for the conformal Bjorken case we have $\tau = 4/3$), the gravity dual in general contains singularities, which are avoided only for $\tau = 4/3$.
- This procedure is general: imposing the absence of singularities selects the physical flow, at all orders in the hydrodynamics expansion.
- The gravity dual of Bjorken flow has an effective horizon that is time-dependent (though it is not quite the usual black hole with the usual horizon).

- One can generalize this procedure, and find a solution to the Einstein equations from a solution to the Navier-Stokes equation, by imposing the absence of singularities for the horizon.

Further Reading

The reviews [23] and [24] give a good introduction to the fluid-gravity correspondence, including the scaling limit for the nonrelativistic limit. The nonrelativistic limit (and the first papers to describe the fluid-gravity correspondence as we know it today) was first described in [98] and [99], though I followed mostly [96]. A good review for the gravity dual of Bjorken flow is [100]. For the Navier-Stokes to Einstein embedding, see [101], [102].

Exercises

(1) Show that the Navier-Stokes equation is obtained in the nonrelativistic limit of the relativistic equation.
(2) Show that at leading order in the nonrelativistic boosted black hole (brane), we obtain $\theta = \partial^i v_i$.
(3) Prove that $\partial_\mu T^{\mu\nu} = 0$ implies the equations

$$\tilde{w}\partial_+ \log P = -\frac{(1 + \tilde{w})^2}{2}\partial_+ y - \frac{\tilde{w}^2 - 1}{2}e^{-2y}\partial_- y$$

$$\tilde{w}\partial_- \log P = +\frac{(1 + \tilde{w})^2}{2}\partial_- y + \frac{\tilde{w}^2 - 1}{2}e^{2y}\partial_+ y \qquad (39.59)$$

(4) Show that the Bjorken flow metric with $s = 4/3$ is consistent with the energy-momentum tensor $T_{\mu\nu}$ sourcing it.
(5) Calculate the Ricci scalar at the effective horizon of the Bjorken flow metric.

PART IV

ADVANCED APPLICATIONS

40

Fermi Gas and Liquid in AdS/CFT

The most important model for electric transport in metals is the free Fermi gas, and its modification, the Fermi liquid theory. In this chapter we will see how we can model them using AdS/CFT. The more challenging cases of the non-Fermi liquids and strange metals, which unlike the cases here do not have a good condensed matter model, will be treated later on.

The free Fermi gas is such a simple case in condensed matter that one could ask why we bother with a gravity dual. First, the embedding for it in a relativistic quantum field theory is nontrivial, as we will see. Second, one could consider all sorts of shapes for the Fermi droplet (the Fermi surface), which leads to different behaviors in metals, and this will correspond to a classification of gravity dual solutions. Thus we can both learn about possible new gravity duals, and reversely, from gravity duals we can learn about free fermion constructions and their properties, though of course we could have learned about that without the gravitational description.

The Fermi liquid on the other hand, despite its successes, is a phenomenological model for a strongly coupled theory (strong electron interaction generating an effective mass). That means that it is ideal for a gravity dual description, and that description can teach us about the details of the Fermi liquid model.

40.1 Gravity Dual for Fermi Gas

Free fermions, with their characteristic Fermi surfaces bounding "droplets" in momentum space, are a very simple yet powerful tool, one that has appeared in string theory in various places, when solving various models through differential equations.

Here, however, we are interested mainly in the application to AdS/CFT, i.e. in obtaining a gravity dual solution that describes arbitrary configurations of free fermions in momentum space, i.e. arbitrary droplets. We will see that the gravity dual has a smooth geometry and no horizons. The constructions, found by Lin, Lunin, and Maldacena, can be done for both 10-dimensional type IIB superstring theory and for the 11-dimensional M-theory, but I will describe only the 10-dimensional type IIB case.

Free Fermions from Matrices

The free fermionic construction appears whenever we try to solve a large N matrix problem with a harmonic potential, as was first understood by Brezin, Itzykson, Parisi, and Zuber in 1978 [103]. The argument roughly goes as follows. Consider that the large N limit system of quantum field theory of Hermitian matrices M with components M_{ij} reduces to the following Schrödinger problem. The Hamiltonian for the Schrödinger problem

$$H\psi = N^2 E(g)\psi \tag{40.1}$$

is a sum of a kinetic part and a potential part,

$$H = -\frac{1}{2}\Delta + V, \tag{40.2}$$

where the Laplacian is

$$\Delta \equiv \sum_i \frac{\partial^2}{\partial M_{ii}^2} + \sum_{i<j} \frac{\partial^2}{\partial \mathrm{Re}M_{ij}^2} + \sum_{i<j} \frac{\partial^2}{\partial \mathrm{Im}M_{ij}^2}, \tag{40.3}$$

and the potential is a sum of a quadratic and a quartic contribution (a fact that needs to be proven in specific models, but we will assume it here)

$$V = \frac{1}{2}\mathrm{Tr}\,M^2 + \frac{g}{N}\mathrm{Tr}\,M^4. \tag{40.4}$$

We consider a ground state wave function ψ symmetric under $U(N)$ (which is a symmetry of the problem), thus becoming a symmetric function of the eigenvalues λ_i of the matrices M, $\psi(\lambda_i)$.

Introduce now the *antisymmetric* function

$$\phi(\lambda_1, \ldots, \lambda_N) = \left[\prod_{i<j}(\lambda_i - \lambda_j)\right]\psi(\lambda_1, \ldots, \lambda_N). \tag{40.5}$$

For it, we obtain the Schrödinger equation

$$\sum_i h_i \phi \equiv \sum_i \left(-\frac{1}{2}\frac{\partial^2}{\partial \lambda_i^2} + \frac{\lambda_i^2}{2} + \frac{g}{N}\lambda_i^4\right)\phi = N^2 E(g)\phi. \tag{40.6}$$

We note that, since ϕ is antisymmetric, this relation is what we expect for a system of N decoupled fermionic harmonic oscillators (since in the large N limit, $g/N \to 0$), which have a Fermi level.

Free Fermions from $\mathcal{N} = 4$ SYM

The 10-dimensional type IIB solution $AdS_5 \times S^5$ is the gravity dual of the 4-dimensional $\mathcal{N} = 4$ SYM theory. Consider then $\mathcal{N} = 4$ SYM theory, but instead of in flat space, on the conformally equivalent $S^3 \times \mathbb{R}$. Indeed, the two spaces, in Euclidean signature, are related by a conformal factor,

$$ds^2 = dx_\mu dx^\mu = dr^2 + r^2 d\Omega_3^2 = r^2\left(\frac{dr^2}{r^2} + d\Omega^2\right) = e^{2\rho}(d\rho^2 + d\Omega_3^2), \tag{40.7}$$

where $\rho = \ln r$. The gravity dual for it is $AdS_5 \times S^5$ in global coordinates.

As we saw in Chapter 28, one can consider the complex scalar $Z = X^1 + iX^2$, and from it construct the ground state $\mathrm{Tr}[Z^J]$. More generally, consider multi-trace operators $\prod_i (\mathrm{Tr}[Z^{n_i}])^{r_i}$.

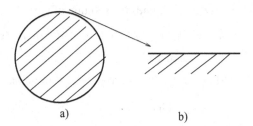

a)   b)

Fig. 40.1 Fermi surfaces for the gravity backgrounds of (a) $AdS_5 \times S^5$ and (b) its Penrose limit, the maximally supersymmetric pp wave (obtained by focusing near a point on the circular surface).

Since we are on $S^3 \times \mathbb{R}$ instead of $\mathbb{R}^4$, we can make a Kaluza-Klein reduction on S^3 of Z. Nonzero KK states would be described in $\mathbb{R}^4$ by acting with covariant derivatives D_i, since these would give "spin" on the angular S^3, like the nontrivial spherical harmonics $Y_{l\bar{m}}(\vec{\Omega})$, and so the states would correspond to $D_i D_j \cdots Z$. But the single Z with no D_i corresponds to the zero (lowest) mode on S^3, which has a harmonic oscillator potential due to the conformal coupling of the scalar Z to the Ricci curvature of the sphere, a term $\mathcal{R}Z^2$ in the Lagrangean, which gives zero in flat space but one (times Z^2) on the unit-radius S^3. The new Hamiltonian after the KK reduction is, as we saw in Chapter 28, $H' = H - J$, where J is the angular momentum for the rotation of directions 1 and 2 that make up Z.

But according to the argument above, the gauge-invariant states of the large N matrix Z with harmonic oscillator potential are N noninteracting fermions in a harmonic oscillator potential, which will form *droplets in momentum space*, bounded by Fermi surfaces. The ground state, which will correspond to $AdS_5 \times S^5$ in the gravity dual, is a *circular droplet*, as in Figure 40.1(a). BPS excitations of this ground state will correspond to particle and hole excitations of the Fermi droplet. We are interested in general BPS states, not necessarily excitations. Droplets of shapes different than the sphere will correspond to gravity duals that are deformed far from $AdS_5 \times S^5$, though we still keep the needed symmetries. In particular, we want to keep the $SO(4) \times \mathbb{R}$ symmetry ($\mathbb{R}$ is the "time" translation corresponding to the new Hamiltonian $H' = H - J$ after the KK reduction) of $S^3 \times \mathbb{R}$, and to have BPS states, one needs to break the $SU(4)$ R-symmetry of $\mathcal{N} = 4$ SYM in flat space, which exchanges the four supersymmetries in four dimensions, to $SO(4) = SU(2) \times SU(2)$ R-symmetry.

Gravity Dual Construction

For the gravity dual of the most general BPS states, we must therefore look for an $SO(4) \times SO(4) \times \mathbb{R}$ invariant ansatz. The $SO(4) \times SO(4)$ part can be satisfied by having 2 S^3 spheres with arbitrary functions multiplying them. Assume then a metric and antisymmetric field strength of the form

$$ds^2 = g_{\mu\nu}dx^\mu dx^\nu + e^{H+G}d\Omega_3^2 + e^{H-G}d\tilde{\Omega}_3^2$$
$$F_{(5)} = F_{\mu\nu}dx^\mu \wedge dx^\nu \wedge d\Omega_3 + \tilde{F}_{\mu\nu}dx^\mu \wedge dx^\nu \wedge d\tilde{\Omega}_3, \tag{40.8}$$

where $\mu = 0, 1, 2, 3$ correspond to the 4-dimensional directions. The rest of the fields are assumed to be trivial.

A rather long analysis finds the solution of type IIB supergravity

$$ds^2 = -h^{-2}(dt + V_i dx^i)^2 + h^2(dy^2 + dx^i dx^i) + ye^G d\Omega_3^2 + ye^{-G} d\tilde{\Omega}_3^2$$

$$h^{-2} = 2y\cosh G$$

$$y\partial_y V_i = \epsilon_{ij}\partial_j z, \qquad y(\partial_i V_j - \partial_j V_i) = \epsilon_{ij}\partial_y z$$

$$z = \frac{1}{2}\tanh G$$

$$F = dB_t \wedge (dt + V) + B_t dV + d\hat{B}$$

$$\tilde{F} = d\tilde{B}_t \wedge (dt + V) + \tilde{B}_t dV + \hat{\tilde{d}B}$$

$$B_t = -\frac{1}{4}y^2 e^{2G} \qquad \tilde{B}_t = -\frac{1}{4}y^2 e^{-2G}$$

$$d\hat{B} = -\frac{1}{4}y^2 *_3 d\left(\frac{z+1/2}{y^2}\right), \qquad \hat{\tilde{d}B} = -\frac{y^3}{4} *_3 d\left(\frac{z-1/2}{y^2}\right). \tag{40.9}$$

Here $i = 1, 2$ and $*_3$ is the flat (y, x_1, x_2) space epsilon tensor.

We see that the solution is parametrized by the single function z, which obeys the equation

$$\partial_i \partial_i z + y\partial_y \left(\frac{\partial_y z}{y}\right) = 0. \tag{40.10}$$

Since from the solution, the product of the radii of the spheres Ω_3 and $\tilde{\Omega}_3$ is y, $y = 0$ would be a singular surface unless something special happens. One can in fact show that the solution is nonsingular if $z = \pm 1/2$ on the $(y = 0, x_1, x_2)$ plane, as could be guessed by looking at the right-hand side of $d\hat{B}$ and $\hat{\tilde{d}B}$ in (40.9).

Defining $\Phi \equiv z/y^2$, (40.10) becomes

$$\partial_i \partial_i \Phi + \partial_y^2 \Phi + \frac{3}{y}\partial_y \Phi = 0, \tag{40.11}$$

which is the Laplace equation in 6 dimensions, with spherical symmetry in four of them, with radius y. As initial data, we need to specify regions of the $(y = 0, x_1, x_2)$ plane where $y = +1/2$ and $y = -1/2$, which correspond to particles and holes, respectively. The general solution with these boundary conditions is given in terms of the Green's function for the above Laplace operator,

$$z(x_1, x_2, y) = y^2 \Phi(x_1, x_2, y) = \frac{y^2}{\pi} \int_D dx_1' dx_2' \frac{z(x_1', x_2', 0)}{[(\vec{x} - \vec{x}')^2 + y^2]^2}$$

$$= -\frac{1}{2\pi} \oint_{\partial D} dl \, n_i \frac{x_i - x_i'}{[(\vec{x} - \vec{x}')^2 + y^2]} + \sigma, \tag{40.12}$$

where $\mathcal{D}$ is a droplet (for $y = +1/2$), $\partial\mathcal{D}$ is its boundary, $\sigma = \pm 1/2$ when we have $z = \pm 1/2$ asymptotically, and n_i is the unit normal vector to the droplet pointing toward the $z = 1/2$ regions. From this z, we can derive V_i by using the third line in (40.9), to give

$$V_i(x_1, x_2, y) = \frac{\epsilon_{ij}}{\pi} \int_D dx_1' dx_2' \frac{z(x_1', x_2', 0)(x_j - x_j')}{[(\vec{x} - \vec{x}')^2 + y^2]^2} = \frac{\epsilon_{ij}}{2\pi} \oint_{\partial D} dx_j' \frac{1}{(\vec{x} - \vec{x}')^2 + y^2}. \tag{40.13}$$

Therefore we see that the solution is given entirely in terms of a choice of droplet on the plane (choice of $z = \pm 1/2$ in domains).

PP Wave Example

The simplest example is the pp wave. As we said, it will turn out that the $AdS_5 \times S^5$ ground state corresponds to a spherical droplet. A pp wave is the Penrose limit of $AdS_5 \times S^5$, which amounts to looking near a null geodesic in the space. In the case of the droplet, it will amount to the limit of looking near the surface of the sphere, when the droplet will appear like a completely filled infinite half plane, as in Figure 40.1(b).

This was not proven, but it was meant as a motivation to use as the droplet boundary condition the half filled plane:

$$z(x_1', x_2', 0) = \frac{1}{2}\text{sign } x_2'. \tag{40.14}$$

With this boundary condition, we can easily calculate z and V_i, obtaining

$$z(x_1, x_2, y) = \frac{1}{2}\frac{x_2}{\sqrt{(x_2)^2 + y^2}}$$

$$V_1 = \frac{1}{2}\frac{1}{\sqrt{(x_2)^2 + y^2}}; \quad V_2 = 0. \tag{40.15}$$

Plugging into the solution (40.9), and changing coordinates by

$$y = r_1 r_2; \quad x_2 = \frac{1}{2}(r_1^2 - r_2^2), \tag{40.16}$$

we obtain the usual form for the (smooth!) metric of the maximally supersymmetric pp wave in 10 dimensions:

$$ds^2 = -2dtdx_1 - (r_1^2 + r_2^2)dt^2 + d\vec{r}_1^2 + d\vec{r}_2^2. \tag{40.17}$$

40.2 Fermi Liquid Construction for Gravity Dual

If we want to describe Fermi liquids, we should have a strong coupling description of interacting fermions at finite temperature and finite chemical potential. Since strongly coupled conformal systems are described by asymptotically AdS gravity duals, we need a gravity theory with a cosmological constant, and since we want finite temperature and finite chemical potential, it should be a charged black hole, with $A \to \mu dt$ at the boundary. Interacting fermions would lead to the existence of fermionic operators, coupling to fermionic sources in the gravity dual, so we need fermions, with a mass corresponding to the dimension of the fermionic operators in the field theory. For concreteness, we will consider 2+1–dimensional field theories, so 3+1–dimensional gravity dual.

Then the action one needs to consider is

$$S_{\text{gravity}} = \frac{1}{2\kappa_{N,4}^2}\int d^4x\sqrt{-g}\left[R + \frac{6}{L^2} + L^2\left(-\frac{F_{\mu\nu}^2}{4} - \bar{\Psi}\Gamma^M D_M\Psi - m\bar{\Psi}\Psi\right)\right]. \tag{40.18}$$

Here $\Gamma^M = e_A^M \Gamma^A$, Γ^A are 4-dimensional gamma matrices, and the covariant derivative is covariant with respect to both gravity (the spin connection ω_μ^{AB}) and the gauge field:

$$D_M \Psi = \left(\partial_M + \frac{1}{8} \omega_M^{AB} [\Gamma^A, \Gamma^B] + ig A_M \right) \Psi. \tag{40.19}$$

One considers a background solution for the gravitational action where the fermion is not turned on, i.e. a charged AdS_4 black hole, with planar horizon. We have considered this solution extensively before. It can be written (in the $A_z = 0$ radial gauge) as

$$\begin{aligned}
ds^2 &= \frac{L^2 \alpha^2}{z^2} (-f(z) dt^2 + dx^2 + dy^2) + \frac{L^2}{f(z)} \frac{dz^2}{z^2} \\
A_0 &= 2q\alpha(z - 1) \\
f(z) &= (1 - z)(z^2 + z + 1 - q^2 z^3).
\end{aligned} \tag{40.20}$$

The temperature of the black hole, equated to the CFT temperature, is $T = \alpha(3 - q^2)/(2\pi)$, and the chemical potential is $\mu = A_0(z = 0) = -2q\alpha$. In this background, we will consider the fermions in the probe approximation, i.e. without back-reaction.

As usual, the CFT partition function equals the gravity partition function, which gives

$$Z_{\text{CFT}}[J] = \langle e^{\int J \cdot \mathcal{O}} \rangle = Z_{\text{gravity}} = \exp\left[i(S_{\text{gravity,bulk}} + S_{\text{gravity,boundary}})[\phi(J)] \right]_{\phi_{\partial AdS} = J}. \tag{40.21}$$

In the case of the fermions, the bulk (Dirac) action vanishes on-shell, so we have only the boundary terms. It is therefore essential to define the correct boundary terms that need to be added to the gravitational action. Defining chirality with respect to Γ^z, $\Gamma^z \Psi_\pm = \pm \Psi_\pm$, we consider as the boundary source $\Psi_+^0 = \Psi_+(z = 0)$. The Dirac equation then relates Ψ_-^0 to Ψ_+^0, which is therefore not an independent source.

It can be shown that the correct boundary term to be added (though I will not explain that here) is

$$S_{\text{gravity,boundary}}[\Psi] = \frac{L^2}{2\kappa_{N,4}^2} \int_{z=z_0 \to 0} d^3 x \sqrt{-g} \bar{\Psi}_+ \Psi_-. \tag{40.22}$$

For studies of the Fermi liquid, one should calculate the retarded single fermion propagator $G^R(\omega, k)$ at finite temperature, and from it the spectral function

$$A(\omega, k) = -\frac{1}{\pi} \operatorname{Im} \operatorname{Tr}(i\gamma^0 G_R(\omega, k)), \tag{40.23}$$

which can be directly compared with experiments. Note that in CFT the normal (time-ordered) propagator $\langle \bar{\Psi} \Psi \rangle = G(\omega, k)$ is completely fixed by conformal symmetry for a fermion of scaling dimension Δ_Ψ to be of the form

$$G_{CFT}^{\Delta_\Psi}(\omega, k) \sim \frac{1}{(\sqrt{-\omega^2 + k^2})^{d - 2\Delta_\Psi}}. \tag{40.24}$$

Unitarity bounds (corresponding to Breitenlohner-Freedman type bounds in AdS space) give the constraint

$$mL = \Delta_\Psi - \frac{d}{2} > -\frac{1}{2},$$

(40.25)

and one obtains *Fermi liquid behavior only in the limit* $mL = -1/2 + \delta$, $\delta \ll 1$, i.e. close to the unitarity limit, which corresponds to a free fermion ($\Delta_\Psi = 1$).

One can calculate the retarded Green's function in the charged AdS black hole background as follows. The on-shell fermionic solutions in AdS space are written in a matrix form in terms of the sources, of the type $\Psi(z) = G(z, z_0)\Psi_0(z_0)$, more precisely

$$\Psi_+^{\rm sol}(z) = F_+(z)F_+^{-1}(z_0)\Psi_+^0(z_0), \qquad \Psi_-^{\rm sol}(z) = F_-(z)F_-^{-1}(z_0)\Psi_-^0(z_0).$$

(40.26)

Then the on-shell action is

$$S^{\rm on-shell} = \frac{L^2}{2\kappa_{N,4}^2} \int_{z=z_0 \to 0} \frac{d\omega d^2k}{(2\pi)^3} \sqrt{-h} \bar{\Psi}_+^0 F_-(z_0)F_+^{-1}(z_0)\Psi_+^0.$$

(40.27)

For the solutions $\Psi_\pm$ one considers infalling boundary conditions at the horizon, $\Psi \sim (1 - z)^{-i\tilde{\omega}-1/4}$, as appropriate for the retarded propagator. Then, by the Son-Starinets prescription, the retarded propagator is (up to a normalization)

$$G^R(\omega, k) \sim F_-(z_0)F_+^{-1}(z_0).$$

(40.28)

From a complicated numerical analysis, one finds expected properties of the Fermi liquid from the spectral function $A(\omega, k)$:

- There is a quasiparticle peak for $A(\omega, k)$ as we approach the Fermi momentum $k \to k_F$, when $A(\omega, k) \propto \delta(\omega - E_F)$. This defines the Fermi surface.
- The spectral function should vanish identically near the peak, at the Fermi energy but away from it for nonzero temperature, i.e. $A(\omega = E_F(T), k) = 0$ independent of k. Here $E_F(T)$ is slightly shifted at $T \neq 0$ from the peak.
- The dispersion relation near the Fermi energy is with $v_F < c = 1$. One finds the maximum frequency $\omega_{\rm max}$ as a function of k, and the plot has a cone shape. One defines the Fermi velocity

$$v_F = \lim_{\omega \to E_F, k \to k_F} \frac{\omega - E_F}{k - k_F},$$

(40.29)

and finds that it is less than one.
- One also finds that the width of the quasiparticle peak grows quadratically with temperature (which is related to the resistivity $\rho \propto T^2$ of Landau's Fermi liquid theory for metals).

There are, however, also some observables, in particular the behavior of the quasiparticle pole as $k_F/T \to 0$, which behave in an unusual manner, more consistent with quantum critical systems like high T_c superconductors. We will not comment further on the possible explanations for it.

More troublesome, it has been argued by Ogawa, Takayanagi, and Ugajin in [132] that a holographic calculation of the entanglement entropy, which will be discussed in Chapter 46, gives a result generically (for any *classical* gravity dual) inconsistent with the Fermi liquid, casting doubt on the interpretation in this section, as well as in the next one, as approximations to a true holographic dual of Fermi liquids. In particular, it is argued that in a true classical gravitational description of a Fermi liquid, the number of Fermi surfaces is of order N^2 (at large N), and a Fermi surface can be defined by the behavior of the entanglement entropy. The description in this section was in terms of a probe approximation with a single Fermi surface, but it becomes unclear if one can modify it to find a true holographic dual.

The construction described in this section can be called "AdS Fermi hair," since "hair" on a black hole would mean long distance charges on a black hole that can be observed from infinity. In this case, it is not quite appropriate, since we simply have a quantum fermion solution in the background of a black hole, and not really a *classical* (backreacted) black hole with Fermi hair. The purpose of the name however, is to distinguish it from the next construction to be analyzed, of the "electron star."

40.3 "Electron Star" Gravity Dual, Its "Neutron Star" Origin, and Interpolation

In the case in the previous section, we considered a single fermion Green's function, as a probe in the charged AdS black hole geometry. An interesting question is: Can we consider a *finite density* for the fermions in AdS space, and what will it correspond to in the boundary field theory?

As we will see, the finite density for the fermions means that the black hole solution actually loses its horizon. The corresponding solution is called the "electron star" gravity dual, by analogy with a "neutron star" gravity dual, which itself is a generalization inside AdS space of the neutron star solution in flat space found in 1939 by Tolman and Oppenheimer and Volkov (TOV). As the name suggests, the neutron star is made up of uncharged fermions (it was devised as a way to describe the astrophysical, and thus electrically neutral, neutron stars), whereas the electron star will describe holographically a field theory at finite chemical potential, which, as we described, requires electrical charge in the gravity dual. Therefore the electron star will be made up of charged fermions. Note that the terms don't refer to gravity duals of neutron or electron stars, but rather gravity duals that have in them a neutron or electron star, and that correspond to some interesting field theory effects on the boundary.

"Neutron Star" Gravity Dual

For completeness, we start with the neutron star gravity dual. Again, we want to describe a system at finite fermion density in the gravity dual, which means that it should correspond to a certain macroscopic number of occupied fermion "states" in the field theory. Since a state corresponds, as we saw for instance in the description of spin chains, to a (gauge-invariant) composite operator (in general, in a conformal field theory, we have a mapping between states and operators), we need to consider the field theory in the presence of "large" operators made up of fermions, with a certain degeneracy.

Consider the single trace fermionic *operator* Ψ in the CFT that is sourced by the fermion ψ *field* in the gravity dual. Ψ has a (conformal) dimension Δ_0, corresponding as Δ_0 becomes large to a mass $m = \Delta_0/L$ (L is the AdS radius) for ψ in the gravity dual. In the large N limit, the single trace operator behaves like a free fermion, since interactions, encoded in $n > 2$-point correlators, are suppressed by inverse powers of N. We can use this "free fermion" state to construct multi-particle fermionic states, out of Ψ's, and derivatives acting on them.

The states (or operators) Φ that we are interested in are products of terms, each with n derivatives acting on Ψ, for each given n existing a number of such terms equal to the binomial coefficient $\binom{n+d-1}{d-1}$. The maximum n is called n_F, and it is clearly associated with a Fermi surface for the (free) fermions Ψ. The total number of Ψ's N and total conformal dimension Δ of Φ is then

$$N = \sum_{n=0}^{n_F} \binom{n+d-1}{d-1}, \quad \Delta = \sum_{n=0}^{n_F}(n+\Delta_0)\binom{n+d-1}{d-1}. \tag{40.30}$$

From them, we can define a total mass M and Fermi energy ϵ_F in AdS space from the total dimension Δ and the dimension Δ_F of the Ψ with the largest number n_F of derivatives on it:

$$M = \frac{\Delta}{L}; \quad \epsilon_F = \frac{\Delta_F}{L} = \frac{n_F + \Delta_0}{L}. \tag{40.31}$$

In the large Δ_0, n_F limit, the sums defining them become integrals written in terms of a density of states $g(\epsilon)$, exactly like for free fermions (which are now *in the bulk*, not on the boundary; they are states added to the gravity dual):

$$N(\epsilon_F) = \int_m^{\epsilon_F} d\epsilon \, g(\epsilon); \quad M(\epsilon_F) = \int_m^{\epsilon_F} d\epsilon \, \epsilon \, g(\epsilon). \tag{40.32}$$

We consider the limit of Δ_0, n_F large (as well as N large), with their ratio fixed, so that ϵ_F/m is fixed.

In the *gravity dual* this gives a hydrodynamic limit with a finite density, when the particles form an ideal Fermi gas. For a gravity dual of the type

$$ds^2 = -A(r)^2 dt^2 + B(r)^2 dr^2 + r^2 d\Omega_{d-1}^2, \tag{40.33}$$

one considers an ideal hydrodynamic energy-momentum tensor,

$$T_{\mu\nu} = (\rho + p)u_\mu u_\nu + pg_{\mu\nu}, \tag{40.34}$$

with a static 4-velocity $u_\mu = (A(r), \vec{0})$, so that $u_\mu u_\nu g^{\mu\nu} = -1$. Conservation of $T_{\mu\nu}$, $\partial^\mu T_{\mu\nu} = 0$, leads to

$$\frac{dp}{dr} + \frac{1}{A}\frac{dA}{dr}(\rho + p) = 0. \tag{40.35}$$

In terms of the number density n, we have the usual relations $d\rho = \mu dn$ and $\rho + p = \mu n$, which imply that (40.35) is solved by

$$\mu(r) = \frac{\epsilon_F}{A(r)}. \tag{40.36}$$

This will drop until $\mu = m$, when we need that ρ and p should drop to zero. This must then happen at $r = R$ defined by

$$A(r = R) = \frac{\epsilon_F}{m}, \tag{40.37}$$

and the "neutron star" will end at $r = R$.

For an asymptotically AdS_{d+1} space, one can write an ansatz for $A(r)$ and $B(r)$ that is inspired by the AdS-Schwarzschild solution of a mass $M(r)$:

$$A^2(r) = e^{2\chi(r)}\left(1 - \frac{2C_d M(r)}{r^{d-2}} + \frac{r^2}{L^2}\right)$$

$$B^2(r) = \left(1 - \frac{2C_d M(r)}{r^{d-2}} + \frac{r^2}{L^2}\right)^{-1}. \tag{40.38}$$

We note that for $\chi = 0$ and $M(r) = M$ we would have the AdS-Schwazschild solution. Here $C_d = 16\pi G_{N,d+1}/((d-1)\Omega_{d-1})$. One can show that then the Einstein equations reduce to

$$M'(r) = \Omega_{d-1}\rho(r)r^{d-1}$$

$$\chi'(r) = \frac{\Omega_{d-1}C_d}{2}(\rho(r) + p(r))rB^2(r), \tag{40.39}$$

which together with (40.35) are the TOV equations in asymptotically AdS space.

We take as boundary conditions $M(0) = 0$, so that $M(r)$ is the contribution of a mass density inside the star. Since $r = R$ is the edge of the star, by the Birkhoff theorem, outside it we must have AdS-Schwarzschild, i.e. we must have

$$\chi(r) = 0, \quad M(r) = M \quad \text{for} \quad r \geq R. \tag{40.40}$$

For small enough M and large enough R, we expect that the solution we find is nonsingular, corresponding to a neutron star, as opposed to a black hole. Indeed, there is an "Oppenheimer-Volkov limit" analogous to the one in flat space, and to the general "Chandrasekhar limit" for the mass, beyond which gravity forces it to collapse and form a black hole.

"Electron Star" Gravity Dual

For the electron star, we introduce also a Maxwell term $\int[-F_{\mu\nu}^2/4]$ in the gravity dual action, as well as a current source J^μ for the Maxwell field, which satisfies

$$\nabla_\mu F^{\mu\nu} = -e^2 J^\nu, \tag{40.41}$$

and we consider a perfect fluid 4-current

$$J^\mu = \sigma u^\mu, \tag{40.42}$$

where σ is a charge density.

One writes a planar ansatz, which therefore would asymptote to the planar AdS-Reissner-Nordstrom (charged black hole) solution:

$$ds^2 = L^2 \left(-F(\tilde{z})^2 dt^2 + G(\tilde{z})^2 d\tilde{z}^2 + \frac{dx^2 + dy^2}{\tilde{z}^2} \right)$$

$$A = \frac{eL}{\kappa_{N,4}} H(\tilde{z}) dt. \tag{40.43}$$

In a "locally flat approximation," we can use the flat space formulae for the energy density ρ of the fermion fluid, and now for the charge density σ, which is just the particle density n times e,

$$\rho = \int_m^\mu d\epsilon \, \epsilon \, g(\epsilon); \quad \sigma = e \int_m^\mu d\epsilon \, g(\epsilon), \tag{40.44}$$

satisfying $\rho + p = \mu\sigma/e$. Here

$$g(\epsilon) = \beta\epsilon\sqrt{\epsilon^2 - m^2}, \tag{40.45}$$

with β of order one, is the density of states for the relativistic fermions.

We can also define a local chemical potential in analogy with what we would have at infinity (where by the AdS/CFT map we have $A \to \mu dt$), namely

$$\mu(\tilde{z}) = \frac{A_t}{LF(\tilde{z})} = \frac{eH(\tilde{z})}{\kappa_{N,4}F(\tilde{z})}, \tag{40.46}$$

and use it as the upper limit of integration in ρ and σ.

This approximation of locally defined thermodynamic quantities for the fermions, in the TOV description, in condensed matter physics would correspond to the Thomas-Fermi approximation, described in Part I.

In the interior of the electron star, which is at $\tilde{z} \to \infty$ ($r \sim 1/\tilde{z} \to 0$ in the neutron star), one finds, numerically solving the Einstein equations, the equivalent of the TOV equations above, Lifshitz scaling

$$F(\tilde{z}) = \frac{1}{\tilde{z}^z}(1 + F_1\tilde{z}^\alpha + \cdots), \quad G(\tilde{z}) = \frac{G_\infty}{\tilde{z}}(1 + G_1\tilde{z}^\alpha + \cdots),$$

$$H(\tilde{z}) = \frac{H_\infty}{\tilde{z}^z}(1 + H_1\tilde{z}^\alpha + \cdots). \tag{40.47}$$

For given parameters m, β, one can solve from the equations of motion for all the coefficients in the above, including for the Lifshitz exponent z, which for instance at $\beta \to \infty$ is

$$z \to \frac{1}{1 - \frac{m^2\kappa_{N,4}^2}{e^2}}. \tag{40.48}$$

This Lifshitz scaling is evidence of strong coupling effects in the field theory that do not follow the Fermi liquid picture, but rather behave like a strange metal.

As in the case of the neutron star, we can determine the radius $z = Z$ where the electron star ends by imposing that $\mu(z) = m$ there, so that $g(\mu) = 0$, and thus $\rho = \sigma = p = 0$

there, from $g(\epsilon) \propto \sqrt{\epsilon^2 - m^2}$ and $\rho, \sigma \sim \int^{\mu(z)} d\epsilon\, g(\epsilon) \times \cdots$. The condition becomes

$$\frac{h(Z)}{F(Z)} = \frac{m\kappa_{N,4}}{e}. \tag{40.49}$$

Note that the boundary of space is $z \to 0$, where $z \to \infty$ is the center of the electron star. Outside the electron star, the solution must become the planar AdS-Reissner-Nordstrom, so like in the neutron star case, we can anticipate that by writing the solution in general as

$$F^2(z) = \frac{c^2(z)}{z^2} - M(z)z + \frac{z^2 Q^2(z)}{2} = c^2(z)\left(\frac{1}{z^2} - \tilde{M}(z)z + \frac{z^2 \tilde{Q}^2(z)}{2}\right)$$

$$G^2(z) = \frac{c^2(z)}{z^4 F^2(z)} = z^{-4}\left(\frac{1}{z^2} - \tilde{M}(z)z + \frac{z^2 \tilde{Q}^2(z)}{2}\right)^{-1}$$

$$H(z) = \mu - zQ(z) = c(\tilde{\mu} - z\tilde{Q}(z)). \tag{40.50}$$

Then, outside the electron star, for $z < Z$, we must have $M(z) = M$, $Q(z) = Q$, and $c(z) = c$ (that can be fixed to 1), but inside it varies, and we can choose $Q(\infty) = M(\infty) = 0$.

Like in the case of the astrophysical neutron stars, there is a limit beyond which the star never forms. However, in this case the limit is found (and numerically confirmed) to depend on the fermion mass m instead, and be at

$$\frac{m\kappa_{N,4}}{e} = 1, \tag{40.51}$$

beyond which we have no star formed. But, since the horizon is planar, the interpretation is more complicated.

If we compactify the Euclidean time, thus putting the theory at finite temperature, we need for regularity of the solution that there is a horizon *inside* the electron star, generating a temperature, so really we have a solution like a black hole surrounded by fermions. For temperatures smaller than a critical temperature $T_c(m)$, we have an electron star, whereas for higher temperatures the star collapses to a usual AdS-Reissner-Nordstrom one, with no fermions. As $m \to e/\kappa_{N,4}$, $T_c(m) \to 0$.

After a complicated analysis, one can compute the electrical conductivity σ of the field theory from this gravity dual, and find that at small frequencies we have

$$\text{Re}(\sigma(\omega)) \propto \delta(\omega) + \omega^2. \tag{40.52}$$

Interpolation to AdS Fermi Hair

One can compute the spectral function in the electron star, the same way as it was done in the last section in the unperturbed solution. One finds that the number of bound states present are a good sign of adiabaticity: for small number, we are close to the AdS Fermi hair description, whereas for large (strictly infinite) number, we are in the electron star description. A good parameter for the interpolation is the "effective charge" $q_{\text{eff}} = eL/\kappa_{N,4}$.

We still need to remember the caveat of the entanglement entropy mentioned in the previous subsection, and therefore we have to consider the descriptions of Fermi liquid available until now as tentative only. In fact, as we saw, the AdS Fermi hair was a probe approximation, whereas the electron star obtained only a non-Fermi liquid. It is unclear if a true gravitational dual of a Fermi liquid exists or not.

Important Concepts to Remember

- Large N matrices in a harmonic potential behave like free fermions in a harmonic potential, with a Fermi surface.
- Gauge-invariant ground states (BPS) in $\mathcal{N} = 4$ SYM on $S^3 \times \mathbb{R}_t$ form free fermions in a harmonic potential, giving droplets in momentum space bounded by a Fermi surface.
- The gravity dual description of the BPS states with droplets in momentum space is in terms of the solution of a differential equation, with boundary conditions on a 2-dimensional plane, where a discrete variable $z = \pm 1/2$ corresponds to particles vs. holes.
- One can describe a Fermi liquid (a system of strongly coupled electrons that behave like a renormalized Fermi gas) by a probe (nonbackreacted) fermion in an AdS-RN black hole (with charge corresponding to chemical potential on the boundary), called AdS Fermi (or Dirac) hair.
- One analyzes the Fermi surfaces through the spectral function, which can be found in the gravity dual through a calculation of the retarded propagator using the Son-Starinets prescription.
- One finds properties of Fermi liquid for AdS fermion masses close to the unitarity limit $mL = -1/2 + \delta$, though there are some possible discrepancies.
- A "neutron star" gravity dual is a gravity state with a finite fermion density, corresponding to some large operators with large degeneracies in the field theory. It satisfies TOV equations and has no horizon for a large enough radius and small enough mass.
- An "electron star" gravity dual has also charge density for the fermions, and at zero temperature interpolates between a solution with Lifshitz scaling in the IR (in the center of the star) and an AdS-RN solution at the boundary.
- Compactifying Euclidean time gives finite temperature solutions with horizons that exist only for low enough fermion masses m and for temperatures smaller than a critical temperature $T_c(m)$.
- One can interpolate between the electron star and AdS Fermi hair solutions, by tuning an effective charge parameter, that increases the number of fermionic states between one and infinity.

Further Reading

The free Fermi gas construction in AdS/CFT presented here was done by Lin, Lunin, and Maldacena in [104]. The Fermi liquid by AdS Fermi hair construction was developed by several people, but one of the more important early papers that I have followed here is [105]. The electron star construction was developed in [106] based on earlier works, one of them being the "neutron star" of [107], and the interpolation between electron stars and AdS Fermi hair was done in [108].

Exercises

(1) Calculate the equation of motion for $F_{(5)}$ on the ansatz in (40.8) and (40.9).

(2) Check explicitly that z in (40.12) and V_i in (40.13) satisfy (40.10) and

$$y\partial_y V_i = \epsilon_{ij}\partial_j z; \quad y(\partial_i V_j - \partial_j V_i) = \epsilon_{ij}\partial_y z. \tag{40.53}$$

(3) Prove that the on-shell action for the fermionic solution in (40.26) is (40.27).

(4) Show that the Einstein equations in the "neutron star" background reduce to (40.39).

(5) Calculate ρ and σ at a large z in the interior of the electron star, using $F(z)$ in (40.47).

41

Quantum Hall Effect from String Theory

As we saw in Part I, the Quantum Hall Effect is one of the most challenging in condensed matter, with many interesting physical effects and theories. As such, there have been many attempts to describe them in string theory, in general, and also holographically. In this chapter, I will attempt to describe only some of them, and some general issues.

As we explained when talking about anyons (in Chapter 12), part of the relevant physics of the fractional quantum Hall effect is encoded in a Chern-Simons action of the type $\sim \int \epsilon^{\cdots}(Ada + ada)$. It is therefore to be expected that if we obtain a similar Chern-Simons action from a string theory construction, then we have a good description of the quantum Hall system. A way to obtain it suggests itself since the D-brane action has a WZ term in it that contains couplings of the type $\int \epsilon^{\cdots} C_n \wedge F \cdots \wedge F$, where $F = dA$ lives on the D-brane, and C_n is part of the background. By considering integer fluxes for the background fields C_n ($F_{n+1} = dC_n \sim N\epsilon_{\cdots}$), which can happen only on compact spaces like spheres, we can obtain, by KK reduction on the compact space, a Chern-Simons term.

In the discussion of the Chern-Simons action for FQHE we will realize that a classification of topological insulators and superconductors, based on the Chern-Simons terms, can be obtained from D-branes. Also we will see that the bulk-edge correspondence for the FQHE can be embedded in AdS/CFT as the holographic duality.

One thing that we will not study in this chapter is a description of anyonic and nonabelian states in the Fractional Quantum Hall Effect. That will be touched on in Chapter 44, when describing non-standard statistics from AdS/CFT.

41.1 Quantum Hall Effect and Chern-Simons Theory in String Theory

Chern-Simons and Noncommutative Chern-Simons for FQHE

A simple explanation for the relevance to the FQHE of not only the usual Chern-Simons theory, but moreover of the noncommutative version for it was given by Susskind.

Consider a fluid (the quantum Hall fluid) in 2+1 dimensions described by *comoving* (intrinsic) coordinates y_1, y_2, i.e. moving with the fluid, and coordinates x_1, x_2 in the physical (Euclidean) space, so that the motion is described by $x_i(y, t)$. Consider also a constant particle density in comoving coordinates, ρ_0. Then the system is invariant under area-preserving diffeomorphisms of y_i:

$$\delta y_i = f_i(y) \quad \text{with} \quad f_i = \epsilon_{ij}\frac{\partial \Lambda(y)}{\partial y_j}. \tag{41.1}$$

The density in physical space is

$$\rho = \rho_0 \left| \frac{\partial y}{\partial x} \right|, \tag{41.2}$$

and we consider a potential that is a function of it, $V = V(\rho)$, which has a minimum at ρ_0. Then the static equilibrium configuration is $x_i = y_i$, and a small perturbation around it is defined by a vector A_i, as

$$x_i = y_i + \epsilon_{ij} \frac{a_j}{2\pi \rho_0}. \tag{41.3}$$

This might seem arbitrary, but it is chosen to be perpendicular to the y_i such that we get an x-space density of

$$\rho = \rho_0 - \frac{1}{2\pi} \vec{\nabla} \times \vec{a} \equiv \rho_0 - \frac{b}{2\pi}, \tag{41.4}$$

just like the density of a Landau level is

$$\rho_B = \frac{e}{m} B, \tag{41.5}$$

and the density equals the filling fraction times the above, $\rho_e = \nu \rho_B$. But more precisely, the reason is that a vortex, which in the FQHE is a Laughlin quasiparticle of charge q with respect to the emergent gauge field a_μ, must have a magnetic flux for a of q times the delta function, so

$$\vec{\nabla} \times \vec{a} = 2\pi \rho_0 q \delta^2(y), \tag{41.6}$$

which modifies the density at the vortex positions by $\rho_0 q$.

Then the area-preserving diffeomorphisms become

$$\delta a_i = 2\pi \rho_0 \frac{\partial \Lambda}{\partial y_i} + \epsilon_{lm} \frac{\partial a_i}{\partial y_l} \frac{\partial \Lambda}{\partial y_m}. \tag{41.7}$$

A particle in a magnetic field has force $\vec{F} = e\vec{v} \times \vec{B}$, which in two dimensions is $F_i = \epsilon_{ij} e v_j B$, and it should equal $\nabla^i V = \delta \mathcal{L}/\delta x^i$, so the point particle Lagrangean gets the extra term

$$\frac{eB}{2} \epsilon_{ab} \dot{x}_a x_b, \tag{41.8}$$

which on the fluid gives

$$L = \frac{eB}{2} \int \rho_0 d^2 y \epsilon_{ab} \dot{x}_a x_b = \frac{eB}{8\pi^2 \rho_0} \int d^2 y \, \epsilon_{ab} \dot{a}_a a_b. \tag{41.9}$$

We also have the constraint that the density of the fluid at a fixed point in comoving coordinates, y, is time independent (conserved), and so can be set to one:

$$\frac{\rho_0}{\rho} = \left| \frac{\partial x}{\partial y} \right| = \frac{1}{2} \epsilon_{ij} \epsilon_{ab} \frac{\partial x_a}{\partial y_i} \frac{\partial x_b}{\partial y_j} = 1. \tag{41.10}$$

This constraint, which in fact is a Gauss-type constraint, can be added to the Lagrangean with a Lagrange multiplier a_0, leading to the Lagrangean

$$L = \frac{eB\rho_0}{2}\epsilon_{ab}\int d^2y\left[\left(\dot{x}_a - \frac{1}{2\pi\rho_0}\{x_a, a_0\}\right)x_b + \frac{\epsilon_{ab}}{2\pi\rho_0}a_0\right], \qquad (41.11)$$

where we have used the Poisson bracket

$$\{F(y)G(y)\} = \epsilon_{ij}\partial_i F \partial_j G. \qquad (41.12)$$

Substituting $x_i = y_i + \epsilon_{ij}a_j/(2\pi\rho_0)$ and writing together $(a_i, a_0) \equiv a_\mu$, we obtain the relativistic Lagrangean

$$L = \frac{1}{4\pi\nu}\epsilon_{\mu\nu\rho}\left[\frac{\partial a_\mu}{\partial y_\rho} - \frac{\theta}{3}\{a_\mu, a_\rho\}\right]a_\nu. \qquad (41.13)$$

Here the filling fraction ν appears since as we saw in Part I, we have

$$\nu = \frac{n_e m}{eB} = \frac{2\pi\rho_0}{eB}, \qquad (41.14)$$

and we have defined

$$\theta = \frac{1}{2\pi\rho_0}. \qquad (41.15)$$

The area-preserving diffeomorphisms (41.7) become now

$$\delta a_a = \frac{\partial\lambda}{\partial y_a} + \theta\{a_a, \lambda\}. \qquad (41.16)$$

In fact, the theory with this Lagrangean and gauge invariance is the first order expansion of a *noncommutative Chern-Simons gauge theory*, with Lagrangean

$$L_{\text{NC}} = \frac{1}{4\pi\nu}\epsilon_{\mu\nu\rho}\left(\hat{A}_\mu * \partial_\nu\hat{A}_\rho + \frac{2i}{3}\hat{A}_\mu * \hat{A}_\nu * \hat{A}_\rho\right), \qquad (41.17)$$

where the star product is the standard Moyal noncommutative product:

$$f * g = e^{i\theta_{ij}\partial_i\partial_j'}f(x)g(x')\Big|_{x=x'}. \qquad (41.18)$$

In this case, $\theta_{12} = \theta$.

Another way to obtain the noncommutative Chern-Simons Lagrangian will be more useful in terms of deriving it from string theory. One needs to replace the classical configuration space of K electrons with positions x_a forming the fluid to a space of $K \times K$ Hermitian matrices X_a, just like in the case of D-branes. Then the Lagrangian (41.11) generalizes to

$$L = \frac{eB}{2}\epsilon_{ab}\,\text{Tr}\left(\dot{X}_a - i[X_a, \hat{A}_0]\right)X_b + eB\theta\hat{A}_0. \qquad (41.19)$$

The hat on A_0 is the notation used for noncommutative fields. Indeed, varying with respect to $\hat{A}_0$, we obtain the noncommutativity constraint

$$[X_a, X_b] = i\theta\epsilon_{ab}. \qquad (41.20)$$

This constraint can be solved only in terms of infinite matrices (there is no finite matrix representation of noncommutative spaces).

Note that, once the constraint is taken into account, the Lagrangian (41.19) becomes simply $(eB/2)\epsilon_{ab}\,\mathrm{Tr}[\dot{X}_a X_b]$, which is a Chern-Simons Lagrangian.

Now replacing as before (just that now this is a matrix relation) $X_a = y_a + \epsilon_{ab}\theta\hat{A}_b$, where y_a are fixed matrices solving the noncommutative constraint (or otherwise representing $y_2 = -i\theta\partial/\partial y_1$), we obtain the noncommutative CS Lagrangean (41.17).

String Theory Realizations

As we mentioned, Chern-Simons gauge theories are relatively easy to obtain on the world-volume of D-branes. Moreover, noncommutative gauge theories were obtained in string theory by Seiberg and Witten in the presence of some background fields.

Consider an M5-brane, an object in 11-dimensional M-theory with a 5+1–dimensional worldvolume, wrapped on the compact lightlike direction x^-, and extending also in x^+, X^1, X^2, X^3, X^4. In M-theory, *on the M5-brane* we have a field $\tilde{B}_{\mu\nu}$ with (six-dimensional) self-dual field $H_{\mu\nu\rho}$. On the lightlike compactification, we consider

$$H_{+12} = H_{+34} = H \neq 0. \qquad (41.21)$$

Note that $\eta^{+-} = 1$ implies that the self-duality condition $H_{\mu\nu\rho} = (1/3!)\epsilon_{\mu\nu\rho\tau\epsilon}H^{\sigma\tau\epsilon}$ becomes $H_{+12} = H_{+34}$. Then one can show that carriers of N units of momentum on x^-, $P_- = N/R$, "blow up" (or "puff up") into membranes that satisfy the condition

$$[X^1, X^2] = i\theta, \qquad (41.22)$$

if we take $H \to \infty$ and the 11-dimensional Planck mass $M_{\mathrm{Pl}} \to \infty$, with the fixed ratio

$$\theta = \frac{H}{RM_{\mathrm{Pl}}^6}. \qquad (41.23)$$

Moreover, the membranes naturally reduce to D2-branes in 10-dimensional type IIA string theory under KK reduction on x^-, and the M5-branes reduce to D4-branes. The D2-branes will have a WZ Chern-Simons term on their worldvolume:

$$L_{2,\mathrm{CS}} = kA \wedge F. \qquad (41.24)$$

Here A is a gauge field living on the D2-brane worldvolume. In the presence of the H field above, we have a noncommutativity θ, so the CS Lagrangian becomes noncommutative.

The construction above is due to Susskind and Hellerman, but one can consider related constructions. A D2-brane of 10-dimensional type IIA string theory will have in general the WZ Chern-Simons term above, regardless of its origin (above it appeared from a dimensional reduction of an 11-dimensional momentum mode puffed up into a membrane). And for noncommutativity, there is a general way to obtain it on D-branes in string theory (so in particular it can be found on D2-branes) that has been defined by Seiberg and Witten.

Seiberg and Witten have found that if we consider a theory of closed strings in a 10-dimensional spacetime interacting with open strings living on D-branes, in the background of a *constant* $B_{\mu\nu}$ field in the directions $i, j = 1, \ldots, r$, then there is a relation between the variables felt by the closed string, or *closed string variables* g_{ij}, B_{ij} and string coupling g_s and the variables felt by the open string, or *open string variables* G_{ij}, θ^{ij} and string coupling G_s, given by

$$\left(G + \frac{\theta}{2\pi\alpha'}\right)^{ij} = \left(\frac{1}{g + 2\pi\alpha'B}\right)^{ij}, \qquad (41.25)$$

where G^{ij} is the symmetric part of the right-hand side, and $\theta^{ij}/(2\pi\alpha')$ is the antisymmetric part, as well as the relation

$$G_s = g_s \sqrt{\frac{\det(g + 2\pi\alpha'B)}{\det g}}. \qquad (41.26)$$

Then G_{ij} is found as the inverse of G^{ij}. Explicitly, one finds

$$G_{ij} = g_{ij} - (2\pi\alpha')^2 (B \cdot g^{-1} \cdot B)_{ij}$$

$$\theta^{ij} = -(2\pi\alpha')^2 \left(\frac{1}{g + 2\pi\alpha'B} \cdot B \cdot \frac{1}{g - 2\pi\alpha'B}\right)^{ij}. \qquad (41.27)$$

Consider next the $\alpha' \to 0$ limit, with

$$\alpha' \sim \sqrt{\epsilon} \to 0; \quad g_{ij} \sim \epsilon \to 0. \qquad (41.28)$$

In this limit, we obtain the finite quantities

$$G^{ij} = -\frac{1}{(2\pi\alpha')^2} \frac{1}{B} \cdot g \cdot \frac{1}{B}$$

$$G_{ij} = -(2\pi\alpha')^2 (B \cdot g^{-1} \cdot B)_{ij}$$

$$\theta^{ij} = \left(\frac{1}{B}\right)^{ij}$$

$$G_s = g_s \sqrt{\det'(2\pi\alpha' B \cdot g^{-1})}, \qquad (41.29)$$

for the $i, j = 1, \ldots, r$ directions and trivial ($G^{ij} = g^{ij}$, $G_{ij} = g_{ij}$, $\theta^{ij} = 0$) in the others. Here $\det' = \det_{r \times r}$. The DBI action is written in the usual way in terms of G_{ij} and G_s, but the gauge fields $\hat{F}_{ij}$ become noncommutative with noncommutativity θ^{ij}, i.e. $[x^i, x^j] = i\theta^{ij}$. The hats on $\hat{F}$ are a reminder that one should use the Moyal product. There is in fact a nonlinear field redefinition (Seiberg-Witten map) that takes the noncommutative fields $\hat{F}_{ij}$ to usual commutative fields F_{ij}, but the action becomes complicated under it.

41.1.1 Topological Insulator and Superconductor Classification from D-Branes

Topological insulators and topological superconductors have been classified (in the case of effective noninteracting fermions) according to K theory, a category theory concept that classifies possible charges and the associated topological field theories. But in string theory it is known that K theory classifies also possible D-brane charges. This identification was used by Ryu and Takayanagi to realize the topological insulator and superconductor classes in string theory, on D-branes, and reobtain the classification in string theory.

As we have seen in Part I, the topological insulators and topological superconductors can both be described by some topological field theory related to Chern-Simons (either

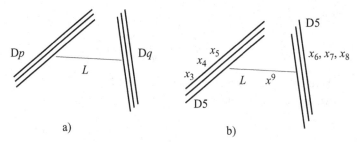

Fig. 41.1 Brane configuration: (a) Dp-branes separated from Dq branes in an overall transverse direction, by a length L. (b) Specifically, for $p = q = 5$, the two sets of D5-branes, one extending in x_3, x_4, x_5, and another in x_6, x_7, x_8, are separated in x^9 by a length L.

directly, or by dimensional reduction). Since the WZ term of the D-brane action contains such Chern-Simons terms, the relation is not surprising.

The basic setup consists of two D-branes of different dimensionalities, and oriented in different directions, i.e. a Dp-brane and a Dq-brane, separated by a finite distance L in an overall transverse direction, as in Figure 41.1(a). One also needs to add in many cases a string theory object called an orientifold plane, or O-plane, that like a D-brane is extended in p spatial directions, but is not dynamical. It creates "mirror images" of the D-branes, with some identification of the string theory (more precisely, it is a combination of an identification of spacetime coordinates, $X \sim -X$, and an inversion of the string worldsheet coordinate σ, i.e. $\sigma \to 2\pi - \sigma$). For simplicity we will ignore these Op-planes in our analysis.

A concrete example (corresponding to the simplest class in the classification, the one described by the 2+1–dimensional abelian topological Chern-Simons theory) that we will consider is a D5-brane in type IIB theory ($p = 5$) in the directions 0, 1, 2 and 3, 4, 5, and another D5-brane ($q = 5$) in the directions 0, 1, 2 and 6, 7, 8, and separated by a distance L in the overall transverse x^9 direction; see Figure 41.1(b). Note that if we make two T-dualities in relative transverse directions, say, directions 4 and 5, the system changes to a D3–D7 system with direction 3 transverse to D7 and parallel to D3. The number of Neumann-Dirichlet direction, i.e. directions that are transverse to one brane (Dirichlet boundary conditions), but parallel to the other (Neumann boundary conditions) is conserved by T-duality, which exchanges Neumann with Dirichlet. Thus no matter the T-dualities, the system we consider will have six ND (Neumann-Dirichlet) directions. The (effective) theory that we are after is in the common 2+1–dimensions of the two D5-branes, x^0, x^1, x^2. Since open strings situated between the two D-branes have a length of (at least) L, their states have masses of $m = \frac{1}{2\pi\alpha'}L$, and their endpoints are (fundamentally) charged under the two gauge groups of the Dp and Dq-branes, i.e. in this case $U(1) \times U(1)$ (gauge fields A_μ and $\tilde{A}_\mu$), with charges $(+1, 0)$ and $(0, +1)$. These states correspond in the 2+1–dimensional theory to massive scalars X^9 and the massive fermions in the same supersymmetry multiplet. In fact, in these constructions, supersymmetry is broken, and actually the scalars have larger masses than the fermions. That means that at low enough energies, we have only massive fermions to consider. Integrating out these charged massive fermions generates a Chern-Simons term for the gauge fields (this is a general property of Chern-Simons-massive fermion theories),

$$\frac{k}{4\pi} \int A \wedge dA \quad \text{and} \quad \frac{k}{4\pi} \int \tilde{A} \wedge d\tilde{A}, \tag{41.30}$$

but it can also be thought of alternatively as follows. Integrating out the massive fermion corresponds from the point of view of the $D(p = 5)$–brane as integrating out the effect of the $D(q = 5)$–brane. But a D5-brane *supergravity solution* (remember that the D-brane = p-brane equality identifies the D-brane with a supergravity solution) is magnetically charged under a $C_2 = C_{\mu\nu}$ RR antisymmetric field, so that $*dC_2 = dC_6$, with C_6 coupling to the 5+1 dimensional worldvolume. Then the $D(q = 5)$–brane generates a C_2 field at the position of the $D(p = 5)$–brane, and hence a nontrivial Chern-Simons (WZ) term,

$$S = \int_{Dp} F \wedge F \wedge C_2 = \int_{\mathcal{M}_3} (dC_2) \int_{\mathbb{R}^{2,1}} A \wedge F \sim k \int_{\mathbb{R}^{2,1}} A \wedge F, \tag{41.31}$$

which gives the required topological field theory in x^0, x^1, x^2.

   Thus we have found that there are no dynamical degrees of freedom in this construction, only topological ones. But in the case of topological insulators and superconductors, there are boundary (edge) states that are massless, in particular fermionic states. We can easily construct a theory with a boundary with massless fermions from a modification of the above setup. The branes were separated by a constant separation L in the x^9 direction, constant as a function of, say, x^2 (one of the coordinates of our effective theory), i.e. parallel in x^2. But consider instead a case where one brane bends at 90° in the (x^2, x^9) plane, i.e. from parallel moves to perpendicular to the other brane, touching it at a given $x^2 = x_0^2$ position. That means that the 2+1–dimensional effective field theory has a boundary at x_0^2. On this boundary lives a massless fermion, the fermion that was massive before, with mass proportional to the distance L. Away from x_0^2 it is still massive, thus can be integrated out, but at x_0^2 it is massless, giving our edge states.

   One can similarly analyze all cases stemming from Dp-Dq-$O(r)$–systems, and find the classification of K theory, matching the one of topological insulators and topological superconductors.

   To obtain a holographic dual description, in principle we would need to consider a large number of branes of a type, thus curving the spacetime and creating a background.

41.2 Quantum Hall Effect and Chern-Simons Models from Holography

We have seen that the description of the quantum Hall effect necessitates a Chern-Simons term, and such a term can be obtained if we have a Dp-brane wrapped on some compact space with flux, KK reduced down to a ("fractional") D2-brane. We have also seen in the case of topological insulators and superconductors that a Dp-Dq system with six ND (Neumann-Dirichlet) coordinates would obtain such a CS term.

   Holographic constructions of the Fractional Quantum Hall Effect are generically variations of this string theory setup, with one type (or maybe an additional type) of branes being a large number, thus being able to create a gravity dual. The FQHE action that one wants to obtain, for a filling fraction $\nu = 1/k$, is of the type

$$S_{\text{FQHE}} = \frac{k}{4\pi} \int a \wedge da + \frac{1}{2\pi} \int a \wedge F_{\text{ext}}, \qquad (41.32)$$

as we have seen in Part I, where a is an emergent gauge field and $F_{\text{ext}} = dA_{\text{ext}}$ is an external electromagnetic field.

ABJM with "Fractional D2-Branes" = Wrapped D4-Branes, and D6-Branes

One model in which we can obtain this is as follows. Note however that the model will contain more fields than the above, so FQHE is only embedded into it. We would like to obtain a model of the D4-D6 type, with six ND directions, i.e. having a common D2-brane, and the other two and four directions be different. Consider the standard toy model for 2+1–dimensional condensed matter, the ABJM model. We have mentioned it in Chapter 27, and it will be described in more detail in the next chapter. The model comes from a limit of N M2-branes in 11–dimensional M-theory, which would dimensionally reduce to a D2-brane in 10–dimensional type IIA string theory. The gravity dual is in type IIA string theory, with background $AdS_4 \times \mathbb{CP}^3$. The Fubini-Study metric on $\mathbb{CP}^3$ is

$$\begin{aligned}
ds^2 &= d\xi^2 + \frac{\cos^2 \xi}{4}(d\theta_1^2 + \sin^2 \theta_1 d\phi_1^2) + \frac{\sin^2 \xi}{4}(d\theta_2^2 + \sin^2 \theta_2 d\phi_2^2) \\
&\quad + \cos^2 \xi \sin^2 \xi (d\psi + A_1 - A_2), \\
A_i &= \frac{1}{2} \cos \theta_i d\phi_i.
\end{aligned} \qquad (41.33)$$

Consider "*fractional D2-branes*" inside this geometry, i.e. D4-branes wrapped on an $\mathbb{CP}^1 = S^2$ space in it, which we can choose, for instance, to be

$$ds^2 = d\theta_1^2 + \sin^2 \theta_1 d\phi_1^2, \qquad (41.34)$$

and the other directions are the field theory directions. The D4-branes have a WZ term in the action that includes the contribution

$$2\pi^2 T_4 \int C_1 \wedge F \wedge F, \qquad (41.35)$$

where $T_4 = 1/(2\pi)^4$ (for $\alpha' = 1$) is the D4-brane tension and C_1 is the RR one-form potential. Such a D-brane will "melt" into the gravity dual, and correspondingly modify the ABJM field theory. We are not interested in describing the modification, since all we are interested in is a $U(1)$ subgroup of the gauge theory on the D4-branes.

But we note that a flux of k units of $F_2 = dC_1$ on $\mathbb{CP}^1$ would give the correct CS term $k \int a \wedge da$ for the FQHE. Such a flux is obtained by also adding D6-branes as "flavor" branes to the theory, i.e. inducing extra fundamental ("flavor" in particle physics language) fields in the ABJM theory. Such D6-branes would be stable if added to the plane

$$\theta_1 = \theta_2 = \frac{\pi}{2}; \quad \psi = \pi. \qquad (41.36)$$

This is a fixed plane for the $\mathbb{Z}_2$ action

$$\phi_1 \to \phi_1 + \pi; \quad \phi_2 \to \phi_2 + \pi; \quad \psi \to \psi + \pi, \qquad (41.37)$$

defining an an $S^3/\mathbb{Z}_2$ space inside $\mathbb{CP}^3$. The induced metric on the space is

$$ds^2 = d\xi^2 + \frac{\cos^2 \xi}{4} d\phi_1^2 + \frac{\sin^2 \xi}{4} d\phi_2^2, \tag{41.38}$$

with coordinate ranges $0 < \xi < \pi/2$ and $0 \le \phi_i \le 2\pi$. One considers D6-branes wrapped on it, and extending in all four AdS_4 directions (including the radial direction, unlike the D4-branes!). Note that considering only D6-branes is an approximation, since then the RR flux lines, transverse to the D6-branes, on the compact space $\mathbb{CP}^3$, do not have a corresponding negative "sink." Such a sink would be an "orientifold" O6-plane, so a correct construction actually involves one O6-plane and four D6-brane, for a vanishing total charge. But we are interested in separating off one D6-brane, such that we have a nonzero flux for $F_2 = dC_1$ transverse to it (the D6-brane couples magnetically to $F_2 = *F_8 = *dC_7$ Ramond-Ramond flux). Note that the system is now of the type we advertized, with six ND directions: the D2-branes are common to the D4-brane and the D6-brane, and the D6 extends in the $S^3/\mathbb{Z}_2$ space and the radial AdS_4 direction, whereas the D4-brane extends in the transverse $\mathbb{CP}^1 = S^2$ directions.

We can actually calculate F_2 and moreover its integral on $\mathbb{CP}^1$, and we find

$$\int_{\mathbb{CP}^1} F_2 = 2\pi k. \tag{41.39}$$

That means that the D4-brane (fractional D2-brane) WZ term in the action becomes

$$\frac{1}{8\pi^2} \int_{\mathbb{CP}^1} F_2 \wedge \int_{\mathbb{R}^{2,1}} A \wedge dA = \frac{k}{4\pi} \int_{\mathbb{R}^{2,1}} A \wedge dA, \tag{41.40}$$

i.e., the required first term in the FQHE action.

To obtain also the second term, we can add a background RR-flux. Since the WZ term of the D4-brane includes a contribution $\int C_3 \wedge F$, we can choose the value

$$C_3 = \frac{4\pi k}{R^3} A_{\text{ext}} \wedge \omega, \tag{41.41}$$

where ω is the same volume form for $\mathbb{CP}^1$ appearing in F_2 as $F_2 = -2k^2\omega/R^3$, and A_{ext} is a 1-form inside AdS_4 that is our wanted external gauge field. Then the WZ term gives

$$\frac{1}{(2\pi)^4} \int_{\mathbb{R}^{2,1} \times \mathbb{CP}^1} 2\pi F \wedge C_3 = \frac{1}{2\pi} \int_{\mathbb{R}^{2,1}} A_{\text{ext}} \wedge F. \tag{41.42}$$

Note that the fractional D2-branes and the above C_3 flux would in principle change the background, but we can neglect this backreaction and treat it in the probe approximation, just like the case of the D6-branes.

D3-D7-Brane Model

We now present a D3-D7-brane model, also with six ND directions. We consider N D3-branes, for large N, creating the $AdS_5 \times S^5$ gravity dual of $\mathcal{N} = 4$ SYM. The D3-branes would be in directions x^0, x^1, x^2, x^3.

The first possibility would be to consider $N_7 \ll N$ D7-branes in directions x^0, x^1, x^2 and $x^4 - x^9$, which can be interpreted as probes, so they would wrap an $S^4 \subset S^5$ and an

$AdS_4 \subset AdS_5$. However, it is found that this is an unstable configuration, and the D7-branes can "slip off" the S^5.

Another related possibility is for the $N_7 \ll N$ D7-branes to be in directions x^0, x^1, x^2 and the radial direction r of AdS_4, as well as in an $S^2 \times S^2$ direction. Thus consider the gravity dual

$$ds^2 = \frac{r^2}{R^2}\left(-dx_0^2 + dx_1^2 + dx_2^2 + dx_3^2\right) + \frac{R^2}{r^2}\left(dr^2 + r^2 d\Omega_5^2\right)$$
$$F_5 = 4R^4(r^3 dx^0 \wedge dx^1 \wedge dx^2 \wedge dx^3 \wedge dr + d\Omega_5), \tag{41.43}$$

with the 5-sphere metric decomposed as a fibration of $S^2 \times S^2$ on the interval $\psi \in [0, \pi/2]$:

$$d\Omega_5^2 = d\psi^2 + \cos^2\psi\left[d\Omega_2^{(1)}\right]^2 + \sin^2\psi\left[d\Omega_2^{(2)}\right]^2$$
$$[d\Omega_5^{(i)}]^2 = d\theta_i^2 + \sin^2\theta_i d\phi_i^2. \tag{41.44}$$

Under this decomposition, the 4-form potential C_4 ($F_5 = dC_4$) becomes

$$C_4 = R^4\left(r^4 dx^0 \wedge dx^1 \wedge dx^2 \wedge dx^3 + \frac{1}{2}c(\psi)d\Omega_2^{(1)} \wedge d\Omega_2^{(2)}\right). \tag{41.45}$$

The D7-brane wraps the $\Omega_2^{(1)} \times \Omega_2^{(2)}$ spheres. One finds again that the embedding is unstable to slipping toward a trivial solution, but now it can be stabilized by introducing a background worldvolume magnetic flux on the D7-brane(s),

$$F_2 = \frac{1}{2}\left(n_1 d\Omega_2^{(1)} + n_2 d\Omega_2^{(2)}\right), \tag{41.46}$$

where n_1, n_2 are integers.

As before, the WZ term on the D7-brane contains a coupling

$$-\frac{(2\pi\alpha')^2 T_7}{2}\int C_4 \wedge F \wedge F, \tag{41.47}$$

that, when we integrate C_4 over $\Omega_2^{(1)} \wedge \Omega_2^{(2)}$, gives a CS term form for the gauge field A, written as a boundary term in r ($\int_0^\infty dr \int dx^0 dx^1 dx^2 (F \wedge F)$, with $F \wedge F = d(A \wedge dA)$).

41.2.1 Hierarchical FQHE, Topological Insulators, and Bulk/Edge Correspondence as a Holographic Duality

We can have models for more general FQHEs, specifically in the case that the filling fraction ν is a continued fraction, called hierarchical FQHE. A topological insulator is in fact an example of an FQHE system, since it has no bulk modes, and only surface (edge) modes. As such, there is a certain correspondence between the bulk and the edge of the topological insulator. We will in fact understand this correspondence as the AdS/CFT correspondence, since we will be able to embed the hierarchical FQHE in string theory in the bulk of a gauge/gravity duality only.

The more general, hierarchical FQHE is described by a set of $U(1)$ emergent gauge fields $a^{(i)}$, coupled to an external electromagnetic gauge field, with a matrix coupling K_{ij} between the gauge fields and an *emergent* charge vector $\vec{q} = (q_1, q_2, \ldots)$, i.e. with an action

$$S = \frac{1}{4\pi} \int_\Sigma \left[\sum_{i,j=1}^r K_{ij} a^{(i)} \wedge da^{(j)} + 2 \sum_{i=1}^r q_i a^{(i)} \wedge dA_{\text{ext}} \right]. \tag{41.48}$$

The equations of motion of this action are

$$K_{ij} \partial_\mu a_\nu^{(j)} + q_i \partial_\mu A_{\text{ext},\nu} \Rightarrow \partial_\mu a_\nu^{(i)} = -K^{ij} q_j \partial_\mu A_{\text{ext},\nu}, \tag{41.49}$$

where K^{ij} is the inverse matrix to K_{ij}.

We can define the electric current by varying the action with respect to the electromagnetic field $A_{\text{ext},\mu}$, and calculate it by replacing the formula for $\partial_\mu a_\nu^{(i)}$ from the equations of motion, to obtain

$$J^\mu \equiv \frac{\delta S}{\delta A_{\text{ext},\mu}} = -\frac{1}{2\pi} \epsilon^{\mu\nu\rho} q_i \partial_\nu a_\rho^{(i)} = \left(\frac{1}{2\pi} q_i K^{ij} q_j \right) \epsilon^{\mu\nu\rho} \partial_\nu A_{\text{ext},\rho}. \tag{41.50}$$

This defines the Hall conductivity, since

$$\sigma_{xy} = \frac{J_x}{E_y} = \frac{J_x}{\partial_0 A_y}. \tag{41.51}$$

The filling fraction is found to be

$$\nu = q_i K^{ij} q_j, \tag{41.52}$$

and then the Hall conductivity is

$$\sigma_{xy} = \frac{1}{2\pi} q_i K^{ij} q_j = \frac{\nu}{2\pi}. \tag{41.53}$$

The equation of motion (41.49) gives a relation between the electromagnetic field $\vec{E}$ and the emergent field $\vec{e}^{(i)}$ given by $K^{ij} q_j$. That means that for a quasiparticle with *emergent* charge vector $\vec{l}$, the *electric* charge $Q(l)$ is defined by

$$l_i \vec{e}^{(i)} = Q(l) \vec{E} \Rightarrow Q(l) = l_i K^{ij} q_j. \tag{41.54}$$

In turn, by the general arguments we gave in Part I for calculating the anyonic phase, that means that we have an anyonic phase factor $e^{i\theta}$ when exchanging a quasiparticle with $\vec{l}$ with a quasiparticle with $\vec{l}'$ of

$$\theta = \pi l_i K^{ij} l_j'. \tag{41.55}$$

In the case of a charge vector only on the first position, $\vec{q} = (1, 0, 0, 0, \ldots)$ and a matrix K_{ij} of components

$$K_{ij} = \begin{pmatrix} a_1 & -1 & & \\ -1 & a_2 & -1 & \\ & -1 & a_3 & -1 \\ & & -1 & \ddots \end{pmatrix}, \tag{41.56}$$

we obtain, after calculating the inverse matrix K^{ij}, a continued fraction ν, of

$$\nu = \cfrac{1}{a_1 - \cfrac{1}{a_2 - \cfrac{1}{a_3 \dots}}}. \tag{41.57}$$

In order to embed the hierarchical FQHE action (41.48) in string theory, we must do so in a gravity dual, where there are background fields. I will not explain much about the background, but I will just define it: consider type IIA string theory on $\mathbb{R}^{1,2} \times S^3 \times \mathbb{C}^2/\mathbb{Z}_{n(p)}$, where in the *orbifold* (note that in general an orbifold is obtained by taking a smooth space and dividing by, i.e. identifying under, the action of a discrete isometry group) $\mathbb{C}^2/\mathbb{Z}_{n(p)}$ we divide by the action $(z_1, z_2) \to (e^{\frac{e\pi i}{n}} z_1, e^{\frac{e\pi i p}{n}} z_2)$, where $-n+1 \le p \le n+1$ and $(n, |p|) = 1$. The background is the near-horizon limit of a large number of NS5-branes of type IIA string theory wrapped on $\mathbb{C}^2/\mathbb{Z}_n$. One also introduces k units of flux of the NS-NS B-field of string theory (with $H = dB$) through S^3,

$$\int_{S^3} H = -4\pi^2 k, \tag{41.58}$$

which has the effect of changing the $\mathbb{R}^{2,1}$ to AdS_3.

One can "blow up" the singularities by replacing them with some 2-cycles, obtaining a space M_4, such that the intersection number of successive 2-cycles in 1, and self-intersection number is $-a_i$, with $\{a_i\}$ given by the Hirzeburch-Jung continued fraction for $n/p = a_1 - 1/(a_2 - 1/(a_3 \dots))$. The intersection matrix of the "blown-up" (resolved) orbifold M_4 is now the same as (41.56), with a minus sign, i.e. $-K_{ij}$.

The result is that a Kaluza-Klein compactification on S^3 of the string theory in the $\mathbb{R}^{2,1} \times S^3 \times \mathbb{C}^2/\mathbb{Z}_3$ space, *at the singularities* of the orbifold, i.e. a theory on $\mathbb{R}^{2,1}$ (or rather on AdS_3), has an RR 3-form field

$$C_{\mu\nu\rho} = (2\pi)^2 \sum_i a_\mu^{(i)} \omega_{\nu\rho}^{(i)}, \tag{41.59}$$

where the 2-forms $\omega^{(i)}$ are normalized according to $\int_{[i]} \omega^{(j)} = \delta_{ij}$, and $K_{ij} = \int_{M_4} \omega_{(i)} \wedge \omega_{(j)}$.

Type IIA supergravity in 10 dimensions has a Chern-Simons term that then becomes

$$\frac{1}{4\kappa_{10}^2} \int_{AdS_3 \times S^3 \times M_4} H_3 \wedge C_3 \wedge dC_3$$

$$= \frac{(2\pi)^4}{4\kappa_{10}^2} \left(\int_{S^3} H \right) \wedge \int_{AdS_3 \times M_4} \left(\sum_i a^{(i)} \wedge \omega_{(i)} \right) \wedge \left(\sum_j da^{(j)} \wedge \omega_{(j)} \right)$$

$$= \frac{k}{4\pi} \int_{AdS_3} \sum_{i,j=1}^{r} K_{ij} a^{(i)} \wedge da^{(j)}. \tag{41.60}$$

To obtain an external gauge field coupling of the Chern-Simons form, add a D4-brane, with $U(1)$ gauge field A_{ext} on it, wrapped on a linear combination of 2-cycles weighted by a charge vector $\vec{q}$. The WZ term on the D4-brane contains a coupling $\frac{1}{(2\pi)^3} \int C_3 \wedge F_{\text{ext}}$

($F_{\text{ext}} = dA_{\text{ext}}$), which becomes

$$\frac{1}{(2\pi)^3} \int_{D4 = AdS_3 \times q_i \omega^{(i)}} C_3 \wedge F_{\text{ext}} = \frac{1}{2\pi} \int_{AdS_3} \sum_i q_i a^{(i)} \wedge F_{\text{ext}}. \qquad (41.61)$$

We have now completely reproduced the hierarchical FQHE action (41.48) in the bulk of a gravity dual. The field theory at the boundary of AdS_3 has a chiral boson, just like the theory at the boundary (edge) of the topological insulator; therefore the AdS/CFT correspondence turns into a (better defined version of) the bulk/edge correspondence.

Important Concepts to Remember

- The FQHE can be described by a noncommutative version of the Chern-Simons theory, with a Moyal product with a $\theta = 1/(2\pi \rho_0)$, ρ_0 being a particle density.
- We can realize it on branes, for instance on M2-branes in M-theory, or on D-branes with B-field on them, which in an $\alpha' \to 0$ limit become noncommutative, as found by Seiberg and Witten.
- Topological superconductors and insulators can be classified according to K-theory, which is the same that classifies D-brane charges, therefore string theory offers a way to not only classify them, but construct them explicitly, from Dp-Dq-brane systems, perhaps with O(r) planes.
- We get the required topological Chern-Simons terms from the WZ terms on the Dp-branes, or from integrating out the massive fermions that live between the branes.
- To obtain the FQHE holographically, we need to obtain the $\int a \wedge da$ emergent CS term and the $\int a \wedge dA_{\text{ext}}$ CS terms on D-branes. One constructs Dp-Dq-brane systems, with six ND directions.
- One model is the ABJM model with fractional D2-branes, i.e. D4-branes wrapped on a $\mathbb{CP}^1 = S^2$ inside $\mathbb{CP}^3$, and D6-branes wrapped on an $S^3/\mathbb{Z}_2$ space in $\mathbb{CP}^3$.
- Another model is an D3-D7 model, with N D3's giving an $AdS_5 \times S^5$ background, and the D7-branes wrapping (as probes) an $S^2 \times S^2$ inside S^5, with a magnetic flux for the gauge field on them for stabilization, and extending in x^0, x^1, x^2 and the radial direction r.
- We can extend the FQHE to a hierarchical one with several gauge fields, and a filling fraction that is a continued fraction.
- The model can be embedded in string theory *inside a gravity dual*, and then the AdS/CFT correspondence becomes the bulk/edge correspondence of topological insulators.

Further Reading

The noncommutative Chern-Simons picture for the FQHE was developed by Susskind in [109], and the string theory picture in M-theory was put forward in [110]. Noncommutative gauge theory on D-branes was obtained by Seiberg and Witten in [111]. The classification of topological insulators and superconductors from D-branes was developed in [112] and [113]. The FQHE models from holography described here were found in [114], [115] (with a correct treatment of the D6-branes with O6-planes added in [116]), and [117].

Exercises

(1) Expand the noncommutative Chern-Simons Lagrangean (41.17), with the Moyal product, to obtain (41.11).

(2) Check that the limit $\alpha' \sim \sqrt{\epsilon} \to 0$, $g_{ij} \sim \epsilon \to 0$ on the open string variables defined by

$$\left(G + \frac{\theta}{2\pi\alpha'} \right)^{ij} = \left(\frac{1}{g + 2\pi\alpha'B} \right)^{ij},$$

$$G_s = g_s \sqrt{\frac{\det(g + 2\pi\alpha'B)}{\det g}}, \qquad (41.62)$$

gives the result (41.29).

(3) Prove that the Fubini-Study metric on $\mathbb{CP}^3$ is obtained from the metric on $S^7/\mathbb{Z}_k$ as $k \to \infty$.

(4) Calculate the DBI action for the D7-brane probe in the background (41.43).

(5) Invert the $N \times N$ matrix K_{ij} in (41.56) in the large N limit to find the filling fraction as a continued fraction

$$\nu = \frac{1}{a_1 - \frac{1}{a_2 - \frac{1}{a_3 \cdots}}}. \qquad (41.63)$$

42 Quantum Critical Systems and AdS/CFT

In this chapter, we aim to describe quantum critical systems using AdS/CFT. In fact, we have seen in Chapter 40 that we can obtain non-Fermi liquid behavior in AdS/CFT models, and we will see in future chapters also how to obtain particular quantum critical systems like strange metals (Chapter 48). But here we want to see how to obtain the generic behavior of quantum critical systems in a simple construction. We have seen in Part I that the quantum critical phase is the phase that appears when we extend to finite temperature the (zero temperature) phase transition in the coupling g, at some g_c. A new phase opens up from the $g = g_c$ point into an interval of couplings for $T > 0$. As with all phase transition points (critical points), the theory at $g = g_c$ and $T = 0$ is conformal, so in effect the quantum critical phase can be thought of as a finite temperature version of a conformal field theory (the temperature, being a scale, breaks the conformal invariance).

Since the phase transition is in the coupling g as opposed to in T, as is more common, the fluctuations that govern the phase are quantum in nature, as opposed to thermal, so the new phase is a quantum critical phase. Since the coupling of the phase at $T = 0$ is $g_c \sim 1$, the system is also at strong coupling. Moreover, we have seen that we can have an *effective* relativistic description in terms of a relativistic Landau-Ginzburg effective theory, even though (quasi-)particles do not travel at the speed of light c.

Therefore, having a strongly coupled effectively relativistic conformal field theory, the problem is very much suited to a description via AdS/CFT in its regular form, as a duality obtained in the decoupling limit of a brane system. Moreover, since the systems of interest for condensed matter are mostly in 2+1–dimensions, we are looking for a 2+1–dimensional theory with a gravity dual. As an example of the quantum critical phase, a type II high T_c superconductor is thought to be of a quantum critical type, and described by the relativistic Landau-Ginzburg theory, since in that case for $g < g_c$ we have a symmetry-breaking theory (superconductor), and for $g > g_c$ we have a disordered state (insulator).

42.1 The ABJM Model and its Vacuum

Then at least as a toy model, we can consider the standard AdS/CFT toy model in 2+1 dimensions, the ABJM model. It is a gauge theory, but unlike the standard 3+1–dimensional toy model, the $\mathcal{N} = 4$ Super Yang-Mills, it is a Chern-Simons (CS) gauge theory instead. The CS gauge theory in 2+1 dimensions has no propagating degrees of freedom, so from that point of view it is almost topological (it is not quite topological, but has no local degrees of freedom). It also has the $\mathcal{N} = 6$ supersymmetry for generic parameters, compared with the maximum of $\mathcal{N} = 8$, though at specific parameters it has the full $\mathcal{N} = 8$. Note that a spinor in two or three dimensions has two components, compared to four for four dimensions, so the maximal $\mathcal{N} = 4$ supersymmetry of 3+1 dimensions translates into maximal $\mathcal{N} = 8$ in 2+1 dimensions.

The Chern-Simons action is

$$S = \frac{k}{4\pi} \int d^{2+1}x \, \mathrm{Tr} \left[A \wedge dA + \frac{2}{3} A \wedge A \wedge A \right], \tag{42.1}$$

where k is an integer, since e^{iS} must be well-defined, and one can prove that

$$\frac{1}{8\pi^2} \int d^{2+1}x \, \mathrm{Tr} \left[A \wedge dA + \frac{2}{3} A \wedge A \wedge A \right] = \frac{S}{2\pi k} \tag{42.2}$$

changes by an integer number under a (large) gauge transformation (that cannot be smoothly deformed to the identity), leading to the quantization of k.

The gauge group for the ABJM model is $SU(N) \times SU(N)$ (or $U(N) \times U(N)$ in another variant of the theory), with gauge fields A_μ and $\hat{A}_\mu$ for the two factors, and we also have bifundamental fields C^I (complex scalars) and ψ_I (spin 1/2 fermions), i.e. fields that belong to the representation $(\mathbf{N}, \bar{\mathbf{N}})$ of the product gauge group. The action is of the type

$$S = \int d^{2+1}x \left[\frac{k}{4\pi} \epsilon^{\mu\nu\rho} \, \mathrm{Tr} \left(A_\mu \partial_\nu A_\rho + \frac{2i}{3} A_\mu A_\nu A_\rho - \hat{A}_\mu \partial_\nu \hat{A}_\rho - \frac{2i}{3} \hat{A}_\mu \hat{A}_\nu \hat{A}_\rho \right) \right.$$
$$\left. - \mathrm{Tr}(D_\mu C_I^\dagger D^\mu C^I) - i \, \mathrm{Tr}(\psi^{I\dagger} \gamma^\mu D_\mu \psi_I) + V_6(C^I) + \mathrm{Tr}(CC^\dagger \psi \psi^\dagger \quad \text{term}) \right], \tag{42.3}$$

where $V_6(C^I)$ is a complicated sextic potential for the scalars and the last term is a Yukawa term in three dimensions. Here $I = 1, 2, 3, 4$ is an index in an $SU(4)$ (=$SO(6)$) global symmetry, called R-symmetry. In fact, the full R-symmetry is $SU(4) \times U(1)$, where the $U(1)$ acts by multiplication with an overall phase of all the matter fields.

The above theory has $\mathcal{N} = 6$ supersymmetry, and in fact the supersymmetry is preserved under a deformation by a mass parameter μ. This is an unusual situation, since, for instance, in 3+1 dimensions there is no maximally supersymmetric mass deformation, and deforming $\mathcal{N} = 4$ SYM by a mass term lowers its supersymmetry. The μ deformation is defined as follows. One splits the four scalars C^I as $C^I = (Q^\alpha, R^{\dot{\alpha}})$, where $\alpha, \dot{\alpha} = 1, 2$, and $Q^\alpha, R^{\dot{\alpha}}$ are treated differently under the mass deformation, which leads to the breaking of the R-symmetry to $SU(2)_A \times SU(2)_B \times U(1)_A \times U(1)_B \times \mathbb{Z}_2$, where the $SU(2)_A$ and $SU(2)_B$ act separately on Q^α and $R^{\dot{\alpha}}$. The μ-deformed potential is of the type

$$V_6 = \mathrm{Tr}[|\mu Q^\alpha + \cdots|^2 + |-\mu R^\alpha + \cdots|^2], \tag{42.4}$$

more specifically

$$V_6 = \mathrm{Tr}[|M^\alpha|^2 + |N^\alpha|^2], \tag{42.5}$$

where

$$M^\alpha = \mu Q^\alpha + \frac{2\pi}{k}\left(2Q^{[\alpha}Q_\beta^\dagger Q^{\beta]} + R^\beta R_\beta^\dagger Q^\alpha - Q^\alpha R_\beta^\dagger R^\beta + 2Q^\beta R_\beta^\dagger R^\alpha - 2R^\alpha R_\beta^\dagger Q^\beta\right),$$

$$N^\alpha = -\mu R^\alpha + \frac{2\pi}{k}\left(2R^{[\alpha}R_\beta^\dagger R^{\beta]} + Q^\beta Q_\beta^\dagger R^\alpha - R^\alpha Q_\beta^\dagger Q^\beta + 2R^\beta Q_\beta^\dagger Q^\alpha - 2Q^\alpha Q_\beta^\dagger R^\beta\right).$$

$$(42.6)$$

The mass deformed theory has a ground state, obtained by putting the quantities in the modulus squared above to zero, that is of the type of a *fuzzy sphere*. The vacua are

$$R^\alpha = cG^\alpha, \quad Q^\alpha = 0 \quad \text{and} \quad Q_\alpha^\dagger = cG^\alpha, \quad R^\alpha = 0, \qquad (42.7)$$

where

$$c = \sqrt{\frac{\mu k}{2\pi}}, \qquad (42.8)$$

and the constant matrices G^α satisfy the "algebra"

$$G^\alpha G_\beta^\dagger G^\beta - G^\beta G_\beta^\dagger G^\alpha = G^\alpha. \qquad (42.9)$$

A fuzzy sphere is a quantum version of a sphere, or rather a matrix approximation to a sphere; i.e. it satisfies something like $\sum_i (X_i)^2 = R^2$, but functions on this sphere, instead of being expanded into an infinite set of spherical harmonics, say

$$f(\theta, \phi) = \sum_{l \in \mathbb{Z}} \sum_{m=-l}^{l} a_{lm} Y_{lm}(\theta, \phi), \qquad (42.10)$$

are expanded only into an independent set of $N \times N$ matrices (a basis), i.e. can be thought of as $N \times N$ matrices expanded into a basis. We say that the fuzzy sphere is a quantum, or "fuzzy" approximation to a sphere.

On the other hand, the massless ($\mu = 0$) ABJM theory doesn't have a nontrivial vacuum solution, but rather it has a nontrivial BPS solution, i.e. a solution that breaks half of the supersymmetry, leaving only $\mathcal{N} = 3$, of the *fuzzy funnel* type. That means that it is a sphere whose radius depends on an extra coordinate, and this radius varies from zero to infinity. Specifically, it is the same as the above fuzzy sphere, except the constant c is replaced by the function

$$c(s) = \sqrt{\frac{k}{4\pi s}}, \qquad (42.11)$$

s being one of the two spatial coordinates of the ABJM theory. Thus the radius of the fuzzy sphere is zero at $s = \infty$, and is infinite at $s = 0$, giving a fuzzy funnel.

The matrices G^α defining the fuzzy sphere satisfy the "algebra" (42.9), which is a sort of generalization of a Lie algebra, $T^a T^b - T^b T^a = f^{ab}{}_c T^c$ (it can be shown in fact to be equivalent to the Lie algebra of $SU(2)$). As such, it admits many representations, but the irreducible matrix representation is

$$(G^1)_{mn} = \sqrt{m-1}\,\delta_{mn} \quad (G^2)_{mn} = \sqrt{N-m}\,\delta_{m+1,n}$$

$$(G_1^\dagger)_{mn} = \sqrt{m-1}\,\delta_{mn} \quad (G_2^\dagger)_{mn} = \sqrt{N-n}\,\delta_{n+1,m}, \qquad (42.12)$$

and reducible representations can be found as block-diagonal matrices with the above matrices in the blocks.

42.2 Gravity Dual of the ABJM Model

The gravity dual of pure (massless) ABJM theory is string theory in the $AdS_4 \times \mathbb{CP}^3$ background. The compact space is obtained as

$$\mathbb{CP}^3 = S^7/\mathbb{Z}_k; \quad k \to \infty. \tag{42.13}$$

The ABJM theory has four complex scalars C^I, which as we saw in the case of the theory on the D-branes, correspond to coordinates transverse to the brane. In this case, the ABJM theory is defined via branes in the 11-dimensional M-theory (string theory at strong coupling, with the coupling being the eleventh dimension), so the four complex C^I are mapped to four complex transverse coordinates Z^i. The 7-sphere is then obtained as the transverse sphere in these coordinates, i.e. as

$$\sum_{i=1}^{4} |Z^i|^2 = 1. \tag{42.14}$$

This S^7 is obtained as a *Hopf fibration* with fiber S^1 over $\mathbb{CP}^3$, which means the space is written identifying an S^1 fiber in the S^7 space whose radius varies with the position in a $\mathbb{CP}^3$ base, forming the total space (which therefore is not of a product type). The S^1 fiber is the overall $U(1)$ phase multiplying the Euclidean coordinates Z^i, i.e. the phase

$$Z^i \to e^{i\alpha} Z^i; \quad \alpha \in \mathbb{R}. \tag{42.15}$$

Then the action of the $\mathbb{Z}_k$ is by a phase which is an k-th order root of unity, i.e. by

$$Z^i \to e^{\frac{2\pi i n}{k}} Z^i; \quad n = 0, 1, \ldots, k-1, \tag{42.16}$$

and one identifies under this action (considers equivalence classes under this transformation by an element of $\mathbb{Z}_k$). Note that $\mathbb{Z}_k$ is the group of k-th order roots of unity, $a^k = 1$, which under multiplication form a cyclical group. Then obviously by identifying with respect to this, as $k \to \infty$, one identifies with an infinitesimal transformation in the $U(1)$ direction, $e^{\frac{2\pi i}{k}}$, thus effectively eliminating the fiber, and remaining with just the $\mathbb{CP}^3$ base. The fact that the remaining space (the base) is $\mathbb{CP}^3$ is understood this way: $\mathbb{CP}^n$ is defined in general as the space of n complex numbers Z^i, identified under a common multiplication by an arbitrary complex number λ, i.e. $Z^i \sim \lambda Z^i$. By identifying $\mathbb{C}^4$ with the modulus $|\lambda| = R$ of a complex number only, where R is the radius of the given Z^i's,

$$\sum_i |Z^i|^2 = R^2, \tag{42.17}$$

we are left with the unit 7-sphere above, $\sum_i |Z^i|^2 = 1$. By further identifying with the phase $e^{i\alpha}$, the $U(1)$ fiber in the Hopf fibration of S^7, such that in total we identify with the arbitrary complex number $\lambda = e^{i\alpha} R$, we are therefore left, by definition, with $\mathbb{CP}^3$.

The gravity dual of the massive ABJM is even more complicated, and will not be described here. To apply our simple AdS/CFT rules, we must dimensionally reduce on the compact space $\mathbb{CP}^3$ in the massless case, on a more complicated space in the massive case, and also to consider string corrections to the supergravity action, that correspond to taking only a finite gauge group in the gauge theory. This would be complicated to do in general, but an analysis of symmetries, possible terms that could appear, and redundancies leads to a simple action.

42.3 Models for Quantum Critical Systems

The action must contain at least the metric $g_{\mu\nu}$, we need an AdS background for it, so we need a cosmological constant term in the action, and we also need a gauge field to couple to the currents of the boundary field theory. The leading supergravity action must then be

$$S_0 = \int d^4x \left[\frac{1}{2\kappa_{N,4}^2} \left(R + \frac{6}{L^2} \right) - \frac{1}{4e^2} F_{ab} F^{ab} \right], \tag{42.18}$$

where the cosmological constant was written in terms of L, the radius of the AdS background solution to the above action. The AdS black hole solution corresponding to a conformal field theory at finite temperature T on the boundary is

$$ds^2 = \frac{L^2}{u^2} \frac{du^2}{f(u)} + \frac{u^2}{L^2}(-f(u)dt^2 + d\vec{x}^2), \tag{42.19}$$

and

$$f(u) = 1 - \frac{R^3}{u^3}. \tag{42.20}$$

The string corrections are encoded in the next order term,

$$S_1 = \int d^4x \sqrt{-g}\gamma \frac{L^2}{e^2} C_{abcd} F^{ab} F^{cd}, \tag{42.21}$$

where C_{abcd} is the Weyl curvature tensor, which is as we saw in Chapter 36 is the Riemann curvature tensor R_{abcd} minus some traces, such that it becomes zero on an AdS background (as opposed to the Riemann tensor, which becomes constant, i.e. proportional to metrics, on an AdS background).

Then the $S_0 + S_1$ action is a kind of toy model action for the gravity dual to the ABJM model, which we can use without needing to do any long calculation in string theory to find the exact result, and we can treat the coefficient γ (which normally would have a calculable value) as a free parameter, which we can vary and use to describe the system.

But the question is why would the ABJM model be a good model to describe quantum critical systems, other than the fact that it is the quintessential toy model in 2+1 dimensions? The answer we will give is that, like in real condensed matter systems when the relativistic Landau-Ginzburg model appears as an effective low-energy field theory that correctly describes the macroscopic behavior, in the ABJM model we also have an effective LG model as an effective low-energy field theory. One issue one could worry about is that the LG model is an abelian ($U(1)$) model, whereas as we saw, for the existence of a good gravity dual model we need a large N (large rank) gauge group. The reason, however, that we will be able to use the LG model is that it appears now in a nontrivial nonabelian embedding in the full ABJM theory, which means that it corresponds simply to a (set of) solution(s) in the gravity dual of the ABJM model.

42.4 Abelian Reduction of the ABJM Model to Landau-Ginzburg

Specifically, consider the nontrivial abelian reduction ansatz

$$
\begin{aligned}
A_\mu &= a_\mu^{(2)} G^1 G_1^\dagger + a_\mu^{(1)} G^2 G_2^\dagger \\
\hat{A}^\mu &= a_\mu^{(2)} G_1^\dagger G^1 + a_\mu^{(2)} G_2^\dagger G^2 \\
Q^\alpha &= \phi_\alpha G^\alpha \\
R^\alpha &= \chi_\alpha G^\alpha,
\end{aligned}
\tag{42.22}
$$

which implies also

$$
\begin{aligned}
F_{\mu\nu} &= f_{\mu\nu}^{(2)} G^1 G_1^\dagger + f_{\mu\nu}^{(1)} G^2 G_2^\dagger \\
\hat{F}_{\mu\nu} &= f_{\mu\nu}^{(2)} G_1^\dagger G^1 + f_{\mu\nu}^{(1)} G_2^\dagger G^2.
\end{aligned}
\tag{42.23}
$$

Note that the matrices G^α, since they appear in the ansatz for the bifundamental fields $Q^\alpha R^\alpha$, are also bifundamental, i.e. in the representation $(\mathbf{N}, \bar{\mathbf{N}})$ of the product gauge group. But that means that $G^1 G_1^\dagger$ and $G^2 G_2^\dagger$ are in the $(\mathbf{N} \otimes \bar{\mathbf{N}}, 1)$ representation (the indices of the second group are summed over in the matrix multiplication), i.e. the adjoint of the first group factor and the singlet of the second, whereas $G_1^\dagger G^1$ and $G_2^\dagger G^2$ are in the $(1, \mathbf{N} \otimes \bar{\mathbf{N}})$ representation, i.e. the singlet of the first factor and the adjoint of the second, which justified the reduction ansatz.

Substituting the reduction ansatz in the ABJM action, after some algebra, one obtains the reduced action,

$$
S = -\frac{N(N-1)}{2} \int d^3x \left[\frac{k}{4\pi} \epsilon^{\mu\nu\rho} (a_\mu^{(2)} f_{\nu\rho}^{(1)} + a_\mu^{(1)} f_{\nu\rho}^{(2)}) + |D_\mu \phi_i|^2 + |D_\mu \chi_i|^2 + U(|\phi_i|, |\chi_i|) \right],
\tag{42.24}
$$

where the covariant derivatives are defined as

$$
D_\mu \phi_i = \left(\partial_\mu - i a_\mu^{(i)} \right) \phi_i; \quad D_\mu \chi_i = \left(\partial_\mu - i a_\mu^{(i)} \right) \chi_i,
\tag{42.25}
$$

and the scalar potential U will not be made explicit. Note that the two CS terms are equivalent by partial integration, but we kept them both to have symmetry under exchange of the one and two indices.

The above truncation is consistent, i.e. equations of motion of the reduced theory solve (are consistent with) the equations of motion of the original theory. We can check that imposing the equations of motion of the reduced theory, on the reduction ansatz, we solve the equations of motion of the original theory. We have encountered the notion of consistent truncation when discussing KK reductions in Part II.

We can make a further consistent truncation to $\phi_1 = \phi_2 = 0$ and $\chi_1 = b$ (b is an arbitrary constant), after which we obtain the action

$$S = -\frac{N(N-1)}{2} \int d^3x \left[\frac{k}{2\pi} \epsilon^{\mu\nu\rho} a_\mu^{(1)} f_{\nu\rho}^{(2)} + [a_\mu^{(1)}]^2 |b|^2 + |D_\mu \chi_2|^2 + V \right], \quad (42.26)$$

where the scalar potential is

$$V = \frac{4\pi^2}{k^2} \left[|b|^2 |\chi_2|^4 + |\chi_2|^2 [(|b|^2 - c^2)^2 - 2|b|^2 c^2] + c^4 |b|^2 \right], \quad (42.27)$$

and as before, $c^2 = \mu k/(2\pi)$. Note that now $a_\mu^{(1,2)}$ do not appear on equal footing anymore, which is why we have written the CS term in a single form. Indeed, now we observe that $a_\mu^{(1)}$ has become auxiliary, and can be eliminated via its equation of motion, to obtain

$$a_\mu^{(1)} = -\frac{k}{4\pi |b|^2} \epsilon^{\mu\nu\rho} f_{\nu\rho}^{(2)}. \quad (42.28)$$

What happens here in this Higgs vacuum, around the VEV $\chi_1 = b$, is an example of the *Mukhi-Papageorgakis Higgs mechanism* in 2+1 dimensions. The usual Higgs mechanism in 3+1 dimensions, which can be generalized to any dimension, is the fact that around the Higgs vacuum, the *Yang-Mills* gauge fields (with two degrees of freedom, or in general dimension d, $d-2$ degrees of freedom) in the broken symmetry directions eat the would-be Goldstone bosons (massless scalars) in the same directions, to become *massive* (with three degrees of freedom, or in a general dimension d, $d-1$ degrees of freedom). The Mukhi-Papageorgakis Higgs mechanism in 2+1 dimensions is that *Chern-Simons* gauge fields (with no degrees of freedom) eat the would-be Goldstone bosons (massless scalars) in the broken symmetry directions, to become *Yang-Mills* (with one degree of freedom).

We see that the CS gauge field $a_\mu^{(1)}$ has become auxiliary, which means still no degrees of freedom, but the CS gauge field $a_\mu^{(2)}$ eats χ_1 and becomes Yang-Mills (or rather, Maxwell). Indeed, by substituting the on-shell value of $a_\mu^{(1)}$ back in the action, one gets

$$S = -\frac{N(N-1)}{2} \int d^3x \left[\frac{k^2}{8\pi^2 |b|^2} (f_{\mu\nu}^{(2)})^2 + |D_\mu \chi_2|^2 + V \right], \quad (42.29)$$

by using $\epsilon^{\mu\nu\rho} \epsilon_{\sigma\nu\rho} = -2\delta_\sigma^\mu$, which can be checked directly using $\epsilon^{012} = +1$ and $\epsilon_{012} = -1$.

We can use a canonical normalization for the fields when $N \to \infty$, defined via the rescaling

$$a_\mu^{(2)} = \frac{2\pi\mu}{Nk} \tilde{a}_\mu^{(2)}; \quad \chi_2 = \frac{\tilde{\chi}_2}{N}, \quad (42.30)$$

after which the action becomes

$$S = \int d^3x \left[-\frac{1}{4} (\tilde{f}_{\mu\nu}^{(2)})^2 - |D_\mu \tilde{\chi}_2|^2 - V \right]. \quad (42.31)$$

To have the usual covariant derivative,

$$D_\mu = \partial_\mu - ig\tilde{a}^{(2)}_\mu, \qquad (42.32)$$

we see that we must define the coupling as

$$g = \frac{2\pi |b|}{Nk}. \qquad (42.33)$$

With this redefinition, the potential becomes

$$V = \frac{g^2}{2}\left[|\tilde{\chi}_2|^4 + \frac{\mu^2 k^2 N^4}{4\pi^2} + |\tilde{\chi}_2|^2 N^2\left(-\frac{4\mu k}{2\pi} + |b|^2 + \frac{\mu^2 k^2}{4\pi^2|b|^2}\right)\right]. \qquad (42.34)$$

This potential contains a standard χ^4 quartic term, an irrelevant constant, and a mass term (quadratic term) that can be either positive or negative, depending on parameters, so it is like the $|\phi|^2(g - g_c)$ of Landau-Ginzburg theory. Roughly, we can think of μ and k as giving g_c, and the arbitrary VEV $|b|$ giving the coupling g, but one can make a more precise identification of g and g_c. In any case, we have shown that the resulting reduced action is really of the LG form.

Moreover, we see that the mass term vanishes for

$$|b|^2 = \frac{\mu k}{2\pi}(2 \pm \sqrt{3}), \qquad (42.35)$$

whereas generically the mass is of order μ. On the other hand, for this special value of $|b|^2$, we find

$$g^2 \sim \frac{2\pi \mu}{N^2 k} \ll \mu \qquad (42.36)$$

if $N \gg 1$ (independent of k).

That means that for the value of b close to the massless one, then also $m^2 \ll \mu^2$, which means that all the other modes decouple from the reduced modes, which are almost massless. In particular, it means that not only is the reduction consistent, but loops of the massive modes are negligible, and the reduced theory is a consistent truncation even at the quantum level, not only at the classical (equations of motion) level. Thus the reduced theory, the relativistic Landau-Ginzburg theory, is an effective low-energy theory, just like in the condensed matter case. This justifies the use of the gravity dual of ABJM theory to describe the quantum critical phase.

42.5 Abelian Reduction to Landau-Ginzburg in Condensed Matter Systems?

But it would still be useful to find a direct derivation, at the microscopic level, for the relevance of ABJM theory to the quantum critical phase of specific systems, in particular for high T_c superconductivity. While it is not clear how to do that rigorously, here I will sketch a physical picture for how the reduction of degrees of freedom from ABJM to LG is similar to the reduction of degrees of freedom in a 2+1–dimensional superconductor to the effective LG theory.

Consider a system of spinless bosons ϕ_{ij}, where i is an index for the 2-dimensional spatial lattice. In particular, for the superconductor, this would correspond to the Cooper pairs matching electrons of opposite spins and momenta at sites i and j:

$$\phi_{ij} = \bar{\psi}_i^{\uparrow}(\vec{k})\psi_j^{\downarrow}(-\vec{k}). \qquad (42.37)$$

In reality, i is a 2-dimensional vector $\vec{i}$, and we describe the pair $(\vec{i}\vec{j})$ as the pair of the midpoint vector $\vec{i'} = (\vec{i} + \vec{j})/2$, thought of as the lattice point where we situate the Cooper pair, and the relative vector $\vec{j} - \vec{i}$, that can be described also as the pair of cartesian lattice coordinates, $\vec{j} - \vec{i} = (ab)$, which characterizes the Cooper pair. Consider a lattice of size $N \times N$, so $a, b \leq N$ and $|\vec{i} - \vec{j}| \lesssim N$. It is known experimentally that Cooper pairs form from electrons that are spread over macroscopic distances, of the order of the sample size, which justifies considering $|\vec{i} - \vec{j}| \lesssim N$.

We therefore describe the Cooper pair through the normalized wavefunctions $\phi_{\vec{i'}}^{ab}$, which give the probability of existence of the pair,

$$\mathcal{P} \propto \phi_{\vec{i'}}^{ab}. \qquad (42.38)$$

By rotational invariance, we diagonalize the $\phi_{\vec{i'}}^{ab}$ to radial modes $\psi(r)$ (where $r = \sqrt{a^2 + b^2}$), or approximately by the diagonal element $\psi(a)$. This matches what happens in the ABJM theory, where we have $N \times N$ matrix fields $\psi^{ab}(\vec{x}) \to \phi_i^{ab}$, if the continuous 2-dimensional coordinate $\vec{x}$ is replaced by the discrete (lattice) vector $\vec{i}$. In the ABJM model, we have a $U(N) \times U(N)$ gauge symmetry, $U^{ab,a'b'} = fU^{aa'}V^{bb'}$, but in the condensed matter system we expect perhaps a more general $U^{ab,a'b'}$ unitary transformation.

A consistent and simple choice for the probability of formation of the Cooper pair on the atomic lattice is

$$|\psi(a)|^2 \propto N - a, \qquad (42.39)$$

such that the probability vanishes at the boundary $a = N$ (since clearly the Cooper pairs cannot form for electrons outside the sample), and is maximum at the center $a = 0$. This is similar to what happens in the ABJM theory, where χ_2 multiplies

$$(G^2)_{mn} = \sqrt{N - m}\delta_{m+1,n} \Rightarrow (G^2 G_2^{\dagger})_{mn} = (N - m)\delta_{mn}, \qquad (42.40)$$

so the nonabelian field $(R^2)_{mn}(\vec{x}) = \chi_2(\vec{x})(G^2)_{mn}$ is mapped to the quantum field (diagonal piece of the matrix field)

$$\psi_{\vec{i'}}^{aa} = \sum_a \psi(a)_{\vec{i}} \hat{b}^{\dagger aa'}. \qquad (42.41)$$

Here the wavefunction $\psi(a)_{\vec{i}}$ multiplies the creation operator $b^{\dagger aa'}$ to form the quantum field (we usually have $\phi(x) = \sum_a \phi_a(x)b_a^{\dagger}$).

We have sketched how the reduction in degrees of freedom from ABJM to LG could be mapped to a real reduction in degrees of freedom inside a superconductor, but of course a precise map is still lacking.

42.6 Nonrelativistic Limit of Duality and the Jackiw-Pi Model

We have seen that we can reduce the massive ABJM model to the bosonic abelian action (42.24), which can be further reduced to the Landau-Ginzburg theory. In fact, one can reduce to a supersymmetric version of the action in (42.24), by including the fermions in the reduction, obtaining

$$
\begin{aligned}
S = -\frac{N(N-1)}{2} \int d^3x \Bigg\{ & \frac{k}{4\pi} \epsilon^{\mu\nu\lambda} \left(a_\mu^{(2)} f_{\nu\lambda}^{(1)} + a_\mu^{(1)} f_{\nu\lambda}^{(2)} \right) + |D_\mu \phi_i|^2 + |D_\mu \chi_i|^2 \\
& + i \sum_{i=1,2} \left[\bar{\eta}_i (\slashed{D} + \mu)\eta_i + \bar{\tilde{\eta}}_i (\slashed{D} - \mu)\tilde{\eta}_i \right] \\
& - \frac{2\pi i}{k} \left[(|\phi_1|^2 + |\chi_1|^2)(\bar{\eta}_2 \eta_2 + \bar{\tilde{\eta}}_2 \tilde{\eta}_2) + (|\phi_2|^2 + |\chi_2|^2)(\bar{\eta}_1 \eta_1 + \bar{\tilde{\eta}}_1 \tilde{\eta}_1) \right] \\
& + \left(\frac{2\pi}{k}\right)^2 \left[(|\phi_1|^2 + |\chi_1|^2)(|\chi_2|^2 - |\phi_2|^2 - c^2)^2 + (|\phi_2|^2 + |\chi_2|^2)(|\chi_1|^2 - |\phi_1|^2 - c^2)^2 \right. \\
& \left. + 4|\phi_1|^2 |\phi_2|^2 (|\chi_1|^2 + |\chi_2|^2) + 4|\chi_1|^2 |\chi_2|^2 (|\phi_1|^2 + |\phi_2|^2) \right] \Bigg\},
\end{aligned}
\tag{42.42}
$$

where $\eta_1, \tilde{\eta}_1$, and $\eta_2, \tilde{\eta}_2$ are complex 2-component Dirac spinors. By the reduction of the supersymmetry transformation rules of the massive ABJM theory, one can find also the supersymmetry transformation rules of the above reduced action:

$$
\begin{aligned}
\delta\phi_1 &= i\bar{\epsilon}\tilde{\eta}_{\hat{1}}, \\
\delta\phi_2 &= -i\bar{\epsilon}\tilde{\eta}_{\hat{2}}, \\
\delta\chi_{\hat{1}} &= -i\bar{\epsilon}\eta_1, \\
\delta\chi_{\hat{2}} &= i\bar{\epsilon}\eta_2, \\
\delta a_\mu^{(1)} &= \frac{2\pi}{k} \left(\bar{\epsilon}\gamma_\mu[\phi_2 \tilde{\eta}_{\hat{2}}^* - \chi_{\hat{2}}\eta_2^*] + \bar{\epsilon}^*\gamma[\phi_2^* \tilde{\eta}_{\hat{2}} - \chi_{\hat{2}}^*\eta_2] \right), \\
\delta a_\mu^{(2)} &= -\frac{2\pi}{k} \left(\bar{\epsilon}\gamma_\mu[\phi_1 \tilde{\eta}_{\hat{1}}^* - \chi_{\hat{1}}\eta_1^*] + \bar{\epsilon}^*\gamma^\mu[\phi_1^* \tilde{\eta}_{\hat{1}} - \chi_{\hat{1}}^*\eta_1] \right), \\
\delta\eta_1 &= \gamma^\mu D_\mu \chi_{\hat{1}} + \frac{2\pi}{k}\epsilon\chi_{\hat{1}}(|\phi_2|^2 + |\chi_{\hat{2}}|^2) - \mu\epsilon\chi_{\hat{1}}, \\
\delta\eta_2 &= -\gamma^\mu \epsilon D_\mu \chi_{\hat{2}} - \frac{2\pi}{k}\epsilon\chi_{\hat{2}}(|\phi_1|^2 + |\chi_{\hat{1}}|^2) + \mu\epsilon\chi_{\hat{2}}, \\
\delta\tilde{\eta}_{\hat{1}} &= -\gamma^\mu \epsilon D_\mu \phi_1 - \frac{2\pi}{k}\epsilon\phi_1(|\phi_2|^2 + |\chi_{\hat{2}}|^2) - \mu\epsilon\phi_1, \\
\delta\tilde{\eta}_{\hat{2}} &= \gamma^\mu \epsilon D_\mu \phi_2 + \frac{2\pi}{k}\epsilon\phi_2(|\phi_1|^2 + |\chi_{\hat{1}}|^2) + \mu\epsilon\phi_2.
\end{aligned}
\tag{42.43}
$$

Here ϵ is complex, which means an $SO(2) = U(1)$ R-symmetry, which in turn means that we have $\mathcal{N} = 2$ supersymmetry in three dimensions. Note that the scalar transformation rules are all standard, like in the simplest example in Chapter 25, as are the leading terms in the fermion transformation rules.

In this section we will consider a nonrelativistic limit of the above action and supersymmetry transformation rules, which will lead to a supersymmetric version of a very important nonrelativistic model that has been used for condensed matter applications, the Jackiw-Pi model.

The first step is to introduce the factors of c and $\hbar$, and separate space and time components. We also write $(\partial_0, A_0) = (\partial_t, A_t)/c$ and replace μ by $mc/\hbar$, obtaining for the Lagrangian

$$
-\frac{2}{N(N-1)}\mathcal{L} = -\frac{1}{c^2}(D_t\tilde{\phi}_j)\overline{(D_t\tilde{\phi}_j)} + (D_i\tilde{\phi}_j)\overline{(D_i\tilde{\phi}_j)} - \frac{1}{c^2}(D_t\tilde{\chi}_j)\overline{(D_t\tilde{\chi}_j)}
$$
$$
+ (D_i\tilde{\chi}_j)\overline{(D_i\tilde{\chi}_j)} + \frac{m^2c^2}{\hbar^2}\left(|\tilde{\phi}_1|^2 + |\tilde{\phi}_2|^2 + |\tilde{\chi}_1|^2 + |\tilde{\chi}_2|^2\right)
$$
$$
- \frac{8\pi}{k}\frac{mc}{\hbar}\frac{1}{\hbar c}\left(|\tilde{\chi}_1|^2|\tilde{\chi}_2|^2 - |\tilde{\phi}_1|^2|\tilde{\phi}_2|^2\right) + \frac{4\pi^2}{(k\hbar c)^2}
$$
$$
\times \left[(|\tilde{\chi}_1|^2 + |\tilde{\phi}_1|^2)(|\tilde{\chi}_2|^2 + |\tilde{\phi}_2|^2)(|\tilde{\chi}_1|^2 + |\tilde{\chi}_2|^2 + |\tilde{\phi}_1|^2 + |\tilde{\phi}_2|^2)\right]
$$
$$
+ \frac{k\hbar}{4\pi}\epsilon^{\mu\nu\lambda}\left(A_\mu^{(2)}F_{\nu\lambda}^{(1)} + A_\mu^{(1)}F_{\nu\lambda}^{(2)}\right)
$$
$$
+ i\sum_{i=1,2}\left[\bar{\eta}_i\left(\slashed{D} + \frac{mc}{\hbar}\right)\eta_i + \bar{\tilde{\eta}}_i\left(\slashed{D} - \frac{mc}{\hbar}\right)\tilde{\eta}_i\right]
$$
$$
- \frac{2\pi i}{k\hbar c}\left[(|\phi_1|^2 + |\chi_1|^2)(\bar{\eta}_2\eta_2 + \bar{\tilde{\eta}}_2\tilde{\eta}_2) + (|\phi_2|^2 + |\chi_2|^2)(\bar{\eta}_1\eta_1 + \bar{\tilde{\eta}}_1\tilde{\eta}_1)\right]
$$
$$
\tag{42.44}
$$

and supersymmetry transformation rules

$$
\delta\phi_1 = i\bar{\epsilon}\tilde{\eta}_1,
$$
$$
\delta\phi_2 = -i\bar{\epsilon}\tilde{\eta}_2,
$$
$$
\delta\chi_1 = -i\bar{\epsilon}\eta_1,
$$
$$
\delta\chi_2 = i\bar{\epsilon}\eta_2,
$$
$$
\delta A_\mu^{(1)} = \frac{2\pi}{kc}\left(\bar{\epsilon}\gamma_\mu[\phi_2\tilde{\eta}_2^* - \chi_2\eta_2^*] + \bar{\epsilon}^*\gamma[\phi_2^*\tilde{\eta}_2 - \chi_2^*\eta_2]\right),
$$
$$
\delta A_\mu^{(2)} = -\frac{2\pi}{kc}\left(\bar{\epsilon}\gamma_\mu[\phi_1\tilde{\eta}_1^* - \chi_1\eta_1^*] + \bar{\epsilon}^*\gamma^\mu[\phi_1^*\tilde{\eta}_1 - \chi_1^*\eta_1]\right), \tag{42.45}
$$
$$
\delta\eta_1 = \gamma^\mu\epsilon D_\mu\chi_1 + \frac{2\pi}{kc}\epsilon\chi_1(|\phi_2|^2 + |\chi_2|^2) - \frac{mc}{\hbar}\epsilon\chi_1,
$$
$$
\delta\eta_2 = -\gamma^\mu\epsilon D_\mu\chi_2 - \frac{2\pi}{kc}\epsilon\chi_2(|\phi_1|^2 + |\chi_1|^2) + \frac{mc}{\hbar}\epsilon\chi_2,
$$
$$
\delta\tilde{\eta}_1 = -\gamma^\mu\epsilon D_\mu\phi_1 - \frac{2\pi}{kc}\epsilon\phi_1(|\phi_2|^2 + |\chi_2|^2) - \frac{mc}{\hbar}\epsilon\phi_1,
$$
$$
\delta\tilde{\eta}_2 = \gamma^\mu\epsilon D_\mu\phi_2 + \frac{2\pi}{kc}\epsilon\phi_2(|\phi_1|^2 + |\chi_1|^2) + \frac{mc}{\hbar}\epsilon\phi_2. \tag{42.46}
$$

Note that we have put tildes on ϕ and χ to distinguish from the corresponding nonrelativistic fields to be defined below. Here ϵ is a 2-component spinor, and we can denote its components by ϵ_1 and ϵ_2, and in the non-relativistic limit these will correspond to independent supersymmetries. I included the supersymmetry rules here since as, we will see below, they will enforce a condition in the nonrelativistic limit that will also affect the bosonic action.

For a massive scalar field, the non-relativistic limit is found by splitting the nonrelativistic time dependence $e^{-i\frac{E_0}{\hbar}t} = e^{-i\frac{mc^2}{\hbar}t}$, multiplying by $\hbar/\sqrt{2m}$ for the purpose of obtaining the right kinetic term for ∂_i (when compared with ∂_t), and splitting in particle ϕ and antiparticle $\hat{\phi}^*$ sectors, separately conserved:

$$\tilde{\phi} = \frac{\hbar}{\sqrt{2m}}\left[\phi e^{-i\frac{mc^2}{\hbar}t} + \hat{\phi}^* e^{+i\frac{mc^2}{\hbar}t}\right]. \qquad (42.47)$$

In condensed matter we are interested only in the zero antiparticle sector, so we drop $\hat{\phi}^*$, meaning that the nonrelativistic scalar fields are defined by

$$(\tilde{\phi}, \tilde{\chi}) \longrightarrow \left(\frac{\hbar}{\sqrt{2m}}\phi(x, t)e^{-imc^2t/\hbar}, \frac{\hbar}{\sqrt{2m}}\chi(x, t)e^{-imc^2t/\hbar}\right). \qquad (42.48)$$

Similarly, for fermions the fields are rescaled by $\sqrt{\hbar c}$, and written as

$$\tilde{\eta} = \sqrt{\hbar c}\left[\psi e^{-i\frac{mc^2}{\hbar}t} + \sigma_2 \hat{\psi}^* e^{+i\frac{mc^2}{\hbar}t}\right], \qquad (42.49)$$

but since in condensed matter we are interested in the zero antiparticle sector, we write

$$\eta_i = \sqrt{\hbar c}\psi_i(x, t)e^{-i\frac{mc^2}{\hbar}t}. \qquad (42.50)$$

More importantly, only half of the components remain dynamical, and we can write the other components in terms of them. One can calculate this relation more precisely, and obtain (I will not show the details here)

$$\tilde{\eta}_i \longrightarrow \sqrt{\hbar c}e^{-i\frac{mc^2}{\hbar}t}\begin{pmatrix}\tilde{\psi}_{i1} \\ \frac{\hbar}{2mc}D_-\tilde{\psi}_{i1}\end{pmatrix} \quad, \quad \eta_i \longrightarrow \sqrt{\hbar c}e^{-i\frac{mc^2}{\hbar}t}\begin{pmatrix}-\frac{\hbar}{2mc}D_+\psi_{i2} \\ \psi_{i2}\end{pmatrix}. \qquad (42.51)$$

The nonrelativistic limit now corresponds to taking $c \to \infty$, while keeping ϕ_i, χ_i, m, k, ψ_{i1}, ψ_{i2} fixed. After a (somewhat long) calculation, one obtains

$$-\frac{2}{N(N-1)}\mathcal{L}_{\text{NR}} = -\bar{\phi}_i\left(i\hbar D_t + \frac{\hbar^2}{2m}D_i^2\right)\phi_i - \bar{\chi}_i\left(i\hbar D_t + \frac{\hbar^2}{2m}D_i^2\right)\chi_i$$

$$+ \frac{2\pi\hbar^2}{mk}\left(|\phi_1|^2|\phi_2|^2 - |\chi_1|^2|\chi_2|^2\right) + \frac{k\hbar}{4\pi}\epsilon^{\mu\nu\lambda}\left(A_\mu^{(2)}F_{\nu\lambda}^{(1)} + A_\mu^{(1)}F_{\nu\lambda}^{(2)}\right)$$

$$+ \sum_{j=1,2}\left[-\psi_j^\dagger\left(i\hbar D_t + \frac{1}{2m}\left(D_i^2 + F_{12}^{(j)}\right)\right)\psi_j\right.$$

$$\left. - \tilde{\psi}_j^\dagger\left(i\hbar D_t + \frac{1}{2m}\left(D_i^2 - F_{12}^{(j)}\right)\right)\tilde{\psi}_j\right]$$

$$- \frac{\pi\hbar^2}{km}\left[(|\phi_1|^2 + |\chi_1|^2)(\psi_2^\dagger\psi_2 - \tilde{\psi}_2^\dagger\tilde{\psi}_2)\right.$$

$$\left. + (|\phi_2|^2 + |\chi_2|^2)(\psi_1^\dagger\psi_1 - \tilde{\psi}_1^\dagger\tilde{\psi}_1)\right]. \qquad (42.52)$$

For the supersymmetry rules, one must also rescale the supersymmetry parameters as well. One finds that if one rescales both ϵ_i in the same way, $\epsilon_i \to \sqrt{\hbar/(2mc)}\epsilon_i$, one obtains a rather trivial theory with two supersymmetries. The other possibility arises if we consider $\delta\phi_1$ in (42.46), expanding ϵ as (ϵ_1, ϵ_2) and $\tilde{\eta}$ from (42.51), which implies that to keep both terms, we need to rescale ϵ_1 and ϵ_2 differently,

$$(\epsilon_1, \epsilon_2) \longrightarrow \left(\sqrt{\frac{\hbar}{2mc}}\epsilon_1, \sqrt{\frac{c}{2m\hbar}}\epsilon_2 \right), \tag{42.53}$$

in which case for consistency of the transformation rules (considering, for instance, $\delta\phi_1$ and $\delta\chi_1$ together) we are forced to consider the further truncation of the model to $\chi_i = \psi_i = 0$. Then finally the truncated action is

$$
\begin{aligned}
S_{\text{NR,trunc.}} = -\frac{N(N-1)}{2} \int d^3x \Bigg\{ & \frac{k\hbar}{4\pi}\epsilon^{\mu\nu\lambda}\left(A_\mu^{(2)}F_{\nu\lambda}^{(1)} + A_\mu^{(1)}F_{\nu\lambda}^{(2)}\right) - \bar{\phi}_i\left(i\hbar D_t + \frac{\hbar^2}{2m}D_j^2\right)\phi_i \\
& - \sum_{j=1,2}\left[\psi_j^\dagger\left(i\hbar D_t + \frac{1}{2m}\left(D_i^2 - F_{12}^{(j)}\right)\right)\psi_j\right] \\
& + \frac{\pi\hbar^2}{km}\left[(|\phi_1|^2)(\psi_2^\dagger\psi_2) + (|\phi_2|^2)(\psi_1^\dagger\psi_1)\right] + \frac{2\pi\hbar^2}{mk}\left(|\phi_1|^2|\phi_2|^2\right) \Bigg\}, \tag{42.54}
\end{aligned}
$$

and the supersymmetry transformation rules are

$$\delta\phi_1 = -\epsilon_1^*\psi_{1,1} + \frac{1}{2m}\epsilon_2^*D_-\psi_{1,1},$$

$$\delta\phi_2 = \epsilon_1^*\psi_{2,1} - \frac{1}{2m}\epsilon_2^*D_-\psi_{2,1},$$

$$\delta A_t^{(1)} = +\frac{\pi\hbar}{mk}(\epsilon_1^*\phi_2\psi_{2,1}^* + \epsilon_1\phi_2^*\psi_{2,1}) + \frac{2\pi\hbar}{(2m)^2 k}(\epsilon_2^*\phi_2 D_+\psi_{2,1}^* + \epsilon_2\phi_2^*D_-\psi_{2,1}),$$

$$\delta A_1^{(1)} = -\frac{i\pi\hbar}{mk}(\epsilon_2^*\phi_2\psi_{2,1}^* + \epsilon_2\phi_2^*\psi_{2,1}),$$

$$\delta A_2^{(1)} = -\frac{\pi\hbar}{mk}(\epsilon_2^*\phi_2\psi_{2,1}^* + \epsilon_2\phi_2^*\psi_{2,1}), \tag{42.55}$$

$$\delta A_t^{(2)} = -\frac{\pi\hbar}{mk}(\epsilon_1^*\phi_1\psi_{1,1}^* + \epsilon_1\phi_1^*\psi_{1,1}) - \frac{2\pi\hbar}{(2m)^2 k}(\epsilon_2^*\phi_1 D_+\psi_{1,1}^* + \epsilon_1\phi_1^*D_-\psi_{1,1}),$$

$$\delta A_1^{(2)} = \frac{i\pi\hbar}{mk}(\epsilon_2^*\phi_1\psi_{1,1}^* + \epsilon_2\phi_1^*\psi_{,11}),$$

$$\delta A_2^{(1)} = \frac{\pi\hbar}{mk}(\epsilon_2^*\phi_1\psi_{1,1}^* + \epsilon_2\phi_1^*\psi_{1,1}),$$

$$\delta\psi_{1,1} = \frac{1}{2m}\epsilon_2 D_-\phi_1 - \epsilon_1\phi_1,$$

$$\delta\psi_{2,1} = -\frac{1}{2m}\epsilon_2 D_-\phi_2 + \epsilon_1\phi_2. \tag{42.56}$$

This abelian, nonrelativistic, supersymmetric model contains an important bosonic sub-model. Considering the further truncation

$$A_\mu^{(1)} = A_\mu^{(2)} = A_\mu, \quad \phi_1 = \phi_2 = \phi, \tag{42.57}$$

leads to the *bosonic* action (dropping the fermions)

$$S_{JP} = -N(N-1)\int d^3x \left\{ \frac{k\hbar}{4\pi}\epsilon^{\mu\nu\lambda}A_\mu F_{\nu\lambda} - \bar\phi\left(i\hbar D_t + \frac{\hbar^2}{2m}D_j^2\right)\phi + \frac{\pi\hbar^2}{mk}(\phi\bar\phi)^2) \right\}. \tag{42.58}$$

This is the action of the *Jackiw-Pi model*. The equation of motion for ϕ reads

$$i\hbar D_t\phi = -\frac{\hbar^2}{2m}D_j^2\phi + \frac{2\pi\hbar^2}{mk}(\bar\phi\phi)\phi, \tag{42.59}$$

and as we can see, it is a gauged nonlinear Schrödinger equation, with very interesting properties. In fact, we have already obtained such a gauged nonlinear Schrödinger equation, in the form of the Landau-Ginzburg equation (11.41) and the Gross-Pitaevskii equations (11.57) and (11.58) after gauging. But the gauge field equation of motion will be different, because of the Chern-Simons coupling. That suggests that the Jackiw-Pi model has vortex solutions, related to the gauged vortices described in the context of the Landau-Ginzburg equation, that correspond to (topological) vortices with flux, but with a different profile. Their properties will be described in the next chapter. Being nonrelativistic, this model of a more general interest to condensed matter than the case of the relativistic Landau-Ginzburg theory from the last section.

Important Concepts to Remember

- Quantum critical systems in 2+1 dimensions, systems with a quantum phase transition at $g = g_c$, can be described via AdS/CFT because they are conformal, strongly coupled and relativistic. They are described by an effective relativistic Landau-Ginzburg model.
- The standard 2+1–dimensional toy model is ABJM theory, a CS gauge theory for the gauge group $SU(N) \times SU(N)$, with $\mathcal{N} = 6$ supersymmetry, and bifundamental scalars and fermions.
- The ABJM theory admits a maximally supersymmetric mass deformation that splits the scalars in two, $C^I = (Q^\alpha, R^{\dot\alpha})$.
- The vacuum of the massive theory is a fuzzy sphere (a quantum version, or approximation, of the classical sphere) of constant radius, $R^\alpha = cG^\alpha$, and the BPS solution of the massless model is a fuzzy funnel, with $c = c(s)$.
- The gravity dual of the massless ABJM is string theory in $AdS_4 \times \mathbb{CP}^3$ background, with $\mathbb{CP}^3$ appearing as $S^7/\mathbb{Z}_k$ for $k \to \infty$, which kills off the fiber in the Hopf fibration of S^7 with S^1 fiber.
- We can describe the string-corrected gravity dual of ABJM by a metric (with cosmological constant) coupled to a gauge field, and a coupling of the Weyl tensor to two field strengths.

- We can write a nontrivial abelian reduction of the massive ABJM model to the relativistic Landau-Ginzburg model. The reduction is a consistent truncation, even at the quantum level, i.e. the LG model is a low-energy effective field theory, like in a condensed matter system.
- The reduction includes a Mukhi-Papageorgakis Higgs mechanism, where a CS gauge field eats a would-be Goldstone boson scalar to become Yang-Mills.
- We can sketch the process of reduction of degrees of freedom in a superconductor to LG as similar to the reduction in degrees of freedom from ABJM to LG.
- The reduction of the massive ABJM to Landau-Ginzburg can be made supersymmetric.
- One can take a nonrelativistic limit of the supersymmetric Landau-Ginzburg model, and obtain, after a necessary truncation, a supersymmetric model that contains as a bosonic subsector the Jackiw-Pi model, with equation of motion the gauged nonlinear Schrödinger equation.

Further Reading

The original papers [118] and [119].

Exercises

(1) Show that the most general *irreducible* solution to the algebra (42.9) is (42.12), up to a gauge transformation.
(2) Calculate the action $S_0 + S_1$ on the ansatz for the AdS black hole in (42.19).
(3) Show that the kinetic terms of the ABJM action reduce to the kinetic terms in (42.24) on the reduction ansatz (42.22).
(4) Show that the abelian reduction of the ABJM model on the ansatz (42.22) is a consistent trunction.
(5) Find the equations of motion of the Jackiw-Pi model (42.58) and show that they solve the equations of motion of the supersymmetric action (42.52).

43

Particle-Vortex Duality and ABJM vs. $AdS_4 \times \mathbb{CP}^3$ Duality

In this chapter we want to explore the consequences of a 2+1–dimensional strong/weak duality, particle-vortex duality, described in Chapter 21. For condensed matter applications, in particular in the area of quantum phase transitions, 2+1–dimensional strongly coupled systems are of interest. Often these systems possess solitons of vortex type, as we have seen, for example, in the case of superconductivity. But in 3+1–dimensional quantum field theories, a Maxwell-type duality between solitons (specifically monopoles) and particles has led to many physical results, in particular to the Seiberg and Witten solution for the low-energy effective action for $\mathcal{N} = 2$ supersymmetric theories.

It is therefore of great interest to analyze particle-vortex duality, its physical consequences, and how to embed it in our standard 2+1–dimensional toy model, the ABJM model. We also would like to understand how that duality manifests itself in the gravity dual: we will see that there it becomes the usual Maxwell duality in 3+1 dimensions, and how to take the nonrelativistic limit of the duality. We will obtain a relation between particles and Jackiw-Pi vortices, which will be of greater interest in the context of the mostly nonrelativistic condensed matter applications.

43.1 Particle-Vortex Dualities in the Path Integral

In Section 21.2 in Part II we have seen how we can define particle-vortex duality in the path integral, relating an abelian-Higgs model with vortex solutions to a theory with an extra Maxwell gauge field and a real scalar. The duality at the level of the fields was roughly for the phase α of the complex scalar mapping to an abelian gauge field b_μ, $\Phi_0^2 \partial_\mu \alpha \sim \epsilon_{\mu\nu\rho} \partial^\nu b^\rho$, while the real modulus of the scalar, Φ_0, remains unchanged. Moreover, the particle current $j_\mu \sim \Phi_0^2 \partial_\mu \alpha$ and vortex current j_{vortex}^μ were related by the duality also, as $\Phi_0^2 j_{\text{vortex}}^\mu \sim \epsilon^{\mu\nu\rho} \partial_\nu j_\rho$.

Symmetric Duality

But we would like to find also a symmetric duality, that exchanges one action with another one formally of the same type, just exchanging the fields and couplings.

Consider then a symmetric starting point, with path integral

$$
Z = \int \mathcal{D}a_\mu \mathcal{D}\Phi_0 \mathcal{D}\chi_0 \mathcal{D}\theta \mathcal{D}\tilde{b}_\mu \exp\left\{ -i \int d^3x \left[\frac{1}{2}|(\partial_\mu - iea_\mu)\Phi_0 e^{-i\theta}|^2 + \frac{1}{2}(\partial_\mu \chi_0)^2 \right. \right.
$$
$$
\left. \left. + \frac{1}{4e^2\chi_0^2} f_{\mu\nu}^{(\tilde{b})} f^{(\tilde{b})\mu\nu} + \epsilon^{\mu\nu\rho} a_\mu \partial_\nu \tilde{b}_\rho - \frac{2\pi}{e} \tilde{b}_\mu \tilde{j}_{\text{vortex}}^\mu(t) + V(\Phi_0^2) + V(\chi_0^2) \right] \right\}. \quad (43.1)
$$

Here j^μ_{vortex} is a current that will be associated with the nontrivial (vortex) boundary conditions for the scalar in the dual description, $\tilde{b}_\mu$ is a gauge field to be dualized, and χ_0 will become the modulus of the dual scalar.

We see that we need to dualize both the gauge field $\tilde{b}_\mu$ and the phase of Φ. As in Chapter 21 we write a first order formulation for $\lambda_\mu = \partial_\mu\theta$, with θ the phase of Φ, and impose that it has no field strength through Lagrange multipliers b_μ. We also do the same procedure in reverse for $\tilde{b}_\mu$, defining

$$(\tilde{\lambda}_{\mu,\text{smooth}} + \tilde{\lambda}_{\mu,\text{vortex}} + ea_\mu)e\chi_0^2 = \epsilon^{\mu\nu\rho}\partial_\nu\tilde{b}_\rho. \tag{43.2}$$

The master path integral for the duality must have it as an equation of motion, so it becomes

$$Z = \int \mathcal{D}a_\mu\mathcal{D}\Phi_0\mathcal{D}\chi_0\mathcal{D}\lambda_\mu\mathcal{D}b_\mu\mathcal{D}\tilde{\lambda}_\mu\mathcal{D}\tilde{b}_\mu$$

$$\times \exp\left\{-i\int d^3x\left[\frac{1}{2}(\partial_\mu\Phi_0)^2 + \frac{1}{2}(\partial_\mu\chi_0)^2 + \frac{1}{e}\epsilon^{\mu\nu\rho}(b_\mu\partial_\nu\lambda_\rho + \tilde{b}_\mu\partial_\nu\tilde{\lambda}_\rho)\right.\right.$$

$$+\frac{1}{2}(\lambda_{\mu,\text{smooth}} + \lambda_{\mu,\text{vortex}} + ea_\mu)^2\Phi_0^2$$

$$\left.\left.+\frac{1}{2}(\tilde{\lambda}_{\mu,\text{smooth}} + \tilde{\lambda}_{\mu,\text{vortex}} + e\tilde{a}_\mu)^2\chi_0^2 + V(\Phi_0^2) + V(\chi_0^2)\right]\right\}. \tag{43.3}$$

We can now do the dualities, i.e. repeat the procedure that led to the master action exchanging tilde and untilde quantities: integrate over λ_μ and $\tilde{b}_\mu$, obtaining $\tilde{\lambda}_\mu = \partial_\mu\tilde{\theta}$. The dual path integral is then

$$Z = \int \mathcal{D}a_\mu\mathcal{D}\Phi_0\mathcal{D}\chi_0\mathcal{D}\tilde{\theta}\mathcal{D}b_\mu\exp\left\{-i\int d^3x\left[\frac{1}{2}|(\partial_\mu - iea_\mu)\chi_0e^{-i\tilde{\theta}}|^2 + \frac{1}{2}(\partial_\mu\Phi_0)^2\right.\right.$$

$$\left.\left.+\frac{1}{4e^2\Phi_0^2}f^{(b)}_{\mu\nu}f^{(b)\mu\nu} + \epsilon^{\mu\nu\rho}a_\mu\partial_\nu b_\rho - \frac{2\pi}{e}b_\mu j^\mu_{\text{vortex}}(t) + V(\Phi_0^2) + V(\chi_0^2)\right]\right\}. \tag{43.4}$$

Mukhi-Papageorgakis Higgs Mechanism

We can now define more precisely the Mukhi-Papageorgakis Higgs mechanism alluded to in the last chapter. As we said there, it is a mechanism specific to 2+1 dimensions, where in analogy with the usual Higgs mechanism (defined originally in 3+1 dimensions, but also valid in 2+1 dimensions), a Chern-Simons gauge field (thus with no propagating degrees of freedom) eats up a dynamical scalar (via a field redefinition) and becomes a Maxwell gauge field (with one propagating degree of freedom in 2+1 dimensions).

Since the Mukhi-Papageorgakis Higgs mechanism will be used in the context of (abelian) particle-vortex duality, we will describe here a simple abelian version for it that will be used subsequently.

Consider an abelian Higgs model, with the gauge field being nondynamical, but coupled in a Chern-Simons term to another gauge field $\tilde{a}_\mu$:

$$S = -\int d^3x\left[\frac{k}{2\pi}\epsilon^{\mu\nu\rho}a_\mu\partial_\nu\tilde{a}_\rho + \frac{1}{2}|(\partial_\mu - iea_\mu)\Psi|^2 + V(|\Psi|^2)\right]. \tag{43.5}$$

Consider also that the vacuum (minimum of the potential) is at $\Psi = b$. Expand around this vacuum like in the usual Higgs mechanism, just that now we separate a possible vortex contribution θ_{vortex}, a singular part of θ that appears for a vortex solution,

$$\Psi = (b + \delta\psi)e^{-i\delta\theta}; \quad \delta\theta = \theta_{\text{smooth}} + \theta_{\text{vortex}}, \tag{43.6}$$

to obtain the perturbative action (dropping higher order terms in the perturbations)

$$S = -\int d^3x \left[\frac{k}{2\pi}\epsilon^{\mu\nu\rho}a_\mu\partial_\nu\tilde{a}_\rho + \frac{1}{2}(\partial_\mu\delta\psi)^2 + \frac{1}{2}(\partial_\mu\theta_{\text{smooth}} + \partial_\mu\theta_{\text{vortex}} + ea_\mu)^2 b^2 + \cdots \right]. \tag{43.7}$$

Exactly like in the usual Higgs mechanism, the "eating" of the one scalar degree of freedom and consequent "transmutation" of the gauge field is achieved through the field redefinition

$$ea_\mu + \partial_\mu\theta_{\text{smooth}} + \partial_\mu\theta_{\text{vortex}} = ea'_\mu. \tag{43.8}$$

Replacing this redefinition in the perturbative action, partially integrating the term linear in $\tilde{a}_\mu$ to obtain the $j^\mu_{\text{vortex}}\tilde{a}_\mu$ term, and solving for a'_μ, which gives

$$a^\mu + \frac{1}{e}\partial^\mu\delta\theta = a'^\mu = -\frac{k}{2\pi b^2}\epsilon^{\mu\nu\rho}\partial_\nu\tilde{a}_\rho, \tag{43.9}$$

we find the perturbative action

$$S = \int d^3x \left[-\frac{k^2}{16\pi^2 b^2}(\tilde{f}_{\mu\nu})^2 - \frac{1}{2}(\partial_\mu\delta\psi)^2 + \frac{k}{e}j^\mu_{\text{vortex}}\tilde{a}_\mu + \cdots \right], \tag{43.10}$$

The relation (43.9) takes a form similar to the particle-vortex duality relation, as does the relation between (43.5) and (43.10), though note that we have now a *perturbative* relation between actions, and also the interpretation of the formal manipulations is different. This will become apparent in the application.

Symmetric Duality for Dynamical Scalar Theory

We apply the Mukhi-Papageorgakis Higgs mechanism to the case of the symmetric duality. Instead of having a duality between a theory with a scalar and a gauge field, and the dual theory with the scalar and the gauge field "interchanged," now we want to define a duality between theories with two scalars (coupled to external gauge fields).

We consider therefore the path integral to be dualized to be

$$Z = \int \mathcal{D}a_\mu\mathcal{D}\Phi_0\mathcal{D}\theta\mathcal{D}\tilde{b}_\mu\mathcal{D}\chi\mathcal{D}\chi^*\mathcal{D}\mathcal{A}_\mu \exp\left\{-i\int d^3x \left[\frac{1}{2}|(\partial_\mu - iea_\mu)\Phi_0 e^{-i\theta}|^2 \right.\right.$$
$$\left.\left. + \frac{1}{2}|(\partial_\mu - ie\mathcal{A}_\mu)\chi_0 e^{-i\phi}|^2 + \epsilon^{\mu\nu\rho}\left(\frac{1}{e}\mathcal{A}_\mu\partial_\nu\tilde{b}_\rho + a_\mu\partial_\nu\tilde{b}_\rho\right) + V(\phi_0^2) + V(\chi_0^2)\right]\right\}. \tag{43.11}$$

To apply the Mukhi-Papageorgakis Higgs mechanism, we redefine $\mathcal{A}_\mu$ to $\mathcal{A}'_\mu$ via (43.8) therefore "eating" ϕ, allowing us to do trivially the path integral over it, since the path integral is now ϕ independent. Since now $\mathcal{A}'_\mu$ appears algebraically in the path integral, we can solve for it by using its equation of motion, thus getting the starting point of the symmetric duality (43.1), which we have shown is dual to (43.4).

We can now "undo" the Mukhi-Papageorgakis Higgs mechanism, by introducing a first order formulation for $\tilde{f}_{\mu\nu}^{(b)}$ in terms of an auxiliary field $\tilde{\mathcal{A}}_{\mu}'$, introduce a trivial path integration over a variable $\tilde{\phi}$, and redefine $\tilde{\mathcal{A}}_{\mu}'$ to "regurgitate" (i.e., make dynamical) the scalar $\tilde{\phi}$ via a redefinition equivalent to the one for $\mathcal{A}_{\mu}'$:

$$e\tilde{\mathcal{A}}_{\mu} + \partial_{\mu}\tilde{\phi} + \partial_{\mu}\tilde{\phi}_{\text{vortex}} \equiv e\tilde{\mathcal{A}}_{\mu}' = \frac{1}{e^2\Phi_0^2}\epsilon^{\mu\nu\rho}\partial_{\nu}b_{\rho}. \qquad (43.12)$$

Finally, we obtain the dual path integral:

$$Z = \int \mathcal{D}a_{\mu}\mathcal{D}\chi_0\mathcal{D}\tilde{\theta}\mathcal{D}b_{\mu}\mathcal{D}\Phi\mathcal{D}\Phi^*\mathcal{D}\tilde{\mathcal{A}}_{\mu}$$
$$\times \exp\left\{-i\int d^3x \left[\frac{1}{2}|(\partial_{\mu} - iea_{\mu})\chi|^2 + \frac{1}{2}|(\partial_{\mu} - ie\tilde{\mathcal{A}}_{\mu})\Phi|^2\right.\right.$$
$$\left.\left. + \epsilon^{\mu\nu\rho}\left(\frac{1}{e}\tilde{\mathcal{A}}_{\mu}\partial_{\nu}b_{\rho} + a_{\mu}\partial_{\nu}b_{\rho}\right) + V(\phi_0^2) + V(\chi_0^2)\right]\right\}. \qquad (43.13)$$

Here $\chi = \chi_0 e^{-i\tilde{\theta}}$ and $\Phi = \Phi_0 e^{-i\tilde{\phi}}$.

43.2 Particle-Vortex Duality Constraints on Condensed Matter Transport

As we saw, exchanging a particle description with a vortex description modifies the Lagrangean, and as a result, when particles and vortices are considered as the charge carriers, it will also modify the electromagnetic response to external fields, quantified in the case of *conductors* by the Ohmic conductivity σ_{xx} and the Hall conductivity σ_{xy}. Constructing the complex conductivity

$$\sigma = \sigma_{xy} + i\sigma_{xx}, \qquad (43.14)$$

duality will amount to a subgroup of the $PSl(2, \mathbb{Z})$ action on σ,

$$\tilde{\sigma} = \frac{a\sigma + b}{c\sigma + d}, \qquad (43.15)$$

where the integers a, b, c, d satisfy $ad - bc = 1$.

In general, the quantum field theory effective action as a function of external (electromagnetic) fields is defined by

$$e^{i\Gamma(A_{\text{ext}})} = \int \mathcal{D}[\text{fund.fields}]e^{i\int \mathcal{L}_{\text{eff}}(\text{fund.fields}, A_{\text{ext}})}, \qquad (43.16)$$

where the quadratic part, defining the linear response, defines a $\Pi^{\mu\nu}$ by

$$\Gamma(A) = -\frac{1}{2}\int d^3x d^3x' A_{\mu}(x)\Pi^{\mu\nu}(x - x')A_{\nu}(x'). \qquad (43.17)$$

A particle action will be a function of $a_\mu + A_\mu^{\text{ext}}$, whereas a vortex action will be a function of b_μ, where a_μ is the statistical gauge field and b_μ is the dual of the scalar field phase.

Indeed, adding a statistical gauge field a_μ on the starting particle action in (21.17), we have

$$
\mathcal{L}_p = -\frac{1}{2}\partial_\mu \Phi_0^2 - V - \frac{F_{\mu\nu}^2}{4} - \frac{\pi}{2\theta}\epsilon^{\mu\nu\rho}a_\mu \partial_\nu a_\rho
$$
$$
-\frac{1}{2}[\partial_\mu \alpha + (a_\mu + A_\mu)]^2, \tag{43.18}
$$

so we obtain instead of (21.20) the dual action

$$
\mathcal{L}_v = -\frac{1}{2}\partial_\mu \Phi_0^2 - V - \frac{F_{\mu\nu}^2}{4} - \frac{\pi}{2\theta}\epsilon^{\mu\nu\rho}a_\mu \partial_\nu a_\rho
$$
$$
-\epsilon^{\mu\nu\rho}b_\mu \partial_\nu(a_\rho + A_\rho) - b_\mu j_{\text{vortex}}^\mu
$$
$$
-\frac{(f_{\mu\nu}^b)^2}{4\Phi_0^2}, \tag{43.19}
$$

where one could neglect the term on the last line, since it is subleading at low energy. Note that with respect to (12.24), the normalization of the statistical gauge field action differs by a factor of $e^2/(2\pi)$.

In analogy with the full quantum effective action, one can write an effective action where one integrates over the particle action, except for the statistical gauge field,

$$
e^{i\Gamma_p(a+A)} = \int \mathcal{D}[\cdots]e^{i\mathcal{L}_p(\ldots,a+A)}, \tag{43.20}
$$

where the quadratic part, defining the (linear) *particle response function $P^{\mu\nu}$* is

$$
\Gamma_p = -\frac{1}{2}\int d^3x d^3x' a_\mu(x)P^{\mu\nu}(x-x')a_\nu(x'). \tag{43.21}
$$

We can also define the effective action when we integrate over everything except for the dual gauge field for vortices b_μ, such that

$$
e^{i\Gamma_v(b)} = \int \mathcal{D}[\cdots]e^{i\mathcal{L}_v(\ldots,b)}, \tag{43.22}
$$

and the quadratic part, defining the (linear) *vortex response function $V^{\mu\nu}$*, is

$$
\Gamma_v = -\frac{1}{2}\int d^3x d^3x' b_\mu(x)V^{\mu\nu}b_\nu(x'). \tag{43.23}
$$

One can write a form factor expansion, isolating allowed momentum structures for $\Pi^{\mu\nu}, P^{\mu\nu}, V^{\mu\nu}$, as

$$\Pi^{\mu\nu} = \Pi_1(p^2)\left(\delta^{\mu\nu} - \frac{p^\mu p^\nu}{p^2}\right) + \Pi_2(p^2)i\epsilon^{\mu\nu\rho}\frac{p_\rho}{\sqrt{p^2}}$$

$$P^{\mu\nu} = P_1(p^2)\left(\delta^{\mu\nu} - \frac{p^\mu p^\nu}{p^2}\right) + P_2(p^2)i\epsilon^{\mu\nu\rho}\frac{p_\rho}{\sqrt{p^2}}$$

$$V^{\mu\nu} = V_1(p^2)\left(\delta^{\mu\nu} - \frac{p^\mu p^\nu}{p^2}\right) + V_2(p^2)i\epsilon^{\mu\nu\rho}\frac{p_\rho}{\sqrt{p^2}}, \tag{43.24}$$

and form the complex combinations $\Pi = \Pi_1 + i\Pi_2$, $P = P_1 + iP_2$, $V = V_1 + iV_2$.

Then one can relate Π to P and V by doing the (linearized) functional integrations in the corresponding effective actions, obtaining (putting $\hbar = 1$)

$$\Pi = \frac{i\sqrt{p^2}\frac{\pi}{\theta}(p^2 + VP)}{p^2 + V\left(P + i\sqrt{p^2}\frac{\pi}{\theta}\right)}. \tag{43.25}$$

In the case of no vortices, we have $V \to \infty$, which from (43.25) gives the *particle-mediated electromagnetic response*:

$$\Pi = i\sqrt{p^2}\frac{\pi}{\theta}\frac{P}{P + i\sqrt{p^2}\frac{\pi}{\theta}}. \tag{43.26}$$

In the case of no particles, we have $P = 0$, which from (43.25) gives the *vortex-mediated electromagnetic response*:

$$\tilde{\Pi} = \frac{p^2}{V - i\sqrt{p^2}\frac{\theta}{\pi}}. \tag{43.27}$$

The statement of duality is to identify $\tilde{V} = P$, i.e. V in (43.27) with P in (43.26), the dual response in terms of vortices equals the original response in terms of particles. This relates the electromagnetic response for the two carriers:

$$\frac{\tilde{\Pi}}{p^2} = \frac{i\sqrt{p^2}\frac{\pi}{\theta} - \Pi}{p^2 + i\sqrt{p^2}\left(\frac{\pi}{\theta} + \frac{\theta}{\pi}\right)\Pi}. \tag{43.28}$$

The case of interest is of conductors, where the low-energy behavior of the response functions is linear in momentum (so that the response is proportional to the electric field $p_{[i}A_{j]}$):

$$\Pi_1(p^2) \to \sigma_{xx}\sqrt{p^2} + \cdots$$
$$\Pi_2(p^2) \to \sigma_{xy}\sqrt{p^2} + \cdots. \tag{43.29}$$

Then forming $\Pi = \Pi_1 + i\Pi_2$ and $\sigma = \sigma_{xy} + i\sigma_{xx}$, we have

$$\Pi = i\sqrt{p^2}\sigma^*, \tag{43.30}$$

so that finally the particle-vortex duality relation *on the conductivity* is

$$\tilde{\sigma} = \frac{\frac{\pi}{\theta} - \sigma}{1 - \left(\frac{\theta}{\pi} + \frac{\pi}{\theta}\right)\sigma} = \frac{1 - \frac{\theta}{\pi}\sigma}{\frac{\theta}{\pi} - \left(1 + \left(\frac{\theta}{\pi}\right)^2\right)\sigma}. \tag{43.31}$$

Note that, since this takes the form (43.15) with $ad - bc = 1$, we indeed have a subgroup of $PSl(2, \mathbb{Z})$ for both fermionic charge carriers $\theta = \pi$, and bosonic charge carriers $\theta = 0$. Indeed, for $\theta/\pi \in \mathbb{Z}$, the coefficients a, b, c, d in (43.15) are all integers.

43.3 Particle-Vortex Duality Embedded in the ABJM Model

We have seen how to construct an abelian particle-vortex duality in the path integral, and to use it to obtain constraints on the transport in conductors. But to use string theory methods, including AdS/CFT, we should embed it in a model with a gravity dual. Our standard 2+1–dimensional toy model is the ABJM model, so we will try to embed the duality in it.

More precisely, in general the duality maps a theory to another, but as we saw, we could make the duality symmetric, such that it is a self-duality, that changes only the interpretations of fields and couplings, but not the model itself.

We want therefore to embed the action in (43.11) into the action of the ABJM model. Therefore we need to identify the two complex scalars Φ and χ, the two gauge fields coupled to them, a_μ and $\mathcal{A}_\mu$, and the neutral gauge field $\tilde{b}_\mu$. Except for the field $\tilde{b}_\mu$ that couples them, the others give identical decoupled actions, which could therefore have similar origins.

To obtain the *abelian* action in (43.11), we choose to split the N-dimesional matrix space into two block diagonal $N/2$-dimensional subspaces. In the first subspace, we write the abelian reduction ansatz:

$$\begin{aligned}
A_\mu &= a_\mu^{(1)} \mathbf{1}_{N/2 \times N/2}, \\
\hat{A}_\mu &= \hat{a}_\mu^{(1)} \mathbf{1}_{N/2 \times N/2}, \\
Q^1 &= \phi G_{N/2 \times N/2}^1, \\
Q^2 &= \phi G_{N/2 \times N/2}^2, \\
R^\alpha &= 0.
\end{aligned} \tag{43.32}$$

Here A_μ and $\hat{A}_\mu$ are the gauge fields corresponding to the two $U(N)$ factors of the $U(N) \times U(N)$ gauge group, under which Q^α and R^α, $\alpha = 1, 2$ are the bifundamental scalars.

We first calculate the covariant derivative on Q^α,

$$D_\mu Q^\alpha = G_{N/2 \times N/2}^\alpha \left(\partial_\mu \phi + i\left(a_\mu^{(1)} - \hat{a}_\mu^{(1)}\right)\phi\right), \tag{43.33}$$

which gives for the scalar kinetic term

$$\text{Tr}[|D_\mu Q^\alpha|^2] = 2\frac{N}{2}\left(\frac{N}{2} - 1\right)|\partial_\mu + i\left(a_\mu^{(1)} - \hat{a}_\mu^{(1)}\right)\phi|^2. \tag{43.34}$$

For the potential we also need the scalar combination

$$M^\alpha = \mu Q^\alpha + \frac{2\pi}{k}(Q^\alpha Q^\dagger_\beta Q^\beta - Q^\beta Q^\dagger_\beta Q^\alpha) = G^\alpha_{N/2 \times N/2}\left(\mu\phi + \frac{2\pi}{k}\phi^3\right), \quad (43.35)$$

which leads to the scalar potential

$$V = \mathrm{Tr}[|M^\alpha|^2] = \frac{N}{2}\left(\frac{N}{2} - 1\right)|\phi|^2\left|\mu + \frac{2\pi}{k}\phi^2\right|^2 = \frac{N}{2}\left(\frac{N}{2} - 1\right)|\phi|^2\left(\mu + \frac{2\pi}{k}|\phi|^2\right)^2.$$
$$(43.36)$$

On the other hand, the Chern-Simons term on the abelian reduction ansatz becomes

$$\frac{k}{4\pi}\frac{N}{2}\epsilon^{\mu\nu\rho}\left(a^{(1)}_\mu \partial_\nu a^{(1)}_\rho - \hat{a}^{(1)}_\mu \partial_\nu \hat{a}^{(1)}_\rho\right) = \frac{k}{4\pi}\frac{N}{2}\epsilon^{\mu\nu\rho}\left(a^{(1)}_\mu + \hat{a}^{(1)}_\mu\right)\partial_\nu\left(a^{(1)}_\rho - \hat{a}^{(1)}_\rho\right). \quad (43.37)$$

Together, between the scalar kinetic term, the scalar potential and the Chern-Simons term, we obtain the first half of the action in (43.11), with the identification $\Phi \to \phi$, $a_\mu \to a^{(1)}_\mu - \hat{a}^{(1)}_\mu$ and $\tilde{b}_\mu \to a^{(1)}_\mu + \hat{a}^{(1)}_\mu$. We obtain the other half, for the fields χ and $\tilde{A}$ from the second $N/2$ subspace, but we must obtain *the same* field $\tilde{b}_\mu$, so the reduction on the two subspaces must be supplemented with the constraint:

$$\tilde{b}_\mu = a^{(1)}_\mu + \hat{a}^{(1)}_\mu = a^{(2)}_\mu + \hat{a}^{(2)}_\mu. \quad (43.38)$$

Therefore, we have obtained the action required for the self-duality from the reduction of the ABJM model.

43.4 Particle-Vortex Duality as Maxwell Duality in AdS_4

Now that we have embedded particle-vortex duality in the ABJM model, we want to understand what it corresponds to in the gravity dual to the ABJM model, $AdS_4 \times \mathbb{CP}^3$. We will see that in fact it corresponds to the usual Maxwell duality in AdS_4.

AdS/CFT Mapping
We start by deriving what the Maxwell duality in AdS_4 corresponds to on the boundary. For the 3-dimensional conformal field theory on the boundary, the partition function in Euclidean signature for a $U(1)$ current operator J_i coupled to a source a_i, boundary value for the AdS_4 bulk field A_μ, is

$$Z_{\mathrm{CFT}}[a_i] = \int \mathcal{D}[\mathrm{fields}]e^{-S[\mathrm{fields}]+\int d^3 x J^i a_i}. \quad (43.39)$$

The AdS/CFT relation equates this with the on-shell supergravity partition function with boundary source a_i. In the case of A_μ, the action is the on-shell Maxwell action, so we have

$$Z_{\mathrm{sugra}}[a_i] = e^{-\int d^4 x \sqrt{-g}\left[+\frac{1}{4g^2}F^2_{\mu\nu}\right]}. \quad (43.40)$$

We consider the *radial gauge* $A_z = 0$, so we simply have $A_i \to a_i$ on the boundary.

Then the 4-dimensional Maxwell duality,

$$\tilde{F}^{\mu\nu} = \frac{1}{2\sqrt{-g}} \epsilon^{\mu\nu\rho\sigma} F_{\rho\sigma}, \tag{43.41}$$

leaves invariant the action, therefore the partition function is written in terms of the dual fields as

$$Z_{sugra}[a_i] = Z_{sugra}[\tilde{a}_i] = e^{-\int d^4x \sqrt{-g}\left[+\frac{1}{4g^2}\tilde{F}^2_{\mu\nu}\right]}. \tag{43.42}$$

We define a particle-vortex duality for the currents J_i on the boundary in the usual way defined in Chapter 21:

$$J^i = \frac{1}{2}\epsilon^{ijk}\partial_j \tilde{J}_k. \tag{43.43}$$

Transforming the CFT partition function using this duality for currents, we obtain

$$Z_{\text{CFT}}[a_i] = \int \mathcal{D}[\text{fields}]e^{-S[\text{fields}]+\int d^3x \frac{1}{2}\epsilon^{ijk}(\partial_j \tilde{J}_k)a_i} = \int \mathcal{D}[\text{fields}]e^{-S[\text{fields}]+\int d^3x \tilde{J}^i(\frac{1}{2}\epsilon^{ijk}\partial_j a_k)}, \tag{43.44}$$

where we have done a partial integration. That means that if the sources a_i would also transform in the same way,

$$\tilde{a}^i = \frac{1}{2}\epsilon^{ijk}\partial_j a_k, \tag{43.45}$$

then $Z_{\text{CFT}}[\tilde{a}_i]$ would be written exactly like the dual of $Z_{sugra}[\tilde{a}_i]$; therefore the particle-vortex duality on the boundary would come from Maxwell duality in the bulk.

Maxwell Duality in the Bulk

For the case of AdS_4 in Poincaré coordinates, with metric

$$ds^2 = \frac{-dt^2 + dx^2 + dy^2 + dz^2}{z^2}, \tag{43.46}$$

the Maxwell duality (43.41) exchanges electric and magnetic fields as

$$\tilde{F}_{01} = -F_{23}; \quad \tilde{F}_{23} = -F_{01}, \cdots \tag{43.47}$$

In the radial gauge $A_3 = \tilde{A}_3 = 0$, near the boundary $z = 0$ of AdS space, we can write the expansion

$$A_i = a_i + z\bar{a}_i + \frac{z^2}{2}a_i^{(2)} + \frac{z^3}{3!}a_i^{(3)} + \cdots$$

$$\tilde{A}_i = \tilde{a}_i + z\tilde{\bar{a}}_i + \frac{z^2}{2}\tilde{a}_i^{(2)} + \frac{z^3}{3!}\tilde{a}_i^{(3)} + \cdots, \tag{43.48}$$

and we have

$$\tilde{F}_{01}(z = 0) = \partial_z A_2(z = 0); \quad F_{01}(z = 0) = \partial_z \tilde{A}_2(z = 0), \tag{43.49}$$

which means that the duality relations on the first two leading terms give

$$\tilde{f}_{ij} = \frac{1}{2}\epsilon_{ijk}\bar{a}_k; \quad f_{ij} = \frac{1}{2}\epsilon_{ijk}\tilde{\bar{a}}_k. \tag{43.50}$$

Here f_{ij} and $\tilde{f}_{ij}$ are boundary field strengths. The Maxwell duality for the higher terms and the Maxwell equations relate the subleading terms in the boundary expansion $a^{(2)}$, $a^{(3)}$, etc. in terms of the first two, as one could verify.

We see that we can specify independently a_i and $\tilde{a}_i$, or a_i and $\bar{a}_i$, and then the solutions for A_i and $\tilde{A}_i$ are fixed. The duality relations exchange $\tilde{a}_i$ with $\bar{a}_i$ and a_i with $\tilde{a}_i$. But this duality corresponds to the particle-vortex duality exchanging currents (43.43), as we saw in (43.45).

43.5 Nonrelativistic Limit and Duality with Jackiw-Pi Vortex

Finally, we consider a nonrelativistic limit of the particle-vortex duality, for it to more easily apply to condensed matter systems. As we saw in the last chapter, we can take a nonrelativistic limit of the abelian reduction of the ABJM model that gives the Lagrangean (42.58), which is the Jackiw-Pi model, whose ϕ equation of motion we had already seen.

Considering the Jackiw-Pi action with a general coupling, usually of opposite sign (putting $\hbar = 1$ and $\kappa \equiv k/(2\pi)$),

$$S_{JP} = \int d^3x \left\{ \frac{\kappa}{2} \epsilon^{\mu\nu\lambda} A_\mu F_{\nu\lambda} + \bar{\phi}\left(iD_t + \frac{D_j^2}{2m}\right)\phi + g\left(\phi\bar{\phi}\right)^2 \right\}, \qquad (43.51)$$

for which the equations of motion are

$$iD_t\phi = -\frac{1}{2m}D_i^2\phi - g\bar{\phi}\phi\phi$$
$$-\epsilon^{\mu\nu\lambda}F_{\nu\lambda} = \frac{1}{\kappa}j^\mu \equiv \frac{1}{\kappa}\frac{\delta S_\phi}{\delta A_\mu}. \qquad (43.52)$$

We have a Bogomolny'i limit $g \to 1/(m|\kappa|)$ for which we can write, as usual for topological solitons, first order equations of motion, obtained either by imposing that in the Hamiltonian the sum of squares vanishes independently, or in a supersymmetric context (the model was embedded in a supersymmetric one), by imposting that half the supersymmetry is preserved. These are

$$D_i\phi = i\epsilon_{ij}D_j\phi,$$
$$\epsilon_{ij}\,\partial_i A_j = -\frac{1}{\kappa}\bar{\phi}\phi, \qquad (43.53)$$

together with the Gauss law for this Chern-Simons case, stating that a solution with electric charge q has also a magnetic flux $\Phi = -q/\kappa$. Consider the ansatz for the vortex solution

$$\phi(x) = \sqrt{\rho(x)}e^{i\omega}, \qquad (43.54)$$

where the phase is given by $\omega = N\theta$. Here θ is the polar angle in the plane (the complex coordinate z is $z = re^{i\theta}$), as we saw that we need for a general vortex. Then on this ansatz, the equations of motion reduce to the equation for ρ:

$$\nabla^2 \ln \rho = -\frac{2}{\kappa}\rho. \tag{43.55}$$

The solution to this equation is given in terms of an arbitrary holomorphic function $f(z)$ as

$$\rho(r) = \kappa \nabla^2 \ln(1 + |f(z)|^2) = \frac{4\kappa|f'(z)|^2}{(1 + |f(z)|^2)^2}. \tag{43.56}$$

For the case $f(z) = c_0 z^{-n}$, we obtain the $(n-1)$–vortex solution (with topological charge $n-1$) with scale parameter r_0, which is explicitly

$$\phi(r, \theta) = \frac{2\sqrt{\kappa}n}{r}\left(\left(\frac{r_0}{r}\right)^n + \left(\frac{r}{r_0}\right)^n\right)^{-1} e^{i(1-n)\theta}. \tag{43.57}$$

Note that another remarkable property of these vortices is that, unlike the usual relativistic ANO vortex, we can now write *exact* solutions, without the need for numerics.

We now try to understand the particle-vortex duality in the nonrelativistic limit. The naive expectation is that it will relate the nonrelativistic limits of the two dual actions. Indeed, we will see that this expectation is correct.

As we have seen in Chapter 6, the kinetic action for the external electromagnetic field, the potential and the kinetic action for the modulus of the scalar are not affected by the duality. The only transformation is between

$$-\frac{1}{2}\int d^3x (\partial_\mu \alpha_{\text{smooth}} + \partial_\mu \alpha_{\text{vortex}} + ea_\mu)^2 \Phi_0^2$$
$$\rightarrow \int d^3x \left[-\frac{(f_{\mu\nu}^b)^2}{4\Phi_0^2} - \epsilon^{\mu\nu\rho}b_\mu \partial_\nu a_\rho - \frac{2\pi}{e}b_\mu j_{\text{vortex}}^\mu\right], \tag{43.58}$$

with the duality relation

$$\partial_\mu \alpha + ea_\mu = -\frac{1}{\Phi_0^2}\epsilon^{\mu\nu\rho}\partial_\nu b_\rho. \tag{43.59}$$

In our case, the original theory does indeed involve the terms above, in the nonrelativistic limit. Since the nonrelativistic limit involves peeling off a factor $e^{-i\frac{mc^2}{\hbar}}$ from ϕ, which corresponds to extracting an addtive factor of $mc^2/\hbar$ from the phase α:

$$\partial_\mu \tilde{\alpha} + ea_\mu + \frac{mc^2}{\hbar}\delta_{\mu 0} = -\frac{1}{\Phi_0^2}\epsilon^{\mu\nu\rho}\partial_\nu b_\rho. \tag{43.60}$$

The left-hand side term squared is what gave the nonrelativistic scalar kinetic term $\phi^* D_t \phi + \phi^2 D^2/(2m)\phi$, whereas the right-hand side squared gives the electromagnetic kinetic term, which is the same in the nonrelativistic limit:

$$\int d^3x\left[\frac{1}{2}\vec{E}^2 - \frac{1}{2}\vec{B}^2\right]. \tag{43.61}$$

The Chern-Simons term $\int d^3x \epsilon^{\mu\nu\rho}b_\mu \partial_\nu a_\rho$ is also unaffected by the nonrelativistic limit, and the terms that are duality invariant have the usual nonrelativistic limit as we saw, so the dual action is

$$
S_{\text{dual,nonrel.}} = \int d^3x \left[\bar{\Phi}_0 \left(i\hbar D_t + \frac{\hbar^2 D_i^2}{2m} \right) \Phi_0 + \frac{1}{2}(\vec{E}^2 - \vec{B}^2) - V(\Phi_0) \right.
$$
$$
\left. + \frac{1}{2}((\vec{E}^{(b)})^2 - (\vec{B}^{(b)})^2) - \hbar\epsilon^{\mu\nu\rho} b_\mu \partial_\nu a_\rho - \frac{2\pi}{e}(b_0 q + \vec{b} \cdot \vec{j}) \right]. \quad (43.62)
$$

Important Concepts to Remember

- One can define a particle-vortex duality in the path integral that acts on the fields as $\Phi_0^2 \partial_\mu \alpha \sim \epsilon_{\mu\nu\rho} \partial^\nu b^\rho$ and on the currents like $\Phi_0^2 j_{\text{vortex}}^\mu \sim \epsilon^{\mu\nu\rho} \partial_\nu j_\rho$, and one can turn it into a symmetric duality by starting with fields Φ (complex), χ_0 (real) and b_μ.
- When added to the Mukhi-Papageorgakis Higgs mechanism, which is that a Chern-Simons gauge fields eats a scalar and becomes Maxwell, the symmetric duality takes a scalar action into a scalar action, when we also have some Chern-Simons gauge fields coupled to them.
- At the level of the conductivity of conductors, particle-vortex duality acts like a certain $Sl(2, \mathbb{Z})$ transformation, depending on the anyonic phase θ of the charge carriers.
- One can embed the particle-vortex self-dual scalar+CS Abelian action into the non-abelian ABJM model.
- In the gravity dual, the boundary particle-vortex duality becomes regular Maxwell duality in AdS_4.
- The Jackiw-Pi model, the nonrelativistic limit of the abelianized ABJM model, has exact vortices.
- One can define a nonrelativistic particle-vortex duality that acts on the Jackiw-Pi model and its vortices.

Further Reading

Particle-vortex duality in the path integral, the embedding in ABJM and the gravity dual interpretation were found in [120]. The constraints on condensed matter transport were found in [121].

Exercises

(1) Describe the Mukhi-Papageorgakis Higgs mechanism for a nonabelian action of the type

$$
S = -\int d^3x \left[\frac{k}{2\pi} \epsilon^{\mu\nu\rho} A_\mu \partial_\nu \tilde{A}_\nu - \frac{1}{2}|(\partial_\mu - ieA_\mu)C^I|^2 + V(C^I) \right], \quad (43.63)
$$

where $A_\mu, \tilde{A}_\mu \in U(N)$, C^I is acted on the left by A_μ, and $V(C^I)$ has a minimum at $V = 0$ for $C^I = b\delta^{I4}, b \in \mathbb{R}$.

(2) Consider the anyonic $\theta = \pi/k$ in the particle-vortex duality relation on the conductivity. What group does it correspond to?

(3) Calculate the duality relation coming from Maxwell duality, all the coefficients in the expansion of A_i, $\tilde{A}_i$ in (43.48).

(4) Write the Hamiltonian for the Jackiw-Pi action (43.51) and show that when $g = 1/(m|\kappa|)$ it can be written as a topological term plus a sum of squares that vanishes on the first order equations (43.53).

(5) Check that (43.56) is a solution to the equation

$$\nabla^2 \ln \rho = -\frac{2}{\kappa}\rho, \tag{43.64}$$

and that for $f(z) = c_0 z^{-n}$ we obtain (43.57).

Topology and Nonstandard Statistics from AdS/CFT

In this chapter we describe how to obtain topology, in the form of the Berry phase, and nonstandard statistics, in the form of anyonic and nonabelian statistics from string theory, in an AdS/CFT context.

We have already seen some signs of it in the chapter on the Quantum Hall Effect, where Chern-Simons terms were obtained from string theory.

44.1 Berry Phase from AdS/CFT

We start with a description of the Berry connection and phase. There are many ways we could obtain such a Berry connection, but we will describe one related to a model we have already used. It is easiest to obtain a Berry phase in the context of paths in some parameter space that form a nontrivial loop. Fortunately, we have studied a model where this is possible. In Chapter 37 we have described how to obtain a holographic Josephson junction. In a model that obtained the AC Josephson effect, we had considered two Dp-brane halves ending at different points in an S^2 inside the Ω_{8-9} transverse to the Dp-branes, which was wrapped by a D$(p+2)-$brane.

We had considered the case when the Dp-brane halves moved on a circle inside S^2, and then we obtained the usual abelian Josephson effect, but if we consider a motion on a nontrivial path, not parametrized by a single angle ϕ on S^2, but depending on both angles ϕ and θ, we obtain a nonabelian Josephson effect. Instead of having only the magnetic potential (coupling with the endpoints of the two Dp-brane halves, i.e. the intersections with the D$(p+2)-$brane where they end) $\mathcal{A}_{\phi 12...p-1} = B \cos(\Omega t)$, which leads to only $\mathcal{F}_{\phi t 12...p}$, we consider the more general fluxes:

$$\mathcal{F}_{\phi t 12...p-1} = \mathcal{B}_\phi = B_\phi \sin(\Omega t); \quad \mathcal{F}_{\theta t 12...p-1} = \mathcal{B}_\theta = B_\theta \cos(\Omega t); \quad \mathcal{F}_{\phi\theta 12...p-1} = E.$$

(44.1)

This will induce a nontrivial motion in the (θ, ϕ) plane for the endpoints of the Dp-brane halves, which can form curved paths, and lead to nonabelian (noncommuting) effects for the condensate of the superfluid.

The parameters of the motion of the Dp-brane endpoints are the magnetic fields $\mathcal{B}_\phi$, $\mathcal{B}_\theta$ and electric field E. Motion along a closed curve in this space, like the circle parametrized by Ω, would imply a nontrivial Berry holonomy for the state of the system. We know that the Berry connection is related to the change of the parameter space of the system, in this case the magnetic field, along the trajectory.

The state of the system is given by the condensate of the superfluid, the position of the endpoint of the Dp-brane, more specifically its nonabelian ($SU(2)$) phase given by the position on the S^2.

But it is easier to describe the motion of the system not in the (θ, ϕ) "plane" (2-sphere), but in the complex plane, by a stereographic projection:

$$z = \tan \frac{\theta}{2} e^{i\phi}. \tag{44.2}$$

We want to define an $SU(2)$ Berry connection (for the symmetry of the 2-sphere), embedded in the $Gl(2, \mathbb{C})$ symmetry of the plane, defined by an element $g = \begin{pmatrix} a & b \\ c & d \end{pmatrix}$ acting on z by

$$z' \equiv \hat{g}z = \frac{az + b}{cz + d}. \tag{44.3}$$

Consider matrices h which are in the $SU(2)$ subgroup of $Gl(2, \mathbb{C})$, i.e. with a, b, c, d restricted by $|a|^2 + |b|^2 = |c|^2 + |d|^2 = 1$ and $ac^* + bd^* = 0$. The trajectory on the plane is related to the trajectory on the sphere, and it takes the form of multiplication by an $SU(2)$ matrix h (in the sense of (44.3)),

$$\dot{z} = i\hat{h}z, \tag{44.4}$$

where the matrix is

$$h(t) = \dot{\phi}\frac{\sigma_3}{2} + \dot{\theta}\frac{\sigma_2}{2} + \dot{\theta}\sin\phi\frac{\sigma_1}{2}. \tag{44.5}$$

We leave this statement as an exercise. But we want to rewrite $\dot{z} = i\hat{h}z$ as the equation defining *parallel transport* along the trajectory in parameter ($\mathcal{B}$) space. The trajectory is in $\mathcal{B} = (\mathcal{B}_\theta, \mathcal{B}_\phi)$ space, thus the unit vector tangent to it (the trajectory being parametrized by time t) is

$$\vec{n} = \frac{1}{\sqrt{(\partial_t \mathcal{B}_\phi)^2 + (\partial_t \mathcal{B}_\theta)^2}}(\partial_t \mathcal{B}_\theta, \partial_t \mathcal{B}_\phi). \tag{44.6}$$

The parallel transport equation defining $\vec{A}$ in parameter space is (this is the definition of parallel transport, since it means to have the connection-covariant derivative always perpendicular to the trajectory along which we transport)

$$\vec{n} \cdot (\vec{\nabla} - \vec{A}) = 0, \tag{44.7}$$

which gives the Berry connection in the tangent direction as

$$\vec{n} \cdot \vec{A} = \frac{h(t)}{\sqrt{(\partial_t \mathcal{B}_\phi)^2 + (\partial_t \mathcal{B}_\theta)^2}}. \tag{44.8}$$

The Berry holonomy, i.e. the change in the state of the system, or motion in the space of condensate phases, or in the z plane, corresponds to an $SU(2)$ group element acting on the coordinate z, specifically

$$g = e^{i \oint_C dx^\mu A_\mu} = e^{i \oint_C \vec{n}\cdot\vec{A}}. \tag{44.9}$$

44.2 Anyons in AdS/CFT

It is straightforward to obtain a description of anyons in AdS/CFT. In fact, as we saw in Part I, all we need is a Chern-Simons term in the action, and we naturally have anyons. This is so, since adding point sources for the particles to the action, coupled to the gauge field A_μ, results in the equation of motion $B \equiv F_{12} \propto q\delta^2(\vec{x} - \vec{x}_0^i)$, i.e. to having a singular magnetic flux tied to the (quasi)particles at x_0^i. In turn, that means that there is an anyonic phase for interchanging two of these (quasi)particles.

Chern-Simons terms are rather easy to obtain in string theory, and also in field theories with AdS/CFT descriptions. In fact, our standard toy model for condensed matter in 2+1 dimensions, the ABJM model, does have nonabelian Chern-Simons terms, so it could be said to describe anyons, and moreover actually nonabelian anyons (nonabelions).

We have also seen that Chern-Simons terms are easily obtained on the worldvolume of Dp-branes, coming from the WZ term in the presence of RR fluxes. We have seen examples that gave the FQHE, and topological insulators and superconductors.

But all these examples were of *relativistic* systems, which do have their uses in condensed matter, like we saw in Chapter 42. It would be nice, however, to describe also nonrelativistic systems with anyons, since those are more standard. As we saw in Chapters 42 and 43, we can take a nonrelativistic limit of the abelian reduction of the ABJM model, and arrive at a model with an abelian Chern-Simons term, (42.54), an extension of the Jackiw-Pi model, that is in fact invariant under the Schrödinger group.

Nonrelativistic Anyons in a Harmonic Trap

Consider a nonrelativistic theory that is invariant under the Schrödinger group, the nonrelativistic analog of the conformal group. In this theory, if we have bosons ϕ_i and fermions ψ_i, with densities $\rho_B = \phi_i^\dagger \phi_i$ and $\rho_F = \psi_i^\dagger \psi_i$, the numbers of bosons and fermions are

$$N_B = \int d^2x \rho_B; \quad N_F = \int d^2x \rho_F, \tag{44.10}$$

which depending on the theory could be individually conserved, and the special conformal transformations are generated by

$$C = \frac{m}{2} \int d^2x |z|^2 (\rho_B + \rho_F), \tag{44.11}$$

which looks like a harmonic trap for the particles.

But then we can define the new Hamiltonian

$$L_0 = H + C, \tag{44.12}$$

which is the Hamiltonian with an added harmonic trap, and we show that when acting on primary operators, it gives the spectrum of the dilatation operator. Therefore the spectrum of the dilatation operator D equals the spectrum of the Hamiltonian placed in a harmonic trap.

Primary operators are defined in the nonrelativistic conformal theory similarly to the relativistic case, as operators satisfying

$$[C, \mathcal{O}] = 0. \tag{44.13}$$

For them, there is a map (again like in the relativistic theory) between operators and states, with the state being

$$|\Psi_\mathcal{O}\rangle = e^{-H}\mathcal{O}(0)|0\rangle, \tag{44.14}$$

and $|0\rangle$ is the vacuum of H.

Part of the Schrödinger algebra (which we have described in Part II) were the commutators

$$[H, -iD] = -2H; \quad [H, C] = -iD. \tag{44.15}$$

But, using

$$e^H C e^{-H} = C + [H, C] + \frac{1}{2}[H, [H, C]] + \frac{1}{3!}[H, [H, [H, C]]] + \cdots, \tag{44.16}$$

we show that

$$L_0|\Psi_\mathcal{O}\rangle = (H + C)e^{-C}\mathcal{O}(0)|0\rangle = e^{-H}(H + C - iD - H + 0)\mathcal{O}(0)|0\rangle. \tag{44.17}$$

Using also the definition of the spectrum of the dilatation operator, namely the fact that the scaling dimensions $\Delta_\mathcal{O}$ appear from

$$[-iD, \mathcal{O}] = \Delta_\mathcal{O}\mathcal{O}, \tag{44.18}$$

we obtain

$$L_0|\Psi_\mathcal{O}\rangle = \Delta_\mathcal{O}|\Psi_\mathcal{O}\rangle. \tag{44.19}$$

Finally, that means that the spectrum of primary operators of different scaling dimensions (under the dilatation operator) equals to the spectrum of states (or wavefunctions) of the Hamiltonian of the theory in a harmonic trap.

For our abelian theory with a Chern-Simons term, the states will have a flux attached to them, so are *anyons*. Moreover, besides the Schrödinger group invariance, there is also an invariance under some supersymmetry, and correspondingly some R-symmetry, which relates the supersymmetries with each other. The commutator of the supersymmetry Q with the superconformal symmetry S,

$$\{Q, S^\dagger\} = \frac{i}{2}(iD - J_0 + N_B - N_F), \tag{44.20}$$

contains the orbital angular momentum

$$J_0 = \int d^2x(izP - i\bar{z}\bar{P}), \tag{44.21}$$

and we want to write it as

$$\{Q, S^\dagger\} = \frac{i}{2}\left(iD - J + \frac{3}{2}R\right). \tag{44.22}$$

The right-hand side is zero for *primary states* that obey ($[J, \mathcal{O}] = j_\mathcal{O}\mathcal{O}$, $[R, \mathcal{O}] = r_\mathcal{O}\mathcal{O}$)

$$\Delta_\mathcal{O} = j_\mathcal{O} - \frac{3}{2}r_\mathcal{O}, \qquad (44.23)$$

which are *chiral* primary states, i.e. that are invariant under Q as well. In fact, when diagonalizing the commutations of Q and S, we find that we can have either sign of Δ, i.e. the right-hand side of the above equation has a $\pm$ in front.

This means that there is some freedom in defining J and R. One choice is

$$J = J_0 - \frac{1}{2k}N_B + \frac{k-1}{2k}N_F$$
$$R = \frac{2k-1}{3k}N_B - \frac{k+1}{3k}N_F, \qquad (44.24)$$

and means the transformation of the bosonic and fermionic fields is ($[J_0, \phi_i^\dagger] = 0$)

$$[J, \phi_i^\dagger] = -\frac{1}{2k}\phi_i^\dagger; \quad [J, \psi_i^\dagger] = \frac{k-1}{2k}\psi_i^\dagger. \qquad (44.25)$$

But ϕ_i and ψ_i couple to the gauge field, which has a Chern-Simons kinetic term, and therefore attaches magnetic flux to particles. To construct gauge-invariant states, we must dress ϕ_i and ψ_i with flux. More specifically, since ϕ transforms by $i\lambda\phi$, the gauge-invariant combination is found by considering the dual photon σ, $F_{ij} = \epsilon_{ijk}\partial^k\sigma$, and by the Gauss constraint (A_0 equation of motion) we have $1/k$ units of flux tied to a quasiparticle. Therefore the gauge-invariant combinations are

$$\Phi_i = e^{-i\sigma/k}\phi_i; \quad \Psi_i = e^{-i\sigma/k}\psi_i. \qquad (44.26)$$

They transform in the same way under J as the ϕ_i and ψ_i, respectively.

One can calculate the value of J_0 for quasiparticle states, and one can show that it is a sum over quasiparticle pairs, each with a $-1/k$ contribution, so for n quasiparticles:

$$J_0 = -\frac{n(n-1)}{2k}. \qquad (44.27)$$

Then the total angular momentum J is

$$J = -\frac{n^2}{2k}. \qquad (44.28)$$

Consider the chiral primary operator

$$\mathcal{O} = (\Phi^\dagger)^n. \qquad (44.29)$$

It is made of n gauge-invariant quasiparticle bosonic units, so we obtain

$$j_\mathcal{O} = -\frac{n^2}{2k}; \quad r_\mathcal{O} = \frac{2k-1}{3k}n. \qquad (44.30)$$

Since it is chiral primary, it has as we saw $\Delta = \pm(j - 3r/2)$, so we find the dimensions

$$\Delta_\mathcal{O} = \mp\left[n + \frac{n(n-1)}{2k}\right]. \qquad (44.31)$$

We said that these dimensions, i.e. eigenvalues of the dilatation operator, are identified as the eigenenergies for the *anyonic* states in a harmonic trap. These states are known and are called "linear" solutions in condensed matter.

The (ground state) wavefunctions corresponding to these operators $(\Phi^\dagger)^n$ can be found based on the symmetries of the quasiparticles that make them and are

$$\Psi_{0,n} = \prod_{i<j=1}^{n} |z_i - z_j|^{1/k}, \tag{44.32}$$

for $k < 0$, when the 4-point $|\phi|^4$ interaction in (42.54) is attractive. The wavefunctions *in the harmonic trap* are

$$\tilde{\Psi}_{0,n} = \prod_{i<j=1}^{n} |z_i - z_j|^{1/k} \exp\left(-\frac{m}{2} \sum_{j=1}^{n} |z_i|^2\right). \tag{44.33}$$

44.3 Nonabelian Statistics in AdS/CFT

For nonabelian statistics we simply need to consider the nonabelian version of the ABJM constructions in the previous section. ABJM theory itself is nonabelian, with a nonabelian Chern-Simons term, so it automatically gives (quasi)particles with nonabelian magnetic flux, therefore with nonabelian anyonic phases, i.e. nonabelions.

So the ABJM model itself contains nonabelions, and its gravity dual is $AdS_4 \times \mathbb{CP}^3$. But in condensed matter we are more interested in nonrelativistic systems, so we can take a nonrelativistic limit of the ABJM model in the same way as we did for the abelianization of ABJM. We will obtain a model that is similar to (42.54), but is nonabelian. One should in principle follow the same construction of the previous section to find the wavefunctions for nonabelions in a harmonic trap, but there are some technical issues that make it more difficult.

There are other constructions of nonabelions in string theory and AdS/CFT. Vafa proposed in 2015 a way to embed the Fractional Quantum Hall Effect in string theory, including the nonabelion states, in a theory on M5-branes in M-theory, wrapped on a squashed 3-sphere S^3 and a 2-dimensional Riemann surface Σ, such that the FHE lives in $\mathbb{R}_t \times \Sigma$, but we will not explain it here, as it is quite complicated.

Important Concepts to Remember

- We can define a Berry phase and connection in the case of a nonabelian Josephson junction, where the nonabelian phase of the condensate follows a nontrivial motion on a sphere, as the parameters of the theory, magnetic fields, are varied on closed loops. This generates a holonomy on the sphere in the form of the Berry phase.

- Anyons can be obtained from Chern-Simons terms, which are relatively easy to obtain in string theory. In particular, they arise in the ABJM model, and on the worldvolume of Dp-branes in RR-flux (perhaps generated by a Dq-brane).
- Nonrelativistic anyons are obtained from nonrelativistic models with Chern-Simons terms, like the nonrelativistic limit of the abelianized ABJM model.
- In nonrelativistic theories invariant under the Schrödinger group (equivalent of the conformal group), with special conformal generator C, the Hamiltonian $H + C$ is the Hamiltonian with a harmonic trap added, and is given on primary operators by the eigenvalues of the dilatation operator. If H contained a CS term, we have anyons in a harmonic trap.
- We can find chiral primary operators, corresponding to wave functions for states of the anyons in a harmonic trap, as gauge-invariant states with required charges and symmetries.
- Nonabelian Chern-Simons leads to nonabelions, so the ABJM model has them, as would a nonrelativistic limit of it.

Further Reading

The Berry phase construction described here was presented in [92]. The construction of abelian and nonabelian anyons was presented in [122].

Exercises

(1) Prove that the matrix $h(t)$ corresponding to the motion on S^2 onto the stereographic plane is given by (44.5).
(2) Calculate the Berry curvature of the connection (44.8).
(3) Consider a kinetic term for bosons and fermions

$$S_k = \frac{1}{2} \int d^2x \, dt \left[\psi_i^\dagger i D_t \psi_i + \phi_i^\dagger i D_t \phi_i - \frac{|D_a \phi_i|^2}{2m} - \frac{|D_a \psi_i|^2}{2m} \right]. \quad (44.34)$$

Show that it is invariant under C, and find the spacetime symmetry it corresponds to.

(4) Show that the addition of C to the Hamiltonian shifts the anyon wavefunctions from (44.32) to (44.33).

(5) Calculate the nonabelian flux associated with an added point particle coupled to the gauge field in the ABJM model, and from it, the nonabelian holonomy $P \exp \left(i \oint_C \vec{A} \cdot d\vec{x} \right)$ acting on the wavefunction of the point particle.

45

DBI Scalar Model for QGP/Black Hole Hydro- and Thermo-Dynamics

In this chapter we will use and generalize a simple scalar model from 1952 by Heisenberg, introduced to describe the asymptotic behavior of the cross section for scattering in QCD with the center of mass energy. The model will give a holographic description of a plasma in terms of a black hole in a gravity dual, and will describe the thermodynamics and hydrodynamics using the simple scalar theory.

45.1 The Heisenberg Model and Quark-Gluon Plasma

The Heisenberg model was developed to describe the asymptotic behavior for the total scattering cross section $\sigma_{tot}(s)$ in high-energy ($s \to \infty$) scattering in QCD (the center of mass energy is $\sqrt{s} \equiv \sqrt{-(p_1 + p_2)^2}$). In QCD, we have the well-known Froissart unitarity bound, i.e. a bound obtained imposing the conservation of probability, or unitarity, which says that

$$\sigma_{tot}(s) \leq C \ln^2 \frac{s}{s_0}; \quad C \leq \frac{\pi}{m^2}. \tag{45.1}$$

The model by Heisenberg was introduced, in 1952, to find $\sigma_{tot}(s \to \infty)$ in high-energy nucleon collisions, and it was found to saturate the Froissart unitarity bound, before Froissart (1961), and before even there was a QCD theory.

Our interest in the model will be not the behavior of $\sigma_{tot}(s)$, which is a particle physics issue, but rather in describing the properties of the quark-gluon plasma forming in the high-energy nucleon-nucleon scattering. Indeed, we know that, *at least in heavy ion collisions* (according to experimental data), but probably even in nucleon-nucleon scattering (at least according to the Heisenberg model), a strongly coupled plasma of quarks and gluons forms. We have described some of the properties of its equivalent in $\mathcal{N} = 4$ SYM in previous chapters. It would be good if we could use a simple model instead of the complicated strongly coupled QCD calculations.

Heisenberg started with the simple but important observation that the classical limit for a field theory corresponds to a large occupation number. For instance, the classical electromagnetic field corresponds to a large number of photons, and for an emitted wave, it encodes the energy radiated by the field at large distances. In the case of nuclear interactions, an effective description, at large enough distances so that the microscopic description in terms of QCD is irrelevant, is in terms of the lightest QCD particle, the pion. The pion is strictly speaking in a triplet representation (for the $SU(2)$ isospin symmetry that interchanges the up u quark and down d quark) (π^+, π^-, π^0), but the simple model of Heisenberg considered a scalar singlet field ϕ.

Then we can use the classical field theory of the scalar pion ϕ to describe nuclear collisions in which many pions are created, like the case of the $s \to \infty$, $t = -(p_2 + p_3)^2$ fixed scattering, or the total cross section in the $s \to \infty$ limit, dominated by the same. The calculation of the (quantum) total cross section is obtained in a semiclassical way: first, one calculates the emitted energy per unit pion momentum $d\mathcal{E}/dk_0$, which is identified with the classically calculated pion field energy per unit momentum dE/dk, and from this, one finds the average per pion emitted energy, $\langle k_0 \rangle = \mathcal{E}/N$. Finally, assuming that the radiated energy is proportional to the center of mass energy $\sqrt{s}$ and the pion wave function overlap e^{-mb}, where b is an impact parameter and that at the collision limit $b_{\max}$ we have $\mathcal{E} = \langle k_0 \rangle$ (a single emitted pion), one can calculate the total scattering cross section $\sigma_{\text{tot}}(s) = \pi b_{\max}^2(s)$.

The classical ϕ field configuration we want is obtained by the following reasoning. If we collide two finite-sized nuclei at $s \to \infty$, the Lorentz contraction in the direction of the collision means that the colliding nuclei are pancake-shaped, or in the strict limit, like a shockwave. Moreover, the pion field surrounding them is also Lorentz contracted, so we are colliding two pion field shockwaves. But the shockwave collision is hard to calculate. Instead, one can consider a single shockwave, and still extract relevant information, since we find that there is information about the radiated energy in the shockwave.

The shockwave is considered as a function of

$$\tilde{s} = t^2 - x^2, \tag{45.2}$$

only, where x is the position in the collision direction, and $\phi = 0$ before the shockwave passes, which means also $\phi(x^+ = 0) = 0$ or $\phi(x^- = 0) = 0$ ($x^\pm = t \pm x$). Then we must have $\phi \propto (x^+)^q$ near $x^+ \sim 0$, which means that for relativistic invariance we need to have $\phi = \phi(x^+ x^-) = \phi(\tilde{s})$.

The second observation of Heisenberg is that we need a classical action for this high-energy, yet nonperturbative (since it involves many quanta of ϕ) regime. The action needs to be nonlinear, to reflect the nonperturbative nature. But we also need to have a finite $(\partial_\mu \phi)^2$ at the shock $\tilde{s} = 0$, for the nonlinearities to actually play a role at the shock. If one would take a polynomial action, $\mathcal{L} = -(\partial_\mu \phi)^2/2 - V(\phi)$, with $V = \sum_n a_n \phi^n$, one would find that $(\partial_\mu \phi)^2$ is not finite at the shock. Moreover, one would find that $\sigma_{\text{tot}}(s)$ is constant, and therefore does not saturate the Froissart bound.

Instead, it is found that the only nonlinear action for ϕ that satisfies the required criteria is the DBI action for the scalar:

$$\mathcal{L} = l^{-4} \left[1 - \sqrt{1 + l^4 [(\partial_\mu \phi)^2 + m^2 \phi^2]} \right]. \tag{45.3}$$

Its equation of motion is

$$-\Box \phi + m^2 \phi + l^4 \frac{[(\partial_\mu \partial_\nu \phi)(\partial_\mu \phi)\partial_\nu \phi + (\partial_\mu \phi)^2 m^2 \phi]}{1 + l^4 [(\partial_\mu \phi)^2 + m^2 \phi^2]} = 0. \tag{45.4}$$

But we are interested in relativistic field configuration $\phi = \phi(\tilde{s})$, for which

$$(\partial_\mu \phi)^2 = -4\tilde{s} \left(\frac{d\phi}{d\tilde{s}} \right)^2, \tag{45.5}$$

so the condition of $(\partial_\mu \phi)^2$ being finite at the shock means $\phi(\tilde{s}) \sim \sqrt{\tilde{s}}$ as $\tilde{s} \to 0$. On these field configurations, the DBI Lagrangean becomes

$$\mathcal{L} = l^{-4} \left[1 - \sqrt{1 + l^4 \left(-4\tilde{s} \left(\frac{d\phi}{d\tilde{s}} \right)^2 + m^2 \phi^2 \right)} \right]. \tag{45.6}$$

The equations of motion become, after some nontrivial manipulations, multiplying with a denominator and canceling terms,

$$4\frac{d}{d\tilde{s}} \left(\tilde{s} \frac{d\phi}{d\tilde{s}} \right) + m^2 \phi = 8\tilde{s} l^4 \left(\frac{d\phi}{d\tilde{s}} \right)^2 \frac{\left[\frac{d\phi}{d\tilde{s}} + m^2 \phi \right]}{1 + l^4 m^2 \phi^2}. \tag{45.7}$$

An exact solution of these equations can be found when $m = 0$,

$$\phi = \frac{1}{a} \log \left(1 + \frac{a^2}{2l^4} \tilde{s} + \frac{a}{2l^4} \sqrt{4l^4 \tilde{s} + a^2 \tilde{s}^2} \right), \quad \tilde{s} \geq 0, \tag{45.8}$$

and $\phi = 0$ for $\tilde{s} < 0$, where a is a parameter.

When $m \neq 0$, we can find a perturbative solution away from the shock as

$$\phi = \frac{\sqrt{\tilde{s}}}{l^2}(1 + a\tilde{s}m^2 + \cdots), \quad 0 \leq \tilde{s} \ll 1/m^2, \tag{45.9}$$

and $\phi = 0$ for $\tilde{s} < 0$.

In some sense, this pion field shockwave has to give a description of the quark-gluon plasma being created in the high-energy collisions. We will explain in more detail in the next sections how that will happen. Since we are not interested in particle physics issues, we will not explain further details about the calculation of $\sigma_{\text{tot}}(s)$.

45.2 DBI Scalar Model for Black Hole

When we create a quark-gluon plasma "fireball" (at a nonzero temperature) in high-energy collisions, it should correspond in the gravity dual to creating a black hole. Indeed, in previous chapters we have described the finite temperature plasmas as corresponding to black holes. But in a paper in 2005, I have proposed that moreover, high-energy collisions that form "fireballs" are dual to gravitational shockwave collisions that create black holes in the gravity dual, more specifically in the IR region of the gravity dual.

A simple model for gravity dual is the so-called "hard-wall model," where we just cut off AdS space,

$$ds^2 = \frac{r^2}{R^2} d\vec{x}^2 + \frac{R^2}{r^2} dr^2 + R^2 ds_X^2, \tag{45.10}$$

at an $r_{\min} \sim R^2 \Lambda_{\text{QCD}}$. We will explain more about it in below.

In such a model, the black hole that is formed will be approximately at the cutoff $r_{\min}$, more so the more energy it has, i.e. the more energy we put into the collision that forms it. At high enough energies, the quantum fluctuations away from this cutoff will be more and more suppressed, and the black hole will be more and more classical.

We can then effectively consider the scattering happening *on the cutoff in the IR*. The gravitational shockwave solution in this case has the asymptotic form at large r:

$$\Phi(r, y = 0) \simeq R_s \sqrt{\frac{2\pi R}{r}} C_1 e^{-M_1 r}; \quad C_1 = \frac{j_{1,1}^{-1/2} J_2(j_{1,1})}{a_{1,1}}; \quad J_1(z) \sim a_{1,1}(z - j_{1,1}); \quad z \to j_{1,1}.$$

$$(45.11)$$

Here $M_1 = j_{1,1}/R$ is the mass of the lightest mode of the graviton KK reduced onto the $r = r_{\min}$ surface and $j_{1,1}$ is the first zero of $J_1(z)$. As such, it should correspond to the mass of the lightest QCD particle in the field theory, i.e. the pion mass. But as we saw, the pion field wavefunction indeed decays as $\phi(r) \sim e^{-m_\pi r}$ at large r.

We come therefore to the realization that the collisions creating black holes in the IR of the gravity dual should correspond to the collisions producing quark-gluon plasma in the gauge theory, and moreover those should be able to be described by the collisions of pion field shockwaves, which should also produce a "fireball."

We should therefore be able to describe the black hole in the IR by a pion field "fireball" that should also emit Hawking-like radiation at a finite temperature. In the gravitational theory, Hawking radiation was thought to lead to "quantum information loss," since the information about the states that fall into the black hole would be lost if the black hole emits perfectly thermal radiation only. But it was shown that in string theory, at least in certain limits, we can describe the Hawking radiation as a unitary process of quantum scattering. In fact, as we argued, the description of such a process in string theory led to the definition of AdS/CFT. If the black hole is dual to a pion field "fireball," it means that in a unitary quantum field theory like the pion scalar field theory from the Heisenberg model, it could also appear that we have a perfectly thermal radiating state, despite the fact that the scattering is unitary.

This expectation is borne out by the explicit construction of a field theory that leads to a black hole–type solution with finite temperature. Consider a scalar field solution of (45.3) of the type $\phi = \Phi(r) + \delta\Phi$. Then the equation of motion for the perturbation $\delta\Phi$ is

$$-\partial_t \frac{1}{\sqrt{1 + (\vec{\nabla}\Phi)^2}} \partial_t \delta\Phi + \partial_i \frac{1}{\sqrt{1 + (\vec{\nabla}\Phi)^2}} \left(\delta^{ij} - \frac{\partial^i\Phi}{\sqrt{1 + (\vec{\nabla}\Phi)^2}} \frac{\partial^j\Phi}{\sqrt{1 + (\vec{\nabla}\Phi)^2}} \right) \partial_j \delta\Phi = 0.$$

$$(45.12)$$

Unruh has considered analogs of black holes in acoustic theory, i.e. sound wave propagation in fluids, where he called them "dumb holes." These arise in ultrasonic hydrodynamic flows ("sonic booms"), where the velocity reaches ultrasonic levels $v > c$. The horizon of the dumb holes corresponds to the boom surface $v = c$. The flow is irrotational ($\vec{\nabla} \times \vec{v} = 0$), so it is described by a potential Φ, with $\vec{v} = \vec{\nabla}\Phi$, with fluctuation equation

$$\frac{1}{\rho}\left(\frac{d}{dt} + \vec{v}\cdot\vec{\nabla} + (\vec{\nabla}\cdot\vec{v})\right)\frac{\rho}{c^2}\left(\frac{d}{dt} + \vec{v}\cdot\vec{\nabla}\right)\delta\Phi - \frac{1}{\rho}\vec{\nabla}(\rho\vec{\nabla}\delta\Phi) = 0. \quad (45.13)$$

This can be identified with the fluctuation equation for a scalar field in a curved spacetime, with $\partial_\mu \sqrt{g} g^{\mu\nu} \partial_\nu \delta\Phi = 0$, if

$$\sqrt{g} g^{\mu\nu} = \rho \begin{pmatrix} \frac{1}{c^2} & \frac{v^i}{c^2} \\ \frac{v^j}{c^2} & \frac{v^i v^j}{c^2} - \delta^{ij} \end{pmatrix}. \tag{45.14}$$

If we define a new time coordinate by $d\tau = dt + v^i dx^i/(c^2 - v^2)$, the metric becomes, in the case of a radial flow,

$$ds^2 = \frac{\rho}{c} \left[(1 - v^2/c^2)c^2 d\tau^2 - \frac{dr^2}{1 - v^2/c^2} - r^2 d\Omega^2 \right], \tag{45.15}$$

whereas the scalar fluctuation equation becomes

$$\partial_0 \frac{\rho/c^2}{1 - v^2/c^2} \partial_0 \delta\Phi + \partial_i \rho \left(\frac{v^i v^j}{c^2} - \delta^{ij} \right) \partial_j \delta\Phi = 0. \tag{45.16}$$

We see that we can identify the scalar field fluctuation in the DBI solution (45.12) with the scalar field fluctuation in the radial flow (45.16), if we identify the Φ's, i.e. the pion field scalar with the velocity potential for the fluid flow, and

$$c^2 = 1 + (\vec{\nabla}\Phi)^2, \quad \rho = \frac{1}{\sqrt{1 + (\vec{\nabla}\Phi)^2}}; \quad v^i = \partial^i \Phi. \tag{45.17}$$

As Unruh described, all we need is for the scalar field propagation to be in an "effective metric" that has a horizon with a nonzero surface gravity κ. Then we can follow Hawking's calculation of the temperature of the black hole and find the usual $T = \kappa/(2\pi)$.

Consider a static, spherically symmetric solution with only $g_{rr}(r)$ and $g_{tt}(r)$ nontrivial, and maybe a nontrivial r-dependent factor in front of the sphere metric $d\Omega^2$. Like we did in the case of the Schwarzschild and Reissner-Nordstrom black hole, we can calculate the surface gravity from

$$(2\kappa)^2 = \lim_{\text{horizon}} \frac{g^{rr}}{g_{tt}} (\partial_r g_{tt})^2. \tag{45.18}$$

In the case of the DBI scalar pion theory, with the metric (45.15), we find

$$2\kappa = \left| \sqrt{1 + \Phi'^2} \frac{d}{dr} \left[\frac{1}{1 + \Phi'^2} \right] \right|_{r = r_{\text{horizon}}}. \tag{45.19}$$

For this to be finite, we need to add a special kind of source term to the DBI scalar action, at $m_\pi \simeq 0$. A source term, linear in ϕ, will modify the shape of $\Phi(r)$, and corresponds to a spread out nucleon, sourcing the pion field. Consider the source term

$$\int d^4x \phi \frac{\alpha}{r^2}. \tag{45.20}$$

A source term will not modify the fluctuation equation (which comes from the quadratic piece in Φ: one for the equation of motion, the other for variation $\delta\Phi$), so it will modify only the static solution $\Phi(r)$. For the above source, the solution is

$$\Phi(r) = \int_r^\infty dx \frac{\bar{C} + \alpha x}{\sqrt{x^4 - (\bar{C} + \alpha x)}}, \qquad (45.21)$$

with $\bar{C} = -\alpha^2 l^2/4$ and $r_{\text{horizon}} = \alpha l^2/2$ for $\alpha > 0$. The horizon of this solution is where $\Phi'(r) \to \infty$, where the square root in the solution vanishes. Near $r \simeq r_{\text{horizon}}$, we find the approximate solution (defined up to a constant ϕ_0)

$$\Phi(r) \simeq (\phi_0+)\frac{\alpha}{2\sqrt{2}} \ln(r - r_0). \qquad (45.22)$$

Plugging this into (45.19) and calculating the temperature, we find

$$T = \frac{\sqrt{2}}{\pi \alpha l^2}. \qquad (45.23)$$

In fact, it can be proven that to have a scalar field solution with a horizon with a finite surface gravity, we need to have the DBI action, plus a source term that decays faster than $1/r$ at infinity, thus the action we considered is in fact unique!

The thermal horizon described here also has another property characteristic for black holes, namely an infinite time delay for information to get out of it. Indeed, the reason that a black hole is black is that no information can escape the horizon, taking an infinite time from the point of view of the observer at infinity. An equivalent explanation would be that (the phase velocity $v_{\text{ph}} = \omega/k$ and) the group velocity $v_{\text{gr}} = d\omega/dk$, describing the propagation of information in a field wave, go to zero at the horizon.

In the case of our scalar field black hole, dubbed "pionless hole," of course information propagating in a light (electromagnetic) wave travels at $v = c$, since we are in a relativistic theory, but the only issue here is what is the speed of propagation of the pion field wave. Considering a spherical wave $\delta\Phi = Ae^{i(\omega t - kr)}$ in the perturbation equation (45.16), we obtain the dispersion relation near the horizon

$$\omega^2(k) \simeq \frac{1}{1+\Phi'^2}\left(k^2 - 3ik\frac{\Phi'\Phi''}{1+\Phi'^2}\right) \simeq \frac{8(r - r_{\text{horizon}})}{\alpha^2}[(r - r_{\text{horizon}})k^2 + 3ik], \quad (45.24)$$

which implies vanishing phase and group velocities at the horizon *for all k*:

$$c_{\text{ph}}^2 = \frac{\omega^2}{k^2} \simeq \frac{8(r - r_{\text{horizon}})}{\alpha^2}\left[r - r_{\text{horizon}} + \frac{3i}{k}\right] \to 0;$$

$$c_{\text{gr}} = \frac{d\omega}{dk} \simeq \frac{\sqrt{i}}{\alpha}\sqrt{\frac{6(r - r_{\text{horizon}})}{k}} \to 0. \qquad (45.25)$$

Note that this is not something so unusual. In fact, we have experimentally obtained even light speeds in specific media that could be slowed even to observable speeds, i.e. to zero for all intents and purposes. The point is that if *all frequencies* are slowed to zero at a surface, the surface should act as a thermal horizon, by the above formalism borrowed from Hawking's calculation.

45.3 Generalizations of the Heisenberg Model

We now want to understand how unique is the Heisenberg model described by the DBI action (45.3), and how can we generalize it. The simplest generalization is to replace inside the square root the mass term $m^2\phi^2$ to a general potential $2V$, obtaining

$$\mathcal{L} = l^{-4}\left[1 - \sqrt{1 + l^4[(\partial_\mu\phi)^2 + 2V(\phi)]}\right]. \qquad (45.26)$$

Of course, we could also add the potential outside the square root, though that is less interesting.

Truncating the action with V inside the square root to the first subleading term,

$$\begin{aligned}
\mathcal{L} &= -\frac{1}{2}(\partial_\mu\phi)^2 - V(\phi) + \frac{l^4}{8}[(\partial_\mu\phi)^2 + 2V(\phi)]^2 \\
&= -\frac{1}{2}(\partial_\mu\phi)^2 - \tilde{V}(\phi) + \frac{l^4}{8}[(\partial_\mu\phi)^2]^2 + \frac{l^4}{2}(\partial_\mu\phi)^2 V(\phi),
\end{aligned} \qquad (45.27)$$

one finds that the equation of motion does not admit a real scalar field solution. We leave this as an exercise.

More generally, we can consider an action with a series in derivatives of ϕ with arbitrary coefficients:

$$\mathcal{L} = -\frac{1}{2}(\partial_\mu\phi)^2 + \sum_{n\geq 2} C_n[(\partial_\mu\phi)^2]^n. \qquad (45.28)$$

Its equations of motion are

$$-\Box\phi + \Box\phi \sum_{n\geq 2} 2nC_n[(\partial_\mu\phi)^2]^{n-1} + (\partial_\mu\phi)(\partial_\nu\phi)(\partial_\mu\partial_\nu\phi)\sum_{n\geq 2} 4n(n-1)[(\partial_\mu\phi)^2]^{n-1} = 0.$$

$$(45.29)$$

We want to check whether $\phi = A\sqrt{\tilde{s}}$, the near shock solution of the DBI action, the only one that has the required property of $(\partial_\mu\phi)^2$ being finite at the shock, is also a solution for the above equations. Therefore we evaluate the equations of motion *on the solution*, and obtain

$$\frac{A}{\sqrt{\tilde{s}}}\left(1 + \sum_{n\geq 2} 2nC_n(-1)^n A^{2(n-1)}\right) + \left(\frac{A}{2\sqrt{\tilde{s}}} - \frac{A}{2\sqrt{\tilde{s}}}\right)\sum_{n\geq 2} 8n(n-1)C_n(-1)^n A^{2(n-1)} = 0.$$

$$(45.30)$$

For arbitrary coefficients, there is no solution, but for the coefficients that come from the expansion of the DBI action, the second sum over n in (45.30) is geometric, and we obtain

$$\frac{A}{\sqrt{\tilde{s}}} + \frac{2l^4 A^2}{1 - A^4 l^2}\left(\frac{A}{2\sqrt{\tilde{s}}} - \frac{A}{2\sqrt{\tilde{s}}}\right) = 0, \qquad (45.31)$$

which does have a solution, based on the fact that now both the numerator and the denominator vanish.

From this analysis, we deduce that it is quite difficult to generalize the Heisenberg model, since it is quite unique, at least in terms of an expansion in $(\partial_\mu \phi)^2$.

A possible generalization appears when we introduce several scalar fields ϕ^i, $i = 1, \ldots, N$, with different masses, corresponding to different scalar mesons. Indeed, in QCD there are many scalar mesons, the lightest of which is the pion. Heisenberg already considered summing up individual DBI Lagrangeans L_i for ϕ^i. However, inspired by the DBI action for a brane in a gravity dual, we consider the generalization with the sum inside the square root, considered also in d dimensions:

$$\mathcal{L} = l^{-(d+1)}\left[h(\phi^i) - f(\phi)\sqrt{1 + l^{d+1}[g_{ij}(\phi^k)(\partial_\mu \phi^i)(\partial_\mu \phi^j) + 2V(\phi^i)]}\right]. \quad (45.32)$$

If we consider the simpler generalization with just sums inside the square root,

$$\mathcal{L} = l^{-4}\left[1 - \sqrt{1 + l^4\left[\sum_a (\partial_\mu \phi^a)(\partial_\mu \phi^a) + \sum_a m_a^2 \phi_a^2 \right]}\right], \quad (45.33)$$

the equations of motion have the solution

$$\phi^a = A^a \sqrt{\tilde{s}}, \quad (45.34)$$

and from the equations we find the constraint

$$\sum_a (A^a)^2 = 1/l^4. \quad (45.35)$$

We leave the proof as an exercise.

Another possible generalization is to introduce also *vector mesons*. Indeed, the DBI action was first written (by Born and Infeld) for electromagnetism, as we said. The DBI action for a brane does indeed includes both scalars and electromagnetic fields, though the electromagnetic fields are massless. But we can introduce masses for the vectors the same way as for the scalar mesons: in the square root. Therefore we arrive at the action

$$\mathcal{L} = l^{-4}\left[1 - \sqrt{\det(\eta_{ab} + l^4 \partial_a \phi \partial_b \phi + l^2 F_{ab}) + m^2\phi^2 + M_V^2 A_a^2}\right]$$

$$= l^{-4}\left[1 - \sqrt{1 + l^4[(\partial_\mu \phi)^2 + m^2\phi^2] + \frac{l^4}{2}F_{ab}F^{ab} - l^8\left(\frac{1}{4}\tilde{F}_{ab}F^{ab}\right)^2 + M_V^2 A_a^2 + \cdots}\right]. \quad (45.36)$$

Here $\tilde{F}_{ab} = \frac{1}{2}\epsilon_{abcd}F^{cd}$ is the dual of the electromagnetic field.

45.4 Gravity Duals and the Heisenberg Model

We now come to the gravity dual description of the Heisenberg model. We have already seen a simple way to think about it, as the "hard-wall model," or cutoff AdS_5; see (45.10). One can calculate the scattering of gravitational shockwaves, corresponding to high-energy nucleons in the field theory. Then the field theory scattering is obtained as a convolution with a wavefunction of the gravitational shockwave scattering, which can be thought of as a weighted integral over the radial direction of AdS_5. The result of the calculation is that the higher the center of mass energy $\sqrt{s}$ of the collision, the more the scattering is situated near the IR cutoff.

In the asymptotic limit then, we have a one-to-one correspondence between the pion field shockwaves, standing in for the effect of the high-energy nucleons, and gravitational shockwaves in the gravity dual, living on the IR cutoff $r_{\min}$. We have in fact seen that the gravitational shockwave profile $\Phi(r)$ takes the form (45.11), which matches the form of the pion field wavefunction $\phi(r)$, which at large r is $\sim e^{-m_\pi r}$.

But more to the point, pure gravity theory in the gravity dual would correspond to just glueballs in the field theory (gauge-invariant combinations of the QCD, or glue, fields). To consider also the scalar pion, we need a scalar in the gravity dual, which is provided by the position X of the IR cutoff of AdS_5, promoted to a dynamical D3-brane. As we already know, the position X transverse to the brane corresponds to a DBI scalar ϕ on the D3-brane worldvolume. If the brane is situated in the IR, with a reflection symmetry with respect to the cutoff, the DBI scalar action takes the approximative form of the one in flat space, namely (45.3) at $m = 0$. If the gravity dual is not simply AdS_5 and there is a stabilization mechanism for the brane, it will generate a potential $V(\phi)$, or at least a mass for the scalar.

In fact, for a gravity dual of higher dimensionality and of a more general form, the DBI action will have the more general form (45.32).

One could also consider a gravity dual (generated by a large number N of Dp-branes) with a Dq-brane probe. This will give a lower supersymmetry, or no supersymmetry at all, and introduce in the field theory fundamental fields ("quarks") from open strings stretching between the Dp's and the Dq. Such a scenario was considered, for instance, by Sakai and Sugimoto, who considered a D4–D8-brane system, with a large number N of D4-branes generating a background, and a probe D8-brane of a U-shape, which is stable in the D4-brane background.

A modified gravity dual construction can be considered, which gives a DBI action of a generalized Heisenberg model. A *nonextremal*, i.e. finite temperature (black brane type), near-horizon D4-brane background is

$$ds^2 = \left(\frac{u}{R_{D4}}\right)^{3/2}\left[-dt^2 + \delta_{ij}dx^i dx^j + f(u)dx_4^2\right] + \left(\frac{R_{D4}}{u}\right)^{3/2}\left[\frac{du^2}{f(u)} + u^2 d\Omega_4^2\right]$$

$$F_4 = \frac{2\pi N_c}{V_4}\epsilon_4, \quad e^{\tilde{\phi}} = g_s\left(\frac{u}{R_{D4}}\right)^{3/4}, \quad R_{D4}^3 = \pi g_s N_c l_s^3, \quad f(u) = 1 - \left(\frac{u_\Lambda}{u}\right)^3. \tag{45.37}$$

In order to get a 3+1–dimensional field theory, we consider x_4 to be compact. Here V_4 is the volume of the unit four sphere Ω_4 and ϵ_4 its corresponding volume form, and we denoted the dilaton of string theory by $\tilde{\phi}$ so as not to confuse with the Heisenberg scalar. Substituting it in the DBI action of the D8-brane,

$$S_{DBI} = -T_8 \int d^9\sigma e^{-\tilde{\phi}}\sqrt{-\det[\partial_\mu X^i \partial_\nu X^j g_{ij}(X) + 2\pi\alpha' F_{\mu\nu}]}, \tag{45.38}$$

with $F_{\mu\nu} = 0$, and integrating over the 4-sphere coordinates, we obtain

$$S_{DBI} = -\tilde{T}_8 \int dt\, d^3x\, dx_4\ \phi^4 \sqrt{f(\phi) + \left(\frac{R_{D4}}{\phi}\right)^3 \left[\partial_\mu\phi\partial^\mu\phi + \frac{1}{f(\phi)}(\partial_{x_4}\phi)^2\right]}. \tag{45.39}$$

Here $\tilde{T}_8 = T_8\Omega_4/g_s$, and we have identified the u coordinate with the scalar ϕ of the Heisenberg model, from the point of view of the DBI action. We still need to deal with the coordinate x_4. In fact, the field $\phi(x^\mu, x_4)$ is expanded in modes on x_4,

$$\phi(x_4, x^\mu) = \phi_{cl}(x_4) + \sum_n \delta\phi_n(x^\mu)\zeta_n(x_4), \tag{45.40}$$

and only the zero mode is identified with the scalar pion $\phi(x^\mu)$ of the Heisenberg model. By integrating over x_4 the $\zeta_n(x_4)$ we will generate mass terms $m_n^2\phi_n^2$ plus higher orders. Assuming that these higher orders organize themselves into the mass inside the DBI square root, and keeping only the zero mode $\phi_0 = \phi(x^\mu)$, we get

$$S_{\mathrm{DBI}} = -\tilde{T}_8 \int dt\, d^3x\, \phi^4 \sqrt{f(\phi) + \left(\frac{R_{D4}}{\phi}\right)^3 \left[\partial_\mu\phi\partial^\mu\phi + m^2\phi^2\right]}. \tag{45.41}$$

This is indeed of the general form in (45.32). A -1 inside the integral appears from a WZ term.

45.5 QGP Hydrodynamics from DBI Scalar

Finally, we want to understand how we can derive a hydrodynamics description from the DBI action in the Heisenberg model. Indeed, the model was introduced to describe the high-energy scattering in QCD, and we know that if we collide heavy ions at high energy, we create a strongly coupled plasma that is well described by hydrodynamics. Therefore it should be possible to derive a hydrodynamics description from the DBI scalar.

The usual energy momentum tensor, derived from coupling with gravity, called the Belinfante tensor, for the DBI action (45.3) is

$$T_{\mu\nu}^B = -\frac{2}{\sqrt{-g}}\frac{\delta S}{\delta g^{\mu\nu}} = \frac{\partial_\mu\phi\partial_\nu\phi - g_{\mu\nu}(\partial_\rho\phi)^2 - g_{\mu\nu}l^{-4}(1 + l^4 m^2\phi^2)}{\sqrt{1 + l^4[(\partial_\mu\phi)^2 + m^2\phi^2]}} + \frac{g_{\mu\nu}}{l^4}. \tag{45.42}$$

We note that T_{00}^B goes to infinity at the shock, since the square root goes to zero there.

The relativistic hydrodynamics expansion of an isotropic fluid in Landau frame, with energy density ρ and pressure P is, to first order, as we saw in Part I,

$$\begin{aligned}
T_{\mu\nu} &= \rho u_\mu u_\nu + P(g_{\mu\nu} + u_\mu u_\nu) + \pi_{\mu\nu} \\
\pi_{\mu\nu} &= -2\eta\left[\frac{\nabla_\mu u_\nu + \nabla_\nu u_\mu}{2} + \frac{a_\mu u_\nu + a_\nu u_\mu}{2} - \frac{1}{3}(\nabla^\rho u_\rho)(g_{\mu\nu} + u_\mu u_\nu)\right] \\
&\quad - \zeta(\nabla^\mu u_\mu)(g_{\mu\nu} + u_\mu u_\nu) + \cdots,
\end{aligned} \tag{45.43}$$

where $a_\mu = u^\rho\nabla_\rho u_\mu$. Here η is the shear viscosity and ζ the bulk viscosity.

We would like to identify the energy-momentum tensor in (45.42) with the relativistic expansion. But there is a subtlety. In $d = 4$ the Belinfante tensor is not traceless for conformal models, one needs to add the total divergence of a current to construct a traceless tensor. For instance, for the free massless scalar,

$$S = \int d^4x \left[-\frac{1}{2}(\partial_\mu \phi)^2 \right] \Rightarrow$$

$$T^{B\mu}{}_\mu \equiv -\frac{2}{\sqrt{-g}} g^{\mu\nu} \frac{\delta S}{\delta g^{\mu\nu}} = \frac{2-d}{2}(\partial_\rho \phi)^2. \tag{45.44}$$

We can check that adding to it the total divergence $-\frac{1}{6}(\partial_\mu \partial_\nu - g_{\mu\nu}\partial^2)\phi^2$ makes the resulting *improved energy-momentum tensor* $T^I_{\mu\nu}$ traceless *on-shell*, i.e. by using $\partial^2 \phi = 0$. We will add a similar term to (45.42).

But to identify (45.42) with the relativistic hydrodynamics expansion, we need first to understand what is ϕ, the DBI scalar, and reversely, where in the DBI action do we find the velocity u^μ?

We will try only to describe ϕ near the shock, which corresponds to the behavior near the collision that forms quark-gluon plasma, as hydrodynamics. Therefore we will concentrate on the $\phi(\tilde{s})$ solution near $\tilde{s} = 0$.

We then need to think of ϕ as a kind of potential for the hydrodynamics. For an irrotational nonrelativistic flow, the velocity comes from a potential Φ as $\vec{v} = \vec{\nabla}\Phi$. Relativistically, if we expand the scalar in momentum (Fourier) modes, we have $\partial_\mu \phi \sim ik_\mu \phi$, and if $\phi(k)$ peaks on a single value, we have $u^\mu = k^\mu/m \propto \partial^\mu \phi$. We then define in general the 4-velocity of the fluid flow described by the scalar ϕ as

$$u^\mu = \frac{\partial^\mu \phi}{\sqrt{-(\partial_\mu \phi)^2}}. \tag{45.45}$$

Indeed, now $u^\mu u_\mu = -1$, if $(\partial_\mu \phi)^2 < 0$, which is true for $\phi(\tilde{s})$. In fact, as we saw, at $\tilde{s} \to 0$, $(\partial_\mu \phi)^2 \to -1$. Moreover, as a verification, for a canonically normalized scalar field with a potential, $\int[-(\partial_\mu \phi)^2/2 - V(\phi)]$, with $\phi = \phi(t)$ only, we obtain $u^\mu = (1, 0, 0, 0)$, and $T_{\mu\nu} = \text{diag}(\rho, p, p, p)$, with

$$\rho_\phi = \frac{1}{2}\dot{\phi}^2 + V(\phi); \quad p_\phi = \frac{1}{2}\dot{\phi}^2 - V(\phi). \tag{45.46}$$

This is the usual definition of ρ and p for a homogenous isotropic ideal fluid described by a scalar.

But not only the velocity, but also the density ρ and pressure p of the fluid, will be described by the scalar field ϕ.

We can check that in the improved energy-momentum tensor of the free massless scalar, $T^I_{\mu\nu}$, written as

$$T^I_{\mu\nu} = \frac{2}{3}\partial_\mu \phi \partial_\nu \phi - \frac{1}{6}g_{\mu\nu}(\partial_\rho \phi)^2 - \frac{\phi}{3}\left(\partial_\mu \partial_\nu \phi - \frac{1}{4}g_{\mu\nu}\partial^2 \phi\right), \tag{45.47}$$

the first two terms can be identified with the ones of the ideal fluid, with

$$\rho = -\frac{1}{2}(\partial_\mu \phi)^2$$

$$P = -\frac{1}{6}(\partial_\mu \phi)^2, \tag{45.48}$$

and the next term with a shear viscosity term, with

$$2\eta = \frac{\phi}{3}\sqrt{-(\partial_\lambda \phi)^2}, \tag{45.49}$$

and $\zeta = 0$. We leave this as an exercise. Then, however, even though η is found to be finite at the shock, what matters is that $\eta/(\rho + P) \to 0$ as $\tilde{s} \to 0$. Indeed, the thermodynamic relation $U + PV - TS = 0$ gives $\rho + P = Ts$, so we get $\eta/s \to 0$.

We consider next the ideal DBI scalar, with $m \to 0$, and an improved energy-momentum tensor of

$$T^I_{\mu\nu} = T^B_{\mu\nu} + \partial^\rho J_{\rho\mu\nu}$$

$$J_{\rho\mu\nu} = -\frac{1}{6}\frac{(g_{\mu\nu}\partial_\rho - g_{\mu\rho}\partial_\nu)\phi^2}{\sqrt{1 + l^4(\partial_\lambda \phi)^2}}. \tag{45.50}$$

By using the DBI equation of motion, it can be rewritten as

$$T^I_{\mu\nu} = \frac{\frac{4}{3}(\partial_\mu \phi \partial_\nu \phi - g_{\mu\nu}(\partial_\rho \phi)^2) - g_{\mu\nu} l^{-4}}{\sqrt{1 + l^4(\partial_\lambda \phi)^2}} + \frac{g_{\mu\nu}}{\beta^2}$$

$$+ \frac{\phi}{3\sqrt{1 + l^4(\partial_\lambda \phi)^2}}\left(\partial_\mu \partial_\nu \phi - \frac{l^4 \partial_\nu \phi (\partial_\mu \partial^\rho \phi) \partial_\rho \phi}{1 + l^4(\partial_\lambda \phi)^2}\right). \tag{45.51}$$

From the terms on the first line, with single derivatives on ϕ, and knowing the definition of the 4-velocity (45.45) we extract the density ρ and pressure P of the fluid as

$$\rho = \frac{1}{l^4\sqrt{1 + l^4(\partial_\lambda \phi)^2}} - \frac{1}{l^4}$$

$$P = \frac{1}{l^4}(1 - \sqrt{1 + l^4(\partial_\lambda \phi)^2}) - \frac{(\partial_\rho \phi)^2}{3\sqrt{1 + l^4(\partial_\lambda \phi)^2}}. \tag{45.52}$$

Near the shock, where $l^4(\partial_\mu \phi)^2 \to -1$, we obtain the conformal relation $3P \simeq \rho \to +\infty$.

The last term (with two derivatives on ϕ, i.e. with one derivative on the velocity) in the energy-momentum tensor gives a shear viscosity, but like in the case of the free massless scalar, we have $\eta/(\rho + P) = \eta/(Ts) \to 0$ at the shock.

To obtain a nontrivial (finite) shear viscosity/entropy density at the shock, we can add a term inside the square root of the DBI action:

$$S = -l^{-4}\int d^4x\left[\sqrt{1 + l^4[(\partial \phi)^2 + m^2 \phi^2] + \alpha\left[\partial^2 \phi - \frac{(\partial_\mu \phi)(\partial^\mu \partial^\rho \phi)(\partial_\rho \phi)}{(\partial_\lambda \phi)^2}\right]} - 1\right]. \tag{45.53}$$

On the shock solution $\phi(\tilde{s})$, it becomes

$$S = -l^{-4} \int d^4x \left[\sqrt{1 + l^4 \left(-4\tilde{s} \left(\frac{d\phi}{d\tilde{s}} \right)^2 + m^2\phi^2 \right) + \alpha \left(-2\frac{d\phi}{d\tilde{s}} \right)} - 1 \right]. \quad (45.54)$$

From it, we obtain a shear viscosity that blows up at the shock,

$$2\eta = \frac{\alpha\sqrt{-(\partial_\lambda\phi)^2}}{l^4 \sqrt{1 + l^4[(\partial\phi)^2 + m^2\phi^2] + \alpha\left[\partial^2\phi - \frac{(\partial_\mu\phi)(\partial^\mu\partial^\rho\phi)(\partial_\rho\phi)}{(\partial_\lambda\phi)^2}\right]}} \rightarrow \infty, \quad (45.55)$$

in such a way that the ratio to the entropy density is finite:

$$\frac{\eta}{s} = T\frac{\alpha}{l^4\sqrt{-(\partial_\lambda\phi)^2}} \rightarrow T\frac{\alpha}{l^2}. \quad (45.56)$$

We also obtain a nonzero bulk viscosity, $\zeta = 2\eta/3$.

That means that the plasma situated near the shock obeys the thermodynamic relation

$$\frac{1}{T}\frac{\eta}{s} = \frac{\alpha}{l^2}. \quad (45.57)$$

Important Concepts to Remember

- The Heisenberg model of high energy nucleon scattering describe the scattering in terms of pion field shockwaves sourced by the nucleons (or hadrons, in general), having a singular surface at $\tilde{s} = 0$ due to Lorentz contraction.
- The classical pion field description works because of the high pion multiplicity, and the action needed is nonperturbative: it is the massive DBI scalar action.
- In a gravity dual description for the collision, gravitational shockwaves in cutoff AdS_5 (the "hard-wall model") collide to form black holes situated near the IR cutoff, or on it in the asymptotic limit.
- Thus there is a one-to-one correspondence between the black hole on the IR cutoff, or brane, and a pion field configuration describing the quark-gluon plasma created in the collision.
- We can find a map between a $\Phi(r)$ static pion field solution and Unruh's "dumb holes," analogs of black holes in ultrasonic acoustic flows, with $\vec{v} = \vec{\nabla}\Phi$, $c^2 = 1 + (\vec{\nabla})^2$, $\rho = 1/\sqrt{1 + (\vec{\nabla}\Phi)^2}$.
- By adding a source term (coming from the hadronic source for the pions) that drops off faster than $1/r$, the (unique) action for the pion gives a finite temperature.
- The thermal horizon appears because scalar (pion) field information propagation is slowed to zero at this surface, just as in the event horizon of a black hole the same thing happens for *all* information.
- The Heisenberg model is unique, since no finite number of terms (with arbitrary coefficients) in a derivative expansion gives the desired result. But it can be generalized with some more functions of Φ (mimicking gravity dual realizations), or by adding more scalars Φ^i, or by adding vector mesons A^i_μ.

- The simplest gravity dual of the Heisenberg model is the hard-wall model, but one can consider more complicated constructions, like a nonextremal (finite temperature) D4-brane metric with a D8-brane probe, a modification of the Sakai-Sugimoto model.
- The energy-momentum tensor from the massive DBI scalar action can be identified with the hydrodynamics expansion of a fluid energy-momentum tensor, with $u^\mu = \partial_\mu \phi / \sqrt{-(\partial_\mu \phi)^2}$ and P and ρ defined as well in terms of ϕ.
- One can add a term inside the square root of the DBI action to define a finite shear viscosity over entropy density η/s.

Further Reading

The mapping of the saturation of the Froissart bound in the Heisenberg model and the one in the gravity dual (in gravitational shockwave collisions) was done in [123], and the correspondence between the quark-gluon plasma fireball created in heavy ion collisions and black holes on the IR brane was proposed in [124]. The map between $\Phi(r)$ static pion field solutions and Unruh's dumb holes, and the corresponding finite temperature horizons, was first defined in [125], and its details and uniqueness were found in [126]. A review of these issues is found in [127]. The identification between the DBI scalar theory and the hydrodynamics expansion of the QGP fluid was done in [128].

Exercises

(1) Show that the equations of motion of the DBI action for $\phi = \phi(\tilde{s})$ are given by (45.7).
(2) Show that the equations of motion of the DBI action truncated to the first subleading term (45.27) have no real scalar field solutions.
(3) Show that the equations of motion for (45.33) have the solution $\phi^a = A^a \sqrt{\tilde{s}}$, with $\sum_a (A^a)^2 = 1/l^4$.
(4) Check the identifications of ρ, P and η, ζ for the (improved) free massless scalar hydrodynamics in the text.
(5) Check that if we add a source term $\int d^4x \phi \alpha / r$ to the DBI action, the temperature of the "horizon" associated with $\Phi(r)$, from the "surface gravity" in (45.19), is not finite.

46

Holographic Entanglement Entropy in Condensed Matter

An observable that has been gaining importance recently in condensed matter physics, spurred by the possibility to measure it in interesting systems, is entanglement entropy. But in 2006, Ryu and Takayanagi have proposed a way to calculate the entanglement entropy holographically, for systems at strong coupling, by a prescription inside a gravity dual. In this chapter we will describe the holographic prescription and some of its recent applications for condensed matter.

46.1 Entanglement Entropy in Quantum Field Theory

Shortly put, the entanglement entropy is the von Neumann entropy of a reduced subsystem, when we trace over the degrees of freedom not part of the subsystem. More precisely, consider a total system in a pure state $|\Psi\rangle$, thus with density matrix

$$\rho_{\text{tot}} = |\Psi\rangle\langle\Psi|. \qquad (46.1)$$

Its total von Neumann entropy is zero, $S_{\text{vN,tot}} = -\operatorname{Tr}\rho_{\text{tot}}\ln\rho_{\text{tot}} = 0$, since the state is pure.

But we can consider the system as being composed of two subsystems A and B, not necessarily by any physical division, but rather as an abstract construction. Then the Hilbert space splits, $\mathcal{H}_{\text{tot}} = \mathcal{H}_A \otimes \mathcal{H}_B$, and so do specific wavefunctions.

The density matrix observed by an observer that has access only to A is traced over the degrees of freedom of B:

$$\rho_A = \operatorname{Tr}\rho_{\text{tot}}. \qquad (46.2)$$

This reduced density matrix has now a nontrivial von Neumann entropy called the *entanglement entropy*:

$$S_A = -\operatorname{Tr}_A \rho_A \log\rho_A. \qquad (46.3)$$

It is a measure of how entangled the original state $|\Psi\rangle$ is.

For a system at finite temperature, the density matrix $\rho_{\text{tot}} = |\Psi\rangle\langle\Psi|$ is replaced by the thermal density matrix $\rho_{\text{tot}} = e^{-\beta\hat{H}}$, but then we can define the finite temperature entanglement entropy $S_A(\beta)$ in the same was as a function of the new ρ_{tot}. If $B = 0$, so A is the total system, then $\rho_A = \rho_{\text{tot}} = e^{-\beta\hat{H}}$, so the entanglement entropy $S_A(\beta) = -\operatorname{Tr}\rho_{\text{tot}}\ln\rho_{\text{tot}}$ equals the usual thermal entropy.

For a pure state ρ_{tot}, the entanglement entropy of the subsystem A is the same as for the system B, $S_A = S_B$, which means that the entanglement entropy is *nonextensive*. More generally, the entanglement entropy of the sum of two subsystems is less than or equal to the sum of entanglement entropies of the individual subsystems:

$$S_{A+B} \leq S_A + S_B. \tag{46.4}$$

The difference between the right-hand side and the left-hand side is called the *mutual information* $I(A, B)$,

$$I(A, B) = S_A + S_B - S_{A+B} \geq 0, \tag{46.5}$$

and is another important property of subsystems.

We will be mainly interested in the case of a quantum field theory defined on a spatial manifold M, times time $\mathbb{R}_t$, with a submanifold A with boundary ∂A, separating from B, the complement of A inside M. The entanglement entropy is always divergent in quantum field theory, so we need to introduce an UV cutoff ϵ to write a formula for it. Note that in physical condensed matter situations, it could be that the UV cutoff is a fixed lattice size a instead of a regulator ϵ that can be removed by taking it to zero.

In $d+1$ spacetime dimensions, the leading term in the entanglement entropy of a system of size L is of order $(L/\epsilon)^{d-1}$. For *conformal field theories* in d even we have only the odd powers of (L/ϵ) in the expansion of S_A, whereas for d odd we have only the even powers, starting with $\log(L/\epsilon)$ standing in for the zeroth power. This was first predicted from the holographic prescription, but has been proven to be true in many cases.

The leading term is more precisely proportional to the area of the boundary ∂A, with a coefficient γ that depends on the system, but not on A,

$$S_A = \gamma \frac{\text{Area}(\partial A)}{\epsilon^{d-1}} + \mathcal{O}\left(\frac{l}{\epsilon}\right)^{d-3}, \tag{46.6}$$

if $d > 1$. This is called the *area law* for entanglement entropy. However, in the special case of a conformal field theory in $d = 1$ ($1 + 1$ dimensions), for an infinitely long total system and a subsystem A of length l, we have instead

$$S_A = \frac{c}{3} \log \frac{l}{\epsilon}, \tag{46.7}$$

as predicted by the general expansion. Here c is the central charge of the conformal field theory.

In the case that we have a conformal field theory (or a quantum field theory with an UV fixed point) defined on a subsystem A which is a strip of width l and length L,

$$A : \left\{ (x_1, \ldots, x_d) \,\bigg|\, -\frac{l}{2} \leq x_1 \leq \frac{l}{2}, \quad 0 \leq x_2, \ldots, x_d \leq L \right\}, \tag{46.8}$$

embedded in a system of size L^d, the entanglement entropy has an area law corrected by a constant term:

$$S_A = \gamma \frac{\text{Area}(\partial A)}{\epsilon^{d-1}} - \alpha \frac{\text{Area}(\partial A)}{l^{d-1}} = \gamma \frac{L^{d-1}}{\epsilon^{d-1}} - \alpha \frac{L^{d-1}}{l^{d-1}}. \tag{46.9}$$

However, for theories with a Fermi surface, the expected modification of the area law with a large subsystem size l is instead

$$S_A = \gamma \frac{L^{d-1}}{\epsilon^{d-1}} + \eta (L k_F)^{d-1} \log(l k_F) + \mathcal{O}(l^0). \qquad (46.10)$$

Here k_F refers to the Fermi momentum. A free fermion calculation done in [131] obtains this result, and it was argued that strong interactions in a Fermi liquid don't modify it, and shown that in three dimensions for massless Dirac fermions even a finite chemical potential doesn't change it.

Note that if one considers instead free fermions on a fixed lattice (UV cutoff a fixed), for a subsystem A of large but finite size L in *all directions*, then the leading term in the entanglement entropy is of the form

$$S_A = \eta \left(\frac{L}{a} \right)^{d-1} \log \frac{L}{a} + \cdots \qquad (46.11)$$

Note that the would-be area law $\gamma (L/a)^{d-1}$ would be subleading with respect to this term.

46.2 The Ryu-Takayanagi Prescription

We want to calculate the entanglement entropy of a quantum field theory in $d+1$ dimensions at strong coupling, so that it has a well-defined gravity dual in $d+2$ dimensions. More precisely, consider a *conformal field theory* defined on the spatial manifold $M = \mathbb{R}^d$, with AdS_{d+2} gravity dual, in coordinates $t, x^i, z, i = 1, \ldots, d$.

Since we need to introduce an UV regulator in the conformal field theory, we consider the corresponding regulator in AdS space, namely a cutoff in z at $z_{UV} = \epsilon$. The spatial manifold $M = \mathbb{R}^d$ (at fixed time) is then situated at the $z = \epsilon$ surface. Consider a subsystem $A \subset M$ at fixed time, bounded by the *closed $d-1$-dimensional spatial surface* ∂A, with B the complement of A in M.

To calculate the entanglement entropy from the bulk, we consider the d-dimensional spatial surface γ_A in the bulk that ends on the same closed surface ∂A, i.e. $\partial \gamma_A = \partial A$. We consider the *minimal surface* (surface of minimal area) $\gamma_{A,\min}$ in AdS_{d+2}, with a given $\partial A = \gamma_A$ on M. Then the Ryu-Takayanagi prescription for the holographic entanglement entropy is defined in a similar way to the way the Wilson loop[*] (defined in Chapter 26) is found holographically in terms of the minimal worldsheet of the string:

$$S_A = \frac{\text{Area}(\gamma_{A,\min})}{4 G_N^{(d+2)}}. \qquad (46.12)$$

---

[*] We have not yet defined the holographic prescription of a Wilson loop. The Wilson loop was defined as $\langle W[C] \rangle = \langle e^{-\oint_C dx^\mu A_\mu} \rangle$ for a closed contour (loop) C, with A_μ being a gauge field. The holographic prescription for it is to consider the *minimal area* surface S in the gravity dual that ends on C, standing in for the worldsheet of a classical string, and then $\langle W[C] \rangle_{\text{holo}} = e^{-S_{\text{string}}(S)} = e^{-\frac{1}{2\pi\alpha'}\text{Area}(S)}$.

Of course, for the comparison with the Wilson loop, the dimensionality works only in $d = 2$, when γ_A is a 2-dimensional surface, and moreover in order to match we would need to be in Euclidean signature, since γ_A is a Euclidean surface, but the string worldsheet is a spacetime, with Minkowski signature.

We also note that the Ryu-Takayanagi formula is very similar to the Bekenstein-Hawking formula for the entropy of the black hole, for which the numerator is the area of the horizon of a black hole, perhaps inside a gravity dual. We can then say that the Ryu-Takayangi formula is sort of generalization of the Bekenstein-Hawking formula, since the event horizon of a black hole is a surface of minimal area, albeit one that doesn't end on the boundary of AdS space.

For the Ryu-Takayanagi prescription a heuristic derivation was given in their original paper, which will not be described here. Also, it was later reduced to a reasonable, though still unproven, assumption by Lewkowycz and Maldacena.

46.3 Holographic Entanglement Entropy in Two Dimensions

For a simple application of the Ryu-Takayanagi formula, consider a $1+1$–dimensional conformal field theory defined on the infinite line, with a subsystem A of finite length l, for concreteness between $x = -l/2$ and $x = +l/2$.

One can calculate the central charge of the conformal field theory holographically, although we will not do it here. For an AdS_3 gravity dual of radius R, the central charge of the dual field theory is found to be

$$c = \frac{3R}{2G_N^{(3)}}. \tag{46.13}$$

If A is the interval from $-l/2$ to $+l/2$, then ∂A refers simply to the points $x = -l/2$ and $x = +l/2$, and thus $\gamma_{A,\text{min}}$ is a geodesic in AdS_3 that ends on the two points on the regulated boundary $z = \epsilon$, i.e. at $(x, z) = (-l/2, \epsilon)$ and $(x, z) = (+l/2, \epsilon)$.

The geodesic that ends on these two points is

$$(x, z) = \frac{l}{2}(\cos s, \sin s), \quad \frac{2\epsilon}{l} \leq s \leq \pi - \frac{2\epsilon}{l}. \tag{46.14}$$

We can check that this satisfies the geodesic equation, though it is left as exercise. The "area" of this geodesic is its total length L is

$$L(\gamma) = 2R \int_{2\epsilon/l}^{\pi/2} \frac{ds}{\sin s} = 2R \log \frac{l}{\epsilon}. \tag{46.15}$$

The entanglement entropy, according to Ryu-Takayanagi, is

$$S_A = \frac{L(\gamma_A)}{4G_N^{(3)}} = \frac{R}{2G_N^{(3)}} \log \frac{l}{\epsilon} = \frac{c}{3} \log \frac{l}{\epsilon}. \tag{46.16}$$

Therefore this indeed matches the expected general formula (46.7).

46.4 Condensed Matter Applications

The first application of the holographic entanglement entropy formalism to condensed matter issues is one that has been previewed in Chapter 40, when talking about Fermi liquids and Fermi surfaces. One can define systems with Fermi surfaces by a logarithmic violation of the area law for the entanglement entropy (46.10).

Fermi Surface Holographic Definition

For 2+1–dimensional systems with Fermi surfaces, the 3+1–dimensional gravity background dual to it, if it is translationally and rotationally symmetric in the boundary spatial directions x and y, should be of the form

$$ds^2 = \frac{R^2}{z^2}(-f(z)dt^2 + g(z)dz^2 + dx^2 + dy^2). \tag{46.17}$$

Asymptotically, near the $z = 0$ boundary, we want to have AdS_4, so that the dual field theory has an UV fixed point.

The subsystem A is defined as before by

$$-\frac{l}{2} \leq x \leq \frac{l}{2}; \quad 0 \leq y \leq L, \tag{46.18}$$

thus its boundary ∂A is a square loop of sides L and l.

In this case, the minimal surface γ_A is worldsheet, and it ends on the loop, being the same construction as for the Wilson loop, as we described above. It can be found using the standard technology derived for the Wilson loops. Since we take $L \gg l$, we assume that the surface is y-independent, and is therefore really a curve $x(z)$ translated into the y direction. The curve has two branches: it drops down to a turning point z_*, then goes back up to $z = 0$. Then

$$\text{Area}(\gamma_A) = 2R^2 L \int_\epsilon^{z_*} \frac{dz}{z^2} \sqrt{g(z) + x'(z)^2}. \tag{46.19}$$

Here ϵ is an UV cutoff for the field theory, which in the gravity dual means that we consider the subsystem A to be situated at $z = \epsilon$ instead of at $z = 0$.

Since the area does not depend explicitly on $x(z)$, only on its derivative, the equation of motion is $\frac{d}{dz}\delta \text{Area}/\delta x'(z) = 0$, leading to

$$\frac{x'(z)}{z^2 \sqrt{g(z) + x'^2}} = \text{const.} \tag{46.20}$$

The solution is

$$x'(z) = \frac{z^2}{z_*^2} \sqrt{\frac{g(z)}{1 - \frac{z^4}{z_*^4}}}, \tag{46.21}$$

and it leads to the formula for the length l in the x direction as

$$l = 2 \int_0^{z_*} dz \frac{z^2}{z_*^2} \sqrt{\frac{g(z)}{1 - \frac{z^4}{z_*^4}}}. \tag{46.22}$$

Replacing this in the formula for the area, and then calculating the entanglement entropy $S_A = \text{Area}(\gamma_{A,\min})/(4G_N^{(4)})$, we get

$$S_A = \frac{R^2}{2G_N^{(4)}} L \int_\epsilon^{z_*} \frac{dz}{z^2} \sqrt{\frac{g(z)}{1 - \frac{z^4}{z_*^4}}}, \tag{46.23}$$

where we needed to introduce a cutoff ϵ since otherwise the result is divergent.

Entanglement Entropy of Conformal Field Theories in d Spatial Dimensions
For conformal field theories, for which the gravity dual is AdS_{d+2}, we have $f(z) = g(z) = 1$. We can then calculate the result as an expansion in ϵ and obtain

$$S_A = \frac{\left(RM_{\text{Pl}}^{(d+2)}\right)^d}{4(d-1)} \left[\left(\frac{L}{\epsilon}\right)^{d-1} - \left(\sqrt{\pi} \frac{\Gamma\left(\frac{d+1}{2d}\right)}{\Gamma\left(\frac{1}{2d}\right)}\right)^d \left(\frac{L}{l}\right)^{d-1} \right], \tag{46.24}$$

where for completeness we have written the result for arbitrary d, which can be obtained in a similar way. Here $(M_{\text{Pl}}^{(d+2)})^d = 1/G_N^{(d+2)}$, and we have ignored subleading terms in the expansion in ϵ. We leave the proof of this formula for $d = 2$ as an exercise.

Entanglement Entropy for Systems with Fermi Surfaces and UV Fixed Points
We want to consider the asymptotically AdS case, i.e. for dual field theory systems with UV fixed points. In that case, since $z = 0$ is the boundary, and defining a certain scale z_F corresponding as we will see shortly to the Fermi momentum, we need to have

$$g(z) \simeq 1 \quad \text{for} \quad z \ll z_F. \tag{46.25}$$

On the other hand, in the deep IR, we assume a general power law:

$$g(z) \simeq \left(\frac{z}{z_F}\right)^{2n} \quad \text{for} \quad z \gg z_F. \tag{46.26}$$

We will be interested in the limit $z_* \gg z_F$, corresponding to the deep IR.

We leave it as an exercise to show that then the entropy is found to be

$$S_A = \frac{R^2}{2G_N^{(4)}} \frac{L}{\epsilon} + k_n \frac{R^2}{G_N^{(4)}} \frac{L}{z_F} \left(\frac{l}{z_F}\right)^{\frac{n-1}{n+1}} + \cdots, \tag{46.27}$$

for $n > 1$, where k_n is a positive constant, and we have neglected terms that are small in the limit $z_* \gg z_F$ or $l \gg z_F$.

But more importantly, for $n = 1$, we find instead

$$S_A = \frac{R^2}{2G_N^{(4)}} \frac{L}{\epsilon} + \kappa_1 \frac{R^2}{G_N^{(4)}} \frac{L}{z_F} \log \frac{l}{z_F} + \mathcal{O}(l^0), \tag{46.28}$$

just like the general prediction for Fermi surfaces (46.10). Note that we need to identify $k_F \sim 1/z_F$, and that to have a Fermi surface we need $n = 1$.

Constraints on the possible gravity duals are obtained from the *null energy condition*,

$$T_{\mu\nu} n^\mu n^\nu \geq 0, \tag{46.29}$$

where n^ν is null, believed to hold for sensible quantum theories. It is a manifestation in general relativity of the idea that the local energy density must be positive. In the case of the ansatz (46.17) with a suitable null vector, it is found that, if also $f(z)$ is a power law at large z, $\propto z^{-2m}$, the null energy condition implies

$$m \geq n. \tag{46.30}$$

Considering the system above at finite temperature, we need to replace

$$f(z) \to z^{-2m} h(z); \quad g(z) \to z^{2n} \tilde{h}(z), \tag{46.31}$$

where now we must have a horizon for a black hole at $z = z_H$, so near it we need to have

$$h(z) \simeq \tilde{h}(z) \sim \frac{z_H - z}{z_H}. \tag{46.32}$$

Far from the horizon, for $z \ll z_H$, but still in the IR $z \gg z_F$, we must still have the z^{-2m} and z^{2n} behavior, i.e. $h(z) \simeq \tilde{h}(z) \simeq 1$. In the same way as we calculated the temperature of gravity duals several times already, we find by requiring that the Euclidean horizon is smooth, that the temperature scales with z_H as

$$T \propto z_H^{-m-n-1}. \tag{46.33}$$

On the other hand, the thermal entropy S is the area of the horizon A_H over 4 in Planck units, so, since the horizon is at $z = z_H, t = t_0$, we have

$$S \propto A_H = \frac{R^2 \Delta x \Delta y}{z_H^2} \propto T^{\frac{2}{m+n+1}}. \tag{46.34}$$

That in turn means that the specific heat is

$$C = \frac{\partial E}{\partial T} = T \frac{\partial S}{\partial T} \propto T^{\frac{2}{m+n+1}} \equiv T^r. \tag{46.35}$$

But we saw that the entanglement entropy requires $n = 1$, and the null energy condition requires $m \geq n = 1$, so

$$r \leq \frac{2}{3}. \tag{46.36}$$

But a Fermi liquid has $C \propto T$, i.e. $r = 1$, hence cannot be obtained by the above kind of holographic dual construction.

Non-Fermi liquids (with Fermi surfaces) on the other hand, can be obtained in this way, and all we need is the correct asymptotics of $g(z)$. We will, however, return to these in the last chapter, when describing strange metals, which are non-Fermi liquids.

Important Concepts to Remember

- The entanglement entropy of a subsystem A is the von Neumann entropy of system M when we trace over the degrees of freedom of the complement B of A in M. It is a measure of how entangled is the state $|\Psi\rangle$ of the system M.
- In a quantum field theory in $d+1$ dimensions, the entanglement entropy has the most divergent term in the UV cutoff ϵ, of order $1/\epsilon^{d-1}$, being proportional to the area of the boundary ∂A of the subsystem, which is known as the area law.
- In $1+1$–dimensional conformal field theories for subsystem A of length l, the entanglement entropy is proportional to the central charge, $S_A = c/3 \log l/\epsilon$.
- In a conformal field theory on a subsystem = strip of width l and length L, the area law is corrected by a constant term of order $(L/l)^{d-1}$, whereas systems with Fermi surfaces have a correction proportional to $(Lk_F)^{d-1} \log(lk_F)$, which can be taken to be a definition of Fermi surfaces.
- The Ryu-Takayanagi holographic prescription for the entanglement entropy of a quantum field theory of a subsystem A of $\mathbb{R}^d$, with closed boundary ∂A, is given in terms of the surface γ_A of minimal area in the gravity dual that ends on ∂A, $\partial \gamma_A = \partial A$, as the area of the surface over 4 in Planck units, $S_A = \text{Area}(\gamma_A)/4G_N$.
- We can find the finite deviation of the area law for conformal field theories from a holographic calculation, as well as the formula for $1+1$–dimensional conformal field theories.
- We can find holographically the entanglement entropy that obeys the deviation from area law characteristic of Fermi surfaces, but the needed geometry, when coupled with the null energy condition, implies a specific heat $C \propto T^r$ with $r \leq 2/3$, as opposed to $r = 1$ for Fermi liquids, hence only non-Fermi liquids can be obtained.

Further Reading

The holographic entanglement entropy calculation was proposed by Ryu and Takayanagi in [129] and [130]. A review of the holographic entanglement entropy developments was done in [131]. Its use for Fermi surfaces was done in [132], and further applications of it were done in [133] and [134].

Exercises

(1) Calculate the entanglement entropy of ρ_A in $\rho_{\text{tot}} = |\Psi\rangle\langle\Psi|$, when

$$|\Psi\rangle = |\Psi\rangle_{A,1} \otimes (|\Psi\rangle_{B,1} + |\Psi\rangle_{B,2}), \tag{46.37}$$

and when

$$|\Psi\rangle = |\Psi\rangle_{A,1} \otimes |\Psi\rangle_{B,1} + |\Psi\rangle_{A,2} \otimes |\Psi\rangle_{B,2}, \tag{46.38}$$

where the various A, B states are orthonormal.

(2) Check that (46.14) satisfies the geodesic equation in AdS_3.

(3) Check that for $d = 2$, the expansion of the entanglement entropy formula (46.23) gives (46.24).

(4) Calculate the holographic Wilson loop,

$$\langle W[C] \rangle = e^{-\frac{1}{2\pi\alpha'}\text{Area}(S)}, \tag{46.39}$$

for a string worldsheet in the gravity dual (46.17), ending on a rectangle of length L in the x direction and of length $T \gg L$ in the time direction.

(5) Calculate the entanglement entropy for the case that $g(z)$ is a power law, deriving (46.27) and (46.28) from (46.23).

47

Holographic Insulators

We have seen in Chapter 13 that the key to describing insulators is momentum dissipation. In a metal, translational invariance is recovered in the deep IR, so we have a Drude peak for $\omega \to 0$: If $1/\tau \to 0$, we have $\sigma(\omega \to 0) \propto \delta(\omega)$. In an insulator on the other hand, momentum dissipation is always important, even in the deep IR, due to the breaking of translational invariance, and we obtain finite DC conductivities.

So to describe insulators holographically, we must describe momentum dissipation. In this chapter, I will present a number of holographic approaches to momentum dissipation, which can lead to holographic descriptions of insulators.

47.1 Momentum Dissipation through Random Disorder: Anderson-Type Insulator?

The first approach to break translational invariance in 2+1 dimensions mimics what Anderson did, by adding impurities as random couplings to the Hamiltonian, of the form

$$\delta H = \int d^2 y V(y) \mathcal{O}(t, y), \tag{47.1}$$

where $V(y)$ is an explicitly space dependent potential, treated statistically, with a Gaussian distribution

$$\langle \cdots \rangle_{\text{imp}} = \int \mathcal{D}V e^{-\int d^2 y (V(y))^2/(2\bar{V}^2)} (\cdots), \tag{47.2}$$

so that

$$\langle V(x) \rangle_{\text{imp}} = 0; \quad \langle V(x) V(y) \rangle_{\text{imp}} = \bar{V}^2 \delta^2(x - y). \tag{47.3}$$

As we explained in Chapter 13, any amount of impurities will localize the wavefunction, so it is enough to consider dilute, weakly coupled impurities, i.e. ones treated as probes. We need to relate (nonholographically) τ_{imp} to something that we can calculate in a holographic dual, but the holographic dual itself will not break translational invariance explicitly.

The operator $\mathcal{O}$ is the most relevant operator that preserves the symmetries of the theory. Since the dimension of V is $[V] = 3 - \Delta_{\mathcal{O}}$, we see that $[\bar{V}] = [V] - 1 = 2 - \Delta_{\mathcal{O}}$ needs be positive for the impurities to be a relevant perturbation.

Considering the retarded propagator for the momentum density $\mathcal{P}$, $G^R_{\mathcal{PP}}(\omega, k)$ and its constant piece

$$\chi_0 \equiv \lim_{\omega \to 0} G^R_{\mathcal{PP}}(\omega, 0) = \epsilon + P, \tag{47.4}$$

the linear response (or susceptibility) of $\mathcal{P}$ to a $\mathcal{P}$ perturbation, as we saw in Chapter 35.

Moreover, consider the commutator of $\mathcal{P}$ with the Hamiltonian $H = H_0 + \delta H$, in the case that H_0 is translationally invariant, so

$$\mathcal{F}(t, x) \equiv [\mathcal{P}, H] = [\mathcal{P}, \delta H] = \int d^2 y V(y) [\mathcal{P}, \mathcal{O}(t, y)] = iV(x)\partial\mathcal{O}(t, x), \tag{47.5}$$

where we have used $[\mathcal{P}(t, x), \mathcal{O}(t, y)] = i\delta^2(x - y)\partial\mathcal{O}(t, y)$.

Like we saw in the general Drude theory in Chapter 13, we can define an impurity scattering time τ_{imp}, and then we have

$$G^R_{\mathcal{PP}}(\omega, 0) = \frac{\chi_0}{1 - i\omega\tau_{\text{imp}}}, \tag{47.6}$$

which means that if we define the *memory function* $M(\omega)$

$$M(\omega) \equiv \frac{\omega G^R_{\mathcal{PP}}(\omega, 0)}{\chi_0 - G^R_{\mathcal{PP}}(\omega, 0)}, \tag{47.7}$$

it will go to i/τ_{imp} as $\omega \to 0$,

$$\frac{i}{\tau_{\text{imp}}} \equiv \lim_{\omega \to 0} M(\omega). \tag{47.8}$$

In fact this $\omega \to 0$ means $\omega \ll T$, i.e. it is part of a *hydrodynamic limit*.

Using the Heisenberg equations of motion $i\partial_t\mathcal{O} = [\mathcal{O}, H]$ twice, as well as the definition of $G^R_{\mathcal{OO}}$ as the Fourier transform of $\langle [\mathcal{O}(x, t), \mathcal{O}(0, 0)] \rangle$, we find

$$\omega^2 G^R_{\mathcal{PP}}(\omega, k) = -G^R_{[\mathcal{P}, H][\mathcal{P}, H]}(\omega, k) + G^R_{[\mathcal{P}, H][\mathcal{P}, H]}(0, k) = -(G^R_{\mathcal{FF}}(\omega, k) - G^R_{\mathcal{FF}}(0, k)). \tag{47.9}$$

But then we can write

$$M(\omega) \simeq \frac{\omega G^R_{\mathcal{PP}}(\omega, 0)}{\chi_0} = -\frac{1}{\chi_0} \frac{G^R_{\mathcal{FF}}(\omega, 0) - G^R_{\mathcal{FF}}(0, 0)}{\omega}, \tag{47.10}$$

and taking the $\omega \to 0$ limit, find

$$\frac{i}{\tau_{\text{imp}}} = -\frac{1}{\chi_0} \lim_{\omega \to 0} \frac{G^R_{\mathcal{FF}}(\omega, 0) - G^R_{\mathcal{FF}}(0, 0)}{\omega}$$

$$= -\frac{i}{\chi_0} \lim_{\omega \to 0} \frac{\text{Im} G^R_{\mathcal{FF}}(\omega, 0)}{\omega}, \tag{47.11}$$

where in the second line we have used the fact that the imaginary part is an odd function of ω, whereas the real part is even, thus its contribution vanishes.

Now with $\mathcal{F} = iV(x)\partial\mathcal{O}$, we can write

$$G_{\mathcal{F}\mathcal{F}}^R(\omega, 0) = i\bar{V}^2 \int_0^\infty dt \, \langle[\partial\mathcal{O}(t, 0), \partial\mathcal{O}(0, 0)]\rangle e^{i\omega t} = -\frac{\bar{V}^2}{2} \int \frac{d^2k}{(2\pi)^2} k^2 G_{\mathcal{O}\mathcal{O}}^R(\omega, k),$$

(47.12)

so finally we find

$$\frac{1}{\tau_{\text{imp}}} = \frac{\bar{V}^2}{2\chi_0} \lim_{\omega \to 0} \int \frac{d^2k}{(2\pi)^2} k^2 \frac{\text{Im} G_{\mathcal{O}\mathcal{O}}^R(\omega, k)}{\omega}.$$

(47.13)

We see then that in the holographic dual all we need is to calculate the retarded correlator of $\mathcal{O}$, without the need to introduce any explicit breaking of translational invariance.

Gravity Dual Calculation

For a gravity dual calculation, one considers a (somewhat) top-down construction. Consider the 4-dimensional $\mathcal{N} = 8$ (maximally) supersymmetric gauged supergravity obtained by KK dimensional reduction on S^7 from 11 dimensions ($AdS_4 \times S^7$ background). The $AdS_4 \times S^7$ background corresponds to the near-horizon limit of N M2-branes in M-theory, so the field theory corresponding to it is the M2-brane field theory. In fact, we know that with some $\mathbb{Z}_k$ identifications on S^7, and after an IR limit, we would obtain the ABJM model.

The details of the supergravity and its relevant truncations will not be shown here. We will just quote the relevant solution, a black hole with electric charge (to consider chemical potential) and magnetic charge (in order to consider magnetic field), i.e. an AdS_4 dyonic black hole, with metric and electromagnetic field of

$$ds^2 = \frac{\alpha^2 L^2}{z^2}[-f(z)dt^2 + dx^2 + dy^2] + \frac{L^2}{z^2}\frac{dz^2}{f(z)}$$
$$F = h\alpha^2 dx \wedge dy + q\alpha dz \wedge dt,$$

(47.14)

where the blackening function f is

$$f(z) = 1 + (h^2 + q^2)z^4 - (1 + h^2 + q^2)z^3,$$

(47.15)

and α, q, h are constants. They define the magnetic field B, charge density ρ, and temperature T by

$$B = h\alpha^2, \quad \rho = -\frac{2q\alpha^2 L^2}{\kappa_{N,4}^2}, \quad T = \frac{\alpha(3 - h^2 - q^2)}{4\pi}.$$

(47.16)

We note that there is a horizon at $z_h = 1$, and the entropy density is given as the area of the horizon over 4 in $G_{N,4} = \kappa_{N,4}^2/(8\pi)$ units:

$$s = \frac{L^2\alpha^2}{4z_h^2 G_{N,4}} = \frac{2\pi L^2\alpha^2}{\kappa_{N,4}^2}.$$

(47.17)

Besides the metric and the gauge field, two scalar fields ϕ and χ are considered, and more precisely their combinations

$$\psi_+ = \frac{h\phi + q\chi}{\sqrt{h^2 + q^2}}; \quad \psi_- = \frac{q\phi - h\chi}{\sqrt{h^2 + q^2}}. \tag{47.18}$$

These fields have masses $m^2L^2 = -2$, which are stable since they are above the Breitenlohner-Freedman bound for AdS_4, of $m^2L^2 \geq -9/4$. Since the AdS/CFT map implies $m^2L^2 = \Delta(\Delta - 3)$, where Δ is the dimension of the field theory operator $\mathcal{O}$ coupling to the boundary value of the scalar, the operators $\mathcal{O}$ we consider can have dimensions $\Delta_{\mathcal{O}} = 1$ and $\Delta_{\mathcal{O}} = 2$.

The general form of the asymptotics for $\psi_\pm$ is (in general $z^{3-\Delta}$ is $z^{D-\Delta}$, with D the spacetime dimension)

$$\psi_\pm = z^{3-\Delta}[A(\omega, k) + \mathcal{O}(z^2)] + z^\Delta[B(\omega, k) + \mathcal{O}(z^2)], \tag{47.19}$$

where A and B are related to the source $\psi_{\mathcal{O}}$ for the operator $\mathcal{O}$ and the expectation value $\langle \mathcal{O} \rangle$, respectively, as

$$\psi_{\mathcal{O}} = \alpha^{3-\Delta_{\mathcal{O}}} A(\omega, k)$$
$$\langle \mathcal{O} \rangle = \frac{\sigma_0}{4} \alpha^{\Delta_{\mathcal{O}}} (2\Delta_{\mathcal{O}} - 3) B(\omega, k), \tag{47.20}$$

and σ_0 is the electric conductivity at $\rho = B = 0$, given by

$$\sigma_0 = \frac{2L^2}{\kappa_{N,4}^2}. \tag{47.21}$$

Since as we know $\langle \mathcal{O} \rangle = G_{\mathcal{O}\mathcal{O}}^R(\omega, k)\phi_{\mathcal{O}}$ (the retarded propagator gives the linear response), we obtain

$$G_{\mathcal{O}\mathcal{O}}^R(\omega, k) = -\frac{\sigma_0}{4\alpha} \frac{A(\omega, k)}{A(\omega, k)}. \tag{47.22}$$

Using (47.17), written as $s = \pi \sigma_0 \alpha^2$, and remembering that $\chi_0 = \epsilon + P$, we have from the general formula (47.13) that

$$\frac{1}{\tau_{\text{imp}}} = -\frac{\bar{V}^2}{T} \frac{sT}{16\pi^2(\epsilon + P)} \lim_{\omega \to 0} \int dk \; k^2 \text{Im} \frac{A(\omega, k)}{B(\omega, k)}. \tag{47.23}$$

One can in fact show that for the AdS_4 black hole, $sT/(\epsilon + P) = \sqrt{\sigma/\sigma_0}$, where σ is the DC conductivity in the absence of impurities. So there is a relation between τ_{imp} responsible for impurity momentum dissipation (leading to a conductivity for insulators) and the DC conductivity that doesn't consider impurities, which seems quite strange. Moreover, one finds that, except for the $\bar{V}^2/T$ prefactor, the rest is a function of $\sqrt{B^2 + \rho^2/\sigma_0^2}/T^2$, and numerically one finds that the function diverges at a value of about 21.

47.2 Momentum Dissipation through Holographic Lattice: Hubbard-Type Insulator?

But as we saw, in the previous case there were no explicit impurities in the holographic dual, and there was also no lattice of ions to dissipate momentum. The first model that dealt with that was the one of Horowitz, Santos, and Tong, which introduced a lattice in the gravity dual. As such, we can think of it as an analog of the Hubbard insulator, in that it is a model where insulation is obtained as a result of a lattice interaction, though as we will see, the properties that will be obtained are different.

The field content of the gravity dual will be an Einstein-Maxwell theory, plus a neutral scalar,

$$S = \frac{1}{16\pi G_{N,4}} \int d^4x \sqrt{-g} \left[R + \frac{6}{L^2} - \frac{1}{2} F_{\mu\nu} F^{\mu\nu} - 2(\nabla_\mu \Phi)^2 - 4V(\Phi) \right], \quad (47.24)$$

and for a potential we take just a mass of $m^2 L^2 = -2$:

$$V(\Phi) = -\frac{\Phi^2}{L^2}. \quad (47.25)$$

Note that the Newton constant has been factored out of the whole action, making A_μ and Φ dimensionless. The scalar field is introduced to provide the lattice. Considering as a background the Poincaré patch of AdS space, with the boundary at $z = 0$, the near boundary behavior of Φ is

$$\Phi \to z\phi_1 + z^2\phi_2 + \mathcal{O}(z^3), \quad (47.26)$$

where ϕ_1 is the source of the operator dual to ϕ, of dimension $\Delta_{\mathcal{O}} = 2$ (since in general $m^2 L^2 = \Delta(\Delta - 3)$), and ϕ_2 is the VEV $\langle \mathcal{O} \rangle$. The lattice is produced through a cosinusoidal source,

$$\phi_1(x) = A \cos(k_0 x). \quad (47.27)$$

Since the stress-energy tensor coming from Φ is quadratic in this expression, the wavenumber of the lattice felt by the dual field theory is $k = 2k_0$.

The ansatz for the gravity dual is

$$ds^2 = \frac{L^2}{z^2} \left[-(1-z)P(z)Q_{tt}dt^2 + \frac{Q_{zz}dz^2}{P(z)(1-z)} + Q_{xx}(dx + z^2 Q_{xz}dz)^2 + Q_{yy}dy^2 \right]$$

$$\Phi = z\phi(x, z)$$

$$A = (1-z)\psi(x, z)dt, \quad (47.28)$$

where, to have a black hole in the gravity dual, we have a factor of $1 - z$ (the horizon thus being at $z = 1$), but the function

$$P(z) = 1 + z + z^2 - \frac{\mu_1 z^3}{2} \quad (47.29)$$

is introduced to define the temperature, which is easily found to be

$$T = \frac{P(1)}{4\pi L} = \frac{6 - \mu_1^2}{8\pi L}. \tag{47.30}$$

From the near boundary behavior of Φ we obtain the one for $\phi(x, z)$,

$$\phi(x, z) = \phi_1(x) + z\phi_2(x) + \mathcal{O}(z^2), \tag{47.31}$$

and from the general case of AdS-Reissner-Nordstrom solutions, we know that

$$\psi(x, z) = \mu + [\mu - \rho(x)]z + \mathcal{O}(z^2), \tag{47.32}$$

where μ is the chemical potential and $\rho(x)$ is the charge density of the field theory.

The rest of the asymptotics (for Q_{ij}) are found from the Einstein's equations, together with a choice of gauge (of coordinates). One finds a numerical solution to the Einstein's equations, and writes perturbations around it,

$$g_{\mu\nu} = g_{\mu\nu}^{(0)} + h_{\mu\nu}; \quad A_\mu = A_\mu^{(0)} + \delta A_\mu; \quad \Phi = \Phi^{(0)} + \delta\Phi, \tag{47.33}$$

using the Lorenz gauge $\nabla^{(0)\mu}\delta A_\mu = 0$ and the de Donder gauge $\nabla^{(0)\mu}\bar{h}_{\mu\nu} = 0$. The field strength of the vector perturbation is $f_{\mu\nu} = \partial_\mu\delta A_\nu - \partial_\nu\delta A_\mu$. The perturbation is taken to be time dependent, with a usual $e^{-i\omega t}$ dependence for all fields, to study ω dependence of the conductivity.

From the general definition of the conductivity of the field theory, it is defined at the boundary as

$$\sigma(\omega, x) \equiv \lim_{z \to 0} \frac{f_{zx}(x, z)}{f_{xt}(x, z)}, \tag{47.34}$$

since f_{xt} is an electric field perturbation $E = i\omega\delta A_x$, whereas $f_{zx} = \partial_z\delta A_x$ is dual to an electric (or particle) current. More precisely, we have the boundary condition

$$\delta A_x = \frac{E}{i\omega} + J_x(x, \omega)z + \mathcal{O}(z^2). \tag{47.35}$$

The homogenous (x-independent) part of the above conductivity, $\sigma(\omega)$, the "optical" conductivity is what we are interested in.

The result of the numerical calculation is that for small ω, specifically $\omega \lesssim T$, the conductivity is well described by the Drude form

$$\sigma(\omega) = \frac{K\tau}{1 - i\omega\tau}, \tag{47.36}$$

whereas for large enough frequencies $\omega \gtrsim T$ (more precisely in the intermediate regime $2 < \omega\tau < 8$), it is well described by the form

$$|\sigma(\omega)| = \frac{B}{\omega^{2/3}} + C. \tag{47.37}$$

If we had $C = 0$, this would be *exactly* the behavior observed for strange metals, in particular in the cuprates (copper oxide high T_c superconductors in the normal, i.e. insulating, phase).

Moreover, we can define a DC resistivity, as $1/\sigma(0) = 1/(K\tau)$, and the numerical data fit an analytical result obtained by Hartnoll and Hofman in the Einstein-Maxwell holographic theory in AdS background[*]:

$$\rho \propto T^{2\nu-1},$$
$$\nu = \frac{1}{2}\sqrt{5 + 2(k/\mu)^2 - 4\sqrt{1 + (k/\mu)^2}}. \tag{47.38}$$

Here $k = 2k_0$ is the wavenumber of the lattice. As an aside, note that for $(k/\mu)^2 = 3/2 + \sqrt{6}$ we would obtain $\rho \propto T$, which is another very characteristic behavior of strange metals, and one that is almost impossible to obtain with usual condensed matter methods.

The thermoelectric conductivity parameter α was also calculated ($J_x = -\alpha\partial_x T$), and in the intermediate frequency regime $2 < \omega\tau < 8$ it was found to be described by

$$|\alpha(\omega)| = \frac{\tilde{B}}{\omega^\eta} + \tilde{C}, \tag{47.39}$$

where $\eta \simeq 5/6$. Moreover, a similar calculation was done for a 4+1–dimensional gravity dual, for a 3+1–dimensional field theory with a holographic lattice, obtaining a conductivity that is well fitted in the intermediate frequencies regime by

$$|\sigma(\omega)| = \frac{\hat{B}}{\omega^\gamma} + \hat{C}, \tag{47.40}$$

where $\gamma \simeq 0.87 \simeq \sqrt{3}/2$.

It was also verified that the type of lattice is not important. An *ionic* holographic lattice, introduced as a varying chemical potential $\mu(x)$, i.e. for a boundary condition for A_t of

$$A_t \to \mu(x) = \mu_0[1 + \cos(k_0 x)], \tag{47.41}$$

gives the same profile for the conductivity $\sigma(\omega)$.

47.3 Effective Massive Gravity Model for Lattice

Modeling the lattice by a sinusoidal energy-momentum tensor in the gravity dual leads to momentum dissipation, but it is very difficult numerical simulation work. Instead, it was realized in work by Vegh, with refinements by Blake and Tong, that there is an *effective description* of the holographic lattice in terms of a theory of massive gravity.

Indeed, to break translational symmetry, we must break the shift symmetry in δA_x, which, however, is found to mix with $g_t^x \equiv g_{xt}g^{xx} \simeq g_{tx}$, so we need to break the shift symmetry in

[*] Specifically, Hartnoll and Hofman found that decoupled gauge-invariant variables $\Phi_\pm = \Psi + \frac{r^2}{\sqrt{6}\kappa^2}(1 \pm \sqrt{1 + (k/\mu)^2})\Phi$, where $\Phi = \delta A_t - \sqrt{\frac{3}{2}}\frac{\delta g_{tt}}{f(r)}$, $\Psi = \delta g_{yy}$, with $f(r) = 1 - r^2/r_+^2$, behave near the boundary $r = 0$ as $\Phi_\pm \sim r^{1/2}(r^{-\nu_\pm} + G_\pm(\omega)r^{\nu_\pm})$, and then the retarded Green's function $G_R(\omega)$ is found from $G_+(\omega)$, as the ratio of subleading to leading terms in Φ_+, leading to a conductivity from a Kubo formula $\sigma \sim \mathrm{Im}G_R(\omega)/\omega \sim \omega^{2\nu--1}$. This in turn leads to the resistivity $\rho(T) \sim T^{2\nu-1}$.

g_{tx}. The simplest option is a mass term for the graviton:

$$\sqrt{-g}m^2(\delta g_{tx})(\delta g^{tx}). \tag{47.42}$$

But since the background is diagonal, $\delta g_{tx} = g_{tx}$, and from the Einstein's equations one finds that also g_{rx} must be nonzero.

Generic massive gravitons have instabilities (such as the Boulware-Deser ghost). De Rham, Gabadadze, and Tolley introduced a theory of massive gravity with higher derivatives, which has two couplings (dimensionless parameters), depending on a fixed rank 2 symmetric tensor $f_{\mu\nu}$, the *reference metric*, and having the usual dynamical metric $g_{\mu\nu}$. The Lagrangean is written using the combination

$$K^\mu{}_\nu \equiv \sqrt{g^{\mu\rho} f_{\rho\nu}}, \tag{47.43}$$

where the square root is taken in the matrix sense:

$$K^\mu{}_\rho K^\rho{}_\nu = g^{\mu\rho} f_{\rho\nu}. \tag{47.44}$$

The Einstein-Maxwell plus higher derivative action of the model is

$$S = \frac{1}{2\kappa_{N,4}^2} \int d^4x \sqrt{-g} \left[R + \Lambda - \frac{2\kappa_{N,4}^2}{4e^2} F_{\mu\nu}^2 + m^2 \sum_{i=1}^4 c_i U_i(g, f) \right]. \tag{47.45}$$

Here U_i are the combinations

$$U_1 = \text{Tr}[K]$$
$$U_2 = \text{Tr}[K]^2 - \text{Tr}[K^2]$$
$$U_3 = \text{Tr}[K]^3 - 3\,\text{Tr}[K]\,\text{Tr}[K^2] + 2\,\text{Tr}[K^3]$$
$$U_4 = \text{Tr}[K]^4 - 6\,\text{Tr}[K^2]\,\text{Tr}[K]^2 + 8\,\text{Tr}[K^3]\,\text{Tr}[K] + 3\,\text{Tr}[K^2]^2 - 6\,\text{Tr}[K^4]. \tag{47.46}$$

If the reference metric is flat, it can be transformed by a coordinate transformation ϕ^a to the form η_{ab}, as

$$f_{\mu\nu} = \partial_\mu \phi^a \partial_\nu \phi^b \eta_{ab}. \tag{47.47}$$

The relevant theory in our case is instead for

$$f_{\mu\nu} = \text{diag}(0, 0, 1, 1), \tag{47.48}$$

which is left invariant by the transformation with ϕ^a given by

$$\phi^{t,r}(t, r); \quad \phi^x = x; \quad \phi^y = y, \tag{47.49}$$

i.e. a general coordinate transformation in t, r. Then the mass term breaks the covariance only in x, y, where $f_{\mu\nu}$ has ones on the diagonal.

One considers only the case $c_1 = \alpha, c_2 = \beta, c_3 = c_4 = 0$, so the graviton mass terms are $m^2\alpha$ and $m^2\beta$. We see that we can put $m = 1$ and use α and β as the only mass parameters. Then one finds an AdS-Reissner-Nordstrom–type solution (the AdS-RN solution can be obtained as a limit if we put back in m and write $f_{\mu\nu} = (0, 0, F, F)$, and take the limit

$m \to 0, F \to 0$)

$$ds^2 = \frac{L^2}{r^2}\left(-f(r)dt^2 + \frac{dr^2}{f(r)} + dx^2 + dy^2\right)$$

$$A_t = \mu\left(1 - \frac{r}{r_h}\right),\tag{47.50}$$

where

$$f(r) = 1 + \frac{\alpha L}{2}r + \beta r^2 - Mr^3 + \frac{\mu^2}{\gamma^2 r_h^2}r^4.\tag{47.51}$$

Here $\gamma^2 = 4e^2 L^2/(2\kappa_{N,4}^2)$. The mass parameter M is defined by the position of the horizon, where $f(r_h) = 0$, to be

$$M = \frac{1}{r_h^3}\left(1 + \frac{\alpha L r_h}{2} + \beta r_h^2 + \frac{\mu^2 r_h^2}{\gamma^2}\right).\tag{47.52}$$

The temperature is found in the usual way to be

$$T = -\frac{f'(r_h)}{4\pi} = \frac{1}{4\pi r_h}\left(3 + \alpha L r_h + \beta r_h^2 - \frac{\mu^2 r_h^2}{\gamma^2}\right),\tag{47.53}$$

and the charge density is given by the coefficient of r in A_t (rescaled by e^2):

$$Q = \frac{\mu}{e^2 r_h}.\tag{47.54}$$

The conductivity is found to obey a membrane paradigm type calculation, i.e. to be determined solely by quantities at the horizon. Now the gauge field A_x mixes in general with the mode

$$\tilde{g}_{rx} = f(r)g_{rx}.\tag{47.55}$$

One finds the system of coupled equations

$$\begin{pmatrix} L_1 & 0 \\ 0 & L_2 \end{pmatrix}\begin{pmatrix} \delta A_x \\ \delta\tilde{g}_{rx} \end{pmatrix} + \frac{\omega^2}{f}\begin{pmatrix} \delta A_x \\ \delta\tilde{g}_{rx} \end{pmatrix} = M\begin{pmatrix} \delta A_x \\ \delta\tilde{g}_{rx} \end{pmatrix},\tag{47.56}$$

where L_1, L_2 are differential operators, and M is an r-dependent mass matrix:

$$M = \begin{pmatrix} 4e^4 Q^2\gamma^{-2}r^2 & e^2 Q m^2(r)(i\omega L^2)^{-1}r^2 \\ 4e^2 QL^2 i\omega\gamma^{-2} & m^2(r) \end{pmatrix}$$

$$m^2(r) = -2\beta - \frac{\alpha L}{r}.\tag{47.57}$$

Note that $\det M = 0$, so there is a massless eigenmode, which can be easily found to be

$$\delta\lambda_1 = \left(1 + \frac{4e^4 Q^2}{\gamma^2 m^2(r)}r^2\right)^{-1}\left[\delta A_x - \frac{e^2 Q r^2}{i\omega L^2}\delta\tilde{g}_{rx}\right].\tag{47.58}$$

We observe that δA_x is not the massless mode (the gauge field) anymore. If it were, the conductivity of the field theory would be defined from the boundary as

$$\sigma(\omega) = \left.\frac{\partial_r \delta A_x}{\partial_t \delta A_x}\right|_{r=0}, \tag{47.59}$$

but now that becomes

$$\sigma(\omega) = \left.\frac{1}{e^2}\frac{\partial_r \delta\lambda_1}{i\omega\delta\lambda_1}\right|_{r=0}. \tag{47.60}$$

One can then define a quantity that is conserved under radial flow for $\omega \to 0$ (limit defining the DC conductivity), and equals the one above at $r = 0$, and therefore it can also be rewritten as a quantity at the horizon:

$$\sigma_{\rm DC} = \frac{1}{e^2}\left(1 + \frac{4e^4 Q^2}{\gamma^2}\frac{r_h^2}{m^2(r_h)}\right). \tag{47.61}$$

When the second term dominates, we get the usual Drude form. Indeed, one can prove that (from a hydrodynamic analysis) in general we have

$$\sigma_{\rm DC} = \frac{1}{e^2} + \frac{Q^2}{\epsilon + P}\tau, \tag{47.62}$$

so if the last term dominates we find

$$\rho_{\rm DC} = \frac{1}{\sigma_{\rm DC}} = \frac{\gamma^2 m^2(r_h)}{4e^2 Q^2 r_h^2} = \frac{sm^2}{4\pi Q^2}, \tag{47.63}$$

where we have used the fact that the entropy density is

$$s = \frac{2\pi L^2}{k^2 r_h^2} = \frac{\pi\gamma^2}{e^2 r_h^2}. \tag{47.64}$$

Holographic Lattice as Effective Graviton Mass

To see that the holographic lattice does indeed generate an effective graviton mass, consider, like in the previous section, the scalar field source

$$\phi_1 = \epsilon\cos(k_0 x), \tag{47.65}$$

with $\epsilon \ll 1$, and then the bulk solution is of the form

$$\phi(r, x, y) = \epsilon\phi_0(r)\cos(k_0 x), \tag{47.66}$$

where $\phi_0(r)$ satisfies the equation of motion:

$$\frac{d}{dr}\left(\frac{f}{r^2}\frac{d\phi_0}{dr}\right) - \frac{k_0^2}{r^2}\phi_0 - \frac{m^2 L^2}{r^4}\phi_0 = 0. \tag{47.67}$$

On top of it, consider a perturbation with the same wavenumber,

$$\delta\phi(r, x, t) = \delta\phi(r, t)\sin(k_0 x) \equiv \epsilon k_0\phi_0(r)\pi(r, t), \tag{47.68}$$

interpreted as a bulk phonon mode, i.e. a mode of oscillation of the lattice. In total, the scalar field is

$$\phi(r, x, t) = \epsilon\phi_0(r)\cos[k_0(x - \pi(r, t))]. \tag{47.69}$$

Note that $\pi(r, t)$ is like a Goldstone boson for the shift symmetry on π. We can of course make a coordinate redefinition $\tilde{x} = x - \pi(r, t)$, where this scalar perturbation is lost, but now δg_{rx} becomes dynamical. This is the gravitational analog of the Higgs mechanism, where a gauge field "eats" the would-be Goldstone boson and becomes massive. Here it is the graviton that eats the would-be Goldstone boson π and becomes massive, thus with a new degree of freedom (g_{rx}). The mass term can be thought of as arising from the $(\partial_x\phi)^2$ term in the action, evaluated on the background solution $\phi = \epsilon\phi_0(r)\cos(k_0x)$. It gives

$$S_m = \frac{1}{2}\int d^4x\sqrt{-g}M^2(r)g^{xx}, \tag{47.70}$$

where the effective mass $M(r)$ is

$$M^2(r) = \frac{1}{2}\epsilon^2 k_0^2 \phi_0(r)^2. \tag{47.71}$$

Expanding the determinant of the metric, $\delta\sqrt{-g} = (1/2)\sqrt{-g}g^{\mu\nu}\delta g_{\mu\nu}$ to second order, we obtain $\sqrt{-g}(\delta g^{rx}\delta g_{rx} + \delta g^{tx}\delta g_{tx})$ terms, i.e. a graviton mass.

47.4 Strong Momentum Dissipation in the IR: New Holographic Insulators and Mott-Type Insulators

Until now we have described momentum dissipation due to impurities and lattices, which could describe insulators. But a question we have not addressed until now is: Can we have a phase transition between a metal and an insulator? It must necessarily be a quantum phase transition, in a strongly coupled system. The fact that the system is strongly coupled means that the phase transition must be driven by interactions, and the insulating phase would likely be of a Mott type.

Concretely, we can differentiate between various systems by their frequency-dependent conductivity $\sigma(\omega)$. For a metal, we expect a Drude peak, i.e. a large maximum at $\omega = 0$, followed by a quick drop that saturates. A Mott insulator on the other hand will have a negligible conductivity at $\omega = 0$, but then would start increasing dramatically at large enough ω's. An intermediate case would be a "bad" (or "incoherent") metal, where the conductivity $\sigma(\omega)$ stays roughly constant, and it would be encountered near the metal-insulator phase transition.

A true insulator would break translationally invariance (and thus dissipate momentum) strongly in the IR (at the largest possible distances), and thus have no Drude peak.

One interesting holographic model that shows that behavior has been put forward by Donos and Hartnoll. It involves a breaking of translational invariance while retaining homogeneity, by introducing a helical lattice of background fields. We consider a 4+1–dimensional action (dual to a 3+1–dimensional field theory) involving two vector fields, A_μ and B_μ, where A_μ is a gauge field (massless), whereas B_μ is a Proca field (massive), with a Chern-Simons interaction.

$$S = \int d^5x \sqrt{-g} \left(R + 12 - \frac{1}{4}F_{\mu\nu}F^{\mu\nu} - \frac{1}{4}W_{\mu\nu}W^{\mu\nu} - \frac{m^2}{2}B_\mu B^\mu \right) - \frac{k}{2}\int B \wedge F \wedge W.$$

(47.72)

Here $F = dA$ and $W = dB$ are the field strengths. The gauge field A introduces chemical potential in the usual way, through $A_t = \mu$, whereas the Proca field B introduces the helical lattice through a boundary source $B^{(0)}$ for the dual operator, in total

$$A^{(0)} = \mu\, dt, \quad B^{(0)} = \lambda\omega_2,$$

(47.73)

where the one-form ω_2 defines the helical structure of pitch p,

$$\omega_2 = \cos(px_1)dx_2 - \sin(px_1)dx_3.$$

(47.74)

Here x_1, x_2, x_3 are the spatial coordinates on the boundary, and we can more generally consider the one-forms:

$$\omega_1 = dx_1, \quad \omega_2 + i\omega_3 = e^{ipx_1}(dx_2 + idx_3).$$

(47.75)

The ansatz for the solution is

$$ds^2 = -U(r)dt^2 + \frac{dr^2}{U(r)} + e^{2v_1(r)}\omega_1^2 + e^{2v_2(r)}\omega_2^2 + e^{2v_3(r)}\omega_3^2$$
$$A = a(r)dt$$
$$B = w(r)\omega_2.$$

(47.76)

At the UV boundary $r \to \infty$, we consider asymptotically AdS_5 space, and the helical sources, so

$$U(r) \to r^2, \quad v_i(r) \to \log r, \quad a(r) \to \mu, \quad w(r) \to \lambda.$$

(47.77)

Then there are three possible IR behaviors, corresponding to the fact that flowing from the UV (i.e., going along geodesics in the radial direction) we can reach any of these IR fixed points (for the dual field theory). One is an $AdS_2 \times \mathbb{R}^3$ "horizon" solution (for the case of zero temperature), with asymptotics as $r \to 0$:

$$U \to 12r^2(1 + u_1 r^\delta); \quad v_i \to v_o(1 + v_{i,1}r^\delta); \quad a \to 2\sqrt{6}(1 + a_1 r^\delta); \quad w \to w_1 r^\delta.$$

(47.78)

For the mass squared m^2 greater than a certain combination of p, k, and v_o, the δ's are either zero or real and positive. Specifically, the condition is

$$m^2 > (2\sqrt{6}k - pe^{-v_o})pe^{-v_o},$$

(47.79)

and then the solution is stable, and the deformations defined by δ are marginal or irrelevant. Then the IR solution is translationally invariant, since $w \to 0$ at $r \to 0$, therefore the helical structure $w(r)$ becomes irrelevant in the IR. This is then a *metallic* state, with a Drude peak.

There is another stable IR geometry possible. For the massless case $m = 0$, as $r \to 0$ we have the solution

$$U \to u_o r^2 + \cdots, \quad e^{v_1} \to e^{v_{1o}} r^{-1/3} + \cdots, \quad e^{v_2} \to e^{v_{2o}} r^{2/3} + \cdots, \quad e^{v_3} \to e^{v_{3o}} r^{1/3} + \cdots$$
$$a \to a_o r^{5/3} + \cdots, \quad w \to w_o + w_1 r^{4/3} + \cdots. \tag{47.80}$$

In this case $w(r)$ does not vanish in the IR, and it still breaks translational invariance through the helical structure. Therefore this is an *insulating* state.

Moreover, mediating the transition between the two possible stable IR geometries, there is an unstable solution, with

$$U \to u_o r^2, \quad v_1 \to v_{1o}, \quad e^{v_2} \to e^{v_{2o}} r^\alpha, \quad e^{v_3} \to e^{v_{3o}} r^\alpha, \quad a \to a_o r, \quad w \to w_o r^\alpha, \tag{47.81}$$

with the various parameters determined numerically from the equations of motion. This solution is unstable to a perturbation by an r^δ, and in the $m = 0$ case this bifurcating solution exists for al $|k| \lesssim 0.57$. This can be thought of as a *bad metal* state.

But in order to really make the identifications of these IR states with various solids, we need to calculate the electrical conductivity $\sigma(\omega)$, by perturbing the solution with

$$\delta A = e^{-i\omega t} A(r) \omega_1$$
$$\delta B = e^{-i\omega t} B(r) \omega_3$$
$$\delta ds^2 = e^{-i\omega t} [C(r) dt \, \omega_1 + D(r) \omega_2 \, \omega_3]. \tag{47.82}$$

One finds that indeed, the metallic phase has a Drude peak with a height that increases as the temperature is lowered. The insulating phase obtains a $\mathrm{Re}(\sigma) \sim \omega^{4/3}$ as $\omega \to 0$ at $T = 0$ without a Drude peak. Correspondingly we find the resistivity as a function of temperature $\rho \propto T^{-4/3}$. At large T in all phases, one has $\rho \propto 1/T$, as expected on dimensional grounds, since we have a conformal field theory in the UV.

In conclusion, this is a new type of insulator, with a phase transition toward a metallic phase, that has some of the properties expected of insulators.

Other kinds of holographic insulators were proposed also, with properties of Mott type in the sense of being due to a strongly coupled electronic phase, with a phase transition to other kinds of phases.

47.5 Hydrodynamic Flow, Dirac Fluid, and Linear Resistivity

Finally, one of the characteristics of the holographic models, besides their strong coupling, is the existence of a hydrodynamical description, related to the existence of an event horizon, as we saw in Part III. But such a hydrodynamical description is quite unexpected from the point of view of condensed matter, where electron transport follows mostly through the collisions of individual electrons with the lattice, or with impurities. The hydrodynamical description is for a strongly interacting system, which loses momentum as a whole (through friction).

Yet recently, in 2016, several experimental groups have presented evidence of hydrodynamical flow in graphene and in other very conductive metals. The fluid under consideration is a fluid associated with Dirac (relativistic!) fermions, i.e. a Dirac fluid. We have seen in the chapter about holographic fermions (Chapter 40) that the "electron star" was a holographic description of a system similar to this.

In fact, graphene has a perfect crystal structure, and the Fermi surface $\epsilon(k)$ for graphene has two opposite cones (almost) touching each other on the tips, thus creating the relativistic dispersion relation of a massless Dirac fermion $\epsilon = \pm k$. The electrons and holes from these bands therefore behave as Dirac particles and antiparticles. In this case, the Fermi surface at strong coupling can describe a Dirac fluid.

In this section we will present a few arguments that a hydrodynamical description, coupled with some reasonable assumptions, can lead to a resistivity linear with temperature, like the one seen in "strange metals," which will be described in more detail in Chapter 48.

In a relativistic system, the diffusion constant in the absence of a chemical potential is related to the shear viscosity over entropy density by

$$\frac{D}{c^2} = \frac{\eta}{Ts} = \frac{\eta}{\epsilon + P}, \tag{47.83}$$

where in the second equality we have used the thermodynamic relation $Ts = \epsilon + P$. In a nonrelativistic limit, $P \simeq 0$ and $\epsilon \simeq m_e n_e$.

On the other hand, $\sigma_{\text{DC}} = (ne^2/m)\tau = \omega_p^2 \tau$, where ω_p is the plasma frequency and τ is the momentum relaxation time scale, so

$$\rho_{\text{DC}} = \frac{1}{\omega_p^2 \tau}. \tag{47.84}$$

The momentum relaxation time scale τ is related to the diffusion coefficient D and the characteristic length scale l over which the translation invariance is broken (the mean free path) by $\tau = l^2/D$, so (for $c = 1$)

$$\rho_{\text{DC}} = \frac{\eta}{\omega_p^2 (\epsilon + P)l^2} = \frac{\eta}{\omega_p^2 l^2 m_e n_e}. \tag{47.85}$$

Then if there is a bound on the shear viscosity over entropy density η/s, like the one proposed by Kovtun, Son, and Starinets (KSS), but perhaps with another coefficient A,

$$\frac{\eta}{s} \geq \frac{A\hbar}{k_B}, \tag{47.86}$$

then *at the saturation of the bound*, i.e. for the least viscous system, considering that $s/n_e = S_e$ is the entropy per electron, we have

$$\rho_{\text{DC}}(T) = \frac{A\hbar}{\omega_p^2 m_e l^2} \frac{S_e(T)}{k_B}. \tag{47.87}$$

If the entropy per electron $S_e(T)$ is linear in temperature, then so is ρ_{DC}.

On the other hand, we saw that in the massive gravity effective action for the holographic lattice we had, according to (47.63), $\rho_{\text{DC}}(T) = s(T)m^2/(4\pi Q^2)$, so again if $s(T)$ is linear in T we have a linear ρ_{DC}. One can find in fact a solution of a massive gravity effective action that has this property, but it appears in the context of strange metals, and will be considered in the next chapter.

Another argument for the linear resistivity goes as follows. The diffusion constant is actually a matrix D_{AB},

$$\frac{d}{dt}n_A = D_{AB}\nabla^2 n_B,$$
(47.88)

with eigenvalues D_+ and D_-, which satisfy

$$D_+D_- = \frac{\sigma}{\chi}\frac{\kappa}{c_p}$$

$$D_+ + D_- = \frac{\sigma}{\chi} + \frac{\kappa}{c_p} + \frac{T(\zeta\sigma - \chi\alpha)^2}{c_p\chi^2\sigma}.$$
(47.89)

Here κ is the thermal conductivity, α is the thermoelectric conductivity, and χ and ζ are susceptibilities defined by

$$\chi = -\frac{\partial^2 f}{\partial\mu^2}; \quad \zeta = -\frac{\partial^2 f}{\partial T\partial\mu},$$
(47.90)

and $f = \epsilon - sT - \mu\rho$ is the Gibbs potential density.

Then the modified KSS bound (47.86), with $D = (\eta/s)(c^2/T)$, and perhaps the speed of light c replaced by the Fermi velocity v_F can be replaced by

$$\frac{\sigma}{\chi} \geq \frac{Av_F^2\hbar}{k_BT}; \quad \frac{\kappa}{c_p} \geq \frac{Av_F^2\hbar}{k_BT}.$$
(47.91)

If the susceptibility χ is almost independent of T, then *at the saturation of the bound,*

$$\rho_{DC} = \frac{1}{\sigma_{DC}} \sim \frac{1}{v_F^2\chi}\frac{k_BT}{A\hbar}$$
(47.92)

is linear in T.

In conclusion, the existence of an ideal (minimal viscosity) fluid for the Fermi surface seems to lead to resistivity linear in temperature.

Important Concepts to Remember

- Insulators have momentum dissipation even in the deep IR, so the lattice breaking of translational invariance stays valid even at large distances.
- By adding random impurities coupling to operators $\mathcal{O}$ in the Hamiltonian, the impurity time scale that defines the (Anderson-type) conductivity can be calculated from a retarded correlator of operators $\mathcal{O}$, which can be calculated holographically. That gives a relation between the momentum dissipation due to impurities and momentum dissipation (conductivity) without impurities, calculated in a similar manner.
- A holographic lattice for momentum dissipation can be introduced by a sinusoidal energy-momentum tensor on the boundary, which can be obtained by a scalar with a boundary sinusoidal source.

- The conductivity of the holographic lattice is of Drude form (like a metal) at small ω and of the form $\sigma(\omega) \sim B\omega^{-2/3} + C$, like a strange metal, at large ω.
- An effective description of the holographic lattice is in terms of a massive gravity action, with general coordinate invariance in t, r and breaking of it in x, y. One can obtain a graviton mass for g_{tx} and g_{rx} from the holographic lattice.
- A helical lattice breaks translational invariance in a homogenous way, and can be introduced as a source for a massive vector (Proca) field.
- The holographic helical lattice obtains a transition between an IR metallic state and an IR insulating state, intermediated by an unstable IR geometry. In the IR insulating state, the helical lattice remains all the way to the IR, thus translational invariance is broken in the IR.
- Hydrodynamic flow for electrons has been found, and one has a fluid of effective Dirac particles.
- In hydrodynamic models, when saturating bounds on viscosity, i.e. for "ideal" fluids, one can obtain $\rho \propto T$, a characteristic of strange metals.

Further Reading

The addition of random impurities for momentum dissipation in the AdS/CFT correspondence was considered in [135]. The dependence $\rho(T) \sim T^{2\nu-1}$ was found in [136]. The holographic lattice was described in [137] and [138]. The massive gravity effective description was developed in [139] and [140], and the fact that the holographic lattice leads to massive gravity was shown in [141]. The holographic helical lattice for the metal-insulator transition was described in [142]. The hydrodynamical model for transport in materials was developed in [143]. A description of the experimental observation of hydrodynamic transport in materials can be found in [144]. For more on the electronic properties of graphene, see [145]. The use of viscosity bounds in hydrodynamic flow to find linear resistivity was described in [146] and [147].

Exercises

(1) Show that the temperature of the gravity dual (47.14) is

$$T = \frac{\alpha(3 - h^2 - q^2)}{4\pi}. \tag{47.93}$$

(2) Calculate explicitly the equation of motion for $\phi(x, z)$ in the "holographic lattice" gravity dual (47.28)), and from it, derive the form of the term of order z^2 in the near boundary expansion as a function of $\phi_1(x)$ and $\phi_2(x)$ (and the expansions of Q_{xx}, Q_{xz}, Q_{zz}).

(3) Calculate explicitly the mass terms with coefficients α and β for the massive gravity, on the ansatz (47.50) with arbitrary $f(r)$. Then, remembering that in the Newtonian

approximation $f(r)$ is related to the Newtonian potential, find that

$$f(r) = 1 + \frac{\alpha L}{2}r + \beta r^2 - Mr^3 + \frac{\mu^2}{\gamma^2 r_h^2}r^4 \tag{47.94}$$

is a solution for small r (Newtonian perturbations).

(4) Diagonalize the matrix equations (47.56) to find the eigenfunctions and eigenvectors.

(5) Calculate the explicit equations of motion for the fields A_μ and B_μ in the gravity dual ansatz (47.76) and find how the first subleading terms in the near-boundary expansion for them are related to the leading sources μ and λ.

Holographic Strange Metals and the Kondo Problem

In this last chapter, we will study how to obtain holographically strange metals, and in this context, we will find a way toward describing holographically the Kondo problem.

The holographic strange metal constructions will be in the general context of some generalized holographic descriptions that exhibit so-called "hyperscaling violation."

48.1 Hyperscaling Violation and Holography

We have already described "Lifshitz scaling," which is the statement that some nonrelativistic field theories in condensed matter scale as $\vec{x} \to \lambda\vec{x}$, $t \to \lambda^z t$, with z the dynamical critical exponent. We have seen how to realize it geometrically, obtaining phenomenological holographic duals for Lifshitz theories.

But there is one more anomalous exponent that can appear in nonrelativistic field theories, the "hyperscaling violation exponent" θ. Hyperscaling is the statement that in the Lifshitz theories, by scaling arguments under Lifshitz scaling, we expect the entropy density of the system to scale with the temperature according to its naive dimension, i.e. $s \propto T^{d/z}$. But in the presence of the hyperscaling violation exponent θ, which appears as an anomalous dimension, we have

$$s \propto T^{\frac{d-\theta}{z}}. \tag{48.1}$$

To realize it geometrically, we consider a metric with a simple generalization of the Lifshitz holographic dual,

$$ds^2 = \frac{1}{r^2}\left(-\frac{dt^2}{r^{2d\frac{z-1}{d-\theta}}} + r^{\frac{2\theta}{d-\theta}}dr^2 + \sum_{i=1}^{d}dx_i^2\right) = \frac{1}{\tilde{r}^{\frac{2(d-\theta)}{d}}}\left(-\frac{dt^2}{\tilde{r}^{2(z-1)}} + d\tilde{r}^2 + \sum_{i=1}^{d}dx_i^2\right), \tag{48.2}$$

where $\tilde{r} = r^{\frac{d}{d-\theta}}$. As usual, the boundary is at $r \to 0$, and the IR at $r \to \infty$. The metric is easily seen to be invariant under the scaling

$$\vec{x} \to \lambda\vec{x}; \quad t \to \lambda^z t$$
$$r \to \lambda^{\frac{d-\theta}{d}}r, \quad \tilde{r} \to \lambda\tilde{r}$$
$$ds \to \lambda^{\theta/d}ds. \tag{48.3}$$

Since under the Lifshitz scaling the proper distance ds is covariant, not invariant, and ds will define the entropy, it is clear that there will be a hyperscaling violation exponent. At nonzero temperature T, there will be a black brane solution with horizon at $r = r_h, t = t_0$, which will have g_{tt} and g_{rr} modified, but not g_{ii}, so the entropy S of the solution will be the horizon area over 4 in Planck units, and thus the entropy density will be

$$s \propto r_h^{-d} \propto t^{-\frac{d-\theta}{z}} \propto T^{\frac{d-\theta}{z}}, \qquad (48.4)$$

so indeed θ is the hyperviolation exponent. Here we have used the fact that time scales as inverse temperature, $t \propto 1/T$.

Constraints on Exponents from the Null Energy Condition

We can obtain constraints on z and θ for the holographic dual (48.2) from the null energy condition. The null energy condition is the weakest bound on the energy momentum tensor in gravity that one can impose, saying that in a certain sense we have locally positive energy density. It states that

$$T_{\mu\nu} n^\mu n^\nu \geq 0 \qquad (48.5)$$

for all null vectors n^μ, i.e. $n_\mu n^\mu = 0$, if the Einstein equations are satisfied. The Einstein equations are $8\pi G_N T_{\mu\nu} = R_{\mu\nu} - \frac{1}{2} g_{\mu\nu} R$, so in effect we need to impose

$$\left(R_{\mu\nu} - \frac{1}{2} g_{\mu\nu} R \right) n^\mu n^\nu \geq 0. \qquad (48.6)$$

It is left as an exercise to show that the metric (48.2) has Ricci tensor components

$$R_{tt} = \frac{(d + z - \theta)(dz - \theta)}{d} \tilde{r}^{-2z}$$

$$R_{rr} = \frac{-d(d + z^2) + (d + z)\theta}{d} \tilde{r}^{-2}$$

$$R_{ij} = -\delta_{ij} \frac{(d - \theta)(d + z - \theta)}{d} \tilde{r}^{-2}. \qquad (48.7)$$

Choosing $n^t = \tilde{r}^{2(z-1)}$ and $n^r = 1, n^i = 0$ or $n^r = 0, n^i = 1$, we obtain the conditions

$$(d - \theta)(d(z - 1) - \theta) \geq 0$$

$$(z - 1)(d + z - \theta) \geq 0. \qquad (48.8)$$

For a Lorentz invariant theory, $z = 1$, we obtain $\theta \leq 0$ or $\theta \geq d$. For no hyperscaling violation $\theta = 0$, we obtain $z \geq 1$ as usual.

Compressible states with fermionic excitations for a $d-1$–dimensional Fermi surface (surface in momentum space), which can disperse along the single transverse direction to the surface, must have $S \propto T^{1/z}$, and thus

$$\theta = d - 1. \qquad (48.9)$$

The null energy condition then implies

$$z \geq 2 - 1/d. \qquad (48.10)$$

For the relevant case of $d = 2$, we obtain $z \geq 3/2$.

One is interested in finding solutions that *in the IR*, i.e. for $r \to \infty$, go to the solution (48.2). Besides the usual Einstein-Maxwell system, we need now also a scalar field Φ, giving a relevant perturbation in the UV, and allowing more IR behaviors. We also introduce two functions, a scalar field potential $V(\Phi)$, and a Maxwell coupling $Z(\Phi)$, for a Lagrangean

$$\mathcal{L} = \frac{1}{2\kappa^2}\left(R - 2(\nabla\Phi)^2 - \frac{V(\Phi)}{L^2}\right) - \frac{Z(\Phi)}{4e^2}F_{\mu\nu}F^{\mu\nu}. \qquad (48.11)$$

Note that the functions $V(\Phi)$ and $Z(\Phi)$ are chosen to be dimensionless. The ansatz for the $(T = 0)$ solution is

$$ds^2 = L^2\left(-f(r)dt^2 + g(r)dr^2 + \frac{dx_i^2}{r^2}\right)$$

$$A = \frac{eL}{\kappa}h(r)$$

$$\Phi = \Phi(r). \qquad (48.12)$$

We have chosen to fix the coefficient of dx_i^2 to be $1/r^2$, which fixes the coordinate r to be the one in the first form of (48.2). We could have chosen to have the same coefficient for dr^2 and dx_i^2, which would have corresponded to the $\tilde{r}$ coordinate form the second form of (48.2).

Moreover, to obtain the correct IR behavior, we must choose exponential forms as $\Phi \to \infty$:

$$Z(\Phi) \to Z_0 e^{\alpha\Phi}$$

$$V(\Phi) \to -V_0 e^{\beta\Phi}. \qquad (48.13)$$

From the equations of motion, one finds the exponents

$$\theta = \frac{d^2\beta}{\alpha + (d-1)\beta}$$

$$z = 1 + \frac{\theta}{d} + \frac{8(d(d-\theta)+\theta)^2}{d^2(d-\theta)\alpha^2}, \qquad (48.14)$$

which can be inverted in the case $\theta = d - 1$ (Fermi surfaces) to give

$$\alpha = \frac{2\sqrt{2}(2d-1)}{d\sqrt{z-2+1/d}}; \quad \beta = \alpha\frac{d-1}{2d-1}. \qquad (48.15)$$

Then the IR solution has $f(r)$ and $g(r)$ given by constants times their values in (48.2), whereas

$$h(r) \to h_1 r^{-d-dz/(d-\theta)}; \quad e^{\Phi(r)} \to e^{\Phi_1} r^{\frac{2d}{\alpha}\left(1+\frac{\theta}{d(d-\theta)}\right)}. \qquad (48.16)$$

The solution depends on a charge parameter Q that is defined as an integration constant from $\delta S/\delta A_t = dQ/dr = 0$.

Moreover, we can consider also solutions at nonzero temperature, obtained by replacing $f(r)$ and $g(r)$ by

$$f_T(r) = f(r)\left(1 - (r/r_h)^{d(1+z/(d-\theta))}\right)$$

$$g_T(r) = g(r)\left(1 - (r/r_h)^{d(1+z/(d-\theta))}\right). \qquad (48.17)$$

From it, one finds indeed an entropy density $s \propto QT^{(d-\theta)/z}$.

48.2 Holographic Strange Metal Constructions

Strange metals are metals that have strongly coupled electrons generating Fermi surfaces, but without Fermi liquid properties, i.e. non-Fermi liquids. As such, we have seen that for systems with hyperscaling violation with $\theta = d - 1$, we can obtain such conditions.

Moreover, the existence of a Fermi surface was defined in Chapter 46, and it could be defined by the logarithmic violation of the area law for entanglement entropy, (46.10). We have also seen that we need a specific behavior for the functions $f(r)$ and $g(r)$ (to compare with Chapter 46, $g(z)/z^2$ there is $g(r)$ here and $f(z)/z^2$ there is $f(r)$ here, with the same coordinate $z \to r$), namely *for $d = 2$*,

$$g(r) \simeq \frac{1}{r^2}\left(\frac{r}{r_F}\right)^2 = \frac{1}{r_F^2}, \quad r \gg r_F$$

$$g(r) \simeq \frac{1}{r^2} \quad r \ll r_F, \tag{48.18}$$

to obtain the needed behavior (46.10) of the entanglement entropy.

Such a behavior in the IR $r \gg r_F$, also with

$$f(r) \simeq kr^{-p-2}, \tag{48.19}$$

can be obtained from the general action for dilaton-Maxwell-Einstein in (48.11) for $d = 2$ space dimensions. In a general dimension, to obtain (46.10) we need to have

$$g(r) \simeq \frac{1}{r^2}\left(\frac{r}{r_F}\right)^{2(d-1)}, \tag{48.20}$$

and from the equations of motion we also find

$$f(r) = kr^{-2d(z-1)-2}, \tag{48.21}$$

which for $d = 2$ means $p = 4(z - 1)$. Comparing with the general hyperscaling violating metric (48.2), we see that this indeed implies that we need $\theta = d - 1$, as advertised.

In the $d = 2$ spatial dimensions case, solving the equations of motion in the IR, we find solutions only for $p > 2$, i.e. for $z > 3/2$ (note that this is the same condition obtained in the previous section from the null energy condition). One finds the solution

$$\Phi(r) = \sqrt{\frac{p-2}{2}}r, \tag{48.22}$$

and also the asymptotic forms of $V(\Phi)$ and $Z(\Phi)$,

$$V(\Phi) = -(8 + 3p + p^2/4)r_F^2 e^{-2\sqrt{\frac{2}{p-2}}\Phi}$$

$$Z(\Phi) = \frac{8A^2}{r_F^2 p(p+8)} e^{6\sqrt{\frac{2}{p-2}}\Phi}, \tag{48.23}$$

where, as we said, $p = 4(z - 1)$. Then indeed, these asymptotic forms agree with the one for general θ and d in (48.13) and (48.16), which for $\theta = d - 1$ become

$$e^{\Phi(r)} = r^{\sqrt{2z-3}}$$
$$V(\Phi) = -V_0 e^{-\frac{2\sqrt{2}(d-1)}{d\sqrt{z-2+1/d}}\Phi}$$
$$Z(\Phi) = Z_0 e^{\frac{2\sqrt{2}(2d-1)}{d\sqrt{z-2+1/d}}\Phi}. \tag{48.24}$$

Also the finite temperature solutions are found form (48.17), with the extra factor being $1 - (r/r_h)^{d(1+z)}$ for $\theta = d - 1$. In $d = 2$ spatial dimensions, we obtain a power of $2(1 + z) = (p + 8)/2$.

Finally, this IR solution can be embedded into a full solution with AdS asymptotics in the UV. We can give the functions $f(r)$ and $g(r)$, and then find the corresponding $\Phi(r)$ solution and the potential $V(\Phi)$ and function $Z(\Phi)$ that gives such a solution. For instance, we can choose

$$g(r) = \frac{1}{r^2}\sqrt{1 + \frac{r^4}{r_F^4}}$$
$$f(r) = \frac{kr^{-p-2}}{1 + kr^{-p-2}}. \tag{48.25}$$

The holographic entanglement entropy of the strip of length L and width l can be calculated using the Ryu-Takayangi construction. The result for the IR metric (48.2) at $z = 1$ (relativistic, but hyperscaling violating) can be found to be, for $\theta \neq d - 1$,

$$S_A = \frac{(M_{\text{Pl}}R)^d}{4(d - \theta - 1)}\left[\left(\frac{\epsilon}{r_F}\right)^{\theta}\left(\frac{L}{\epsilon}\right)^{d-1} - \left(\frac{\sqrt{\pi}\,\Gamma\left(\frac{1+d-\theta}{2(d-\theta)}\right)}{\Gamma\left(\frac{1}{2(d-\theta)}\right)}\right)^{d-\theta}\left(\frac{l}{r_F}\right)^{\theta}\left(\frac{L}{l}\right)^{d-1}\right], \tag{48.26}$$

where r_F is, as before, the crossover scale between the hyperscaling violating metric and the AdS metric. When $\theta = d - 1$, i.e. for Fermi surfaces, however, we find (note that this can also be obtained as a limit of the previous equation, where the power laws tend to 1, and the square bracket vanishes, but also the denominator vanishes)

$$S_A = \frac{(M_{\text{Pl}}R)^d}{4}\left(\frac{L}{r_F}\right)^{d-1}\log\frac{2l}{\epsilon}, \tag{48.27}$$

i.e., the expected logarithmic violation of the area law.

Furthermore, considering the system at finite temperature, in the small temperature regime, we can use the hyperscaling violating metric (48.2) in the holographic entanglement entropy formula of Ryu and Takayanagi, obtaining the formula above, but at large temperature, one effectively finds

$$s_A = \frac{S_A}{L^{d-1}l} \propto T^{\frac{d-\theta}{z}}, \tag{48.28}$$

i.e., the same formula as the thermal entropy. Indeed, this is expected, since in this limit, the Ryu-Takayanagi minimal surface has the maximum value for the area on the horizon, and so effectively measures the horizon area, that appears in the thermal entropy formula.

String Theory Realization

Hyperscaling violating metrics with $\theta \neq 0$, but still relativistic, i.e. with $z = 1$, can be easily obtained in string theory. In fact, they are the near-horizon limits of general (black) Dp-brane metrics.

The black Dp-branes have string frame metric and dilaton

$$ds^2_{\text{string}} = H_p(u)^{-1/2} \left(-f(u)dt^2 + \sum_{i=1}^{p} dx_i^2 \right) + H_p(u)^{1/2} \left(\frac{du^2}{f(u)} + u^2 d\Omega^2_{8-p} \right)$$

$$e^{\phi(u)} = g_s H_p(u)^{\frac{3-p}{4}}, \tag{48.29}$$

where the functions are

$$H_p(u) = 1 + \sinh^2 \beta \, \frac{u_h^{7-p}}{u^{7-p}}$$

$$f(u) = 1 - \frac{u_h^{7-p}}{u^{7-p}}. \tag{48.30}$$

The Einstein frame metric is $ds^2_{\text{Einstein}} = (e^{\phi}/g_s)^{-1/2} ds^2_{\text{string}}$, and this supergravity solution is valid when there are small curvatures and dilaton, which amounts to a condition on the effective (dimensionless) coupling,

$$1 \ll g^2_{\text{eff}} = \frac{g_s N}{u^{3-p}} \ll N^{\frac{4}{7-p}}, \tag{48.31}$$

which gives a restriction on the coordinates.

KK reducing onto S^{8-p} and going to the Einstein frame in the remaining $p+2$ dimensions, one obtains the metric

$$ds^2_{p+2} = u^{\frac{16-2p}{p}} H(u)^{1/p} \left(-f(u)dt^2 + \sum_{i=1}^{p} dx_i^2 + H(u) \frac{du^2}{f(u)} \right), \tag{48.32}$$

and after the change of coordinates

$$ds = \sqrt{H(u)} du, \tag{48.33}$$

we obtain the hyperscaling violating metric (48.2), with exponent (changing notation from p to d, previously used for the number of spatial dimensions)

$$\theta = d - \frac{9-d}{5-d}. \tag{48.34}$$

Note, however, that this doesn't cover the very important case of Fermi surfaces, which has $\theta = d - 1$.

48.2.1 Hydrodynamic Electron Flow and Strange Metals

In the previous chapter we have seen that momentum dissipation through a holographic lattice can be described in an effective theory with a massive gravity. We have also previewed the fact that if we have a strange metal construction, we can obtain a resistivity linear in temperature through an entropy density linear in temperature. This was related to an electron transport that can be described hydrodynamically.

The specific model we must then take is a model with hyperscaling violation, so that it can describe strange metals, and with massive gravity with mass m, so it can have holographic momentum dissipation. We choose

$$Z(\Phi) = e^{\frac{2}{\sqrt{3}}\Phi}$$
$$V(\Phi) = -6\cosh\left(\frac{2}{\sqrt{3}}\Phi\right) \to -3e^{\frac{2}{\sqrt{3}}\Phi} \quad \text{for} \quad \Phi \to \infty, \tag{48.35}$$

which has both an AdS background, since $V(\Phi = 0) = -6$ is a negative cosmological constant, and the required IR ($\Phi \to \infty$) form to give the holographic hyperscaling violating metric. By comparing with (48.14), we see that, since we have $\alpha = \beta = 2/\sqrt{3}$, we would obtain

$$\theta = d = 2, \quad z \to \infty, \tag{48.36}$$

except that of course the introduction of the graviton mass, which changes the IR physics, means that we don't have anymore the IR solution (48.2). The action is then

$$S = \frac{1}{2\kappa_{N,4}^2} \int d^4x \sqrt{-g} \left[R - \frac{1}{4}e^{\frac{2}{\sqrt{3}}\Phi}F_{\mu\nu}F^{\mu\nu} - 2\partial_\mu\Phi\partial^\mu\Phi \right.$$
$$\left. + \frac{6}{L^2}\cosh\left(\frac{2}{\sqrt{3}}\Phi\right) - \frac{m^2}{2}[\text{Tr}(K)^2 - \text{Tr}(K^2)] \right], \tag{48.37}$$

and it can be shown that it has the black hole solution

$$ds^2 = \frac{r^2 g(r)}{L^2}(-h(r)dt^2 + dx^2 + dy^2) + \frac{L^2}{r^2 g(r)h(r)}dr^2$$
$$A_t(r) - \sqrt{\frac{3Q(Q+r_0)}{L^2}\left(1 - \frac{m^2 L^4}{2(Q+r_0)^2}\right)}\left(1 - \frac{Q+r_0}{Q+r}\right) \equiv L\mu\left(1 - \frac{Q+r_0}{Q+r}\right)$$
$$\phi(r) = \frac{1}{3}\log(g(r)), \tag{48.38}$$

where we have defined the functions

$$g(r) = \left(1 + \frac{Q}{r}\right)^{3/2}$$
$$h(r) = 1 - \frac{m^2 L^4}{2(Q+r)^2} - \frac{(Q+r_0)^3}{(Q+r)^3}\left(1 - \frac{m^2 L^4}{2(Q+r_0)^2}\right). \tag{48.39}$$

From the metric we can read off the temperature in the usual way, obtaining

$$T = \frac{r_0 \left(6(1 + Q/r_0)^2 - \frac{m^2 L^4}{r_0^2}\right)}{8\pi L^2 (1 + Q/r_0)^{3/2}}. \tag{48.40}$$

We observe then that $T/\mu \propto \sqrt{r_0/Q}$ at small T/μ (i.e. small r_0/Q).

The entropy is the area of the horizon over 4 in Planck units, which gives for the entropy density divided by the chemical potential squared

$$\frac{s}{\mu^2} = \frac{2\pi L^2}{3\kappa_{N,4}^2} \sqrt{r_0/Q} \sqrt{1 + r_0/Q} \left(1 + \frac{3m^2/\mu^2}{2(1 + r_0/Q)}\right) \propto T/\mu, \quad \text{for small } T. \tag{48.41}$$

As we saw in Chapter 47, this means that then $\rho_{DC} = sm^2/(4\pi Q^2)$ is proportional to T at small T, provided σ/μ^2 is approximately T-independent at small T, as we can check it is.

Electron Stars and Hydrodynamic Electron Flow

We have seen when discussing holographic Fermi surfaces that we can describe the presence of a Fermi fluid in the gravity dual as corresponding to finite density in the field theory, and that naturally one obtained strange metals. We can understand the presence of electron stars as a Thomas-Fermi–type approximation, where the continuous fluid obeys the equation of state of free Dirac fermions in a local chemical potential $\mu(r) = A_t(r)/\sqrt{-g_{tt}(r)}$. The contribution of the Dirac fluid to the Lagrangean can be given as simply the pressure of the fluid in $\mu(r)$.

This electron star has to be considered in the context of the hyperscaling violating Einstein-Maxwell-dilaton action, which is the one we saw gives a better description of the strange metal. Moreover, one should consider a graviton mass to account for momentum dissipation.

In the field theory, this electron star will give rise to a hydrodynamic electron flow for a strange metal at finite chemical potential.

48.3 Supersymmetric Kondo Effect via Holography

We have seen in Chapter 14 that the Kondo model is the interaction between an impurity spin with the conduction electrons, but it led to a diverging zero temperature resistivity. It was solved by Anderson and then Wilson in a renormalization group procedure for the strong coupling effect. In fact, the Kondo model was essential to Wilson's development of the renormalization group approach: the running of the coupling J with the temperature T is such that we have asymptotic freedom (small coupling at high temperatures) and IR slavery (strong coupling at small temperatures). The dynamically generated scale, the Kondo temperature, is the equivalent of the similar QCD scale. The Kondo lattice is the generalization to a lattice of impurity spins, and it led to two scales: the Kondo temperature and the RKKY temperature.

Noticing that the same holographic physics describes the interaction of an impurity lattice (added one at a time) with a CFT, and the conduction of electrons in a strange metal,

Sachdev argued that the physics of the Kondo lattice should be relevant to the holographic strange metal.

A maximally supersymmetric Kondo model was developed, and shown to have a simple holographic description that we will show here. Basically, the defect spin of the Kondo model will arise by an intersection of M D5-branes treated as probes, with the N D3-branes that curve the space to form an $AdS_5 \times S^5$ geometry. The D5-branes wrap an $AdS_2 \times S^4$ submanifold.

In the Kondo Hamiltonian for interaction of an impurity spin $S_{\alpha\beta}$ in a representation R of $SU(N)$ ($N = 2$ for the usual spin) situated at the origin, we can integrate out either the impurity, to find an effective theory, or the electrons, to provide an effective quantum mechanical theory for the impurity, though we will not describe the latter.

Given that at low temperature the Kondo coupling flows to strong coupling, integrating out the impurity corresponds to saying that the interaction of the impurity with the electrons is minimized by the binding of one electron with the spin, to form a singlet and thus screen the impurity, with the only effect of removing the electron field at the position of the impurity. Thus integrating out the impurity amounts to a boundary condition $\psi(r = 0) = 0$ for the electron wavefunction, and otherwise having free electrons.

The fixed point of the Kondo model can be described by a conformal field theory, coupled to the localized impurity, described by a fermion χ with kinetic term, via an operator $\mathcal{O}$ in the CFT coupled to $\bar{\chi}\chi$:

$$S = S_{\text{CFT}} + \int dt \; \bar{\chi}(i\partial_t + \mathcal{O}(x = 0))\chi. \tag{48.42}$$

To use AdS/CFT, we consider the $\mathcal{N} = 4$ $SU(N)$ SYM CFT, and couple it with M 0+1–dimensional impurity fermions χ_i^I, $i = 1, \ldots, N$, $I = 1, \ldots, M$, with fermion number k, i.e. with the constraint $\bar{\chi}\chi = k$. The action is then

$$S = S_{\mathcal{N}=4\text{SYM}} + \int dt \; i[\bar{\chi}\partial_t\chi + \bar{\chi}T_A(A_0^A(t, 0) + \vec{n} \cdot \vec{\phi}^A(t, 0))\chi] + S_k, \tag{48.43}$$

where T_A are the generators of the adjoint of $SU(N)$, $\vec{n}$ is a unit vector in the transverse scalar space $\mathbb{R}^6$, and S_k is the action for the fermion number constraint, imposed with an $M \times M$ matrix Lagrange multiplier $\tilde{A}_0{}^I{}_J$:

$$S_k = \int dt[\bar{\chi}\tilde{A}_0\chi - k\,\text{Tr}_I\,\tilde{A}_0]. \tag{48.44}$$

Note that we have added a coupling to the scalars $\vec{n} \cdot \vec{\phi}$ in the impurity action, needed for supersymmetry. Indeed, the CFT action has $\mathcal{N} = 4$ supersymmetry, but the addition of the impurity action still leaves half of it unbroken, the one described by supersymmetry parameters ϵ satisfying $\gamma_0 n^a \gamma_a \epsilon = \epsilon$.

The action (48.43) is understood like the action of $\mathcal{N} = 4$ SYM coupled to external "quarks" (impurities) that are infinitely heavy, since their positions are *fixed*. Then integrating these quarks (impurities) out results in a Wilson loop insertion, as we are familiar from QCD. The Wilson loop is a gauge-invariant observable in gauge theory, already described in Chapters 26 and 46. The normal one, appearing in QCD, was defined as we saw for a closed loop C by

$$W_R[C] = \text{Tr}_R \, P \exp\left(i \oint_C dx^\mu A_\mu\right), \qquad (48.45)$$

but in $\mathcal{N} = 4$ SYM the relevant Wilson loop is one that is invariant under supersymmetry, obtained by adding a scalar field term in the $\vec{n} \cdot \vec{\phi}$. This is the one obtained by integrating out the impurities in (48.43). The contour C needs in principle to be closed. But one can consider a loop C that is an infinitely long rectangle in the time direction, so it becomes effectively two straight lines from $t = -\infty$ to $t = +\infty$. The Wilson loop obtained by integrating out (48.43) has only one such line, thus it is

$$W_R = \text{Tr}_R \, P \exp\left(i \int dt \, (A_0 + \vec{n} \cdot \vec{\phi})\right). \qquad (48.46)$$

In general, for arbitrary M, the representation R is more complicated. But for $M = 1$, the representation R obtained is the k-th antisymmetric representation of $SU(N)$. It arises because the partition function with k fermions has a product of k phases $e^{i \int dt m_{i_k}}$, which organize into the Wilson line with this representation.

Holographic Description

The first clue about the holographic description of this Kondo model comes from the Wilson loops, which are described holographically through macroscopic (long) string worldsheets ending on the contour C at the boundary of AdS space. In that case, the holographic description is found by considering that one of the N D3-branes that form the AdS background is split off, and the string ends on it. So the impurities should correspond to some strings ending on D-branes in the gravity dual.

The impurities are 1+0–dimensional, so we should consider strings extending between the N D3-branes and some M D5-branes, in the case that the D3-branes and D5-branes share only the time direction, the spatial ones being different: x^1, x^2, x^3 for D3 and $x^4, \ldots, x^8$ for D5. They should also be separated in x_9, which will be spanned by k strings extending between D3-branes and D5-branes. The strings between the two D-brane groups, each with its own gauge group, i.e. $SU(N) \times SU(M)$, will be bifundamental (each end is charged under a different gauge group). Call the two gauge fields A_μ and $\tilde{A}_\mu$. If we treat the strings as impurities and the D-branes as fixed, only the bifundamental fermionic degree of freedom remains, the scalar one being massive, to the D3–D5 string action becomes the impurity action

$$S_{\text{imp}} = \int \left[i\bar{\chi}\partial_t \chi + \bar{\chi}_i^I (A_0(t,0) + \vec{n} \cdot \vec{\phi})^i_{\ j} \chi_I^j + \bar{\chi}_i^I (\tilde{A}_0)^I_{\ J} \chi_J^i - k(\tilde{A}_0)^I_{\ I}\right]. \qquad (48.47)$$

In the decoupling limit for the N D3-branes, we are left with the 3-5 strings, fundamental with respect to $SU(N)$ and $SU(M)$, with the $SU(M)$ group decoupled, i.e. $SU(M)$ becomes a global ("flavor") symmetry.

In the probe approximation for the D5-branes, they are simply classical branes wrapping a cycle in $AdS_5 \times S^5$, with a worldvolume flux for the gauge field A on the D5-branes. The cycle can be calculated from the spacetime configuration described above.

Writing the $AdS_5 \times S^5$ metric as

$$ds^2 = R^2(du^2 + \cosh^2 u \, ds^2_{AdS_2} + \sinh^2 u \, d\Omega_2^2 + d\theta^2 + \sin^2\theta \, d\Omega_4^2), \quad (48.48)$$

the embedding surface is $AdS_2 \times S^4$, defined by the embedding

$$u = 0, \quad \theta = \theta_k, \quad (48.49)$$

where θ_k is defined in terms of N and k by

$$\frac{k}{N} = \frac{1}{\pi}\left(\theta_k - \frac{1}{2}\sin 2\theta_k\right). \quad (48.50)$$

The description above corresponds to the Kondo fixed point (for the renormalization group), and one can show that it is stable under perturbations.

A study of the specific heat C and magnetic susceptibility χ shows that the defect has properties different than the Fermi liquid, and also in general different than the usual strange metals. This was to be expected, since we have here a supersymmetric version of the Kondo model. One can also consider a backreacted solution for the D5-branes and the corresponding Wilson loop, which is of the general form of Lunin, Lin, and Maldacena studied in the case of holographic free fermions.

48.4 Toward the Real Kondo Problem

I will now describe a recent attempt at describing the real Kondo problem.

For a Kondo lattice, the RKKY interactions occur via conduction electrons scattering off the impurities, and are (anti)ferromagnetic, whereas the Kondo effect, as we saw in the last section for the case of a single impurity, tends to screen the spins and produce a nonmagnetic ground states. The competition between the two leads to a quantum phase transition as parameters like the impurity concentration are changed.

One of the ingredients in a good description of the Kondo effect then is the renormalization of the coupling, and in view of the above, it would also be advisable to obtain a phase transition between the (anti)ferromagnetic ground state and a nonmagnetic ground state. We should introduce the spin impurity as an impurity in the holographic model.

One example of such a model is obtained as follows. One considers the AdS_3 black hole (AdS "BTZ" black hole in three dimensions),

$$ds^2 = \frac{R^2}{z^2}\left(-f(z)dt^2 + dx^2 + \frac{dz^2}{f(z)}\right)$$

$$f(z) = 1 - \frac{z^2}{z_h^2}, \quad (48.51)$$

with temperature $T = 1/(2\pi z_h)$. The boundary of the gravity dual is at $z = 0$. In it, the holographic impurity is a 1+1–dimensional defect situated at $x = 0$ (so on the boundary, it is a 0+1–dimensional impurity, as needed), all along the z domain, from $z = 0$ to $z = z_h$. On the defect live a scalar Φ dual to the operator $\mathcal{O}$ describing the impurity spin via $2\vec{S}\cdot\vec{J} = \mathcal{O}\mathcal{O} - (q/N)\bar{\psi}\psi$, i.e. $\mathcal{O} = \bar{\psi}\chi$ (the impurity spin is written in terms of "slave fermions" χ as $S^a = \bar{\chi}T^a\chi$) and a gauge field a_m, dual to the charge Q of the slave fermions. The defect action is

$$S_{\text{imp}} \simeq -N \int_{AdS_3} \delta(x)\left[\frac{1}{4}f_{mn}f^{mn} + (D_m\Phi)^2 + V(\Phi^\dagger\Phi)\right]. \tag{48.52}$$

With an expansion of the gauge field defining the charge Q and chemical potential μ of the slave fermions, $a_t \simeq \frac{Q}{z} + \mu$, and a potential in the form of a mass term $V = M^2\Phi^\dagger\Phi$, one finds from the equations of motion the expansion

$$\Phi(z) \simeq \sqrt{z}(\alpha_\Lambda \log(\Lambda z) + \beta_\Lambda). \tag{48.53}$$

Here as usual α_Λ is dual to the source for $\mathcal{O}$, and β_Λ is dual to the VEV of $\mathcal{O}$. But then the ratio $\alpha_\Lambda/\beta_\Lambda = \kappa_\Lambda$ defines (up to a constant) the coupling of the double trace $\mathcal{O}\mathcal{O}$, i.e. the Kondo coupling.

The scale Λ is a dynamically generated scale, so $\Phi(z)$ should be independent of it. That means that $\Phi(z)$ for Λ should equal $\Phi(z)$ for the scale $2\pi T = 1/z_h$, so

$$\frac{\alpha_\Lambda}{\kappa_\Lambda}(1 + \kappa \log(\Lambda z)) = \frac{\alpha_T}{\kappa_T}(1 + \kappa_T \log(2\pi T z)). \tag{48.54}$$

Considering the equation at $z = z_h = 1/(2\pi T)$, and assuming that the coefficient of the leading term, α_Λ, does not change, so $\alpha_\Lambda = \alpha_T$, we obtain

$$\kappa_T = \frac{\kappa_\Lambda}{1 + \kappa_\Lambda \log\left(\frac{\Lambda}{2\pi T}\right)}. \tag{48.55}$$

This is the form of the running of the Kondo coupling, and from it we obtain the Kondo temperature, when the coupling diverges in perturbation theory, as

$$T_K = \frac{\Lambda}{2\pi}e^{1/\kappa_\Lambda}. \tag{48.56}$$

One also obtains a phase transition, depending on $\Phi(z)$: at $\Phi = 0$ we have $\langle\mathcal{O}\rangle = 0$, but at $\Phi(z)$, we have a nonzero VEV.

But obtaining the full phase diagram expected for the Kondo model proves challenging, as we need among other facts to consider a Kondo lattice of impurities, and there are many observables that one should verify have the right behavior.

Important Concepts to Remember

- Hyperscaling violation is an anomalous dimension (the hyperscaling violation exponent θ) in the scaling with T of the entropy density, $s \propto T^{\frac{d-\theta}{z}}$.
- It is obtained holographically (geometrized) by a scaling $ds \to \lambda^{\theta/d}ds$ of the line element.
- It can be obtained from an Einstein-Maxwell-dilaton theory with a scalar potential $V(\Phi)$ and an electromagnetic kinetic function $Z(\Phi)$.
- A holographic strange metal, with a Fermi surface characterized by a logarithmic breaking of the entanglement entropy area law, needs to have $\theta = d - 1$. This is obtained if as $\Phi \to \infty$, $V(\Phi)$ and $Z(\Phi)$ are well-defined exponentials.

- In string theory, near horizon black p-branes give relativistic ($z = 1$) hyperscaling violation, but with $\theta \neq d - 1$.
- Hydrodynamic electronic flow and $\rho \propto T$ can be obtained for holographic strange metals coming from hyperscaling violating models with a massive gravity (for momentum dissipation).
- A supersymmetric holographic Kondo model can be obtained, since the impurities can be realized as Wilson loops. The dual impurities are fermionic D3-D5 strings, stretching between D3-branes, forming an $AdS_5 \times S^5$ gravity background, and probe D5-branes, on $AdS_2 \times S^4$.
- In a real holographic Kondo model, we want to obtain the scaling of the Kondo coupling and a phase transition between magnetic and nonmagnetic ground states.

Further Reading

Holographic hyperscaling violation was introduced in [148], and further developed in [133] and [134], where the strange metal description in terms of entanglement entropy, started by [132], also was developed. The holographic description of conductivity in strange metals was done in [146]. The connection between strange metals and the Kondo model and Kondo lattices was argued for in [149]. The holographic supersymmetric Kondo model was described in [150]. A recent attempt at a holographic dual of the real Kondo model was done in [151].

Exercises

(1) Show that the metric (48.2) has Ricci tensor (48.7).
(2) Find $V(\Phi)$ and $Z(\Phi)$ that give $f(r)$ and $g(r)$ in (48.25), and

$$h(r) = h_1 e^{-d - dz/(d-\theta)} \tag{48.57}$$

by using the equations of motion for Φ and A_μ coming from the action (48.11).
(3) Show that the ansatz (48.38) solves the equations of motion for Φ coming from the action in (48.37).
(4) Write down the DBI+WZ action for the D5-brane in the metric (48.48), and show that $u = 0, \theta = \theta_k$ is a solution of its equations of motion.
(5) Write down the equations of motion for Φ on the solution (48.51), and show that

$$\Phi(z) \simeq \sqrt{z}(\alpha_\Lambda \log(\Lambda z) + \beta_\Lambda) \tag{48.58}$$

is a solution near the boundary.

References

[1] S. A. Hartnoll, "Lectures on holographic methods for condensed matter physics," Class. Quant. Grav. **26**, 224002 (2009), arXiv:0903.3246 [hep-th].

[2] N. Iqbal, H. Liu, and M. Mezei, "Lectures on holographic non-Fermi liquids and quantum phase transitions," arXiv:1110.3814 [hep-th].

[3] C. P. Herzog, "Lectures on holographic superfluidity and superconductivity," *J. Phys. A* **42**, 343001 (2009), arXiv:0904.1975 [hep-th].

[4] J. McGreevy, "Holographic duality with a view toward many-body physics," *Adv. High Energy Phys.* **2010**, 723105 (2010), arXiv:0909.0518 [hep-th].

[5] J. Zaanen, Y. W. Sun, Y. Liu, and K. Schalm, *Holographic Duality in Condensed Matter Physics*, Cambridge University Press, 2016.

[6] C. Kittel, *Introduction to Solid State Physics*, John Wiley and Sons, 2005.

[7] P. Phillips, *Advanced Solid State Physics*, Westview Press, 2003.

[8] H. Năstase, *Introduction to the AdS/CFT Correspondence*, Cambridge University Press, 2015.

[9] L. D. Faddeev, "How algebraic Bethe ansatz works for integrable model," arXiv:hep-th/9605187.

[10] T. R. Klassen and E. Melzer, "The thermodynamics of purely elastic scattering theories and conformal perturbation theory," *Nucl. Phys. B* **350**, 635 (1991).

[11] J. Polchinski, *String Theory*, vol. I, Cambridge University Press, 2000.

[12] S. Sachdev, *Quantum Phase Transitions*, Cambridge University Press, 2011.

[13] G. V. Dunne, "Aspects of Chern-Simons theory," arXiv:hep-th/9902115.

[14] A. Stern, "Anyons and the quantum Hall effect – A pedagogical review," *Ann. Phys.* **323**, 204 (2008).

[15] S. Rao, "An Anyon primer," arXiv:hep-th/9209066.

[16] E. Witten, "Three lectures on topological phases of matter," arXiv:1510.07698 [cond-mat.mes-hall].

[17] G. W. Moore and N. Read, "Nonabelions in the fractional quantum Hall effect," *Nucl. Phys. B* **360**, 362 (1991).

[18] X.-L. Qi and S.-C. Zhang, "Topological insulators and superconductors," *Rev. Mod. Phys.* **83** (2011) 1057, arXiv:1008.2026 [cond-mat.mes-hall].

[19] X. L. Qi, E. Witten, and S. C. Zhang, "Axion topological field theory of topological superconductors," *Phys. Rev. B* **87**, 134519 (2013), arXiv:1206.1407 [cond-mat.supr-con].

[20] X. L. Qi, T. Hughes, and S. C. Zhang, "Topological field theory of time-reversal invariant insulators," *Phys. Rev. B* **78**, 195424 (2008), arXiv:0802.3537 [cond-mat.mes-hall].

[21] P. Coleman, "Heavy fermions and the Kondo lattice: A 21st century perspective," arXiv:1509.05769 [cond-mat.str-el].

[22] L. D. Landau and E. M. Lifshitz, *Course of Theoretical Physics*, vol. 6, *Fluid Mechanics*, 2nd ed., Elsevier, 1987.

[23] M. Rangamani, "Gravity and hydrodynamics: lectures on the fluid-gravity correspondence," *Class. Quant. Grav.* **26**, 224003 (2009), arXiv:0905.4352 [hep-th].

[24] V. E. Hubeny, S. Minwalla, and M. Rangamani, "The fluid-gravity correspondence," arXiv:1107.5780 [hep-th].

[25] P. J. E. Peebles, *Principles of Physical Cosmology*, Princeton University Press, 1993.

[26] R. M. Wald, *General Relativity*, University of Chicago Press, 1984.

[27] C. W. Misner, K. S. Thorne, and J. A. Wheeler, *Gravitation*, Freeman and Co., 1973.

[28] S. W. Hawking and G. F. R. Ellis, *The Large Scale Structure of Space-time*, Cambridge University Press, 1973.

[29] S. W. Hawking, "Particle creation by black holes," *Commun. Math. Phys.* **43**, 199 (1975), *Commun. Math. Phys.* **46**, 206 (1976).

[30] J. M. Bardeen, B. Carter, and S. W. Hawking, "The four laws of black hole mechanics," *Commun. Math. Phys.* **31**, 161 (1973).

[31] M. J. Duff, B. E. W. Nilsson, and C. N. Pope, "Kaluza-Klein supergravity," *Phys. Rept.* **130**, 1 (1986).

[32] H. Năstase, D. Vaman, and P. van Nieuwenhuizen, "Consistency of the AdS(7) × S(4) reduction and the origin of selfduality in odd dimensions," *Nucl. Phys. B* **581**, 179 (2000), arXiv:hep-th/9911238.

[33] R. I. Nepomechie, "Magnetic monopoles from antisymmetric tensor gauge fields," *Phys. Rev. D* **31**, 1921 (1985).

[34] C. Teitelboim, "Gauge invariance for extended objects," *Phys. Lett. B* **167**, 63 (1986).

[35] C. Teitelboim, "Monopoles of higher rank," *Phys. Lett. B* **167**, 69 (1986).

[36] M. J. Duff, R. R. Khuri, and J. X. Lu, "String solitons," *Phys. Rept.* **259**, 213 (1995), arXiv:hep-th/9412184.

[37] A. A. Tseytlin, "Harmonic superpositions of M-branes," *Nucl. Phys. B* **475**, 149 (1996), arXiv:hep-th/9604035.

[38] M. Cvetic and A. A. Tseytlin, "Nonextreme black holes from nonextreme intersecting M-branes," *Nucl. Phys. B* **478**, 181 (1996), arXiv:hep-th/9606033.

[39] C. P. Burgess and F. Quevedo, "Bosonization as duality," *Nucl. Phys. B* **421**, 373 (1994), arXiv:hep-th/9401105.

[40] J. Murugan and H. Năstase, "A nonabelian particle-vortex duality," *Phys. Lett. B* **753**, 401 (2016), arXiv:1506.04090 [hep-th].

[41] L. Alvarez-Gaume and S. F. Hassan, "Introduction to S duality in N=2 supersymmetric gauge theories: A pedagogical review of the work of Seiberg and Witten," *Fortsch. Phys.* **45**, 159 (1997), arXiv:hep-th/9701069.

[42] M. B. Green, J. H. Schwarz, and E. Witten, *Superstring Theory*, Cambridge University Press, 1987.

[43] B. Zwiebach, *A First Course in String Theory*, Cambridge University Press, 2009.

[44] C. Johnson, *D-Branes*, Cambridge University Press, 2003.

[45] K. Becker, M. Becker, and J. H. Schwarz, *String Theory and M-Theory*, Cambridge University Press, 2007.

[46] M. Ammon and J. Erdmenger, *Gauge/Gravity Duality: Foundations and Applications*, Cambridge University Press, 2015.

[47] O. Aharony, S. S. Gubser, J. M. Maldacena, H. Ooguri, and Y. Oz, "Large N field theories, string theory and gravity," *Phys. Rept.* **323**, 183 (2000), arXiv:hep-th/9905111.

[48] J. M. Maldacena, "The large N limit of superconformal field theories and supergravity," *Int. J. Theor. Phys.* **38**, 1113 (1999) [*Adv. Theor. Math. Phys.* **2**, 231 (1998)], arXiv:hep-th/9711200.

[49] E. Witten, "Anti-de Sitter space and holography," *Adv. Theor. Math. Phys.* **2**, 253 (1998), arXiv:hep-th/9802150.

[50] S. S. Gubser, I. R. Klebanov, and A. M. Polyakov, "Gauge theory correlators from noncritical string theory," *Phys. Lett. B* **428**, 105 (1998), arXiv:hep-th/9802109.

[51] D. E. Berenstein, J. M. Maldacena, and H. S. Năstase, "Strings in flat space and pp waves from N=4 superYang-Mills," *JHEP* **0204**, 013 (2002), arXiv:hep-th/0202021.

[52] J. C. Plefka, "Lectures on the plane wave string gauge theory duality," *Fortsch. Phys.* **52**, 264 (2004), arXiv:hep-th/0307171.

[53] J. Kowalski-Glikman, "Vacuum states in Supersymmetric Kaluza-Klein theory," *Phys. Lett. B* **134**, 194 (1984).

[54] R. Penrose, "Any spacetime has a plane wave as a limit," *Differential Geometry and Relativity*, Reidel, 1974, pp. 271–275.

[55] M. Blau, J. M. Figueroa-O'Farrill, C. Hull, and G. Papadopoulos, "A new maximally supersymmetric background of IIB superstring theory," *JHEP* **0201**, 047 (2002), arXiv:hep-th/0110242.

[56] M. Blau, J. M. Figueroa-O'Farrill, C. Hull, and G. Papadopoulos, "Penrose limits and maximal supersymmetry," *Class. Quant. Grav.* **19**, L87 (2002), arXiv:hep-th/0201081.

[57] P. C. Aichelburg and R. U. Sexl, "On the gravitational field of a massless particle," *Gen. Rel. Grav.* **2**, 303 (1971).

[58] G. T. Horowitz and A. R. Steif, "Space-time singularities in string theory," *Phys. Rev. Lett.* **64**, 260 (1990).

[59] J. A. Minahan and K. Zarembo, "The Bethe ansatz for N = 4 superYang-Mills," *JHEP* **0303**, 013 (2003), arXiv:hep-th/0212208.

[60] J. Plefka, "Spinning strings and integrable spin chains in the AdS/CFT correspondence," *Living Rev. Rel.* **8**, 9 (2005), arXiv:0507136 [hep-th].

[61] K. Zarembo, "Semiclassical Bethe ansatz and AdS/CFT," *Comptes Rendus Physique* **5**, 1081 (2004) [*Fortsch. Phys.* **53**, 647 (2005)], arXiv:hep-th/0411191.

[62] M. F. Paulos, J. Penedones, J. Toledo, B. C. van Rees, and P. Vieira, "The S-matrix bootstrap I: QFT in AdS," arXiv:1607.06109 [hep-th].

[63] D. Bernard, "An introduction to Yangian symmetries," *Int. J. Mod. Phys. B* **7**, 3517 (1993), arXiv:hep-th/9211133.

[64] L. Dolan, C. R. Nappi, and E. Witten, "Yangian symmetry in D = 4 superconformal Yang-Mills theory," arXiv:hep-th/0401243.

[65] N. Beisert, V. Dippel, and M. Staudacher, "A novel long range spin chain and planar N = 4 super Yang-Mills," *JHEP* **0407**, 075 (2004), arXiv:hep-th/0405001.

[66] N. Beisert, "The SU(2–2) dynamic S-matrix," *Adv. Theor. Math. Phys.* **12**, 948 (2008), arXiv:hep-th/0511082.

[67] N. Beisert, B. Eden, and M. Staudacher, "Transcendentality and crossing," *J. Stat. Mech.* **0701**, P01021 (2007), arXiv:hep-th/0610251.

[68] S. Kachru, X. Liu, and M. Mulligan, "Gravity duals of Lifshitz-like fixed points," *Phys. Rev. D* **78**, 106005 (2008), arXiv:0808.1725 [hep-th].

[69] M. Taylor, "Non-relativistic holography," arXiv:0812.0530 [hep-th].

[70] T. Griffin, P. Hořava, and C. M. Melby-Thompson, "Lifshitz gravity for Lifshitz holography," *Phys. Rev. Lett.* **110**, no. 8, 081602 (2013), arXiv:1211.4872 [hep-th].

[71] C. P. Herzog, M. Rangamani, and S. F. Ross, "Heating up Galilean holography," *JHEP* **0811**, 080 (2008), arXiv:0807.1099 [hep-th].

[72] A. Adams, K. Balasubramanian, and J. McGreevy, "Hot spacetimes for cold atoms," *JHEP* **0811**, 059 (2008), arXiv:0807.1111 [hep-th].

[73] J. Maldacena, D. Martelli, and Y. Tachikawa, "Comments on string theory backgrounds with non-relativistic conformal symmetry," *JHEP* **0810**, 072 (2008), arXiv:0807.1100 [hep-th].

[74] D. T. Son, "Toward an AdS/cold atoms correspondence: A geometric realization of the Schrödinger symmetry," *Phys. Rev. D* **78**, 046003 (2008), arXiv:0804.3972 [hep-th].

[75] K. Balasubramanian and J. McGreevy, "Gravity duals for non-relativistic CFTs," *Phys. Rev. Lett.* **101**, 061601 (2008), arXiv:0804.4053 [hep-th].

[76] E. Witten, "Anti-de Sitter space, thermal phase transition, and confinement in gauge theories," *Adv. Theor. Math. Phys.* **2**, 505 (1998), arXiv:hep-th/9803131.

[77] J. Casalderrey-Solana, H. Liu, D. Mateos, K. Rajagopal, and U. A. Wiedemann, "Gauge/string duality, hot QCD and heavy ion collisions," arXiv:1101.0618 [hep-th].

[78] S. A. Hartnoll and P. Kovtun, "Hall conductivity from dyonic black holes," *Phys. Rev. D* **76**, 066001 (2007), arXiv:0704.1160 [hep-th].

[79] S. S. Gubser, A. Nellore, S. S. Pufu, and F. D. Rocha, "Thermodynamics and bulk viscosity of approximate black hole duals to finite temperature quantum chromodynamics," *Phys. Rev. Lett.* **101**, 131601 (2008), arXiv:0804.1950 [hep-th].

[80] O. DeWolfe, S. S. Gubser, and C. Rosen, "A holographic critical point," *Phys. Rev. D* **83**, 086005 (2011), arXiv:1012.1864 [hep-th].

[81] H. A. Chamblin and H. S. Reall, "Dynamic dilatonic domain walls," *Nucl. Phys. B* **562**, 133 (1999), arXiv:hep-th/9903225.

[82] D. T. Son and A. O. Starinets, "Minkowski space correlators in AdSCFT correspondence: Recipe and applications," *JHEP* **0209**, 042 (2002), arXiv:hep-th/0205051.

[83] N. Iqbal and H. Liu, "Universality of the hydrodynamic limit in AdS/CFT and the membrane paradigm," *Phys. Rev. D* **79**, 025023 (2009), arXiv:0809.3808 [hep-th].

[84] C. Lopez-Arcos, H. Nastase, F. Rojas, and J. Murugan, "Conductivity in the gravity dual to massive ABJM and the membrane paradigm," *JHEP* **1401**, 036 (2014), arXiv:1306.1263 [hep-th].

[85] S. S. Gubser, "Momentum fluctuations of heavy quarks in the gauge-string duality," *Nucl. Phys. B* **790**, 175 (2008), arXiv:hep-th/0612143.

[86] J. Casalderrey-Solana and D. Teaney, "Heavy quark diffusion in strongly coupled N = 4 Yang-Mills," *Phys. Rev. D* **74**, 085012 (2006), arXiv:hep-th/0605199.

[87] U. Gursoy, E. Kiritsis, L. Mazzanti, and F. Nitti, "Langevin diffusion of heavy quarks in non-conformal holographic backgrounds," *JHEP* **1012**, 088 (2010), arXiv:1006.3261 [hep-th].

[88] G. T. Horowitz and V. E. Hubeny, "Quasinormal modes of AdS black holes and the approach to thermal equilibrium," *Phys. Rev. D* **62**, 024027 (2000), arXiv:hep-th/9909056.

[89] S. S. Gubser, "Breaking an Abelian gauge symmetry near a black hole horizon," *Phys. Rev. D* **78**, 065034 (2008), arXiv:0801.2977 [hep-th].

[90] S. A. Hartnoll, C. P. Herzog, and G. T. Horowitz, "Holographic superconductors," *JHEP* **0812**, 015 (2008), arXiv:0810.1563 [hep-th].

[91] G. T. Horowitz, J. E. Santos, and B. Way, "A holographic Josephson Junction," *Phys. Rev. Lett.* **106**, 221601 (2011), arXiv:1101.3326 [hep-th].

[92] S. K. Domokos, C. Hoyos, and J. Sonnenschein, "Holographic Josephson Junctions and Berry holonomy from D-branes," *JHEP* **1210**, 073 (2012), arXiv:1207.2182 [hep-th].

[93] K. S. Thorne, R. H. Price, and D. A. MacDonald, *Black Holes: The Membrane Paradigm*, Yale University Press, 1986.

[94] C. Eling, I. Fouxon, and Y. Oz, "The incompressible Navier-Stokes equations from membrane dynamics," *Phys. Lett. B* **680**, 496 (2009), arXiv:0905.3638 [hep-th].

[95] C. Eling and Y. Oz, "Relativistic CFT Hydrodynamics from the membrane paradigm," *JHEP* **1002**, 069 (2010), arXiv:0906.4999 [hep-th].

[96] C. Eling, I. Fouxon, and Y. Oz, "Gravity and a geometrization of turbulence: an intriguing correspondence," arXiv:1004.2632 [hep-th].

[97] C. Eling, Y. Neiman, and Y. Oz, "Membrane paradigm and holographic hydrodynamics," *J. Phys. Conf. Ser.* **314**, 012032 (2011), arXiv:1012.2572 [hep-th].

[98] S. Bhattacharyya, S. Minwalla, and S. R. Wadia, "The incompressible non-relativistic Navier-Stokes equation from gravity," *JHEP* **0908**, 059 (2009), arXiv:0810.1545 [hep-th].

[99] S. Bhattacharyya, V. E. Hubeny, S. Minwalla, and M. Rangamani, "Nonlinear fluid dynamics from gravity," *JHEP* **0802**, 045 (2008), arXiv:0712.2456 [hep-th].

[100] M. P. Heller, R. A. Janik, and R. Peschanski, "Hydrodynamic flow of the quark-gluon plasma and gauge/gravity correspondence," *Acta Phys. Polon. B* **39**, 3183 (2008), arXiv:0811.3113 [hep-th].

[101] I. Bredberg, C. Keeler, V. Lysov, and A. Strominger, "Wilsonian approach to fluid/gravity duality," *JHEP* **1103**, 141 (2011), arXiv:1006.1902 [hep-th].

[102] I. Bredberg, C. Keeler, V. Lysov, and A. Strominger, "From Navier-Stokes to Einstein," *JHEP* **1207**, 146 (2012), arXiv:1101.2451 [hep-th].

[103] E. Brezin, C. Itzykson, G. Parisi, and J. B. Zuber, "Planar diagrams," *Commun. Math. Phys.* **59**, 35 (1978).

[104] H. Lin, O. Lunin, and J. M. Maldacena, "Bubbling AdS space and 1/2 BPS geometries," *JHEP* **0410**, 025 (2004), arXiv:hep-th/0409174.

[105] M. Cubrovic, J. Zaanen, and K. Schalm, "String theory, quantum phase transitions and the emergent Fermi-Liquid," *Science* **325**, 439 (2009), arXiv:0904.1993 [hep-th].

[106] S. A. Hartnoll and A. Tavanfar, "Electron stars for holographic metallic criticality," *Phys. Rev. D* **83**, 046003 (2011), arXiv:1008.2828 [hep-th].

[107] J. de Boer, K. Papadodimas, and E. Verlinde, "Holographic neutron stars," *JHEP* **1010**, 020 (2010), arXiv:0907.2695 [hep-th].

[108] M. Cubrovic, Y. Liu, K. Schalm, Y. W. Sun, and J. Zaanen, "Spectral probes of the holographic Fermi groundstate: Dialing between the electron star and AdS Dirac hair," *Phys. Rev. D* **84**, 086002 (2011), arXiv:1106.1798 [hep-th].

[109] L. Susskind, "The Quantum Hall fluid and noncommutative Chern-Simons theory," arXiv:hep-th/0101029.

[110] S. Hellerman and L. Susskind, "Realizing the quantum Hall system in string theory," arXiv:hep-th/0107200.

[111] N. Seiberg and E. Witten, "String theory and noncommutative geometry," *JHEP* **9909**, 032 (1999), arXiv:hep-th/9908142.

[112] S. Ryu and T. Takayanagi, "Topological insulators and superconductors from D-branes," *Phys. Lett. B* **693**, 175 (2010), arXiv:1001.0763 [hep-th].

[113] S. Ryu and T. Takayanagi, "Topological insulators and superconductors from string theory," *Phys. Rev. D* **82**, 086014 (2010), arXiv:1007.4234 [hep-th].

[114] M. Fujita, W. Li, S. Ryu, and T. Takayanagi, "Fractional quantum Hall effect via holography: Chern-Simons, edge states, and hierarchy," *JHEP* **0906**, 066 (2009), arXiv:0901.0924 [hep-th].

[115] Y. Hikida, W. Li, and T. Takayanagi, "ABJM with flavors and FQHE," *JHEP* **0907**, 065 (2009), arXiv:0903.2194 [hep-th].

[116] J. Murugan and H. Năstase, "On abelianizations of the ABJM model and applications to condensed matter," *Braz. J. Phys.* **45**, no. 4, 481 (2015), arXiv:1301.0229 [hep-th].

[117] O. Bergman, N. Jokela, G. Lifschytz, and M. Lippert, "Quantum Hall effect in a holographic model," *JHEP* **1010**, 063 (2010), arXiv:1003.4965 [hep-th].

[118] A. Mohammed, J. Murugan, and H. Năstase, "Abelian-Higgs and vortices from ABJM: Towards a string realization of AdS/CMT," *JHEP* **1211**, 073 (2012), arXiv:1206.7058 [hep-th].

[119] A. Mohammed, J. Murugan, and H. Nastase, "Towards a realization of the condensed-matter/gravity correspondence in string theory via consistent Abelian truncation," *Phys. Rev. Lett.* **109**, 181601 (2012), arXiv:1205.5833 [hep-th].

[120] J. Murugan, H. Năstase, N. Rughoonauth, and J. P. Shock, "Particle-vortex and Maxwell duality in the $AdS_4 \times \mathbb{CP}^3$/ABJM correspondence," *JHEP* **1410**, 51 (2014), arXiv:1404.5926 [hep-th].

[121] C. P. Burgess and B. P. Dolan, "Particle vortex duality and the modular group: Applications to the quantum Hall effect and other 2-D systems," *Phys. Rev. B* **63**, 155309 (2001), arXiv:hep-th/0010246.

[122] N. Doroud, D. Tong, and C. Turner, "On superconformal anyons," *JHEP* **1601**, 138 (2016), arXiv:1511.01491 [hep-th].

[123] K. Kang and H. Năstase, "Heisenberg saturation of the Froissart bound from AdS-CFT," *Phys. Lett. B* **624**, 125 (2005), arXiv:hep-th/0501038.

[124] H. Năstase, "The RHIC fireball as a dual black hole," arXiv:hep-th/0501068.

[125] H. Năstase, "DBI skyrmion, high energy (large s) scattering and fireball production," arXiv:hep-th/0512171.

[126] H. Năstase, "A black hole solution of scalar field theory," arXiv:hep-th/0702037.

[127] H. Năstase, "AdS-CFT and the RHIC fireball," *Prog. Theor. Phys. Suppl.* **174**, 274 (2008), arXiv:0805.3579 [hep-th].

[128] H. Năstase, "DBI scalar field theory for QGP hydrodynamics," arXiv:1512.05257 [hep-th].

[129] S. Ryu and T. Takayanagi, "Holographic derivation of entanglement entropy from AdS/CFT," *Phys. Rev. Lett.* **96**, 181602 (2006), arXiv:hep-th/0603001.

[130] S. Ryu and T. Takayanagi, "Aspects of holographic entanglement entropy," *JHEP* **0608**, 045 (2006), arXiv:hep-th/0605073.

[131] T. Nishioka, S. Ryu, and T. Takayanagi, "Holographic entanglement entropy: an overview," *J. Phys. A* **42**, 504008 (2009), arXiv:0905.0932 [hep-th].

[132] N. Ogawa, T. Takayanagi, and T. Ugajin, "Holographic Fermi surfaces and entanglement entropy," *JHEP* **1201**, 125 (2012), arXiv:1111.1023 [hep-th].

[133] L. Huijse, S. Sachdev, and B. Swingle, "Hidden Fermi surfaces in compressible states of gauge-gravity duality," *Phys. Rev. B* **85**, 035121 (2012), arXiv:1112.0573 [cond-mat.str-el].

[134] X. Dong, S. Harrison, S. Kachru, G. Torroba, and H. Wang, "Aspects of holography for theories with hyperscaling violation," *JHEP* **1206**, 041 (2012), arXiv:1201.1905 [hep-th].

[135] S. A. Hartnoll and C. P. Herzog, "Impure AdS/CFT correspondence," *Phys. Rev. D* **77**, 106009 (2008), arXiv:0801.1693 [hep-th].

[136] S. A. Hartnoll and D. M. Hofman, "Locally critical resistivities from Umklapp scattering," *Phys. Rev. Lett.* **108**, 241601 (2012), arXiv:1201.3917 [hep-th].

[137] G. T. Horowitz, J. E. Santos, and D. Tong, "Optical conductivity with holographic lattices," *JHEP* **1207**, 168 (2012), arXiv:1204.0519 [hep-th].

[138] G. T. Horowitz, J. E. Santos, and D. Tong, "Further evidence for lattice-induced scaling," *JHEP* **1211**, 102 (2012), arXiv:1209.1098 [hep-th].

[139] D. Vegh, "Holography without translational symmetry," arXiv:1301.0537 [hep-th].

[140] M. Blake and D. Tong, "Universal resistivity from holographic massive gravity," *Phys. Rev. D* **88**, no. 10, 106004 (2013), arXiv:1308.4970 [hep-th].

[141] M. Blake, D. Tong, and D. Vegh, "Holographic lattices give the graviton an effective mass," *Phys. Rev. Lett.* **112**, no. 7, 071602 (2014), arXiv:1310.3832 [hep-th].

[142] A. Donos and S. A. Hartnoll, "Interaction-driven localization in holography," *Nature Phys.* **9**, 649 (2013), arXiv:1212.2998 [hep-th].

[143] S. A. Hartnoll, P. K. Kovtun, M. Muller, and S. Sachdev, "Theory of the Nernst effect near quantum phase transitions in condensed matter, and in dyonic black holes," *Phys. Rev. B* **76**, 144502 (2007), arXiv:0706.3215 [cond-mat.str-el].

[144] J. Zaanen, "Electrons go with the flow in exotic material systems," *Science* **351**, 1026 (2016).

[145] A. H. Castro Neto, F. Guinea, N. M. R. Peres, K. S. Novoselov, and A. K. Geim, "The electronic properties of graphene," *Rev. Mod. Phys.* **81**, 109 (2009).

[146] R. A. Davison, K. Schalm, and J. Zaanen, "Holographic duality and the resistivity of strange metals," *Phys. Rev. B* **89**, no. 24, 245116 (2014), arXiv:1311.2451 [hep-th].

[147] S. A. Hartnoll, "Theory of universal incoherent metallic transport," *Nature Phys.* **11**, 54 (2015), arXiv:1405.3651 [cond-mat.str-el].

[148] C. Charmousis, B. Gouteraux, B. S. Kim, E. Kiritsis, and R. Meyer, "Effective holographic Theories for low-temperature condensed matter systems," *JHEP* **1011**, 151 (2010), arXiv:1005.4690 [hep-th].

[149] S. Sachdev, "Strange metals and the AdS/CFT correspondence," *J. Stat. Mech.* **1011**, P11022 (2010), arXiv:1010.0682 [cond-mat.str-el].

[150] S. Harrison, S. Kachru, and G. Torroba, "A maximally supersymmetric Kondo model," *Class. Quant. Grav.* **29**, 194005 (2012), arXiv:1110.5325 [hep-th].

[151] J. Erdmenger, M. Flory, C. Hoyos, M. N. Newrzella, A. O'Bannon, and J. Wu, "Holographic impurities and Kondo effect," arXiv:1511.09362 [hep-th].

Index

Printed in the United States
by Baker & Taylor Publisher Services